The Digital Consumer Technology Handbook

The Digital Consumer Technology Handbook

A Comprehensive Guide to Devices, Standards, Future Directions, and Programmable Logic Solutions

by Amit Dhir
Xilinx, Inc.

ELSEVIER

AMSTERDAM • BOSTON • HEIDELBERG • LONDON
NEW YORK • OXFORD • PARIS • SAN DIEGO
SAN FRANCISCO • SINGAPORE • SYDNEY • TOKYO
Newnes is an imprint of Elsevier

Newnes

Newnes is an imprint of Elsevier
200 Wheeler Road, Burlington, MA 01803, USA
Linacre House, Jordan Hill, Oxford OX2 8DP, UK

 Recognizing the importance of preserving what has been written, Elsevier prints its books on acid-free paper whenever possible.

Library of Congress Cataloging-in-Publication Data

Dhir, Amit.
  Consumer electronics handbook : a comprehensive guide to digital technology, applications and future directions / by Amit Dhir.
    p. cm.
  ISBN 0-7506-7815-1
    1. Digital electronics--Handbooks, manuals, etc. 2. Household electronics--Handbooks, manuals, etc. I. Title.

TK7868.D5D478 2004
621.381--dc22

                                                                    2003068632

British Library Cataloguing-in-Publication Data
A catalogue record for this book is available from the British Library.

For information on all Newnes publications
visit our website at www.newnespress.com

04 05 06 07 08 10 9 8 7 6 5 4 3 2 1

Printed in the United States of America

Dedications and Acknowledgments

This book is dedicated to my loving wife, Rita, for her undying patience, support, and understanding. For the countless hours she sat beside me giving me the strength to complete this effort and to always believe in myself, I owe this book to her. My special thanks to my parents, family, and friends who have supported me through the completion of this book.

I would especially like to acknowledge the efforts of Tom Pyles and Robert Bielby in getting this book published. My special thanks to Tom who went beyond the call of his job at Xilinx to help me edit and format the book in a brief two months. I thank Robert for providing me motivation to complete this book and for his efforts in reviewing and overseeing its development. He was also instrumental in guiding my professional development over the last four years.

About the Author

Amit Dhir is a Senior Manager in the Strategic Solutions Marketing group at Xilinx, Inc. He has published over 90 articles in business and technical publications dealing with the role of programmable logic in wired and wireless communications and in embedded and consumer applications. Amit is the author of the popular *The Home Networking Revolution: A Designer's Guide*, a book describing convergence, broadband access, residential gateways, home networking technologies, information appliances, and middleware. As a Xilinx representative, he has presented at numerous networking, home networking, wireless LAN, and consumer conferences. Until recently he served as marketing chairman of the HiperLAN2 committee. He holds a BSEE from Purdue University (1997) and a MSEE from San Jose State University (1999). Amit may be contacted at amit.dhir@xilinx.com.

Foreword

Studying the history of consumer electronics is nothing short of fascinating. The landscape is filled with countless stories of product successes and failures— fickle consumer adoptions, clever marketing campaigns that outsmart the best technologies, better packaging winning over better technology, and products that are simply ahead of their time.

This book was not written to trace the history of consumer electronics. Rather, it discusses the current state of the art of digital consumer devices. However, it is almost certain that what is considered leading edge today will eventually become another obsolete product in a landfill—a casualty of continued technological advances and progress. But make no mistake, although technological advances may render today's products obsolete, they are the lifeblood of the digital consumer revolution.

Pioneers and visionaries such as Boole, Nyquist, and Shannon well understood the benefits of digital technologies decades before digital circuits could be manufactured in a practical and cost-effective manner. Only through advances in semiconductor technology can the benefits of digital technology be realized. The role of semiconductor technology in driving digital technologies into consumer's hands is shown by looking at computer history.

The first computers were built of vacuum tubes and filled an entire room. They cost millions of dollars, required thousands of watts to power, and had mean time between failures measured in minutes. Today, that same level of computing power, using semiconductor technology, fits in the palm of your hand. It takes the form of a hand-held calculator that can operate from available light and, in many cases, is given away for free. Semiconductor advances enable products to be built that are significantly more powerful than their predecessors and sell at a fraction of the price.

But semiconductor technology is only the basic fabric upon which the digital consumer product is built. As you read this book, it's important to realize that there are many elements that factor into the success of a product. Building and delivering a successful consumer product requires the alignment of dynamics such as infrastructure, media, content, "killer" applications, technologies, government regulation/deregulation, quality, cost, consumer value, and great marketing.

Just as the availability of "killer" software applications was key to the growth of the personal computer, the Internet is a main driver behind the success of many of today's—and tomorrow's—digital consumer devices. The ability to exchange digital media (i.e., music, data, pictures, or video) between information appliances and personal computers has become a key component of today's consumer revolution. The success of today's digital consumer revolution is based on the infrastructure that was created by yesterday's consumer successes. This dynamic is sometimes referred to as the "virtuous cycle"—the logical antithesis of the vicious cycle.

In the case of the Internet, the virtuous cycle means that more bandwidth drives greater capability, which drives new applications, which drive an increased demand for more bandwidth. Mathematically, the numerator of the fraction—capabilities—continues to grow while the denominator—cost—continues to shrink as semiconductor technology advancements drive costs down. The result of this equation is a solution that tends towards infinity. This is why consumer digital electronics continues to grow at such an explosive rate.

However, there are key elements such as consumer acceptance and governmental regulation/deregulation that can negatively affect this explosive growth. A good example of government regulation is illustrated by personal digital audio recording.

Over 20 years ago Digital Audio Tape (DAT) was developed and introduced into the consumer market. DAT allowed consumers to make high-quality digital recordings in the home using 4mm DAT tape cartridges. The Recording Institute Association of America, RIAA, was opposed to this development. They claimed that this new format would enable rampant bootleg recordings that would cause irreversible financial damage to the music industry.

Perhaps the argument was somewhat melodramatic since no consumer recording technology up until that time had any real measurable impact on music sales. However, the impact of that argument stalled the widespread introduction of DAT for almost 10 years. The result is that DAT never realized anything near its market potential. Today, over 20 years later, CD RD/WR has become the premier consumer digital audio recording technology. And history continues to prove that this format has had no measurable effect on media sales.

Another phenomenon that has impacted product success is the delivery of consumer technology ahead of available content. This was the case with the introduction of color television. It was available years before there was any widespread availability of color television programming. The added cost of color television without color programming almost killed the product.

The most recent case of this phenomenon has been seen in the slow introduction of HDTV—high definition television. Here, the absence of high definition programming and the high cost of HDTV television sets have significantly impacted the adoption rate. HDTV is just now seeing some growth resulting from wide-range acceptance of the DVD and the gradual transition to HDTV programming format.

The term "digital consumer devices" covers a wide range of topics and products that make up today's consumer technology suite. Through Mr. Dhir's insights and expertise in this market, you are provided with a comprehensive review of an exciting and dynamic topic that most of us rely upon daily. This book will help you navigate the vast landscape of new consumer technologies and gain a better understanding of their market drivers.

Robert Bielby
Senior Director, Strategic Solutions Marketing, Xilinx, Inc.

Contents

Contents

Contents

Contents

Preface

Introduction

The consumer electronics market is flooded with new products. In fact, the number of new consumer devices that has been introduced has necessitated entirely new business models for the production and distribution of goods. It has also introduced major changes in the way we access information and interact with each other.

The digitization of technology—both products and media—has led to leaps in product development. It has enabled easier exchange of media, cheaper and more reliable products, and convenient services. It has not only improved existing products, it has also created whole new classes of products.

For example, improvements in devices such as DVRs (digital video recorders) have revolutionized the way we watch TV. Digital modems brought faster Internet access to the home. MP3 players completely changed portable digital music. Digital cameras enabled instant access to photographs that can be emailed within seconds of taking them. And eBooks allowed consumers to download entire books and magazines on their eBook tablets.

At the same time, digital technology is generating completely new breeds of products. For example, devices such as wireless phones, personal digital assistants (PDAs) and other pocketsize communicators are on the brink of a new era where mobile computing and telephony converge.

All this has been made possible through the availability of cheaper and more powerful semiconductor components. With advancements in semiconductor technology, components such as microprocessors, memories, and logic devices are cheaper, faster, and more powerful with every process migration. Similar advances in programmable logic devices have enabled manufacturers of new consumer electronics devices to use FPGAs and CPLDs to add continuous flexibility and maintain market leadership.

The Internet has brought about similar advantages to the consumer. Communication tasks such as the exchange of emails, Internet data, MP3 files, digital photographs, and streaming video are readily available today to everyone.

This book is a survey of the common thinking about digital consumer devices and their related applications. It provides a broad synthesis of information so that you can make informed decisions on the future of digital appliances.

Organization and Topics

With so many exciting new digital devices available, it is difficult to narrow the selections to a short list of cutting edge technologies. In this book I have included over twenty emerging product categories that typify current digital electronic trends. The chapters offer exclusive insight into digital consumer products that have enormous potential to enhance our lifestyles and workstyles. The text in each chapter follows a logical progression from a general overview of a device, to its market dynamics, to the core technologies that make up the product.

Chapter 1 is an introduction to the world of digital technologies. It outlines some of the key market drivers behind the success of digital consumer products.

Chapter 2 reviews the fundamentals of digitization of electronic devices. It presents an in-depth understanding of the market dynamics affecting the digital consumer electronics industry.

Digital TV and video are growing at very fast pace. *Chapter 3* deals extensively with these products and includes in-depth coverage on set-top boxes, digital video recorders and home-entertainment systems.

Chapter 4 discusses the emerging digital audio technologies such as MP3, DVD-Audio and Super Audio CD (SACD).

Chapter 5 explains next generation cellular technologies and how cell phones will evolve in the future.

The popularity of gaming consoles is growing because of the availability of higher processing power and a large number of games in recent years. *Chapter 6* overviews the main competitive products—Xbox, PlayStation 2 and GameCube.

The digital videodisc (DVD) player continues to be the fastest growing new product in consumer electronics history. *Chapter 7* describes how a DVD player works and provides an insight into the three application formats of DVD.

While consumer devices perform a single function very well, the PC remains the backbone for information and communication exchange. *Chapters 8 and 9* provide extensive coverage of PC and PC peripheral technologies.

Chapter 10 includes an overview of the various alternative display technologies including digital CRT, Flat panel displays, LCD's, PDP and DLP technologies.

Chapter 11 presents details on digital imaging—one of the fastest growing digital applications. It describes the market and the technology beneath digital cameras and camcorders.

Chapter 12 describes market, application, technology, and component trends in the web terminal and web pad products areas.

PDAs and other handheld products constitute one of the highest growth segments in the consumer market. *Chapter 13* describes the market dynamics and technology details of these devices.

Chapter 14 focuses on the developments in screenphone and videophone products.

Telematics technology has made a broad change to the facilities available to us while we travel. *Chapter 15* discusses the market dynamics and challenges of the evolving telematics industry.

Chapter 16 provides details on the emerging category of eBook products which promise to change the way we access, store and exchange documents.

Changes occurring in consumer devices such as VCRs, pagers, wireless email devices, email terminals, and NetTVs are discussed in *Chapter 17*.

Chapter 18 covers key technologies that continue to change the consumer landscape such as set-top boxes which distribute voice, video, and data from the high-speed Internet to other consumer devices.

Chapter 19 provides a synopsis of the book and highlights some of its key trends and takeaways.

Who Should Read this Book

This book is directed to anyone seeking a broad-based familiarity with the issues of digital consumer devices. It assumes you have some degree of knowledge of systems, general networking and Internet concepts. It provides details of the consumer market and will assist you in making design decisions that encompass the broad spectrum of home electronics. It will also help you design effective applications for the home networking and digital consumer devices industry.

Generation D—The Digital Decade

The Urge to Be Connected

Humans have an innate urge to stay connected. We like to share information and news to enhance our understanding of the world. Because of this, the Internet has enjoyed the fastest adoption rate of any communication medium. In a brief span of 5 years, more than 50 million people connected to the web. Today we have over 400 million people connected and sharing information. In fact, in 2003 we exchanged 6.9 trillion emails! Thanks to the Internet, we've become a "connected society."

The Internet has created a revolution in the way we communicate, share information, and perform business. It is an excellent medium for exchanging content such as online shopping information, MP3 files, digitized photographs, and video-on-demand (VoD). And its effect on business has been substantial. For example, a banking transaction in person costs a bank $1.07 to process, while the same transaction costs just a cent over the Internet. This translates into reduced time, lower cost, lower investment in labor, and increased conveniences to consumers.

The dynamics of the networking industry are also playing into homes. We are demanding faster Internet access, networked appliances, and improved entertainment. Hence, there is an invasion of smart appliances such as set-top boxes, MP3 (MPEG type III) players, and digital TVs (televisions) into today's homes.

The digitization of data, voice, video, and communications is driving a consumer product convergence, and this convergence has given birth to the need for networked appliances. The ability to distribute data, voice, and video within the home has brought new excitement to the consumer industry.

The Internet, combined with in-home networking and "smart," connected appliances, has resulted in Generation D—the Digital Decade. Our personal and business lives have come to depend on this digital revolution.

Our Daily Lives

Ten years ago cell phones were a luxury and personal digital assistants (PDAs) and web pads were unheard of. Now our lives are filled with digital consumer devices.

Remember the post-it notes we left everywhere to remind us of important tasks? They've been replaced by the PDA which tracks all our activities. The digital camera helps us save pictures in digital format on our PCs and on CDs (compact disks). And we use the Internet to share these pictures with friends and family around the world. Our music CDs are being replaced with MP3 audio files that we can store, exchange with friends, and play in a variety of places.

Thanks to our cell phones, the Internet is constantly available to us. Renting cars and finding directions are easy with the use of auto PCs now embedded in most cars. No more looking at maps and asking directions to make it from the airport to the hotel. And we no longer need to stand in line at the bank. Through Internet banking from our PCs, we can complete our financial transactions with the click of a few buttons.

While email was one step in revolutionizing our lives, applications such as Yahoo! Messenger, MSN Messenger, and AOL Instant Messenger provide the ability to communicate real-time around the world. Applications such as Microsoft NetMeeting allow us to see, hear, and write to each other in real time, thus saving us substantial travel money. Digital consumer devices such as notebook PCs, DVDs (digital versatile disk), digital TVs, gaming consoles, digital cameras, PDAs, telematics devices, digital modems, and MP3 players have become a critical part of our everyday lives.

So what has enabled us to have this lifestyle? It is a mix of:

- Digital consumer devices
- Broadband networking technologies
- Access technologies
- Software applications
- Communication services.

These items provide us with the communication, convenience, and entertainment that we've come to expect.

The Digitization of Consumer Products

In the first 50 years of the 20th century, the mechanical calculator was developed to process digital information faster. It was supercharged when the first electronic computers were born. In the scant 50 years that followed we discovered the transistor, developed integrated circuits, and then learned how to build them faster and cheaper.

With the advent of commodity electronics in the 1980s, digital utility began to proliferate beyond the world of science and industry. Mass digital emergence was at hand and the consumer landscape began a transition from an industrial-based analog economy to an information-based digital economy. Thus was born the PC, the video game machine, the audio CD, graphical computing, the Internet, digital TV, and the first generation of dozens of new digital consumer electronic products.

Today digital data permeates virtually every aspect of our lives. Our music is encoded in MP3 format and stored on our hard drives. TV signals are beamed from satellites or wired through cable into our homes in MPEG (Moving Pictures Experts Group) streams. The change from an analog CRT (cathode-ray tube) TV to digital TV provides us with better, clearer content through high-definition TV. The move from photographs to digital pictures is making the sharing and storage of image content much easier. The move from analog to digital cellular phones provides clearer, more reliable calls. And it enables features such as call waiting, call forwarding, global roaming, gaming, video communications, and Internet access from our cellular phones. We monitor news and investments over the Internet. We navigate the planet with GPS (global positioning satellites) guided graphical moving maps. We record pictures and movies in Compact FLASH. We buy goods, coordinate our schedules, and send and receive email from the palms of our hands. All these applications have been enabled by the move from analog to digital formats.

Why Digitization?

What's driving this digital firestorm? There are three answers to this question: utility, quality, and affordability.

By its nature digital data can be replicated exactly and distributed easily, which gives it much more utility. (With analog formatted data, every subsequent generation deteriorates.) For example, through digital technology you can quickly and easily send a copy of a picture you've just taken to all your friends.

Digital content has several quality advantages as well. It exhibits less noise as illustrated by the lack of background hiss in audio content. It also enables superior image manipulation as seen in shadow enhancement, sharpness control, and color manipulations in medical imaging.

Economically, digital data manipulation is much less expensive since it is able to leverage semiconductor economies of scale. It becomes better and more affordable every year. For example, an uncompressed, photographic quality digital picture requires less than 20 cents of hard drive space to store.

The digital revolution is growing rapidly. The digitization of consumer products has led not only to the improvement of existing products, but to the creation of whole new classes of products such as PVRs (personal video recorders), digital modems, and MP3 players.

Digital products are synonymous with quality, accuracy, reliability, speed, power, and low cost. Simply stated, anything that is digital is superior.

Converging Media

Communication is not about data or voice alone. It includes voice, data, and video. Since digital media offers superior storage, transport and replication qualities, the benefits of analog to digital migration are driving a demand for digital convergence. The different media types are converging and benefiting from the advantages of digital media. We find a demand for its capabilities in an ever-expanding range of devices and applications:

- Digital data (the world wide web)
- Digital audio (CDs, MP3)
- Digital video (DVD, satellite TV)
- HDTV (high definition TV)
- Digital communications (the Internet, Ethernet, cellular networks, wireless LANs, etc.)

For example, the circuit-switched network was the primary provider of voice communication. However, the far cheaper data network (Internet backbone) accommodates voice, data and video communication over a single network. Digital convergence has allowed availability of all these media types through a host of consumer devices.

Faster and Cheaper Components

The availability of faster and cheaper components is fueling the digital revolution. Everyday consumer appliances are now embedded with incredible computing power. This is due to Moore's law, which predicts that the number of transistors on a microprocessor will approximately double every 12 to 18 months. Hence, every generation of new products yields higher computing power at lower costs.

Moore's law, coupled with process migration, enables certain component types to provide more gates, more features, and higher performance at lower costs:

- Memory
- FPGAs (field programmable gate arrays)
- ASICs (application specific integrated circuits)

- ASSPs (application specific standard products)
- Microprocessors

The microprocessor, for example, has seen no boundaries and continues to expand beyond speeds of 3 GHz. While it remains to be seen what computing power is truly needed by the average consumer, this development has enabled the embedding of smart processors into every electronic device—from a toaster to an oven to a web tablet. It has allowed software developers to provide the highest quality in communication, entertainment, and information. And it enables programmable logic devices such as FPGAs and CPLDs, memory and hard disk drives to offer more power at lower costs.

Broadband Access—The Fat Internet Pipe

What started as a healthy communication vehicle among students, researchers, and university professors is today the most pervasive communications technology available. The Internet has positively affected millions of people around the world in areas such as communication, finance, engineering, operations, business, and daily living. Perhaps the Internet's best-known feature is the World Wide Web (the Web), which provides unlimited resources for text, graphics, video, audio, and animation applications.

Unfortunately, many people who access the Web experience the frustration of endless waiting for information to download to their computers. The need for fast access to the Internet is pushing the demand for broadband access. We are moving from phone line dial-up to new and improved connectivity platforms. Demand for high-speed Internet access solutions has fueled the proliferation of a number of rival technologies that are competing with each other to deliver high-speed connections to the home. The bulk of these connections are delivered by technologies based on delivery platforms such as:

- Cable
- DSL
- ISDN
- Wireless local loop
- Ethernet
- Fiber
- Satellite or Powerline

Telecommunication, satellite, and cable companies are looking for ways to enhance their revenue sources by providing high-speed Internet access. While the average phone call lasts three minutes, an Internet transaction using an analog modem averages over three hours and keeps the circuits busy. This increased traffic requires more circuits, but comes with no incremental revenue gain for the service provider.

Phone companies had to look for new techniques for Internet access. One such technology is DSL, which offers several advantages. While phone lines are limited to speeds up to 56 Kb/s (kilobits per seconds) that many find inadequate, broadband communications offer much higher transfer rates. And broadband connections are always on, which means users don't have to go through a slow log-on process each time they access the Internet.

Home Networking

Many of us don't realize that there are already several computers in our homes—we just don't call them computers. There are microchips in refrigerators, microwaves, TVs, VCRs (videocassette recorders), and stereo systems. With all of these processors, the concept of *home networking* has emerged. Home networking is the distribution of Internet, audio, video, and data between different appliances in the home. It enables communications, control, entertainment, and information between consumer devices.

The digitization of the different smart appliances is pushing the need for the Internet which, in turn, is fueling the demand for networked appliances. Several existing networking technologies such as phonelines, Ethernet, and wireless are currently used to network home appliances. Home networking will enable communications, control, entertainment, and information exchange between consumer devices.

The Day of the Digital Consumer Device

While many believe that the day of the PC is over, this is not the case. The PC will survive due to its legacy presence and to improvements in PC technology. The PC of the future will have increased hard disk space to store audio, image, data and video files. It will provide access to the Internet and control the home network. It will be a single device that can provide multiple services as opposed to many recently introduced appliances that provide one unique benefit. In general, hardware and software platforms will become more reliable.

The digitization of the consumer industry is leading to convergence and allowing consumers to enjoy the benefits of the Internet. The digital decade introduces three types of product usage that shape the market for new electronics devices—wireless, home networking and entertainment. Wireless products will be smart products that enable wireless, on-the-go communications at all times for voice, video and data. Home networking products such as gateways will extend PC computing power to all smart devices in the home. In the entertainment area, digital technology is enabling tremendous strides in compression and storage techniques. This will enable us to download, store, and playback virtually any size content.

Digital technologies will continue to enable wireless convenience through consumer products and provide better communication, information, and entertainment.

Bringing it All Together

This book examines the different consumer devices that are emerging to populate our homes. It previews the digital home of the future and describes the market and technology dynamics of appliances such as digital TVs, audio players, cellular phones, DVD players, PCs, digital cameras, web terminals, screen phones, and eBooks.

This book also provides information, including block diagrams, on designing digital consumer devices. And it explains how solutions based on programmable logic can provide key advantages to the consumer in flexibility, performance, power, and cost.

CHAPTER **2**

Digital Consumer Devices

Introduction

The average home is being invaded—by faster and smarter appliances. Traditional appliances such as toasters, dishwashers, TVs, and automobiles are gaining intelligence through embedded processors. Our cars have had them for years. Now, they appear in our children's toys, our kitchen appliances, and our living room entertainment systems. And their computing power rivals that of previous generation desktop computers.

In addition, drastic price reductions are bringing multiple PCs into our homes. This has created the need for high-speed, always-on Internet access in the home. The PC/Internet combination not only serves as a simple communication tool, it also enables vital applications like travel, education, medicine, and finance. This combination also demands resource sharing. Data, PC peripherals (i.e., printers, scanners) and the Internet access medium require device networking to interact effectively. These requirements have helped spawn a new industry called *home networking*.

This chapter provides a peek into the consumer home and shows how the landscape of digital consumer devices is evolving. It describes the types of digital consumer devices that are emerging and how they interact with each other.

The Era of Digital Consumer Devices

The home product industry is buzzing with new digital consumer devices. They are also called information appliances, Internet appliances, intelligent appliances, or iAppliances. These smart products can be located in the home or carried on the person. They are made intelligent by embedded semiconductors and their connection to the Internet. They are lightweight and reliable and provide special-purpose access to the Internet. Their advanced computational capabilities add more value and convenience when they are networked.

Although the PC has long dominated the home, it has been error-prone, which has hindered its penetration into less educated homes. This compares with the television or the telephone, which have seen much wider adoption not only among the educated, but also the uneducated and lower-income households. This has led to the introduction of digital consumer devices. They are instant-on devices that are more stable and cheaper than PCs when developed for a specific application. They can also be more physically appropriate than a large, bulky PC for certain applications.

Since they are focused on functionality, digital consumer devices generally perform better than a multi-purpose PC for single, specific functions. Web surfing is the most popular application, with other applications such as streaming media and online gaming growing in popularity. These devices are easier to use than PCs, simple to operate, and require little or no local storage.

Digital consumer devices encompass hardware, software, and services. They are designed specifically to enable the management, exchange, and manipulation of information. They are enabled by technologies such as:

- Internet convergence and integration
- Wired and wireless communications
- Software applications
- Semiconductors technology
- Personal computing
- Networking
- Consumer electronics

Digital consumer devices also enable a broader range of infotainment (information and entertainment) by providing for services such as accessing email, checking directions when on the road, managing appointments, and playing video games.

Examples of key digital consumer devices

The following list includes some of the more common digital consumer devices.

- Audio players – MP3 (Internet audio), CD, SACD, DVD-Audio player
- Digital cameras and camcorders
- Digital displays (PDP, LCD, TFT)
- Digital TVs (SDTV, HDTV)
- Digital VCRs or DVRs (digital video recorder) or personal video recorders (PVR)
- DVD players
- EBooks
- Email terminals
- Energy management units, automated meter reading (AMR) and RF metering
- Gaming consoles
- Internet-enabled picture frames
- Mobile phones
- NetTVs or iTV-enabled devices
- PDAs and smart handheld devices
- Robot animals
- Screen phones or video phones
- Security units
- Set-top boxes
- Telematic or auto PCs
- Web pads, fridge pads, Web tablets
- Web terminals
- White Goods (dish washer, dryer, washing machine)

While PCs and notebook PCs are not traditionally considered digital consumer devices, they have one of the highest penetration rates of any consumer product. Hence, this book will cover digital consumer devices as well as PCs and PC peripherals.

Market Forecast

The worldwide acceptance and shipment of digital consumer devices is growing rapidly. While market forecasts vary in numbers, market data from several research firms show exploding growth in this market.

Market researcher IDC (International Data Corporation) reports that in 2003, 18.5 million digital consumer device units shipped compared with 15.7 million units of home PCs. Dataquest-GartnerGroup reports that the worldwide production of digital consumer devices has exploded from 1.8 million units in 1999 to 391 million units in 2003. Complementing the unit shipment, the worldwide revenue forecast for digital consumer devices was predicted to grow from $497 million in 1999 to $91 billion in 2003.

The digital consumer device segment is still emerging but represents exponential growth potential as more devices include embedded intelligence and are connected to each other wirelessly. The digital consumer device space includes:

- Internet-enabling set-top boxes
- Screen phones
- Internet gaming consoles
- Consumer Internet clients
- Digital cameras
- Smart handheld devices (personal and PC companions, PDAs, and vertical industry devices)
- High-end and low-end smart phones
- Alphanumeric pagers

Bear Sterns believes that the market for these products will grow to 475 million by 2004 from roughly 88 million in 2000. The market for digital consumer devices will grow from 100 million-plus users and $460 billion in the mid-1990s to over 1 billion users and $3 trillion by 2005.

The high-tech market research firm Cahners In-Stat finds that the digital consumer devices market will heat up over the next several years with sales growing over 40% per year between 2000 and 2004. Predicted worldwide unit shipments for some key digital consumer devices are:

- DVD players growing to over 92 million units by 2005.
- Digital cameras growing to 41 million units and digital camcorders growing to 20 million units by 2005.
- Digital TV growing to 10 million units by 2005.
- Smart handheld devices growing to 43 million units by 2004.
- Gaming consoles reaching 38 million units by 2005.
- DVRs growing to 18 million units by 2004.
- Internet audio players growing to 17 million units by 2004.
- NetTVs growing to 16 million units by 2004.
- Email terminals, web terminals and screen phones reaching 12 million units by 2004.
- PDAs growing to 7.8 million units by 2004.
- Digital set-top boxes growing to over 61 million units by 2005.
- Mobile handsets reaching 450 million units by 2005.

The Consumer Electronics Association (CEA) reports that the sales of consumer electronics goods from manufacturers to dealers surpassed $100 billion in 2002 and was expected to top $105 billion in 2003. This will set new annual sales records and represent the eleventh consecutive year of growth for the industry. The spectacular growth in sales of consumer electronics is due in large part to the wide variety of products made possible by digital technology. One of the categories most affected by the digital revolution has been the in-home appliance, home information, and mobile electronics product categories. CEA predicts strong growth in the video appliance category in 2004, with products such as digital televisions, camcorders, set-top boxes, personal video recorders, and DVD players leading the charge. By 2005, worldwide home-network installations are expected to exceed 40 million households and the total revenues for such installations are expected to exceed $4.75 billion.

Certainly the digital consumer device is seeing significant market penetration. But a critical issue plaguing the market is the widespread perception that digital consumer devices are focused on failed products such as Netpliance i-opener and 3Com Audrey. However, the market is not considering the success of handheld (PDA) or TV-based devices (set-top boxes, digital TVs, etc.). Internet handheld devices, TV-based devices and Internet gaming consoles are the most popular consumer devices based on both unit shipments and revenues generated.

Market Drivers

The most influential demand driver for digital consumer devices is infotainment acceleration. This is the ability to access, manage, and manipulate actionable information at any time and any place.

Most people feel that the PC takes too long to turn on and establish an Internet connection. Users want instant-on devices with always-on Internet access. Moreover, people want the ability to take their information with them. For example, the Walkman allows us to take our music anywhere we go. The pager makes us reachable at all times. And the mobile phone enables two-way communication from anywhere Interconnected, intelligent devices yield a massive acceleration in communication and data sharing. They also drastically increase our productivity and personal convenience.

Digital consumer devices are only now coming into their own because the *cost* of information must be reasonable compared to the *value* of that information. Technology is finally tipping the balance. Cost can be measured in terms of ease-of-use, price, and reliability. Value translates into timeliness of information delivery, usefulness, and flexibility. The convergence of several technologies enables vendors to introduce appliances that are simple-on-the-outside and complex-on-the-inside. Buyers can now obtain value at an affordable price.

The Integrated Internet

The development of the Internet has been key to the increases in information acceleration. This is best illustrated by the exponential growth in Internet traffic. The ongoing development of Internet technologies is enabling the connection of every intelligent device, thus speeding the flow of information for the individual. This linkage is also making possible the introduction of Internet-based services. For example, consumers are realizing the benefit of integrated online travel planning and management services. They can make real-time changes based on unforeseen circumstances and send that information automatically to service providers over the Internet. If a flight is early, this information is passed from the airplane's onboard tracking system to the airline's scheduling system. From there, it is sent to the passenger's car service which automatically diverts a driver to the airport 20 minutes early to make the pickup.

Today, many find it difficult to imagine life without the Internet. Soon, it will be hard to imagine life without seamless access to information and services that doesn't require us to be "on-line" for Internet access.

Semiconductors

Advances in semiconductor technology have produced low-power silicon devices that are small, powerful, flexible, and cheap. Such innovation has created advances such as system-on-a-chip (SoC) technology. Here, multiple functions such as micro-processing, logic, signal processing, and memory are integrated on a single chip. In the past, each of these functions was handled by a discrete chip.

Smaller chips that execute all required functions have several advantages. They allow smaller product form factors, use less power, reduce costs for manufacturers, and offer multiple features. In addition, many can be used for multiple applications and products depending on the software placed on top. There are several companies specifically designing chips for digital consumer devices including Xilinx, Motorola, ARM, MIPS, Hitachi, Transmeta, and Intel. This trend in semiconductor technology allows new generations of consumer products to be brought to market much faster while lowering costs. With process technologies at 90-nm (0.09 micron) and the move to 12-inch (300-mm) wafers, the cost of the devices continues to see drastic reduction. In addition, integration of specific features allows processors to consume less power and enhance battery life. This provides longer operation time in applications such as laptop PCs, web tablets, PDAs, and mobile phones.

Software and Middleware

New software architectures like Java and applications like the wireless application protocol (WAP) micro-browser are proving key to enabling digital consumer devices. For example, the Java platform and its applications allow disparate devices using different operating systems to connect over-wire and wireless networks. They enable the development of compact applications or applets for use on large (e.g., enterprise server) or small (e.g., smart phone) devices. These applets can also be used within Web pages that support personal information management (PIM) functions such as e-mail and calendars. Other middleware technologies such as Jini, UPnP, HAVi, and OSGi also contribute to increased production of digital consumer devices. And the move toward software standards in communications (3G wireless, Bluetooth, wireless LANs) and copyrights makes it easier to produce and distribute digital devices.

Storage

The demand for low-cost storage is another key market driver for digital consumer devices. Here, the declining cost per megabyte (MB) lowers the cost of delivering a given product. The lower cost results in new storage products for both hard drives and solid state memory such as DRAM, SRAM, and flash memory. Also hard drive technology continues to embrace lower costs and this is fueling new products such as set-top boxes and MP3 players.

Communications

Wired and wireless communications are gradually becoming standardized and less expensive. They are driven by market deregulation, acceptance of industry standards, and development of new technologies. They provide advantages for rapid, convenient communication. For example, the advent of 2.5G and 3G wireless will enable greater transfer rates. And new wireless technologies such as IEEE 802.11b/a/g and HiperLAN2 could be used to connect public areas (hotels and airports), enterprises, and homes. These LAN technologies evolved from enterprise data networks, so their focus has been on increasing data rates since over two-thirds of network traffic is data. To that end, newer wireless LAN technologies are designed to provide higher quality of service.

Phases of Market Acceptance

While huge growth is predicted for digital consumer devices, the acceptance of these products will be in phases.

In the pre-market acceptance phase there is:

- Hype and debate
- Few actual products
- Limited business models
- Limited Web pervasiveness
- Less technology ammunition

The acceptance of these products will lead to:

- Rise of the digital home
- Broader range of products
- Diversification of business models
- Heightened industry support
- Increased bandwidth for home networking and wireless
- Maturation of the Internet base

Success Factors and Challenges

Some of the factors contributing to the success of digital consumer devices are:

- *Services* are key to the business model. The success of some products such as web tablets requires the development of thorough business models to provide real value. The key is to offer low-cost solutions backed by partnerships and sustainable services.
- Product design must achieve *elegance.* Consumers often buy products that are functionally appealing and possess design elegance. For example, thinner PDAs will achieve higher success than bigger ones.
- *Branding and channels* are important because customers hesitate to buy products that do not have an established name and supply channel.
- *Industry standards* must be adopted in each area to enable adjacent industries to develop products. For example, wired and wireless data communications are becoming standardized and less expensive. They are driven by market deregulation, acceptance of industry standards, and development of new broadband and networking technologies. Critical technologies such as broadband (high-speed Internet) access, wireless, and home networking must hit their strides.
- Industry and venture capital *investment* must be heightened.
- New product concepts must gain significant *consumer awareness.* (Consumers who do not understand the value and function of a tablet PC, for example, will not be encouraged to buy one.)
- The *Internet* must be developed as a medium for information access, dissemination, and manipulation.
- Advances in *semiconductor* technology that enabled development of inexpensive SoC technologies must continue.
- Development of *software platforms* such as Java must continue.
- The functionality of multiple devices must be *bundled.* Multiple-function products can be very compelling. (Caveat: Bundling functionality is no guarantee of market acceptance, even if the purchase price is low. Because there is no formula for what users want in terms of function, form factor, and price, an opportunity exists for new product types in the market.)

Some of the major outstanding issues affecting the growth of the market are:

- **Single-use versus Multi-use** – The three design philosophies for digital consumer device are:
  - Single-purpose device, like a watch that just tells time.
  - Moderately integrated device such as pagers with email, phones with Internet access, and DVD players that play multiple audio formats.
  - All-purpose device such as the PC or the home entertainment gateway.

  Note that there is no clear way to assess each category before the fact. One must see the product, the needs served, and the customer reaction.
- **Content Integration** – How closely tied is the product solution to the desired data? Can users get the data they want, when they want it, and in the form they want it? Can they act on it in a timely fashion? Digital consumer devices allow this, but not without tight integration with the information sources.
- **Replacement or Incremental** – Are the products incremental or do they replace existing products? Most products represent incremental opportunities similar to the impact of the fractional horsepower engine. Some legacy product categories provide benefits over new categories like the speed of a landline connection and clarity of full-screen documents. Also, how many devices will users carry? Will users carry a pager, a cell phone, a wireless e-mail device, a PDA, a Walkman tape, *and* a portable DVD player?
- **How Much Complexity** – There is no rule on how much complexity to put into a device. But there is a tradeoff, given power and processing speeds, between ease-of-use and power. While most devices today have niche audiences, ease-of-use is key to mass audience acceptance.
- **Access Speeds, Reliability, and Cost** – Issues facing Internet access via mobile digital consumer devices include:
  - Establishing standards
  - Developing successful pricing models
  - Improving reliability and user interfaces
  - Increasing geographical coverage.
- **Business Model** – What is the right business model for a digital consumer devices company? A number of business models such as service subscription, on-demand subscription, technology licensing, and up-front product purchase can be successful. Investors should focus on whether a company is enabling a product solution and whether the company is unlocking that value.
- **Industry Standards** – Industry standards have not yet been determined in many areas such as software platform, communication protocol, and media format. This increases the risk for vendors investing in new products based on emerging technologies. And it raises the risk for users that the product they purchase may lose support as in the case of the DIVX DVD standard. It also raises incompatibility issues across geographies. In most areas, however, vendors are finding ways to manage ompatibility problems and standards bodies are gaining momentum.

Functional Requirements

Some of the key requirements of digital consumer devices to gain market acceptance include:

- **Ubiquity** – The prevalence of network access points.
- **Reliability** – Operational consistency in the face of environmental fluctuation such as noise interference and multi-path.

- **Cost** – Affordable for the mass market.
- **Speed** – Support of high-speed distribution of media rich content (>10 Mb/s)
- **Mobility/portability** – Support of untethered devices. The ability to access and manipulate actionable information at any time and in any place is a must.
- **QoS** (Quality of Service) – Scalable QoS levels for application requirements of individual devices.
- **Security** – User authentication, encryption, and remote access protection.
- **Remote management** – Enabled for external network management (queries, configuration, upgrades).
- **Ease-of-use** – Operational complexity must be similar to existing technologies such as TVs and telephones.

What About the Personal Computer?

Technology, not economics, is the Achilles' heel of the PC. Until the arrival of the digital consumer device, access to the Web and e-mail was the exclusive domain of the PC. Digital consumer devices are clearly being marketed now as an alternative to the PC to provide network services. Previously, volume economics gave the PC an edge. But it has weaknesses such as:

- Long boot-up time
- Long Internet connection time
- Inability to be instant-on
- Not always available
- Not easy to use
- Not truly portable
- Failure-prone
- Expensive
- Complicated software installations

This has created an opportunity for a new wave of consumer products. Users want instant-on appliances that leverage their connection to the Internet and to other devices.

Digital consumer devices provide Internet access and they are low cost, consumer focused, and easy to use. The acceptance of these appliances is causing PC manufacturers to reduce prices. However, the PC will not go away because of the infrastructure surrounding it and its productivity. We have entered the era of *PC-plus* where the PC's underlying digital technology extends into new areas. For example, the notebook PC continues to survive because of its valuable enhancements and its plummeting prices. It is the ideal home gateway given its capabilities and features such as storage space and processor speeds.

Digital Home

With the invasion of so many digital consumer devices and multiple PCs into the home, there is a big push to network all the devices. Data and peripherals must be shared. The growing ubiquity of the Internet is raising consumer demand for home Internet access. While many digital home devices provide broadband access and home networking, consumers are not willing to tolerate multiple broadband access technologies because of cost and inconvenience. Rather, consumers want a single broadband technology that networks their appliances for file sharing and Internet access.

Home networking is not new to the home. Over the past few years, appliances have been networked into "islands of technologies." For example, the PC island includes multiple PCs, printer, scanner, and PDA. It networks through USB 1.1, parallel, and serial interfaces. The multimedia island consists of multiple TVs, VCR players, DVD players, receivers/amplifiers, and speakers. This island has been connected using IEEE 1394 cables. Home networking technologies connect the different devices and the networked islands.

The digital home consists of:

- **Broadband Access** – Cable, xDSL, satellite, fixed wireless (IEEE 802.16), powerline, Fiber-to-the-home (FTTH), ISDN, Long Reach Ethernet (LRE), T1
- **Residential Gateways** – Set-top boxes, gaming consoles, digital modems, PCs, entertainment gateways
- **Home Networking Technologies** –
  - No new wires – phonelines, powerlines
  - New wires – IEEE 1394, Ethernet, USB 2.0, SCSI, optic fiber, IEEE 1355
  - Wireless – HomeRF, Bluetooth, DECT, IEEE 802.11b/a/g, HiperLAN2, IrDA
- **Middleware Technologies** – OSGi, Jini, UPnP, VESA, HAVi
- **Digital Consumer Devices** – Digital TV, mobile phones, PDAs, web pads, set-top box, digital VCRs, gaming consoles, screen phones, auto PCs/telematics, home automation, home security, NetTV, PCs, PC peripherals, digital cameras, digital camcorders, audio players (MP3, other Internet, CD, DVD-Audio, SACD), email terminals, etc.

King of All—The Single All-Encompassing Consumer Device

Will there ever be a single device that provides everything we need? Or will there continue to be disparate devices providing different functionality? As ideal as a single-device solution would be, the feasibility is remote because of these issues:

- **Picking the Best in Class** – Users often want to have more features added to a particular device. But there is a tendency to prefer the best-in-class of each category to an all-in-one device because of the tradeoffs involved. For example, a phone that does everything gives up something, either in battery life or size.
- **Social Issues** – How do consumers use devices? Even though there is a clock function in a PDA, a wireless e-mail device, and a phone, we still use the wristwatch to find the time. This is due to information acceleration and the fact that we do not want to deal with the tradeoffs of a single device. But we do want all devices to communicate with each other and have Internet protocol (IP) addresses so that we see the same times on all the clock sources.
- **Size versus Location** – There is a strong correlation between the size of a device and the information it provides. For example, we don't want to watch a movie on our watch, or learn the time from our big-screen TV. Also, we would prefer a wired connection for speed, but we will access stock quotes on a PDA when there is no alternative.
- **Moore's Law Implications** – We never really get the device we want because once we get it, they come out with a better, faster, and sleeker one. For example, when we finally got the modem of all modems—at 14.4K—they introduced the 28.8K version. The key issue is not the device, but the continuation of Moore's and Metcalfe's Laws.

However, having the king of all devices—also known as the *gateway*—does provide a benefit. It introduces broadband access and distributes it to multiple devices via home networking. It is also has a unique platform which can provide massive storage for access by these devices. And it provides features such as security at the source of broadband access.

Summary

The era of digital convergence is upon us. From pictures to e-mail, from music to news, the world has gone digital. This digital explosion and media convergence has given birth to several digital consumer devices which provide communication, connectivity and conveniences. The landscape of tomorrow's home will look very different as consumer devices grow into smarter appliances that are networked to increase productivity and provide further conveniences. Market researchers predict that digital consumer devices will out-ship consumer PCs by 2005 in the U.S. While there are several new-product issues that need to be addressed, the high-volume digital consumer devices will be PDAs, set-top boxes, digital TVs, gaming consoles, DVD players, and digital cameras.

Digital Television and Video

Introduction

Digital television (TV) offers a way for every home, school, and business to join the Information Society. It is the most significant advancement in television technology since the medium was created almost 120 years ago. Digital TV offers more choices and makes the viewing experience more interactive.

The analog system of broadcasting television has been in place for well over 60 years. During this period viewers saw the transition from black and white to color TV technology. This migration required viewers to purchase new TV sets, and TV stations had to acquire new broadcast equipment. Today the industry is again going through a profound transition as it migrates from conventional TV to digital technology. TV operators are upgrading their existing networks and deploying advanced digital platforms.

While the old analog-based system has served the global community very well, it has reached its limits. Picture quality is as good as it can get. The conventional system cannot accommodate new data services. And, an analog signal is subject to degradation and interference from things such as low-flying airplanes and household electrical appliances.

When a digital television signal is broadcast, images and sound are transmitted using the same code found in computers—ones and zeros. This provides several benefits:

- Cinema-quality pictures
- CD-quality sound
- More available channels
- The ability to switch camera angles
- Improved access to new entertainment services

Many of the flaws present in analog systems are absent from the digital environment. For example, both analog and digital signals get weaker with distance. While a conventional TV picture slowly degrades for viewers living a long distance from the broadcaster, a digital TV picture stays perfect until the signal becomes too weak to receive.

Service providers can deliver more information on a digital system than on an analog system. For example, a digital TV movie occupies just 2% of the bandwidth normally required by an analog system. The remaining bandwidth can be filled with programming or data services such as:

- Video on demand (VoD)
- Email and Internet services
- Interactive education
- Interactive TV commerce

Eventually, all analog systems will be replaced by digital TV. But the move will be gradual to allow service providers to upgrade their transmission networks and manufacturers to mass-produce sufficient digital products.

History of Television

Television started in 1884 when Paul Gottlieb patented the first mechanical television system. It worked by illuminating an image via a lens and a rotating disc (Nipkow disc). Square apertures were cut out of the disc which traced out lines of the image until the full image had been scanned. The more apertures there were, the more lines were traced, producing greater detail.

In 1923 Vladimir Kosma Zworykin replaced the Nipkow disc with an electronic component. This allowed the image to be split into more lines, which produced greater detail without increasing the number of scans per second. Images could also be stored between electronic scans. This system was named the *Iconoscope* and patented in 1925.

J.L. Baird demonstrated the first mechanical color television in 1928. It used a Nipkow disc with three spirals, one for each primary color—red, green and blue. However, few people had television sets at that time so the viewing experience was less than impressive. The small audience of viewers watched a blurry picture on a 2- or 3-inch screen.

In 1935 the first electronic television system was demonstrated by the Electric Musical Industries (EMI) company. By late 1939, sixteen companies were making or planning to make electronic television sets in the U.S.

In 1941, the National Television System Committee (NTSC) developed a set of guidelines for the transmission of electronic television. The Federal Communications Commission (FCC) adopted the new guidelines and TV broadcasts began in the United States. In subsequent years, television benefited from World War II in that much of the work done on radar was transferred directly to television set design. Advances in cathode ray tube technology are a good example of this.

The 1950s heralded the golden age of television. The era of black and white television commenced in 1956 and prices of TV sets gradually dropped. Towards the end of the decade, U.S. manufacturers were experimenting with a wide range of features and designs.

The sixties began with the Japanese adoption of the NTSC standards. Towards the end of the sixties Europe introduced two new television transmission standards:

- Systeme Electronique Couleur Avec Memoire (SECAM) is a television broadcast standard in France, the Middle East, and most of Eastern Europe. SECAM broadcasts 819 lines of resolution per second.
- Phase Alternating Line (PAL) is the dominant television standard in Europe. PAL delivers 625 lines at 50 half-frames per second.

The first color televisions with integrated digital signal processing technologies were marketed in 1983. In 1993 the Moving Picture Experts Group (MPEG) completed definition of MPEG-2 Video, MPEG-2 Audio, and MPEG-2 Systems. In 1993 the European Digital Video Broadcasting (DVB) project was born. In 1996 the FCC established digital television transmission standards in the United States by adopting the Advanced Television Systems Committee (ATSC) digital standard. By 1999 many communication mediums had transitioned to digital technology.

Components of a Digital TV System

Behind a simple digital TV is a series of powerful components that provide digital TV technology. These components include video processing, security and transmission networks.

A TV operator normally receives content from sources such as local video, cable, and satellite channels. The content is prepared for transmission by passing the signal through a digital broadcasting system. The following diagram illustrates this process.

Note that the components shown are logical units and do not necessarily correspond to the number of physical devices deployed in the total solution. The role of each component shown in Figure 3.1 is outlined in the following category descriptions.

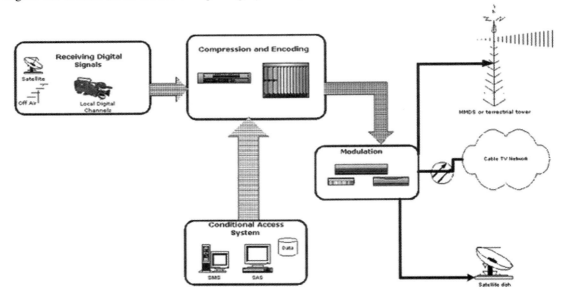

Figure 3.1: Basic Building Blocks of a Digital Broadcasting System

Compression and Encoding

The compression system delivers high quality video and audio using a small amount of network bandwidth by minimizing information storage capacity. This is particularly useful for service providers who want to 'squeeze' as many digital channels as possible into a digital stream.

A video compression system consists of encoders and multiplexers. Encoders digitize, compress and scramble a range of audio, video and data channels. They allow TV operators to broadcast several digital video programs over the same bandwidth that was formerly used to broadcast a single analog video program. Once the signal is encoded and compressed, an MPEG-2 data stream is transmitted to the multiplexer. The multiplexer combines the outputs from the various encoders with security and program information and data into a single digital stream.

Modulation

Once the digital signal has been processed by the multiplexer, the video, audio, and data are modulated (amalgamated) with the carrier signal. The unmodulated digital signal from the multiplexer has only two possible states—zero or one. By passing the signal through a modulation process, a number of states are added that increase the data transfer rate. The modulation technique used by TV operators depends on the geography of the franchise area and the overall network architecture.

Conditional access system

Broadcasters and TV operators interact with viewers on many levels, offering a greater program choice than ever before. The deployment of a security system, called conditional access, provides viewers with unprecedented control over what they watch and when. A conditional access system is best described as a gateway that allows viewers to access a virtual palette of digital services.

A conditional access system controls subscribers' access to digital TV pay services and secures the operator's revenue streams. Consequently, only customers who have a valid contract with the network operator can access a particular service. Using today's conditional access systems, network operators can directly target programming, advertisements, and promotions to subscribers. They can do this by geographical area, market segment, or viewers' personal preferences. Hence, the conditional access system is a vital aspect of the digital TV business.

Network Transmission Technologies

Digital transmissions are broadcast through one of three systems:
- Cable network
- TV aerial
- Satellite dish

Different service providers operate each system. If cable television is available in a particular area, viewers access digital TV from a network based on Hybrid fiber/coax (HFC) technology. This refers to any network configuration of fiber-optic and coaxial cable that redistributes digital TV services. Most cable television companies are already using it. HFC networks have many advantages for handling next-generation communication services—such as the ability to simultaneously transmit analog and digital services. This is extremely important for network operators who are introducing digital TV to their subscribers on a phased basis.

Additionally, HFC meets the expandable capacity and reliability requirements of a new digital TV system. The expandable capacity feature of HFC-based systems allows network operators to add services incrementally without major changes to the overall network infrastructure. HFC is essentially a 'pay as you go' architecture. It matches infrastructure investment with new revenue streams, operational savings, and reliability enhancements.
An end-to-end HFC network is illustrated in Figure 3.2.

HFC network architecture is comprised of fiber transmitters, optical nodes, fiber and coaxial cables, taps, amplifiers and distribution hubs. The digital TV signal is transmitted from the head-end in a star-like fashion to the fiber nodes using fiber-optic feeders. The fiber node, in turn, distributes the signals over coaxial cable, amplifiers and taps throughout the customer service area.

Customers who have aerials or antennas on their roofs are able to receive digital TV through one of the following wireless technologies:
- **Multi-channel multi-point distribution system (MMDS)** – MMDS is a relatively new service used to broadcast TV signals at microwave frequencies from a central point or head-end to small, rooftop antennas. An MMDS digital TV system consists of a head-end that receives signals from satellites, off-the-air TV stations, and local programming. At the head-

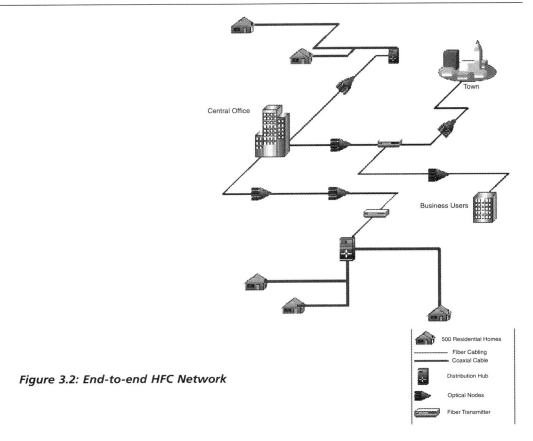

Figure 3.2: End-to-end HFC Network

end the signals are mixed with commercials and other inserts, encrypted, and broadcasted. The signals are then re-broadcast from low-powered base stations in a diameter of 35 miles from the subscriber's home. The receiving rooftop antennas are 18 to 36 inches wide and have a clear line of site to the transmitting station. A down converter, usually a part of the antenna, converts the microwave signals into standard cable channel frequencies. From the antenna the signal travels to a special set-top box where it is decrypted and passed into the television. Today there are MMDS-based digital TV systems in use all around the U.S. and in many other countries including Australia, South Africa, South America, Europe, and Canada.

■ **Local multi-point distribution system (LMDS)** – LMDS uses microwave frequencies in the 28 GHz frequency range to send and receive broadband signals which are suitable for transmitting video, voice and multimedia data. It is capable of delivering a plethora of Internet- and telephony-based services. The reception and processing of programming and other head-end functions are the same as with the MMDS system. The signals are then re-broadcast from low-powered base stations in a 4 to 6 mile radius of the subscribers' homes. Signals are received using a six-square-inch antenna mounted inside or outside of the home. As with MMDS, the signal travels to the set-top box where it is decrypted and formatted for display on the television.

- **Terrestrial** – Terrestrial communications, or DTT, can be used to broadcast a range of digital services. This system uses a rooftop aerial in the same way that most people receive television programs. A modern aerial should not need to be replaced to receive the DTT service. However, if the aerial is out of date, updating may be necessary. Additionally, DTT requires the purchase of a new digital set-top box to receive and decode the digital signal. DTT uses the Coded Orthogonal Frequency Division Multiplexing (COFDM) modulation scheme. COFDM makes the terrestrial signal immune to multi-path reflections. That is, the signal must be robust enough to traverse geographical areas that include mountains, trees, and large buildings.
- **Satellite** – Digital television is available through direct broadcast satellite (DBS) which can provide higher bandwidth than terrestrial, MMDS or LMDS transmission. This technology requires a new set-top box and a satellite dish because existing analog satellite dishes are unable to receive digital TV transmissions. The fact that satellite signals can be received anywhere makes service deployment as easy as installing a receiver dish and pointing it in the right direction. Satellite programmers broadcast, or uplink, signals to a satellite. The signals are often encrypted to prevent unauthorized reception. The are then sent to a set-top box which converts them to a TV-compatible format.

Digital TV Standards

The standard for broadcasting analog television in most of North America is NTSC. The standards for video in other parts of the world are PAL and SECAM. Note that NTSC, PAL and SECAM will all be replaced over the next ten years with a new suite of standards associated with digital television.

International organizations that contribute to standardizing digital television include:
- **Advanced Television Systems Committee (ATSC)**
- **Digital Video Broadcasting (DVB)**

The Advanced Television Systems Committee was formed to establish a set of technical standards for broadcasting television signals in the United States. ATSC digital TV standards include high-definition television, standard definition television, and satellite direct-to-home broadcasting. ATSC has been formally adopted in the United States where an aggressive implementation of digital TV has already begun. Additionally, Canada, South Korea, Taiwan, and Argentina have agreed to use the formats and transmission methods recommended by the group.

DVB is a consortium of about 300 companies in the fields of broadcasting, manufacturing, network operation and regulatory matters. They have established common international standards for the move from analog to digital broadcasting. DVB has produced a comprehensive list of standards and specifications that describe solutions for implementing digital television in areas such as transmission, interfacing, security, and interactivity for audio, video, and data.

Because DVB standards are open, all compliant manufacturers can guarantee that their digital TV equipment will work with other manufacturers' equipment. There are numerous broadcast services around the world using DVB standards and hundreds of manufacturers offering DVB-compliant equipment. While the DVB has had its greatest success in Europe, the standard also has implementations in North and South America, China, Africa, Asia, and Australia.

SDTV and HDTV Technologies

DTV is an umbrella term that refers to both high-definition television (HDTV) and standard definition television (SDTV). DTV can produce HDTV pictures with more image information than conventional analog signals. The effect rivals that of a movie theater for clarity and color purity. HDTV signals also contain multi-channel surround sound to complete the home theatre experience. Eventually, some broadcasters will transmit multiple channels of standard definition video in the same spectrum amount now used for analog broadcasting. SDTV is also a digital TV broadcast system that provides better pictures and richer colors than analog TV systems. Although both systems are based on digital technologies, HDTV generally offers nearly six times the sharpness of SDTV-based broadcasts.

Consumers who want to receive SDTV on their existing television sets must acquire a set-top box. These devices transform digital signals into a format suitable for viewing on an analog-based television. However, consumers who want to watch a program in HDTV format must purchase a new digital television set.

Commonly called wide-screen, HDTVs are nearly twice as wide as they are high. They present an expanded image without sacrificing quality. Standard analog televisions use squarer screens that are 3 units high and 4 units wide. For example, if the width of the screen is 20 inches, its height is about 15 inches. HDTV screens, however, are about one third wider than standard televisions, with screens that are 9 units high by 16 wide. Current analog TV screens contain approximately 300,000 pixels while HDTV televisions have up to 2 million pixels.

In addition to classifying digital TV signals into HDTV and SDTV, each has subtype classifications for *lines of resolution* and *progressive interlaced scanning*.

A television picture is made up of horizontal lines called lines of resolution, where resolution is the amount of detail contained in an image displayed on the screen. In general, more lines mean clearer and sharper pictures. Clearer pictures are available on new digital televisions because they can display 1080 lines of resolution versus 525 lines on an ordinary television.

Progressive scanning is when a television screen refreshes itself line-by-line. It is popular in HDTVs and computer monitors. Historically, early TV tubes could not draw a video display on the screen before the top of the display began to fade. Interlaced scanning overcomes this problem by the way it refreshes lines. It is popular in analog televisions and low-resolution computer monitors.

Digital Set-top Boxes

Most consumers want to take advantage of the crystal-clear sound and picture quality of DTV, but many cannot afford a new digital television set. To solve the problem, these viewers can use a set-top box that translates digital signals into a format displayable on their analog TVs. In many countries, service providers are retrofitting subscribers' analog set-top boxes with new digital set-top boxes. Additionally, some countries are pushing second-generation set-top boxes that support a range of new services.

A set-top box is a type of computer that translates signals into a format that can be viewed on a television screen. Similar to a VCR, it takes input from an antenna or cable service and outputs it to a television set. The local cable, terrestrial, or satellite service provider normally installs set-top boxes.

The digital TV market can be broadly classified into these categories:

■ **Analog set-top boxes** – Analog set-top boxes receive, tune and de-scramble incoming television signals.

■ **Dial-up set-top boxes** – These set-top boxes allow subscribers to access the Internet through the television. An excellent example is the NetGem Netbox.

■ **Entry-level digital set-top boxes** – These digital boxes can provide traditional broadcast television complemented by a Pay Per View system and a very basic navigation tool. They are low cost and have limited memory, interface ports, and processing power. They are reserved for markets that demand exceptionally low prices and lack interactivity via telephone networks.

■ **Mid-range digital set-top boxes** – These are the most popular set-top boxes offered by TV operators. They normally include a return path, or back channel, which provides communication with a server located at the head-end. They have double the processing power and storage capabilities of entry-level types. For example, while a basic set-top box needs approximately 1 to 2MB of flash memory for coding and data storage, mid-range set-top boxes normally include between 4MB and 8MB. The mid-range set-top box is ideal for consumers who want to simultaneously access a varied range of multimedia and Internet applications from the home.

■ **Advanced digital set-top boxes** – Advanced digital set-top boxes closely resemble a multimedia desktop computer. They can contain more than ten times the processing power of a low-level broadcast TV set-top box. They offer enhanced storage capabilities of between 16-32 MB of flash memory in conjunction with a high-speed return path that can run a variety of advanced services such as:
- Video teleconferencing
- Home networking
- IP telephony
- Video-on-demand (VoD)
- High-speed Internet TV

Advanced set-top box subscribers can use enhanced graphical capabilities to receive high-definition TV signals. Additionally, they can add a hard drive if required. This type of receiver also comes with a range of high-speed interface ports which allows it to serve as a home gateway.

There is uncertainty about the development of the digital set-top box market in the coming years. Most analysts predict that set-top boxes will evolve into a residential gateway that will be the primary access point for subscribers connecting to the Internet.

Components of a Digital Set-top Box

The set-top box is built around traditional PC hardware technologies. It contains silicon chips that process digital video and audio services. The chips are connected to the system board and communicate with other chips via buses. Advanced digital set-top boxes are comprised of four separate hardware subsystems: TV, conditional access (CA), storage, and home networking interfaces (see Figure 3.3).

The TV subsystem is comprised of components such as:

■ Tuner

■ Channel modulator

■ Demultiplexer

■ MPEG-2 audio and video decoders

- Graphics hardware
- CPU
- Memory

The tuner in the box receives a digital signal from a cable, satellite, or terrestrial network and converts it to base-band. Tuners can be divided into three broad categories:

1. **Broadcast In-band (IB)** – Once a signal arrives from the physical transmission media, the IB tuner isolates a physical channel from a multiplex and converts it to baseband.
2. **Out Of Band (OOB)** – This tuner facilitates the transfer of data between the head-end systems and the set-top box. OOB tuners are widely used in cable set-top boxes to provide a medley of interactive services. OOB tuners tend to operate within the 100 - 350 MHz frequency band.
3. **Return path** – This tuner enables a subscriber to activate the return path and send data back to an interactive service provider. It operates within the 5 to 60 MHz frequency band.

The baseband signal is then sampled to create a digital representation of the signal. From this digital representation, the demodulator performs error correction and recovers a digital transport layer bitstream.

A transport stream consists of a sequence of fixed-sized transport packets of 188 bytes each. Each packet contains 184 bytes of payload and a 4-byte header. The header field is a critical 13-bit program identifier (PID) that identifies the transport stream in which the packet belongs. The transport demultiplexer selects and de-packetizes the audio and video packets of the desired program.

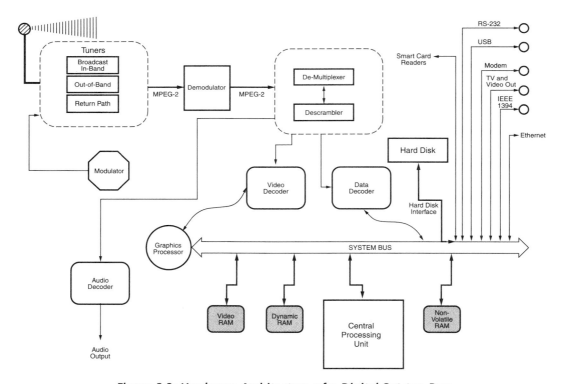

Figure 3.3: Hardware Architecture of a Digital Set-top Box

Once the demultiplexer finishes its work, the signal is forwarded to three different types of decoders:

1. **Video** – Changes the video signal into a sequence of pictures which are displayed on the television screen. Video decoders support still pictures and are capable of formatting pictures for TV screens with different resolutions.
2. **Audio** – Decompresses the audio signal and sends to the speakers.
3. **Data** – Decodes interactive service and channel data stored in the signal.

The graphics engine processes a range of Internet file formats and proprietary interactive TV file formats. Once rendered by the graphics engine, the graphics file is often used to overlay the TV's standard video display. The power of these processors will continue to increase as TV providers attempt to differentiate themselves by offering new applications such as 3D games to their subscriber base.

The following modem options are available to set-top box manufacturers:

- **Standard telephone modem** for terrestrial, satellite and MMDS environments.
- **Cable modem** for a standard cable network.

The system board's CPU typically provides these functions:

- Initializing set-top box hardware components.
- Processing a range of Internet and interactive TV applications.
- Monitoring hardware components.
- Providing processing capabilities for graphic interfaces, electronic program guides, and user interfaces.

Processing chipsets contain electronic circuits that manage data transfer within the set-top box and perform necessary computations. They are available in different shapes, pin structures, architectures, and speeds. The more electronic circuits a processor contains, the faster it can process data. The MHz value of the set-top box processor gives a rough estimate of how fast a processor processes digital TV signals. 300-MHz chips are now becoming standard as demand from service providers increases.

A set-top box requires memory to store and manipulate subscriber instructions. In addition, set-top box elements like the graphics engine, video decoder, and demultiplexer require memory to execute their functions. Set-top boxes use both RAM and ROM. Most functions performed by a set-top box require RAM for temporary storage of data flowing between the CPU and the various hardware components. Video, dynamic, and non-volatile RAM are also used for data storage.

The primary difference between RAM and ROM is volatility. Standard RAM loses its contents when the set-top box is powered off; ROM does not. Once data has been written onto a ROM chip, the network operator or the digital subscriber cannot remove it. Most set-top boxes also contain EPROMs and Flash ROM.

A digital set-top box also includes a CA system. This subsystem provides Multiple Service Operators (MSOs) with control over what their subscribers watch and when. The CA system is comprised of a de-encryption chip, a secure processor, and hardware drivers. The decryption chip holds the algorithm section of the CA. The secure processor contains the keys necessary to decrypt digital services. The condition access system drivers are vendor-dependent.

The ability to locally store and retrieve information is expected to become very important for digital TV customers. While storage in first-generation set-top boxes was limited to ROM and RAM, high capacity hard drives are beginning to appear. Their presence is being driven by the popularity of

DVR technology, among other things. They allow subscribers to download and store personal documents, digital movies, favorite Internet sites, and e-mails. They also provide for:

1. Storing the set-top box's software code.
2. Storing system and user data (such as user profiles, configuration, the system registry and updateable system files).
3. Storing video streams (PVR functionality).

Some set-top boxes support home networking technology While Ethernet is the most popular technology, we can expect to see Bluetooth, DECT, IEEE 1394, USB 2.0, and HomePNA in future boxes.

Economically, set-top box manufacturers must design to a network operator's unique requirements, so costs are high. To overcome this, several manufacturers are beginning to use highly integrated chips that incorporate most required functionality in a single chip.

Market Outlook

Digital set-top boxes continue to be one of the fastest-growing products in the consumer electronics marketplace. A report from Gartner Dataquest estimates that the digital set-top box market will grow to 61 million units by 2005. They represent a huge revenue-earning potential for content owners, service providers, equipment manufacturers, and consumer-oriented semiconductor vendors.

Many industry experts are predicting that set-top boxes will evolve into home multimedia centers. They could become the hub of the home network system and the primary access to the Internet. The set-top market is very dynamic and will change dramatically over the next few years. We are already seeing a convergence of technology and equipment in TV systems such as the addition of a hard drive to store and retrieve programs. Such changes will continue to occur as the market expands.

Integrated Digital Televisions

An integrated digital television contains a receiver, decoder and display device. It combines the functions of a set-top box with a digital-compatible screen. While the digital set-top box market is increasing steadily, the integrated TV set market remains small. The main reason is that the business model adopted by digital services providers is one of subsidizing the cost of set-top boxes in return for the customer's one- or two-year TV programming subscription. This is consistent with the model adopted by mobile phone companies. The subsidized model favors customers in that their up-front cost is minimal.

Additionally, integrated TVs lag behind digital set-top boxes in production and penetration of worldwide TV households due to the following facts:

■ Worldwide production of analog TV sets is still growing.
■ Each analog TV set has an estimated life span of 10 to 15 years. Consumers are unlikely to throw away a useable TV set in order to move to digital TV.
■ Low-cost digital set-top boxes offer an easier route to the world of digital TV than high-cost digital TV sets.
■ Set-top boxes will remain the primary reception devices for satellite, terrestrial, and cable TV services.

Digital Home Theater Systems

A typical home theater includes:

- Large-screen TV (25 inches or larger)
- Hi-fi VCR
- VCR or DVD player
- Audio/video receiver or integrated amplifier
- Four or five speakers

According to CEA market research, nearly 25 million households—about 23% of all U.S. households—own home theater systems. Perhaps the most important aspect of home theater is the surround sound. This feature places listeners in the middle of the concert, film, TV show, or sporting event. The growing interest in home theater is driven in large part by an interest in high-quality audio. High-resolution audio formats such as multi-channel DVD—Audio and Super Audio CD (SACD)—have emerged. Both offer higher resolution than CD, multi-channel playback, and extras like liner notes, lyrics and photos.

Digital Video Recorders

While the potential for personalized television has been discussed for years, it has failed to capture viewers' imaginations. This is beginning to change because some of the larger broadcasting companies have begun to develop business models and technology platforms for personal TV.

The concept of personalizing the TV viewing experience has become more attractive with offerings from companies like TiVo that sell digital video recorders (DVRs). Similar to a VCR, a DVR can record, play, and pause shows and movies. However, a DVR uses a hard disk to record television streams. DVRs also use sophisticated video compression and decompression hardware. Hard disk recording offers a number of benefits including:

- **Storing data in digital format** – Because the hard drive stores content digitally, it maintains a higher quality copy of content than a VCR based system. While an analog tape suffers some loss with each recording and viewing, the quality of a digital recording is determined by the DVR's file format, compression, and decompression mechanisms.
- **'Always-On recording'** – The storage mechanism is always available to record; you don't have to hunt for a blank tape. Like a PC, a DVR can instantly determine and inform you of disk space available for recording. In addition, a DVR can prompt you to delete older material on the hard disk to free up space.
- **Content management** – A DVR file management system lets you categorize and search for particular programs and video files. Rather than trying to maintain labels on tapes, you can instantly browse DVR contents. (This is especially helpful for people who do not label tapes and have to play the first few minutes of every tape to recall its contents!)

DVRs are also optimized for simultaneous recording and playback. This lets the viewer pause live television, do an instant replay, and quick-jump to any location on the recording media. To do this on a VCR, rewind and fast forward must be used excessively. Coupled with a convenient electronic program guide and a high-speed online service, DVRs can pause live television shows, recommend shows based on your preferences, and automatically record your favorite shows. Taking this a step further, some DVRs can preemptively record shows that might be of interest to you. They do so by monitoring the shows you frequently watch and identifying their characteristics. It can then record other shows with similar characteristics.

Core Technologies

A DVR is like a stripped-down PC with a large hard disk. A remote is used to select a program to record, then the DVR captures and stores these programs on its hard disk for future viewing. A DVR box is comprised of both hardware and software components that provide video processing and storage functions. The software system performs these functions:

- Manage multiple video streams.
- Filter video and data content.
- Learn viewers' program preferences
- Maintain programming, scheduling, and dial-up information.

Current DVR software providers include TiVo, SonicBlue, Microsoft, and Metabyte.

The DVR's tuner receives a digital signal and isolates a particular channel. The signal is then:

- Encoded to MPEG-2 for storage on a local hard drive.
- Streamed off the hard drive.
- Transformed from MPEG-2 into a format viewable on your television.

The process of storing and streaming video streams from the hard disk allows viewers to pause live television. The length of the pause is determined by the amount of hard drive storage. While most DVRs use standard hard disks, some manufacturers are optimizing their products for storing and streaming MPEG-2 video streams. For instance, Seagate technology is using Metabyte's MbTV software suite to control the DVR's disc drive recording features. This enables optimization of video stream delivery and management of acoustic and power levels.

The DVR's CPU controls the encoding and streaming of the MPEG-2 signal. Companies like IBM are positioning products such as the PowerPC 400 series processors for the DVR market. Eventually, CPUs will include software applications that encode and decode the MPEG-2 video stream.

Figure 3.4 shows a detailed block diagram of a digital video recorder.

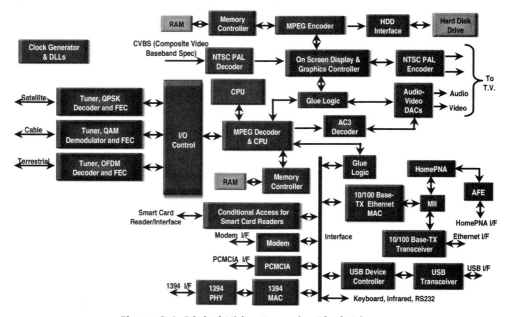

Figure 3.4: Digital Video Recorder Block Diagram

Market Outlook

Currently, there are only a few million homes worldwide that own a DVR. With all the benefits of DVR technology, why is the number so low? The answer lies the technology and pricing structure for the components required to manufacture DVRs. The DVR is expensive when compared to a $99 DVD player or $69 VCR. At present, prices start at about $300 and go up to $2000 for a model that can store hundreds of hours of TV shows. And many of them incur a $10 monthly service fee on top of the purchase price—the last straw for many consumers.

For starters, a DVR requires a large hard disk to store real-time video. Large hard disks have been available for some time but they have been far too expensive for use in digital consumer products. However, this situation is changing as hard disk capacity continues to increase while costs decrease, making high-capacity disks inexpensive enough for DVRs.

Another key DVR component is the video processing subsystem. This system is responsible for simultaneously processing an incoming digital stream from a cable or satellite network, and streaming video streams off a hard disk. Due to the complexity of this system, many manufacturers find it difficult to design cost-effective video processing chipsets. This too has begun to change, as manufacturers are incorporating all required functionality into a single chip. Multi-purpose chips foretell a steep price decline for DVR technology in the future

Two other factors are also acting as a catalyst to promote these types of digital devices. The first is rapid consumer adoption of DVD and home theater systems, indicating high consumer interest in quality digital video. Analog VCRs are unable to offer these capabilities. Second, digital set-top box manufacturers have started to integrate DVR capabilities into their products. These boxes are providing an ideal platform for DVR features. Sheer market size is creating a whole new wave of growth for DVR-enabled systems.

Informa Media forecasts that by 2010, nearly 500 million homes globally will have DVR functionality; 62% will be digital cable customers, 18% digital terrestrial, 13% digital satellite and 7% ADSL.

Geographically, the United States is the initial focus for companies bringing DVRs to market. This is changing, however, with the announcement by TiVo and British Sky Broadcasting Group to bring the firm's recording technology to the United Kingdom for the first time.

While the benefits of the DVR are very compelling, it is a new category of consumer device and there are issues that could limit its success. For instance, if the DVR disk drive fails, the subscriber can't access services. This would lead to loss of revenue and customer loyalty for the operator.

To get around the high prices and recurring fees, some consumers are building their own DVRs. Basically, creation of a DVR from a PC requires the addition of a circuit board and some software to receive and process TV signals. The hard drive from a PC can be used for program storage. And personalized TV listings are available for free from several Internet sites such as tvguide.com and zap2it.com.

Summary

In its simplest form, television is the conversion of moving images and sounds into an electrical signal. This signal is then transmitted to a receiver and reconverted into visible and audible images for viewing. The majority of current television systems are based on analog signals. As viewers demand higher quality levels and new services, the analog format is proving inefficient. Digital TV offers fundamental improvements over analog services. For instance, where an analog signal degrades with distance, the digital signal remains constant and perfect as long as it can be received. The advent of digital TV offers the general public crystal clear pictures, CD quality sound, and access to new entertainment services.

A set-top box can be used to access digital TV. Once a relatively passive device, it is now capable of handling traditional computing and multimedia applications. A typical set-top box is built around traditional PC hardware technologies.

The biggest advance in video in recent years has been the DVR. Similar to a VCR, a DVR stores programs on a hard drive rather than on recording tape. It can pause and instant-replay live TV. It can also let the consumer search through the electronic program guides for specific shows to watch and record. After considerable hype in the early DVR market, expectations have cooled. But analysts still believe that DVR will be the main attraction in interactive TV.

Audio Players

Introduction

Today's music formats are very different from those of a few years ago. In the past, the vinyl record format lasted a relatively long time. In comparison, the life of the cassette format was shorter. Today the popular CD (compact disc) format is facing extinction in even less time. The late 1990s saw the introduction of the MP3 (MPEG Layer III) format, which has gained prominence because of its ease-of-use and portability. MP3 music files can be stored in the hard drive of a PC or flash memory of a portable player. However, the number of competing recording standards has increased as the number of music vendors vying for market prominence has risen. Some of the emerging audio technologies are MP3, DVD-A (DVD-Audio), and SACD (Super Audio CD).

In general, music producers favor two types of media format—physical and Internet-based. Physical media formats include CD,- MiniDisc, SACD, and DVD-Audio. Internet media formats include MP3, Secure MP3, MPEG2 AAC, and SDMI. These will be explained in later sections of this chapter.

The Need for Digital Audio—Strengths of the Digital Domain

The first recording techniques by Thomas Edison and others were based on a common process—the reproduction of an audio signal from a mechanical or electrical contact with the recording media. After nearly 100 years, analog audio has reached a mature state and nearly all of its shortcomings have been addressed. It can offer no further improvements without being cost-prohibitive.

The very nature of the analog signal leads to its own shortcomings. In the analog domain the playback mechanism has no means to differentiate noise and distortion from the original signal. Further, every copy made introduces incrementally more noise because the playback/recording mechanism must physically contact the media, further damaging it on every pass. Furthermore, total system noise is the summation of all distortion and noise from each component in the signal path. Finally, analog equipment has limited performance, exhibiting an uneven frequency response (which requires extensive equalization), a limited 60-dB dynamic range, and a 30-dB channel separation that affects stereo imaging and staging.

Digital technology has stepped in to fill the performance gap. With digital technology, noise and distortion can be separated from the audio signal. And a digital audio signal's quality is not a function of the reading mechanism or the media. Performance parameters such as frequency response, linearity, and noise are only functions of the digital-to-analog converter (DAC). Digital audio parameters include a full audio band frequency response of 5 Hz to 22 kHz and a dynamic range of over 90-dB with flat response across the entire band.

The circuitry upon which digital audio is built offers several advantages. Due to circuit integration, digital circuits do not degrade with time as analog circuits do. Further, a digital signal suffers no degradation until distortion and noise become so great that the signal is out of its voltage threshold. However, this threshold has been made intentionally large for this reason. The high level of circuit integration means that digital circuitry costs far less than its analog counterpart for the same task. Storage, distribution, and playback are also superior in digital media technology. The only theoretical limitation of digital signal accuracy is the quantity of numbers in the signal representation and the accuracy of those numbers, which are known, controllable design parameters.

Principles of Digital Audio

Digital audio originates from analog audio signals. Some key principles of digital audio are sampling, aliasing, quantization, dither, and jitter as described in the following sections.

Sampling

Sampling converts an analog signal into the digital domain. This process is dictated by the Nyquist sampling theorem, which dictates how quickly samples of the signal must be taken to ensure an accurate representation of an analog signal. Furthermore, it states that the sampling frequency must be greater than or equal to the highest frequency in the original analog signal. The sampling theorem requires two constraints to be observed. First, the original signal must be band-limited to half the sampling frequency by being passed through an ideal low-pass filter. Second, the output signal must again be passed through an ideal low-pass filter to reproduce the analog signal. These constraints are crucial and will lead to aliasing if not observed.

Aliasing

Aliasing is audible distortion. For a frequency that is exactly half the sampling frequency, only two samples are generated, since this is the minimum required to represent a waveform. The modulation creates image frequencies centered on integer multiples of Fs. These newly generated frequencies are then imaged, or aliased, back into the audible band. For a sampling rate of 44.1 kHz, a signal of 23 kHz will be aliased to 21.1 kHz. More precisely, the frequency will be folded back across half the sampling frequency by the amount it exceeds half the sampling frequency—in this case, 950 Hz.

Quantization

Once sampling is complete, the infinitely varying voltage amplitude of the analog signal must be assigned a discrete value. This process is called quantization. It is important to note that quantization and sampling are complementary processes. If we sample the time axis, then we must quantize the amplitude axis and vice versa. Unfortunately, sampling and quantization are commonly referred to as just quantization, which is not correct. The combined process is more correctly referred to as digitization.

In a 16-bit audio format, a sinusoidal varying voltage audio signal can be represented by 216 or 65,536 discrete levels. Quantization limits performance by the number of bits allowed by the quantizing system. The system designer must determine how many bits create a sufficient model of the original signal. Because of this, quantizing is imperfect in its signal representation whereas sampling is theoretically perfect. Hence, there is an error inherent in the quantization process regardless of how well the rest of the system performs.

Dither

Dither is the process of adding low-level analog noise to a signal to randomize, or confuse, the quantizer's small-signal behavior. Dither specifically addresses two problems in quantization. The first is that a reverberating, decaying signal can fall below the lower limit of the system's resolution. That is, an attempt to encode a signal below the LSB (Least Significant Bit) results in encoding failure and information is lost. The second problem is that system distortion increases as a percent of a decreasing input signal. However, dither removes the quantization error from the signal by relying on some special behavior of the human ear. The ear can detect a signal masked by broadband noise. In some cases, the ear can detect a midrange signal buried as much as 10 to 12 dB below the broadband noise level.

Jitter

Jitter is defined as time instability. It occurs in both analog-to-digital and digital-to-analog conversion. For example, jitter can occur in a compact disc player when samples are being read off the disc. Here, readings are controlled by crystal oscillator pulses. Reading time may vary and can induce noise and distortion if:

- The system clock pulses inaccurately (unlikely)
- There is a glitch in the digital hardware
- There is noise on a signal control line

To combat the fear of jitter, many questionable products have been introduced. These include disc stabilizer rings to reduce rotational variations and highly damped rubber feet for players. However, a system can be designed to read samples off the disc into a RAM buffer. As the buffer becomes full, a local crystal oscillator can then clock-out the samples in a reliable manner, independent of the transport and reading mechanisms. This process—referred to as time-base correction—is implemented in quality products and eliminates the jitter problem.

Digital Physical Media Formats

Physical media formats include vinyl records, cassettes, and CDs. Today CD technology dominates the industry. In 2001 alone over 1 billion prerecorded CDs were sold in the U.S., accounting for 81% of prerecorded music sales. It took 10 years for the CD format to overtake the cassette in terms of unit shipments. Popular digital physical music media formats include CD, MD, DVD-A, and SACD as discussed in the following sections.

Compact Disc

The compact disc player has become hugely popular, with tens of millions sold to date. CD players have replaced cassette players in the home and car and they have also replaced floppy drives in PCs. Whether they are used to hold music, data, photos, or computer software, they have become the standard medium for reliable distribution of information.

Understanding the CD

A CD can store up to 74 minutes of music or 783,216,000 bytes of data. It is a piece of injection-molded, clear polycarbonate plastic about 4/100 of an inch (1.2 mm) thick. During manufacturing, the plastic is impressed with microscopic bumps arranged as a single, continuous spiral track. Once the piece of polycarbonate is formed, a thin, reflective aluminum layer is sputtered onto the disc, covering the bumps. Then a thin acrylic layer is sprayed over the aluminum to protect it.

Cross-section of a CD

A CD's spiral data track circles from the inside of the disc to the outside. Since the track starts at the center, the CD can be smaller than 4.8 inches (12 cm). In fact, there are now plastic baseball cards and business cards that can be read in a CD player. CD business cards can hold about 2 MB of data before the size and shape of the card cuts off the spiral.

The data track is approximately 0.5 microns wide with a 1.6 micron track separation. The elongated bumps that make up the track are 0.5 microns wide, a minimum of 0.83 microns long, and 125 nm high. They appear as pits on the aluminum side and as bumps on the side read by the laser. Laid end-to-end, the bump spiral would be 0.5 microns wide and almost 3.5 miles long.

CD Player

A CD drive consists of three fundamental components:

1. A drive motor to spin the disc. This motor is precisely controlled to rotate between 200 and 500 rpm depending on the track being read.
2. A laser and lens system to read the bumps.
3. A tracking mechanism to move the laser assembly at micron resolutions so the beam follows the spiral track.

The CD player formats data into blocks and sends them to the DAC (for an audio CD) or to the computer (for a CD-ROM drive). As the CD player focuses the laser on the track of bumps, the laser beam passes through the polycarbonate layer, reflects off the aluminum layer, and hits an optoelectronic sensor that detects changes in light. The bumps reflect light differently than the "lands" (the rest of the aluminum layer) and the sensor detects that change in reflectivity. To read the bits that make up the bytes, the drive interprets the changes in reflectivity.

The CD's tracking system keeps the laser beam centered on the data track. It continually moves the laser outward as the CD spins. As the laser moves outward from the center of the disc, the bumps move faster past the laser. This is because the linear, or tangential, speed of the bumps is equal to the radius times the speed at which the disc is revolving. As the laser moves outward, the spindle motor slows the CD so the bumps travel past the laser at a constant speed and the data is read at a constant rate.

Because the laser tracks the data spiral using the bumps, there cannot be extended gaps where there are no bumps in the track. To solve this problem, data is encoded using EFM (eight-fourteen modulation). In EFM, 8-bit bytes are converted to 14 bits and EFM modulation guarantees that some of those bits will be 1s.

Because the laser moves between songs, data must be encoded into the music to tell the drive its location on the disc. This is done with subcode data. Along with the song title, subcode data encodes the absolute and relative position of the laser in the track.

The laser may misread a bump, so there must be error-correcting codes to handle single-bit errors. Extra data bits are added to allow the drive to detect single-bit errors and correct them. If a scratch on the CD causes an entire packet of bytes to be misread (known as burst error), the drive must be able to recover. By interleaving the data on the disc so it is stored non-sequentially around one of the disc's circuits, the problem is solved. The drive reads data one revolution at a time and un-interleaves the data to play it. If a few bytes are misread, a little fuzz occurs during playback.

When data instead of music is stored on a CD, any data error can be catastrophic. Here, additional error correction codes are used to insure data integrity.

CD Specifications

Sony, Philips, and Polygram jointly developed the specifications for the CD and CD players. Table 4.1 contains a summary of those spec standards.

Table 4.1: Specifications for the Compact Disc System

Functions	Specifications
Playing time	74 minutes, 33 seconds maximum
Rotation	Counter-clockwise when viewed from readout surface
Rotational speed	1.2 – 1.4 m/sec. (constant linear velocity)
Track pitch	1.6 µm
Diameter	120 mm
Thickness	1.2 mm
Center hole diameter	15 mm
Recording area	46 mm - 117 mm
Signal area	50 mm - 116 mm
Minimum pit length	0.833 µm (1.2 m/sec) to 0.972 µm (1.4 m/sec)
Maximum pit length	3.05 µm (1.2 m/sec) to 3.56 µm (1.4 m/sec)
Pit depth	~0.11 µm
Pit width	~0.5 µm
Material	Any acceptable medium with a refraction index of 1.55
Standard wavelength	780 nm (7,800 Å)
Focal depth	± 2 µm
Number of channels	2 channels (4 channel recording possible)
Quantization	16-bit linear
Quantizing timing	Concurrent for all channels
Sampling frequency	44.1 KHz
Channel bit rate	4.3218 Mb/s
Data bit rate	2.0338 Mb/s
Data-to-channel bit ratio	8:17
Error correction code	Cross Interleave Reed-Solomon Code (with 25% redundancy)
Modulation system	Eight-to-fourteen Modulation (EFM)

MiniDisc

MiniDisc™ (MD) technology was pioneered and launched by Sony in 1992. It is a small-format optical storage medium with read/write capabilities. It is positioned as a new consumer recording format with smaller size and better fidelity than audio. MiniDisc was never meant to compete with the CD; it was designed as a replacement for the cassette tape. Its coding system is called ATRAC (Adaptive Transform Acoustic Coding for MiniDisc).

Based on psychoacoustic principles, the coder in a MiniDisc divides the input signal into three sub-bands. It then makes transformations into the frequency domain using a variable block length. The transform coefficients are grouped into non-uniform bands according to the human auditory system. They are then quantized on the basis of dynamics and masking characteristics. While maintaining the original 16-bit, 44.1-kHz signal, the final coded signal is compressed by an approximate ratio of 1:5.

In 1996, interest in the MiniDisc from the consumer market increased. While MD technology has been much slower to take off in the U.S., it enjoys a strong presence in Japan because:

- Most mini-component systems sold in Japan include an MD drive
- Portable MD players are offered by several manufacturers in a variety of designs
- Sony has advertised and promoted this technology through heavy advertisements and low prices for players and media.

In 2003 the forecast for MiniDisc player shipments for the home, portable, and automobile market was expected to exceed 3.7 million units in the U.S.

DVD-Audio

The first popular optical storage medium was the audio CD. Since then the fields of digital audio and digital data have been intertwined in a symbiotic relationship. One industry makes use of the other's technology to their mutual benefit. It took several years for the computer industry to realize that the CD was the perfect medium for storing and distributing digital data. In fact, it was not until well into the 1990s that CD-ROMs became standard pieces of equipment on PCs. With the latest developments in optical storage, the record industry is now looking to use this technology to re-issue album collections, for example. The quest for higher fidelity CDs has spawned a number of standards that are battling with the CD to become the next accepted standard. Among these are DVD-A, SACD, and DAD (digital audio disc).

DVD-A is a new DVD (digital versatile disc) format providing multi-channel audio in a lossless format. It offers high-quality two-channel or 5.1 channel surround sound. The winning technology could produce discs with 24-bit resolution at a 96-kHz sampling rate, as opposed to the current 16-bit/44.1-kHz format.

When DVD was released in 1996, it did not include a DVD-Audio format. Following efforts by the DVD Forum to collaborate with key music industry players, a draft standard was released in early 1998. The DVD-Audio 1.0 specification (minus copy protection) was subsequently released in 1999. DVD-A is positioned as a replacement for the CD and brings new capabilities to audio by providing the opportunity for additional content such as video and lyrics. A DVD-Audio disc looks similar to a normal CD but it can deliver much better sound quality. It allows sampling rates of 44.1, 88.2, 176.4, 48, 96, and 192 kHz with a resolution of 16, 20, or 24 bits. While the two best sampling rates can only be applied to a stereo signal, the others can be used for 5.1 surround sound channels with quadrupled capacity. Even though a DVD-Audio disc has a storage capacity of up to 5 GB, the original signal takes even more space. To account for this, DVD-Audio uses a type of lossless packing called Meridian Lossless Packing (MLP) applied to the PCM (pulse code modulation) bit stream.

DVD-Audio includes the option of PCM (pulse code modulation) digital audio with sampling sizes and rates higher than those of audio CD. Alternatively, audio for most movies is stored as discrete, multi-channel surround sound. It uses Dolby Digital or Digital Theatre Systems (DTS) Digital Surround audio compression similar to the sound formats used in theatres. DTS is an audio encoding format similar to Dolby Digital that requires a decoder in the player or in an external receiver. It accommodates channels for 5.1 speakers—a subwoofer plus five other speakers (front left, front center, front right, rear left, and rear right). Some say that DTS sounds better than Dolby because of its lower compression level. As with video, audio quality depends on how well the processing and encoding was done. In spite of compression, Dolby Digital and DTS can be close to or better than CD quality.

Like DVD-Video titles, we can expect DVD-Audio discs to carry video. They will also be able to carry high-quality audio files and include limited interactivity (Internet, PCs). The capacity of a dual layer DVD-Audio will be up to at least two hours for full surround sound audio and four hours for stereo audio. Single-layer capacity will be about half these times.

As part of the general DVD spec, DVD-Audio is the audio application format of new DVD-Audio/Video players. It includes:

- Pulse code modulation
- High-resolution stereo (2-channel) audio
- Multi-channel audio
- Lossless data compression
- Extra materials, which include still images, lyrics, etc.

Pulse Code Modulation (PCM)

Digitizing is a process that represents an analog audio signal as a stream of digital 1's and 0's. PCM is the most common method of digitizing an analog signal. It samples an analog signal at regular intervals and encodes the amplitude value of the signal in a digital word. The analog signal is then represented by a stream of digital words.

According to digital sampling theory, samples must be taken at a rate at least twice as fast as the frequency of the analog signal to be reproduced. We can hear sounds with frequencies from 20 Hz to 20,000 Hz (20 kHz), so sampling must be performed at least at 40,000 Hz (or 40,000 times per second) to reproduce frequencies up to 20 kHz. CD format has a sampling frequency of 44.1 kHz, which is slightly better than twice the highest frequency that we can hear. While this method is a minimum requirement, it can be argued mathematically that twice is not fast enough. Hence, higher sampling frequencies in PCM offer better accuracy in reproducing high-frequency audio information. Compared to PCM, the CD format's sampling frequency is barely adequate for reproducing the higher frequencies in the range of human hearing.

Another critical parameter in PCM is word length. Each sample of the analog signal is characterized by its magnitude. The magnitude is represented by the voltage value in an analog signal and by a data word in a digital signal. A data word is a series of bits, each having a value of 1 or 0. The longer the data word, the wider the range (and the finer the gradations of range values) of analog voltages that can be represented. Hence, longer word length enables a wider dynamic range of sounds and finer sound nuances to be recorded. The CD format has a word length of 16 bits, which is enough to reproduce sounds with about 96 dB (decibels) in dynamic range.

While most people think that the 44.1-kHz sampling rate and the 16-bit word length of audio CDs are adequate, audiophiles disagree. High fidelity music fans say that audio CD sounds "cold" and exhibits occasional ringing at higher frequencies. That is why consumer electronics manufacturers have designed the DVD-Audio format. Even the average listener using DVD-Audio format on a properly calibrated, high-quality system can hear the differences and the improvement over CD format.

High-Resolution Stereo (2-Channel) Audio

DVD-Audio offers much higher resolution PCM. DVD-Audio supports sampling rates up to 192 kHz (or more than four times the sampling rate of audio CD). It also supports up to a 24-bit word length. The higher sampling rate means better reproduction of the higher frequencies. The 192-kHz sampling rate is over nine times the highest frequency of the human hearing range. One can hear the quality improvement on a high- quality, well-balanced music system. Though DVD-Audio supports up to 192-kHz sampling, not all audio program material has to be recorded using the highest rate. Other sampling rates supported are 44.1, 48, 88.2 , and 96 kHz.

With a word length up to 24 bits, a theoretical dynamic range of 144 dB can be achieved. However, it is not possible to achieve such high dynamic ranges yet, even with the best components. The limiting factor is the noise level inherent in the electronics. The best signal-to-noise ratio that can be achieved in today's state-of-the-art components is about 120 dB. Hence, a 24-bit word length should be more than enough for the foreseeable future. Though DVD-A can support word lengths of 16 and 20 bits, high-resolution stereo usually uses a word length of 24 bits.

Multi-Channel Audio

Another characteristic of DVD-Audio format is its capability for multichannel discrete audio reproduction. Up to six full-range, independent audio channels can be recorded. This allows us not just to hear the music, but to experience the performance as if it were live in our living room. No other stereo music programs can approach this feeling.

Usually, the sixth channel serves as the low frequency effects (LFE) channel to drive the subwoofer. But it is also a full frequency channel. It can serve as a center surround channel (placed behind the listener) or as an overhead channel (placed above the listener) for added dimensionality to the soundstage. The application and placement of the six audio channels are limited only by the imagination of the artist and the recording engineer.

Note that multichannel DVD-Audio does not always mean six channels or 5.1 channels. Sometimes it uses only four channels (left front, right front, left surround, and right surround), or three channels (left front, center, right front). Multichannel DVD-Audio can use up to 192 kHz and up to a 24-bit word length to reproduce music. Practically speaking, it usually uses 96 kHz sampling due to the capacity limitation of a DVD-Audio disc. (Six-channel audio uses three times the data capacity of a two-channel stereo when both use the same sampling rate and word length.) DVD-Audio uses data compression to accommodate the high-resolution stereo and/or multi-channel digital information.

Lossless Data Compression

There are two types of compression:
- **Lossy** – data is lost at the compression stage, such as MPEG-2, Dolby Digital, and DTS.
- **Lossless** – data is preserved bit-for-bit through the compression and decoding processes.

To efficiently store massive quantities of digital audio information, the DVD Forum has approved the use of Meridian's Lossless Packing (MLP) algorithm as part of the DVD-Audio format. Hence, no digital information is lost in the encoding and decoding process, and the original digital bitstream is re-created bit-for-bit from the decoder.

Extra Materials

The DVD-Audio format supports tables of contents, lyrics, liner notes, and still pictures. Additionally, many DVD-Audio titles are actually a combination of DVD-Audio and DVD-Video—called DVD-Audio/Video. Record labels can use the DVD-Video portion to include artist interviews, music videos, and other bonus video programming. Similarly, DVD-Audio discs may actually include DVD-ROM content that can be used interactively on a PC with a DVD-ROM drive.

Backward Compatibility with DVD-Video Players

A DVD-Audio player is required to hear a high-resolution stereo (2-channel) or a multichannel PCM program on the DVD-Audio disc. The Video portion on an Audio/Video disc contains a multichannel soundtrack using Dolby Digital and/or optionally DTS surround sound which can be played back by existing DVD-Video players. The DVD-Video player looks for the DVD-Video portion of the disc and plays the Dolby Digital or DTS soundtracks. Dolby Digital and DTS soundtracks use lossy compression and do not feature high-resolution stereo and multichannel information. However, consumers are pleased with how the music sounds in DVD-Video audio formats, particularly with DTS.

DVD-Audio Market

The DVD-Audio market will grow quickly as manufacturers scramble to add features to keep their DVD offerings at mid-range prices. By 2004, over 27 million DVD-Audio/Video players will ship. Most manufacturers of mid-range players will offer DVD-Audio/Video players, not DVD Video-only players. Very few DVD-Audio-only players will ship during this period since these products will be popular only among audiophiles and in the automotive sound arena. To date, not many music titles have been released to DVD-Audio since the format's launch in the summer of 2000. Record labels have just recently entered the DVD-Audio market with some releases.

Sales of DVD-Audio player shipments in 2003 for the home, portable, and automobile market are expected to be 5.7 million units in the U.S. With competition from the SACD camp, consumers will hesitate buying into the DVD-Audio format now. However, we might see universal players that can play both DVD-Video and DVD-Audio discs. However, these should not be expected for some time. Some affordable universal players promise the ability to play SACD, DVD-Audio, DVD-Video, CD, CD-R, and CD-RW. These players will offer high-resolution stereo and multichannel music with multiple formats.

Super Audio Compact Disc (SACD)

Super Audio CD is a new format that promises high-resolution audio in either two-channel stereo or multi-channel audio. It represents an interesting and compelling alternative to the DVD-Audio format. But unlike DVD-Audio, it does not use PCM. Instead, SACD technology is based on Digital Stream Direct (DSD), whose proponents claim is far superior to PCM technology.

SACD is said to offer unsurpassed frequency response, sonic transparency, and more analog-like sound reproduction. As much promise as SACD holds, the technology is still new and has not yet gained mainstream status. Very few stereo and multichannel titles are available in SACD. It is fully compatible with the CD format and today's CDs can be played in next-generation SACD players. While DVD-Audio is positioned as a mass-market option, SACD is considered an audiophile format. SACD technology was jointly developed by Sony and Philips to compete with DVD-Audio. The format is the same size as CD and DVD media, but offers potential for higher sampling rates.

Using DSD, the bit stream of the SACD system is recorded directly to the disc, without converting to PCM. Unlike PCM, DSD technology uses a 1-bit sample at very high sampling rates (up to 2,822,400 samples per second), which is 64 times faster than the audio CD standard. Using noise shaping, the final signal has a bandwidth of more than 100 kHz and a dynamic range of 120 dB. Since this technique is much more efficient than PCM, it will allow for up to 6 independent, full bandwidth channels with lossless packing. The implication is that DSD will be able to better reproduce the original analog audio signal.

The Disc

The SACD disc looks like an audio CD disc and resembles a DVD in physical characteristics and data capacity. A single-layer SACD disc has a single, high-density layer for high-resolution stereo and multi-channel DSD recording. A dual-layer SACD disc has two high-density layers for longer play times for stereo or multi-channel DSD recordings. There is also a hybrid SACD disc that features a CD layer and a high-density layer for stereo and multi-channel DSD recording. The hybrid SACD looks like an ordinary audio CD to existing CD players and can be played in any CD player. However, the CD player would only reproduce CD-quality stereo sound, not the high-resolution, multichannel DSD tracks of the high-density layer.

Backward Compatibility

When present, the hybrid CD layer makes the SACD disc CD player-compatible. If you buy a hybrid SACD title, you could play it in any CD player or in your PC's CD-ROM drive. However, this cannot be done for DVD-Audio albums. Conversely, SACD discs cannot contain video content as do DVD-Audio/Video discs. Nevertheless, the SACD format is a strong contender for the new high-resolution audio format.

Not all SACD titles or discs will be pressed with the hybrid CD layer option since they are significantly more expensive to make. Those that feature the hybrid CD layer construction are clearly marked as "Hybrid SACD" or "compatible with all CD players." All SACD players can play regular audio CDs so current CD collections will be SACD-compatible.

SACD Players

The first stereo-only Super Audio CD players made their debut in 1999. It was not until late 2000 that multichannel SACD discs and players appeared. Initially, SACD technology was marketed to audiophiles since the prices of the first SACD players were quite high. Now, entry-level, multichannel SACD players can be purchased for as low as $250.

A stereo-capable audio system is all that's needed for stereo DSD playback. For multichannel DSD playback, a multichannel-capable SACD player is required. In addition, a 5.1-channel receiver or preamplifier with 5.1-channel analog audio inputs and an 'analog direct' mode is necessary. To prevent signal degradation, the analog direct mode allows the analog audio signals to pass without additional analog-to-digital (A/D) and digital-to-analog (D/A) conversions in the receiver or preamplifier.

In 2003, Super Audio CD player shipments for the home, portable and automobile markets are expected to exceed 1 million units in the U.S.

Internet Audio Formats

An Internet audio player is a device or program that plays music compressed using an audio compression algorithm. The best-known compression algorithm is MP3. Internet audio formats are gaining popularity because they remove the middlemen and lead to higher revenues and market share. Hence, major music labels are looking to leverage this format. Internet audio players are available as Macintosh and PC applications and as dedicated hardware players including in-dash automotive players.

The most popular player is the portable MP3 player, which resemble a Walkman™ or a pager. Portable MP3 players store music files in flash memory. They typically come with 32 or 64 MB of flash memory and can be expanded to 64 MB and beyond via flash card. For PC usage, music is transferred to the player using a parallel or USB cable.

Analysts predict tremendous growth in the market for portable Internet audio players. They expect online music sales to rise to $1.65 billion, or nearly 9% of the total market and over 10 million users by 2005. This compares to a mere $88 million in annual domestic sales of retail music, or less than 1% of the total music market in 1999.

The Internet audio market is seeing increased support from consumer electronics vendors because of the growing popularity of digital downloads. In spite of the growth forecasts, the following issues will impact product definition and limit market growth rate:

- **Music file format** – While MP3 is the dominant downloading format, other formats are emerging.
- **Copyright protection** – New formats were designed to provide copyright protection for the music industry.
- **Feature set supported by players** –

Internet audio formats include MP3, Secure MP3, MPEG2 AAC, Liquid Audio, Windows Media Player, a2b, EPAC, TwinVQ, MPEG-4, Qdesign Music Codec, and SDMI. The following sections provide details on these formats.

MPEG-1

The Moving Pictures Experts Group (MPEG) was set up under the International Organization for Standardization (ISO) to provide sound and video compression standards and the linkage between them.

The audio part of MPEG-1 describes three layers of compression:

1. **Layer I** – 1:4
2. **Layer II** – 1:6 to 1:8
3. **Layer III** – 1:10 to 1:12

These layers are hierarchically compatible. Layer III decoders can play all three layers, while Layer II decoders can play Layer I and Layer II bitstreams. A Layer I decoder plays only Layer I bitstreams. MPEG specifies the bitstream format and the decoder for each layer, but not the encoder. This was done to give more freedom to the implementers and prevent participating companies from having to reveal their business strategies. Nevertheless, the MPEG-group has submitted some C-language source for explanation purposes.

All three layers are built on the same perceptual noise-shaping standard and use the same analysis filter bank. To ensure compatibility, all the compressed packets have the same structure—a header explaining the compression followed by the sound signal. Hence, every sequence of audio frames can be used separately since they provide all necessary information for decoding. Unfortunately, this increases the file size. The ability to insert program-related information into the coded packets is also described. With this feature, items such as multimedia applications can be linked in.

MP3 (MPEG-1 Layer III Audio Coding)

The three layers of MPEG-1 have different applications depending on the bit rate and desired compression ratio. The most popular has been Layer III, called MP3. The name MP3 was implemented when making file extensions on the Windows platform. Since the typical extension consists of three letters, "MPEG-1 Layer III" became "MP3". Unfortunately, the name has resulted in confusion since people tend to mix the different MPEG standards with the corresponding layers. And, ironically, an MPEG-3 spec does not exist. Table 4.2 shows some of the different Layer III parameters.

Table 4.2: Typical Performance Data of MPEG-1 Layer III

Quality	Bandwidth	Mode	Bit Rate	Compression Ratio
Telephone	2.5 KHz	Mono	8 Kb/s	1:96
Short-wave	2.5 KHz	Mono	16 Kb/s	1:48
AM radio	7.5 KHz	Mono	32 Kb/s	1:24
FM radio	11 KHz	Stereo	56-64 Kb/s	1:24-26
Near-CD	15 KHz	Stereo	96 Kb/s	1:16
CD	>15 KHz	Stereo	112-128 Kb/s	1:12-14

Layer III enhancements over Layers I and II include:
- Nonuniform quantization
- Usage of a bit reservoir
- Hoffman entropy coding
- Noise allocation instead of bit allocation

These enhancements are powerful and require better encoders than do the other layers. But today, even the cheapest computer easily manages to process such files. Each layer supports decoding audio sampled at 48, 44.1, or 32 kHz. MPEG 2 extends this family of codes by adding support for 24, 22.05, or 16 kHz sampling rates as well as additional channels for surround sound and multilingual applications.

MP3 supports audio sampling rates ranging from 16 to 48 kHz and up to five main audio channels. It is a variable-bit codec where users can determine the sampling rate for encoded audio. Higher sampling rates mean better fidelity to the original audio, but result in less compression. The better the quality, the larger the resulting files, and vice versa. For most listeners, MP3 files encoded at reasonable rates (96 or 120 Kb/s) are indistinguishable from CDs.

Secure MP3

Note that MP3 has no built-in security and no safeguards or security policies to govern its use. However, secure MP3 is working to add security to the format. Some of the security approaches are:

- Packing the MP3 files in a secure container that must be opened with a key. The key can be associated with a particular system or sent to the user separately. However, issues such as key tracking and lack of consumer flexibility remain problematic.
- Using encryption technology combined with the encoded file. The key could be the drive ID. Information held in the encryption code ensures copyright protection, establishment of new business models, and the specific uses of songs.

Music label companies are not MP3 fans due to the free distribution of MP3 music files by ventures such as Napster and Gnutella. Hence, secure MP3 may offer a solution for distributing music over the Internet securely with revenue protection.

MPEG-2

MPEG-2 BC became an official standard in 1995. Carrying the tag BC (Backward-Compatible), it was never intended to replace the schemes presented in MPEG-1. Rather, it was intended to supply new features. It supports sampling frequencies from 16 kHz to 22.05 kHz and 24 kHz at bit rates from 32 to 256 Kb/s for Layer I. For Layers II and III, it supports sampling frequencies from 8 to 160 Kb/s. For the coding process, it includes more tables to the MPEG-1 audio encoder.

MPEG-2 is also called MPEG-2 multichannel because it includes the addition of multichannel sound. MPEG-1 only supports mono and stereo signals, and it was necessary to design support for 5.1 surround sound for coding movies. This includes five full bandwidth channels and one low frequent enhancement (LFE) channel operating from 8 kHz to 100 kHz. For backward compatibility, it was necessary to mix all six channels down to a stereo signal. Hence, the decoder can reproduce a full stereo picture.

MPEG-2 AAC (Advanced Audio Coding)

Dolby Laboratories argued that MPEG-II surround sound was not adequate as a new consumer format and that it was limited by the backward compatibility issues. So MPEG started designing a new audio compression standard, originally thought to be MPEG-3. Since the video part of the new standard could easily be implemented in MPEG-2, the audio part was named MPEG-2 AAC. Issued in 1997, the standard features Advanced Audio Coding (AAC) which represents sound differently than PCM does. Note that AAC was called NBC (Non-Backward-Compatible) because it is not compatible with MPEG-1 audio formats.

AAC is a state-of-the-art audio compression technology. It is more efficient than MP3. Formal listening tests have demonstrated its ability to provide slightly better audio quality. The essential benefits of AAC are:

- A wider range of sampling rates (from 8 kHz to 96 kHz)
- Enhanced multichannel support (up to 48 channels)
- Better quality
- Three different complexity profiles

Instead of the filter bank used by previous standards, AAC uses Modified Discrete Cosine Transform (MDCT). Leveraging Temporal Noise Shaping, this method shapes the distribution of quantization noise in time by prediction in the frequency domain. Together with a window length of 2048 lines per transformation, it yields compression that is approximately 30% more efficient than that of MPEG-2 BC.

Since MPEG-2 AAC was never designed to be backward-compatible, it solved the MPEG-2 BC limitation when processing surround sound. In addition, MPEG changed the highly criticized transport syntax. It left the encoding process to decide whether to send a separate header with all audio frames. Hence, AAC provides a much better compression ratio relative to former standards. It is appropriate in situations where backward compatibility is not required or can be accomplished with simulcast. Formal listening tests have shown that MPEG-2 AAC provides slightly better audio quality at 320 Kb/s than MPEG-2 BC provides at 640 Kb/s. At this point, it is expected that MPEG-2 AAC will be the sound compression system of choice in the future.

With time, AAC will probably be the successor of MP3. AAC can deliver equivalent quality to MP3 at 70% of the bit rate (70% of the size at a rate of 128 Kb/s). And it can deliver significantly better audio at the same bit rate. Like MPEG-1 audio encoding standards, AAC supports three levels of encoding complexity. Perhaps the best indication of its significance is its use as the core audio encoding technology for AT&T's a2b, Microsoft's WMA, and Liquid Audio.

MPEG-4

MPEG-4 was developed by the same group that supported MPEG-1 and MPEG-2. It has better compression capabilities than previous standards and it adds interactive support. With this standard MPEG wants to provide a universal framework that integrates tools, profiles, and levels. It not only integrates bitstream syntax and compression algorithms, but it also offers a framework for synthesis, rendering, transport, and integration of audio and video.

The audio portion is based on MPEG-2 AAC standards. It includes Perceptual Noise Substitution (PNS) which saves transmission bandwidth for noise-like signals. Instead of coding these signals, it transmits the total noise-power together with a noise-flag. The noise is re-synthesized in the decoder during the decoding process. It also provides scalability, which gives the encoder the ability to adjust the bit rate according to the signal's complexity.

Many developers are interested in synthesizing sound based on structured descriptions. MPEG-4 does not standardize a synthesis method; it standardizes only the description of the synthesis. Hence, any known or unknown sound synthesis method can be described. Since a great deal of sounds and music are already made through synthesis methods, the final audio conversion can be left for the end computer by using MPEG-4.

Text-To-Speech Interfaces (TTSI) have existed since the advent of personal computers. MPEG-4 will standardize a decoder capable of producing intelligible synthetic speech at bit rates from 200 b/s to 1.2 Kb/s. It will be possible to apply information such as pitch contour, phoneme duration, language, dialect, age, gender and speech rate. MPEG-4 can also provide sound synchronization in animations. For example, the animation of a person's lips could easily be synchronized to his/her lips so that they correspond regardless of the person's language or rate of speech.

An MPEG-4 frame can be built up by separated elements. Hence, each visual element in a picture and each individual instrument in an orchestral sound can be controlled individually. Imagine you're listening to a quintet playing Beethoven and you turn off one of the instruments and play that part yourself. Or you choose which language each actor speaks in your favorite movie. This concept of hypertextuality offers unlimited possibilities.

MPEG-7

By 1996 the MPEG consortium found that people were finding it difficult to locate audiovisual digital content on worldwide networks because the web lacked a logical description of media files. The consortium remedied this problem with the Multimedia Content Description Interface, or MPEG-7 for short. MPEG-7 describes media content. For example, if you hum lines of a melody into a microphone connected to your computer, MPEG-7 will search the web and produce a list of matching sound files. Or you can input musical instrument sounds and MPEG-7 will search for and display sound files with similar characteristics. MPEG-7 can also be used with Automatic Speech Recognition (ASR) for similar purposes. In this capacity, MPEG-7 provides the tools for accessing all content defined within an MPEG-4 frame.

RealAudio G2

RealAudio 1.0 was introduced in 1995 to provide fast downloads with conventional modems. The latest version is called RealAudio G2 and features up to 80% faster downloads than its predecessors. It offers improved handling of data loss while streaming. The available bandwidth on the web may vary and this can result in empty spaces in the sound being played. With RealAudio G2, data packets are built up by parts of neighboring frames so they overlap each other. One package may contain parts of several seconds of music. If some packets are lost, the possible gap will be filled in by an interpolation scheme, similar to interlaced GIF-pictures. And even if packets are lost, the engine will still produce a satisfactory result

RealAudio G2 is optimized for Internet speeds of 16 Kb/s to 32 Kb/s, with support for rates from 6 Kb/s to 96 Kb/s. This allows a wide range of bit rates as well as the ability to constantly change bit rate while streaming. Due to its success, RealNetworks has expanded the scope offerings for sound and video transfer as well as for multimedia platforms such as VRML and Flash. They are also working on a light version of MPEG-7 to describe the content of the media being played. However, RealNetworks products suffer from a lack of public source code and limitations in the free coding tools. Consumers might reject it as too expensive. And even big companies would rather use tools such as AAC or Microsoft Audio that are free and easily available

Microsoft—Audio v4.0 and Windows Media Player

Microsoft has been intimately involved with audio and has incorporated multimedia in its operating systems since Windows 95. It provides basic support for CDs and WAV files and for the recently-introduced MS Audio format. Windows Media Player, for example, is a proprietary multimedia platform shipped with Windows that has a default front end for playing audio and video files. Currently, Microsoft is aggressively developing new technologies for the secure download of music.

WMA (Windows Media Audio) recently developed a codec that is twice as effective as MP3. It is a variable-rate technology with new compression technology that provides high quality in very small files. It has licensed its codec to other software providers and has incorporated WMA on server-side technology in Windows NT Server OS (royalty-free). Microsoft is also working on support for other Microsoft products such as PocketPC OS for hand-held devices and set-top boxes.

Liquid Audio

The Liquid Audio Company was formed in 1996. It developed an end-to-end solution to provide high-quality, secure music. It enables music to be encoded in a secure format, compressed to a reasonable file size, purchased online, downloaded, and played. It also included back-end solutions to manage and track payments—a sort of clearinghouse for downloaded music.

The company worked with Dolby Labs to create a compression format that retains high fidelity to the original. The compression format encodes music better than MP3 does, resulting in files that average about 0.75 MB per minute while retaining very high quality. As part of the security technology, content owners can encode business rules and other information into the songs. Options include artwork and promotional information that appears on the player while the music plays. It can also include lyrics or other song information, a time limit for playback, and even a URL for making a purchase. One possible application is the presentation of a Web site to promote CDs. The site could distribute a full song for download with a time limit for playback. While the song plays the Web site name can be featured prominently to invite the listener to place an order for the full CD.

The Liquid Music Network includes over 200 partners and affiliates including Sony, Hewlett-Packard, Iomega, S3/Diamond Multimedia, Sanyo, Texas Instruments, and Toshiba. The network has been hindered by its PC-only playback capabilities, but has recently gained support from silicon vendors and consumer electronics companies.

a2b

a2b format is a compression and security technology supported by AT&T. The web site promotes the a2b codec with free players and available content. It has a limited installed base of players and little momentum in the market, thus preventing a2b from becoming a widely supported format.
EPAC (Enhanced Perceptual Audio Codec)
EPAC is a codec developed by Lucent Technologies (Bell Labs). It compresses audio at a rate of 1:11. It is supported by Real Networks in its popular G2 player, but few content owners and distributors support it.

TwinVQ (Transform-domain Weighted Interleave Vector Quantization)

This compression technology, targeted at download applications, was originally developed by Yamaha. It has been incorporated, along with AAC, into the MPEG-4 specification. The underlying algorithms are significantly different from the algorithm used in MP3. It has attracted high industry interest due to its quality and compression capabilities. The codec can compress audio at rates of 1:18 to 1:96 that implies a near CD-quality file size of about 0.55 MB per minute.

Qdesign Music Codec

Based in British Columbia, Qdesign developed a high-quality, streaming audio codec. It is distributed by Apple Computer as part of its QuickTime media architecture. It gives excellent quality at dialup data rates.

SDMI

The Secure Digital Music Initiative (SDMI) is a forum of more than 180 companies and organizations representing information technology, music, consumer electronics, security technology, the worldwide recording industry, and Internet service providers. SDMI's charter is to develop open technology specifications that protect the playing, storing, and distributing of digital music.

In June of 1999 SDMI announced that they had reached a consensus on specifications for a new portable music player. This spec would limit digital music consumers to two options:

- Users could transfer tracks from CDs they purchased onto their own players.
- Users could pay for and download music from the Internet from its legitimate publisher.

The proposed version 1.0 specification provides for a two-phase system. Phase I commences with the adoption of the SDMI Specification. Phase II begins when a screening technology is available to identify pirated songs from new music releases and refuses playback.

During Phase I SDMI-compliant portable devices may accept music in all current formats, whether protected or unprotected. In early 2000 record companies started imprinting CD content with a digital watermark that secured music against illegal copying. In Phase II, consumers wishing to download new music releases that include new SDMI technology would be prompted to upgrade their Phase I device to Phase II in order to play or copy that music. As an incentive to upgrade—with music now secured against casual pirating—music companies may finally be ready to put their music libraries on line.

The new proposal's impact on MP3 is far less than it might have been. The Big 5—Sony Music Entertainment, EMI Recorded Music, Universal Music Group, BMG Entertainment and Warner Music Group—had initially advocated making SDMI-compliant players incompatible with MP3 files. However, they may agree to a security scheme that is backward-compatible with the MP3 format. If so, consumers will be able to copy songs from their CDs and download unprotected music, just as they do now.

The open technology specifications released by SDMI will ultimately:

- Provide consumers with convenient access to music available online and in new, emerging digital distribution systems.
- Enable copyright protection for artists' works.
- Promote the development of new music-related business and technologies.

Components of MP3 Portable Players

Portable MP3 players are small and at present cost between $100-$300. They leverage a PC for content storage, encoding, and downloading. MP3 home systems must support multiple output formats and have robust storage capabilities. Automotive versions require broad operating environment support, specific industrial design specifications, and multiple format support including radio and CD. Product differentiation trends include adding Bluetooth capability, user interface, video, games, and day-timer features. Figure 4.1 shows a block diagram of a typical MP3 player.

The most expensive components in an MP3 player are the microprocessor/digital signal processor (DSP) and the flash memory card. The MP3 songs are downloaded into flash memory via the PC parallel port, USB port, or Bluetooth. The user interface controls are interpreted by the main control logic. The song data is manipulated to play, rewind, skip, or stop. The main control logic interfaces directly to the MP3 decoder to transfer MP3 data from the flash memory to the MP3 decoder.

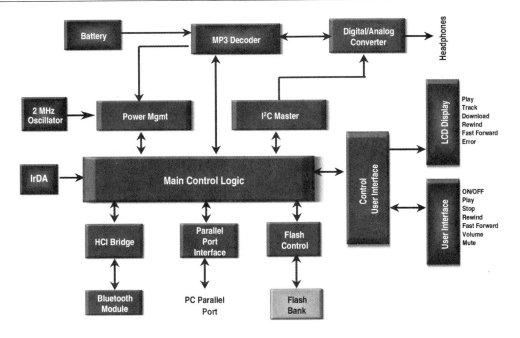

Figure 4.1: MP3 Block Diagram

Flash Memory

Flash memory has enabled the digital revolution by providing for the transfer and storage data. Flash memory cards have become the dominant media storage technology for mobile devices requiring medium storage capacities. The largest application markets for flash cards are digital audio (MP3) players, digital still cameras, digital camcorders, PDAs, and mobile cellular handsets. Flash memory remains the de facto choice for MP3 storage.

The five most popular flash standards are:

- CompactFlash
- SmartMedia
- Memory Stick
- MultiMediaCard (MMC)
- Secure Digital Card

Recently, shortages of flash cards and rising prices may have stunted the growth of the MP3 player market. Manufacturers are beginning to turn to hard disk drives and to Iomega's Clik drives to increase storage capacity.

Even with the increased availability of these alternatives, flash cards will continue to dominate the market. The density of flash cards in MP3 players will rise steadily over the next few years. By 2004 most players shipped will contain 256 MB or 512 MB cards.

Internet Audio Players – Market Data and Trends

Sales of portable digital music players will drastically increase over the next few years. MP3 player sales soared to $1.25 billion by the end of 2002. By 2004 the worldwide market for portable digital audio players (MP3 and other formats) will grow to more than 10 million units and to just under $2 billion in sales. This compares to a mere $125 million for MP3 player sales in 1999. The sales will top more than 25 million units for digital music players (portable, home, automotive) by 2005.

With the arrival of portable MP3 players, the market for portable CD players has for the first time started to see a decrease in unit sales in the U.S. The MP3 player market is concentrated in North America with over 90% of worldwide shipments last year occurring in the United States and Canada. Over the next few years MP3 player sales will climb with increasing worldwide Internet access penetration levels.

Some market trends in the Internet audio player industry are:

- There is tremendous consolidation at every level, thus increasing the market power for the remaining large music labels. This consolidation occurs because corporations are seeking economies of scale and profitability in a mature industry. This, in turn, prompts artists and independent labels to search for new ways to promote and distribute music. Efficient systems provide lower costs and higher profits for the labels. The new medium provides the opportunity for promotion and exposure to artists.

- Communication, entertainment, and information are transitioning from analog to digital. The shift has been occurring in the music industry for decades via CD formats as consumers expect greater fidelity, depth, and range in music. Also, digitization makes it easier to copy without losing quality. PCs aid this transition by making digital files easy to store and replicate.

- The Internet has become a superb distribution channel for digital data. It is serving as a new medium for music promotion and distribution.

- Digital compression benefits the Internet audio player industry. Reducing music files to a fraction of their size makes them easier to manage and distribute without sacrificing quality.

- Nontraditional and experimental business models are being implemented. Knowledge of consumer information reduces the distance between the fans and the artists. Focused marketing allows closer relationships while increasing profits. Electronic distribution allows per-play fees, limited-play samples, and subscription models.

- Broadband (high-speed) access to the home will promote electronic music distribution over the Internet.

- The Internet audio market will probably not converge on a single standard any time soon. In fact, it will probably get more fragmented as it grows. Each standard will be optimized for different applications, bit rates, and business agendas of the media providers.

- Users will want players that support multiple standards. An essential component of a player is the range of music that it makes available. This trend has begun with recent announce-ments of products that support multiple standards and have the ability to download support for new standards. It will be important for audio players to support physical formats such as CD, DVD-Audio, and SACD since consumers cannot afford three players for three formats.

- Another emerging trend is support for metadata in music standards. Metadata is non-music data included in the music files. It includes data such as track information and cover art. A potential use for metadata is advertisements.

- Copyright protection is the biggest issue hindering the growth of Internet music distribution. MP3 files can be easily distributed across the Internet using web page downloads or email. Since MP3 has no inherent copy protection, once it is made available on the web everyone has access to it at no charge. The recording industry represented by the Recording Industry Association of America (RIAA) sees the combination of MP3 and the Internet as a

Pandora's box that will result in widespread piracy of copyrighted material. The RIAA believes the threat is significant enough that it took legal action to block the sale of the Diamond Rio in late 1998. The fear of piracy has limited the availability of legitimate MP3 material and tracks from emerging and mainstream artists.

Several trends will spur growth of the portable digital music player market:

- **More available hardware** – Over the past 12 months more than 50 manufacturers announced new portable music players.
- **Greater storage capacity** – The density of flash cards, which are the most common storage media for MP3 players, is steadily increasing. Most players now offer 64 MB of storage—about one hour's worth of music.
- **More storage options** – Cheaper alternatives to flash cards (hard disk drives, Iomega's Clik drives) are becoming available.
- **More digital music files** – Legal or not, Napster, Gnutella, and other file sharing sites have increased awareness and availability of digital music.

Other Portable Audio Products

Sales of traditional cassette and CD headphone stereos, Internet audio headphone portables, and boomboxes are expected to exceed $3.2 billion in 2003. Sales continue to grow because, at prices as low as $29, headphone CD players have become viable gifts for teens and pre-teens. Sales also grew because anti-shock technologies have reduced player mis-track, making them more practical when walking, hiking or working out.

Another growth booster is the launch of the first headphone CD players capable of playing back MP3-encoded CDs. In the boombox category, factory-level sales of CD-equipped models will continue to grow at 4.6%. Cassette boomboxes will continue their downward spiral, dropping 34%. Within the CD-boombox segment, sales of higher priced three-piece units (featuring detachable speakers) will fall 25% as consumers opt increasingly for more powerful tabletop shelf systems whose prices have grown increasingly affordable.

Convergence of MP3 Functionality in Other Digital Consumer Devices

MP3 technology will proliferate as manufacturers integrate it into other electronics devices. Several manufacturers are combining MP3 functionality into consumer devices such as cellular phones, digital jukeboxes, PDAs, wristwatches, automotive entertainment devices, digital cameras, and even toasters. Convergence between CD and DVD players has already been popular for several years. This convergence will move further to players that incorporate DVD-Video, CD, DVD-Audio, and SACD. And next generation portable players will support multiple Internet audio formats.

PC-Based Audio

Though PC-based audio has existed for over 20 years, only during the past two years have personal computers become widely used to play and record popular music at the mainstream level. Sound card and computer speaker sales are strong. Many fans are abandoning their home stereos in favor of PCs with high storage capacity because they are ideal for storing music for portable MP3 players.

The home audio convergence and the shift to the PC will continue as new home audio networking products hit the market. These devices will call up any digital audio file stored on a home PC and play it in any room of a house—through an existing home theater system or through a separate pair of powered speakers. If the PC has a cable modem or DSL connection, the home networking products can even access Internet streamed audio.

It is easy to store CDs on a computer hard drive. Software for CD "ripping" has been available for several years. The process is as simple as placing a disc in a CD-ROM drive and pressing start. Some titles even let users tap into online databases containing song information and cover art. After the CDs are stored, users can use a remote control to access custom play-lists or single tracks. Add the ability to retrieve Internet streamed audio - a radio broadcast or an overseas sporting event, for example - and the benefits of convergence emerge.

Digital entertainment is migrating from the computer room to the living room. PCs of the future will provide for the distribution of broadband signals to the homes through residential gateways - cable or DSL connections. The technology will quickly move beyond audio to provide whole-house digital video access to movies and shows that are stored on a PC or on the Internet.

"Thin" operating systems such as PocketPC are also appearing in home theater components. While some fear domination by a single OS manufacturer, standardization will permit interaction among different component brands. For example, with a touch of a remote button devices such as the television, stereo and lighting systems will synchronize. And this will not even require custom programming.

In the coming years, the PC as we know it will disappear. Instead of a multi-function box, the home computer will become a powerful information server for everyday devices. Consumers must prepare to experience a new dimension in home audio access, one that will dramatically change our media consumption habits.

Internet Radio

The Internet also extended its reach to influence the design of home radios, turning AM/FM radios into AM/FM/Internet radios. Internet radios are available that enable consumers to stream music from thousands of websites without having to sit near an Internet-enabled PC. Today the selection of Internet radios is expanded to include devices from mainstream audio suppliers and computer-industry suppliers. Internet radios are showing up in tabletop music systems and in audio components intended for a home's audio-video system. Some connect to super fast broadband modems to achieve live streaming with the highest possible sound quality.

Digital Audio Radio

The digital revolution has also come to radio. In 2001 Sirius Satellite Radio and XM Satellite Radio delivered up to 100 channels of coast-to-coast broadcasting via satellite. The first terrestrial AM and FM radio stations could soon begin commercial digital broadcasting if the Federal Communications Commission authorizes digital conversion.

Online Music Distribution

Record companies have always been paranoid about piracy. This was rarely justified until the digital age made high-fidelity copying a ubiquitous reality. First came the MP3 compression format. Then came the Internet which made electronic content distribution worldwide viable. Next, portable players appeared. Finally, sites such as Napster, AudioGalaxy, Gnutella, and LimeWire appeared as venues for easy file sharing.

These technologies and services disrupted the long-established business model for the $50-billion-a-year recording industry. The system was threatened and shareholders worried that stock values would collapse. Companies replied by suing. Shortly after Diamond Multimedia Systems launched the Rio 300 MP3 player in the fall of 1998, the RIAA sued to have the device taken off the market, alleging that it was an illegal digital recording device. The Rio 300 wasn't the first MP3 player, but it was the first one marketed and sold heavily in the United States. The suit triggered a media blitz that portrayed the RIAA as the bad guy. Ironically, the media displayed pictures of the match-box-size Rio everywhere, bringing it to the attention of many who didn't know about MP3 players.

The RIAA dropped its suit after a federal circuit court agreed with Diamond that Rio did not fit the law's definition of an illegal digital recording device. That news triggered another sales surge with major consumer electronics companies such as Creative Technology, Thomson Multimedia, and Samsung Electronics jumping on the bandwagon. MP3 file-sharing Web sites, including Napster, became favorite stops for music lovers.

In late 1999 the RIAA sued Napster on copyright infringement, bringing more media attention and spurring sales of even more portable players. A federal court ruling stopped Napster, but not before three billion files were swapped illegally. Although MP3 player sales slowed a bit, other websites picked up where Napster left off. During all the years that people were downloading free MP3 files, only 2% of U.S. households ever paid for downloaded music. The record companies are fearful of losing control over content. They are looking at SDMI and other secure schemes as a possible solution for protecting the playing, storing, and distributing of digital music.

The copyright infringement lawsuits have been settled, with Napster and MP3.com being purchased by separate labels. These types of deals and partnerships between consumer electronics vendors, music labels, Internet portals, retailers, and online music players will enable legitimate online music distribution channels. The key ingredients in a successful online music service will be an easily negotiable interface with a broad array of content accessible from any device at a reasonable price. It would be comprised of a paid service that approximates the convenience and ubiquity of a Napster, but with better file quality and a broader range of services. Forecasts predict that the number of music service provider users in the U.S. will exceed 10 million by 2005 and revenues will exceed $1.6 billion.

Summary

By 1981 the vinyl LP was more than 30 years old and the phonograph was more than 100 years old. It was obvious that the public was ready for a new, advanced format. Sony, Philips, and Polygram collaborated on a new technology that offered unparalleled sound reproduction—the CD. Instead of mechanical analog recording, the discs were digital. Music was encoded in binary onto a five-inch disc covered with a protective clear plastic coating and read by a laser.

Unlike fragile vinyl records, the CD would not deteriorate with continued play, was less vulnerable to scratching and handling damage, held twice as much music, and did not need to be flipped over. It was an immediate sensation when it was introduced in 1982. By 1988 sales of CDs surpassed vinyl as the home playback medium of choice. It passed the pre-recorded cassette in sales in 1996. But the CD could only play back digital music. Now accustomed to the perfection of digital music playback, consumers demanded the ability to record digitally as well. In 1986 Sony introduced Digital Audio Tape (DAT), followed by MiniDisc in 1992. Philips countered that same year with DCC (Digital Compact Cassette). None of these formats caught on simply because none of them were CD. The recordable CD was unveiled in 1990, but the first consumer CD-R devices would not be introduced until the mid-1990s.

With the success of the DVD player and the definition of a DVD format audio specification, DVD-Audio is one of the most popular of emerging formats. A similar audio format—Super Audio CD—is also being introduced. The arrival of MP3 and similar formats has introduced several exciting trends, one of which is the distribution of music over the Internet. Changes in the multibillion dollar audio music market will spell success for some technologies and failure for others. At the very least, we are guaranteed to see improved sound quality and more secure and easy distribution of music in the coming years.

Cellular/Mobile Phones

Introduction

The wireless industry includes electronics systems such as pagers, GPS, cordless telephones, notebook PCs with wireless LAN functionality, and cellular phones. Cellular handset phones represent the largest and most dynamic portion of the wireless communication market. No product has become as dominant in unit shipments as the cellular phone. It has evolved from being a luxury for urgent communications to a platform that provides voice, video, and data services. It's now a necessity for both business and personal environments.

Definition

The term cellular comes from a description of the network in which cellular or mobile phones operate. The cellular communications system utilizes numerous low-power transmitters, all interconnected to form a grid of "cells." Each ground-based transmitter (base station) manages and controls communications to and from cellular phones in its geographic area, or cell. If a cellular phone user is traveling over a significant distance, the currently transmitting cell transfers the communications signal to the next adjacent cell. Hence, a cellular phone user can roam freely and still enjoy uninterrupted communication. Since cellular communications rely on radio waves instead of on a fixed-point wired connection, a cellular phone can be described as a radiophone. The dependence on radio waves is the common denominator for wireless communications systems.

Wireless and cellular roots go back to the 1940s when commercial mobile telephony began. The cellular communications network was first envisioned at Bell Laboratories in 1947. In the United States cellular planning began in the mid-1940s, but trial service did not begin until 1978. It was not until 1981 in Sweden that the first cellular services were offered. Full deployment in America did not occur until 1984. Why the delay? The reasons are limited technology, Bell System ambivalence, cautiousness, and governmental regulation. The vacuum tube and the transistor made possible the early telephone network. But the wireless revolution began only after low-cost microprocessors, miniature circuit boards, and digital switching became available. Thus, the cellular phone industry can still be considered a "young" electronic system segment—much younger than the PC industry, which began in the mid-1970s.

Landscape—Migration to Digital and 3G

History—The Wireless Beginning

Although CB (Citizens Band) radio and pagers provided some mobile communications solutions, demand existed for a completely mobile telephone. Experiments with radio telephones began as far back as the turn of the century, but most of these attempts required the transport of bulky radio transmitters or they required tapping into local overhead telephone wires with long poles. The first practical mobile radio telephone service, MTS (Mobile Telephone Service), began in 1946 in St. Louis. This system was more like a radio walkie-talkie - operators handled the calls and only one person at a time could talk.

The idea of permanent "cells" was introduced in 1947, the same year radio telephone service was initiated between Boston and New York by AT&T. Automatic dialing—making a call without the use of an operator—began in 1948 in Indiana. But it would be another 16 years before the innovation was adopted by the Bell system.

The Bell Labs system (part of AT&T then) moved slowly and with seeming lack of interest at times toward wireless. AT&T products had to work reliably with the rest of their network and they had to make economic sense. This was not possible for them with the few customers permitted by the limited frequencies available at the time. The U.S. FCC was the biggest contributor to the delay, stalling for decades on granting more frequency space. This delayed wireless technology in America by perhaps 10 years. It also limited the number of mobile customers and prevented any new service from developing. But in Europe, Scandinavia, Britain, and Japan where state-run telephone companies operated without competition and regulatory interference, cellular came more quickly. Japanese manufacturers equipped some of the first car-mounted mobile telephone services, their technology being equal to what America was producing.

During the next 15 years the introduction of the transistor and an increase in available frequencies improved radio telephone service. By 1964 AT&T developed a second-generation cell phone system. It improved Mobile Telephone Service to have more of the hallmarks of a standard telephone, though it still allowed only a limited number of subscribers. In most metropolitan service areas there was a long waiting list. The idea of a mobile phone was popularized by Secret Agent 86, Maxwell Smart, who used a shoe phone on the TV spy spoof, "Get Smart." In 1973 Motorola filed a patent for a radio telephone system and built the first modern cell phone. But the technology would not reach consumers until 1978.

First There Was Analog

The first cellular communications services (first generation, or 1G) were analog systems. Analog systems are based on frequency modulation (FM) using bandwidths of 25 kHz to 30 kHz. They use a constant phase variable frequency modulation technique to transmit analog signals.

Among the most popular of analog wireless technologies is AMPS (Advanced Mobile Phone System), developed by Bell Labs in the 1970s. It operates in the 800-MHz band and uses a range of frequencies between 824 MHz and 894 MHz for analog cell phones. To encourage competition and keep prices low, the U. S. government required the presence of two carriers, known as carrier A and B, in every market. One of the carriers was normally the local phone company. Carriers A and B are each assigned 832 frequencies—790 for voice and 42 for data. A pair of frequencies, one for transmit and one for receive, makes up one channel. The frequencies used are typically 30 kHz wide. 30 kHz was chosen because it provides voice quality comparable to a wired telephone.

The transmit and receive frequencies of each voice channel are separated by 45 MHz to keep them from cross-interference. Each carrier has 395 voice channels as well as 21 data channels for housekeeping activities such as registration and paging. A version of AMPS known as Narrowband AMPS (NAMPS) incorporates some digital technology, which allows the system to carry about three times as many calls as the original version. Though it uses digital technology, it is still considered analog. AMPS and NAMPS operate in the 800-MHz band only and do not offer features such as e-mail and Web browsing.

When AMPS service was first fully initiated in 1983, there were a half million subscribers across the U.S. By the end of the decade there were two million cell phone subscribers. But demand far outstripped the supply of frequency bands and cell phone numbers.

First-generation analog networks are gradually being phased out. They are limited in that there can only be one user at a time per channel and they can't provide expanded digital functionality (e.g., data services). Also, these systems could not contain all the potential wireless customers. Companies began researching digital cellular systems in the 1970s to address these limitations. It was not until 1991, however, that the first digital cellular phone networks (second generation, or 2G) were established.

Along Comes Digital

Digital cell phones use analog radio technology, but they use it in a different way. Analog systems do not fully utilize the signal between the phone and the cellular network because analog signals cannot be compressed and manipulated as easily as digital. This is why many companies are switching to digital—they can fit more channels within a given bandwidth.

Digital phones convert a voice into binary information (1s and 0s) and compress it. With this compression, between three and 10 digital cell phone calls occupy the space of a single analog call. Digital cellular telephony uses constant frequency variable phase modulation techniques to transmit its analog signals. With this technology digital cellular can handle up to six users at a time per channel compared to one user at a time with analog. This is especially critical in densely populated urban areas. Thus, digital technology helped cellular service providers put more cellular subscribers on a given piece of transmission bandwidth. The cost to the service provider of supplying digital cellular services to users can be as little as one-third the cost of providing analog services. Besides reduced costs to the service provider, digital cellular technology also offers significant benefits to the user. These benefits include encryption capability (for enhanced security), better voice quality, longer battery life, and functionality such as paging, caller ID, e-mail, and FAX.

In the 1970s digital cellular research concentrated on narrow-band frequency division multiple access (FDMA) technology. In the early 1980s the focus switched to time division multiple access (TDMA) techniques. In 1987 narrow-band TDMA with 200 kHz channel spacing was adopted as the technology of choice for the pan-European GSM digital cellular standard. In 1989 the U.S. and Japan also adopted the narrow-band TDMA. More recently cellular networks are migrating to CDMA (code division multiple access) technology.

The three common technologies used to transmit information are described in the following section:

- **Frequency division multiple access (FDMA)** – Puts each call on a separate frequency. FDMA separates the spectrum into distinct voice channels by splitting it into uniform chunks of bandwidth. This is similar to radio broadcasting where each station sends its signal at a different frequency within the available band. FDMA is used mainly for analog transmission. While it is capable of carrying digital information, it is not considered to be an efficient method for digital transmission.
- **Time division multiple access (TDMA)** – Assigns each call a certain portion of time on a designated frequency. TDMA is used by the Electronics Industry Alliance and the Telecom-

munications Industry Association for Interim Standard 54 (IS-54) and Interim Standard 136 (IS-136). Using TDMA, a narrow band that is 30 kHz wide and 6.7 milliseconds long is split into three time slots. Narrow band means "channels" in the traditional sense. Each conversation gets the radio for one-third of the time. This is possible because voice data that has been converted to digital information is compressed to take up significantly less transmission space. Hence, TDMA has three times the capacity of an analog system using the same number of channels. TDMA systems operate in either the 800 MHz (IS-54) or the 1900 MHz (IS-136) frequency band. TDMA is also used as the access technology for Global System for Mobile communications (GSM).

- **Code division multiple access (CDMA)** – Assigns a unique code to each call and spreads it over the available frequencies. Multiple calls are overlaid on each other on the channel. Data is sent in small pieces over a number of discrete frequencies available for use at any time in the specified range. All users transmit in the same wide-band chunk of spectrum but each phone's data has a unique code. The code is used to recover the signal at the receiver. Because CDMA systems must put an accurate time-stamp on each piece of a signal, they reference the GPS system for this information. Between eight and ten separate calls can be carried in the same channel space as one analog AMPS call. CDMA technology is the basis for Interim Standard 95 (IS-95) and operates in both the 800 MHz and 1900 MHz frequency bands. Ideally, TDMA and CDMA are transparent to each other. In practice, high-power CDMA signals raise the noise floor for TDMA receivers and high-power TDMA signals can cause overloading and jamming of CDMA receivers.

With 1G (analog) cellular networks nearing their end and 2G (digital) ramping up, there is much discussion about third-generation (3G) cellular services. 3G is not expected to have a significant impact on the cellular handset market before 2002-03. While 3G deployments are currently underway, 2.5G is being considered as the steppingstone for things to come. The evolution of the digital network from 2G, to 2.5G, to 3G is described next.

2G

Second-generation digital systems can provide voice/data/fax transfer as well as other value-added services. They are still evolving with ever-increasing data rates via new technologies such as HSCSD and GPRS. 2G systems include GSM, US-TDMA (IS_136), cdmaOne (IS-95), and PDC. US-TDMA/PDC have been structured atop existing 1G analog technology and are premised on compatibility and parallel operation with analog networks. GSM/IS-95, however, are based on an entirely new concept and are being increasingly adopted worldwide.

2.5G

2.5G technologies are those offering data rates higher than 14.4 Kb/s and less than 384 Kb/s. Though these cellular data services may seem like a steppingstone to things to come, they may be with us much longer than some cellular carriers want to believe.

2.5G technologies accommodate cellular nicely. First, they are packet-based as opposed to 2G data services, which were generally connection based. This allows for always-on services. And since no real connection needs to be established, latency in sending data is greatly reduced. Second, since 2.5G services don't require new spectrum or 3G licenses, the carrier's cost to deploy these services is modest. This might make them cheaper for consumers. Third, enabling handsets for 2.5G services are fairly inexpensive, typically adding less than $10 to the handset cost.

3G

3G (third generation) mobile communications systems include technologies such as cdma2000, UMTS, GPRS, WCDMA, and EDGE. The vision of a 3G cellular system was originally articulated by the International Telecommunications Union (ITU) in 1985. 3G is a family of air-interface standards for wireless access to the global telecommunications infrastructure. The standards are capable of supporting a wide range of voice, data, and multimedia services over a variety of mobile and fixed networks. 3G combines high-speed mobile access with enhanced Internet Protocol (IP)-based services such as voice, text, and data. 3G will enable new ways to communicate, access information, conduct business, and learn.

3G systems will integrate different service coverage zones including macrocell, microcell, and picocell terrestrial cellular systems. In addition, they support cordless telephone systems, wireless access systems, and satellite systems. These systems are intended to be a global platform and provide the infrastructure necessary for the distribution of converged services. Examples are:

- Mobile or fixed communications
- Voice or data
- Telecommunications
- Content
- Computing

To enhance the project, the ITU has allocated global frequency ranges to facilitate global roaming and has identified key air-interface standards for the 3G networks. It has also identified principal objectives and attributes for 3G networks including:

- increasing network efficiency and capacity.
- anytime, anywhere connectivity.
- high data transmission rates: 44 Kb/s while driving, 384 Kb/s for pedestrians, and 2 Mb/s for stationary wireless connections.
- interoperability with fixed line networks and integration of satellite and fixed-wireless access services into the cellular network.
- worldwide seamless roaming across dissimilar networks.
- bandwidth on demand and the ability to support high-quality multimedia services.
- increased flexibility to accommodate multiple air-interface standards and frequency bands and backward compatibility to 2G networks.

The ITU World Radio Conference (WRC) in 1992 identified 230 MHz on a global basis for IMT-2000, including both satellite and terrestrial components. However, following unexpectedly rapid growth in both the number of mobile subscribers and mobile services in the 1990s, the ITU is considering the need for additional spectrum since WRC 2000. The currently identified frequency bands are:

- 806 – 960 MHz
- 1,710 – 1,885 MHz,
- 1,885 – 2,025 MHz
- 2,110 – 2,200 MHz
- 2,500 – 2,690 MHz

Of the five 3G air-interface standards approved by the ITU, only two—cdma2000 and Wideband CDMA—have gained serious market acceptance.

Digital technologies are organized into these categories:

- **2G** – GSM, TDMA IS-136, Digital-AMPS, CDMA, cdmaOne, PDC, PHS, and iDEN, PCS
- **2.5G** – GSM, HSCSD (High Speed Circuit Switched Data), GPRS (General Packet Radio Service), EDGE (Enhanced Data Rate for GSM Evolution), and cdma2000 1XRTT
- **3G** – cdma2000 1XEV, W-CDMA, TDMA-EDGE, and cdma2000 3XRTT

Some of the key digital wireless technologies are:

- **CDMA** – Code Division Multiple Access, known in the U.S. as Inc. and as IS-95. The Telecommunications Industry Association (TIA) adopted the CDMA standard in 1993.

  It was developed by Qualcomm and characterized by high capacity and small cell radius. It uses the same frequency bands and supports AMPS, employing spread-spectrum technology and a special coding scheme.

- **cdmaOne** – This is considered a 2G technology mobile wireless technology. It is used by member companies of the CDMA Development Group to describe wireless systems complying with standards including the IS-95 CDMA air interface and the ANSI-41 network standard for switch interconnection. CdmaOne describes a complete wireless system that incorporates the ANSI-41 network standard for switch interconnection. The IS-95A protocol employs a 1.25-MHz carrier, operates in radio-frequency bands at either 800 MHz or 1.9 GHz, and supports data speeds up to 14.4 Kb/s. IS-95B can support data speeds up to 115 Kb/s by bundling up to eight channels.

- **cdma2000** – 3G wireless standard that is an evolutionary outgrowth of cdmaOne. It provides a migration path for current cellular and PCS operators and has been submitted to the ITU for inclusion in IMT-2000. Cdma2000 offers operators who have deployed a 2G cdmaOne system a seamless migration path to 3G features and services within existing spectrum allocations for both cellular and PCS operators. Cdma2000 supports the 2G network aspect of all existing operators regardless of technology (cdmaOne, IS-136 TDMA, or GSM). This standard is also known by its ITU name IMT-CDMA Multi-Carrier (1X/3X). Cdma2000 has been divided into 2 phases. First phase capabilities are defined in the 1X standard which introduces 144 Kb/s packet data in a mobile environment and faster speeds in a fixed environment. Cdma2000 phase two, known as 3X, incorporates the capabilities of 1X and:
  - supports all channel sizes (5 MHz, 10 MHz, etc.).
  - provides circuit and packet data rates up to 2 Mb/s.
  - incorporates advance multimedia capabilities.
  - includes a framework for advanced 3G voice services and vocoders (voice compression algorithm codecs), including voice over packet and circuit data.

- **CDPD** – Cellular Digital Packet Data. It refers to a technology that allows data packets to be sent along idle channels of CDPD cellular voice networks at very high-speeds during pauses in conversations. CDPD is similar to packet radio technology in that it moves data in small packets that can be checked for errors and retransmitted.

- **CT-2** – A 2G digital cordless telephone standard that specifies 40 voice channels (40 carriers times one duplex bearer per carrier).

- **CT-3** – A 3G digital cordless telephone standard that is a precursor to DECT.

- **D-AMPS** – Digital AMPS, or D-AMPS, is an upgrade to the analog AMPS. Designed to address the problem of using existing channels more efficiently, D-AMPS (IS-54) employs the same 30 KHz channel spacing and frequency bands (824 - 849 MHz and 869 - 894 MHz) as AMPS. By using TDMA, IS-54 increases the number of users from one to three per channel (up to 10 with enhanced TDMA). An AMPS/D-AMPS infrastructure can support analog AMPS or digital D-AMPS phones. Both operate in the 800 MHz band.

- **DECT** – Initially, this was Ericsson's CT-3, but it grew into ETSI's Digital European Cordless Standard. It is intended to be a far more flexible standard than CT-2 in that it has 120 duplex voice channels (10 RF carriers times 12 duplex bearers per carrier). It also has a better multimedia performance since 32 Kb/s bearers can be concatenated. Ericsson is developing a dual GSM/DECT handset that will be piloted by Deutsche Telekom.

- **EDGE** – Enhanced Data for GSM Evolution. EDGE is an evolution of the US-TDMA systems and represents the final evolution of data communications within the GSM standard. EDGE uses a new enhanced modulation scheme to enable network capacity and data throughput speeds of up to 384 Kb/s using existing GSM infrastructure.

- **FSK** – Frequency Shift Keying. Many digital cellular systems rely on FSK to send data back and forth over AMPS. FSK uses two frequencies, one for 1 and the other for 0. It alternates rapidly between the two to send digital information between the cell tower and the phone. Clever modulation and encoding schemes are required to convert the analog information to digital, compress it, and convert it back again while maintaining an acceptable level of voice quality.

- **GPRS** – General Packet Radio Service. It is the first implementation of packet-switched data primarily for GSM-based 2G networks. Rather than sending a continuous stream of data over a permanent connection, packet switching utilizes the network only when there is data to be sent. GPRS can send and receive data at speeds up to 115 Kb/s. GPRS network elements consists of two main elements - SGSN (Service GPRS Support Node) and GGSN (Gateway GPRS Support Node).

- **GSM** – Global System for Mobile Communications. The first digital standard developed to establish cellular compatibility throughout Europe. Uses narrow-band TDMA to support eight simultaneous calls on the same radio frequency. It operates at 900, 1800, and 1900 MHz. GSM is a technology based on TDMA which is the predominant system in Europe, but is also used worldwide. It operates in the 900 MHz and 1.8 GHz bands in Europe and the 1.9 GHz PCS band in the U.S. It defines the entire cellular system, not just the air interface (TDMA, CDMA, etc.). GSM provides a short messaging service (SMS) that enables text messages up to 160 characters to be sent to and from a GSM phone. It also supports data transfer at 9.6 Kb/s to packet networks, ISDN, and POTS users. GSM is a circuit-switched system that divides each 200 KHz channel into eight 25 KHz time slots. It has been the backbone of the success in mobile telecoms over the last decade and it continues to evolve to meet new demands. One of GSM's great strengths is its international roaming capability, giving a potential 500 million consumers a seamless service in about 160 countries. The imminent arrival of 3G services is challenging operators to provide consumer access to high-speed, multimedia data services and seamless integration with the Internet. For operators now offering 2G services, GSM provides a clear way to make the most of this transition to 3G.

- **HSCSD** – High Speed Circuit Switched Data. Introduced in 1999, HSCSD is the final evolution of circuit-switched data within the GSM environment. It enables data transmission over a GSM link at rates up to 57.6 Kb/s. This is achieved by concatenating consecutive GSM timeslots, each of which is capable of supporting 14.4 Kb/s. Up to four GSM timeslots are needed for the transmission of HSCSD.

- **IDEN** – Integrated Digital Enhanced Network.

- **IMT-2000** – International Mobile Telecommunication 2000. This is an ITU (International Telecommunications Union) standards initiative for 3G wireless telecommunications services. It is designed to provide wireless access to global telecommunication infrastructure through satellite and terrestrial systems, serving fixed and mobile phone users via public and private telephone networks. IMT-2000 offers speeds of 144 Kb/s to 2 Mb/s. It allows operators to access methods and core networks to openly implement their technologies, depending on regulatory, market and business requisites. IMT-2000 provides high-quality, worldwide roaming capability on a small terminal. It also offers a facility for multimedia applications (Internet browsing, e-commerce, e-mail, video conferencing, etc.) and access to information stored on PC desktops.

- **IS–54** – TDMA-based technology used by the D-AMPS system at 800 MHz.
- **IS–95** – CDMA-based technology used at 800 MHz.
- **IS–136** – The wireless operating standard for TDMA over AMPS. It was previously known as D-AMPS (Digital AMPS). It was also known as US TDMA/IS-136 which was the first digital 2G system adopted in the U.S.
- **PCS** – Personal Communications Service. The PCS is a digital mobile phone network very similar to cellular phone service but with an emphasis on personal service and extended mobility. This interconnect protocol network was implemented to allow cellular handset access to the public switched telephone network (PSTN). While cellular was originally created for use in cars, PCS was designed for greater user mobility. It has smaller cells and therefore requires a larger number of antennas to cover a geographic area. The term "PCS" is often used in place of "digital cellular," but true PCS means that other services like messaging (paging, fax, email), database access, call forwarding, caller ID, and call waiting are bundled into the service.

  In 1994 the FCC auctioned large blocks of spectra in the 1900-MHz band. This frequency band is dedicated to offering PCS service access and is intended to be technology-neutral. Technically, cellular systems in the U.S. operate in the 824-MHz to 894-MHz frequency bands. PCS operates in the 1850-MHz to 1990-MHz bands. While it is based on TDMA, PCS has 200-kHz channel spacing and eight time slots instead of the typical 30-kHz channel spacing and three time slots found in digital cellular. PCS networks are already operating throughout the U.S.

  GSM 1900 is one of the technologies used in building PCS networks—also referred to as PCS 1900, or DCS 1900. Such networks employ a range of technologies including GSM, TDMA, and cdmaOne. Like digital cellular, there are several incompatible standards using PCS technology. CDPD, GSM, CDMA, and TDMA cellular handsets can all access the PCS network provided they are capable of operating in the 1900-MHz band. Single-band GSM 900 phones cannot be used on PCS networks.

- **PDC** – Personal Digital Cellular. This is a TDMA-based Japanese digital cellular standard operating in the 800-MHz and 1500-MHz bands. To avoid the lack of compatibility between differing analog mobile phone types in Japan (i.e., the NTT type and the U.S.-developed TACS type), digital mobile phones have been standardized under PDC. In the PDC standard, primarily six-channel TDMA (Time Division Multiple Access) technology is applied. PDC, however, is a standard unique to Japan and renders such phone units incompatible with devices that adopt the more prevalent GSM standard. Nevertheless, digitalization under the standard makes possible ever-smaller and lighter mobile phones which, in turn, has spurred market expansion. As a result, over 93% of all mobile phones in Japan are now digital.

- **PHS** – Personal Handy Phone System. Soon after PDC was developed, Japan developed PHS. It is considered to be a low-tier TDMA technology.

- **TDMA** – Time Division Multiple Access was the first U.S. digital standard to be developed. It was adopted by the TIA in 1992 and the first TDMA commercial system began operating in 1993. It operates at 800 MHz and 1900 MHz and is used in current PDC mobile phones. It breaks voice signals into sequential, defined lengths and places each length into an information conduit at specific intervals. It reconstructs the lengths at the end of the conduit. GSM and US-TDMA standards apply this technique. Compared to the FDMA (Frequency Division Multiple Access) applied in earlier analog mobile phones, TDMA accommodates a much larger number of users. It does so by more finely dividing a radio frequency into time slots and allocating those slots to multiple calls. However, a shortage in available channels is anticipated in the near future so a more efficient system adopting CDMA is currently being developed under IMT-2000.

- **UMTS** – Universal Mobile Telephone Standard. It is the next generation of global cellular which should be in place by 2004. UMTS proposes data rates of < 2Mb/s using a combination of TDMA and W-CDMA. IT operates at about 2 GHz. UMTS is the European member of the IMT-2000 family of 3G cellular mobile standards. The goal of UMTS is to enable networks that offer true global roaming and can support a wide range of voice, data, and multimedia services. Data rates offered by UMTS are 144 Kb/s for vehicles, 384 Kb/s for pedestrians, and 2 Mb/s for stationary users.

- **W-CDMA** – Wideband Code Division Multiple Access standard. Also known as UMTS, it was submitted to the ITU for IMT-2000. W-CDMA includes an air interface that uses CDMA but isn't compatible for air and network interfaces with cdmaOne, cdma2000, or IS-136. W-CDMA identifies the IMT-2000 CDMA Direct Spread (DS) standard. W-CDMA is a 3G mobile services platform based on modern, layered network-protocol structure, similar to GSM protocol. It was designed for high-speed data services and for internet-based packet-data offering up to 2 Mb/s in stationary or office environments. It also supports up to 384 Kb/s in wide area or mobile environments. It has been developed with no requirements on backward compatibility with 2G technology. In the radio base station infrastructure-level, W-CDMA makes efficient use of radio spectrum to provide considerably more capacity and coverage than current air interfaces. As a radio interface for UMTS, it is characterized by use of a wider band than CDMA. It also offers high transfer rate and increased system capacity and communication quality by statistical multiplexing. W-CDMA efficiently uses the radio spectrum to provide a maximum data rate of 2 Mb/s.

- **WLL** – Wireless Local Loop. It is usually found in remote areas where fixed-line usage is impossible. Modern WLL systems use CDMA technology.

- **WP-CDMA** – Wideband Packet CDMA is a technical proposal from Golden Bridge Technology that combines WCDMA and cdma2000 into one standard.

- **UWC-136** – UWC-136 represents an evolutionary path for both the old analog Advanced Mobile Phone System (AMPS) and the 2G TIA/EIA-136 technologies. Both were designed specifically for compatibility with AMPS. UWC-136 radio transmission technology proposes a low cost, incremental, evolutionary deployment path for both AMPS and TIA/EIA operators. This technology is tolerant of its frequency band of 500 MHz to 2.5 GHz.

Standards and Consortia

Success of a standard is based on industry-wide support of that standard. Open standards are also key for a vital communications industry. Some of the key standards bodies promoting next-generation cellular technologies are discussed next.

3GPP

The Third Generation Partnership Project (3GPP) is a global collaboration agreement that brings together several telecommunications standards bodies known as "organizational partners." The current organizational partners are ARIB, CWTS, ETSI, T1, TTA, and TTC.

Originally, 3GPP was to produce technical specs and reports for a globally applicable 3G mobile system. This system would be based on evolved GSM core networks and radio access technologies supported by the partners. In this case, Universal Terrestrial Radio Access (UTRA) for both Frequency Division Duplex (FDD) and Time Division Duplex (TDD) modes. The scope was subsequently amended to include the maintenance and development of the GSM technical specifications and reports including evolved radio access technologies such as GPRS and EDGE.

3GPP2

The Third Generation Partnership Project 2 (3GPP2) is a 3G telecommunications standards-setting project comprised of North American and Asian interests. It is developing global specifications for ANSI/TIA/EIA-41.

3GPP2 was born out of the International Telecommunication Union's IMT-2000 initiative which covered high speed, broadband, and Internet Protocol (IP)-based mobile systems. These systems featured network-to-network interconnection, feature/service transparency, global roaming, and seamless services independent of location. IMT-2000 is intended to bring high-quality mobile multimedia telecommunications to a worldwide market by:
- increasing the speed and ease of wireless communications.
- responding to problems caused by increased demand to pass data via telecommunications.
- providing "anytime, anywhere" services.

The UMTS Forum

The UMTS Forum is the only body uniquely committed to the successful introduction and development of UMTS/IMT-2000 3G mobile communications systems. It is a cross-industry organization that has more than 260 member organizations drawn from mobile operator, supplier, regulatory, consultant, IT, and media/content communities worldwide.

Some UMTS objectives are:
- Promote global success for UMTS/3G services delivered on all 3G system technologies recognized by the ITU.
- Forge dialogue between operators and other market players to ensure commercial success for all.
- Present market knowledge to aid the rapid development of new service.

Geography-Dependent Standards Bodies

Standards bodies govern the technical and marketing specifications of cellular and telecommunication interests. The European Telecommunications Standards Institute (ETSI) is defining a standard for 3G called the UMTS. The Japan Association of Radio Industries and Business (ARIB) primarily focuses on WCDMA for IMT-2000. In Canada, the primary standards development organization is the Telecommunications Standards Advisory Council of Canada (TSACC).

The American National Standards Institute (ANSI) is a U.S. repository for standards considered to be semi-permanent. The United States Telecommunications Industry Association (TIA) and T1 have presented several technology proposals on WCDMA, TDMA UWC-136 (based upon D-AMPS IS-136), and cdma2000 (based upon IS-95). ANSI accredits both TIA and T1. The primary standards working groups are TR45 (Mobile and Personal Communications 900 and 1800 Standards) and TR46 (Mobile and Personal Communications 1800 only standards).

Standards development organizations in Asia include the Korean Telecommunications Technology Association (TTA) and the China Wireless Telecommunications Standards Group (CWTS).

Market Data

The Yankee Group predicted that global handset unit sales would exceed 498.5 million in 2003, 542.8 million in 2004, and 596 million in 2005. New network technologies will drive the introduction of new services and applications that will create demand among consumers for the latest cell phone models. The introduction of new network features will drive sales for devices and personalized applications, and new data services will drive increased demand.

Other analysts predict 1.4 billion subscribers by the end of 2004, and 1.8 billion by the end of 2006. This compares to the estimate of 60 million cellular subscribers in 1996.

Some of the developing countries are seeing faster growth in the wireless subscriber base than developed countries. This is because the digital cellular access provides better quality and cheaper service compared to landline phones.

Today the market is dominated by 2G technologies such as GSM, TDMA IS-136, cdmaOne, and PDC. This is changing to a market that will be dominated by 2.5G and 3G technologies. It is believed that by 2006, 56% of subscribers—nearly 1 billion—will be using 2.5G or 3G technologies. GPRS terminals are forecast to be the largest 2.5G market—nearly 625 million terminals will be shipped by 2004. For the same year cellular terminals based on 3G will ship nearly 38 million terminals.

Market Trends

Cell phones provide an incredible array of functions and new ones are being added at a breakneck pace. You can store contact information, make task lists, schedule appointments, exchange e-mail, access the Internet, and play games. You can also integrate other devices such as PDAs, MP3 players, and GPS receivers. Cell phone technology will evolve to include some of the trends discussed next.

Smart Phones—Converging Cellular Phones, Pagers and PDAs

A smart phone combines the functions of a cellular phone, pager, and PDA into a single, compact, lightweight device—a wireless phone with text and Internet capabilities. It can handle wireless phone calls, hold addresses, and take voice mail. It can also access the Internet and send and receive e-mail, page, and fax transmissions. Using a small screen located on the device, users can access information services, task lists, phone numbers, or schedules.

The smart phone market is growing at a phenomenal rate. As the mobile workforce evolves, a premium is being placed on mobility and "anytime, anywhere" data access. In 2002, analysts predicted nearly 400 million cell phone subscribers worldwide. Since that time, smart phone production rose 88% to meet increased demand.

Although smart phones have been around since 1996, they are just coming of age. As wireless technology takes off, manufacturers have begun vying for a slice of the smart phone pie. Telecom majors Nokia, Qualcomm, and Motorola were among the earliest contenders offering Web-enabled phones. Now, new companies are dedicating themselves wholly to the production of smart phones.

Smart phone technology has also led to a spate of joint ventures. The latest is a merger between software giant Microsoft and telecomm company LM Ericsson. The alliance calls for joint development of mobile phones and terminals equipped with Microsoft's Mobile Explorer micro-browser and mobile e-mail for network operators. Microsoft will also develop a more sophisticated Mobile Explorer for smart phones based on an optimized version of Windows PocketPC.

So far, the primary obstacle to wide adoption of smart phones has been a lack of content designed for their tiny screens. There is also a lack of support for color or graphics. However, over the past year several content providers (AOL-Time Warner, Reuters, etc.) and Internet portals have announced efforts to close the gap. Early last year, CNN and Reuters both announced content delivery plans aimed at smart phones. Soon web sites with location- and topic-specific information will be available from any WAP (Wireless Application Protocol)-enabled smart phone. The sites will deliver proximity-based listings of real-world stores. This will provide smart phone users with relevant, local merchant information such as store hours, specials, and parking information.

Conflicting wireless standards have so far limited the growth of the smart phone segment in the U.S. Many vendors still employ analog cellular technology which does not work with smart phones. Although vendors are trying to move to digital technology, they are unable to reach a consensus on which specific technology to use. In Europe and Asia, GSM is the digital cellular technology of choice. However, in the U.S. GSM is just one of several technologies vying for acceptance. Until standardization is achieved, smart phones may be slow to catch on.

Video or Media Phones

Following e-mail, fax-mail, and voice-mail, video-mail is seen as the evolutionary step for cellular terminals. Sending and receiving short text messages is currently one of the fastest growing uses for mobile phones. Within the next few years it will also be used to send video messages as 3G mobile phones become available

First generation videophones are just coming to the market. These PDA-cell phone convergence devices incorporate a small camera in the terminal that can record short visual images to be sent to other compliant wireless devices. Handsets in the future will include voice, video, and data communications. They will have full-PDA capabilities including Internet browsing, e-mail, and handwriting recognition.

Imaging Phones

The imaging phone can carry strong emotional attachment because it allows the sharing of experiences through communication. It facilitates person-to-person or person-to-group multimedia communication based on self-created content. It includes excellent messaging and imaging, an easy-to-use interface, and a compact and attractive design. It is fun and can be personalized for individual use.

Bluetooth-Enabled Cell Phones

Bluetooth is a personal area networking standard that is inexpensive and power-friendly. It offers short-range radio transmission for voice and data. It uses frequency-hopping technology on the free, 2.4 GHz industrial, scientific, and medical (ISM) band. It was developed to allow a user to walk into a room and have a seamless, automatic connection with all other devices in that room. These devices could include cellular phones, printers, cameras, audio-visual equipment, and mobile PCs. Bluetooth enables mobile users to easily connect a wide range of computing and telecommunications devices without the need for cables. Also, it connects without concern for vendor interoperability. Other Bluetooth applications include synchronization, email, Internet and Intranet access, wireless headsets, and automobile kits for hands-free communication while driving. Bluetooth radios, which theoretically are placed inside every connected device, are about the size of a quarter and can transfer data at 728 Kb/s through walls and around corners. The Bluetooth standard was established by the Bluetooth special interest group (SIG) in July 1999 by founding companies Intel, IBM, Toshiba, Ericsson, and Nokia. Today it is supported by over 2000 companies.

Cellular terminals represent the biggest single opportunity for Bluetooth in its development. Initial development of Bluetooth technology allows for a transmission range of only 10 m. But its next stage of development will allow transmission over a distance up to 100 m. The penetration of Bluetooth into cellular terminals is forecast to rise from (1% in 2000) to 85% by 2005. This corresponds to an increase in shipments of Bluetooth-enabled cellular terminals from 4 million in 2000 to over 900 million in 2005.

Mobile phones represent the single largest application for Bluetooth technology. They allow Bluetooth-enabled headsets and handsets to communicate without the need for wires. This will enable drivers to safely use their phones while driving, for example. And from a health perspective, the phones will be located away from the brain, a subject which has generated much discussion over the last few years.

Mobile workers are a key target market for all Bluetooth products. Bluetooth will enable mobile workers to link their PDAs and portable computers to the Internet and to access their mobile terminals without the need for a cable link. However, mobile workers today represent only a small portion of the total cellular subscriber base. Although mobile workers may drive early sales of Bluetooth cellular phones, this application alone will not drive massive Bluetooth cellular terminal application volumes. For high volumes to be achieved, users must switch to communication with headsets or by a number of the other applications.

Cordless terminals could be replaced by data-function terminals that can link to cellular networks. They could interface with an access point in the home to route calls via the PSTN network. Bluetooth could also link smart phones into the PSTN via conveniently located public access points for accessing email and other Internet services. Use of Bluetooth local radio standard and infrared links will allow mobile phones to communicate with shop terminals. This could lead to a revolution in shopping, going far beyond the replacement of credit cards.

Wireless LANs—Friend or Foe

Wireless LANs (WLANs) provide high-bandwidth, wireless connectivity for a corporate network. They seamlessly provide a connection to the Internet and the ability to access data anywhere, anytime for the business user. Wireless LANs combine data connectivity with user mobility to provide a good connectivity alternative to business customers. WLANs enjoy strong popularity in vertical markets such as healthcare, retail, manufacturing, and warehousing. These businesses enjoy productivity gains by using handheld terminals and notebook PCs to transmit information to centralized hosts for processing. With their growing popularity, cellular phone manufacturers and operators wonder if WLANs will eat into the profit potential of flashier 3G mobile carriers.

The growing ubiquity of WLANs will likely cause wireless carriers to lose nearly a third of 3G revenue as more corporate users begin using WLANs. The growth of public WLAN "hotspots" in airports, hotels, libraries, and coffee shops will increase WLAN revenue to over $6 billion by 2005 with over 15 million users. This compares to over $1 million earned by WLANs in 2000. Over 20 million notebook PCs and PDAs are shipped every year, showing the potential for WLAN networks. They pose a real threat to cellular carriers. Fueling the increased interest in WLANs is the growing availability of fast Internet access and the increasingly common use of wireless networks in homes and office. In comparison to the rise of WLANs, the handheld market is in a free-fall and mobile carriers are stumbling. In addition, the slow introduction of intermediate 2.5G mobile technology is aiding a 10% decline in cell phone sales.

WLANs are an attractive alternative to mobile 3G because of their convenient installation and ease of use. And they enable consumers to use the larger screens on their notebook PCs for better viewing. In contrast to the reported $650 billion spent worldwide by carriers to get ready for 3G, setting up a WLAN hotspot requires very little. An inexpensive base station, a broadband connection, and an interface card using the WLAN networking standard are all that's needed. And WLAN cards are readily available for laptops, PDAs, and smart phones. However, the two technologies use different radio frequencies and are targeting different markets. Where 3G is mostly phone-based and handles both voice and data, WLAN is purely data-driven.

While carriers have invested heavily in 3G, they are more financially sound. WLANs do not have the capital to compete with Cingular, Sprint, NTT DoCoMo, British Telecom, or AT&T. Examples such as the failure of WLAN provider MobileStar (VoiceStream recently purchased MobileStar) do not bode well for the technology. It is unlikely that WLANs will expand outdoors beyond a limited area, such as a corporate campus. WLANs are a potentially disruptive technology. While network equipment providers will see a huge demand of WLAN products, WLAN network operators must determine ways to charge subscribers.

WLANs still face technical and regulatory issues. For instance, there are already potential interference issues between IEEE 802.11b and Bluetooth. These issues will only worsen as the technologies become more widely deployed. Also, WLAN security needs improvement. The Wired Equivalency Privacy (WEP) standard is the encryption scheme behind 802.11 technology. However, WEP encryption must be installed manually and has been broken. Hence, next-generation WLAN specifications are implementing higher security standards. While carriers will survive WLAN competition, the cost of recent auctions of 3G airwaves may lead to WLAN encroachment on potential profit from 3G services.

Rather than being viewed as a threat, WLAN could help introduce consumers to the wireless world. Today's WLANs are about five times faster and more accurate than 3G will be. It is unlikely that a 3G network could ever be cost-effectively upgraded to speeds on par with WLANs. To serve demanding applications, 3G wireless network operators need public wireless LANs. 3G operators and public WLAN operators must start working together to ensure their mutual success. This cooperation will enable public WLANs to obtain sufficient coverage to be truly useful. Wireless operators need the WLANs to offload heavy indoor traffic. Like wireless operators, WLAN operators need subscription control, roaming agreements, and centralized network management. There are significant e-commerce opportunities for WLAN operators, most notably location-dependant, targeted promotions. Public WLANs could serve as a distribution point for multimedia. In the future, the market will see an introduction of products that switch between 3G and WLAN technologies to provide seamless coverage.

From a cost benefit viewpoint the use of the combined technologies could gain traction. But from the technology and market perspective the implementation of such convergence seems unlikely. To begin with, 3G offers wide area mobility and coverage. Each base station typically covers a 5-10 mile radius and a single wide area cell is typically equivalent to 10,000 WLAN cells. In contrast, WLAN offers mobility and coverage across shorter distances—typically less then 150 feet. The two technologies address different market needs and are more likely to share the nest than to compete for it. While WLANs offer many benefits at a local level, they will never match the benefits of wireless access to the Internet since they were never intended to. There is also the question of air interfaces—while Bluetooth and some WLAN technologies share the 2.4 GHz and 5 GHz frequency bands, 3G does not. The question of access arises. Unless wireless manufacturers develop devices that unite 3G and WLAN and combine their air interfaces, convergence will never happen.

Whatever the result, WLANs have a bright future. They offer a range of productivity, convenience, and cost advantages over traditional wired networks. Key players in areas such as finance, education, healthcare, and retail are already benefiting from these advantages. With the corporate success of IEEE 802.11b or WiFi, the WLAN market is slowly advancing. WiFi operates at 2.4 GHz and offers as much as 11 Mb/s data speed. With the arrival of next-generation 5 GHz technologies such as HiperLAN2 and IEEE 802.11a, faster (up to 54 Mb/s) and better service will be addressed. Consumers, corporate travelers, and vertical market employees will see a better alternative in services at home and in public areas. WLANs have the capability to offer a true mobile broadband experience for Internet access, entertainment, and voice. Mobile operators may not be able to offer this for a while. 3G mobile phone operators will need public WLANs to offload heavy indoor traffic from their lower speed, wide area networks. Though WLANs may pose a threat to 3G revenue, 3G is still some years away and WLANs are already available. The longer the delay in 3G introduction, the greater will be the usage of WLANs.

A comparison of wireless technologies such as Bluetooth, WLANs, and cellular reveals room for each standard. Bluetooth provides a low-cost, low-power, wireless cable replacement technology. WLANs provide a good solution for bandwidth-intensive distribution of data and video. However, WLANs will require too many base stations or access points to provide similar coverage as cellular. Cellular technologies will focus on broad geographical coverage with moderate transfer rates for voice and data services. Personal, local, and wide area networks are well covered by the three technologies.

Touch Screen and Voice Recognition as Dominant User Interfaces

The alphanumeric keypad has been the most dominant user interface. With the emergence of the wireless Internet and the issue of cellular handset size, manufacturers must provide a way to incorporate maximum screen display with minimal terminal size.

It is predicted that voice recognition will become the dominant user interface in the future. In the interim a touch screen is projected to co-exist with voice to provide functions that are not currently possible with voice recognition alone. Touch screen has been widely available for a few years, but has only recently been used on mobile phones. With the right software interface it allows a novice to easily interact with a system.

A number of products are coming to market that have a touch screen display and a voice recognition address book. This will allow cellular handset manufacturers to eliminate alphanumeric keypads. Inputs to the handset will be via touch screen, voice recognition, or a combination of both. This will maximize the screen area for data applications and optimize the handset size.

Once reserved for specialized applications, voice input technology is now being developed for the mobile and commercial markets. Voice activation is projected to become standard in the next generation of terminals. As phones shrink in size their keyboarding capabilities are becoming too limited for useful web surfing. Voice recognition solves this problem.

Dual Band vs. Dual Mode

Travelers will probably want to look for phones that offer dual band, dual mode, or both

- **Dual band** – A dual band phone can switch frequencies. It can operate in both the 800 MHz and 1900 MHz bands. For example, a dual band TDMA phone could use TDMA services in either an 800-MHz or a 1900-MHz system.
- **Dual mode** – "Mode" refers to the type of transmission technology used. A phone that supported AMPS and TDMA could switch back and forth as needed. One of the modes must be AMPS to give you analog service where digital support is unavailable.
- **Dual band/Dual mode** – This allows you to switch between frequency bands and transmission modes as needed.

Changing bands or modes is done automatically by phones that support these options. Usually the phone will have a default option set, such as 1900 MHz TDMA. It will try to connect at that frequency with that technology first. If it supports dual bands, it will switch to 800 MHz if it cannot connect at 1900 MHz. If the phone supports more than one mode, it will try the digital mode(s) first, and then switch to analog.

Some tri-mode phones are available; however, the term can be deceptive. It may mean that the phone supports two digital technologies such as CDMA and TDMA, as well as analog. It can also mean that it supports one digital technology in two bands and also offers analog support. A popular version of the tri-mode phone has GSM service in the 900-MHz band for Europe and Asia, and the 1900-MHz band for the United States. It also has analog service.

Multi-mode/band

A multi-mode terminal is a cellular phone that operates with different air interfaces. It offers wider roaming capabilities and it can use different networks under different conditions for different applications.

There is no single cellular standard available in every country worldwide, and certain standards are only used in a small number of countries. When travelling to foreign countries, cellular subscribers often find that there are no cellular networks on which their terminals work. For example, a cdmaOne user from the U.S. could not use a cdmaOne terminal on any networks within the European Union. In the EU there are no cdmaOne networks. Similarly, a Japanese user could not use a PDC terminal except in Japan because there are no PDC networks outside Japan. In many countries such as China, the U.S., and Russia, individual digital cellular networks do not provide nationwide coverage. However, combinations of networks using different standards do give nationwide cellular coverage. Terminals that can operate under a number of different standards have to be used to address this problem.

The next generation of networks will reveal the availability of services and features in specific areas. Multi-mode terminals will allow adoption of higher data rates where they can be accessed. When operators begin to upgrade their networks for faster data rates and increased services, this may not be done across the operators' whole service area. The initial investment to do so would be too high. Operators will probably offer the new network technologies in areas of high cellular subscriber penetration first. This will enable them to offer the advanced services to as wide a customer base as possible with the smallest investment. This is expected to occur with the change from 2G to 2.5G and from 2.5G to 3G services.

The ability to offer connection to different networks under different conditions for different applications is the driving force for multi-mode terminals. It will allow users from one network (e.g., 2G) to have higher data rates, extra features, and more services in another network (e.g., 3G).

Multi-mode combinations already exist in the mass market. CdmaOne/AMPS and TDMA/AMPS combinations are present in the U.S. and PDC/GSM combinations are present in Japan. The European market has not seen many multi-mode terminals because of the omnipresence of GSM between European nations. With the launch of 2.5G and 3G networks, future dual mode phone markets are projected to consist of 2G/2.5G, 2G/3G, and 2.5G/3G technologies.

Positioning Technologies
Mobile Internet location-based services are key to unlocking the potential of non-voice services. Without location, web services are more suited to the PC than a mobile phone. Location awareness and accurate positioning are expected to open up a vast new market in mobile data services. It it has been demonstrated with GPS that future localization of information with up to four meters of accuracy will be possible with handsets. Positioning systems will enable m-commerce and mobile Internet. It will also enable the location of 911 calls from a mobile phone. At the same time, it is unclear how to define a coherent strategy for adding location to consumer and business services. Positioning technologies are not new, but they have not been quickly accepted by the cellular industry. This has been due mainly to limitations with infrastructure technology or the cost of deployment. Also, there has not been high public demand for such technologies considering the high price for terminals that adopt such technology.

Wireless Data
Wireless data capability is slowly being added to mobile appliances. But the wireless networks to which the devices are connecting and the speeds at which those connections are made will determine what services may be offered.

Today's wireless networks can be divided into analog and digital systems. Digital systems provide cheaper and more advanced voice and data transmission. Currently the world's wireless networks are divided into a number of incompatible standards such as GSM, CDMA, and TDMA. As a result there is no unified global network. Network carriers will not be able to implement a network standard compatible across geographies before the introduction of 3G wireless (1-2 Mb/s through-put). This technology is expected to roll out in the next 2-3 years in Europe and Asia and in the next 3-4 years in the United States.

Digital systems can carry data using circuit-switched technology or packet-switched technology. Circuit-switched technology establishes a dedicated link between sender and receiver to provide better data transmission, but cannot be shared by others. Packet-switched technology is always online and sends and receives data in small packets over a shared link. Because of competitive issues, U.S. wireless carriers have rolled out incompatible wireless networks. These make it difficult for users to roam and for device makers to offer globally compatible devices.

The following sections show how networks have approached wireless data services.

- Carriers using CDMA and TDMA networks who want to leverage existing wireless network assets offer wireless data with speeds between 9.6 Kb/s and 14.4 Kb/s
- Carriers with GSM networks have developed their own standard, GPRS, for wireless packet-switched data transmission with speeds of 115 Kb/s
- Terrestrial data networks such as American Mobile's DataTAC system, BellSouth Wireless Data's Mobitex system, and Metricom's Ricochet system use packet data for wireless data transmission between 19.2 Kb/s and 128 Kb/s.

Packet data service in the U.S. currently offers the best wireless data coverage and speed. Packet data carriers include American Mobile Satellite and BellSouth Mobile. On the horizon is the availability of evolved 2G and 3G wireless data service that offers broadband speeds over wireless networks. 3G would require the rollout of a completely new network. This might be possible in small regions in Europe and Asia, but it would be an enormous undertaking in the U.S. A number of 2.5G services may be rolled out over the next three years in the U.S. that allow 128 Kb/s transfer speeds. This would enable wireless streaming audio, video, and multimedia content.

The GPRS, mentioned earlier, is of special interest. It is a 2.5G technology that effectively upgrades mobile data transmission speeds of GSM to as fast as 115 Kb/s from current speeds of 9.6 Kb/s to 38.4 Kb/s. The higher speed is expected to be especially useful for emerging PDA applications. The full commercial launch of GPRS services occurred in mid-2001, which was concurrent with the launch of a new range of GPRS-enabled phones and PDAs.

Connecting Wireless Networks and the Internet

Wireless data transmission connects the mobile user to the carrier's network. To connect that network to data located on the Internet or with an enterprise, data networking protocols must be integrated over the wireless networks, or a gateway connecting the wireless and external networks must be built. Most carriers have chosen to use Internet protocols (TCP/IP) as the basis for connecting wireless devices with the Internet and with other external networks.

There are currently three ways to connect mobile devices to data residing on the Internet or to external networks:

- **One-Way Wireless Access** – Data such as brief messages and alerts can be sent out in one direction from the network to the device.
- **Serial Port or USB** – Data are downloaded via serial port or USB to a personal consumer device during synchronization with a connected device. The user can then access, manipulate, and repost to the network or Internet upon the next synchronization.
- **Two-Way Wireless Messaging** – Data can be sent wirelessly between a user's mobile consumer device and another device via the Internet or a network. This allows users to seamlessly access and manipulate data residing on the network via their mobile devices.

While the first two are more commonly used, two-way wireless will be the most compelling method over the next few years and will drive the creation of many applications and services that leverage this mobile connectivity.

WAP

The Wireless Application Protocol (WAP) is an open, global specification that empowers wireless device users to instantly access information and services. It will enable easy, fast delivery of information and services to mobile users on high-end and low-end devices such as mobile phones, pagers, two-way radios, and smart phones. WAP works with most wireless networks such as:

- CDPD
- CDMA
- GSM
- PDC
- PHS
- TDMA
- FLEX
- ReFLEX
- IDEN
- TETRA
- DECT
- DataTAC
- Mobitex

WAP is both a communications protocol and an application environment. It can be built on any operating system including PalmOS, EPOC, Windows PocketPC, FLEXOS, OS/9, and JavaOS. It can also provide service interoperability between different device families.

WAP includes a micro browser and a network server. With minimal risk and investment, operators can use WAP to decrease churn, cut costs, and increase revenues. Technology adoption has been slowed by the industry's lack of standards that would make handheld products Internet compatible. WAP provides a de facto global open standard for wireless data services.

WAP is similar to HTML as it transformed the Internet into the World Wide Web. The protocols are similar and will allow wireless devices to be a simple extension of the Internet. WAP was defined primarily by wireless equipment manufacturers and has support from all major, worldwide standards bodies. It defines both the application environment and the wireless protocol stack.

The WAP standard allows Internet and database information to be sent direct to mobile phone screens. It is an open standard so information can be accessed from any WAP-compliant phone. It uses the Wireless Mark-up Language (WML) to prepare information for presentation on the screen. The graphics are simplified and the phones can call up existing Internet content. And like HTML, WAP includes links to further information.

M-Commerce

Cellular operators are increasingly looking to m-Commerce (Mobile Electronic Commerce) to attract new business and maintain existing business. M-Commerce is forecast to progress from the e-commerce services available today (via fixed data line services such as the Internet) to mobile data services such as WAP.

Alliances among operators, content providers, financial institutions, and service administrators will build this market. Examples of initial services being offered are links to cinema Internet sites for purchasing movie tickets and a platform allowing purchases from any site on the Internet. Mobile devices will also be used to handle short-distance transactions to point-of-sale machines.

i-mode

Japan is the first country to offer mobile Internet through its i-mode service. I-mode is a packet-based mobile Internet service which allows users to connect to the Internet without having to dial up an Internet service provider. Considering its over 25 million subscribers, it is surprising that this has been achieved at the slow network speed of 9.6 Kb/s. It is text-based with limited capabilities to send low-quality still images. The service is fairly inexpensive considering that the network is packet based, meaning charges are not based on airtime but on volume of data transmitted or received.

In addition to voice and e-mail, there is a plethora of information and content services available such as retrieving summarized newspaper texts, making travel arrangements, finding an apartment or a job, playing games, or downloading ringing tones for mobile phone. Mobile e-mail is a hit with Japanese teenagers who can use truncated text to converse with each other while in earshot of parents or teachers!

The service is based on compact HTML which has successfully promoted and differentiated i-mode from WAP-based services. Since it is a subtext of HTML, users with a homepage can reformat their web pages for i-mode access.

There are a number of factors that contribute to i-mode's success in comparison to other services in Japan. It is easier to use and has better quality content. NTT DoCoMo, i-mode's service provider, provides aggressive advertising. And market factors such as the demand for internet access services, timing, and the proliferation of short message service (SMS) and e-mails.

I-mode is also very entertainment orientated—entertainment content accounts for 55% of its total access. This includes limited-functionality games, musical tunes for the phone's incoming ring, daily horoscopes, and cartoon characters for the phone's display.

Comparatively low value-added content accounts for over 50% of total access with the majority of users in their teens and 20's. This indicates that i-mode demand is still in a boom and has yet to stabilize.

As other functionality such as file attachments, faster network speeds, and e-commerce transactions are added, new services and applications will be developed. If prices stay stable and usage remains simple, i-mode will be used for more value-added applications by a broader group of users.

Technologically, i-mode is not particularly revolutionary but its marketing strategy is. By using a subtext of HTML, NTT DoCoMo has enabled content providers to readily reformat their web sites for i-mode access. This has undoubtedly contributed to rapid i-mode content development. Also, NTT DoCoMo offers billing and collection services to content providers. And, Internet access is easy and the packet-based fee structure is reasonable. Currently, NTT DoCoMo is in discussion with U.S. operators with a view toward launching i-mode mobile Internet services in America.

E-mail and SMS (Short Messaging Service)

SMS allows short text messages to be exchanged between cellular phones. E-mail service basically comes standard with a web-enabled cellular terminal. To access e-mail the terminal and network must be able to connect and deliver e-mail from a standard POP account. Although SMS has been available for a few years, the service available with this technique is very limited. It has a 160-character text limit and transfers only data between cellular terminals. E-mail via a cellular phone, however, enables communication with other cellular phone users (as with SMS) and with anyone who has Internet access via cellular phone, personal computer, television, PDA, etc.

Initial deployment of the service in Japan has proved extremely popular. If this trend continues in Europe, the demand for e-mail through cellular phones will be huge. And as e-mail becomes more sophisticated with the ability to send pictures and video clips via handsets, cellular phone email service will continue to gain in popularity.

Other cellular phone trends

- Color LCDs have started to appear in the market.
- With handset prices continuing to fall, the disposable cellular handset market is slowly becoming a reality.
- Handsets have reached the ideal size.
- Battery life has continued to improve with higher standby times.
- Integration of an MP3 player
- FM and digital radio capability
- Embedded digital camera capability
- Smart card readers for online purchases and verification
- TV reception
- Enhanced multimedia messaging
- Entertainment (ring tones, logos, etc.)
- Vertical markets (logistics and sales force automation)
- Satellite handset phone systems could impact the cellular handset market. Satellite phone systems can transmit at very high rates which will offer satellite phone users similar capabilities as 3G cellular—video, conferencing, etc. The failure of Iridium LLC, which filed for bankruptcy in 1999 with only 50,000 subscribers, has seen limited success and bad press for satellite phones. However, other satellite phone service operators are looking to reach about 10 million users by 2006.

Components of a Mobile/Cellular Handset

Cell phones are extremely intricate devices. Modern digital cell phones can process millions of calculations per second to compress and decompress the voice stream.

The typical handset contains the subsystems described in the next sections.

- **DSP/Baseband Processor and Analog codec subsection** – The digital baseband and analog section includes:
  - Channel and speech encoding/decoding
  - Modulation/demodulation
  - Clock
  - D/A and A/D conversion
  - Display
  - RF interfacing.

  This section performs high-speed encoding and decoding of the digital signal and manages the keypad and other system-level functions. For analog voice and radio signals to become digital signals that the DSP can process, they must be converted in the analog baseband section. The voice and RF codecs perform analog-to-digital and digital-to-analog conversion and filtering. The handset processor is responsible for functions such as keyboard entry, display updates, phonebook processing, setup functions, and time functions. These functions are typically handled by an 8-bit or 16-bit micro-controller, but increasingly the processor is being integrated into the baseband chip.

- **Memory** – This includes SRAM, EEPROM and flash. Flash is a non-volatile memory that provides random access to large amounts of program information or code. There are two types of flash memories—NAND and NOR. NAND is better suited for serial access of large amounts of data. SRAM is much faster than flash and is used as low-power/low-density

RAM for cache. ROM and flash memory provide storage for the phone's operating system and customizable features such as the phone directory.

- **RF section** – The RF section includes the RF/IF transmitter and receiver ICs, RF power amplifiers, PLLs, synthesizers, and VCOs. These components receive and transmit signals through the antenna. RF functions include modulation, synthesis, up and down conversion, filtering, and power amplification.

- **Battery and/or power management** – This section supplies and controls power distribution in the system since different components often operate at different voltages. The power management unit also controls power-down and stand-by modes within the handset and regulates the transmitter power of the handset's power amp. Phones spend the majority of their time waiting for a call, when much of the phone's circuitry is not required and can be placed in a low-power standby state. When a call comes through, the power management system powers-up the sleeping components to enable the call. During the call, the power management section continuously adjusts the phone's transmitted power to the level required to maintain the connection to the base station. Should the handset get closer to the base station, the handset's output power is reduced to save battery power. The power management function also regulates the battery charging process and feeds battery charge state information to the phone's processor for display. The charging and monitoring components for Lithium-ion batteries are typically the most complex so these types of batteries usually contain power management chips embedded in the battery housing. Conversely, Nickel-Metal-Hydride and Nickel-Cadmium batteries typically have control chips embedded within the phone itself.

- **Display** – Cell phone displays can be color or monochrome. Some are backlit, some reflective, while others are organic electroluminescent displays.

- **Operating system** – Several vendors have announced operating systems for the cell phones. Microsoft has introduced the PocketPC 2002 Phone Edition which is a PDA operating system based on Windows CE 3.0.

- Keypad/keyboard

- Microphone and speaker

A mobile handset block diagram is shown in Figure 5.1.

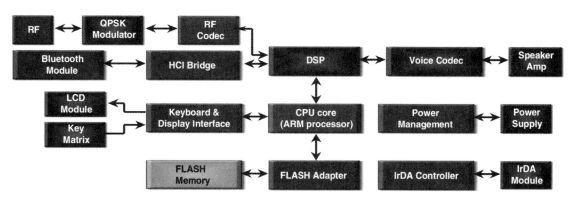

Figure 5.1: Mobile Handset Block Diagram

Summary

People with talking devices can now be seen everywhere. Cellular phones are the most convenient consumer device available. And mobile phones are really just computers—network terminals—linked together by a gigantic cellular network.

Cellular technology started as an analog (1G) network but is migrating to digital. Just as the 2G digital cellular technology is ramping up, there is much discussion about third-generation (3G) cellular services. While 3G deployments are currently underway, 2.5G is being seen as the steppingstone for things to come. 2G, 2.5G, and 3G networks are all made up of combinations of technologies that prevail in different geographies. The vision of 3G includes global and high-speed mobile access with enhanced IP-based services such as voice, video, and text.

The cell phone was initially a basic voice access system with a monochrome display, but is evolving to include voice, video, text, image access with color display, Internet access, and integrated digital camera. Other evolving wireless technologies such as Bluetooth and wireless LANs are looking to complement next-generation cellular technologies. The future for wireless and mobile communications remains very bright and undoubtedly a host of new applications will be generated in the years to come.

Gaming Consoles

Gaming consoles are among the most popular consumer devices to come on the market in the last few years. While they have existed for a while, their popularity is growing because of higher processing power and a larger number of available games. While several companies have introduced gaming consoles, the current market has three main hardware providers—Microsoft, Nintendo, and Sony.

Definition

Gaming consoles deliver electronic game-based entertainment. They feature proprietary hardware designs and software operating environments. They include a significant base of removable, first and/or third party software libraries. Gaming consoles are marketed primarily for stationary household use. They rely on AC power for their primary energy source and they must typically be plugged into an external video display such as a television. Examples of popular gaming consoles include the Nintendo GameCube, Sony PlayStation2, and Microsoft Xbox. They can also provide Internet and email access. Gaming consoles are platform independent from PCs, video-game consoles, and NetTVs. They are also media independent from Internet, cable television, and DBS.

Note that certain consoles are excluded from this definition:

- Machines with embedded software only (i.e., containing no removable software)
- Those without a significant library of external first and/or third party title support. (Examples: Milton Bradley's Pocket Yahtzee and embedded LCD pocket toys from Tiger Electronics)
- Systems geared for development, hobbyist, or other non-mass-entertainment use (Example: Sony's Yaroze)
- Those not targeted primarily as a gaming device (e.g., interactive DVD players)

Handheld gaming devices are designed for mobile household use. They rely on DC (battery) power and they include an embedded video display such as a LCD. Examples include the Nintendo GameBoy Color and Nintendo GameBoy Advance.

Market Data and Trends

Currently there are about 29 million console players, 11 million PC game players, and 7 million who play both. There are clear differences in content so there is limited customer overlap. And the consumer pool is large enough to allow for multiple companies to do well.

IDC predicts that for 2003 shipments of and revenue from game consoles in the U.S. will exceed 21.2 million units and U.S. $2.3 billion, respectively. This is down from a peak of $3.2 billion in 2001. The worldwide shipment of game consoles is forecast to exceed 40 million units for the same year.

According to the Informa Media Group, the worldwide games industry is expected to grow 71% from $49.9 billion in 2001 to $85.7 billion in 2006. This includes everything from arcades to game consoles to PC games, set-top box games, and cell phone games. Analysis by the investment bank Gerard Klauer Mattison predicts that worldwide console sales will double in the next five years to a total of 200 million units. There should also be new opportunities for game startups within this fast-growing market, especially for wireless games and devices.

Market Trends

Some of the market trends for gaming consoles are:

- Generation Y is entering the teenage years with a keen interest in games and interactive entertainment
- The growing pervasiveness of the Internet will help online gaming.
- Online gaming with high-speed access and in-home gaming between consoles and PCs will increase demand for home networking and broadband access solutions.

Market Accelerators

- **Market momentum** – The momentum driving today's video game market cannot be underestimated. Growth beyond the hardcore gamer has given the market a much-needed boost in exposure and revenue. While the hardcore gamer is still important, continual market growth will be driven by the casual gamer. This has led to tremendous momentum, priming the market for the launch of next-generation hardware and games as well as related products including videos, toys, and gaming paraphernalia.
- **Games** – The availability of high-quality games along with the launch of a new platform can significantly accelerate video game platform sales. Conversely, it can also inhibit growth. The importance of games as a purchase factor cannot be underestimated. In many cases the games, not the platform, sell the new technology. For instance, the long-term popularity of older-generation platforms such as PlayStation and Nintendo 64 has been sustained in part by the continued introduction of quality and original game titles. Efforts to expand into other demographic segments will result in new and innovative games that will, in turn, spur market growth.
- **New functionality** – Better games will be a primary driver for console sales. Over the long term these new platforms will have greater appeal because of:
  - Additional features such as DVD-video and online capabilities.
  - The potential for the console to exist as a set-top box and residential gateway.

Market Inhibitors

- **Limited success of on-line, console-based gaming** – Quality of service (QoS) and high-bandwidth connection to the Internet are required to enable real-time, high-speed video exchange online gaming. The slow growth of high-speed broadband limits online gaming.
- **Low-cost PCs** – With the prices of PCs plummeting, consumers are encouraged to own multiple PCs. Moore's Law allows availability of features, performance improvements, and lower prices—all of which make the PC a gaming platform.
- **Higher prices** – With prices of consoles and games remaining high, consumers must consider the number of years they will be keeping their gaming console. In comparison, a PC provides more benefits and functionality than a gaming device.
- **Lack of games** – Improved hardware technology has been and will continue to be an attractive selling point for next-generation consoles. But the availability of new and original

titles will be just as important, if not a more important. Gaming consoles will come under fire for a lack of games that take advantage of improved hardware. A lack of games could inhibit growth—a problem that has historically been encountered by the release of new products.

- **Poor reputation** – Although much of the negativity surrounding the video game industry after the school shootings in 1999 has subsided, the threat of negative publicity remains problematic. The Interactive Digital Software Association (IDSA) has taken many steps to mitigate negative publicity through special studies and lobbying efforts. Despite evidence that video games are not linked to violent behavior, the link remains and does not promise to go away anytime soon. If and when acts of violence do surface, fingers are sure to point to the video game industry.
- **Market confusion** – The availability of three mainstream video game console platforms and the continued availability of older platforms give consumers much choice. However, this choice could cause confusion and lead to buyer hesitation. This may be particularly problematic because the software industry hesitates to introduce games for specific platforms while awaiting public response. And consumers could delay purchasing existing products while waiting for the introduction of new, powerful competitive systems and features such as broadband.

Key Players

Microsoft Corporation

While relatively new to the market, Microsoft has a number of products in the home consumer space including its Xbox gaming platform. Xbox hardware is built with an impressive Intel processor, a custom-designed Microsoft graphics processor with Nvidia memory, and a hard disk drive. It supports HDTV and includes a DVD, four I/O ports for the game controller, and an Ethernet port.

Nintendo

Based in Japan, Nintendo was the original developer of the Nintendo 64 and the more recent Nintendo GameCube systems. The company has stated that it is not pursuing a home gateway strategy and that it will stick to games only.

Sega

Based in Japan, Sega brought to market the Dreamcast, a second-generation 128-bit gaming console with:

- CD player
- Built-in 56K modem for Internet play
- Internet access and e-mail
- Expansion slots for home networking.

Although Sega has exited the market to concentrate on gaming software, it was first to market with its second-generation game console. The company also provides accessories such as a keyboard, extra memory, and various controllers.

After Sega ceased production of the Dreamcast gaming console, the company refashioned itself as a platform-agnostic game developer. By the end of the year Sega will bring its games to handheld devices (such as PDAs and cell phones), set-top boxes, and even rival consoles. In addition Sega agreed to deliver Dreamcast chip technology to set-top box manufacturers to enable cable subscribers to play Dreamcast games. Sega has been the number three video game console supplier and has been hemorrhaging money since the introduction of Dreamcast in 1999. The release of Microsoft's Xbox most likely influenced Sega's move to exit the game console market.

Sony

Based in Japan, Sony Computer Entertainment Inc. is positioning itself as the consumer's one-stop shop for home consumer devices ranging from entertainment to personal computing. Sony has overcome a number of challenges to create a unified network of consumer devices and digital services. It brought together diverse parts of the company to work together on a single, cohesive Internet strategy. Sony is positioning its 128-bit PlayStation2 as a gateway into the home from which it can offer Internet access, value-added services, and home networking. Its home networking would use USB or iLink (also called IEEE 1394 or FireWire) ports to connect devices. PlayStation2 contains a DVD player that could enable digital video playback and could, in time, include an integrated digital video recorder. With its added functionality the Playstation2 is leaving the "game console" designation behind.

Game of War

Strategic differences between the superpowers of the video game console industry are surfacing as they launch the next versions of their products. In the battle for control of the living room, the contenders are:

- Nintendo, who has a lock on the market for kids who think games are just another toy.
- Sony, who is pushing the convergence of different forms of entertainment like DVDs and games.
- Microsoft who is making its pitch to game artists and aficionado players.

The Xbox management team has used the "game is art" statement for entering the console market, even though Sony sold 20 million PlayStation2 units before Microsoft sold its first Xbox. Sony drew its line in the sand with the PlayStation2 launch in March 2000 while Microsoft launched the Xbox in November 2001. Meanwhile, Nintendo had a double-barreled blast since it launched its GameBoy Advance and GameCube console in 2001 in the U.S.

The gaming titan most likely to prevail will have to do more than pontificate about its philosophy. The three companies have likely spent a combined $1.5 billion worldwide on ferocious marketing campaigns. Not only will they fight for customers, they also must woo developers and publishers who make the console games.

To date, console makers have been embroiled in a battle of perception with lots of saber rattling before the ground war begins in retail stores worldwide. Microsoft has signed up 200 companies to make games for its Xbox console. But most industry observers declared Nintendo the winner. The seasoned Japanese company is believed to have the best games and the strongest game brands of any contender. And most believe they have the lowest-price hardware as well.

The Incumbents

The philosophy behind the design of the Sony game console was to combine a step up in graphics quality. It offers a new graphics synthesizer with unprecedented ability to create characters, behavior, and complex physical simulations in real time via massive floating-point processing. Sony calls this concept "Emotion Synthesis." It goes well beyond simulating how images look by depicting how characters and objects think and act.

The key to achieving this capability is the Emotion Engine, a 300-MHz, 128-bit microprocessor based on MIPS RISC CPU architecture. Developed by MIPS licensee Toshiba Corp. and fabricated on a 0.18 micron CMOS process, the CPU combines two 64-bit integer units with a 128-bit SIMD multimedia command unit. It includes two independent floating-point, vector-calculation units, an MPEG-2 decoder, and high-performance DMA controllers, all on a single chip.

The processor combines these functions across a 128-bit data bus. As a result, it can perform complicated physical calculations, curved surface generation, and 3-dimensional (3D) geometric transformations that are typically difficult to perform in real time with high-speed PCs. The Emotion Engine delivers floating-point calculation performance of 6.2 Gflops/s. When applied to the processing of geometric and perspective transformations normally used in 3D graphics calculations, it can reach 66 million polygons per second.

Although the Emotion Engine is the most striking innovation in the PlayStation2, equally important are the platform's extensive I/O capabilities. The new platform includes the entire original PlayStation CPU and features a highly integrated I/O chip developed by LSI Logic that combines IEEE 1394 and USB support.

It is projected to have a long life because it includes:
- Extensive I/O capability via IEEE 1394 ports with data rates between 100 and 400 Mb/s.
- USB 1.1 which can handle data rates of 1.5 to 12 Mb/s.
- Broadband hard-drive communications expansion of the system.

The IEEE 1394 interface allows connectivity to a wide range of systems including VCRs, set-top boxes, digital cameras, printers, joysticks, and other input devices.

As creator of the original PlayStation CPU, LSI Logic Corp. was the natural choice to develop this high-performance I/O chip. They boast a strong track record in ASIC and communications IC design. And being backward compatible is the only way to guarantee that the game can be played on the exact same processor on which it was originally played.

While the PlayStation2 represents the high mark in video-game performance, rival Nintendo has introduced its next-generation console called GameCube. Nintendo officials argue that their system will offer game developers a better platform than PlayStation2 because it focuses exclusively on those applications. Unlike PlayStation2 which offers extensive multimedia and Internet capabilities, Nintendo designers did not try to add such home-information, terminal-type features. For example, GameCube does not add DVD movie playback features. Instead, it uses proprietary 3-inch optical disks as a storage medium. Nintendo is planning to add DVD capability to a higher-end GameCube.

The new Nintendo platform gets high marks for its performance. It features a new graphics processor designed by ArtX Inc. The platform, code-named Flipper, runs at 202 MHz and provides an external bus that supports 1.6 GB bandwidth.

One key distinction between the GameCube and its predecessor lies in its main system processor. While earlier Nintendo machines used processors based on the MIPS architecture, Nintendo has turned to IBM Microelectronics and the PowerPC. Code-named Gekko, the new custom version of the PowerPC will run at a 405 MHz clock rate and sport 256 Kbytes of on-board L2 cache. Fabricated on IBM's 0.18-micron copper wire technology, the chip will feature 32-bit integer and 64-bit floating-point performance. It benchmarks at 925 DMIPS using the Dhrystone 2.1 benchmark.

The RISC-based computing punch of the PowerPC will help accelerate key functions like geometry and lighting calculations typically performed by the system processor in a game console.

The 925 DMIPS figure is important because it adds what the game developers call "artificial intelligence." This is the game-scripting interaction of multiple bodies and other items that may not be very glamorous, but are incredibly important. It allows the creativity of the game developer to come into play.

New Kid on the Block

Microsoft Xbox was launched in November 2001. It was designed to compete with consoles from Nintendo and Sony and with with books, television and the Internet. While the PlayStation2 renders beautiful images and the Nintendo Game Cube offers superb 3D animation, what is under the hood of the Xbox makes the true difference. The Xbox sports a 733-MHz Intel Pentium III processor, about the same level of power as today's low-end computers. It also includes a 233-MHz graphics chip designed jointly by Microsoft and Nvidia Corp., a top player in the PC video adapter market. This combination offers more power than both its competitors. Like the PlayStation2, the Xbox will play DVD movies with the addition of a $30 remote control. The Xbox also plays audio CDs and supports Dolby Digital 5.1 Surround Sound. Nintendo's Game Cube uses proprietary 1.5-GB mini-discs that hold only a fifth as much information as regular discs, so they will not play DVDs.

The front panel of the Xbox provides access to four controller ports (for multi-player gaming), power switch, and eject tray for the slide-out disk drive. The Xbox has 64 MB of RAM and an 8-GB hard drive, making it much more like a PC than the other consoles. The design speeds up play because the Xbox can store games temporarily on its fast hard disk instead of reading from a DVD which accesses data far slower. It also allows users to store saved games internally instead of buying memory cards.

It offers another unique feature. If you do not like a game's soundtrack, you can "rip" music from your favorite audio CD to the hard drive. Then you can substitute those tunes for the original—if the game publisher activates the feature. Of course, all this does not mean much if you cannot see the results, and Microsoft can be proud of what's on the screen. Thanks to crisp, clean images and super-smooth animation, watching the backgrounds in some games can be awe-inspiring.

On the downside, Microsoft's bulky controller is less than ideal with two analog joysticks, triggers, a D-pad, and other buttons. For average and small-sized hands, the sleeker PlayStation2 and smaller GameCube controllers feel better. However, third party substitutes are available for all gaming stations.

Microsoft hopes the Xbox's expandability will make a serious dent in PlayStation2 sales. The Xbox's Ethernet connection which is designed for cable and DSL modem hookups should allow for Internet play and the ability to download new characters and missions for games. It will also provide the ability to network different appliances in the home. Microsoft has also announced that the new version on the Xbox should be expected in 2006.

Meanwhile, Sony has not made any date commitment on the launch of the successor to the PlayStation2 (most likely to be named PlayStation3). However, in May 2003 it introduced the PSX, a mid-life follow-up to the PlayStation 2. Microsoft touted it as a device that creates a new home entertainment category. In addition to the basic features of a game console, PSX will offer a DVD recorder, a 120 GB hard drive, a TV tuner, an Ethernet port, a USB 2.0 port and a Memory Stick slot. The PSX shares a number of components with PlayStation 2 including the Emotion Engine processor and the operating system. Few details have been released about the PlayStation3 which is expected to be launched in late 2005 or 2006. The new "Cell" processor chip is touted to be a thousand times more powerful than the processor in the current PlayStation 2. This is a multi-core architecture in which a single chip may contain several stacked processor cores. It will be based on industry-leading circuitry widths of 65 nanometers. This reduction in circuitry widths allows more transistors (read: more processing power) to be squeezed into the core. It will be manufactured on 300 mm wafers for further cost reduction.

The new multimedia processor, touted as a "supercomputer on a chip," is being created by a collaboration of IBM, Sony, and Toshiba. It is termed a "cell" due to the number of personalities it can take and, hence, the number of applications it can address. While the processor's design is still under wraps, the companies say Cell's capabilities will allow it to deliver one trillion calculations per second (teraflop), or more floating-point calculations. It will be able to do more than 1 trillion mathematical calculations per second, roughly 100 times more than a single Pentium 4 chip running at 2.5 GHz. Cell will likely use between four and 16 general-purpose processor cores per chip. A game console might use a chip with 16 cores while a less complicated device like a set-top box would have a processor with fewer cores.

While Cell's hardware design might be difficult, software is being created for the chip that may be difficult to establish in the market. Creating an operating system and set of applications that can take advantage of its multiprocessing and peer-to-peer computing capabilities will determine if Cell will be successful. Toshiba hinted that it aims to use the chip in next-generation consumer devices. These would include set-top boxes, digital broadcast decoders, high-definition TVs, hard-disk recorders, and mobile phones. Elements of the cell processor design are expected in future server chips from IBM.

According to analysts about eight million units of the Xbox console had been sold globally by June of 2003, with less than a million going to Asia. Despite deposing Nintendo as the number two in units shipped worldwide, Microsoft remains number three in Japan. Meanwhile, Sony's PlayStation console is number 1 with more than 50 million sold worldwide.

As a new kid on the block the Xbox has one major disadvantage compared to the PlayStation2. Sony's console has over 175 games available, compared to 15 or 20 Xbox titles at launch time. This has increased to a little over 50 games since. Because games are the key point in attracting players, the Xbox has a long way to go. However, Microsoft is expanding its Asian territories with localization efforts for China and Korea. It is also bringing in its Xbox Live online game product.

The GameCube seems to be geared toward gamers 10 years old or younger with a growing selection of more age-appropriate titles. It's a toss-up for older children and adults. If you want an excellent machine with plenty of games available right now, the PlayStation2 is your best bet. If you want the hottest hardware with potential for the future, buy the Xbox.

Support System

Developer support is crucial to the success of any video console. Sega's demise can be attributed to its failure to attract Electronic Arts or Japan's Square to create the branded games it needed to compete with Sony. Today, however, many of the large game publishers spread their bets evenly by creating games that can be played on each console. Smaller developers don't have the resources to make multiple versions of their titles or to create branded games for more than one platform. They gravitate to the console that makes their life easiest.

Naturally, the console makers are jockeying for both large publishers and small developers. And in a few cases makers have even funded game developers. Microsoft, which owes much of its overall success to its loyal software developers, started the Xbox in part because it foresaw PC developers moving to the Sony PlayStation and its successor, PlayStation2. Sony has lured a total of 300 developers into its fold, mostly because it has the largest installed base and offers the best potential return on investment for the developers.

In its previous consoles Nintendo's cartridge format made it hard for developers and publishers to generate large profits on anything but the biggest hit titles. This was because publishers had to order cartridges far in advance to accommodate a manufacturing process that took up to ten weeks. Nintendo's latest console has a disk drive instead and the company claims it is embracing third-party developers.

Nintendo launched its console with only 8 titles and it had 17 titles in North America by the end of the year. By contrast, Microsoft had 15 to 20 titles at launch because it distributed its development kits early and it actively courted developers with a system that was easy to program. (In comparison, Sony's system is considered difficult to program.)

The importance of external developer loyalty will wane over time. The major publishers will continue to support each console platform, leaving it up to the console makers' in-house game developers to distinguish each platform with exclusive titles. From most game developers' perspectives the three video consoles are created equal.

Featured Attractions

Sony's machine uses parallel processing to generate images. This allows the console to run compute-intensive games—a boon for gamers who delight in vivid graphics, but long considered a curse by some developers.

PC game developers believe that the Xbox has a clear edge over older machines like the PlayStation2. Microsoft's console uses PC-like tools so game development for it is relatively easy. It also has several times the computing performance of the Sony box and it comes with a hard drive. The question remains whether game developers will use these features to make their games stand apart from Sony's.

Meanwhile, the Nintendo console offers decent performance, but to save costs, it left off a high-cost DVD player and a hard drive—features that rivals have.

Driving down production costs will be a determining factor in profitability over the next five years. According to most estimates Sony's PlayStation2 cost the company $450 per unit upon initial production in early 2000. The company had at first sold the machine as a loss leader for $360 in Japan and for $300 in the United States and Europe. The strategy paid off with the first Play Station because Sony was able to reduce the product's cost from $480 in 1994 to about $80 now. (It was initially priced at $299 and sells at about $99 today). Meanwhile, the company sold about nine games for every console. That model allowed Sony to make billions of dollars over the life of the PlayStation, even if it lost money at first. Sony says its PlayStation2 is selling three times as fast as the original PlayStation.

Part of the reason Sony's game division has yet to report a profit is the disarray in its game software unit. Sony combined two separate game divisions before the PlayStation2 launch. This resulted in big talent defections and a poor showing of Sony-owned titles at the launch of the

PlayStation2 in the United States. With its slow launch of internally produced games, Sony may have squandered its first-mover advantage. The good news for Sony is that its original PlayStation titles are continuing to sell.

Microsoft's Xbox is believed to have initially cost the company about $425 thanks to components like a hard disk drive, extra memory, and ports for four players. The company added these extras to snare consumers who want more "gee-whiz" technology than the PlayStation2 can deliver. The console's launch price was approximately $299. Microsoft's home and entertainment division which operates the Xbox business continues to generate losses.

Perhaps the most significant difference between the Xbox and PlayStation2 is a hard drive. An 8-GB hard drive is built into the Xbox but the PlayStation2 does not have a hard drive for the non-PSX version. The hard drive provides for a richer game experience by giving Xbox gamers more realism, speed, expandability, and storage. Fans of sports games will no longer have to wait for their console to catch up to the action and deliver in real time. Microsoft wants to take those technology advances into the living room where 86% of U.S. families with teens have one or more game consoles.

Microsoft's original manufacturing goal was to take advantage of the PC industry's economies of scale by using standardized PC components. (Sony, however, relied on costly internally built components.) But this may not work. At best, the Xbox cost per unit may fall to $200 over time, but not much lower. Many of the key components in the Xbox already have low costs because they're used in computers. According to the market research firm Disk/Trend, hard drive costs will not fall much lower than the estimated $50 to $60 that Microsoft is paying now. The Intel microprocessor inside the Xbox already costs less than $25.

Heavy Betting

To break even, Microsoft must sell about eight or nine games for every hardware unit, and at least three of those titles must be produced in-house. Typically, games sell for about $50 with profit margins of about 50 to 70% for in-house titles. Third-party titles bring in royalties of about 10%, or $5 to $10 for the console maker. Microsoft says it has more than 80 exclusive titles in the works. And to shore up its business model Microsoft is betting heavily on online gaming services. It hopes that gamers will pay a small fee, say $10 a month, to play other gamers on the Internet and to download new games. But today that is uncharted territory. The more Xbox units Microsoft sells, the more money it loses on hardware. Some sources estimate that the company will lose $800 million in the Xbox's first four years and a total of $2 billion over eight years. The only chance for an upside is if the online plan, untested though it is, pulls off a miracle.

Nintendo says that the company's games should turn a profit on hardware alone. The GameCube console sold at $199 at launch. Over time Nintendo should be able to reduce the cost of its system from $275 to less than $100. Nintendo currently generates the highest profits on its consoles because anywhere from 50 to 70% of its games are internally produced. In the past Nintendo lost out to Sony in part because it had to sell its costlier cartridge games at $50 to $60 when Sony was selling games at $40. This time Nintendo has a 3-inch disk that is smaller and cheaper than the DVDs that Sony is using—and far cheaper than the cartridges.

Nintendo's upper hand in profitability is strengthened because it completely dominates the portable-games market. GameBoy Advance titles are dominating the best-seller lists. And the device sells for about $90 per unit. Nintendo can use portable device and game revenues to make up for any losses in the console hardware market. Nintendo is expecting to sell 4 million GameCube units and 10 million games in addition to 23.5 million Game Boy handhelds and 77.5 million portable games.

Self Promotion

Marketing is another success factor. Microsoft has the No. 2 brand in the world according to a survey by Interbrand, Citicorp, and BusinessWeek. Sony ranked No. 20 while Nintendo was 29. But Microsoft's brand has a lot less power in the games market. That is one reason that Microsoft spent $500 million on marketing worldwide over the first 18 months after the introduction of the Xbox. Microsoft got a lot of attention when it announced that figure. However, this comes out to $111 million a year in the U.S. market which is much less than the $250 million Sony plans to spend promoting both the PlayStation2 and PlayStation One in the North American market.

Targeting will also matter. Sony's dominance was achieved by targeting older, 18- to 34-year-old male game players. As role models for younger players, these players are key influencers. By contrast, Nintendo targeted young players. While younger players are more likely to grow up and adopt the tastes of older players, older players are not likely to adopt "their younger brother's machine." Microsoft targets the same age group as Sony.

Console makers will also have to delight their audiences. Nintendo has shown that with a franchise like Pokemon it can make just as much money as Sony, and with fewer titles. By contrast, Microsoft will follow the Sony model of producing scores of titles for more variety. Sony has a raft of titles slated for this fall but Microsoft's games are untested and lack an obvious blockbuster.

The online-games market is a wild card. Since Microsoft has all the hardware for online games built into the machine, it can more easily recruit developers. Sega plans to exploit the Xbox's online advantage and develop specific online games.

Game Over

Sony will come out ahead of the game. According to a forecast by the European Leisure Software Publishers Association, Sony is expected to have 42% of the worldwide video console market by 2004. According to the same forecast Nintendo and Microsoft are expected to take about 29% of the market each. Microsoft is expected to do its best in Europe and the United States but poorly in Japan. Although Sony grabs the largest market share, it does not necessarily garner the biggest profits.

Nintendo will occupy second place in the worldwide console market. But it will garner the most overall profits, thanks to its portable empire. The Xbox is winning over consumers with its superior technology. Microsoft is trying to supplement its profits with online revenues. And like Sony, Microsoft benefits most if the demographics of gaming expand to increase the market size. If Microsoft is committed to a ten-year battle, it can stay in the game despite heavy initial losses. But five years may be too short a horizon for it to catch up to market leader Sony.

Components of a Gaming Console

A typical gaming console is made up of several key blocks that include a CPU, graphics processor, memory, storage mediums, and Internet connectivity. A typical gaming platform has a CPU, a graphics chip capable of several billion operations per second, and a 3D audio processor with 32 or 64 audio channels. It may support a multi-gigabyte hard drive, memory card, DVD, and on-board SRAM, DRAM or flash memory. The MPEG processor also supports a display controller. While upcoming boxes will support an analog modem, future boxes will provide broadband connectivity for online game play. The graphic synthesizer connects to NTSC, PAL, DTV, or VESA via a NTSC/PAL decoder. The I/O processing unit contains a CPU with bus controller to provide an interface for HomePNA, Ethernet, USB, wireless, and IEEE 1394. These connections make it possible to use the existing PC peripherals and create a networked home.

Figure 6.1 shows the block diagram and components of a typical gaming console.

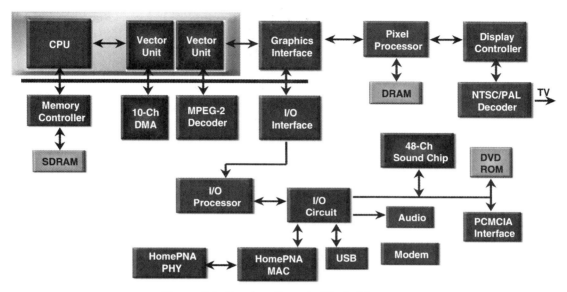

Figure 6.1: Gaming Console Block Diagram

Broadband Access and Online Gaming

To succeed in this market gaming companies must recognize that the video game business is all about the games. From Atari to Nintendo to Sega to PlayStation it has historically been shown time and again that gamers are loyal to the games—not to the hardware. Broadband access will enable users to play with thousands of other gamers on the Internet and will provide the ability to download the latest updates to games. For example, in a basketball video game a user will be able to download the most current status of teams in the NBA so the game will accurately reflect that information.

Online gaming is interactive game play involving another human opponent or an offsite PC. Online gaming capabilities are promoting gaming consoles into potential home gateways. Playing between different players across the Internet has been gaining prominence with the coming of broadband access and the ubiquity of the Internet. Most gaming consoles have a 56K (analog) modem embedded to allow this. The addition of home-networking capabilities within the gaming console will make the evolution into a residential gateway a realistic possibility in the near future.

Online capabilities in next-generation game consoles, whether through a built-in or add-on narrowband or broadband modem, can substantially expand the console's gaming capabilities and open up new areas of entertainment. However, well-defined plans must be laid out for the developer and the consumer in order to use these communication capabilities.

Several console developers have already announced the addition of online functionality. This functionality comes in several guises ranging from looking up instructions to downloading new characters and levels. The latter approach has already been used for PC games in a monthly subscription scheme with some level of success.

Some of the major video game hardware players have introduced a comprehensive online program. When Sega introduced its Dreamcast it provided the Sega.net Internet service, used Dreamcast's built-in narrowband modem, and rolled out Sega.com, a site optimized for narrowband Dreamcast online gaming. Games are being introduced to take advantage of this online functionality. In addition, the web sites of these companies will feature optimized online games.

Microsoft's Xbox contains an internal Ethernet port. Sony announced that the PlayStation2 will feature an expansion unit for network interface and broadband connectivity. Nintendo has announced that both Dolphin and GameBoy Advance will have online connectivity.

Broadband access capabilities are very promising and will eventually become a reality for most households. IDC believes that only 21.2 million U.S. households will have broadband access by 2004. This low number serves as a sobering reminder that those video game households without broadband access could be left out of online console gaming unless plans are made to support narrowband access.

The difficulties many households experience with the installation of broadband can significantly affect the adoption of broadband via the gaming console. Logistically, technology needs to advance the capability of more households to adopt broadband Internet access. In addition, ensuring that the broadband capabilities interface properly with the PC is often a trying experience. Once broadband is fully working, will a household tempt fate by trying to extend that connection to the console? The solution appears to be the creation of a home network. This can be a daunting task for many, often requiring financial outlays and a great deal of self-education. Households that choose to use the broadband functionality of the console will confront these and many other issues.

IDC predicts that by 2004 the number of consoles being used for online gaming will be 27.2 million units, or 33% of the installed base.

Gaming Consoles—More Than Just Gaming Machines

Current 128-bit gaming consoles such as the Xbox, GameCube, and PlayStation2 promise an ultra-realistic gaming experience. They will include Internet access, interactive television capabilities, and the ability to network with other devices in the home through expansion ports (PC card slots). Traditional gaming console vendors are positioning their devices as the nerve center, or gateway, of the home. This will make them true home consumer devices. Products in development will enable:

- Broadband access
- Traditional Internet access
- E-mail
- Internet-enabled value-added services such as video on demand
- Home networking.

Residential (home or media) gateways will network the different home appliances to distribute audio, video, data, and Internet content among them. In this way consumer devices can share digital content and a single broadband access point. This will also be a platform for utility companies to provide value-added services such as video-on-demand and Internet connectivity.

PC Gaming

Does the introduction of the gaming console mean the end of PC-based gaming? The PC and the gaming console are complementary devices. Each has very distinct audiences. PC games are more cerebral while console games are more visceral. A comparison of the top 10 game lists for these platforms shows that the games don't really match up. The most popular PC games of 1999 included Age of Empires II, Half-Life, and SimCity 3000. The most popular console games of 1999 included Pokemon Snap, Gran Turismo Racing, and Final Fantasy VIII. The gaming console is expected to offer the most advanced graphics, the most flexibility in Internet gaming, and the most realistic play of any game console on the market.

Games have always been recognized as a popular form of home entertainment. As a category, games represent 50% of consumer PC software sales. Consumers are becoming more comfortable with PC technology—more than half of U.S. households now own a PC.

Growing Convergence of DVD Players and Gaming Consoles

Few doubt that DVD (digital versatile/video disk) is becoming the video format of choice for the next couple of decades. Its significant capacity, digital quality, and low cost of media make it far superior to magnetic tape. And growing vendor acceptance and market presence have already created an installed base of over a million players in the U.S. alone. As consumer awareness improves and content availability increases, the demand for DVD players will grow.

Manufacturers can drive prices down as volumes begin to expand. In fact, some vendors are already bringing very low priced players to market. They expect to see second-tier consumer electronic companies try to pick up volumes with even lower priced systems during the coming year. These lower prices, dipping below $100, should further boost volumes for DVD players. However, with low prices comes margin pressure. Traditionally, top vendors were able to differentiate their products from second-tier vendors' offerings. They did this through a strong brand and separation of products within their line through feature sets. However, due to the digital nature of the DVD player, the opportunity to further expand differentiation through the addition of separate functionality is now possible. Already DVD players will play audio CDs. And it is possible to add gaming, Internet access, or even a full PC within the same system.

This opportunity creates a conundrum for consumer electronics companies. If consumers are drawn to the combination products, it will allow vendors to maintain higher product prices, greater product differentiation, and potentially create new lines of revenue. However, there is risk of turning away potential sales if the additional functions are not appealing to buyers, or if the perceived incremental cost of that functionality is too high. Further, maintaining a higher basic system price can result in a lower total market due to the proven price sensitivity of demand for consumer electronics.

There are a number of potential add-on functions that could meaningfully enhance a DVD player. These include Internet connectivity, video recording capabilities, satellite or cable set-tops, and gaming. The integration of gaming is a very interesting concept for a number of reasons. The shared components and potentially common media make the incremental functionality cheap for a vendor. And the combined product would be conceptually appealing for consumers. The following sections describe factors that would make it easy for a vendor to integrate gaming into a DVD player.

Shared Components

A game console and a DVD player actually have a number of common components such as video and audio processors, memory, TV-out encoders, and a host of connectors and other silicon. There is also potential for both functions to use the same drive mechanism which is one of the most expensive parts of the system. Today many of the components are designed specifically for one or the other, but with a little design effort it would be easy to integrate high-quality gaming into a DVD player at little incremental cost.

Performance Overhead

As CPUs, graphics engines, and audio processors become more powerful, the performance overhead inherent in many DVD components is also increasing. The excess performance capabilities make it even easier to integrate additional functionality into a gaming console.

Shared Media

A DVD as the medium would be natural for a combination DVD player/game console. Game consoles are already moving quickly towards higher capacity and lower-cost media, and DVD-based software represents the best of both worlds. Although cartridges have benefits in the area of piracy and platform control, they are expensive and have limited a number of manufacturers. Both Sony and Microsoft gaming consoles have DVD support.

Many DVD video producers are incorporating additional interactive functions into their products. For example, a movie about snowboarding may include additional features for a PC such as a game or information on resorts and equipment. The incremental gaming capabilities of a combination DVD/game console could easily take advantage of these features and enable consumers to enjoy them without a PC.

Potential Enhancements

The combination of a DVD and a gaming console has the potential to enhance the use of both. Besides allowing users to take advantage of the interactive elements discussed above, it would enable game producers to add high-quality MPEG2 video to games. This represents an opportunity to enhance the consumer experience and differentiate products.

This also creates new opportunities for cross-media development and marketing. Historically, games have been based on movies and vice versa, but this would enable a DVD with both. It would differentiate the DVD over a VHS movie or a cartridge game while lowering media and distribution costs for both content owners.

One of the most significant drawbacks to such a product for DVD vendors would be the development and marketing of a gaming platform. To take gaming functionality beyond that of limited, casual games that are built into the product, a substantial effort must be undertaken. However, several companies are already working on adapting gaming technology to other platforms. Partnering with a third party or adopting an outside initiative would avoid these issues. Since the DVD manufacturer is relieved of creating the game technology, cultivating developer relationships, and marketing the platform, the investment and risk to the DVD vendor is minimized. Examples of third party initiatives include VM Labs which is developing game technology for a variety of platforms. And Sony has developed a small footprint version of its PlayStation platform that could be integrated into a number of products. Most gaming hardware suppliers have invested research into gaming integration.

Although the ability to easily integrate functionality is a key part of the equation, consumer interest in such a combination is the critical factor. And the consumer factor is also the most difficult to understand or forecast. Although there is no definitive way to estimate consumer interest without actual market experience, there are factors indicating that consumers may be interested in a gaming-equipped DVD player.

Both the DVD player and the gaming console are entering a significant growth phase where vendor leadership is up for grabs and product attributes are uncertain. This situation could provide success for a product that excels in both categories. During the next few years both product categories will shake out. In the meantime there is opportunity for change.

A significant indicator of consumer interest in a DVD player with gaming capabilities is overlapping consumer interest. If done well, the gaming functionality could enhance the DVD player experience. A potential enhancement is the ability for consumers to take advantage of the interactive elements and games that are added to DVD movies for PC-based DVD drives. Given the common components for DVD players and game consoles, there is also potential for high-quality performance of both functions and significant cost savings to the consumer who is interested in both products. And since the game console and the DVD player will sit near each other and use a TV for display, an opportunity for space savings is created. All things considered, the addition of game playing into a DVD player could be successful.

Gaming Trends

In addition to those specifically interested in a game console, there may be other who would value gaming capabilities in a DVD player. Currently, household penetration of game consoles in the United States is almost 40% and PCs are in 46% of U.S. homes. Electronic entertainment is far more pervasive than just consoles; games are appearing in PCs, handhelds, and even cellular phones.

While there are many positive indicators for a DVD player with gaming functionality, the potential success of such a product is far from guaranteed. There are numerous challenges that such a product would face in the market. And any one of them could derail consumer interest and potentially damage the success of a vendor during this important DVD/game console growth phase.

It's all About the Games

As anyone in the gaming business knows, the hardware is the easy part. To drive demand for a gaming platform, developers have to support it fully, creating blockbuster software titles that are either better or unique to the platform. Without plenty of high-demand games, few game enthusiasts will even consider a product. And casual gamers will see little use for the additional features on a DVD player.

Building support from developers without a track record of success or a large installed base is notoriously difficult. Even successful game companies with great technology are challenged in this area. Any one vendor may have difficulty driving volumes of a combination DVD/gaming player to make a significant enough platform to woo developers. This would especially be true if multiple manufacturers brought competing technologies to market.

Marketing is Key

Marketing the platform and its games to consumers is critical. Hardware and software companies spend millions to create demand for their products, using everything from TV ads to sports endorsements. During this critical game-console transition phase, marketing noise will only increase as the incumbents try to sustain software sales on older platforms while building momentum for new technology. In this environment it will be very difficult to build demand for a new platform. Even making a realistic attempt will require a significant investment.

New Gaming Opportunities

While a bulk of household gaming is focused on traditional gaming platforms such as the gaming console and the PC, the gaming industry is looking for opportunities in emerging or nontraditional gaming platforms. These include PDAs, cellular phones, and interactive TV. These emerging devices will offer the industry several potential channels through which to target new gaming audiences, including those that have not traditionally been gamers. It will also enable vendors to take advantage of new and emerging revenue streams.

However, the industry must resolve several issues to build a successful and profitable business model. Vendors must form partnerships, deliver the appropriate services, and derive revenues from sources such as advertisements, sponsorships, subscriptions, and one-time payment fees.

The following sections present brief descriptions of gaming opportunities for emerging devices.

Wireless Connectivity

In Japan multi-player online gaming has already become extremely popular. Sony plans to market its original PlayStation console as a platform for mobile networks. The company has introduced a version of its popular game platform in Japan that offers the same functionality of the original at one-third the size.

Called PSone, the new unit will feature a controller port that attaches the PlayStation console to mobile phone networks. Players will be able to carry the game console from room to room or use it in a car. Slightly bigger than a portable CD player, PSone will feature a 4-inch TFT display.

As part of this new strategy Sony has announced a partnership with NTT DoCoMo Inc., Tokyo. They will develop services that combine NTT DoCoMo's 10-million i-mode cellular phones with the 20 million PlayStation consoles in Japan. The cellular provider uses a 9.6 Kb/s packet communications scheme to enable access to Internet sites written in Compact-HTML. The company will launch a W-CDMA service that will support 38 Kb/s communications and will eventually expand coverage to other carriers.

The future for mobile gaming beyond this service looks bright. Sony has talked extensively about integrating its PSone with cell phones in Japan. Ericsson, Motorola, and Siemens are working to develop an industry standard for a universal mobile game platform. And several publishers including Sega and THQ have begun to develop content for the platform. For most users the concept is still relatively new. But trials have shown that once offered a gaming service, mobile users tend to become avid gamers. The vast majority of mobile phone gaming will be easy-to-play casual games such as trivia and quick puzzle games.

Wireless connectivity, both to broadband networks and between game consoles and controllers, promises to bring new capabilities to game users. It will enable a whole range of new products and uses in the home. Intel has introduced the first PC game pad in the U.S. that offers wireless connectivity. Part of a family of wireless peripherals, the new game pads use a 900 MHz digital frequency-hopping spread-spectrum (FHSS) RF technology. It supports up to four players simultaneously at a range of up to 10 feet. And it will offer the gamer and the Internet user much more flexibility. It will unleash the whole gaming experience by removing the cord clutter and eliminating the need for the PC user to be a certain distance from the PC.

PDAs

The portable and ubiquitous nature of PDAs such as Palm Pilot, Handspring Visor, and Compaq iPaq makes them perfect as handheld gaming platforms. The open Palm operating system has encouraged the development of many single-player games such as Tetris and minesweeper. Competing platforms such as the Compaq iPaq (with Microsoft's PocketPC operating system) have made specific game cartridges available for single-player gaming. The integration of online connectivity into PDAs has enabled the development of online gaming. Factors such as screen size, operating system, and connectivity speeds will limit the scope and type of online games. However, online games that evolve to these platforms will not be substandard or meaningless to the gamer. Rather, the online game pallet has the potential to include a broad mixture of casual and low-bandwidth games (such as those currently played online) that will appeal to a wide spectrum of individuals.

Interactive TV (iTV)

Digital TV (digital cable and satellite) service providers are deploying digital set-top boxes with two-way communication and new middleware that can support interactive TV. iTV has the potential to bring gaming applications to the masses. Already deployed in Europe, the French-based Canal+ has been extremely successful with iTV online games. Significant iTV opportunities exist in the U.S. as well. The rapid deployment of digital set-top boxes will contribute to significant growth of new online and interactive gaming applications.

Home Networking

Today, home networking is the connection of two or more PCs for sharing broadband Internet access, files, and peripherals within a local area network (LAN). A number of households that have established home networks use the high-speed access to engage in multi-player gaming. The establishment of feature-rich home networks capable of delivering new services and products into the home will facilitate the delivery of online gaming applications. It is envisioned that these gaming applications will be delivered to multiple platforms including PCs, game consoles, PDAs, Internet appliances, and interactive TV.

Summary

The leading game console manufacturers (Microsoft, Nintendo, and Sony) are not playing games. The stakes are much higher than a simple fight for the hearts and minds of little Johnny and Jill. There is a growing belief that products of this type will eventually take over the low-end PC/home-entertainment market because there is more than enough processing power in these new boxes. For the casual Web surfer or for someone who is watching TV and needs more information about a product on a Web site, this kind of box is desirable. It can replace what we do with PCs and TVs today and provide a much wider range of functionality. Although the game industry is forecast to be in the tens of billions, the impact may be much greater than the revenue it generates.

Game systems are likely to become hybrid devices used for many forms of digital entertainment including music, movies, Web access, and interactive television. The first generations of machines to offer these functions hold the potential to drive greater broadband adoption into the consumer household.

Digital Video/Versatile Disc (DVD)

Introduction

The digital versatile disc, or DVD, is a relatively new optical disc technology that is rapidly replacing VHS tapes, laser discs, video game cartridges, audio CDs, and CD-ROMs. The DVD-Video format is doing for movies what the compact disc (CD) did for music. Originally the DVD was known as the digital video disc. But with the addition of applications such as PCs, gaming consoles, and audio applied to the disc, DVD became known as digital versatile disc. Today the DVD is making its way into home entertainment and PCs with support from consumer electronics companies, computer hardware companies, and music and movie studios.

The Birth of the DVD

The DVD format came about in December 1995 when two rival camps ended their struggle over the format of next-generation compact discs (CDs). Sony and Philips promoted their MultiMedia Compact Disc format while Toshiba and Matsushita offered their Super Density format. Eventually the two camps agreed on the DVD format. Together they mapped out how digitally stored information could work in both computer-based and consumer electronic products.

Although it will serve as an eventual replacement for CDs and VCRs, DVD is much more than the next-generation compact disc or VHS tape player. DVD not only builds upon many of the advances in CD technology, it sounds a wake-up call for the motion picture and music industries to prepare their intellectual property for a new era of digital distribution and content.

DVD Format Types

The three application formats of DVD include DVD-Video, DVD-Audio, and DVD-ROM.

The DVD-Video format (commonly called "DVD") is by far the most widely known. DVD-Video is principally a video and audio format used for movies, music concert videos, and other video-based programming. It was developed with significant input from Hollywood studios and is intended to be a long-term replacement for the VHS videocassette as a means for delivering films into the home. DVD-Video discs are played in a machine that looks like a CD player connected to a TV set. This format first emerged in the spring of 1997 and is now considered mainstream, having passed the 10% milestone adoption rate in North America by late 2000.

The DVD-Audio format features high-resolution, two-channel stereo and multi-channel (up to six discrete channels) audio. The format made its debut in the summer of 2000 after copy protection issues were resolved. DVD-Audio titles are still very few in number and have not reached mainstream status, even though DVD-Audio and DVD-Video players are widely available. This is due primarily to the existence of several competing audio formats in the market.

DVD-ROM is a data storage format developed with significant input from the computer industry. It may be viewed as a fast, large-capacity CD-ROM. It is played back in a computer's DVD-ROM drive. It allows for data archival and mass storage as well as interactive and/or web-based content. DVD-ROM is a superset of DVD-Video. If implemented according to the specifications, DVD-Video discs will play with all the features in a DVD-ROM drive, but DVD-ROM discs will not play in a DVD-Video player. (No harm will occur. The discs will either not play, or will only play the video portions of the DVD-ROM disc.) The DVD-ROM specification includes recordable versions - either one time (DVD-R), or many times (DVD-RAM).

At the introduction of DVD in early 1997 it was predicted that DVD-ROM would be more successful than DVD-Video. However, by mid-1998 there were more DVD-Video players being sold and more DVD-Video titles are available than DVD-ROM. DVD-ROM as implemented so far has been an unstable device, difficult to install as an add-on and not always able to play all DVD-Video titles without glitches. It seems to be awaiting the legendary "killer application." Few DVD-ROM titles are available and most of those are simply CD-ROM titles that previously required multiple discs (e.g., telephone books, encyclopedias, large games).

A DVD disc may contain any combination of DVD-Video, DVD-Audio, and/or DVD-ROM applications. For example, some DVD movie titles contain DVD-ROM content portion on the same disc as the movie. This DVD-ROM content provides additional interactive and web-based content that can be accessed when using a computer with a DVD-ROM drive. And some DVD-Audio titles are actually DVD-Audio/Video discs that have additional DVD-Video content. This content can provide video-based bonus programming such as artist interviews, music videos, or a Dolby Digital and/or DTS surround soundtrack. The soundtrack can be played back by any DVD-Video player in conjunction with a 5.1-channel surround sound home theater system.

The DVD specification also includes these recordable formats:
- **DVD-R** – DVD-R can record data once, and only in sequential order. It is compatible with all DVD drives and players. The capacity is 4.7 GB.
- **DVD-RW** – The rewritable/erasable version of DVD-R. It is compatible with all DVD drives and players.
- **DVD+R and DVD+RW** – The rewritable/erasable version of DVD+R.
- **DVD-RAM** – Rewritable/erasable by definition.

The last three erasable (or rewritable) DVD formats—DVD-RW, DVD-RAM, and DVD+RW—are slightly different. Their differences have created mutual incompatibility issues and have led to competition among the standards. That is, one recordable format cannot be used interchangeably with the other two recordable formats. And one of these recordable formats is not even compatible with most of the 17 million existing DVD-Video players. This three-way format war is similar to the VHS vs. Betamax videocassette format war of the early 1980s. This incompatibility along with the high cost of owning a DVD recordable drive has limited the success of the DVD recordable market.

Regional Codes

Motion picture studios want to control the home release of movies in different countries because cinema releases are not simultaneous worldwide. Movie studios have divided the world into six geographic regions. In this way, they can control the release of motion pictures and home videos into different countries at different times. A movie may be released onto the screens in Europe later than in the United States, thereby overlapping with the home video release in the U.S. Studios fear that copies of DVD discs from the U.S. would reach Europe and cut into theatrical sales. Also, studios sell distribution rights to different foreign distributors and would like to guarantee an exclusive market. Therefore, they have required that the DVD standard include codes that can be used to prevent playback of certain discs in certain geographical regions. Hence, DVD player is given a code for the region in which it is sold. It will not play discs that are not allowed in that region. Discs bought in one country may not play on players bought in another country.

A further subdivision of regional codes occurs because of differing worldwide video standards. For example, Japan is region 2 but uses NTSC video compatible with that of North America (region 1). Europe is also region 2 but uses PAL, a video system not compatible with NTSC. Many European home video devices including DVD players are multi-standard and can reproduce both PAL and NTSC video signals.

Regional codes are entirely optional for the maker of a disc (the studio) or distributor. The code division is based on nine regions, or "locales." The discs are identified by the region number superimposed on a world globe. If a disc plays in more than one region it will have more than one number on the globe. Discs without codes will play on any player in any country in the world. Some discs have been released with no codes, but so far there are none from major studios. It is not an encryption system; it is just one byte of information on the disc which recognizes nine different DVD worldwide regions. The regions are:

- **Region 0** – World-wide; no specific region encoded
- **Region 1** – North America (Canada, U.S., U.S. Territories)
- **Region 2** – Japan, Western Europe, South Africa, Middle East (including Egypt)
- **Region 3** – Southeast Asia, East Asia (including Hong Kong)
- **Region 4** – Australia, New Zealand, Pacific Islands, Central America, South America, Caribbean
- **Region 5** – Former Soviet Union, Eastern Europe, Russia, Indian Subcontinent, Africa (also North Korea, Mongolia)
- **Region 6** – China
- **Region 7** – Reserved
- **Region 8** – Special international venues (airplanes, cruise ships, etc.)

In hindsight, the attempt at regional segregation was probably doomed to failure from the start. Some of the region standards proved more complicated to finalize than was originally expected. There were huge variations in censorship laws and in the number of different languages spoken across a region. This was one of the reasons why DVD took so long to become established. For example, it is impossible to include films coded for every country in Region-2 on a single disc. This led the DVD forum to split the region into several sub-regions. And this, in turn, caused delays in the availability of Region-2 discs. By the autumn of 1998 barely a dozen Region-2 discs had been released compared to the hundreds of titles available in the U.S. This situation led to many companies selling DVD players that had been reconfigured to play discs from any region.

For several years now games console manufacturers (Nintendo, Sega and Sony) have been trying to stop owners from playing games imported from other countries. Generally, whenever such regional standards were implemented it took someone only a few weeks to find a way around it, either through a machine modification or use of a cartridge adapter. In real terms, regional DVD coding has cost the DVD Forum a lot of money, delayed market up-take, and allowed third-party companies to make a great deal of money bypassing it.

How Does the DVD Work?

DVD technology is faster and provides storage capacity that is about 6 to 7 times greater than CD technology. Both have the same aerial space—4.75 inches and 3.1 inches, 1.2mm thick. DVD technology provides multiple languages on movies with multiple language subtitles. Since a beam of laser light touches the data portion of a DVD disc, it is never touched by a mechanical part when played, thus eliminating wear.

Some key DVD features include:
- MPEG2 video compression
- Digital Theatre Systems (DTS), Dolby Digital Sound, Dolby AC-3 Surround Sound
- Up to eight audio tracks
- 133 minutes/side video running time (at minimum)
- Disc changers
- Backward compatibility with CD and/or CD-ROM
- Still motion, slow motion, freeze frame, jump-to-scene finding
- Interactive/programming-capable (story lines, subtitles)

A DVD system consists of a master (original) copy, a DVD disc, and a player.

The Master

A master disc must be manufactured before a DVD disc can be duplicated. The backbone of any DVD is the master copy. Once a movie has been transferred from photographic film to videotape, it must be properly formatted before it can be distributed on a DVD. The mastering process is quite complicated and consists of these steps:

1. Scan the videotape to identify scene changes, enter scan codes, insert closed-caption information, and tag objectionable sequences that would be likely targets for parental lockout.
2. Use variable-bit-rate encoding to compress the video into MPEG-2 format.
3. Compress audio tracks into Dolby AC-3 Surround Sound format.
4. In a process called multiplexing, combine the compressed video and audio into a single data stream.
5. Simulate the playback of the disc (emulation).
6. Create a data tape with the image of the DVD.
7. Manufacture a glass master which duplicators use to "press discs."
8. With the exception of disc copying and reproduction (step 7 above), all the mastering process steps are completed on high-end workstations.

One of the key steps to making a master DVD disc from videotape is the encoding process (step 2 above). The encoding process uses compression to eliminate repetitive information. For instance, much of a movie picture remains nearly the same from frame to frame. It may have a redundant background (a cloudless sky, a wall in a room, etc.) and the same foreground. If left in its raw state, capturing and coding these repetitive scenes on digital video would be prohibitive. It would be so extensive and expansive that a feature-length movie would require 4.7 GB of storage capacity. That's where encoding comes in.

Encoding is a complicated process that identifies and removes redundant segments in a movie frame. Using video encoder ICs, chipsets, and sophisticated software, the process makes several passes of the video to analyze, compare, and remove repetitive sequences. The process can eliminate more than 97% of the data needed to accurately represent the video without affecting the quality of the picture.

A DVD encoder uses more data to store complex scenes and less data for simple scenes. Depending on the complexity of the picture, the encoder constantly varies the amount of data needed to store different sequences throughout the length of a movie. The average data rate for video on DVD is about 3.7 Mb/s.

The DVD Disc

DVD discs can be manufactured with information stored on both sides of the disc. Each side can have one or two layers of data. A double-sided, double-layered disc effectively quadruples the standard DVD storage density to 17 GB. A double-sided DVD disc holds 17 GB of data (compared with a CD's 680 MB) because it is actually two tightly packed, dual-layered discs glued back-to-back.

With 17 GB, the DVD provides 26 times the storage capacity of a typical audio CD that has 680 MB. This is the equivalent of four full-length feature movies, or 30 hours of CD quality audio, all on one disc. The DVD format is a nearly perfect medium for storing and delivering movies, video, audio, and extremely large databases.

In late 1999 Pioneer Electronics took optical disc density storage one step further when it demonstrated a 27.4-GB disc that used a 405-nm violet laser. The high-density disc has a track pitch about half that of a 17-GB disc. Several vendors are developing a new technology that uses blue laser technology to provide superior quality video. This technology will be capable of achieving higher data densities with 12- to 30-GB capacity.

DVD and CD discs are developed and replicated in a similar manner due to their physical similarities. A DVD disc is made of a reflective aluminum foil encased in clear plastic. Stamping the foil with a glass master forms the tightly wound spiral of tiny data pits. In the case of a single-sided disc, the stamped disc is backed by a dummy that may show advertisements, information, or entertainment. Two halves, each stamped with information, are attached to the back of a double-sided disc. The stamping technique is a well-understood process, derived in part from CD-ROM manufacturing and in part from the production of laser discs.

Replication and mastering are jobs most often done by large manufacturing plants that also replicate CDs. These companies estimate that the cost of a single-sided, single-layer DVD will eventually be about the same as that of a CD—approximately $0.50. This low cost compares favorably to the roughly $2.50 cost to make and distribute a VHS tape and the $8.00 cost to replicate a laser disc.

There are three reasons for the greater data capacity of a DVD disc:

1. **Smaller pit size** – DVDs have a smaller pit size than CDs. Pits are the slight depressions, or dimples, on the surface of the disc that allow the laser pickup to distinguish between the digital 1s and 0s.
2. **Tighter track spacing** – DVDs feature tighter track spacing (i.e., track pitch) between the spirals of pits. In order for a DVD player to read the smaller pit size and tighter track spacing of the DVD format, a different type of laser with a smaller beam of light is required. This is one of the major reasons why CD players cannot read DVDs, while DVD players can read audio CDs.
3. **Multiple layer capability** – DVDs may have up to 4 layers of information, with two layers on each side. To read information on the second layer (on the same side), the laser focuses deeper into the DVD and reads the pits on the second layer. When the laser switches from

one layer to another, it is referred to as the "layer switch," or the "RSDL (reverse spiral dual layer) switch". To read information from the other side of the DVD, almost all DVD players require the disc to be flipped.

Based on DVD's dual-layer and double-sided options, there are four disc construction formats:

1. **Single-Sided, Single-Layered** – Also known as DVD-5, this simplest construction format holds 4.7 GB of digital data. The "5" in "DVD-5" signifies the nearly 5 GB worth of data capacity. Compared to 650 MB of data on CD, the basic DVD-5 has over seven times the data capacity of a CD. That's enough information for approximately two hours of digital video and audio for DVD-Video or 74 minutes of high-resolution music for DVD-Audio.

2. **Single-Sided, Dual-Layered** – DVD-9 construction holds about 8.5 GB. DVD-9s do not require manual flipping; the DVD player automatically switches to the second layer in a fraction of a second by re-focusing the laser pickup on the deeper second layer. This capability allows for uninterrupted playback of long movies up to four hours. DVD-9 is frequently used to put a movie and its bonus materials or its optional DTS Surround Sound track on the same DVD-Video disc.

3. **Double-Sided, Single-Layered** – Known as DVD-10, this construction features a capacity of 9.4 GB. DVD-10s are commonly used to put a widescreen version of a movie on one side and a full frame version of the same movie on the other side. Almost all DVD players require you to manually flip the DVD which is why the DVD-10 is called the "flipper" disc. (There are a few DVD players that can perform flipping automatically.)

4. **Double-Sided, Dual-Layered** – The DVD-18 construction can hold approximately 17 GB (almost 26 times the data capacity of a CD), or about 8 hours of video and audio as a DVD-Video. Think of DVD-18 as a double-sided DVD-9 where up to four hours of uninterrupted video and audio can be stored on one side. Content providers usually choose two DVD-9s rather than a single DVD-18 because DVD-18s cost far more to produce.

A DVD disc can be used for data storage in a PC by using a DVD ROM drive. Using a similar format, each DVD can store up to 17 GB of data compared to a CD-ROM disc which stores 650 MB of data. While a single layer of a DVD disc can hold 4.7 GB of data, greatly increased storage capacity is accomplished by using both sides of the disc and storing two layers of data on each side. The amount of video a disc holds depends on how much audio accompanies it and how heavily the video and audio are compressed. The data is read using 635-nm and 650-nm wavelengths of red laser.

The DVD Player

A DVD player makes the DVD system complete. The player reads data from a disc, decodes audio and video portions, and sends out those signals to be viewed and heard.

DVD units are built as individual players that work with a television or as drives that are part of a computer. DVD-video players that work with television face the same standards problem (NTSC versus PAL) as videotape and laser disc players. Although MPEG video on DVD is stored in digital format, it is formatted for one of these two mutually incompatible television systems. Therefore, some players can play only NTSC discs, some can only play PAL discs, but some will play both. All DVD players sold in PAL countries play both. A very small number of NTSC players can play PAL discs.

Most DVD PC software and hardware can play both NTSC and PAL video. Some PCs can only display the converted video on the computer monitor, but others can output it as a video signal for a TV.

Most players and drives support a standard set of features including:

- Language choice
- Special effects playback (freeze, step, slow, fast, scan)
- Parental lock-out
- Programmability
- Random play and repeat play
- Digital audio output (Dolby Digital)
- Compatibility with audio CDs

Established standards also require that all DVD players and drives read dual-layer DVD discs.

All players and drives will play double-sided discs, but the discs must be manually flipped over to see and/or hear side two. No manufacturer has yet announced a model that will play both sides. The added cost is probably not justifiable at this point in DVD's history.

Components of a DVD Player

DVD players are built around the same fundamental architecture as CD-ROM drivers/players. Major components of a DVD player include:

- **Disc reader mechanism** – Includes the motor that spins the disc and the laser that reads the information from it.
- **DSP (digital signal processor) IC** – Translates the laser pulses back to electrical form that other parts of the decoder can use.
- **Digital audio/video decoder IC** – Decodes and formats the compressed data on the disc and converts data into superior-quality audio and video for output to TVs and stereo systems. This includes the MPEG video decoder and audio decoder. It also includes a DAC for audio output and an NTSC/PAL encoder for video output.
- **8-, 16-, 32-bit Microcontroller** – Controls operation of the player and translates user inputs from remote control or front panel into commands for audio/video decoder and disc reader mechanism.
- **Copy protection descrambler**
- **Network Interface ICs** – To connect to other electronics devices in the home, the DVD player supports a number of network interfaces such as IEEE 1394, Ethernet, HomePNA, etc.
- **Memory** – SRAM, DRAM, flash

Figure 7.1 shows the different components of a typical DVD player.

The DVD player functions when a visible laser diode (VLD) beams a 635- or 650-nm wavelength red light at a DVD disc spun via a spindle motor at roughly eight times the speed of a typical CD-ROM. An optical pickup head governed by a spindle-control chip feeds data to a DSP which feeds it to an audio/video decoder.

The position of the reading point is controlled by the microcontroller. From the audio/video decoder, MPEG-2 data is separated and synchronized (demultiplexed) into audio and video streams. The stream of video data passes through an MPEG-2 decoder where it is formatted for display on a monitor. The audio stream passes through either an MPEG-2 or Dolby Digital AC-3 decoder where it is formatted for home audio systems. The decoder also formats on-screen displays for graphics and performs a host of other features.

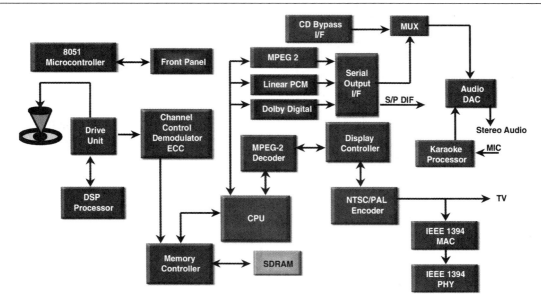

Figure 7.1: DVD Player Block Diagram

First-generation DVD players contained many IC and discrete components to accomplish the coding, decoding, and other data transmission functions. Chipsets in second- and third-generation DVD players have greatly reduced the component count on board these systems. Function integration in fewer integrated circuits is driving prices down in all DVD units. The first DVD players used up to 12 discrete chips to enable DVD playback, mostly to ensure quality by using time-tested chips already on the market. So far, function integration has reduced the number of chips in some players to six. Audio and video decoding, video processing, NTSC/PAL encoding, and content scrambling system (CSS) functions are now typically handled by one chip. Some back-end solutions have also been integrated into the CPU. In addition, front-end functions such as the channel controller, servo control, and DSP are also being integrated into a single chip.

With the integration of front- and back-end functions, single chip DVD playback solutions are in the works, which only require an additional preamplifier, memory, and audio DAC. This component integration is essential for saving space in mini-DVD systems and portable devices. After those chips become widely available, costs will drop. In 2004 an average DVD player bill of materials will fall below $75.

DVD Applications

DVD discs and DVD players/drives serve two major markets - consumer (movie sales and rentals) and PCs (games, database information, etc). In each respective market DVD has the potential to replace older VCR and CD-ROM technologies. DVD improves video and audio and provides wide-screen format, interactivity, and parental control. Because of these advantages the DVD is likely to replace VHS tapes and laser discs. With only a slightly higher cost than CD-ROMs, and because audio CDs and CD-ROMs can be played in DVD-ROM drives, the DVD-ROM will gain presence in PCs. DVD-ROMs will provide high-memory storage in PCs.

DVDs deliver brilliance, clarity, range of colors, and detailed resolution—up to 600 horizontal lines across a TV screen vs. only 300 lines provided by broadcast standards. They also remove fuzziness from the edge of the screen. DVD systems are optimized for watching movies and playing games on a wide-screen format, with aspect ratio of 16:9. They provide superior audio effects by incorporating theatre-quality Dolby Digital AC-3 Surround Sound or MPEG-2 audio. And DVDs provide the interactivity that videotapes couldn't. Because of the storage capacity and the digital signals, consumers can randomly access different segments of the disc. DVDs provide a multi-angle function for movies, concerts, and sporting events with replay capabilities. They allow storage of multiple soundtracks and closed-captioning in multiple languages. And DVDs enable parents to control what their children watch. Parents can select appropriate versions of movies, depending on the ratings.

DVD Market Numbers, Drivers and Challenges

DVD-Audio players and DVD recorders represent the fastest growing consumer electronics product in history. Their recent introduction will propel sales of DVD players to new heights over the next several years. According to the Cahners In-Stat Group, the DVD market grew from nothing in 1996 to more than 28 million units shipped worldwide in 2001. Shipments are expected to exceed 60 million units in 2004.

In 2004 IDC predicts:

- DVD ROM (PC) drive shipments will reach about 27.8 million units
- DVD-R/RW will reach 3.2 million units
- DVD-Audio will reach 4 million units
- DVD game console will reach 9.5 million units
- DVD players will reach 21 million units

Adoption of DVD players is quickly growing and analysts state that unit shipments of DVD players exceeded the total shipments of VCR shipments in 2003. In the U.S. alone in 2003 shipments for DVD players exceeded 20 million units, compared to 18 million VCR players. The installed base for DVD players will grow to 63.5 million units by the end of 2004. This represents the rise of the installed base from 5.5% of the estimated 100 million total U.S. households in 1999 to 63.5% in 2004. Worldwide shipments for DVD players exceeded 51 million units in 2003 and are predicted to exceed 75 million units in 2004. The total market for DVD ROM drives in PCs reached 190 million units in 2003 as consumers started replacing CD drives. Some 400 million DVDs were sold in 2003 compared with roughly 227 million in 2000.

The market is currently experiencing high volume, declining prices, broadening vendor support, and more interactive features (gaming, Internet, etc.) which will help the DVD market.

Key market drivers for the success of DVD players include:

- **DVD-Video players are now mainstream because of low cost** – The adoption rate for DVD-Video players surpasses that of any consumer electronics device to date and has long since passed the "early adopter" stage. With prices as low as $100 for a stand-alone DVD-Video player, DVD-Video is now a mainstream format.
- **High consumer awareness**
- **Pure digital format** – The video and audio information stored on a DVD-Video is pure digital for a crystal clear picture and CD-quality sound. It is the ideal format for movie viewing, collecting, and distribution.

- **Improved picture quality and color** – The DVD format provides 480 (up to 600) horizontal lines of resolution. This is a significant improvement over 260 horizontal lines of resolution of standard VHS. The color is brilliant, rich, and saturated, accurately rendering skin tones. With the right equipment and set-up you can enjoy a picture that approaches the quality of film.
- **Aspect ratio** – True to its promise of delivering the cinematic experience, DVD-Video can reproduce the original widescreen theatrical formats of movies as they are shown in movie theaters. DVD-Video can deliver the 1.85:1, 16:9, or the 2.35:1 aspect ratio. Of course, DVD-Video can also provide the "full-frame" 1.33:1 or 4:3 aspect ratio that represents the standard NTSC television screen (the standard TV format for the U.S. and Canada).
- **State-of-the-art surround sound** – All DVD-Videos include Dolby Digital surround sound consisting of up to six and seven channels of surround sound (i.e., Dolby Digital 5.1-channel and 6.1-channel surround sound). The DVD specification requires Dolby Digital 2.0, 2-channel audio to be encoded on every DVD-Video disc for countries using the NTSC TV standard. This 2-channel soundtrack allows Dolby Surround Pro-Logic to be encoded in the stereo audio channels for backward compatibility with pre-existing Dolby Surround Pro-Logic sound systems. Additionally, some DVDs contain an additional alternative surround sound format called DTS Digital Surround. They can also support the newer extended surround formats such as THX Surround EX, DTS-ES matrix, and DTS-ES discrete 6.1.
- **Multiple language dialogues and soundtracks** - Many DVD movies are distributed with multiple language options (up to a maximum of 8), each with its own dialogue. Closed captioning and/or subtitles are also supported with up to 32 separate closed caption and/or subtitle/karaoke tracks encoded into the DVD disc.
- **Multiple angles option** – DVDs can support the director's use of simultaneous multiple camera angles (up to 9) to put a new spin on the plot.
- **Bonus materials** – Many DVD-Video movie releases come with bonus materials that are normally not included in the VHS version. These might include:
  - Segments on the making of a movie
  - Cast and crew interviews
  - Theatrical trailers
  - TV spots
  - Director's audio commentary
  - Music videos
  - Cast and crew biographies
  - Filmographies

  Some bonus features are actually DVD-ROM features where the same disc features DVD-ROM application content. Such DVD-ROM content (e.g., full screenplay text cross-referenced with video playback, web access, and games) requires a computer with a DVD-ROM drive for viewing. Overall, bonus features make movie collecting on the DVD-Video format all the more rewarding. This material can contain multilingual identifying text for title name, album name, song name, actors, etc. Also there are menus varying from a simple chapter listing to multiple levels with interactive features.
- **Standard audio CDs may be played in DVD players.**
- **Random access to scenes** – Movies on DVDs are organized into chapters similar to how songs are recorded on tracks of an audio CD. You can jump to your favorite scenes directly by using the "skip chapter" button on the DVD player, by entering the chapter number, or by

using the DVD disc's menu feature. And there is no more rewinding of videotapes. DVD-Video has a unique feature called "seamless branching" where different video segments can be pre-programmed to combine in various combinations (reorder the sequence of scene playback). This allows for the same DVD-Video disc to contain different versions of the same film, like an original theatrical release version and a director's cut version of the same film. For example, if you chose the "original theatrical release" version from the main menu, the DVD-Video disc will play the original version of the movie by playing the same scenes as shown in the movie theaters. If you chose the "director's cut" version from the main menu, the DVD-Video disc will play back the director's cut of the movie. This may skip to scenes that were previously unreleased during certain segments, then automatically branch back to the common scenes shared with the theatrical version. These scene transitions are nearly instantaneous and transparent to the viewer.

- **Parental control** – The DVD format offers parents the ability to lock out viewing of certain materials by using the branching function. And different versions of the same movie with different MPAA ratings (e.g., G, PG, PG-13, R, and NC-17) can be stored on the same DVD.

- **Durable disc format** – The DVD disc format offers durability and longevity similar to that of audio CDs. With proper handling and care the DVD disc should last a very long time. There is no wear and tear to worry about since there is no contact between the laser pickup and the DVD disc. Unlike VHS videotapes, there is virtually no deterioration or physical wear with repeated use. With its durability and small size, DVD is a great format in which to collect movies and other video titles. The discs are resistant to heat and are not affected by magnetic fields.

- **Content availability** – With over 10,000 titles available on DVD-Video, there is a wide selection of DVD movies available at national, local and independent video rental stores. There are even on-line merchants that rent DVDs such as NetFlix.com. DVDs are here to stay and have become the new medium of choice for home viewing and movie collecting.

- **High (and multiple) storage capacities** – Over 130 minutes of high-quality digital video are available on a single-layered, single-sided disc. Over four hours are available on a single-layered, dual-sided disc or dual-layered, single-sided disc. And over eight hours are theoretically possible with the dual-layered, dual-sided format.

Some of the challenges limiting the success of DVD products are:

- **Content (copy) protection and encryption** – Copy protection is a primary obstacle that the DVD industry must overcome for rapid acceptance and growth. Movie studios do not want to see highly valuable content copied illegally and distributed via PCs and the Internet—like what happened with MP3 music. The industry has developed CSS as a deterrent to illegal copying. A CSS license gives a company permission to build a device with decrypt or anti-scrambling code. Built-in copy protection can also interfere with some display devices, line doublers, etc.

- **The failure of DIVX** – DIVX was essentially a limited-use, pay-per-view DVD technology. It was marketed as a more affordable DVD format. With the financial backing of Circuit City retail stores, it allowed a user to purchase a DIVX disc at a minimal cost and view its contents for an unlimited number of times within a 48-hour period. Once the 48 hours were up, the user was charged for each additional use. The DIVX machines had a built-in modem which automatically called the central billing server about twice a month to report player usage. Users had the option to purchase the right to unlimited viewing for a sum equal to the cost of a DVD-Video disc. Given that a DIVX player was basically a DVD-Video player with additional features to enable a pay-per-view, it is not surprising that it was capable of playing standard DVD-Video discs. Obviously a standard DVD player did not allow viewing of a DIVX disc.

In addition to the built-in modem, the typical DIVX player contained decrypting circuitry to read the DIVX discs which were encoded with a state-of-the-art algorithm [Triple-DES]. Also, the player read the unique serial number off the disc which is recorded on an area known as the Burst Cutting Area (BCA) located in the inner most region of the disc. Essentially, this could be used to record up to 188 bytes of data after the disc has been manufactured. DIVX uses this number to keep track of the viewing period.

Some consumers balked at the idea of having two different standards for digital discs. Others objected to having to keep paying for something they had already purchased. Still, DIVX appeared to be gaining acceptance among consumers with sales of the enhanced players reportedly matching those of standard DVD units. Then its backers pulled the plug on the format in mid-1999, blaming the format's demise on inadequate support from studios and other retailers. Its fate was effectively sealed when companies, including U.S. retail chain Blockbuster, announced plans to rent DVDs to consumers instead of DIVX discs.

Though it didn't last long, DIVX played a useful role in creating a viable rental market essential for DVD-Video to become as popular as VHS. Furthermore, its BCA feature offered some interesting possibilities for future distribution of software on DVD-ROM discs. For example, it could mean an end to forcing consumers to manually enter a long string of characters representing a product's serial number during software installation. A unique vendor ID, product ID, and serial number can be stored as BCA data and automatically read back during the installation process. Storing a product's serial number as BCA data could also deter pirating by making it almost impossible to install a software product without possessing an authentic copy of the disc.

DVD does have its limitations:
- There is consumer confusion due to DVD rewritable format incompatibility
- Increasing DVD capabilities are converging with other devices such as gaming consoles.
- There are region code restrictions. By encoding each DVD-Video disc and DVD players with region codes, only similarly coded software can be played back on DVD hardware. Hence, a DVD-Video coded "region 1" can be played back only by a DVD player that is compatible with region 1. This allows movie studios to release a DVD-Video of a movie while preparing the same movie for theatrical release of that movie overseas. Regional lockout prevents playback of foreign titles by restricting playback of titles to the region supported by the player hardware.
- Poor compression of video and audio can result in compromised performance. When 6-channel discrete audio is automatically down-mixed for stereo/Dolby Surround, the result may not be optimal.
- Not HDTV standard
- Reverse play at normal speed is not yet possible.
- The number of titles will be limited in the early years.

Some emerging DVD player technology trends are:
- DVD-Video players are combining recordability features.
- DVD players are combining DVD-Audio capabilities, enabling DVD players to achieve higher-quality audio. DVD-Audio provides a major advance in audio performance. It

enables the listener to have advanced resolution stereo (2 channels), multi-channel surround sound (6 channels), or both.

- The interactive access to the Internet feature is gaining prominence in DVD players and gaming consoles. DVD players contain almost all of the circuitry required to support an Internet browser, enabling many new opportunities for DVD. The connection to the Internet would allow the customer to get current prices, order merchandise, and communicate via email with a personal shopper.
- PCs are adopting DVD read and write drives.
- A networked DVD player allows the user to play DVD movies across the home network in multiple rooms.
- Gaming functionality is being added.

Convergence of Multiple Services

The functionality provided by DVD players and DVD-ROM devices will converge and be incorporated within existing consumer electronic appliances and PCs. Several companies are producing a DVD player/Internet appliance in set-top boxes and recent video game consoles from Sony and Microsoft allow users to play DVD movies on their units. TV/DVD combinations are becoming as prevalent as car entertainment systems were. DVD is being combined with CD recorders, VCRs, or hard disc drives to form new products. Several companies have introduced DVD home cinema systems that integrate a DVD Video player with items such as a Dolby digital decoder, CD changer, tuner, and cassette deck. Some consumer device manufacturers are integrating DVD functionality within existing set-top boxes and gaming consoles. DVD-Video players are combining recordability features. And a few consumer electronic manufacturers have introduced integrated DVD players and receivers/amplifiers. This device provides a one-box gateway that will provide excellent quality surround sound and digital video and audio playing.

Convergence of DVD-ROM Devices and PCs

Many mid- and high-priced new desktop and notebook computers are being delivered with DVD-ROM drives that will also play DVD-Videos—either on the computer monitor or on a separate TV set. In order to play DVD-Videos a computer must be equipped to decode the compressed audio and video bitstream from a DVD. While there have been attempts to do this in software with high-speed processors, the best implementations currently use hardware, either a separate decoder card or a decoder chip incorporated into a video card.

An unexpected byproduct of the ability of computers to play DVD videos has been the emergence of higher quality video digital displays suitable for large-screen projection in home theaters. Computers display in a progressive scan format where all the scan lines are displayed in one sweep, usually in a minimum of 1/60th of a second. Consumer video displays in an interlaced scan format where only half the scan lines (alternating odd/even) are displayed on a screen every 1/60th of a second (one "field"). The complete image frame on a TV takes 1/30th of a second to be displayed.

The progressive scan display is referred to as "line-doubled" in home theater parlance because the lines of a single field become twice as many. It has been very expensive to perform this trick. The best line doublers cost $10,000 and up. However, movie films are stored on DVD in a format that makes it simple to generate progressive scan output, typically at three times the film rate, or 72 Hz. Currently, only expensive data-grade CRT front projectors can display the higher scan rates required.

But less expensive rear projectors that can do this have been announced. Convergence of computers and home entertainment appears poised to explode.

This progressive scan display from DVD-Video is technically referred to as 720x480p. That is, 720 pixels of horizontal resolution and 480 vertical scan lines displayed progressively rather than interlaced. It is very close to the entry-level format of digital TV. The image quality and resolution exceed anything else currently available in consumer TV.

Convergence of DVD Players and Gaming Devices

As described in Chapter 6, DVD players will converge with gaming consoles. The significant storage capacity, digital quality, features, and low cost of media make it far superior to game cartridges. A game console and a DVD player actually have a number of components in common such as video and audio processors, memory, TV-out encoders, and a host of connectors and other silicon. There is also potential for both functions to use the same drive mechanism which is one of the most expensive parts of the system. Today, many of the components are designed specifically for one or the other, but with a little design effort it would be easy to integrate high-quality gaming into a DVD player at little incremental cost.

Game consoles are already moving quickly towards higher-capacity and lower-cost media, and DVD-based software represents the best of both worlds. Although cartridges have benefits in the area of piracy and platform control, they are expensive and have limited a number of manufacturers. Gaming consoles announced by both Sony and Microsoft have DVD support.

Convergence of DVD Players and Set-top Boxes

Consumers will want plenty of versatility out of their DVD systems. Instead of receiving satellite, terrestrial, and cable signals via a set-top box or reading a disc from their DVD player, they will expect to download video and audio from a satellite stream and record it using their DVD-RAM system.

Because current DVD players, satellite receivers, and cable set-top boxes use similar electronic components, a converged player may not be far away. DVD is based on the MPEG-2 standard for video distribution as are most satellite and wireless cable systems. The audio/video decoder chip at the heart of a DVD player is an enhanced version of the video and audio decoder chips necessary to build a home satellite receiver or a wireless cable set-top box. With many similarities, it is likely that most of these appliances will eventually converge into a single unit— a convergence box that blends the features of three separate systems into one unit.

Convergence of DVD-Video Players and DVD-Audio Players

The audio market is seeing several technologies compete to become the next-generation audio technology that proliferates into consumer homes. Some of these include CD, MP3, SACD, and DVD-Audio. The DVD specification includes room for a separate DVD-Audio format. This allows the high-end audio industry to take advantage of the existing DVD specification. DVD-Audio is releasing music in a PCM (pulse code modulation) format that uses 96 KHz and 24 bits as opposed to the CD PCM format of 44.1 KHz and 16 bits.

All current DVD-Video players can play standard audio CDs (although not required by the specification) as well as DVDs with 48 or 96 KHz/16-, 20- or 24-bit PCM format. Most second-generation DVD-Video players will actually use 96 KHz digital to analog converters, whereas some of the first generation players initially down-converted the 96 KHz bitstream to 48 KHz before decoding. Hence, several next-generation DVD players will automatically add DVD-Audio capability. However, they may support multiple other audio formats and standards as well.

DVD-A combined into a DVD video player will help prevent severe price deflation by maintaining the value of the DVD video player. Because it will require only a small premium for DVD-A to be incorporated into a DVD video player, players capable of DVD-A will quickly win market share due to the mainstream price point of this advanced player.

Since DVD players can play audio CDs, sales of standalone CD players have shrunk. Consumer electronics manufacturers are looking to integrate audio functionality within DVD players for added revenues. DVD video players combined with DVD-A functionality will eventually replace DVD video players with CD player capabilities.

The "convergence box" concept with its features and functions is not for everyone. For example, high-quality 3D-imaging for video games may be great for some, but not for all. Likewise, reception of 180-plus television stations via satellite is not a necessity for everyone. Though convergence seems inevitable, convergence boxes will probably be packaged next to stand-alone products (DVD, satellite-dish, cable system, etc.) for consumers who will not want to pay extra for features they will not use.

Summary

The DVD format is doing for movies what the CD format did for music. Because of its unique advantages, the DVD has become the most successful consumer technology and product.

The three application formats of DVD include DVD-Video for video entertainment, DVD-Audio for audio entertainment, and DVD-ROM for PCs. In the future DVD will provide higher storage capacities, but it will change the way consumers use audio, gaming, and recording capabilities.

DVD technology will see rapid market growth as it adds features and integrates components to obtain lower prices. In addition, sales will increase as new and existing digital consumer devices and PCs incorporate DVD technology.

Desktop and Notebook Personal Computers (PCs)

Introduction

Over the last 10 years the PC has become omnipresent in the work environment. Today it is found in almost every home, with a sizeable percentage of homes having multiple PCs. Dataquest estimates that PC penetration exceeds 50% of U.S. households—out of 102 million U.S. households, 52 million own a PC. They also find that households with multiple PCs have grown from 15 million in 1998 to over 26 million in 2003.

A plethora of new consumer devices are available today with a lot more to come. Some consumer devices such as MP3 players, set-top boxes, PDAs, digital cameras, gaming stations and digital VCRs are gaining increased popularity. Dataquest predicts that the worldwide unit production of digital consumer devices will explode from 1.8 million in 1999 to 391 million in 2004. The next consumer wave is predicted to transition from a PC-centric home to a consumer device-centric home.

Despite growth in the shipments of digital consumer devices, the PC will continue to penetrate homes for several years. While most consumer devices perform a single function very well, the PC will remain the backbone for information and communication. In the future it will grow to become the gateway of the home network. A large number of consumer devices such as Internet-enabled cell phones and portable MP3 players will not compete directly with PCs. Rather, a set-top box with a hard-drive or a game console may provide similar functionality as the PC and may appeal to a certain market segment. However, the PC in both desktop and notebook variations will continue to be used for a variety of tasks.

Definition of the Personal Computer

Since the PC is so prominent, defining it seems unnecessary. However, it is interesting to look at how the PC has evolved and how its definition is changing.

The PC is a computer designed for use by one person at a time. Prior to the PC, computers were designed for (and only affordable by) companies who attached multiple terminals to a single large computer whose resources were shared among all users. Beginning in the late 1980s technology advances made it feasible to build a small computer that an individual could own and use. The term "PC" is also commonly used to describe an IBM-compatible personal computer in contrast to an Apple Macintosh computer. The distinction is both technical and cultural. The IBM-compatible PC contains Intel microprocessor architecture and an operating system such as Microsoft Windows (or DOS previously) that is written to use the Intel microprocessor. The Apple Macintosh uses Motorola microprocessor architecture and a proprietary operating system—although market economics are now forcing Apple Computer to use Intel microprocessors as well. The IBM-compatible PC is associated with business and home use. The "Mac," known for its more intuitive user interface, is associated with graphic design and desktop publishing. However, recent progress in operating systems from Microsoft provides similar user experiences.

For this discussion, the PC is a personal computer—both desktop and laptop - that is relatively inexpensive and is used at home for computation, Internet access, communications, etc. By definition, the PC has the following attributes:

- Digital computer
- Largely automatic
- Programmable by the user
- Accessible as a commercially manufactured product, a commercially available kit, or in widely published kit plans
- Transportable by an average person
- Affordable by the average professional
- Simple enough to use that it requires no special training beyond an instruction manual

Competing to be the Head of the Household

The PC is the most important and widely used device for computing, Internet access, on-line gaming, data storage, etc. It is also viewed as a strong potential candidate for a residential gateway. In this capacity, it would network the home, provide broadband access, control the functionality of other appliances, and deploy other value-added services. The rapid growth in multi-PC households is creating the need for sharing broadband access, files and data, and peripherals (e.g., printers and scanners) among the multiple PCs in different rooms of a house. This is leading to the networking of PCs, PC peripherals, and other appliances.

With worldwide PC shipments totaling 134.7 million units in 2000, and predicted to exceed 200 million in 2005, the PC (desktop and notebook) is seeing healthy demand. This is due to ongoing price per performance improvements and the position of the PC as a productivity tool. However, there are still several weaknesses with the PC. It is often viewed as being complex, buggy, and confusing. But with the recent introduction of operating systems that have a faster boot-up time and are less buggy, the PC has become friendlier. Standardization of components is also addressing some of the weaknesses.

PC market opportunities include:

- Growth in multiple PC households
- Evolution of new business models
- Introduction of new and innovative PC designs
- The growing ubiquity of the Internet

PC Market threats include:

- Digital consumer devices
- Web-based services and applications
- Saturation in key markets
- Low margins

With this in mind, an emerging category of digital consumer electronics device is coming to market. This category includes these features:

- Low-cost
- Easy-to-use
- Instant-on
- Lightweight

- Reliable
- Special-purpose access to the Internet.

Devices which pose a threat to the existence of the PC include set-top boxes, digital TV, gaming consoles, Internet screen phones, Web pads, PDAs, mobile phones, and Web terminals. Smarter chips and increased intelligence are being embedded into everyday consumer devices. The digitization of data, voice, and video is enabling the convergence of consumer devices. Applications such as email, Web shopping, remote monitoring, MP3 files, and streaming video are pushing the need for Internet access. These devices have advanced computational capabilities that provide more value and convenience when networked.

Over the last couple of years the Internet has grown ubiquitous to a large number of consumers. Consumers are demanding high-speed Internet access to PCs and other home appliances such as Web pads, Web terminals, digital TV, and set-top boxes. Using a single, broadband access point to the Internet provides cost savings, but requires the networking of PCs and digital consumer devices.

Shipments for PCs exceeded 11.6 million units and revenue exceeded $21 billion in the U.S. in 1998. Market analysts project an increase in shipments to 25.2 million units in the year 2001 for the U.S. The growth in PC revenues has stalled over the last few years. Non-PC vendors are seeking an opportunity to capture the $21 billion market in U.S. alone. Predictions for 2001 in the U.S. include sales of 22 million home digital consumer devices (excluding Internet-enabled mobile phones and telematics systems) compared with 18 million home PCs. In 2004 digital consumer appliance revenues will rise above falling PC revenues.

Digital consumer appliances include:
- Internet-connected TVs
- Consumer network computers
- Web tablets
- E-mail-only devices
- Screen phones
- Internet game consoles
- Handheld PCs/personal digital assistants (PDAs)
- Internet-enabled mobile phones
- Automotive telematics systems

The firm forecasts that total revenues from all digital consumer appliances (including Internet-enabled mobile phones and telematics systems) will reach $33.7 billion by the end of 2005. Market numbers for digital consumer devices are largely skewed between research firms because some of the numbers include appliances such as cell-phones and telematics systems that do not directly compete with the PC.

This slower growth in revenues is due to the increase in the number of digital consumer devices and falling ASPs (average selling prices) of PCs. In 2001 shipments of digital consumer devices exceeded the unit shipments for PCs in the U.S.

Factors driving digital consumer devices are:
- Aggressive vendor pursuit
- Meeting consumer demands (ease-of-use, providing specific functionality, requiring Internet connectivity)
- Advancing bandwidth capacity
- Lower product costs

Digital consumer devices are targeting three specific areas:

- Replacing PCs and providing robust Web browsing, email, and interactivity (e.g., email terminals, Web terminals, Web pads).
- Supplementing PCs and coexisting with them (e.g., PDAs, printers, scanners).
- Sidestepping PCs with appliances that provide functionality quite different from the PC and are not a significant threat to its existence (e.g., set-top boxes, cellular phones.)

Some factors that are driving the success of digital consumer devices are:

- Aggressive vendor pursuit
- Consumer market demands
- Advancing bandwidth capacity
- Lower product costs
- Consumer needs
- Device distribution subsidized by service contracts rather than via retail or direct sales, like for PCs. An example would be the digital video recorder (DVR).

Because digital consumer devices are based on the idea that the PC is too complex, comparisons between digital consumer devices and PCs have always been made. It is this constant comparison that has cast the Web terminal segment in such a difficult light. Many in the industry had wrongly expected that shipments of the Web terminal segment (without the help of handheld- or TV-based devices) would single-handedly exceed and replace the PC market. In light of these misinterpretations, comparing shipments of digital consumer devices with shipments of PCs is not completely correct.

Also, there are multiple definitions for digital consumer devices and some include devices that don't directly compete with PCs. The more commonly perceived definition includes only Web terminals, Web tablets, and email terminals. These products barely even register on the radar screen in comparison with PCs. Once the definition of digital consumer devices is expanded to include handheld, gaming, and iTV-enabled devices, the change is obvious. A very healthy spike in digital consumer device shipments in 2001—largely attributable to next-generation gaming consoles becoming Internet-enabled—pushes the definition of digital consumer devices beyond just shipments for PCs. Even with high volumes of handhelds and TV-based devices, it is important to remember that not one single category is expected to out-ship PCs on its own.

Note that this does not mean PCs are going away. Despite certain fluctuations, the PC market is still expected to remain solid. In fact, the function and usage could be completely different (e.g., web surfing on a PC versus online gaming on a console). This may happen even though digital consumer devices represent an alternative to PCs as an Internet access method.

Additionally, many vendors approaching the post-PC market see the PC continuing as a central point (or launching pad) for the other devices that peacefully coexist in the networked home. Hence, many in the industry further argue that the term "post-PC era" is better replaced with the term "PC-plus era." Finally, keep in mind that this comparison with PCs is primarily applicable to the United States where PCs are currently seeing over 55% household penetration. In international markets PC penetration is much lower. In those instances the more appropriate comparison might be with mobile phones.

The PC Fights Back

The PC is not going away soon because of its:

- **Compatibility** – Interoperability between documents for business, education, and government.
- **Flexibility in the PC platform** – Video editing, music authoring, Web hosting, gaming, etc., can all be done on one PC. Comparatively, digital consumer devices are dedicated to one function.
- **Momentum** – PCs have a huge installed base, high revenue, and annual shipments.
- **Awareness** – More than 50% of consumer homes have PCs.
- **Investment protection and reluctance to discard**
- **Pace of improvement** – Faster processors, bigger hard drives, better communication.
- **Being an established industry** – Corporate momentum will continue due to PC vendors, software developers, stores, and support networks.

Some of the market trends for PCs include:

- **Lower price points** – Rapidly moving below $500 which causes blurred lines between PCs and digital consumer devices.
- **New downstream revenue opportunities** – Internet services, financing/leasing, and e-commerce options. Higher intelligence is making the PC an ideal platform for the residential gateway of the future. PCs are already shipped with analog or digital modems, allowing broadband Internet access. This platform is ideal to provide home networking capabilities to multiple consumer devices.
- **Super-efficient distribution** – File data can be distributed effectively via the Internet.
- **Offsetting-profit products** – Products such as servers, services (Internet, high-speed access), and workstations enable aggressive prices to consumers.

Most consumer devices focus on a single value-add to the customer and provide it very well. Hence, some focus on the unique negatives of a PC and show how the PC will never survive. For example, MP3 players focus on portability and ease-of-use to customers. However, today there is no better device than the PC to download and mass-store music in a library .

There are applications that digital consumer devices perform well. For example, a Web pad lets you perform tasks such as scheduling, ordering groceries, and sending email—all with portability. While very convenient within the home, it is incapable beyond this and requires an access point and a gateway to provide a high-speed Internet connection. The PC can provide all the above functions and more—such as video editing, gaming, and Word editing. Hence, PCs will continue to maintain a stronghold on homes. Some digital consumer devices have unique value propositions and are making their way into consumer lives. These include gaming consoles, Web pad, set-top boxes, and telematics. Both PCs and consumer devices will coexist in tomorrow's homes.

Components of a PC

Figure 8.1 shows a generic PC that includes processor, memory, hard-disk drive, operating system, and other components. These make up a unique platform that is ideal for providing multiple functions such as Internet and e-commerce services. Future PCs will provide digital modems and home networking chipsets. Broadband access will provide high-speed Internet access to the PC and other digital consumer devices. Home networking will network multiple PCs, PC peripherals, and other consumer devices.

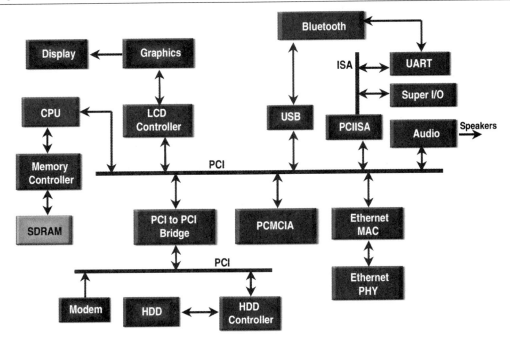

Figure 8.1: The PC

The following sections describe some key components that make up the PC.

Processor

The processor (short for microprocessor, also called the CPU, or central processing unit) is the central component of the PC. This vital component is responsible for almost everything the PC does. It determines which operating systems can be used, which software packages can be run, how much energy the PC uses, and how stable the system will be. It is also a major determinant of overall system cost. The newer, faster, and more powerful the processor, the more expensive the PC.

Today, more than half a century after John Von Neumann first suggested storing a sequence of instructions (program) in the same memory as the data, nearly all processors have a "Von Neumann" architecture. It consists of the central arithmetical unit, the central control unit, the memory, and the Input/Output devices.

Principles

The underlying principles of all computer processors are the same. They take signals in the form of 0s and 1s (binary signals), manipulate them according to a set of instructions, and produce output in the form of 0s and 1s. The voltage on the line at the time a signal is sent determines whether the signal is a 0 or a 1. On a 2.5V system, an application of 2.5 volts means that it is a 1, while an application of 0V means it is a 0.

Processors execute algorithms by reacting in a specific way to an input of 0s and 1s, then return an output based on the decision. The decision itself happens in a circuit called a logic gate. Each gate requires at least one transistor, with the inputs and outputs arranged differently by different operations. The fact that today's processors contain millions of transistors offers a clue as to how complex the logic system is.

The processor's logic gates work together to make decisions using Boolean logic. The main Boolean operators are AND, OR, NOT, NAND, NOR, with many combinations of these as well. In addition, the processor uses gate combinations to perform arithmetic functions and trigger storage of data in memory. Modern day microprocessors contain tens of millions of microscopic transistors. Used in combination with resistors, capacitors, and diodes, these make up logic gates. And logic gates make up integrated circuits which make up electronic systems.

Intel's first claim to fame was the Intel 4004 released in late 1971. It was the integration of all the processor's logic gates into a single complex processor chip. The 4004 was a 4-bit microprocessor intended for use in a calculator. It processed data in 4 bits, but its instructions were 8 bits long. Program and data memory were separate, 1 KB and 4 KB respectively. There were also sixteen 4-bit (or eight 8-bit) general-purpose registers. The 4004 had 46 instructions, using only 2,300 transistors. It ran at a clock rate of 740 KHz (eight-clock cycles per CPU cycle of 10.8 microseconds).

Two families of microprocessor have dominated the PC industry for some years—Intel's Pentium and Motorola's PowerPC. These CPUs are also prime examples of the two competing CPU architectures of the last two decades—the former being a CISC chip and the latter a RISC chip.

CISC (complex instruction set computer) is a traditional computer architecture where the CPU uses microcode to execute very comprehensive instruction sets. These may be variable in length and use all addressing modes, requiring complex circuitry to decode them. For many years the tendency among computer manufacturers was to build increasingly complex CPUs that had ever-larger sets of instructions. In 1974 IBM decided to try an approach that dramatically reduced the number of instructions a chip performed. By the mid-1980s this led to a trend reversal where computer manufacturers were building CPUs capable of executing only a very limited set of instructions.

RISC (reduced instruction set computer) CPUs:

- Keep instruction size constant.
- Ban the indirect addressing mode.
- Retain only those instructions that can be overlapped and made to execute in one machine cycle or less.

One advantage of RISC CPUs is that they execute instructions very fast because the instructions are so simple. Another advantage is that RISC chips require fewer transistors, which makes them cheaper to design and produce.

There is still considerable controversy among experts about the ultimate value of RISC architectures. Its proponents argue that RISC machines are both cheaper and faster, making them the machines of the future. Conversely, skeptics note that by making the hardware simpler, RISC architectures put a greater burden on the software. For example, RISC compilers have to generate software routines to perform the complex instructions that are performed in hardware by CISC computers. They argue that this is not worth the trouble because conventional microprocessors are becoming increasingly fast and cheap anyway. To some extent the argument is becoming moot because CISC and RISC implementations are becoming more and more alike. Even the CISC champion, Intel, used RISC techniques in its 486 chip and has done so increasingly in its Pentium processor family.

Basic Structure

A processor's major functional components are:

- **Core** – The heart of a modern processor is the execution unit. The Pentium has two parallel integer pipelines enabling it to read, interpret, execute, and dispatch two instructions simultaneously.

- **Branch Predictor** – The branch prediction unit tries to guess which sequence will be executed each time the program contains a conditional jump. In this way, the pre-fetch and decode unit can prepare the instructions in advance.

- **Floating Point Unit** – This is the third execution unit in a processor where non-integer calculations are performed.

- **Primary Cache** – The Pentium has two on-chip caches of 8 KB each, one for code and one for data. Primary cache is far quicker than the larger external secondary cache.

- **Bus Interface** – This brings a mixture of code and data into the CPU, separates them for use, then recombines them and sends them back out.

All the elements of the processor are synchronized by a "clock" which dictates how fast the processor operates. The very first microprocessor had a 100-kHz clock. The Pentium Pro used a 200-MHz clock, which is to say it "ticks" 200 million times per second. Programs are executed as the clock ticks. The Program Counter is an internal memory location that contains the address of the next instruction to be executed. When the time comes for it to be executed, the Control Unit transfers the instruction from memory into its Instruction Register (IR).

At the same time, the Program Counter increments to point to the next instruction in sequence. The processor then executes the instruction in the IR. The Control Unit itself handles some instructions. Here, if the instruction says "jump to location 213," the value of 213 is written to the Program Counter so that the processor executes that instruction next.

Many instructions involve the arithmetic and logic unit (ALU). The ALU works in conjunction with the General Purpose Registers (GPR) which provides temporary storage areas that can be loaded from memory or written to memory. A typical ALU instruction might be to add the contents of a memory location to a GPR. As each instruction is executed, the ALU also alters the bits in the Status Register (SR) which holds information on the result of the previous instruction. Typically, SR bits indicate a zero result, an overflow, a carry, and so forth. The control unit uses the information in the SR to execute conditional instructions such as "jump to address 740 if the previous instruction overflowed."

Architectural Advances

According to Moore's Law, CPUs double their capacity and capability every 18-24 months. In recent years Intel has managed to doggedly follow this law and stay ahead of the competition by releasing more powerful chips for the PC than any other company. In 1978 the 8086 ran at 4.77 MHz and had under a million transistors. By the end of 1995 their Pentium Pro had a staggering 21 million on-chip transistors and ran at 200 MHz. Today's processors run in excess of 3 GHz.

The laws of physics limit designers from increasing the clock speed indefinitely. Although clock rates go up every year, this alone wouldn't give the performance gains we're used to. This is the reason why engineers are constantly looking for ways to get the processor to undertake more work in each tick of the clock. One approach is to widen the data bus and registers. Even a 4-bit processor can add two 32-bit numbers, but this takes lots of instructions. A 32-bit processor could do the task in a single instruction. Most of today's processors have a 32-bit architecture, and 64-bit variants are on the way.

In the early days processors could only deal with integers (whole numbers). It was possible to write a program using simple instructions to deal with fractional numbers, but it was slow. Virtually all processors today have instructions to handle floating-point numbers directly. To say that "things happen with each tick of the clock" underestimates how long it actually takes to execute an instruction. Traditionally, it took five ticks—one to load the instruction, one to decode it, one to get the data, one to execute it, and one to write the result. In this case a 100-MHz processor could only execute 20 million instructions per second.

Most processors now employ pipelining, a process resembling a factory production line. One stage in the pipeline is dedicated to each of the stages needed to execute an instruction. Each stage passes the instruction on to the next when it is finished. Hence, at any one time one instruction is being loaded, another is being decoded, data is being fetched for a third, a fourth is being executed, and the result is being written for a fifth. One instruction per clock cycle can be achieved with current technology.

Furthermore, many processors now have a superscalar architecture. Here, the circuitry for each stage of the pipeline is duplicated so that multiple instructions can pass through in parallel. For example, the Pentium Pro can execute up to five instructions per clock cycle.

The 4004 used a 10-micron process—the smallest feature was 10 millionths of a meter across. This is huge by today's standards. Under these constraints a Pentium Pro would measure about 5.5 inches × 7.5 inches and would be very slow. (To be fast, a transistor must be small.) By 2003 most processors used a 0.13-micron process with 90 nanometers being the goal in late 2003.

Software Compatibility

In the early days of computing many people wrote their own software so the exact set of instructions a processor could execute was of little importance. However, today people want to use off-the-shelf software so the instruction set is paramount. Although there's nothing magical about the Intel 80x86 architecture from a technical standpoint, it has long since become the industry standard.

If a third party makes a processor which has different instructions, it will not run industry standard software. So in the days of 386s and 486s, companies like AMD cloned Intel processors, which meant they were always about a generation behind. The Cyrix 6x86 and the AMD K5 were competitors to Intel's Pentium, but they weren't carbon copies. The K5 has its own instruction set and translates 80x86 instructions into its native instructions as they are loaded. In this way, AMD didn't have to wait for the Pentium before designing the K5. Much of it was actually designed in parallel—only the translation circuitry was held back. When the K5 eventually appeared, it leap-frogged the Pentium in performance (if clock speeds were equal).

The use of standard busses is another way processors with different architectures are given some uniformity to the outside world. The PCI bus has been one of the most important standards in this respect since its introduction in 1994. PCI defines signals that enable the processor to communicate with other parts of a PC. It includes address and data buses as well as a number of control signals. Processors have their own proprietary buses so a chipset is used to convert from a "private" bus to the "public" PCI bus.

Historical Perspective

The 400 was the first Intel microprocessor. To date, all PC processors have been based on the original Intel designs. The first chip used in an IBM PC was Intel's 8088. At the time it was chosen, it was not the best CPU available. In fact, Intel's own 8086 was more powerful and had been released earlier. The 8088 was chosen for economic reasons—its 8-bit data bus required a less costly motherboard than the 16-bit 8086. Also, most of the interface chips available were intended for use in 8-bit designs. These early processors would have nowhere near the performance to run today's software. Today the microprocessor market for PCs primarily belongs to two companies—Intel and AMD. Over the last 15-20 years, the introduction of PC architectures and families (with the Intel 80286, 80386, and 80486) has been more often followed by the Pentium family series.

Third generation chips based on Intel's 80386 processors were the first 32-bit processors to appear in PCs. Fourth generation processors (the 80486 family) were also 32-bit and offered a number of enhancements which made them more than twice as fast. They all had 8K of cache memory on the chip itself, right beside the processor logic. These cached data that was transferred from the main memory. Hence, on average the processor needed to wait for data from the motherboard for only 4% of the time because it was usually able to get required information from the cache. The 486 model brought the mathematics co-processor on board as well. This was a separate processor designed to take over floating-point calculations. Though it had little impact on everyday applications, it transformed the performance of spreadsheet, statistical analysis, and CAD packages.

Clock doubling was another important innovation introduced on the 486. With clock doubling, the circuits inside the chip ran at twice the speed of the external electronics. Data was transferred between the processor, the internal cache, and the math co-processor at twice the speed, considerably enhancing performance. Versions following the base 486 took these techniques even further - tripling the clock speed to run internally at 75 or 100 MHz and doubling the primary cache to 16 KB. The Intel Pentium is the defining processor of the fifth generation. It provides greatly increased performance over a 486 due to architectural changes that include a doubling of the data bus width to 64 bits. Also, the processor doubled on-board primary cache to 32 KB and increased the instruction set to optimize multimedia function handling.

The word Pentium™ does not mean anything. It contains the syllable pent—the Greek root for five—symbolizing the 5th generation of processors. Originally, Intel was going to call the Pentium the 80586 in keeping with the chip's 80x86 predecessors. But the company did not like the idea that AMD, Cyrix, and other clone makers could use the name 80x86 as well. So they decided to use 'Pentium'—a trademarkable name. The introduction of the Pentium in 1993 revolutionized the PC market by putting more power into the average PC than NASA had in its air-conditioned computer rooms of the early 1960s. The Pentium's CISC-based architecture represented a leap forward from the 486. The 120-MHz and above versions had over 3.3 million transistors, fabricated on a 0.35-micron process. Internally, the processor uses a 32-bit bus, but externally the data bus is 64 bits wide. The external bus required a different motherboard. To support this, Intel also released a special chipset for linking the Pentium to a 64-bit external cache and to the PCI bus.

The Pentium has a dual-pipelined superscalar design which allows it to execute more instructions per clock cycle. Like a 486, there are still five stages of integer instruction execution—pre-fetch, instruction decode, address generate, execute and write back. But the Pentium has two parallel integer pipelines that enable it to read, interpret, execute, and dispatch two operations simultaneously. These handle only integer calculations; a separate Floating-Point Unit handles real numbers.

The Pentium also uses two 8-KB, two-way set, associative buffers known as primary, or Level 1, cache. One is used for instructions, the other for data. This is twice the amount of its predecessor, the 486. These caches increase performance by acting as temporary storage for data instructions obtained from the slower main memory. A Branch Target Buffer (BTB) provides dynamic branch prediction. It enhances instruction execution by remembering which way an instruction branched, then applying the same branch the next time the instruction is used. Performance is improved when the BTB makes a correct prediction. An 80-point Floating-Point Unit provides the arithmetic engine to handle real numbers. A System Management Mode (SMM) for controlling the power use of the processor and peripherals rounds out the design.

The Pentium Pro was introduced in 1995 as the successor to the Pentium. It was the first of the sixth generation of processors and introduced several unique architectural features that had never been seen in a PC processor before. The Pentium Pro was the first mainstream CPU to radically change how it executes instructions. It translates them into RISC-like microinstructions and executes them on a highly advanced internal core. It also featured a dramatically higher-performance secondary cache compared to all earlier processors. Instead of using motherboard-based cache running at the speed of the memory bus, it used an integrated Level 2 cache with its own bus running at full processor speed, typically three times the speed of a Pentium cache.

Intel's Pentium Pro had a CPU core consisting of 5.5 million transistors, with 15.5 million transistors in the Level 2 cache. It was initially aimed at the server and high-end workstation markets. It is a superscalar processor incorporating high-order processor features and is optimized for 32-bit operation. The Pentium Pro differs from the Pentium in that it has an on-chip Level 2 cache of between 256 KB and 1 MB operating at the internal clock speed. The positioning of the secondary cache on the chip, rather than on the motherboard, enables signals to pass between the two on a 64-bit data path rather than on the 32-bit Pentium system bus path. Their physical proximity also adds to the performance gain. The combination is so powerful that Intel claims the 256 KB of cache on the chip is equivalent to over 2 MB of motherboard cache.

An even bigger factor in the Pentium Pro's performance improvement is the combination of technologies known as "dynamic execution." This includes branch prediction, data flow analysis, and speculative execution. These combine to allow the processor to use otherwise-wasted clock cycles. Dynamic execution makes predictions about the program flow to execute instructions in advance.

The Pentium Pro was also the first processor in the x86 family to employ super-pipelining, its pipeline comprising 14 stages, divided into three sections. The in-order, front-end section, which handles the decoding and issuing of instructions, consists of eight stages. An out-of-order core which executes the instructions has three stages, and the in-order retirement consists of a final three stages.

The other, more critical distinction of the Pentium Pro is its handling of instructions. It takes the CISC x86 instructions and converts them into internal RISC micro-ops. The conversion helps avoid some of the limitations inherent in the x86 instruction set such as irregular instruction encoding and register-to-memory arithmetic operations. The micro-ops are then passed into an out-of-order execution engine that determines whether instructions are ready for execution. If not, they are shuffled around to prevent pipeline stalls.

There are drawbacks to using the RISC approach. The first is that converting instructions takes time, even if calculated in nano or microseconds. As a result the Pentium Pro inevitably takes a performance hit when processing instructions. A second drawback is that the out-of-order design can be particularly affected by 16-bit code, resulting in stalls. These tend to be caused by partial register updates that occur before full register reads, imposing severe performance penalties of up to seven clock cycles.

The Pentium II proved to be an evolutionary step from the Pentium Pro. Architecturally, the Pentium II is not very different from the Pentium Pro, with a similar x86 emulation core and most of the same features. The Pentium II improved on the Pentium Pro by doubling the size of the Level 1 cache to 32 KB. Intel used special caches to improve the efficiency of 16-bit code processing. (The Pentium Pro was optimized for 32-bit processing and did not deal quite as well with 16-bit code.) Intel also increased the size of the write buffers. However, its packaging was the most talked-about aspect of the new Pentium II. The integrated Pentium Pro secondary cache running at full processor speed was replaced with a special circuit board containing the processor and 512 KB of secondary cache running at half the processor's speed.

The Pentium III, successor to the Pentium II, came to market in the spring of 1999. The Single Instruction Multiple Data (SIMD) process came about with the introduction of MMX—multimedia extensions. SIMD enables one instruction to perform the same function on several pieces of data simultaneously. This improves the speed at which sets of data requiring the same operations can be processed. The new processor introduces 70 new Streaming SIMD Extensions, but does not make any other architecture improvements. Fifty of the new SIMD Extensions improve floating-point performance to assist data manipulation. There are 12 new multimedia instructions to complement the existing 57 integer MMX instructions. They provide further support for multimedia data processing. The final eight instructions, referred to by Intel as cache-ability instructions, improve the efficiency of the CPU's Level 1 cache. They allow sophisticated software developers to boost performance of their applications or games. Other than this, the Pentium III makes no other architecture improvements. It still fits into slot 1 motherboards and still has 32 KB of Level 1 cache.

Close on the heels of the Pentium III came the Pentium Xeon with the new Streaming SIMD Extensions (SSE) instruction set. Targeted at the server and workstation markets, the Pentium Xeon shipped as a 700-MHz processor with 512 KB, 1 MB, or 2 MB of Level 2 cache. Since then Intel has announced processors with 0.18-micron process technology that provides smaller die sizes and lower operating voltages. This facilitates more compact and power-efficient system designs, and makes possible clock speeds of 1 GHz and beyond.

Intel and Hewlett Packard announced their joint research and development project aimed at providing advanced technologies for workstation, server, and enterprise-computing products. The product, called Itanium, uses a 64-bit computer architecture that is capable of addressing an enormous 16 TB of memory. That is 4 billion times more than 32-bit platforms can handle. In huge databases, 64-bit platforms reduce the time it takes to access storage devices and load data into virtual memory, and this has a significant impact on performance.

The Instruction Set Architecture (ISA) uses the Explicitly Parallel Instruction Computing (EPIC) technology which represents Itanium's biggest technological advance. EPIC—incorporating an innovative combination of speculation, prediction, and explicit parallelism—advances state-of-art in processor technologies. Specifically, it addresses performance limitations found in RISC and CISC technologies. Both architectures already use various internal techniques to process more than one instruction at once where possible. But the degree of parallelism in the code is only determined at run-time by parts of the processor that attempt to analyze and re-order instructions on the fly. This approach takes time and wastes die space that could be devoted to executing instructions rather than organizing them. EPIC breaks through the sequential nature of conventional processor architectures by allowing software to communicate explicitly to the processor when operations can be performed in parallel.

The result is that the processor can grab as large a chunk of instructions as possible and execute them simultaneously with minimal pre-processing. Increased performance is realized by reducing the number of branches and branch mis-predicts, and reducing the effects of memory-to-processor latency. The IA-64 Instruction Set Architecture applies EPIC technology to deliver massive resources with inherent scalability not possible with previous processor architectures. For example, systems can be designed to slot in new execution units whenever an upgrade is required, similar to plugging in more memory modules on existing systems. According to Intel the IA-64 ISA represents the most significant advancement in microprocessor architecture since the introduction of its 80386 chip in 1985.

IA-64 processors will have massive computing resources including 128 integer registers, 128 floating-point registers, 64 predicate registers, and a number of special-purpose registers. Instructions will be bundled in groups for parallel execution by the various functional units. The instruction set has been optimized to address the needs of cryptography, video encoding, and other functions required by the next generation of servers and workstations. Support for Intel's MMX technology and Internet Streaming SIMD Extensions is maintained and extended in IA-64 processors.

The Pentium 4 is Intel's first new IA-32 core aimed at the desktop market since the introduction of the Pentium Pro in 1995. It represents the biggest change to Intel's 32-bit architecture since the Pentium Pro. Its increased performance is largely due to architectural changes that allow the device to operate at higher clock speeds. And it incorporates logic changes that allow more instructions to be processed per clock cycle.

Foremost among these innovations is the processor's internal pipeline—referred to as the Hyper Pipeline—which is comprised of 20 pipeline stages versus the ten for P6 micro-architecture. A typical pipeline has a fixed amount of work that is required to decode and execute an instruction. This work is performed by individual logical operations called "gates." Each logic gate consists of multiple transistors. By increasing the stages in a pipeline, fewer gates are required per stage. Because each gate requires some amount of time (delay) to provide a result, decreasing the number of gates in each stage allows the clock rate to be increased. It allows more instructions to be "in flight" (be at various stages of decode and execution) in the pipeline. Although these benefits are offset somewhat by the overhead of additional gates required to manage the added stages, the overall effect of increasing the number of pipeline stages is a reduction in the number of gates per stage. And this allows a higher core frequency and enhances scalability.

Other new features introduced by the Pentium 4's micro-architecture—dubbed NetBurst—include:

- An innovative Level 1 cache implementation comprised of an 8-KB data cache and an Execution Trace Cache that stores up to 12K of decoded x86 instructions (micro-ops). This arrangement removes the latency associated with the instruction decoder from the main execution loops.

- A Rapid Execution Engine that pushes the processor's ALUs to twice the core frequency. This results in higher execution throughput and reduced latency of execution. The chip actually uses three separate clocks: the core frequency, the ALU frequency and the bus frequency. The Advanced Dynamic—a very deep, out-of-order speculative execution engine—avoids stalls that can occur while instructions are waiting for dependencies to resolve by providing a large window of memory from which units choose.
- A 256-KB Level 2 Advanced Transfer Cache that provides a 256-bit (32-byte) interface to transfer data on each core clock.
- A much higher data throughput channel – 44.8 GB/s for a 1.4-GHz Pentium 4 processor.
- SIMD Extensions 2 (SSE2) – The latest iteration of Intel's Single Instruction Multiple Data technology. It integrates 76 new SIMD instructions and improves 68 integer instructions. These changes allow the chip to grab 128 bits at a time in both floating-point and integer
- CPU-intensive encoding and decoding operations such as streaming video, speech, 3D rendering, and other multimedia procedures.
- The industry's first 400-MHz system bus, providing a three-fold increase in throughput compared with Intel's current 133-MHz bus.

The first Pentium 4 units with speeds of 1.4 GHz and 1.5 GHz showed performance improvements on 3D applications (games) and on graphics intensive applications (video encoding). The performance gain appeared much less pronounced on everyday office applications such as word processing, spreadsheets, Web browsing, and e-mail. More recent Pentium 4 shipments show speeds in excess of 3 GHz.

Intel's roadmap shows future processor developments that shrink in process utilized (using copper interconnects), increase the number of gates, boost performance, and include very large on-chip caches.

Intel has enjoyed its position as the premier PC processor manufacturer in recent years. Dating from the 486 line of processors in 1989, Cyrix and long-time Intel cloner Advanced Micro Devices (AMD) have posed the most serious threat to Intel's dominance. While advancements in microprocessors have come primarily from Intel, the last few years have seen a turn of events with Intel facing some embarrassments. It had to recall all of its shipped 1.13-GHz processors after it was discovered that the chip caused systems to hang when running certain applications. Many linked the problem to increasing competition from rival chipmaker AMD who succeeded in beating Intel to the 1-GHz barrier a few weeks earlier. Competitive pressure may have forced Intel into introducing faster chips earlier than it had originally planned.

AMD's involvement in personal computing spans the entire history of the industry. The company supplied every generation of PC processor from the 8088 used in the first IBM PCs, to the new, seventh-generation AMD Athlon processor. Some believe that the Athlon represents the first time in the history of x86 CPU architecture that Intel surrendered the technological lead to a rival chip manufacturer. However, this is not strictly true. A decade earlier AMD's 386 CPU bettered Intel's 486SX chip in speed, performance, and cost.

In the early 1990s, both AMD and Cyrix made their own versions of the 486. But their products became better known with their 486 clones, one copying the 486DX2-66 (introduced by Intel in 1992) and the other increasing internal speed to 80 MHz.

Although Intel stopped improving the 486 with the DX4-100, AMD and Cyrix kept going. In 1995 AMD offered the clock-quadrupled 5x86, a 33-MHz 486DX that ran internally at 133 MHz.

AMD marketed the chip as comparable in performance to Intel's new Pentium/75. But it was a 486DX in all respects, including the addition of the 16-K Level 1 cache (built into the processor) that Intel introduced with the DX4. Cyrix followed suit with its own 5x86 called the M1sc, but this chip was much different from AMD's. In fact, the M1sc offered Pentium-like features even though it was designed for use on 486 motherboards. Running at 100 MHz and 120 MHz, the chip included a 64-bit internal bus, a six-stage pipeline (as opposed to the DX4's five-stage pipeline), and branch-prediction technology to improve the speed of instruction execution. However, it is important to remember that the Cyrix 5x86 appeared after Intel had introduced the Pentium, so these features were more useful in upgrading 486s than in pioneering new systems.

In the post-Pentium era, designs from both manufacturers have met with reasonable levels of market acceptance, especially in the low-cost, basic PC market segment. With Intel now concentrating on its Slot 1 and Slot 2 designs, its competitors want to match the performance of Intel's new designs as they emerge, without having to adopt the new processor interface technologies.

However, Cyrix finally bowed out of the PC desktop business when National Semiconductor sold the rights to its x86 CPUs to Taiwan-based chipset manufacturer VIA Technologies. The highly integrated MediaGX product range remained with National Semiconductor. It became part of their new Geode family of system-on-a-chip solutions being developed for the client devices market. VIA Technologies has also purchased IDT's Centaur Technology subsidiary which was responsible for the design and production of its WinChip x86 range of processors. It is unclear if these moves signal VIA's intention to become a serious competitor in the CPU market, or whether its ultimate goal is to compete with National Semiconductor in the system-on-a-chip market. Traditionally, the chipset makers have lacked the x86 design technology needed to take the trend for low-cost chipsets incorporating increasing levels of functionality on a single chip to its logical conclusion. The other significant development was AMD seizing the technological lead from Intel with the launch of its new Athlon processor.

Components and Interfaces

The PC's adaptability—its ability to evolve many different interfaces allowing the connection of many different classes of add-on component and peripheral device—has been one of the key reasons for its success. In essence, today's PC system is little different than IBM's original design—a collection of internal and external components, interconnected by a series of electrical data highways over which data travels as it completes the processing cycle. These "buses," as they are called, connect all the PC's internal components and external devices and peripherals to its CPU and main memory (RAM).

The fastest PC bus, located within the CPU chip, connects the processor to its primary cache. On the next level down, the system bus links the processor with memory. This memory consists of the small amount of static RAM (SRAM) secondary cache and the far larger main banks of dynamic RAM (DRAM). The system bus is 64 bits wide and 66 MHz—raised to 100 MHz. The CPU does not communicate directly with the memory. Rather, it communicates through the system controller chip which acts as an intermediary to manages the host bus and bridges it to the PCI bus (in modern PCs).

Bus Terminology

A modern PC system includes two classes of bus—a System Bus that connects the CPU to main memory, and a Level 2 cache. A PC system also includes a number of I/O Busses that connect various peripheral devices to the CPU. The latter is connected to the system bus via a "bridge," implemented in the processor's chipset.

In a Dual Independent Bus (DIB) architecture, a front-side bus replaces the single system bus. It shuttles data between the CPU and main memory, and between the CPU and peripheral buses. DIB architecture also includes a backside bus for accessing level 2 cache. The use of dual independent buses boosts performance, enabling the CPU to access data from either of its buses simultaneously and in parallel.

The evolution of PC bus systems has generated a profusion of terminology, much of it confusing, redundant, or obsolete. The system bus is often referred to as the "main bus," "processor bus" or "local bus." Alternative generic terminology for an I/O bus includes "expansion bus," "external bus," "host bus," as well as "local bus."

A given system can use a number of different I/O bus systems. A typical arrangement is for the following to be implemented concurrently:

- ISA Bus – The oldest, slowest, and soon-to-be- obsolete I/O Bus system.
- PCI Bus – Present on Pentium-class systems since the mid-1990s.
- PCI-X Bus
- AGP Bus
- USB Bus – The replacement for the PC's serial port. Allows up to 127 devices to be connected by using a hub device or by implementing a daisy chain..
- IEEE 1394

ISA (Industry Standard Architecture) Bus

When it first appeared on the PC, the 8-bit ISA bus ran at a modest 4.77 MHz - the same speed as the processor. It was improved over the years, eventually becoming the ISA bus in 1982 (with the advent of the IBM PC/AT using the Intel 80286 processor and 16-bit data bus). At this stage it kept up with the speed of the system bus, first at 6 MHz and later at 8 MHz.

The ISA bus specifies a 16-bit connection driven by an 8-MHz clock which seems primitive compared to the speed of today's processors. It has a theoretical data transfer rate of up to 16 MB/s. Functionally, this rate would reduce by a half to 8 MB/s since one bus cycle is required for addressing and another bus cycle is required for the 16 data bits. In the real world it is capable of more like 5 MB/s—still sufficient for many peripherals. A huge number of ISA expansion cards ensured its continued presence into the late 1990s.

As processors became faster and gained wider data paths, the basic ISA design was not able to change to keep pace. As recently as the late 1990s most ISA cards remained 8-bit technology. The few types with 16-bit data paths (hard disk controllers, graphics adapters, and some network adapters) are constrained by the low throughput levels of the ISA bus. But expansion cards in faster bus slots can handle these processes better.

However, there are areas where a higher transfer rate is essential. High-resolution graphic displays need massive amounts of data, particularly to display animation or full-motion video. Modern hard disks and network interfaces are certainly capable of higher rates. The first attempt to establish a new standard was the Micro Channel Architecture (MCA) introduced by IBM. This was closely followed by Extended ISA (EISA), developed by a consortium of IBM's major competitors. Although both these systems operate at clock rates of 10 MHz and 8 MHz respectively, they are 32-bit and capable of transfer rates well over 20 MB/s. As its name suggests, an EISA slot can also take a conventional ISA card. However, MCA is not compatible with ISA. Neither system flourished because they were too expensive to merit support on all but the most powerful file servers.

Local Bus

Intel 80286 motherboards could run expansion slots and the processor at different speeds over the same bus. However, dating from the introduction of the 386 chip in 1987, motherboards provided two bus systems. In addition to the "official" bus—whether ISA, EISA, or MCA—there was also a 32-bit "system bus" connecting the processor to the main memory. The rise in popularity of the Graphical User Interface (GUI)—such as Microsoft Windows—and the consequent need for faster graphics drove the concept of local bus peripherals. The bus connecting them was commonly referred to as the "local bus." It functions only over short distances due to its high speed and to the delicate nature of the processor.

Initial efforts to boost speed were proprietary. Manufacturers integrated the graphics and hard disk controller into the system bus. This achieved significant performance improvements but limited the upgrade potential of the system. As a result, a group of graphics chipset and adapter manufacturers—the Video Electronics Standards Association (VESA)—established a non-proprietary, high-performance bus standard in the early 1990s. Essentially, this extended the electronics of the 486 system bus to include two or three expansion slots—the VESA Local Bus (VL-Bus). The VL-Bus worked well and many cards became available, predominately graphics and IDE controllers.

The main problem with VL-Bus was its close coupling with the main processor. Connecting too many devices risked interfering with the processor itself, particularly if the signals went through a slot. VESA recommended that only two slots be used at clock frequencies up to 33 MHz, or three if they are electrically buffered from the bus. At higher frequencies, no more than two devices should be connected. And at 50 MHz or above they should both be built into the motherboard.

The fact that the VL-Bus ran at the same clock frequency as the host CPU became a problem as processor speeds increased. The faster the peripherals are required to run, the more expensive they are due to the difficulties associated with manufacturing high-speed components. Consequently, the difficulties in implementing the VL-Bus on newer chips (such as the 40-MHz and 50-MHz 486s and the new 60/66-MHz Pentium) created the perfect conditions for Intel's PCI (Peripheral Component Interconnect).

PCI Bus

Intel's original work on the PCI standard was published as revision 1.0 and handed over to the PCI SIG (Special Interest Group). The SIG produced the PCI Local Bus Revision 2.0 specification in May 1993. It took in engineering requests from members and gave a complete component and expansion connector definition which could be used to produce production-ready systems based on 5 volt technology.

Beyond the need for performance, PCI sought to make expansion easier to implement by offering plug and play (PnP) hardware. This type of system enables the PC to adjust automatically to new cards as they are plugged in, obviating the need to check jumper settings and interrupt levels. Windows 95 provided operating system software support for plug and play, and all current motherboards incorporate BIOS that is designed to work with the PnP capabilities it provides.

By 1994 PCI was established as the dominant Local Bus standard. The VL-Bus was essentially an extension of the bus the CPU uses to access main memory. PCI is a separate bus isolated from the

CPU, but having access to main memory. As such, PCI is more robust and higher performance than VL-Bus. Unlike the latter which was designed to run at system bus speeds, the PCI bus links to the system bus through special bridge circuitry and runs at a fixed speed, regardless of the processor clock. PCI is limited to five connectors, although two devices built into the motherboard can replace each. It is also possible for a processor to support more than one bridge chip. It is more tightly specified than VL-Bus and offers a number of additional features. In particular, it can support cards running from both 5 volt and 3.3 volt supplies using different "key slots" to prevent the wrong card being plugged into the wrong slot.

In its original implementation PCI ran at 33 MHz. This was raised to 66 MHz by the later PCI 2.1 specification. This effectively doubled the theoretical throughput to 266 MB/s which is 33 times faster than the ISA bus. It can be configured as both a 32-bit and a 64-bit bus. Both 32-bit and 64-bit cards can be used in either. 64-bit implementations running at 66 MHz—still rare by mid-1999—increase bandwidth to a theoretical 524 MB/s. PCI is also much smarter than its ISA predecessor, allowing interrupt requests (IRQs) to be shared. This is useful because well-featured, high-end systems can quickly run out of IRQs. Also, PCI bus mastering reduces latency, resulting in improved system speeds.

Since mid-1995, the main performance-critical components of the PC have communicated with each other across the PCI bus. The most common PCI devices are the disk and graphics controllers which are mounted directly onto the motherboard or onto expansion cards in PCI slots.

PCI-X Bus

PCI-X v1.0 is a high performance addendum to the PCI Local Bus specification. It was co-developed by IBM, Hewlett-Packard, and Compaq—normally competitors in the PC server market—and was unanimously approved by the PCI SIG. Fully backward compatible with standard PCI, PCI-X is seen as an immediate solution to the increased I/O requirements for high-bandwidth enterprise applications. These applications include Gigabit Ethernet, Fibre Channel, Ultra3 SCSI (Small Computer System Interface), and high-performance graphics.

PCI-X increases both the speed of the PCI bus and the number of high-speed slots. With the current design, PCI slots run at 33 MHz and one slot can run at 66 MHz. PCI-X doubles the current performance of standard PCI, supporting one 64-bit slot at 133 MHz for an aggregate throughput of 1 GB/s. The new specification also features an enhanced protocol to increase the efficiency of data transfer and to simplify electrical timing requirements, an important factor at higher clock frequencies.

For all its performance gains, PCI-X is being positioned as an interim technology while the same three vendors develop a more long-term I/O bus architecture. While it has potential use throughout the computer industry, its initial application is expected to be in server and workstation products, embedded systems, and data communication environments.

The symbolism of a cartel of manufacturers making architectural changes to the PC server without consulting Intel is seen as a significant development. At the heart of the dispute is who gets control over future server I/O technology. The PCI-X faction, already wary of Intel's growing dominance in the hardware business, hopes to wrest some control. They want to develop and define the next generation of I/O standards which they hope Intel will eventually support. Whether this will succeed—or generate a standards war—is a moot point. The immediate effect is that it has provoked Intel into leading another group of vendors in the development of rival I/O technology that they refer to as "Next Generation I/O."

Next Generation High-Speed Interfaces

Intel has announced its support for PCI Express (formerly known as 3GIO and Arapahoe) which will potentially make its way into desktop PCs in the next few years. Meanwhile, Motorola has pledged support for RapidIO and serial RapidIO, and AMD is pushing HyperTransport (formerly known as LDT—Lightning Data Transport). Each consortium has a few supporters for the technology specification and market promotion, and they will eventually build products supporting each technology. These, and several other future high-speed interface technologies, are based on serial IOs with clock and data recovery (the clock is encoded with the data). These interfaces take far fewer pins to parallel interfaces, resulting in much higher speeds, higher reliability, and better EMI management.

AGP

As fast and wide as the PCI bus was, one task threatened to consume all its bandwidth—displaying graphics. Early in the era of the ISA bus, monitors were driven by Monochrome Display adapter (MDA) and Color Graphics Array (CGA) cards. A CGA graphics display could show four colors (two bits of data) with 320×200 pixels of screen resolution at 60 Hz. This required 128,000 bits of data per screen, or just over 937 KB/s. An XGA image at a 16-bit color depth requires 1.5 MB of data for every image. And at a vertical refresh rate of 75 Hz, this amount of data is required 75 times each second. Thanks to modern graphics adapters, not all of this data has to be transferred across the expansion bus. But 3D imaging technology created new problems.

3D graphics have made it possible to model fantastic, realistic worlds on-screen in enormous detail. Texture mapping and object hiding require huge amounts of data. The graphics adapter needs to have fast access to this data to prevent the frame rate from dropping, resulting in jerky action. It was beginning to look as though the PCI peak bandwidth of 132 MB/s was not up to the job.

Intel's solution was to develop the Accelerated Graphics Port (AGP) as a separate connector that operates off the processor bus. The AGP chipset acts as the intermediary between the processor and:

- Level 2 cache contained in the Pentium II's Single Edge Contact Cartridge
- System memory
- The graphics card
- The PCI bus

This is called Quad Port acceleration.

AGP operates at the speed of the processor bus, now known as the frontside bus. At a clock rate of 66 MHz—double the PCI clock speed—the peak base throughput is 264 MB/s.

For graphics cards specifically designed to support it, AGP allows data to be sent during both the up and down clock cycle. This effectively doubles the clock rate to 133 MHz and the peak transfer rate to 528 MB/s—a process knows as X2. To improve the length of time that AGP can maintain this peak transfer, the bus supports pipelining which is another improvement over PCI. A pipelining X2 graphics card can sustain throughput at 80% of the peak. AGP also supports queuing of up to 32 commands via Sideband Addressing (SBA) where the commands are being sent while data is being received. According to Intel, this allows the bus to sustain peak performance 95% of the time.

AGP's four-fold bandwidth improvement and graphics-only nature ensures that large transfers of 3D graphics data don't slow up the action on screen. Nor will graphics data transfers be interrupted by other PCI devices. AGP is primarily intended to boost 3D performance, so it also provides other improvements that are specifically aimed at this function.

With its increased access speed to system memory over the PCI bus, AGP can use system memory as if it is actually on the graphics card. This is called Direct Memory Execute (DIME). A device called a Graphics Aperture Remapping Table (GART) handles the RAM addresses. GART enables them to be distributed in small chunks throughout system memory rather than hijacking one large section. And it presents them to a DIME-enabled graphics card as if they're part of on-board memory. DIME allows much larger textures to be used because the graphics card can have a much larger memory space in which to load the bitmaps used.

AGP was initially only available in Pentium II systems based on Intel's 440LX chipset. But despite no Intel support, and thanks to the efforts of other chipset manufacturers such as VIA, by early 1998 it found its way onto motherboards designed for Pentium-class processors.

Intel's release of version 2.0 of the AGP specification, combined with the AGP Pro extensions to this specification, mark an attempt to have AGP taken seriously in the 3D graphics workstation market. AGP 2.0 defines a new 4X-transfer mode that allows four data transfers per clock cycle on the 66-MHz AGP interface. This delivers a maximum theoretical bandwidth of 1.0 GB/s between the AGP device and system memory. The new 4X mode has a much higher potential throughput than 100-MHz SDRAM (800 MB/s). So the full benefit was not seen until the implementation of 133-MHz SDRAM and Direct Rambus DRAM (DRDRAM) in the second half of 1999. AGP 2.0 was supported by chipsets launched early in 1999 to provide support for Intel's Pentium III processor.

AGP Pro is a physical specification aimed at satisfying the needs of high-end graphics card manufacturers who are currently limited by the maximum electrical power that can be drawn by an AGP card (about 25 watts). AGP Pro caters to cards that draw up to 100 watts. It uses a slightly longer AGP slot that will also take current AGP cards.

Table 8.1 shows the burst rates, typical applications, and the outlook for the most popular PC interfaces—ISA, EISA, PCI, and AGP.

Table 8.1: Interfaces Summary

Standard	Typical Applications	Burst Data Rates	Outlook
ISA	Sound cards, modems	2 MB/s to 8.33 MB/s	Expected to be phased out soon
EISA	Network, SCSI adapters	33 MB/s	Nearly phased out, superseded by PCI
PCI	Graphics cards, SCSI adapters, new generation sound cards	266 MB/s	Standard add-in peripheral bus with most market penetration
AGP	Graphics cards	528 MB/s	Standard in all Intel-based PCs from the Pentium II; coexists with PCI

USB (Universal Serial Bus)

Developed jointly by Compaq, Digital, IBM, Intel, Microsoft, NEC, and Nortel, USB offers a standardized connector for attaching common I/O devices to a single port, thus simplifying today's multiplicity of ports and connectors. Significant impetus behind the USB standard was created in

September of 1995 with the announcement of a broad industry initiative to create an open host controller interface (HCI) standard for USB. The initiative wanted to make it easier for companies including PC manufacturers, component vendors, and peripheral suppliers to more quickly develop USB-compliant products. Key to this was the definition of a nonproprietary host interface, left undefined by the USB specification itself, that enabled connection to the USB bus.

Up to 127 devices can be connected by daisy chaining or by using a USB hub (which has a number of USB sockets and plugs for a PC or for other device). Seven peripherals can be attached to each USB hub device. This can include a second hub to which up to another seven peripherals can be connected, and so on. Along with the signal, USB carries a 5V power supply so that small devices such as handheld scanners or speakers do not have to have their own power cable.

Devices are plugged directly into a four-pin socket on the PC or hub using a rectangular Type A socket. All cables permanently attached to the device have a Type A plug. Devices that use a separate cable have a square Type B socket, and the cable that connects them has a Type A and Type B plug.

USB overcomes the speed limitations of UART (Universal Asynchronous Receiver-Transmitters)-based serial ports. USB runs at a staggering 12 Mb/s which is as fast as networking technologies such as Ethernet and Token Ring. It provides enough bandwidth for all of today's peripherals and many foreseeable ones. For example, the USB bandwidth will support devices such as external CD-ROM drives and tape units as well as ISDN and PABX interfaces. It is also sufficient to carry digital audio directly to loudspeakers equipped with digital-to-analog converters, eliminating the need for a soundcard. However, USB is not intended to replace networks. To keep costs down, its range is limited to 5 meters between devices. A lower communication rate of 1.5 Mb/s can be set up for lower-bit-rate devices like keyboards and mice, saving space for things that really need it.

USB was designed to be user-friendly and is truly plug-and-play. It eliminates the need to install expansion cards inside the PC and then reconfigure the system. Instead, the bus allows peripherals to be attached, configured, used, and detached while the host and other peripherals are in operation. There is no need to install drivers, figure out which serial or parallel port to choose, or worry about IRQ settings, DMA channels, and I/O addresses. USB achieves this by managing connected peripherals in a host controller mounted on the PC's motherboard or on a PCI add-in card. The host controller and subsidiary controllers in hubs manage USB peripherals. This helps reduce the load on the PC's CPU time and improves overall system performance. In turn, USB system software installed in the operating system manages the host controller.

Data on the USB flows through a bidirectional pipe regulated by the host controller and by subsidiary hub controllers. An improved version of bus mastering called isochronous data transfer allows portions of the total bus bandwidth to be permanently reserved for specific peripherals. The USB interface contains two main modules—the Serial Interface Engine (SiE), responsible for the bus protocol, and the Root Hub, used to expand the number of USB ports.

The USB bus distributes 0.5 amps of power through each port. Thus, low-power devices that might normally require a separate AC adapter can be powered through the cable. (USB lets the PC automatically sense the power that is required and deliver it to the device.) Hubs may derive all power from the USB bus (bus powered), or they may be powered from their own AC adapter. Powered hubs with at least 0.5 amps per port provide the most flexibility for future downstream devices. Port switching hubs isolate all ports from each other so that a shorted device will not bring down the others.

The promise of USB is a PC with a single USB port instead of the current four or five different connectors. One large powered device, like a monitor or a printer, would be connected onto this and would act as a hub. It would link all the other smaller devices such as mouse, keyboard, modem, document scanner, digital camera, and so on. Since many USB device drivers did not become available until its release, this promise was never going to be realized before the availability of Windows 98.

USB architecture is complex, and a consequence of having to support so many different types of peripheral is an unwieldy protocol stack. However, the hub concept merely shifts expense and complexity from the system unit to the keyboard or monitor. The biggest impediment to USB's success is probably the IEEE 1394 standard. Today most desktop and laptop PCs come standard with a USB interface.

USB 2.0

Compaq, Hewlett-Packard, Intel, Lucent, Microsoft, NEC, and Philips are jointly leading the development of a next-generation USB 2.0 specification. It will dramatically extend performance to the levels necessary to provide support for future classes of high-performance peripherals. At the time of the February 1999 IDF (Intel Developers Forum), the projected performance hike was on the order of 10 to 20 times over existing capabilities. By the time of the next IDF in September 1999, these estimates had been increased to 30 to 40 times the performance of USB 1.1, based on the results of engineering studies and test silicon. At these levels of performance, the danger that rival IEEE 1394 bus would marginalize USB would appear to have been banished forever. Indeed, proponents of USB maintain that the two standards address differing requirements. The aim of USB 2.0 is to provide support for the full range of popular PC peripherals, while IEEE 1394 targets connection to audiovisual consumer electronic devices such as digital camcorders, digital VCRs, DVDs, and digital televisions.

USB 2.0 will extend the capabilities of the interface from 12 Mb/s, which is available on USB 1.1, to between 360-480 Mb/s on USB 2.0. It will provide a connection point for next-generation peripherals which complement higher performance PCs. USB 2.0 is expected to be both forward and backward compatible with USB 1.1. It is also expected to result in a seamless transition process for the end user.

USB 1.1's data rate of 12 Mb/s is sufficient for PC peripherals such as:
- Telephones
- Digital cameras
- Keyboards
- Mice
- Digital joysticks
- Tablets
- Wireless base stations
- Cartridge, tape, and floppy drives
- Digital speakers
- Scanners
- Printers.

USB 2.0's higher bandwidth will permit higher-functionality PC peripherals that include higher resolution video conferencing cameras and next-generation scanners, printers, and fast storage units.

Existing USB peripherals will operate with no change in a USB 2.0 system. Devices such as mice, keyboards, and game pads will operate as USB 1.1 devices and not require the additional performance that USB 2.0 offers. All USB devices are expected to co-exist in a USB 2.0 system. The higher speed of USB 2.0 will greatly broaden the range of peripherals that may be attached to the PC. This increased performance will also allow a greater number of USB devices to share the available bus bandwidth, up to the architectural limits of USB. Given USB's wide installed base, USB 2.0's backward compatibility could prove a key benefit in the battle with IEEE 1394 to be the consumer interface of the future.

IEEE 1394

Commonly referred to as FireWire, IEEE 1394 was approved by the Institute of Electrical and Electronics Engineers (IEEE) in 1995. It was originally conceived by Apple who currently receives $1 royalty per port. Several leading IT companies including Microsoft, Philips, National Semiconductor, and Texas Instruments have since joined the 1394 Trade Association (1394ta).

IEEE 1394 is similar to USB in many ways, but it is much faster. Both are hot-swappable serial interfaces, but IEEE 1394 provides high-bandwidth, high-speed data transfers significantly in excess of what USB offers. There are two levels of interface in IEEE 1394—one for the backplane bus within the computer and another for the point-to-point interface between device and computer on the serial cable. A simple bridge connects the two environments. The backplane bus supports data-transfer speeds of 12.5, 25, or 50 Mb/s. The cable interface supports speeds of 100, 200 and 400 Mb/s —roughly four times as fast as a 100BaseT Ethernet connection and far faster than USB's 1.5 or 12 Mb/s speeds. A 1394b specification aims to adopt a different coding and data-transfer scheme that will scale to 800 Mb/s, 1.6 Gb/s, and beyond. Its high-speed capability makes IEEE 1394 viable for connecting digital cameras, camcorders, printers, TVs, network cards, and mass storage devices to a PC.

IEEE 1394 cable connectors are constructed with the electrical contacts located inside the structure of the connector, thus preventing any shock to the user or contamination to the contacts by the user's hands. These connectors are derived from the Nintendo GameBoy connector. Field-tested by children of all ages, this small and flexible connector is very durable. These connectors are easy to use even when the user must blindly insert them into the back of machines. Terminators are not required and manual IDs need not be set.

IEEE 1394 uses a six-conductor cable up to 4.5 meters long. It contains two pairs of wires for data transport and one pair for device power. The design resembles a standard 10BaseT Ethernet cable. Each signal pair as well as the entire cable is shielded. As the standard evolves, new cable designs are expected that will allow more bandwidth and longer distances without the need for repeaters.

An IEEE 1394 connection includes a physical layer and a link layer semiconductor chip, and IEEE 1394 needs two chips per device. The physical interface (PHY) is a mixed-signal device that connects to the other device's PHY. It includes the logic needed to perform arbitration and bus initialization functions. The Link interface connects the PHY and the device internals. It transmits and receives 1394-formatted data packets and supports asynchronous or isochronous data transfers. Both asynchronous and isochronous formats are included on the same interface. This allows both non-real-time critical applications (printers, scanners) and real-time critical applications (video, audio) to operate on the same bus. All PHY chips use the same technology, whereas the Link is device-specific. This approach allows IEEE 1394 to act as a peer-to-peer system as opposed to USB's client-server design. As a consequence, an IEEE 1394 system needs neither a serving host nor a PC.

Asynchronous transport is the traditional method of transmitting data between computers and peripherals. Data is sent in one direction followed by an acknowledgement to the requester. Asynchronous data transfers place emphasis on delivery rather than timing. The data transmission is guaranteed and retries are supported. Isochronous data transfer ensures that data flows at a pre-set rate so that an application can handle it in a timed way. This is especially important for time-critical multimedia data where just-in-time delivery eliminates the need for costly buffering. Isochronous data transfers operate in a broadcast manner where one, or many, 1394 devices can "listen" to the data being transmitted. Multiple channels (up to 63) of isochronous data can be transferred simultaneously on a 1394 bus. Since isochronous transfers can only take up a maximum of 80% of the 1394 bus bandwidth, there is enough bandwidth left over for additional asynchronous transfers.

IEEE 1394's scaleable architecture and flexible peer-to-peer topology make it ideal for connecting high-speed devices such as computers, hard drives, and digital audio and video hardware. Devices can be connected in a daisy chain or a tree topology scheme. Each IEEE 1394 bus segment may have up to 63 devices attached to it. Currently, each device may be up to 4.5 meters apart. Longer distances are possible with and without repeater hardware. Improvements to current cabling are being specified to allow longer distance cables. Bridges may connect over 1000 bus segments, thus providing growth potential for a large network. An additional feature is the ability of transactions at different speeds to occur on a single device medium. For example, some devices can communicate at 100 Mb/s while others communicate at 200 Mb/s and 400 Mb/s. IEEE 1394 devices may be hot-plugged—added to or removed from the bus—even with the bus in full operation. Topology changes are automatically recognized when the bus configuration is altered. This "plug and play" feature eliminates the need for address switches or other user intervention to reconfigure the bus.

As a transaction-based packet technology, 1394 can be organized as if it were memory space connected between devices, or as if devices resided in slots on the main backplane. Device addressing is 64 bits wide, partitioned as 10 bits for network Ids, six bits for node Ids, and 48 bits for memory addresses. The result is the capability to address 1023 networks of 63 nodes, each with 281TB of memory. Memory-based addressing rather than channel addressing views resources as registers or memory that can be accessed with processor-to-memory transactions. This results in easy networking. For example, a digital camera can easily send pictures directly to a digital printer without a computer in the middle. With IEEE 1394, it is easy to see how the PC could lose its position of dominance in the interconnectivity environment—it could be relegated to being just a very intelligent peer on the network.

The need for two pieces of silicon instead of one will make IEEE 1394 peripherals more expensive than SCSI, IDE, or USB devices. Hence, it is inappropriate for low-speed peripherals. But its applicability to higher-end applications such as digital video editing is obvious, and it is clear that the standard is destined to become a mainstream consumer electronics interface used for connecting handy-cams, VCRs, set-top boxes, and televisions. However, its implementation to date has been largely confined to digital camcorders where is it known as iLink.

Components/Chipsets

A chipset or "PCI set" is a group of microcircuits that orchestrate the flow of data to and from key components of a PC. This includes the CPU itself, the main memory, the secondary cache, and any devices on the ISA and PCI buses. The chipset also controls data flow to and from hard disks and other devices connected to the IDE channels. While new microprocessor technologies and speed improvements tend to receive all the attention, chipset innovations are equally important.

Although there have always been other chipset manufacturers such as SIS, VIA Technologies, and Opti, Intel's Triton chipsets were by far the most popular for many years. Indeed, the introduction of the Intel Triton chipset caused something of a revolution in the motherboard market, with just about every manufacturer using it in preference to anything else. Much of this was due to the ability of the Triton to get the best out of both the Pentium processor and the PCI bus, and it also offered built-in master EIDE support, enhanced ISA bridge, and the ability to handle new memory technologies like EDO (Extended Data Output) and SDRAM. However, the potential performance improvements in the new PCI chipsets will only be realized when used in conjunction with BIOS capable of taking full advantage of the new technologies.

During the late 1990s things became far more competitive. Acer Laboratories (ALI), SIS, and VIA all developed chipsets designed to operate with Intel, AMD, and Cyrix processors. 1998 was a particularly important year in chipset development when an unacceptable bottleneck—the PC's 66-MHz system bus—was finally overcome. Interestingly, it was not Intel but rival chipmakers that made the first move, pushing Socket 7 chipsets to 100 MHz. Intel responded with its 440BX, one of many chipsets to use the ubiquitous North-bridge/South-bridge architecture. It was not long before Intel's hold on the chipset market loosened further still and the company had no one but itself to blame. In 1999 its single-minded commitment to Direct Rambus DRAM (DRDRAM) left it in the embarrassing position of not having a chipset that supported the 133-MHz system bus speed that its latest range of processors were capable of. This was another situation its rivals were able to exploit to gain market share.

Motherboard

The motherboard is the main circuit board inside the PC. It holds the processor and the memory and expansion slots. It connects directly or indirectly to every part of the PC. It is made up of a chipset (known as the "glue logic"), some code in ROM, and various interconnections, or buses. PC designs today use many different buses to link their various components. Wide, high-speed buses are difficult and expensive to produce. The signals travel at such a rate that even distances of just a few centimeters cause timing problems. And the metal tracks on the circuit board act as miniature radio antennae transmitting electromagnetic noise that introduces interference with signals elsewhere in the system. For these reasons PC design engineers try to keep the fastest buses confined to the smallest area of the motherboard and use slower, more robust buses for other parts.

The original PC had minimum integrated devices, just ports for a keyboard and a cassette deck (for storage). Everything else, including a display adapter and floppy or hard disk controllers, was an add-in component connected via expansion slot. Over time more devices were integrated into the motherboard. It is a slow trend, though. I/O ports and disk controllers were often mounted on expansion cards as recently as 1995. Other components (graphics, networking, SCSI, sound) usually remain separate. Many manufacturers have experimented with different levels of integration, building in some, or even all of these components. But there are drawbacks. It is harder to upgrade the specification if integrated components can't be removed, and highly integrated motherboards often require nonstandard cases. Furthermore, replacing a single faulty component may mean buying an entire new motherboard. Consequently, those parts of the system whose specification changes fastest —RAM, CPU, and graphics—tend to remain in sockets or slots for easy replacement. Similarly, parts that not all users need such as networking or SCSI are usually left out of the base specification to keep costs down.

Motherboard development consists largely of isolating performance-critical components from slower ones. As higher-speed devices become available, they are linked by faster buses, and lower-speed buses are relegated to supporting roles. In the late 1990s there was also a trend towards putting peripherals designed as integrated chips directly onto the motherboard. Initially this was confined to audio and video chips, obviating the need for separate sound or graphics adapter cards. But in time the peripherals integrated in this way became more diverse and included items such as SCSI, LAN, and even RAID controllers. While there are cost benefits to this approach, the biggest downside is the restriction of future upgrade options.

BIOS (Basic Input/Output System)

All motherboards include a small block of Read Only Memory (ROM) which is separate from the main system memory used for loading and running software. The ROM contains the PC's BIOS and this offers two advantages. The code and data in the ROM BIOS need not be reloaded each time the computer is started, and they cannot be corrupted by wayward applications that write into the wrong part of memory. A flash-upgradeable BIOS may be updated via a floppy diskette to ensure future compatibility with new chips, add-on cards etc.

The BIOS is comprised of several separate routines serving different functions. The first part runs as soon as the machine is powered on. It inspects the computer to determine what hardware is fitted. Then it conducts some simple tests to check that everything is functioning normally—a process called the power-on self-test. If any of the peripherals are plug-and-play devices, the BIOS assigns the resources. There's also an option to enter the Setup program. This allows the user to tell the PC what hardware is fitted, but thanks to automatic self-configuring BIOS, this is not used so much now.

If all the tests are passed, the ROM tries to boot the machine from the hard disk. Failing that, it will try the CD-ROM drive and then the floppy drive, finally displaying a message that it needs a system disk. Once the machine has booted, the BIOS presents DOS with a standardized API for the PC hardware. In the days before Windows, this was a vital function. But 32-bit "protect mode" software doesn't use the BIOS, so it is of less benefit today.

Most PCs ship with the BIOS set to check for the presence of an operating system in the floppy disk drive first, then on the primary hard disk drive. Any modern BIOS will allow the floppy drive to be moved down the list so as to reduce normal boot time by a few seconds. To accommodate PCs that ship with a bootable CD-ROM, most BIOS allow the CD-ROM drive to be assigned as the boot drive. BIOS may also allow booting from a hard disk drive other than the primary IDE drive. In this case, it would be possible to have different operating systems or separate instances of the same OS on different drives.

Windows 98 (and later) provides multiple display support. Since most PCs have only a single AGP slot, users wishing to take advantage of this will generally install a second graphics card in a PCI slot. In such cases, most BIOS will treat the PCI card as the main graphics card by default. However, some allow either the AGP card or the PCI card to be designated as the primary graphics card.

While the PCI interface has helped by allowing IRQs to be shared more easily, the limited number of IRQ settings available to a PC remains a problem for many users. For this reason, most BIOS allow ports that are not in use to be disabled. It will often be possible to get by without needing either a serial or a parallel port because of the increasing popularity of cable and ADSL Internet connections and the ever-increasing availability of peripherals that use the USB interface.

CMOS RAM

Motherboards also include a separate block of memory made from very low-power consumption CMOS (complementary metal oxide silicon) RAM chips. This memory is kept "alive" by a battery even when the PC's power is off. This stores basic information about the PC's configuration—number and type of hard and floppy drives, how much memory and what kind, and so on. All this used to be entered manually but modern auto-configuring BIOS does much of this work. In this case, the more important settings are advanced settings such as DRAM timings. The other important data kept in CMOS memory are the time and date which are updated by a Real Time Clock (RTC). The clock, CMOS RAM, and battery are usually all integrated into a single chip. The PC reads the time from the RTC when it boots up. After that, the CPU keeps time, which is why system clocks are sometimes out of sync. Rebooting the PC causes the RTC to be reread, increasing their accuracy.

System Memory

System memory is where the computer holds programs and data that are in use. Because of the demands made by increasingly powerful software, system memory requirements have been accelerating at an alarming pace over the last few years. The result is that modern computers have significantly more memory than the first PCs of the early 1980s, and this has had an effect on development of the PC's architecture. It takes more time to store and retrieve data from a large block of memory than from a small block. With a large amount of memory, the difference in time between a register access and a memory access is very great, and this has resulted in extra layers of "cache" in the storage hierarchy.

Processors are currently outstripping memory chips in access speed by an ever-increasing margin. Increasingly, processors have to wait for data going in and out of main memory. One solution is to use cache memory between the main memory and the processor. Another solution is to use clever electronics to ensure that the data the processor needs next is already in cache.

Primary Cache or Level 1 Cache

Level 1 cache is located on the CPU. It provides temporary storage for instructions and data that are organized into blocks of 32 bytes. Primary cache is the fastest form of storage. It is limited in size because it is built into the chip with a zero wait-state (delay) interface to the processor's execution unit.

Primary cache is implemented using static RAM (SRAM). It was traditionally 16KB in size until recently. SRAM is manufactured similarly to processors—highly integrated transistor patterns are photo-etched into silicon. Each SRAM bit is comprised of between four and six transistors, which is why SRAM takes up much more space than DRAM (which uses only one transistor plus a capacitor). This, plus the fact that SRAM is also several times the cost of DRAM, explains why it is not used more extensively in PC systems.

Launched at the start of 1997, Intel's P55 MMX processor was noteworthy for the increase in size of its Level 1 cache—32 KB. The AMD K6 and Cyrix M2 chips launched later that year upped the ante further by providing Level 1 caches of 64 KB. The control logic of the primary cache keeps the most frequently used data and code in the cache. It updates external memory only when the CPU hands over control to other bus masters, or during direct memory access (DMA) by peripherals such as floppy drives and sound cards.

Some chipsets support a "write-back" cache rather than a "write-through" cache. Write-through happens when a processor writes data simultaneously into cache and into main memory (to assure coherency). Write-back occurs when the processor writes to the cache and then proceeds to the next instruction. The cache holds the write-back data and writes it into main memory when that data line in cache is to be replaced. Write-back offers about 10% higher performance than write-through, but this type of cache is more costly. A third type of write mode writes through with buffer, and gives similar performance to write-back.

Secondary (external) Cache or Level 2 Cache

Most PCs are offered with a secondary cache to bridge the processor/memory performance gap. Level 2 cache is implemented in SRAM and uses the same control logic as primary cache.

Secondary cache typically comes in two sizes—256 KB or 512 KB. It can be found on the motherboard. The Pentium Pro deviated from this arrangement, placing the Level 2 cache on the processor chip itself. Secondary cache supplies stored information to the processor without any delay (wait-state). For this purpose, the bus interface of the processor has a special transfer protocol called burst mode. A burst cycle consists of four data transfers where only the addresses of the first 64 are output on the address bus. Synchronous pipeline burst is the most common secondary cache.

To have a synchronous cache, a chipset must support it. It can provide a 3-5% increase in PC performance because it is timed to a clock cycle. This is achieved by specialized SRAM technology which has been developed to allow zero wait-state access for consecutive burst read cycles. Pipelined Burst Static RAM (PB SRAM) has an access time from 4.5 to 8 nanoseconds and allows a transfer timing of 3-1-1-1 for bus speeds up to 133 MHz. These numbers refer to the number of clock cycles for each access of a burst mode memory read. For example, 3-1-1-1 refers to three clock cycles for the first word and one cycle for each subsequent word.

For bus speeds up to 66 MHz, Synchronous Burst Static RAM (Sync SRAM) offers even faster performance—up to 2-1-1-1 burst cycles. However, with bus speeds above 66 MHz its performance drops to 3-2-2-2, significantly slower than PB SRAM. There is also asynchronous cache, which is cheaper and slower because it is not timed to a clock cycle. With asynchronous SRAM, available in speeds between 12 and 20 nanoseconds, all burst read cycles have a timing of 3-2-2-2 on a 50- to 66-MHz CPU bus. This means there are two wait-states for the leadoff cycle and one wait-state for the following three burst-cycle transfers.

Main Memory/RAM

A PC's third and principal level of system memory is main memory, or Random Access Memory (RAM). It is an impermanent source of data, but is the main memory area accessed by the hard disk.

It acts as a staging post between the hard disk and the processor. The more data available in the RAM, the faster the PC runs.

Main memory is attached to the processor via its address and data buses. Each bus consists of a number of electrical circuits, or bits. The width of the address bus dictates how many different memory locations can be accessed. And the width of the data bus dictates how much information is stored at each location. Every time a bit is added to the width of the address bus, the address range doubles. In 1985 Intel's 386 processor had a 32-bit address bus, enabling it to access up to 4 GB of memory. The Pentium processor increased the data bus width to 64 bits, enabling it to access 8 bytes of data at a time.

Each transaction between the CPU and memory is called a bus cycle. The number of data bits a CPU can transfer during a single bus cycle affects a computer's performance and dictates what type of memory the computer requires. By the late 1990s most desktop computers were using 168-pin DIMMs, which supported 64-bit data paths.

Main PC memory is built up using DRAM (dynamic RAM) chips. DRAM chips are large, rectangular arrays of memory cells with support logic used for reading and writing data in the arrays. It also includes refresh circuitry to maintain the integrity of stored data. Memory arrays are arranged in rows and columns of memory cells called word-lines and bit-lines, respectively. Each memory cell has a unique location, or address, defined by the intersection of a row and a column.

DRAM is manufactured similarly to processors—a silicon substrate is etched with the patterns that make the transistors and capacitors (and support structures) that comprise each bit. It costs much less than a processor because it is a series of simple, repeated structures. It lacks the complexity of a single chip with several million individually located transistors. And DRAM uses half as many transistors so it is cheaper than SRAM. Over the years several different structures have been used to create the memory cells on a chip. In today's technologies, the support circuitry generally includes:

- Sense amplifiers to amplify the signal or charge detected on a memory cell.
- Address logic to select rows and columns.
- Row Address Select (/RAS) and Column Address Select (/CAS) logic to latch and resolve the row and column addresses, and to initiate and terminate read and write operations.
- Read and write circuitry to store information in the memory's cells or read that which is stored there.
- Internal counters or registers to keep track of the refresh sequence, or to initiate refresh cycles as needed.
- Output Enable logic to prevent data from appearing at the outputs unless specifically desired.

A transistor is effectively an on/off switch that controls the flow of current. In DRAM each transistor holds a single bit. If the transistor is "open" (1), current can flow; if it's "closed" (0), current cannot flow. A capacitor is used to hold the charge, but it soon escapes, losing the data. To overcome this problem, other circuitry refreshes the memory (reading the value before it disappears), and writes back a pristine version. This refreshing action is why the memory is called dynamic. The refresh speed is expressed in nanoseconds, and it is this figure that represents the speed of the RAM. Most Pentium-based PCs use 60- or 70-ns RAM.

The process of refreshing actually interrupts/slows data access, but clever cache design minimizes this. However, as processor speeds passed the 200-MHz mark, no amount of caching could compensate for the inherent slowness of DRAM. Other, faster memory technologies have largely replaced it.

The most difficult aspect of working with DRAM devices is resolving the timing requirements. DRAMs are generally asynchronous, responding to input signals whenever they occur. As long as the signals are applied in the proper sequence, with signal duration and delays between signals that meet the specified limits, DRAM will work properly.

Popular types of DRAM used in PCs are:

- Fast Page Mode (FPM)
- EDO (Extended Data Out)
- Burst EDO
- Synchronous (SDRAM)
- Direct Rambus (DRDRAM)
- PC133
- Double Data Rate (DDR DRAM).

While much more cost effective than SRAM per megabit, traditional DRAM has always suffered speed and latency penalties making it unsuitable for some applications. Consequently, product manufacturers have often been forced to opt for the more expensive but faster SRAM technology. However, by 2000 system designers had another option available to them. It offered advantages of both worlds - fast speed, low cost, high density, and lower power consumption. Though 1T-SRAM— Monolithic System Technology (MoSys)—calls its design an SRAM, it is based on single-transistor DRAM cells. Like any DRAM, data must be periodically refreshed to prevent loss. What makes the 1T-SRAM unique is that it offers a true SRAM-style interface that hides all refresh operations from the memory controller. Traditionally, SRAMs have been built using a bulky four- or six- transistor (4T, 6T) cell. The MoSys 1T-SRAM device is built on a single transistor (1T) DRAM cell, allowing a reduction in die size of between 50 and 8% compared to SRAMs of similar density.

Regardless of technology type, most desktop PCs for the home are configured with 256 MB of RAM. But it is not atypical to find systems with 512-MB or even 1-GB configurations.

SIMMs (Single Inline Memory Module)

Memory chips are generally packaged into small plastic or ceramic, dual inline packages (DIPs) which are assembled into a memory module. The SIMM is a small circuit board designed to accommodate surface-mount memory chips. SIMMs use less board space and are more compact than previous memory-mounting hardware.

By the early 1990s the original 30-pin SIMM was superseded by the 72-pin variety. These supported 32-bit data paths and were originally used with 32-bit CPUs. A typical motherboard of the time offered four SIMM sockets. These were capable of accepting either single-sided or double-sided SIMMs with module sizes of 4, 8, 16, 32 or even 64 MB. With the introduction of the Pentium processor in 1993, the width of the data bus was increased to 64 bits. When 32-bit SIMMs were used with these processors, they had to be installed in pairs with each pair of modules making up a memory bank. The CPU communicated with the bank of memory as one logical unit.

DIMMs (Dual Inline Memory Module)

By the end of the millennium memory subsystems standardized around an 8-byte data interface. The DIMM had replaced the SIMM as the module standard for the PC industry. DIMMs have 168 pins in two (or dual) rows of contacts, one on each side of the card. With the additional pins a computer can retrieve information from DIMMs, 64 bits at a time instead of 32- or 16-bit transfers that are usual with SIMMs.

Some of the physical differences between 168-pin DIMMs and 72-pin SIMMs include the length of module, the number of notches on the module, and the way the module installs in the socket. Another difference is that many 72-pin SIMMs install at a slight angle, whereas 168-pin DIMMs install straight into the memory socket and remain vertical to the system motherboard. Unlike SIMMs, DIMMs can be used singly and it is typical for a modern PC to provide just one or two DIMM slots.

Presence Detect

When a computer system boots up, it must detect the configuration of the memory modules in order to run properly. For a number of years parallel-presence detect (PPD) was the traditional method of relaying the required information by using a number of resistors. PPD used a separate pin for each bit of information and was used by SIMMs and some DIMMs to identify themselves. However, PPD was not flexible enough to support newer memory technologies. This led the JEDEC (Joint Electron Device Engineering Council) to define a new standard—serial-presence detect (SPD). SPD has been in use since the emergence of SDRAM technology.

The Serial Presence Detect function is implemented by using an 8-pin serial EEPROM chip. This stores information about the memory module's size, speed, voltage, drive strength, and number of row and column addresses. These parameters are read by the BIOS during POST. The SPD also contains manufacturing data such as date codes and part numbers.

Parity Memory

Memory modules have traditionally been available in two basic flavors—non-parity and parity. Parity checking uses a ninth memory chip to hold checksum data on the contents of the other eight chips in that memory bank. If the predicted value of the checksum matches the actual value, all is well. If it does not, memory contents are corrupted and unreliable. In this event a non-maskable interrupt (NMI) is generated to instruct the system to shut down and avoid any potential data corruption.

Parity checking is quite limited—only odd numbers of bit errors are detected (two parity errors in the same byte will cancel themselves out). Also, there is no way of identifying the offending bits or fixing them. In recent years the more sophisticated and more costly Error Check Code (ECC) memory has gained in popularity. Unlike parity memory, which uses a single bit to provide protection to eight bits, ECC uses larger groupings. Five ECC bits are needed to protect each eight-bit word, six for 16-bit words, seven for 32-bit words, and eight for 64-bit words.

Additional code is needed for ECC protection and the firmware that generates and checks the ECC can be in the motherboard itself or built into the motherboard chipsets. Most Intel chips now include ECC code. The downside is that ECC memory is relatively slow. It requires more overhead than parity memory for storing data and causes around a 3% performance loss in the memory subsystem. Generally, use of ECC memory is limited to mission-critical applications and is more commonly found on servers than on desktop systems.

What the firmware does when it detects an error can differ considerably. Modern systems will automatically correct single-bit errors, which account for most RAM errors, without halting the system. Many can also fix multi-bit errors on the fly. Where that's not possible, they automatically reboot and map out the bad memory.

Flash Memory

Flash memory is a solid-state, nonvolatile, rewritable memory that works like RAM combined with a hard disk. It resembles conventional memory, coming in the form of discrete chips, modules, or memory cards. Just like with DRAM and SRAM, bits of electronic data are stored in memory cells. And like a hard disk drive, flash memory is nonvolatile, retaining its data even when the power is turned off.

Although flash memory has advantages over RAM (its nonvolatility) and hard disks (it has no moving parts), there are reasons why it is not a viable replacement for either. Because of its design, flash memory must be erased in blocks of data rather than single bytes like RAM. It has significantly higher cost, and flash memory cells have a limited life span of around 100,000 write cycles. And while "flash drives" are smaller, faster, consume less energy, and are capable of withstanding shocks up to 2000 Gs (equivalent to a 10-foot drop onto concrete) without losing data, their limited capacity (around 100 MB) make them an inappropriate alternative to a PC's hard disk drive. Even if capacity were not a problem, flash memory cannot compete with hard disks in price.

Since its inception in the mid-1980s, flash memory has evolved into a versatile and practical storage solution. Several different implementations exist. NOR flash is a random access device appropriate for code storage applications. NAND flash—optimized for mass storage applications—is the most common form. Its high speed, durability, and low voltage requirements have made it ideal for use in applications such as digital cameras, cell phones, printers, handheld computers, pagers, and audio recorders.

Some popular flash memory types include SmartMedia and CompactFlash. Flash memory capacities used in consumer applications include 128 MB, 256 MB, 512 MB, and 1 GB.

Hard-Disk Drive

When the power to a PC is switched off, the contents of the main memory (RAM) are lost. It is the PC's hard disk that serves as a nonvolatile, bulk storage medium and as the repository for documents, files, and applications. It is interesting to recall that back in 1954 when IBM first invented the hard disk, capacity was a mere 5 MB stored across fifty 24-in platters. It was 25 years later that Seagate Technology introduced the first hard disk drive for personal computers, boasting a capacity of up to 40 MB and data transfer rate of 625 Kb/s. Even as recently as the late 1980s 100 MB of hard disk space was considered generous! Today this would be totally inadequate since it is hardly enough to install the operating system, let alone a huge application.

The PC's upgradability has led software companies to believe that it doesn't matter how large their applications are. As a result, the average size of the hard disk rose from 100 MB to 1.2 GB in just a few years. By the start of the new millennium a typical desktop hard drive stored 18 GB across three 3.5-in platters. Thankfully, as capacity has gone up, prices have come down. Improved area density levels are the dominant reason for the reduction in price per megabyte.

It is not just the size of hard disks that has increased. The performance of fixed disk media has also evolved considerably. Users enjoy high-performance and high-capacity data storage without paying a premium for a SCSI-based system.

CD-ROMs

When Sony and Philips invented the Compact Disc (CD) in the early 1980s, even they could not have imagined what a versatile carrier of information it would become. Launched in 1982, the audio CD's durability, random access features, and audio quality made it incredibly successful, capturing the majority of the market within a few years. CD-ROM followed in 1984, but it took a few years longer to gain the widespread acceptance enjoyed by the audio CD. This consumer reluctance was mainly due to a lack of compelling content during the first few years the technology was available. However, there are now countless games, software applications, encyclopedias, presentations, and other multimedia programs available on CD-ROM. What was originally designed to carry 74 minutes of high-quality digital audio can now hold up to 650 MB of computer data, 100 publishable photographic scans, or even 74 minutes of VHS-quality full-motion video and audio. Many discs offer a combination of all three, along with other information.

Today's mass-produced CD-ROM drives are faster and cheaper than they have ever been. Consequently, not only is a vast range of software now routinely delivered on CD-ROM, but many programs (databases, multimedia titles, games, movies) are also run directly from CD-ROM—often over a network. The CD-ROM market now embraces internal, external and portable drives, caddy- and tray-loading mechanisms, single-disc and multi-changer units, SCSI and EIDE interfaces, and a plethora of standards.

In order to understand what discs do and which machine reads which CD, it is necessary to identify the different formats. The information describing a CD standard is written in books with colored covers. A given standard is known by the color of its book cover:

- Red Book – The most widespread CD standard; it describes the physical properties of the compact disc and the digital audio encoding.
- Yellow Book – Written in 1984 to describe the storage of computer data on CD, i.e., CD-ROM (Read Only Memory).
- Green Book – Describes the CD-interactive (CD-i) disc, player, and operating system.
- Orange Book – Defines CD-Recordable discs with multi-session capability. Part I defines CD-MO (Magneto Optical) rewritable discs; Part II defines CD-WO (Write Once) discs; Part III defines CD-RW (Rewritable) discs.
- White Book – Finalized in 1993; defines the VideoCD specification.
- Blue Book – Defines the Enhanced Music CD specification for multi-session pressed disc (i.e., not recordable) containing audio and data sessions. These discs are intended to be played on any CD audio player, on PCs, and on future custom designed players.

CD-I Bridge is a Philips/Sony specification for discs intended to play on CD-i players and platforms such as the PC. Photo CD has been specified by Kodak and Philips based on the CD-i Bridge specification.

CD-ROM Applications

Most multimedia titles are optimized for double or, at best, quad-speed drives. If video is recorded to play back in real time at a 300 KB/s sustained transfer rate, anything faster than double-speed is unnecessary. In some cases a faster drive may be able to read off the information quickly into a buffer cache from where it is subsequently played, freeing the drive for further work. However, this is rare.

Pulling off large images from a PhotoCD would be a perfect application for a faster CD-ROM drive. But decompressing these images as they are read off the disc results in a performance ceiling of quad-speed. In fact, just about the only application that truly needs fast data transfer rates is copying sequential data onto a hard disc. In other words, installing software.

Fast CD-ROM drives are only fast for sustained data transfer, not random access. An ideal application for high-sustained data transfer is high-quality digital video, recorded at a suitably high rate. MPEG-2 video as implemented on DVDs requires a sustained data transfer of around 580 KB/s. This compares to MPEG-1's 170 KB/s found on existing White Book VideoCDs. However, a standard 650-MB CD-ROM disc would last less than 20 minutes at those high rates. Hence, high-quality video will only be practical on DVD discs which have a much higher capacity.

Normal music CDs and CD-ROMs are made from pre-pressed discs and encased in indentations on the silver surface of the internal disc. To read the disc, the drive shines a laser onto the CD-ROM's surface, and by interpreting the way in which the laser light is reflected from the disc, it can tell whether the area under the laser is indented.

Thanks to sophisticated laser focusing and error detection routines, this process is pretty close to ideal. However, there is no way the laser can change the indentations of the silver disc. This means there's no way of writing new data to the disc once it has been created. Thus, the technological developments to enable CD-ROMs to be written or rewritten have necessitated changes to the disc media as well as to the read/write mechanisms in the associated CD-R and CD-RW drives.

At the start of 1997 it appeared likely that CD-R and CD-RW drives would be superseded by DVD technology, almost before they had gotten off the ground. During that year DVD Forum members turned on each other, triggering a DVD standards war and delaying product shipment.

Consequently, the writable and rewritable CD formats were given a new lease on life. For professional users, developers, small businesses, presenters, multimedia designers, and home recording artists the recordable CD formats offer a range of powerful storage applications. CD media compatibility is their big advantage over alternative removable storage technologies such as MO, LIMDOW, and PD. CD-R and CD-RW drives can read nearly all the existing flavors of CD-ROMs, and discs made by CD-R and CD-RW devices can be read on both (MultiRead-capable) CD-ROM drives and DVD-ROM drive. Due to their wide compatibility, a further advantage is the low cost of media. CD-RW media is cheap and CD-R media even cheaper. Their principal disadvantage is that there are limitations to their rewriteability. CD-R is not rewritable at all. And until recently, CD-RW discs had to be reformatted to recover the space taken by deleted files when a disc becomes full. This is unlike competing technologies which all offer true drag-and-drop functionality with no such limitation. Even now, CD-RW rewriteability is less than perfect, resulting in a reduction of a CD-RW disc's storage capacity.

DVD

After a life span of ten years, during which time the capacity of hard disks increased a hundredfold, the CD-ROM finally got the facelift it needed to take it into the next century. In 1996 a standard for DVD (initially called digital videodisc, but eventually known as digital versatile disc) was finally agreed upon.

The movie companies immediately saw a big CD as a way of stimulating the video market, producing better quality sound and pictures on a disc that costs considerably less to produce than a VHS tape. Using MPEG-2 video compression—the same system that will be used for digital TV, satellite, and cable transmissions—it is possible to fit a full-length movie onto one side of a DVD disc. The picture quality is as good as live TV and the DVD-Video disc can carry multichannel digital sound.

For computer users, however, DVD means more than just movies. While DVD-Video has been grabbing the most headlines, DVD-ROM is going to be much bigger for a long time to come. Over the next few years computer-based DVD drives are likely to outsell home DVD-Video machines by a ratio of at least 5:1. With the enthusiastic backing of the computer industry in general, and CD-ROM drive manufacturers in particular, DVD-ROM drives are being used more than CD-ROM drives.

Initially, movies will be the principal application to make use of DVD's greater capacity. However, the need for more capacity in the computer world is obvious to anyone who already has multi-CD games and software packages. With modern-day programs fast outgrowing the CD, the prospect of a return to multiple disc sets (which appeared to have disappeared when CD-ROM took over from floppy disc) was looming ever closer. The unprecedented storage capacity provided by DVD lets application vendors fit multiple CD titles (phone databases, map programs, encyclopaedias) on a single disc, making them more convenient to use. Developers of edutainment and reference titles are also free to use video and audio clips more liberally. And game developers can script interactive games with full-motion video and surround-sound audio with less fear of running out of space.

History

When Philips and Sony got together to develop CD, there were just the two companies talking primarily about a replacement for the LP. Engineers carried out decisions about how the system would work and all went smoothly. However, the specification for the CD's successor went entirely the other way, with arguments, confusions, half-truths, and Machiavellian intrigue behind the scenes.

It all started badly with Matsushita Electric, Toshiba, and moviemaker Time/Warner in one corner with their Super Disc (SD) technology. In the other corner were Sony and Philips, pushing their Multimedia CD (MMCD) technology. The two disc formats were totally incompatible, creating the possibility of a VHS/Betamax-type battle. Under pressure from the computer industry, the major manufacturers formed a DVD Consortium to develop a single standard. The DVD-ROM standard that resulted at the end of 1995 was a compromise between the two technologies, but relied heavily on SD. The likes of Microsoft, Intel, Apple, and IBM gave both sides a simple ultimatum—produce a single standard quickly, or don't expect any support from the computer world. The major developers, 11 in all, created an uneasy alliance under what later became known as the DVD Forum. They continued to bicker over each element of technology being incorporated in the final specification.

The reasons for the continued rearguard actions were simple. For every item of original technology put into DVD, a license fee has to be paid to the owners of the technology. These license fees may only be a few cents per drive, but when the market amounts to millions of drives a year, it is well worth arguing over. If this didn't make matters bad enough, in waded the movie industry.

Paranoid about losing all its DVD-Video material to universal pirating, Hollywood first decided it wanted an anti-copying system along the same lines as the SCMS system introduced for DAT tapes. Just as that was being sorted out, Hollywood became aware of the possibility of a computer being used for bit-for-bit file copying from a DVD disc to some other medium. The consequence was an attempt to have the U.S. Congress pass legislation similar to the Audio Home Recording Act (the draft was called "Digital Video Recording Act") and to insist that the computer industry be covered by the proposed new law.

While their efforts to force legislation failed, the movie studios did succeed in forcing a deeper copy protection requirement into the DVD-Video standard. The resultant Content Scrambling System

(CSS) was finalized toward the end of 1996. Further copy-protection systems have been developed subsequent to this.

Formats

There are five physical formats, or books, of DVD:

- DVD-ROM is a high-capacity data storage medium.
- DVD-Video is a digital storage medium for feature-length motion pictures.
- DVD-Audio is an audio-only storage format similar to CD-Audio.
- DVD-R offers a write-once, read-many storage format akin to CD-R.
- DVD-RAM was the first rewritable (erasable) flavor of DVD to come to market and has subsequently found competition in the rival DVD-RW and DVD+RW format.

DVD discs have the same overall size as a standard 120-mm diameter, 1.2-mm thick CD, and provide up to 17 GB of storage with higher than CD-ROM transfer rates. They have access rates similar to CD-ROM and come in four versions:

- DVD-5 is a single-sided, single-layered disc boosting capacity seven-fold to 4.7 GB.
- DVD-9 is a single-sided, double-layered disc offering 8.5 G.
- DVD-10 is a 9.4 GB dual-sided single-layered disc.
- DVD-18 will increase capacity to a huge 17 GB on a dual-sided, dual-layered disc.

DVD-ROM

Like DVD discs, there is little to distinguish a DVD-ROM drive from an ordinary CD-ROM drive since the only giveaway is the DVD logo on the front. Even inside the drive there are more similarities than differences. The interface is ATAPI (AT Attachment Packet Interface) or SCSI for the more up-market drives and the transport is much like any other CD-ROM drive. CD-ROM data is recorded near the top surface of a disc. DVDs data layer is right in the middle so that the disc can be double-sided. Hence, the laser assembly of a DVD-ROM drive needs to be more complex than its CD-ROM counterpart so it can read from both CD and DVD media. An early solution entailed having a pair of lenses on a swivel—one to focus the beam onto the DVD data layers and the other to read ordinary CDs. Subsequently, more sophisticated designs have emerged that eliminate the need for lens switching. For example, Sony's "dual discrete optical pickup" design has separate lasers optimized for CD (780-nm wavelength) and DVD (650-nm wavelength). Many Panasonic drives employ an even more elegant solution that avoids the need to switch either lenses or laser beams. They use a holographic optical element capable of focusing a laser beam at two discrete points.

DVD-ROM drives spin the disk a lot slower than their CD-ROM counterparts. But since the data is packed much closer together on DVD discs, the throughput is substantially better than a CD-ROM drive at equivalent spin speed. While a 1x CD-ROM drive has a maximum data rate of only 150 KB/s, a 1x DVD-ROM drive can transfer data at 1,250 KB/s—just over the speed of an 8x CD-ROM drive.

DVD-ROM drives became generally available in early 1997. These early 1x devices were capable of reading CD-ROM discs at 12x speed—sufficient for full-screen video playback. As with CD-ROM, higher speed drives appeared as the technology matured. By the beginning of 1998 multi-speed DVD-ROM drives had already reached the market. They were capable of reading DVD media at double-speed, producing a sustained transfer rate of 2,700 KB/s. They were also capable of spinning CDs at 24-speed, and by the end of that year DVD read performance had been increased to 5-speed. A year later, DVD media reading had improved to six-speed (8,100 KB/s) and CD-ROM reading to 32-speed.

There is no standard terminology to describe the various generations of DVD-ROM drives. However, second generation (DVD II) is usually used to refer to 2x drives capable of reading CD-R/CD-RW media. Third generation (DVD III) usually means 5x, or sometimes, 4.8x or 6x drives—some of which can read DVD-RAM media.

Removable Storage

Back in the mid-1980s when a PC had a 20-MB hard disk, a 1.2-MB floppy was a capacious device capable of backing up the entire drive on a mere 17 disks. By early 1999 the standard PC hard disk had a capacity of between 3 GB and 4 GB—a 200-fold increase. In the same period the floppy's capacity increased by less than 20%. As a result, it is now at a disadvantage when used in conjunction with any modern large hard disks. For most users, the standard floppy disk just isn't big enough anymore.

In the past this problem only affected a tiny proportion of users, and solutions were available for those that did require high-capacity removable disks. For example, SyQuest's 5.25-in, 44- or 88-MB devices have been the industry standard in publishing. They are used for transferring large DTP or graphics files from the desktop to remote printers. They are quick and easy to use but are reasonably expensive.

Times have changed and today everybody needs high-capacity removable storage. These days, applications do not come on single floppies, they come on CD-ROMs. Thanks to Windows and the impact of multimedia, file sizes have gone through the ceiling. A Word document with a few embedded graphics results in a multi-megabyte data file, quite incapable of being shoehorned into a floppy disk. There is nogetting around the fact that a PC just has to have some sort of removable, writable storage with a capacity in tune with current storage requirements. It must be removable for several reasons—to transport files between PCs, to back up personal data, to act as an over-spill for the hard disk, to provide (in theory) unlimited storage. It is much easier to swap removable disks than to install another hard disk to obtain extra storage capacity.

Phase-change Technology

Panasonic's PD system, boasting the company's own patented phase-change technology, has been around since late 1995. Considered innovative at the time, the PD drive combines an optical disk drive capable of handling 650-MB capacity disks along with a quad-speed CD-ROM drive. It is the only erasable optical solution that has direct overwrite capability. Phase-change uses a purely optical technology that relies only on a laser. It is able to write new data with just a single pass of the read/write head.

With Panasonic's system the active layer is made of a material with reversible properties. A very high-powered laser heats that portion of the active layer where data is to be recorded. This area cools rapidly, forming an amorphous spot of low reflectivity. A low-powered laser beam detects the difference between these spots and the more reflective, untouched, crystalline areas thus identify a binary "0" or "1."

By reheating a spot, re-crystallization occurs, resulting in a return to its original highly reflective state. Laser temperature alone changes the active layer to crystalline or amorphous, according to the data required, in a single pass. Panasonic's single-pass phase-change system is quicker than the two-pass process employed by early MO (Magneto Optical) devices. However, modern day single-pass MO devices surpass its level of performance.

Floppy Replacements

Today's hard disks are measured in gigabytes, and multimedia and graphics file sizes are often measured in tens of megabytes. A capacity of 100 MB to 250 MB is just right for performing the traditional functions of a floppy disk (moving files between systems, archiving, backing up files or directories, sending files by mail, etc.). It is not surprising that drives in this range are bidding to be the next-generation replacement for floppy disk drives. They all use flexible magnetic media and employ traditional magnetic storage technology.

Without doubt, the most popular device in this category is Iomega's Zip drive, launched in 1995. The secret of the Zip's good performance (apart from its high, 3,000-rpm spin rate) is a technology pioneered by Iomega. Based on the Bernoulli aerodynamic principle, it sucks the flexible disk up towards the read/write head rather than vice-versa. The disks are soft and flexible like floppy disks which makes them cheap to make and less susceptible to shock.

The Zip has a capacity of 94 MB and is available in both internal and external versions. The internal units are fast, fit a 3.5-inch bay, and come with a choice of SCSI or ATAPI interface. They have an average 29-ms seek time and a data transfer rate of 1.4 KB/s. External units originally came in SCSI or parallel port versions only. However, the Zip 100 Plus version, launched in early 1998, offered additional versatility. It was capable of automatically detecting which interface applied and operating accordingly. The range was further extended in the spring of 1999 when Iomega brought a USB version to market. In addition to the obvious motivation of Windows 98 with its properly integrated USB support, the success of the Apple iMac was another key factor behind the USB variant.

Any sacrifice the external version makes in terms of performance is more than offset by the advantage of portability. It makes the transfer of reasonable-sized volumes of data between PCs a truly simple task. The main disadvantage of Zip drives is that they are not backward compatible with 3.5in floppies.

The end of 1996 saw the long-awaited appearance of OR Technology's LS-120 drive. The technology behind the LS-120 had originally been developed by Iomega, but was abandoned and sold on to 3M. The launch had been much delayed, allowing the rival Zip drive plenty of time to become established. Even then the LS-120 was hampered by a low-profile and somewhat muddled marketing campaign. Originally launched under the somewhat confusing brand name "a:DRIVE," the LS-120 was promoted by Matsushita, 3M, and Compaq. It was initially available only pre-installed on the latter's new range of Deskpro PCs. Subsequently, OR Technology offered licenses to third party manufacturers in the hope they would fit the a:DRIVE instead of a standard floppy to their PCs.

However, it was not until 1998 that Imation Corporation (a spin-off of 3M's data storage and imaging businesses) launched yet another marketing offensive. Under the brand name "SuperDisk," the product began to gain serious success in the marketplace. A SuperDisk diskette looks very similar to a common 1.44-MB, 3.5-inch disk. But it uses a refinement of the old 21-MB floptical technology to deliver much greater capacity and speed. Named after the "laser servo" technology it employs, an LS-120 disk has optical reference tracks on its surface that are both written and read by a laser system. These "servo tracks" are much narrower and can be laid closer together on the disk. For example, an LS-120 disk has a track density of 2,490 tpi compared with 135 tpi on a standard 1.44-MB floppy. As a result, the LS-120 can hold 120 MB of data.

The SuperDisk LS-120 drive uses an IDE interface (rather than the usual floppy lead) which uses up valuable IDE connections. This represents a potential problem with an IDE controller which supports only two devices rather than an EIDE controller which supports four. While its 450-KB/s data transfer rate and 70-ms seek time make it 5 times faster than a standard 3.5-in.floppy drive, it's comparatively slow spin rate of 720 rpm mean that it is not as fast as a Zip drive. However, there are two key points in the LS-120 specification that represent its principal advantages over the Zip. First, there is backward compatibility. In addition to 120-MB SuperDisk diskettes, the LS-120 can accommodate standard 1.44-MB and 720-KB floppy disks. And these are handled with a 3-fold speed improvement over standard floppy drives. Second, compatible BIOS allows the LS-120 to act as a fail-safe start-up drive in the event of a hard disk crash. Taken together, these make the LS-120 a viable alternative to a standard floppy drive.

Early 1999 saw the entry of a third device in this category with the launch of Sony's HiFD drive. With a capacity of 200 MB per disk, the HiFD provides considerably greater storage capacity than either Iomega's 100-MB Zip or Imation's 120-MB SuperDisk. It was initially released as an external model with a parallel port connector and a pass-through connector for a printer. Equipping the HiFD with a dual-head mechanism provides compatibility with conventional 1.44-MB floppy disks. When reading 1.44-MB floppy disks, a conventional floppy-disk head is used. This comes into direct contact with the media surface which rotates at just 300 rpm. The separate HiFD head works more like a hard disk, gliding over the surface of the disk without touching it. This allows the HiFD disk to rotate at 3,600 rpm with a level of performance that is significantly better than either of its rivals. However, the HiFD suffered a major setback in the summer of 1999 when read/write head misalignment problems resulted in major retailers withdrawing the device from the market.

Super-floppies

The 200-MB to 300-MB range is best understood as super-floppy territory. This is about double the capacity of the would-be floppy replacements with performance more akin to a hard disk than to a floppy disk drive. Drives in this group use either magnetic storage or magneto-optical technology. The magnetic media drives offer better performance. But even MO drive performance, at least for the SCSI versions, is good enough to allow video clips to be played directly from the drives.

In the summer of 1999 Iomega altered the landscape of the super-floppy market with the launch of the 250-MB version of its Zip drive. Like its predecessor, this is available in SCSI and parallel port versions. The parallel port version offers sustained read performance around twice the speed of the 100-MB device and sustained write speed about 50% faster. The actual usable capacity of a formatted 250-MB disk is 237 MB This is because, like most hard drive and removable media manufacturers, Iomega's capacity ratings assume that 1 MB equals 1,000,000 bytes rather than the strictly correct 1,048,576 bytes. The Zip 250 media is backward compatible with the 100-MB disks. The only downside is the poor performance of the drive when writing to the older disks.

By the new millennium the SuperDisk format had already pushed super-floppy territory by doubling its capacity to 240 MB. In 2001 Matsushita gave the format a further boost with the announcement of its FD32MB technology which gives high-density 1.44-MB floppy users the option of reformatting the media to provide a storage capacity of 32 MB per disk.

The technology increases the density of each track on the HD floppy by using the SuperDisk magnetic head for reading and the conventional magnetic head for writing. FD32MB takes a conventional floppy with 80 circumference-shaped tracks, increases that number to 777, and reduces the track pitch from the floppy's normal 187.5 microns to as little as 18.8 microns.

Hard Disk Complement

The next step up in capacity, 500 MB to 1 GB, is enough to back-up a reasonably large disk partition. Most such devices also offer good enough performance to function as secondary, if slow, hard disks. Magnetic and MO technology again predominates, but in this category they come up against competition from a number of Phase-change devices.

Above 1 GB the most commonly used removable drive technology is derived from conventional hard disks. They not only allow high capacities, but also provide fast performance, pretty close to that of conventional fixed hard disks. These drives behave just like small, fast hard disks, with recent hard disk complements exceeding 100 GB storage spaces.

Magnetic storage and MO are again the dominant technologies. Generally, the former offers better performance and the latter larger storage capacity. However, MO disks are two-sided and only half the capacity is available on-line at any given time.

Right from its launch in mid-1996 the Iomega Jaz drive was perceived to be a groundbreaking product. The idea of having 1GB removable hard disks with excellent performance was one that many high-power PC users had been dreaming of. When the Jaz appeared on the market there was little or no competition to what it could do. It allowed users to construct audio and video presentations and transport them between machines. In addition, such presentations could be executed directly from the Jaz media with no need to transfer the data to a fixed disk. In a Jaz, the twin platters sit in a cartridge protected by a dust-proof shutter which springs open on insertion to provide access to well-tried Winchester read/write heads. The drive is affordably priced, offers a fast 12-ms seek time coupled with a data transfer rate of 5.4 MB/s, and comes with a choice of IDE or SCSI-2 interfaces. It is a good choice for audiovisual work, capable of holding an entire MPEG movie.

SyQuest's much-delayed riposte to the Jaz, the 1.5-GB SyJet, finally came to market in the summer of 1997. It was faster than the Jaz, offering a data transfer rate of 6.9 MB/s. It came in parallel-port and SCSI external versions as well as an IDE internal version. Despite its larger capacity and improved performance, the SyJet failed to achieve the same level of success as the Jaz.

When first launched in the spring of 1998, the aggressively-priced 1-GB SparQ appeared to have more prospect of turning the tables on Iomega. Available in both external parallel port or internal IDE models, the SparQ achieved significantly improved performance by using high-density single platter disks and a spindle speed of 5,400 rpm. It turned out to be too little too late, though. On 2 November 1998, SyQuest was forced to suspend trading and subsequently file for bankruptcy under Chapter 11 of U.S. law. A few months later, in early 1999, the company was bought by archrival Iomega.

However, it was not long before Iomega's apparently unchangeable position in the removable storage arena came under threat from another quarter. Castlewood Systems had been founded in 1996 by one of the founders of SyQuest, and in 1999 its removable media hard drives began to make the sort of waves that had been made by the ground-breaking Iomega Zip drive some years previously.

The Castlewood ORB is the first universal storage system built using cutting-edge magnetoresistive (MR) head technology. This makes it very different from other removable media drives that use 20-year-old hard drive technology based on thin film inductive heads. MR hard drive technology—first developed by IBM—permits a much larger concentration of data on the storage medium and is expected to allow real densities to grow at a compound annual rate of 60% in the next decade. The Castle ORB drive uses 3.5-inch removable media that is virtually identical to that used in a fixed hard drive. With a capacity of 2.2 GB and a claimed maximum sustained data transfer rate of 12.2 MB/s, it represents a significant step forward in removable drive performance, making it capable of recording streaming video and audio. One of the benefits conferred by MR technology is a reduced component count. This supports two of the ORB's other advantages over its competition—cooler operation and an estimated MTBF rating 50% better than other removable cartridge products. Both will be important factors as ORB drives are adapted to mobile computers and a broad range of consumer products such as digital VCRs (which use batteries or have limited space for cooling fans). In addition, in mid-1999 the ORB drive and media were available at costs that were a factor of 2 and 3, respectively, less than competitive products.

The Blue Laser

The blue laser represents the best chance that optical storage technology has for achieving a significant increase in capacity over the next ten years. The laser is critical to development because the wavelength of the drive's laser light limits the size of the pit that can be read from the disc. If a future-generation disc with pit sizes of 0.1 micrometer were placed in today's CD-ROM drives, the beam from its laser would seem more like a floodlight covering several tracks instead of a spotlight focusing on a single dot.

The race is now on to perfect the blue laser in a form that can be used inside an optical disc layer. They have a smaller wavelength, so the narrower beam can read smaller dots. But blue lasers are proving a difficult nut to crack. They already exist for big systems and it is likely that they will be used extensively in the future for making the masters for DVD discs. However, this will require special laser-beam recorders the size of a wardrobe. They cost thousands and need a super-clean, vibration-free environment in which to work properly.

The challenge now is to build an affordable blue laser that fits into a PC-ROM drive. Solid-state blue lasers work in the laboratory, but not for long. The amount of power the laser needs to produce is a lot for a device hardly bigger than a match-head. Getting the laser to fire out of one end while not simultaneously punching a hole in its other end has been creating headaches for developers.

Techniques moved forward during 1997 and DVD discs and drives have become a big success. The current target is to use blue lasers for discs that contain 15 GB per layer. Using current DVD disc design could automatically produce a double-sided, 30-GB disc, or a dual-layer disc that could store 25 GB on a side.

From a PC data file point of view, this is an unnecessarily large capacity since most users are still producing documents that fit onto a single floppy. The main opportunity is the ability to store multimedia, video, and audio. Even though the desktop PC-based mass storage market is big, moving optical disc into the video recorder/player market would create a massive industry for the 5-inch silver disc.

Graphics

Video or graphics circuitry is usually fitted to a card or built on the motherboard and is responsible for creating the picture displayed on a monitor. On early text-based PCs this was a fairly mundane task. However, the advent of graphical operating systems dramatically increased the amount of information to be displayed. It became impractical for the main process or handle all this information. The solution was to off-load screen activity processing to a more intelligent generation of graphics card.

As multimedia and 3D graphics use has increased, the role of the graphics card has become ever more important. It has evolved into a highly efficient processing engine that can be viewed as a highly specialized co-processor. By the late 1990s the rate of development in graphics chips had reached levels unsurpassed in any other area of PC technology. Major manufacturers such as 3dfx, ATI, Matrox, nVidia, and S3 were working to a barely believable six-month product life cycle! One of the consequences of this has been the consolidation of major chip vendors and graphics card manufacturers.

Chip maker 3dfx started the trend in 1998 with its acquisition of board manufacturer STB systems. This gave 3dfx a more direct route to market with retail product and the ability to manufacture and distribute boards bearing its own brand. Rival S3 followed suit in the summer of 1999 by buying Diamond Multimedia, thereby acquiring its graphics and sound card, modem, and MP3 technologies. Weeks later 16-year veteran Number Nine announced the abandonment of its chip development business in favor of board manufacturing.

All this maneuvering left nVidia as the last of the major graphics chip vendors without its own manufacturing facility. Many speculated that it would merge with close partner, Creative Labs. While there had been no developments on this front by mid-2000, nVidia's position had been significantly strengthened by S3's sale of its graphics business to VIA Technologies in April of that year. S3 portrayed the move as an important step in transforming the company from a graphics-focused semiconductor supplier to a more broadly based Internet appliance company. It left nVidia as sole remaining big player in the graphics chip business. In any event, it was not long before S3's move would be seen as a recognition of the inevitable.

In an earnings announcement at the end of 2000, 3dfx announced the transfer of all patents, patents pending, the Voodoo brand name, and major assets to bitter rivals nVidia. In effect, they recommended the dissolution of the company. In hindsight, it could be argued that 3dfx's acquisition of STB in 1998 had simply hastened the company's demise. It was at this point that many of its board manufacturer partners switched their allegiance to nVidia. At the same time, nVidia sought to bring some stability to the graphics arena by making a commitment about future product cycles. They promised to release a new chip every autumn, and a tweaked and optimized version of that chip each following spring. To date, they have delivered on their promise, and deservedly retained their position of dominance.

Resolution

Resolution is a term often used interchangeably with addressability, but it more properly refers to the sharpness, or detail, of a visual image. It is primarily a function of the monitor and is determined by the beam size and dot pitch (sometimes called "line pitch"). An image is created when a beam of electrons strikes phosphors which coat the base of the monitor's screen. A pixel is a group of one red, one green, and one blue phosphor. A pixel represents the smallest piece of the screen that can be controlled individually, and each pixel can be set to a different color and intensity. A complete screen image is composed of thousands of pixels, and the screen's resolution (specified by a row by column figure) is the maximum number of displayable pixels. The higher the resolution, the more pixels that can be displayed, and the more information the screen can display at any given time.

Resolutions generally fall into predefined sets. Table 8.2 below shows the video standards used since CGA which was the first to support color/graphics capability:

Table 8.2: Summary of Video Standards

Date	Standard	Description	Resolution	No. of Colors
1981	CGA	Color Graphics Adapter	640 x 200 160 x 200	None 16
1984	EGA	Enhanced Graphics Adapter	640 x 350	16 from 64
1987	VGA	Video Graphics Array	640 x 480 320 x 200	16 from 262,144 256
1990	XGA	Extended Graphics Array	800 x 600 1024 x 768	16.7 million 65,536
1992	SXGA	Super Extended Graphics Array	1280 x 1024	65,536
1992	UXGA	Ultra XGA	1600 x 1200	65,536

Pixel addressing that is better than the VGA standard lacks a widely accepted standard. This presents a problem for everyone—manufacturers, system builders, programmers, and end users. As a result, each vendor must provide specific drivers for each supported operating system for each of their cards. The XGA, configurable with 500 KB or 1 MB, was the first IBM display adapter to use VRAM. SXGA and UXGA are subsequent IBM standards, but neither has been widely adopted.

Typically, an SVGA display can support a palette of up to 16.7 million colors. But the amount of video memory in a particular computer may limit the actual number of displayed colors to something less than that. Image-resolution specifications vary. In general, the larger the diagonal screen measure of an SVGA monitor, the more pixels it can display horizontally and vertically. Small SVGA monitors (14 in. diagonal) usually use a resolution of 800×600. The largest (20 in. or greater, diagonal) can display 1280×1024, or even 1600×1200 pixels.

Pixels are smaller at higher resolutions. Prior to Windows 95—and the introduction of scaleable screen objects—Windows icons and title bars were always the same number of pixels in size whatever the resolution. Consequently, the higher the screen resolution, the smaller these objects appeared. The result was that higher resolutions worked much better on larger monitors where the pixels are correspondingly larger. These days the ability to scale Windows objects, coupled with the option to use smaller or larger fonts, yields far greater flexibility. This makes it possible to use many 15-in. monitors at screen resolutions up to 1024×768 pixels, and 17-in. monitors at resolutions up to 1600×1200.

Color depth

Each pixel of a screen image is displayed using a combination of three different color signals—red, green, and blue. The precise appearance of each pixel is controlled by the intensity of these three beams of light. The amount of information that is stored about a pixel determines its color depth. The more bits that are used per pixel ("bit depth"), the finer the color detail of the image.

For a display to fool the eye into seeing full color, 256 shades of red, green, and blue are required. That is eight bits for each of the three primary colors, or 24 bits in total. However, some graphics cards actually require 32 bits for each pixel to display true color due to the way they use the video memory. Here, the extra eight bits are generally used for an alpha channel (transparencies).

High color uses two bytes of information to store the intensity values for the three colors—five bits for blue, five for red, and six for green. This results in 32 different intensities for blue and red, and 64 for green. Though this results in a very slight loss of visible image quality, it offers the advantages of lower video memory requirements and faster performance.

256-color mode uses a level of indirection by introducing the concept of a "palette" of colors, selectable from a range of 16.7 million colors. Each color in the 256-color palette is described by the standard 3-byte color definition used in true color. Hence, red, blue, and green each get 256 possible intensities. Any given image can then use any color from its associated palette.

The palette approach is an excellent compromise solution. For example, it enables far greater precision in an image than would be possible by using the eight available bits to assign each pixel a 2-bit value for blue and 3-bit values each for green and red. The 256-color mode is a widely used standard, especially for business, because of its relatively low demands on video memory.

Dithering

Dithering substitutes combinations of colors that a graphics card can generate for colors that it cannot produce. For example, if a graphics subsystem is capable of handling 256 colors, and an image that uses 65,000 colors is displayed, colors that are not available are replaced by colors created from combinations of colors that are available. The color quality of a dithered image is inferior to a non-dithered image.

Dithering also refers to using two colors to create the appearance of a third, giving a smoother appearance to otherwise abrupt transitions. That is, it uses patterns to simulate gradations of gray or color shades, or of anti-aliasing.

Components of a PC Graphics Card

The modern PC graphics card consists of four main components: graphics processor, video memory, a random access memory digital-to-analog converter (RAMDAC), and driver software.

Graphics Processor

Instead of sending a raw screen image to the frame buffer, the CPU's dedicated graphics processing chip sends a smaller set of drawing instructions. These are interpreted by the graphics card's proprietary driver and executed by the card's on-board processor.

Operations including bitmap transfers and painting, window resizing and repositioning, line drawing, font scaling, and polygon drawing can be handled by the card's graphics processor. It is designed to process these tasks in hardware at far greater speeds than the software running on the system's CPU could process them. The graphics processor then writes the frame data to the frame buffer. As there is less data to transfer, there is less congestion on the system bus, and CPU workload is greatly reduced.

Video Memory

The memory that holds the video image is also referred to as the frame buffer and is usually implemented on the graphics card itself. Early systems implemented video memory in standard DRAM. However, this requires continual refreshing of the data to prevent it from being lost, and data cannot be modified during this refresh process. Performance is badly degraded at the very fast clock speeds demanded by modern graphics cards.

An advantage of implementing video memory on the graphics board itself is that it can be customized for its specific task. And this has resulted in a proliferation of new memory technologies:

- Video RAM (VRAM) – A special type of dual-ported DRAM which can be written to and read from at the same time. Hence, it performs much better since it requires far less frequent refreshing.
- Windows RAM (WRAM) – Used by the Matrox Millennium card, it is dual-ported and can run slightly faster than conventional VRAM.
- EDO DRAM – Provides a higher bandwidth than normal DRAM, can be clocked higher than normal DRAM, and manages read/write cycles more efficiently.
- SDRAM – Similar to EDO RAM except that the memory and graphics chips run on a common clock used to latch data. This allows SDRAM to run faster than regular EDO RAM
- SGRAM – Same as SDRAM but also supports block writes and write-per-bit which yield better performance on graphics chips that support these enhanced features.
- DRDRAM – Direct RDRAM is a totally new, general-purpose memory architecture that promises a 20-fold performance improvement over conventional DRAM.

Some designs integrate the graphics circuitry into the motherboard itself and use a portion of the system's RAM for the frame buffer. This is called unified memory architecture and is used only to reduce costs. Since such implementations cannot take advantage of specialized video memory technologies, they will always result in inferior graphics performance.

The information in the video memory frame buffer is an image of what appears on the screen, stored as a digital bitmap. But while the video memory contains digital information its output medium, the monitor, uses analogue signals. The analogue signal requires more than just an on or off signal since it is used to determine where, when, and with what intensity the electron guns should be fired as they scan across and down the front of the monitor. This is where the RAMDAC comes in.

RAMDAC

The RAMDAC reads the contents of the video memory many times per second, converts it into an analog RGB signal, and sends it over the video cable to the monitor. It does this by using a look-up table to convert the digital signal to a voltage level for each color. There is one digital-to-analog converter (DAC) for each of the three primary colors used by the CRT to create a complete spectrum of colors. The intended result is the right mix needed to create the color of a single pixel. The rate at which the RAMDAC can convert the information, and the design of the graphics processor itself, dictates the range of refresh rates that the graphics card can support. The RAMDAC also dictates the number of colors available in a given resolution, depending on its internal architecture.

Driver Software

Modern graphics card driver software is vitally important when it comes to performance and features. For most applications, the drivers translate what the application wants to display on the screen into instructions that the graphics processor can use. The way the drivers translate these instructions is very important. Modern graphics processors do more than change single pixels at a time; they have sophisticated line and shape drawing capabilities and they can move large blocks of information around. It is the driver's job to decide on the most efficient way to use these graphics processor features, depending on what the application requires to be displayed.

In most cases a separate driver is used for each resolution or color depth. Taking into account the different overheads associated with different resolutions and colors, a graphics card can have markedly different performance at different resolutions, depending on how well a particular driver has been written and optimized.

Sound Cards

Sound is a relatively new capability for PCs because no one really considered it when the PC was first designed. The original IBM-compatible PC was designed as a business tool, not as a multimedia machine. Hence, it is hardly surprising that nobody thought of including a dedicated sound chip in its architecture. Computers were seen as calculating machines; the only kind of sound necessary was the beep that served as a warning signal. For years, the Macintosh has had built-in sound capabilities far beyond beeps and clicks, but PCs with integrated sound are still few and far between. That's why PCs continue to require an add-in board or sound card to produce decent quality sound.

The popularity of multimedia applications over the past few years has accelerated the development of the sound card. The increased competition between manufacturers has led to these devices becoming cheaper and more sophisticated. Today's cards not only make games and multimedia applications sound great, but with the right software, users can also compose, edit, and print their own music. And they can learn to play the piano, record and edit digital audio, and play audio CDs.

Components

The modern PC sound card contains several hardware systems relating to the production and capture of audio. The two main audio subsystems are for digital audio capture and replay and music synthesis. The replay and music synthesis subsystem produces sound waves in one of two ways—through an internal FM synthesizer, and by playing a digitized, or sampled, sound.

The digital audio section of a sound card consists of a matched pair of 16-bit digital-to-analog (DAC) and analog-to-digital (ADC) converters and a programmable sample rate generator. The computer reads the sample data to or from the converters. The sample rate generator clocks the converters and is controlled by the PC. While it can be any frequency above 5 kHz, it is usually a fraction of 44.1 kHz.

Most cards use one or more Direct Memory Access (DMA) channels to read and write the digital audio data to and from the audio hardware. DMA-based cards that implement simultaneous recording and playback (or full duplex operation) use two channels. This increases the complexity of installation and the potential for DMA clashes with other hardware. Some cards also provide a direct digital output using an optical or coaxial S/PDIF connection.

A card's sound generator is based on a custom DSP (digital signal processor) that replays the required musical notes. It multiplexes reads from different areas of the wave-table memory at differing speeds to give the required pitches. The maximum number of notes available is related to the processing power available in the DSP and is referred to as the card's "polyphony."

DSPs use complex algorithms to create effects such as reverb, chorus, and delay. Reverb gives the impression that the instruments are being played in large concert halls. Chorus gives the impression that many instruments are playing at once, when in fact there's only one. Adding a stereo delay to a guitar part, for example, can "thicken" the texture and give it a spacious stereo presence.

Portable/Mobile Computing

No area of personal computing has changed more rapidly than portable technology. With software programs getting bigger all the time and portable PCs being used for a greater variety of applications, manufacturers have their work cut out for them. They must attempt to match the level of functionality of a PC in a package that can be used on the road. This has led to a number of rapid advancements in both size and power. By mid-1998 the various mobile computing technologies had reached a level where it was possible to buy a portable computer that was as fast as a desktop machine, yet capable of being used in the absence of a mains electricity supply for over five hours.

CPU

In mid-1995 Intel's processor of choice for notebook PCs was the 75-MHz version. This was available in a special thin-film package - the Tape Carrier Package (TCP)—designed to ease heat dissipation in the close confines of a notebook. It also incorporated Voltage Reduction Technology which allowed the processor to "talk" to industry standard 3.3-volt components, while its inner core, operating at 2.5 volts, consumed less power to promote a longer battery life. In combination, these features allowed system manufacturers to offer high-performance, feature-rich notebook computers with extended battery lives.

The processors used in notebook PCs have continued to gain performance features, and use less power. They have traditionally been a generation behind desktop PCs. The new CPU's low power consumption, coupled with multimedia extension technology, provide mini-notebook PC users with the performance and functionality to effectively run business and communication applications "on the road."

Expansion Devices

Many notebook PCs are proprietary designs, sharing few common, standard parts. A consequence of this is that their expansion potential is often limited and the cost of upgrading them is high. While most use standard CPUs and RAM, these components generally fit in unique motherboard designs, housed in unique casing designs. Size precludes the incorporation of items like standard PCI slots or drive bays, or even relatively small features like SIMM slots. Generally, the only cheap way to upgrade a notebook is via its native PC Card slots.

A port replicator is not so much an expansion device; it's more a means to facilitate easier connectivity between the notebook PC and external peripherals and devices. The main reason for the existence of port replicators is the fragility of PC connectors, which are designed for only so many insertions. The replicator remains permanently plugged into a desktop PC. It makes the repeated connections necessary to maintain synchronization between mobile computing devices (notebook, PDA, desktop PC) via the replicator's more robust connections.

A desk-based docking station takes the concept a stage further, adding desktop PC-like expansion opportunities to mobile computing devices. A full docking station will generally feature expansion slots and drive bays to which ordinary expansion cards and external peripheral devices may be fit. It may also come complete with an integrated monitor stand and provide additional higher-performance interfaces such as SCSI.

PC Card

In the early 90's the rapid growth of mobile computing drove the development of smaller, lighter, more portable tools for information processing. PC Card technology was one of the most exciting innovations. The power and versatility of PC Cards quickly made them standard equipment in mobile computers.

PCMCIA

The rapid development and worldwide adoption of PC Card technology has been due in large part to the standards efforts of the Personal Computer Memory Card International Association (PCMCIA).

First released in 1990, the PC Card Standard defines a 68-pin interface between the peripheral card and the socket into which it is inserted. The standard defines three standard PC Card form factors—Type I, Type II, and Type III. The only difference between the card types is thickness. Type I is 3.3 mm, Type II is 5.0 mm, and Type III is 10.5 mm. Because they differ only in thickness, a thinner card can be used in a thicker slot, but a thicker card can not be used in a thinner slot. The card types each have features that fit the needs of different applications. Type I PC Cards are typically used for memory devices such as RAM, Flash, and SRAM cards. Type II PC Cards are typically used for I/O devices such as data/fax modems, LANs, and mass storage devices. Type III PC Cards are used for devices whose components are thicker, such as rotating mass storage devices. Extended cards allow the addition of components that must remain outside the system for proper operation, such as antennas for wireless applications.

In addition to electrical and physical specifications, the PC Card Standard defines a software architecture to provide "plug and play" capability across the widest product range. This software is made up of Socket Services and Card Services which allow for interoperability of PC Cards.

Socket Services is a BIOS-level software layer that isolates PC Card software from system hardware and detects the insertion and removal of PC Cards. Card Services describes an Application Programming Interface (API) which allows PC Cards and sockets to be shared by multiple clients such as devices drivers, configuration utilities, or application programs. It manages the automatic allocation of system resources such as memory, and interrupts once Socket Services detects that a card has been inserted.

CardBus

Derived from the Peripheral Component Interconnect (PCI) Local Bus signaling protocol, CardBus is the 32-bit version of PC Card technology. Enabled in the February 1995 release of the PC Card Standard, CardBus features and capabilities include 32 bits of address and data, 33-MHz operation, and bus master operation.

Battery Technology

Historically, the technology of the batteries that power notebook PCs has developed at a somewhat slower rate than other aspects of mobile computing technology. Furthermore, as batteries get better the power advantage is generally negated by increased consumption from higher performance PCs, with the result that average battery life remains fairly constant.

Nickel-Cadmium (NiCad) was a common chemical formulation up to 1996. As well as being environmentally unfriendly, NiCad batteries were heavy, low in capacity, and prone to "memory effect." The latter was a consequence of recharging a battery before it was fully discharged. The memory effect caused batteries to "forget" their true charge. Instead, they would provide a much smaller charge, often equivalent to the average usage time prior to its recharge. Fortunately, NiCad batteries are almost impossible to find in notebook PCs these days.

The replacement for Nickel-Cadmium was Nickel-Metal Hydride (NiMH), a more environmentally friendly formulation. NiMH has a higher energy density and is less prone to the memory effect. By 1998 the most advanced commercially available formulation was Lithium Ion (Li-Ion). Lithium Ion has a longer battery life (typically around three hours) than NiMH and does not have to be fully discharged before recharging. It is also lighter to carry. Although the price differential is narrowing, Li-Ion continues to carry a price premium over NiMH.

The main ingredient of a Li-Ion battery is lithium. It does not appear as pure lithium metal, but as:

- Atoms within the graphite of the battery's anode (positive terminal)
- Lithium-cobalt oxide or lithium-mangan oxide in the cathode (negative terminal)
- Lithium salt in the battery's electrolyte

It relies on reversible action. When being charged, some of the lithium ions usually stored in the cathode are released. They make their way to the graphite (carbon) anode where they combine into its crystalline structure. When connected to a load such as a notebook PC, the electrochemical imbalance within the battery created by the charging process is reversed, delivering a current to the load. Li-Ion is used in all sorts of high-end electronic equipment such as mobile phones and digital cameras. One downside is that it self-discharges over a few months if left on the shelf, so it requires care to keep it in optimum condition.

1999 saw Lithium polymer (Li-polymer) emerge as the most likely battery technology of the future. Using lithium—the lightest metal on earth - this technology offers potentially greater energy densities than Li-Ion. Instead of using a liquid electrolyte, as in conventional battery technologies, Li-polymer uses a solid or gel material impregnated with the electrolyte. Hence, batteries to be made in almost any shape, allowing them to be placed in parts of a notebook case that would normally be filled with air. Cells are constructed from a flexible, multi-layered 100-micron thick film laminate that does not require a hard leak-proof case. The laminate comprises five layers—a current-collecting metal foil, a cathode, an electrolyte, a lithium foil anode, and an insulator.

Early lithium polymer battery prototypes proved unable to produce currents high enough to be practical. Bell core technology (named after the laboratory that invented it) is a variation on the lithium polymer theme. It addresses the problem by using a mixture of liquid and polymer electrolytes that produce higher levels of current. A number of manufacturers have licensed this technology. Initially, the principal market for these devices will be PDAs and mobile phones where space and weight are at a premium and value is high. This will be followed by the increasing use in notebook PCs.

Zinc-Air technology is another emerging technology and possible competitor to Li-polymer. As the name suggests, Zinc-Air batteries use oxygen in the chemical reaction to produce electricity, thus removing the need to include metal reaction elements. Energy density by weight is reduced, but volume increases due to the need for air-breathing chambers in the battery. This reduces its appeal for small form factor devices and may relegate it to being a niche technology.

Requirements Overdose

The PC is facing lackluster revenue growth. In several geographies, revenues and earnings are declining. Thanks to Moore's Law, as process technologies continue to shrink and component manufacturers derive benefits in performance, density, power, and cost, they can pass these on to consumers. Until just a few years ago, applications running on PCs demanded processing power and memory densities that the components could never satisfy. This caused consumers to upgrade their PCs every few years. However, today's PCs seem adequate in performance and functionality with current applications. Combined with a large population base (in the U.S.) that already owns at least one PC, this is causing a slowdown in overall PC demand. Hence, revenues and earnings are falling for several PC vendors and their component suppliers. These companies are trying to diversify revenues through new streams such as set-top boxes, gaming consoles, and networking equipment.

New PC Demand Drivers

To enhance demand for PCs, new demand drivers will have to be created. Some of these include support for new interfaces and the introduction of new and entertaining products and software applications.

Incremental Changes to the Operating System may not be Enough

While Microsoft Corporation has released its next-generation operating system, Windows XP, with great fanfare, the move will not be enough to pull the PC industry out of its downturn. Beyond XP, a road map of stepwise refinements planned for desktops over the next couple of years may help the industry struggle back to growth. But no major architectural leap can be predicted until 2004, and the massive PC industry will lumber forward for a time without a strong driver.

Windows XP shrinks PC boot time to about 35 seconds while boosting stability and ease-of-use, especially for wireless networking. Meatier hardware advances will beef up next year's computers with USB 2.0, Gigabit Ethernet, Bluetooth, and a greater variety of core logic chip sets. Gigabit Ethernet's arrival to the desktop motherboard will rapidly bridge the cost gap between 10/100 and 10/100/1000 Ethernet silicon. In 2004, there is a good possibility of new product concepts when the Arapahoe replacement to the PCI bus arrives.

153 million PCs shipped worldwide in 2001, down 6% from 2000's record high. However, a 13% rebound in unit shipments in 2002 was based on the strength of cost reductions in Intel Pentium 4-class systems, Advanced Micro Device's move to 0.13-micron and silicon-on-insulator technology, and a fresh set of graphics and chip set offerings. In addition, Windows XP will have a new version that will include native support for the 480-Mb/s USB 2.0 interface as well as the short-range wireless Bluetooth link.

Some companies will craft their own software to roll out interfaces (e.g., Bluetooth) earlier, but such moves come with a potential cost in development and compatibility. Working on their own drivers for Bluetooth will cause compatibility problems. This will also raise software development costs which are already significantly higher than hardware development costs for some high-volume PC subsystems. Further, Bluetooth user profiles have yet to be written, and that could delay broad deployment of Bluetooth for as much as two years.

XP is less prone to crashes because it is based on the more stable Windows NT code base. That will lower OEM support costs dramatically. And XP's improved support for networking, particularly for IEEE 802.11b wireless LANs, is one of its greatest strengths. Dell Computer, IBM, and other PC manufacturers have been shipping notebooks with optional 802.11b internal MiniPCI cards. XP adds the ability to recognize access points automatically for users on the move among office, home, and public WLANs without requiring manual reconfiguration of their cards. This is much better in XP than in Windows 2000. It will build mass appeal and prompt people to start using wireless LANs more often.

This shift could open the door to new devices and services that complement the PC. These would include new systems like the tablet PCs and improved versions of MP3 players and Internet juke-boxes. The next step is introducing pieces of Microsoft's Universal Plug and Play (UPnP) initiative for automatically discovering systems on a network. Microsoft's separate .Net and a parallel Web services initiative (announced by Sun Microsystems' initiative One™) aims to enable Web-based services that mix and match component-based applications to anticipate and automate user wants.

The bottom line is that PCs are moving beyond hardware to software. The real challenges are no longer in the processor, memory, and graphics. As the hardware becomes more commoditized, it requires less effort. The effort lies in helping users manage how they deploy, transition, and protect their data.

More radical hardware changes will come in 2004 with the deployment of the Arapahoe interface, anticipated as a faster serial replacement for PCI.

Arapahoe will put video on the same I/O bus instead of on a dedicated bus, which is a big advantage. Users will also be able to use cable versions so they can have split designs. Here, they could place something on and something under the desk as an option. The desktop could be broken into multiple pieces, something that was not possible before because fast enough data transfer rates over an external cable were not available. Users could also have smaller desktops because Arapahoe has a smaller physical interface than PCI. There will be a significant architectural shift with the rise of fast interfaces like Arapahoe and Gigabit Ethernet.

The hard disk drive will be eliminated from the PC in the next three to five years, leading to an era of network-based storage. But with hard drives being so cheap, it might be better to have a local cache for data, even though users may want to mirror and manage a lot of that from a corporate network storage system.

Tablet PC

Microsoft has unveiled a new technology to help bring the PC back to its former leading role. Called a Tablet PC, it is a pen-driven, fully functional computer. It is built on the premise that next-generation hardware and software takes advantage of the full power of today's desktop computers. Even sleek portables need the power of a desktop PC.

Companies such as Sun and Oracle envision a more varied future, with most of the Internet's complexity placed on central servers. Here, everything from pagers to cell phones will be used to access the Internet via these servers. However, Microsoft believes that intelligence and computing power will reside at the consumer end.

The browser-based era of the Internet, where servers do the heavy computations and PCs merely display Web pages, is in decline. The browser model which has been the focus for the last five years is really showing its age.

While the computer world has historically been either centrally or locally based, the need now is for a more balanced approach. Vast amounts of data will reside on the server and powerful desktops will be needed to sort and interpret the data. However, peer-to-peer computing will not render the server obsolete.

Server software will enable "information agents" that can filter and prioritize all the different messages being sent to a person. The world will be filled with lots of medium-powered servers rather than a smaller number of very powerful ones. A profusion of smart client devices (e.g., tablet PCs, gaming consoles, PCs in automotives, handhelds, and smart cell phones) will be able to deliver a richer Internet experience.

The Tablet PC has a 500- to 600-MHz CPU, 128 MB of RAM, a 10-GB hard disk, and universal serial bus (USB) ports for keyboard and mouse. It is also based on Microsoft Windows. The Tablet PC also features Microsoft's ClearType, a technology that makes text on LCD (liquid crystal display) screens more readable.

While it will cost more than the equivalent portable model when it comes out, it presents an opportunity to both grow the portable PC market and become a significant part of that market. Among the technologies behind the Tablet PC is what Microsoft calls "rich ink." This feature allows a person to take notes on the screen as if it were an infinite pad of paper, albeit a pricey one. The Tablet PC converts handwritten pen strokes into graphics that can be handled similarly to ordinary computer type. Although the computer doesn't recognize specific words, it distinguishes between words and pictures. It also lets people perform some word processing-like tasks such as inserting space between lines, copying text, or boldfacing pen strokes.

The tablet PC platform is a next-generation wireless device charged with merging the computing power of a PC with the mobility of a digital consumer appliance. Both Intel and Transmeta micro-processors will power the devices which may come in a wide range of shapes and sizes. They will be manufactured initially by Acer, Compaq, Fujitsu, Sony, and Toshiba. A full-function, slate-like computer, the Tablet, will feature the newly announced Windows XP operating system. It will take advantage of pen-based input in the form of digital ink.

The Tablet PC will be an even more important advance for PCs than notebook computers were. It combines the simplicity of paper with the power of the PC. It abandons the traditional keyboard and mouse for the more ergonomic pen. Microsoft positions this device as more of a PC than as a consumer appliance. This is due to the fallout in "information appliances" as a commercial viability, with both 3Com and Gateway pulling products from both roadmaps and shelves.

It is a real PC with all of the power and ability of a desktop system, and with the portability to replace laptops. Apple Computer's Newton and Pocket Pad were ambitious precursors to PDA-like devices, and the tablet market has been limited to the vertical marketplace. Hence, the success of a product like the tablet PC in the mainstream market requires an industry-wide effort. Form factors, price points, and features for Tablet PC devices will be left up to the OEM, with the only unbending stipulation that the device is a full-function PC.

Given the current state of the PC industry, it's important to think of new ways to expand the market. Many of the tablet PC's components are leveraged off the notebook platform, but improve-ments such as power consumption and cost are on the way.

Next Generation Microsoft Office—Microsoft Office XP
The next version of the dominant PC office suite will allow people to control the program and edit text with their voices via speech-recognition technology. Another new feature known as "smart tags"

allows Office users to dynamically include information from the Internet into documents. Smart tags pop up after different actions, offering additional suggests and items. For example, a formatting options list might appear after a user pastes information into a document. A new version of Office also will include a Web-based collaboration program known as SharePoint.

Instant Messaging—A New Form of Entertainment

Instant messaging technology is viewed as entertainment as well as a productivity enhancement tool and could become a significant growth driver for the Internet-enabled devices and PCs. First introduced by AOL in 1997, IM technology allows people to communicate with each other real-time via the Internet. Now several companies including Microsoft and Yahoo! have introduced instant messaging service.

There are three ways commonly used to establish the connection—the centralized network method, the peer-to-peer method, and a combination of the two.

- Centralized Networks – In a centralized network, users are connected through a series of connected servers. When a message is sent from a user, the server directs it through the network to the recipient's server which, in turn, forwards the message to the recipient. The message travels throughout the network. This method is currently used by MSN messaging service.
- Peer-to-Peer – With this method, a unique IP address is assigned to each user on the network. When you log on to the network, you are assigned an IP address. You also receive the IP addresses of all the people on your buddy list that are on the network, establishing direct connections among them. Thus, when a message is sent it is directly sent to the recipient. Consequently, the message transfer rates are much quicker than in the centralized network method. ICQ uses the peer-to-peer methodology.
- Combination Method – The combination method uses either the centralized network method or peer-to-peer method, depending on the type of message sent. Currently, AOL Instant Messaging service (AIM) uses the combination method. When a text is sent, AIM uses the centralized network method. When a data intensive message such as a picture, a voice, or a file is sent, it uses the peer-to-peer connection.

Notebook/Portable/Laptop PC

The desktop PC may be headed for the geriatric ward, but svelte and brawny notebooks are ready to take its place as the most significant computer product. Market data paints a dismal future for PCs, with flat rather than double-digit sales growth. But notebooks are showing surprising resilience. In fact, after years of being stalled at about 20% of the overall PC market, notebooks are widening their share at the expense of desktop systems.

Portable PC sales are doing so well that many computer companies are shifting their sales focus to notebooks and away from desktop PCs to drive market growth. There is a widening gulf between the two product categories. Worldwide, notebook shipments grew 21% year over year, compared with paltry desktop PC growth of 1.6%. In the United States notebook shipments increased 6% over fourth quarter 1999, while desktop PC shipments stalled at 0.1% growth. While desktops may account for the lion's share of PC revenue, notebooks are expected to continue closing the gap.

The notebook PC market is also less saturated than that for PCs. For example, only 20% of computer users in the corporate segment rely on notebooks. Every employee who needs a PC already has one. But of those desktop users, 80% are potential notebook buyers. One way for vendors to generate sales is to offer notebooks while narrowing the price gap between notebooks and desktops. In the United States and Europe desktop PCs have shifted to a replacement market from a growth market. Companies with strong service organizations will use the growing notebook market to bridge the gap with the more mature desktop PC segment. Instead of selling 2,000 notebooks, a vendor could build a relationship by offering an upgrade to a new operating system. Even if customers are not buying as frequently, the vendor could establish relationships for the next time they buy.

Falling component prices, particularly for LCDs (liquid-crystal displays), are also expected to lead to lower notebook prices. LCD panels typically account for as much as 40% of a laptop's cost. Lower prices are expected to have the most impact on consumer sales. But the major factor fueling notebook sales is increasingly better perceived value over desktop PCs. While there have been no recent compelling changes to desktop PCs, notebooks are benefiting from advances in wireless communications, mainly in the form of IEEE 802.11b and Bluetooth networking. Wireless networking is a clear driver of notebook sales, particularly with 802.11b terminals available in 30 airports around the country and in major hotel chains. The next step is wireless networking appearing in other public places like restaurants and coffee shops. The value proposition for wireless LAN in public places is really strong. Several manufacturers including Apple Computer, Dell Computer, Compaq, and IBM are among the major manufacturers offering wireless networking antennas integrated inside their notebooks.

IBM developed a full-function portable that can also interpret data and drawings written on notepads. It is also testing a wearable PC as well as a new mobile companion. Apple and Dell have opted for putting the most performance in the smallest size possible, rivaling some workstations in terms of power and features. The notebook market will significantly grow as a percentage of the total PC market.

PocketPC Operating System (OS) Platform

Microsoft Corporation is taking sharper aim at the handheld market by introducing a new PocketPC 2002 platform. It supports fewer processors and promises to serve enterprise-level customers. This will play a big role in its grand plan to chip away at Palm Inc.'s leadership in the handheld market. The rollout signals a subtle shift in direction for Microsoft, which until now has tried to build a handheld platform that appealed to all potential users, from business executives to soccer moms. However, with the new release the software giant is clearly homing in on customers who can buy the devices in volume as productivity tools for their companies. The introduction of PocketPC will help boost the use of the PC as a consumer device. At the same time, Microsoft surprised industry experts by narrowing the list of processors that the PocketPC will support. Microsoft's strategy is a solid one, especially as the economic downturn continues to discourage consumers from buying handheld computers for personal applications. The corporate market for handhelds is just starting to kick in now and it is believed that the PocketPC will eventually overtake the Palm OS.

Both PocketPC 2002 and the PocketPC operating system introduced in April 2000 are built on Windows CE. They act as so-called "supersets" to add a variety of applications and software components to the CE foundation. The PocketPC system is an attempt to right the wrongs of Windows CE which had been spectacularly unsuccessful in its early attempts to enter the palmtop computing market. With the newest version of the PocketPC, software will come closer to finding its niche in the handheld marketplace.

Like the earlier version, PocketPC 2002 is aimed at the high end of the handheld computing market. Products that employ it are generally larger and have color screens, longer battery life, higher price tags, and greater functionality than the electronic organizers popularized by Palm. Pocket PCs also offer far more computing power than the lower-end palm tops. Using Intel's SA-1110 processor, for example, HP/Compaq's iPaq handheld can operate at 206-MHz clock speeds. In contrast, low-end Palm systems employ a 33-MHz Motorola Dragonball processor.

However, the newest version of the PocketPC operating system promises even more functionality. Microsoft added support for Windows Media Video, including streaming video and improved electronic reading capabilities via a new Microsoft Reader. More important, the company has targeted enterprise customers by introducing the ability to connect to corporate information via a virtual private network. It has also made major development investments in connectivity by adding options ranging from local-area networks such as 802.11b and personal-area networks such as Bluetooth, to wide-area networks. With the Pocket PC 2002 Microsoft is laying the groundwork for more powerful wireless e-mail capabilities that could make PocketPC handhelds appeal to large corporations with hundreds or even thousands of traveling employees. Marrying the organizer features to wireless e-mail capabilities will allow a lot of senior employees to sign up for the service.

Paring down the hardware support from three processor cores to one allows Microsoft and the developers to streamline their efforts. Using only ARM-based processors simplifies outside application development, cuts costs, and still provides hardware makers with a choice of chip set suppliers. The company's engineers are already working with Intel on its SA-1110 and X-Scale processors, as well as with Texas Instruments on the ARM 920 processor core. Additional ARM-based chip set suppliers include Parthus, Agilent, SD Microelectronics, and Linkup. The industry analysts are expecting the vast majority of Pocket PC hardware makers to opt for Intel's StrongARM processors, even though the ARM-based processors have been much more successful. Microsoft selected ARM-based processors over Hitachi- or MIPS-based processors based on performance and power efficiency—the Intel StrongARM typically operates at about 1.3 to 1.5 MIPS per mW, while much of the market ranges between 0.5 and 0.9 MIPS per mW.

MIPS and Hitachi have both emphasized that their processor cores will continue to be supported in Windows CE. Even though major manufacturers such as Casio and Hewlett-Packard now use other processors, Microsoft's decision will not have a profound effect on sales of Pocket PCs. Microsoft's processor choice would have a near-term impact on consumers. HP/Compaq iPaq already uses the ARM processor. Existing iPaq users will be able to upgrade to Pocket PC 2002, or later to Talisker, whereas others will not.

Ultimately, Microsoft's strategies will help the company continue to close the longstanding gap with Palm. In the PDA market, Microsoft Windows CE and Pocket PC platforms have 30% of the market in 2003, up from an earlier figure in "the teens" during the first quarter of 2001. In 2003, Palm still holds a 49% share. That figure for the Palm-based PDA market share has dropped from a steady level in excess of 70% during the last several years. In 2006 it is predicted that Palm and Microsoft-based platforms will both have 47% market shares. For the same reason, Microsoft's enterprise focus may continue to be the best way to chip away at Palm's dominance. No one has hit on the feature that would be compelling for soccer moms. When they do, the consumer space will be tremendous for handhelds. Until then, Microsoft is doing the best thing by hammering away at the enterprise focus.

Summary

The growing ubiquity of the Internet and the convergence of voice, data, and video are bringing interesting applications to the home. Lower PC prices have brought multiple PCs into the home. However, newer digital consumer appliances that use the Internet to provide particular functions are arriving in the market. The big question is whether the digital consumer device will replace the PC in its entirety. Digital consumer appliances provide dedicated functionality, while the PC is an ideal platform to become the residential gateway of the future. It could provide home networking capabilities to digital consumer devices, PC peripherals, and other PCs. Hence, while digital consumer devices will continue to penetrate homes and provide the convenience of the Internet, they will not replace PCs. Both will continue to coexist.

PC Peripherals

Introduction

A peripheral device is an external consumer device that attaches to the PC to provide a specific functionality. Peripherals include printers, scanners, smart card readers, keyboards, mice, displays, and digital cameras. Some devices such as printers, scanners, cameras, and smart card readers serve as more than just PC peripherals, and devices such as the keyboard, mouse, and display are required to operate the PC. This chapter discusses these peripherals as well because of their importance and because of developments in these products. Note that displays are covered in Chapter 10 and digital cameras are discussed in Chapter 11.

Printers

Printers have always been a mainstream output technology for the computing industry. In earlier times impact and thermal printers were the cornerstone printer technologies most frequently used. Within the last decade this has dramatically shifted to ink jet and laser technologies, which are extremely price sensitive. Also, note that most of the printer business is held by a relatively few large players. In 1998 the worldwide laser printer market was almost 10 million units and the inkjet market was over 45 million units.

While printers are largely thought off as PC peripherals, leading manufacturers have announced that next-generation printer products will connect to set-top boxes and media centers to allow consumers to print from these gateways. This allows consumers to print images and documents stored on the hard-drive of the set-top box. Sleek and slim printer models are being introduced. These devices will no longer be hidden in a home office but rather are a part of the entertainment center.

Due to the widespread use of small office/home office (SOHO) operations, and the corresponding need for consolidation of the desktop area, another extension exists—the multifunction peripheral (MFP). MFPs combine FAX machine, copier, scanner, and printer all into a single unit. An ink jet or laser printer resides at the heart of an MFP output device.

Laser Printer

In the 1980s dot matrix and laser printers were dominant. Inkjet technology did not emerge in any significant way until the 1990s. In 1984 Hewlett Packard introduced a laser printer based on technology developed by Canon. It worked like a photocopier, but with a different light source. A photocopier page is scanned with a bright light, while a laser printer page is scanned with a laser. In a laser printer the light creates an electrostatic image of the page on a charged photoreceptor, and the photoreceptor attracts toner in the shape of an electrostatic charge.

Laser printers quickly became popular due to their high print quality and their relatively low running costs. As the market for lasers developed, competition among manufacturers became increasingly fierce, especially in the production of budget models. Prices dropped as manufacturers found new ways of cutting costs. Output quality has improved with 600-dpi resolution becoming more standard. And form factor has become smaller, making them more suited to home use.

Laser printers have a number of advantages over rival inkjet technology. They produce better quality black text documents than inkjets do, and they are designed more for longevity. That is, they turn out more pages per month at a lower cost per page than inkjets do. The laser printer fits the bill as an office workhorse. The handling of envelopes, cards, and other non-regular media is another important factor for home and business users. Here, lasers once again have the edge over inkjets.

Considering what goes into a laser printer, it is amazing they can be produced for so little money. In many ways, laser printer components are far more sophisticated than computer components. For example:

- The RIP (raster image processor) might use an advanced RISC processor.
- The engineering which goes into the bearings for the mirrors is very advanced.
- The choice of chemicals for the drum and toner is a complex issue.
- Getting the image from a PC's screen to paper requires an interesting mix of coding, electronics, optics, mechanics, and chemistry.

Communication

A laser printer needs to have all the information about a page in its memory before it can start printing. The type of printer being used determines how an image is transferred from PC memory to the laser printer. The crudest arrangement is the transfer of a bitmap image. In this case, there is not much the computer can do to improve on the quality. It can only send a dot for a dot.

However, if the system knows more about the image than it can display on the screen, there are better ways to communicate the data. A standard A4 sheet is 8.5 inches across and 11inches deep. At 300 dpi, that is more than eight million dots, compared with the eight hundred thousand pixels on a 1024 by 768 screen. There is obviously scope for a much sharper image on paper–even more so at 600 dpi where a page can have 33 million dots.

A major way to improve quality is to send a page description consisting of outline/vector information, and to let the printer make the best possible use of this information. For example, if the printer is told to draw a line from one point to another, it can use the basic geometric principle that a line has length but not width. Here, it can draw that line one dot wide. The same holds true for curves that can be as fine as the printer's resolution allows. This principle is called device-independent printing.

Text characters can be handled in the same way since they are made up of lines and curves. But a better solution is to use a pre-described font shape such as TrueType or Type-1 formats. Along with precise placement, the page description language (PDL) may take a font shape and scale it, rotate it, or generally manipulate it as necessary. And there's the added advantage of requiring only one file per font as opposed to one file for each point size. Having predefined outlines for fonts allows the computer to send a tiny amount of information—one byte per character—and produce text in any of several different font styles and sizes.

Operation

An image to be printed is sent to a printer via page description language instructions. The printer's first job is to convert those instructions into a bitmap, an action performed by the printer's internal processor. The result is an image (in memory) where every dot is placed on the paper. Models designated "Windows printers" don't have their own processors, so the host PC creates the bitmap, writing it directly to the printer's memory.

At the heart of the laser printer is a small rotating drum called the organic photo-conducting cartridge (OPC). It has a coating that allows it to hold an electrostatic charge. A laser beam scans across the surface of the drum, selectively imparting points of positive charge onto the drum's surface. They will ultimately represent the output image. The area of the drum is the same as that of the paper onto which the image will eventually appear. Every point on the drum corresponds to a point on the sheet of paper. In the meantime, the paper is passed through an electrically charged wire that deposits a negative charge on it.

On true laser printers, selective charging is accomplished by turning the laser on and off as it scans the rotating drum. A complex arrangement of spinning mirrors and lenses does this. The principle is the same as that of a disco mirror ball. The lights bounce off the ball onto the floor, track across the floor, and disappear as the ball revolves. In a laser printer, the mirror drum spins extremely rapidly and is synchronized with the on/off switching of the laser. A typical laser printer performs millions of on/off switches every second.

The drum rotates inside the printer to build one horizontal line at a time. Clearly, this has to be done very accurately. The smaller the rotation, the higher the resolution down the page. The step rotation on a modern laser printer is typically 1/600th of an inch, giving a 600-dpi vertical resolution rating. Similarly, the faster the laser beam is switched on and off, the higher the resolution across the page.

As the drum rotates to present the next area for laser treatment, the area written on moves into the laser toner. Toner is a very fine black powder negatively charged so that it attracts to the positively charged points on the drum surface. So after a full rotation, the drum's surface contains the whole of the required black image.

A sheet of paper now comes into contact with the drum, fed in by a set of rubber rollers. As it completes its rotation, it lifts the toner from the drum by magnetic attraction. This is what transfers the image to the paper. Negatively charged areas of the drum don't attract toner, so they result in white areas on the paper.

Toner is specially designed to melt very quickly. A fusing system now applies heat and pressure to the imaged paper to adhere the toner permanently. Wax is the toner ingredient that makes it more amenable to the fusion process. It is the fusing rollers that cause the paper to emerge from a laser printer warm to the touch.

The final stage is to clean the drum of any remnants of toner, getting it ready for the cycle to start again. There are two forms of cleaning—physical and electrical. With the first, the toner that was not transferred to the paper is mechanically scraped off the drum and collected in a bin. With electrical cleaning, the drum is covered with an even electrical charge so the laser can write on it again. An electrical element called a corona wire performs this charging. Both the felt pad which cleans the drum and the corona wire need to be changed regularly.

Inkjet Printers

Although inkjets were available in the 1980s, it was only in the 1990s that prices dropped enough to bring the technology to the high street. Canon claims to have invented what it terms "bubble jet" technology in 1977 when a researcher accidentally touched an ink-filled syringe with a hot soldering iron. The heat forced a drop of ink out of the needle and so began the development of a new printing method.

Inkjet printers have made rapid technological advances in recent years. The three-color printer has been around for several years now and has succeeded in making color inkjet printing an affordable option. But as the superior four-color model became cheaper to produce, the swappable cartridge model was gradually phased out.

Traditionally, inkjets have had one massive attraction over laser printers—their ability to produce color—and that is what makes them so popular with home users. Since the late 1990s when the price of color laser printers began to reach levels that made them viable for home users, this advantage has been less definitive. However, in that time the development of inkjets capable of photographic-quality output has done much to help them retain their advantage in the realm of color.

The down side is that, although inkjets are generally cheaper to buy than lasers, they are more expensive to maintain. Cartridges need to be changed more frequently and the special coated paper required to produce high-quality output is very expensive. When it comes to comparing the cost per page, inkjets work out about ten times more expensive than laser printers.

Since the invention of the inkjet, color printing has become immensely popular. Research in inkjet technology is making continual advances with each new product on the market showing improvements in performance, usability, and output quality. As the process of refinement continues, inkjet printer prices continue to fall.

Operation

Like laser printing, Inkjet printing is a non-impact method. Ink is emitted from nozzles as they pass over a variety of possible media. Liquid ink in various colors is squirted at the paper to build up an image. A print head scans the page in horizontal strips, using a motor assembly to move it from left to right and back. Another motor assembly rolls the paper in vertical steps. A strip of the image is printed and then the paper moves on, ready for the next strip. To speed things up, the print head doesn't print just a single row of pixels in each pass, but a vertical row of pixels at a time.

On ordinary inkjets the print head takes about half a second to print a strip across a page. Since A4 paper is about 8.5 inches wide, and inkjets operate at a minimum of 300 dpi, this means there are at least 2,475 dots across the page. Therefore, the print head has about 1/5000th of a second to respond to whether a dot needs printing. In the future, fabrication advances will allow bigger print heads with more nozzles firing at faster frequencies. They will deliver native resolutions of up to 1200 dpi and print speeds approaching those of current color laser printers (3 to 4 ppm in color, 12 to 14 ppm in monochrome).

Drop-on-demand (DOD) is the most common type of inkjet technology. It works by squirting small droplets of ink onto paper through tiny nozzles. It's like turning a hose on and off 5,000 times a second. The amount of ink propelled onto the page is determined by the driver software which dictates which nozzles shoot droplets, and when.

The nozzles used in inkjet printers are hair-fine. On early models they clogged easily. On modern inkjet printers this is rarely a problem, but changing cartridges can still be messy on some machines. Another problem with inkjet technology is a tendency for the ink to smudge immediately after printing. But this has improved drastically during the past few years with the development of new ink compositions.

Thermal Technology

Most inkjets use thermal technology where heat is used to fire ink onto the paper. Here, the squirt is initiated by heating the ink to create a bubble until the pressure forces it to burst and hit the paper. The bubble then collapses as the element cools and the resulting vacuum draws ink from the reservoir to replace the ink that was ejected. This is the method favored by Canon and Hewlett Packard. Thermal technology imposes certain limitations on the printing process. For example, whatever type of ink is used must be resistant to heat because the firing process is heat-based. The use of heat in thermal printers creates a need for a cooling process as well, which levies a small time overhead on the printing process.

Tiny heating elements are used to eject ink droplets from the print head's nozzles. Today's thermal inkjets have print heads containing between 300 and 600 nozzles, each about the diameter of a human hair (approx. 70 microns). These deliver drop volumes of around 8 – 10 pico-liters (a pico-liter is a million millionth of a liter). Dot sizes are between 50 and 60 microns in diameter. By comparison, the smallest dot size visible to the naked eye is around 30 microns. Dye-based cyan, magenta, and yellow inks are normally delivered via a combined CMY print head. Several small color ink drops—typically between four and eight—can be combined to deliver a variable dot size. Black ink, which is generally based on bigger pigment molecules, is delivered from a separate print head in larger drop volumes of around 35 Pl.

Nozzle density that corresponds to the printer's native resolution varies between 300 and 600 dpi, with enhanced resolutions of 1200 dpi increasingly available. Print speed is chiefly a function of the frequency with which the nozzles can be made to fire ink drops. It's also a function of the width of the swath printed by the print head. Typically this is around 12 MHz and half an inch, respectively. This yields print speeds of between 4 to 8 ppm (pages per minute) for monochrome text and 2 to 4 ppm for color text and graphics.

Piezo-electric Technology

Epson's proprietary inkjet technology uses a piezo crystal at the back of the ink reservoir, similar to a loudspeaker cone. It flexes when an electric current flows through it. Whenever a dot is required, a current is applied to the piezo element. The element flexes and forces a drop of ink out of the nozzle.

There are several advantages to the piezo method. The process allows more control over the shape and size of ink droplet release. The tiny fluctuations in the crystal allow for smaller droplet sizes and higher nozzle density. Unlike thermal technology, the ink does not have to be heated and cooled between cycles, which saves time. The ink is tailored more for its absorption properties than it is for its ability to withstand high temperatures. This allows more freedom for developing new chemical properties in inks.

Epson's mainstream inkjets at the time of this writing have black print heads with 128 nozzles, and color (CMY) print heads with 192 nozzles (64 for each color). This arrangement addresses a resolution of 720 by 720 dpi. Because the piezo process can deliver small and perfectly formed dots with high accuracy, Epson can offer an enhanced resolution of 1440 by 720 dpi. However, this is achieved by the print head making two passes which results in a print speed reduction. For use with its piezo technology, Epson has developed tailored inks which are solvent-based and extremely quick-drying. They penetrate the paper and maintain their shape rather than spreading out on the surface and causing dots to interact with one another. The result is extremely good print quality, especially on coated or glossy paper.

Color Perception

Visible light is sandwiched between ultraviolet and infrared. It falls between 380 nm (violet) and 780 nm (red) on the electromagnetic spectrum. White light is composed of approximately equal proportions of all the visible wavelengths. When these shine on or through an object, some wavelengths are absorbed while others are reflected or transmitted. It is the reflected or transmitted light that gives the object its perceived color. For example, leaves are their familiar colors because chlorophyll absorbs light at the blue and red ends of the spectrum and reflects the green part in the middle.

The "temperature" of the light source, measured in Kelvin (K), affects an object's perceived color. White light as emitted by the fluorescent lamps in a viewing box or by a photographer's flashlight has an even distribution of wavelengths. It corresponds to a temperature of around 6,000 K and does not distort colors. Standard light bulbs, however, emit less light from the blue end of the spectrum. This corresponds to a temperature of around 3,000 K and causes objects to appear more yellow.

Humans perceive color via the retina—a layer of light-sensitive cells on the back of the eye. Key retinal cells contain photo-pigments that render them sensitive to red, green, or blue light. (The other light sensitive cells are called rods and are only activated in dim light.) Light passing through the eye is regulated by the iris and focused by the lens onto the retina where cones are stimulated by the relevant wavelengths. Signals from the millions of cones are passed via the optic nerve to the brain which assembles them into a color image.

Creating Color

Creating color accurately on paper has been one of the major areas of research in color printing. Like monitors, printers closely position different amounts of key primary colors which, from a distance, merge to form any color. This process is known as dithering.

Monitors and printers do this slightly differently because monitors are light sources while the output from printers reflects light. Monitors mix the light from phosphors made of the primary additive colors—red, green, and blue (RGB). Printers use inks made of the primary subtractive colors—cyan, magenta, and yellow (CMY). White light is absorbed by the colored inks, reflecting the desired color. In each case the basic primary colors are dithered to form the entire spectrum. Dithering breaks a color pixel into an array of dots so that each dot is made up of one of the basic colors, or left blank.

The reproduction of color from the monitor to the printer output is a major area of research known as color matching. Colors vary from monitor to monitor and the colors on the printed page do not always match up with what is displayed on-screen. The color generated on the printed page is dependent on the color system used and on the model of printer used. It is not generated by the colors shown on the monitor. Printer manufacturers have put lots of money into the research of accurate monitor/printer color matching.

Modern inkjets can print in color and black and white, but the way they switch between the two varies between different models. The number of ink types in the machine determines the basic design. Printers containing four colors—cyan, magenta, yellow, and black (CMYK)—can switch between black and white text and color images on the same page with no problem. Printers equipped with only three colors cannot.

Some of the cheaper inkjet models have room for only one cartridge. You can set them up with a black ink cartridge for monochrome printing or with a three-color cartridge (CMY) for color printing, but you cannot set them up for both at the same time. This makes a big difference to the operation of the printer. Each time you want to change from black and white to color, you must physically swap the cartridges. When you use black on a color page, it will be made up of the three colors. These result in an unsatisfactory dark green or gray color, usually referred to as composite black. However, the composite black produced by current inkjet printers is much better than it was a few years ago due to the continual advancements in ink chemistry.

Print Quality

The two main determinants of color print quality are resolution (measured in dots per inch–dpi), and the number of levels or graduations that can be printed per dot. Generally, the higher the resolution and the more levels per dot, the better the overall print quality.

In practice, most printers make a trade-off. Some opt for higher resolution and others settle for more levels per dot, the best solution depending on the printer's intended use. For example, graphic arts professionals are interested in maximizing the number of levels per dot to deliver photographic image quality. But general business users require reasonably high resolution so as to achieve good text quality as well as good image quality.

The simplest type of color printer is a binary device in which the cyan, magenta, yellow, and black dots are either on (printed) or off (not printed). Here, no intermediate levels are possible. If ink (or toner) dots can be mixed together to make intermediate colors, then a binary CMYK printer can only print eight "solid" colors (cyan, magenta, yellow, red, green and blue, black and white). Clearly, this isn't a big enough palette to deliver good color print quality, which is where half-toning comes in.

Half-toning algorithms divide a printer's native dot resolution into a grid of halftone cells. Then they turn on varying numbers of dots within these cells to mimic a variable dot size. By carefully combining cells containing different proportions of CMYK dots, a half-toning printer can fool the human eye into seeing a palette of millions of colors rather than just a few.

There is an unlimited palette of solid colors in continuous tone printing. In practice, "unlimited" means 16.7 million colors, which is more than the human eye can distinguish. To achieve this, the printer must be able to create and overlay 256 shades per dot per color, which requires precise control over dot creation and placement. Continuous tone printing is largely the province of dye sublimation printers. However, all mainstream-printing technologies can produce multiple shades (usually between 4 and 16) per dot, allowing them to deliver a richer palette of solid colors and smoother halftones. Such devices are referred to as "con-tone" printers.

Recently, six-color inkjet printers have appeared on the market targeted at delivering photographic-quality output. These devices add two further inks—light cyan and light magenta. These make up for current inkjet technology's inability to create very small (and therefore light) dots. Six-color inkjets produce more subtle flesh tones and finer color graduations than standard CMYK devices. However, they are likely to become unnecessary in the future when ink drop volumes are expected to shrink to around 2 to 4 pico-liters. Smaller drop sizes will also reduce the amount of half-toning required because a wider range of tiny drops can be combined to create a larger palette of solid colors.

Rather than simply increasing dpi, long-time market leader Hewlett Packard has consistently supported increasing the number of colors that can be printed on an individual dot to improve print quality. They argue that increasing dpi sacrifices speed and causes problems arising from excess ink, especially on plain paper. In 1996 HP manufactured the DeskJet 850C, the first inkjet printer to print more than eight colors (i.e., two drops of ink) on a dot. Over the years, it has progressively refined its PhotoREt color layering technology to the point where, by late 1999, it could produce an extremely small 5-pl drop size and up to 29 ink drops per dot. This was sufficient to represent over 3,500 printable colors per dot.

Color Management

The human eye can distinguish about one million colors, the precise number depending on the individual observer and viewing conditions. Color devices create colors in different ways, resulting in different color gamut. Color can be described conceptually by a three-dimensional HSB model:

- **Hue (H)** – Describes the basic color in terms of one or two dominant primary colors (red or blue-green, for example). It is measured as a position on the standard color wheel and is described as an angle in degrees between 0 to 360.
- **Saturation (S)** – Referred to as chroma. It describes the intensity of the dominant colors and is measured as a percentage from 0 to 100%. At 0% the color would contain no hue and would be gray. At 100% the color is fully saturated.
- **Brightness (B)** – Describes the color's proximity to white or black, which is a function of the amplitude of the light that stimulates the eye's receptors. It is measured as a percentage. If any hue has a brightness of 0%, it becomes black. At 100% it becomes fully light.

RGB (Red, Green, Blue) and CMYK (Cyan, Magenta, Yellow, Black) are other common color models. CRT monitors use the former, creating color by causing red, green, and blue phosphors to glow. This system is called additive color. Mixing different amounts each of red, green, or blue creates different colors. Each color can be measured from 0 to 255. If red, green, and blue are set to 0, the color is black. If all are set to 255, the color is white.

Applying inks or toner to white paper creates printed material. The pigments in the ink absorb light selectively so that only parts of the spectrum are reflected back to the viewer's eye. This is called subtractive color. The basic printing ink colors are cyan, magenta, and yellow. A fourth ink, black, is usually added to create purer, deeper shadows and a wider range of shades. By using varying amounts of these "process colors" a large number of different colors can be produced. Here the level of ink is measured from 0% to 100%. For example, orange is represented by 0% cyan, 50% magenta, 100% yellow, and 0% black.

The CIE (Commission Internationale de l'Eclairage) was formed early in this century to develop standards for the specification of light and illumination. It was responsible for the first color space model. This defined color as a combination of three axes—x, y, and z. In broad terms, x represents the amount of redness in a color, y the amount of greenness and lightness (bright-to-dark), and z the amount of blueness. In 1931 this system was adopted as the CIE x*y*z model and is the basis for most other color space models. The most familiar refinement is the Yxy model in which the near triangular xy planes represent colors with the same lightness, with lightness varying along the Y-axis. Subsequent developments such as the L*a*b and L*u*v models released in 1978 map the distances between color coordinates more accurately to the human color perception system.

For color to be an effective tool, it must be possible to create and enforce consistent, predictable color in a production chain. The production chain includes scanners, software, monitors, desktop printers, external PostScript output devices, prepress service bureau, and printing presses. The dilemma is that different devices simply can't create the same range of colors. All of this color modeling effort comes into its own in the field of color management. Color management uses the device-independent CIE color space to mediate between the color gamut of the various devices. Color management systems are based on generic profiles of different color devices that describe their imaging technologies, gamut, and operational methods. These profiles are then fine-tuned by calibrating actual devices to measure and correct any deviations from ideal performance. Finally, colors are translated from one device to another, with mapping algorithms choosing the optimal replacements for out-of-gamut colors that cannot be handled.

Until Apple introduced ColorSync as a part of its System 7.x operating system in 1992, color management was left to specific applications. These high-end systems have produced impressive results, but they are computationally intensive and mutually incompatible. Recognizing the problems of cross-platform color, the ICC (International Color Consortium, originally named the ColorSync Profile Consortium) was formed in March 1994 to establish a common device profile format. The founding companies included Adobe, Agfa, Apple, Kodak, Microsoft, Silicon Graphics, Sun Microsystems, and Taligent.

The goal of the ICC is to provide true portable color that will work in all hardware and software environments. It published its first standard—version 3 of the ICC Profile Format—in June 1994. There are two parts to the ICC profile. The first contains information about the profile itself such as what device created the profile, and when. The second is color-metric device characterization which explains how the device renders color. The following year Windows 95 became the first Microsoft operating environment to include color management and support for ICC-compliant profiles via the ICM (Image Color Management) system.

Ink

Whatever technology is applied to printer hardware, the final product consists of ink on paper. These two elements are vitally important to producing quality results. The quality of output from inkjet printers ranges from poor, with dull colors and visible banding, to excellent, near-photographic quality.

Two entirely different types of ink are used in inkjet printers. One is slow and penetrating and takes about ten seconds to dry. The other is a fast drying ink that dries about 100 times faster. The former is better suited to straightforward monochrome printing, while the latter is used for color. Different inks are mixed in color printing so they need to dry as quickly as possible to avoid blurring. If slow-drying ink is used for color printing, the colors tend to bleed into one another before they've dried.

The ink used in inkjet technology is water-based and this poses other problems. The results from some of the earlier inkjet printers were prone to smudging and running, but over the past few years there have been enormous improvements in ink chemistry. Oil-based ink is not really a solution to the problem because it would impose a far higher maintenance cost on the hardware. Printer manufacturers are making continual progress in the development of water-resistant inks, but the results from inkjet printers are still weak compared to lasers.

One of the major goals of inkjet manufacturers is to develop the ability to print on almost any media. The secret to this is ink chemistry, and most inkjet manufacturers will jealously protect their own formulas. Companies like Hewlett Packard, Canon and Epson invest large sums of money in research to make continual advancements in ink pigments, qualities of light-fastness and water-fastness, and suitability for printing on a wide variety of media.

Today's inkjets use dyes based on small molecules (<50 nm) for cyan, magenta, and yellow inks. These have high brilliance and wide color gamut, but aren't sufficiently light-fast or water-fast. Pigments based on larger (50- to 100-nm) molecules are more waterproof and fade-resistant. But they cannot yet deliver the range of colors that dyes do, and they aren't transparent. This means that pigments are currently only used for the black ink. Current and future developments are concentrating on creating water-fast and light-fast CMY inks based on smaller pigment-type molecules.

Paper

Most current inkjet printers require high-quality coated or glossy paper for the production of photo-realistic output, but this can be very expensive. One of the aims of inkjet printer manufacturers is to make color printing media-independent. The attainment of this goal is generally measured by the output quality achieved on plain copier paper. This has vastly improved over the past few years, but coated or glossy paper is still needed to achieve full-color photographic quality. Some printer manufacturers, like Epson, even have their own proprietary paper that is optimized for use with piezo technology.

Inkjet printers can become expensive when printer manufacturers tie you to their proprietary consumables. Paper produced by independent companies is much cheaper than that supplied directly by printer manufacturers. But it tends to rely on its universal properties and rarely takes advantage of the idiosyncratic features of particular printer models.

A great deal of research has gone into the production of universal paper types that are optimized specifically for color inkjet printers. PLUS Color Jet paper, produced by Wiggins Teape, is a coated paper produced specifically for color inkjet technology. Conqueror CX22 is designed for black ink and spot-color business documents, and is optimized both for inkjet and laser printers.

Paper pre-conditioning seeks to improve inkjet quality on plain paper by priming the media to receive ink with an agent that binds pigment to the paper. This reduces dot gain and smearing. A great deal of effort is brought to bear on trying to achieve this without incurring a dramatic performance hit. If this yields results, one of the major barriers to widespread use of inkjet technology will have been removed.

Manageability and Costs

There is no doubt that the inkjet printer has been one of desktop computing's success stories. Its first phase of development was the monochrome inkjet of the late 1980s–a low-cost alternative to the laser printer. The second phase spanned the arrival of color and its development to the point of effective photographic quality. It gave the inkjet an all-round capability unmatched by any other printer technology. But when it comes to manageability and running costs, inkjet trails laser technology by some distance. Inkjet's third phase of development will focus on improving these aspects of the technology.

Hewlett Packard's HP2000C inkjet signaled encouraging progress in this direction. Most inkjet printers combine the ink reservoir and the print head in one unit. When the ink runs out, it's necessary to replace both, even though print heads can have a lifetime many times that of ink reservoirs. The HP2000C differs radically from traditional designs by using a modular system in which the ink cartridges and print heads are kept as separate units. The printer uses four pressurized cartridges which hold 8 cm³ of ink each. They remain static underneath a hinged cover at the front of the printer. These are connected by tubes that are integrated with a standard ribbon-style cable that runs to the print head carriage. Internal smart chips monitor the supply and activate a plunger on the relevant cartridge when it requires a refill. Each ink cartridge can keep track of how much ink it has used and how much remains, even if it is moved between printers. The print heads are also self-monitoring. They trigger an alert when they need to be replaced. The whole system can look at the requirements for a particular print job and start only if there is sufficient ink to complete it.

Wasted ink is also a problem that adversely affects running costs. Printers that combine cyan, yellow, and magenta inks from a single tri-color cartridge require the replacement of a whole cartridge if just one reservoir empties. This must be done regardless of how much ink is left in the other two reservoirs. The solution deployed by a number of printers is to employ a separate, independently replaceable ink cartridge for each color. The downside to this is increased maintenance effort. An inkjet printer that uses four cartridges typically requires twice the attention of a printer that combines the three colors.

The HP2000C includes another innovative feature for manageability. It incorporates a second paper tray so that two different paper types can be kept in the printer to minimize user attention. Like the ability to warn of impending ink depletion, this is essential in a networked environment. Print capacities must also improve. Print speeds have exceeded 10 ppm, and increased cartridge capacities have come with these increased print speeds. Inkjet manufacturers are expected to introduce workgroup color printers with much larger secondary ink containers linked to small primary ink reservoirs located close to or in the print head. These printers will automatically replenish the small primary reservoir from the secondary as needed.

Paper is another area where running cost reductions can be made. Expectations are that the preoccupation with obtaining photographic quality on high-cost glossy paper will diminish. Inkjet technologies will focus on obtaining better results from plain paper for the next generations of inkjet printers.

Other Printers

While lasers and inkjets dominate market share, there are a number of other important print technologies. Solid ink has a significant market presence because it's capable of producing high-quality output on a wide range of media. And thermal wax transfer and dye sublimation play an important role in more specialized printing fields. Dot matrix technology remains relevant in situations where a fast impact printer is required, but it is at a big disadvantage in that it does not support color printing.

Solid Ink

Marketed almost exclusively by Tektronix, solid ink printers are page printers that use solid wax ink sticks in a "phase-change" process. They liquefy wax ink sticks into reservoirs. Then they squirt the ink onto a transfer drum from where it is cold-fused onto the paper in a single pass.

Once they've been warmed up, thermal wax devices should not be moved or wax damage may occur. They are intended to remain powered up in a secure area. They are designed to be shared over a network so they are equipped with Ethernet, parallel, and SCSI ports that allow for comprehensive connectivity.

Solid ink printers are generally cheaper than color lasers. They are economical to run considering their low component count and the Tektronix policy of giving black ink away free. Output quality is good with multi-level dots supported by high-end models. But output quality is generally not as good as the best color lasers for text and graphics. And it's not as good as the best inkjets for photographs. Resolution starts at a native 300 dpi, rising to a maximum of around 850 by 450 dpi. Color print speed is typically 4 ppm in standard mode, rising to 6 ppm in a reduced resolution mode.

Solid ink printers are well suited for general business use. They are also good for specialized tasks such as large-format printing and delivering color transparencies at high speeds. They offer connectivity and relatively low running costs, and they accept the widest range of media of any color printing technology.

Dye-sublimation

Dye-sublimation printers are specialized devices used widely in graphic arts and photographic applications. True dye-subs work by heating the ink so that it turns from a solid into a gas. The heating element can be set to different temperatures to control the amount of ink laid down in one spot. This means that color is applied as a continuous tone rather than in dots as with an inkjet. One color is laid over the whole of one sheet at a time, starting with yellow and ending with black. The ink is on large rolls of film that contain sheets of each color. For example, for an A4 print the film roll contains an A4-size sheet of yellow, followed by a sheet of cyan, and so on. Dye sublimation requires particularly expensive special paper. This is because the dyes are designed to diffuse into the paper surface, mixing to create precise color shades. Print speeds are low, typically between 0.25 and 0.5 ppm.

There are now some inkjet printers on the market that actually deploy dye-sublimation techniques. This type of printer uses the technology differently than a true dye-sub. Its inks are in cartridges that can only cover the page one strip at a time. It uses a heating element to heat the inks to form a gas. The heating element can reach temperatures up to 500° C—higher than the average dye sublimation printer can produce. The proprietary Micro Dry technique employed in Alps' printers is an example of this hybrid technology. These devices operate at 600 to 1200 dpi. In some, the standard cartridges can be swapped for special "photo ink" units to produce photographic-quality output.

Thermo Autochrome

The thermo autochrome (TA) print process is considerably more complex than either inkjet or laser technology. It has emerged recently in printers marketed as companion devices for use with digital cameras. TA paper contains three layers of pigment—cyan, magenta, and yellow—each of which is sensitive to a particular temperature. Of these pigments, yellow has the lowest temperature sensitivity, then magenta, followed by cyan. The printer is equipped with both thermal and ultraviolet heads and the paper is passed beneath these three times. For the first pass, the paper is selectively heated at the temperature necessary to activate the yellow pigment. This is then fixed by the ultraviolet before passing onto the next color (magenta). Although the last pass (cyan) isn't followed by an ultraviolet fix, the end results are claimed to be far more permanent than those obtained with dye-sublimation.

Thermal Wax

Thermal wax is another specialized technology. It is very similar to dye-sublimation and is well suited to transparency printing. It uses CMY or CMYK rolls containing page-sized panels of plastic film coated with wax-based colorants. Thermal wax printers works by melting ink dots. The printers are generally binary, though some higher-end models are capable of producing multi-level dots on special thermal paper. Resolution and print speeds are low—typically 300 dpi and around 1 ppm—which makes them suitable for specialized applications only.

Dot Matrix

Dot matrix was the dominant print technology in the home computing market before inkjet technology emerged. Dot matrix printers produce characters and illustrations by striking pins against an ink ribbon to print closely spaced dots in the appropriate shape. They are relatively expensive and do not produce high-quality output. However, they can print on continuous stationary multi-page forms, something laser and inkjet printers cannot do.

Print speeds, specified in characters per second (cps), vary from about 50 to over 500 cps. Most dot-matrix printers offer different speeds depending on the desired print quality. Print quality is determined by the number of pins (the mechanisms that print the dots). Typically, this varies from between 9 to 24. The best dot-matrix printers (24 pins) are capable of near letter-quality type.

Figure 9.1 illustrates a typical printer, and Figure 9.2 shows a multi-function peripheral.

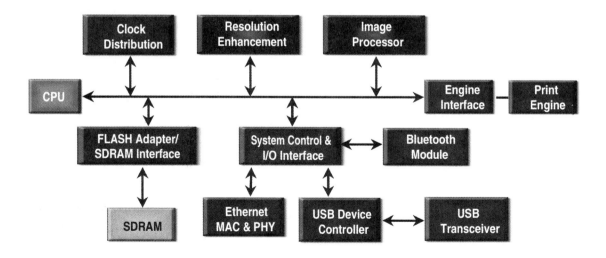

Figure 9.1: Printer Block Diagram

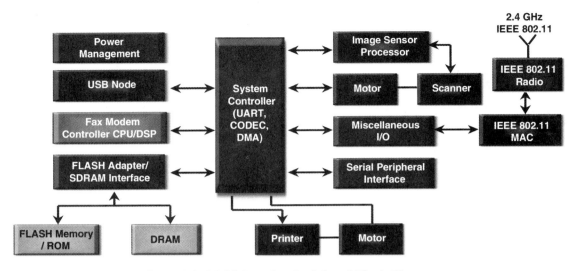

Figure 9.2: Multi-Function Peripheral Block Diagram

Scanners

The scanner should be accessible by any PC in the home, enabling the transfer of scanned images and other files across the home network. Fundamentally, a scanner works like a digital camera. An image is scanned through a lens and onto either a CMOS sensor or a charge-coupled device (CCD). A CCD is an array of light-sensitive diodes. The sensor chip is typically housed on a daughter card along with numerous A/D converters. The CCD and its circuitry create a digital reproduction of the image through a series of photodiodes—each containing red, green, and blue filters—which respond to different ranges of the optical spectrum. Once the picture is scanned, the DSP and pixel co-processor produces a JPEG image that can be displayed on a screen. Most of this work is done by the DSP processing power. Scanners are differentiated by their scanning quality

Digital imaging has come of age. Equipment that was once reserved for expensive applications is now commonplace on the desktop. The powerful PCs required to manipulate digital images are now considered entry level, so it comes as no surprise to learn that scanners are one of the fastest growing markets today.

The list of scanner applications is almost endless. This has resulted in the development of products to meet specialized requirements:

- High-end drum scanners capable of scanning both reflective art and transparencies, from 35- mm slides to 16-foot x 20-in material at high (over 10,000 dpi) resolutions.
- Compact document scanners designed exclusively for OCR and document management.
- Dedicated photo scanners that move a photo over a stationary light source.
- Slide/transparency scanners that pass light through an image rather than reflecting light off of it.
- Handheld scanners for the budget end of the market, or for those with little desk space.

Flatbed scanners are the most versatile and popular format. They are capable of capturing color pictures, documents, and pages from books and magazine. With the right attachments, they can even scan transparent photographic film.

Operation

On the simplest level, a scanner is a device that converts light into 0s and 1s (a computer-readable format). That is, scanners convert analog data into digital data.

All scanners work on the same principle of reflectance or transmission. The image is placed before the carriage, which consists of a light source and sensor. In the case of a digital camera, the light source could be the sun or artificial lighting. When desktop scanners were first introduced, many manufacturers used fluorescent bulbs as light sources, but these have two distinct weaknesses. They rarely emit consistent white light for long. And while they are on, they emit heat which can distort the other optical components. For these reasons, most manufacturers have moved to "cold-cathode" bulbs. These differ from standard fluorescent bulbs in that they have no filament. They operate at much lower temperatures and are more reliable. Standard fluorescent bulbs are now found primarily on low-cost units and older models.

By late 2000, Xenon bulbs emerged as an alternative light source. Xenon produces a very stable, full-spectrum light source that's long lasting and quick to initiate. However, xenon light sources do consume power at a higher rate than cold cathode tubes.

To direct the light from the bulb onto the sensors that read light values, CCD scanners use prisms, lenses, and other optical components. Like eyeglasses and magnifying glasses, these items can vary quite a bit in quality. A high-quality scanner uses high-quality glass optics that are color-corrected and coated for minimum diffusion. Lower-end models typically skimp in this area, using plastic components to reduce costs.

The amount of light reflected by or transmitted through the image and picked up by the sensor is converted to a voltage proportional to the light intensity. The brighter the part of the image, the more light is reflected or transmitted, resulting in a higher voltage. This analog-to-digital conversion (ADC) is a sensitive process that is susceptible to electrical interference and noise in the system. In order to protect against image degradation, the best scanners on the market today use an electrically isolated analog-to-digital converter that processes data away from the main circuitry of the scanner. However, this introduces additional manufacturing costs, so many low-end models include integrated analog-to-digital converters built into the scanner's primary circuit board.

The sensor component is implemented using one of three different technologies:

- **PMT (photo-multiplier tube)** – A device inherited from drum scanner technology.
- **CCD (charge-coupled device)** – The type of sensor used in desktop scanners.
- **CIS (contact image sensor)** – A newer technology which integrates scanning functions into fewer components, allowing scanners to be more compact in size.

Figure 9.3 shows the block diagram of a scanner.

179

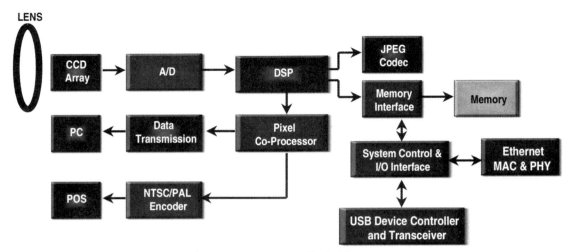

Figure 9.3: Scanner Block Diagram

Smart Card Readers

Instead of fumbling for coins, imagine buying the morning paper using a card charged with small denominations of money. The same card could be used to pay for a ride on public transportation. And after arriving at work, you could use that card to unlock the security door, enter the office, and boot up your PC with your personal configuration. In fact, everything you purchase, whether direct or through the Internet, would be made possible by the technology in this card. It may seem far-fetched but the rapid advancements of semiconductor technologies make this type of card a reality. In some parts of the world, the "smart card" has already started to obsolete cash, coins, and multiple cards. An essential part of the smart card system is the card reader, which is used to exchange or transfer information.

Why is the smart card replacing the magnetic strip card? Because the smart card can hold up to a 100 times more information and data than a traditional magnetic strip card. The smart card is classified as an integrated circuit (IC) card. There are actually two types of IC card—memory cards and smart cards. Memory cards contain a device that allows the card to store various types of data. However, they do not have the ability to manipulate this data. A typical application for memory type cards is a pre-paid telephone card. These cards hold typically between 1 KB and 4 KB of data. A memory card becomes a smart card with the addition of a microprocessor. The key advantage of smart cards is that they are easy to use, convenient, and can be used in several applications. They provide benefits to both consumers and merchants in many different industries by making data portable, secure, and convenient to access.

Components of a Smart Card

A smart card resembles an ordinary credit card, but it has embedded IC's (memory and microcontroller) and contacts for the IC's on one side. It may also include a magnetic strip for conventional transactions. The embedded microcontroller enables the card to make computations and decisions, and to manipulate data. Figure 9.4 shows a block diagram of the smart card internal circuitry.

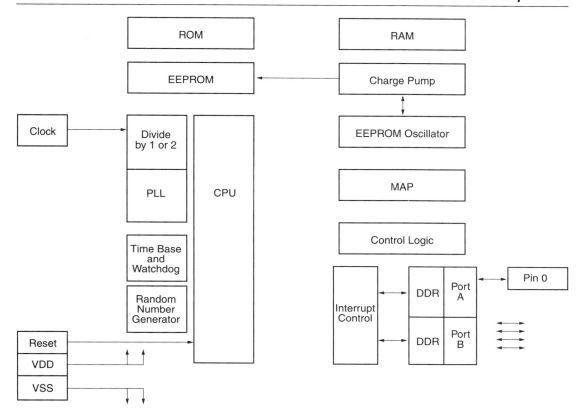

Figure 9.4: Block Diagram of Smart Card Circuitry

A typical smart card consists of an 8-bit microcontroller (MCU), 16 KB of ROM, 512 bytes of RAM, and up to 16 KB of EEPROM or flash memory, all on a single device. There are many types of memory within the card. For temporary data storage, it contains RAM which is only used when power is applied (usually when the card is in contact with the reader). It also contains ROM which stores fixed data and the operating system. The use of non-volatile memory such as EEPROM or Flash memory is ideal for storing data that changes, such as an account PIN or transaction data. This type of data must remain stored once power is removed.

Manufacturers of smart cards are moving to a 32-bit microprocessor to increase processing power and handle more applications.

There are two types of smart card:

■ **Contact smart cards** – These require insertion into a smart card reader.
■ **Contact-less smart cards** – These require only close proximity to an antenna.

The contact smart card has a small gold chip about one-half inch in diameter on the front (instead of a magnetic strip on the back like a credit card). When the card is inserted into a smart card reader, it makes contact with the electrical connectors that read information from the chip and write to the chip.

A contact-less smart card looks like a typical credit card, but it has a built-in microprocessor and an antenna coil that enables it to communicate with an external antenna. Contact-less smart cards are used when transactions must be processed quickly, as in mass-transit toll collection. The "combicard" is a single card that functions as both a contact and contact-less card.

Smart cards have several advantages over traditional magnetic strip cards:

- Proven to be more reliable than the magnetic strip card.
- Can store up to 100 times more information than the magnetic strip card.
- Reduce tampering and counterfeiting through high security mechanisms.
- Can be reusable.
- Have a wide range of applications (banking, transportation, health care, etc.).
- Are compatible with portable electronics (PCs, telephones, PDAs, etc.).
- Can store many types of information (finger print data, credit, debit and loyalty card details, self-authorization data, access control information, etc.).

History of Smart Cards

Bull CP8 and Motorola developed the first "smart card" in 1977. It was a two-chip solution consisting of a microcontroller and a memory device. Motorola produced a single chip card called the SPOM 01.

Smart cards have taken off at a phenomenal rate in Europe by replacing traditional credit cards. The key to smart card success has been its ability to authorize transactions off-line. A smart card stores the "charge" of cash, enabling a purchase up to the amount of money stored in the card. Motorola's single chip solution was quickly accepted into the French banking system. It served as a means of storing the cardholder's account number and personal identification numbers (PIN) as well as transaction details. By 1993 the French banking industry completely replaced all bankcards with smart cards.

In 1989 Bull CP8 licensed its smart card technology for use outside the French banking system. The technology was then incorporated into a variety of applications such as Subscriber Identification Modules (SIM cards) in GSM digital mobile phones. In 1996 the first combined modem/smart card reader was introduced. We will probably soon see the first generation of computers that read smart cards as a standard function.

In May 1996 five major computer companies (IBM, Apple, Oracle, Netscape, and Sun) proposed a standard for a "network computer" designed to interface directly with the Internet, and it has the ability to use smart cards. Also in 1996 the alliance between Hewlett Packard, Informix, and Gemplus was launched to develop and promote the use of smart cards for payment and security on all open networks.

Besides e-commerce, some smart card applications are:

- Transferring favorite addresses from a PC to a network computer
- Downloading airline ticket and boarding pass
- Booking facilities and appointments via Websites
- Storing log-on information for using any work computer or terminal

Smart Card Market Potential

Smart card usage worldwide is increasing at an extraordinary rate. There are already 2 billion smart card units in circulation. A leading smart card reader manufacturer (Gemplus) predicts that by 2018 there will be 5 billion phone cards alone. And this does not include smart cards for medical information, personal ID, loyalty information, etc. All of these smart cards in circulation require smart card readers.

To date, Europe has dominated the smart card industry in both production and usage. The region produces as much as 90% of the world's smart cards and consumes about two thirds. However, Europe's share of the smart card market has been declining as the cards have started gaining popularity in other parts of the world. By the end of the decade, smart card usage is expected to be evenly split Europe, the Americas, and Asia.

College campus smart cards have been a major success in the U.S. These multi-application smart card systems provide students with services such as

- ID
- Parking
- ATM access
- Library check out
- Dormitory access
- Payment services at vending machines, laundry, telephones, and book stores

The number of students carrying smart cards in the U.S. has grown to more than 1 million since 1996. This represents approximately one in 17 students. The growth in campus cards is producing a generation of people who already understand the benefits of smart card technology and who may be more inclined to use them in larger, open system applications.

Smart Card Applications

- Smart cards are being used in applications ranging from stored value cards (SVCs) to transportation, medical, and identification cards. As they become cheaper to produce, disposable smart cards will become available alongside long-term use cards, such as multi-function credit cards. The low-cost disposable smart card will become commonplace in applications such as one-day travel cards, flight tickets, and even concert tickets. Smart card applications are increasing daily in this rapidly growing area. The following list describes some of the ways smart cards are being used today.

- **Stored Value Cards (SVCs)** – Also known as electronic purses, they are being championed by companies such as Mondex International (Mondex), Banksys (Proton), and Chipper International (Chipper). They allow small denominations of money to be stored on the card in various currencies. SVCs can be used for small value purchases where it is inappropriate to use a credit card. The SVC needs to be 'charged' with cash. As each transaction is completed, the appropriate amount is deducted until the card is empty. Some companies are producing small key-ring type readers with a small display that can read the amount left on a SVC. For example, a single card could be used as a credit card, debit card, SVC, access card, video rental card, and medical record file. Other applications for SVCs include vending machines, parking meters, pay TV, cinemas, and convenience stores.

- **Phone Cards** – Public pay phones are beginning to replace the coin slot with a smart-card reader. Using smart cards to pay for calls reduces the need to carry cash and prevents theft from the phone company. This type of stored value card counts down the money spent on each call. It can then be re-charged or disposed of when it's empty.

- **Health Care Cards** – Health care cards can store pertinent information such as:
  - Cardholder's doctor
  - Blood type
  - Allergic reactions

- Medications
- Next of kin
- Emergency telephone numbers
- Dental records
- Health card details
- Scheduled medical visits

This type of card has proved to be very popular in Germany, with most of the population carrying one.

■ **Transportation Cards** – It is estimated that there are 20 billion commuter transactions worldwide. All these transactions take time, so the need for smart card technology is increasing, especially for contact-less transactions. For example, smart cards could drastically reduce the time it takes to pay for a subway ticket and pass through the security barrier. As you pass through a turnstile, you could hold your card next to the reader. The card would be read, money would be deducted, and the entrance barrier would be opened.

■ **Pre-pay Utility Meter Cards** – This type of card is very popular in regions with a lot of seasonal worker movement. They are also good for short-term tenant agreements where utilities (water, electricity, and gas) must be paid for in advance or as-used. The card is "charged" with money and then inserted into the utility card reader. Money is deducted for a certain amount of gas, electricity, or water. The meter usually shows as a countdown of how much fuel is left. This type of utility payment has proved very popular in South Africa.

■ **Personal ATM** – Public and private telephones and PCs with smart card readers could make personal ATMs possible. This would allow users to load funds onto their smart cards from their bank accounts. Or they could top up the limit of a pre-authorized debit card. Financial institutions can make these facilities available wherever there is a phone, without the need for costly traditional ATMs. These telephone transactions can enable funds to be transferred from person to retailer, bank to account holder, and even person to person.

Smart Card System Architecture

The complete smart card system consists of the card, the card reader, and the operating software. The smart card contains the appropriate IC's to store and manipulate data. When the card is inserted into a reader, the reader communicates with the smart card microcontroller to perform authentication functions. The card then performs the requested function such as a deduction of a stored value. The card is removed when the transaction is complete, and the reader is ready to accept the next card transaction.

Standardization plays an important role in the interaction between the card and the reader. For example, the need to carry a separate smart card for each merchant purchased from is inefficient. Consumers already carry several different cards including video rental cards, discount shopping cards, and credit and debit cards. Standardization in cards and card readers would allow for one card to be accepted and read by different readers.

A multi-functional operating system is the key to a multi-application card. It would prevent interference among the programs. JavaCard (Sun Microsystems) and MULTOS™ are two operating systems that let cards perform multiple functions. Some companies support Java and some support Multos, but fortunately for users, they are compatible with each other. Microsoft Smart Card for Windows is another provider of OS software for smart cards.

MULTOS™ is a multi-application operating system for smart cards; the MAOSCO consortium controls its specifications. The main elements of the MULTOS™ architecture are a virtual machine (MEL interpreter) and an application loader. Smart card applications are coded in MEL (MULTOS Executable Language) and are hardware independent. In contrast to the JavaCard, the MULTOS™ virtual machine (VM) is completely realized inside the card. Generally, there is a distinction between off-card and on-card VMs. MULTOS™ offers a 100% on-card VM with firewalls between the applications. With this design, MULTOS™ meets the highest security criteria.

The application loader uses a process called dynamic loading and unloading. This process ensures secure smart card application loading and deletion, to and from the EEPROM. This also applies to cards already issued. The application loader, the loading procedures, and data formats are all part of the MULTOS™ specifications.

Recently, many companies have teamed up to provide smart card users with extra benefits. Beenz.com, creators of the Web currency beenz, and Mondex International have teamed up to work on the development of a Multos smart card capable of carrying Mondex e-cash, beenz, and complimentary e-commerce services. The companies envision that the card will be used with PCs, wireless devices, and digital TV, as well as in the high street. Mondex is cash in electronic form and is particularly suited to high volume, low value payments in the real and virtual worlds. Beenz is a universal web currency. Beenz cannot be bought directly by consumers, but earned on-line by consumers visiting, interacting with, or shopping at web sites. It can be spent with participating merchants on thousands of products. Beenz now have the potential to be earned in the real world as well as on-line in the virtual world.

American Express Blue Card has launched the first mainstream U.S. credit card that has both a magnetic strip and a smart card chip. The company is offering a free "on-line wallet." It stores personal information that can be secured with a free smart card reader that connects to a user's PC. A PIN unlocks the wallet on-line when the smart card is inserted into the reader. A major benefit is that users do not have to type in their account details every time they wish to purchase on-line.

The loyalty card is the predicted killer-application for smart cards. This application is really taking off in the U.S. with drug chain Rite-Aid and movie rental chain Blockbuster leading the way. The smart card loyalty scheme is transacted at the point of sale (POS) terminal. It offers discounts and other incentives according to loyalty points earned. Rite-Aid also sells a smart card gift certificate which is redeemable (with stored value programmed into the chip) for merchandise in the store.

What is a Smart Card Reader?

An essential part of the smart card system is the card reader. The card reader exchanges or transfers information. There are several types of card readers, each specific to a particular application. One type connects to a PC to purchase via the Internet (e-commerce) or to load money onto a smart card through on-line banking. Another type is a handheld wireless terminal used by taxi drivers and in restaurants.

There are two main categories of smart card reader—contact and contact-less. The difference between them is that the contact-less version contains an antenna coil. It sends and receives data without the need to make contact with the smart card. The basic functionality of both types of smart card reader is the same. With the introduction of the combicard (both contact and contact-less functions on one card), the card reader may accept both types of card interrogation and interaction.

The contact smart card tends to be used for cash transactions, whereas the contact-less type is used more for access control applications.

Smart card readers for use with PCs would use a PC for data input and a monitor for viewing card status. Smart card reader functions may also be integrated into cash registers, vending machines, public payphones, set top boxes, mobile phones, utility meters, and so forth.

Figure 9.5 shows the basic components of a smart card reader.

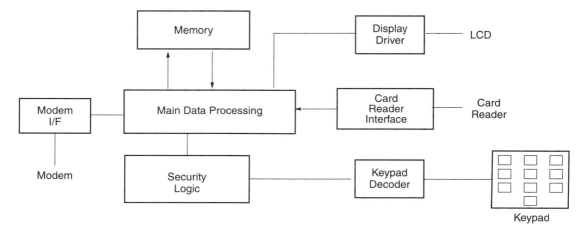

Figure 9.5: Smart Card Reader Block Diagram

The functional blocks that make up the system are:
- **Main data processing** – This is typically a 16- or 32-bit microprocessor for computational functions.
- **Memory** – Stores data (operating system and variable/data storage) and microprocessor boot code.
- **Security logic** – Aids data encryption.
- **Card reader interface** – Supports a smart card reader (contact and contact-less) and a magnetic card reader.
- **Keypad and keypad decoder** – Enters PINs, data, and the associated logic necessary to decode character input.
- LCD display driver
- **Modem and modem interface** – Interfaces to wireless, cellular, and radio modems (usually PCMCIA type).

To describe the functions of the smart card reader, let's review a typical consumer transaction. This is shown pictorially in Figure 9.6.

The merchant inserts the smart card into the card reader and power is applied to the card. The reader communicates with the smart card MCU to perform the card authentication cycle. During the initial read function, the smart card interface logic passes the data to the card reader microprocessor via the security logic. The reader then instructs the user to enter a PIN via a message on the LCD. The user enters his/her PIN via the keypad. This is authenticated by the reader microprocessor. The PIN is verified by the microcontroller in the card which compares the entered PIN to the PIN stored

Figure 9.6: Typical Customer Transaction

in its RAM. If the comparison is negative, the CPU refuses to work. The smart card keeps track of how many wrong PINs are entered. If it is over a predetermined number, the card blocks itself against any future use.

If the PIN is verified, the merchant enters the amount to be deducted. If the smart card has a stored value matching the amount, the deduction of the stored value occurs. If the transaction is not a SVC transaction, the amount to be debited from the bank account is verified using the modem (wireless, cellular, or radio). When the transaction is complete, the card is ejected and removed. The smart card reader is then ready for the next transaction.

The flow of information between an interface device and a smart card occurs via transport protocols in the form of command-response pairs. In most cases the card reader has the role of master. That is, the card reader generates and processes the commands. Smart cards are making high technology applications of the 21st century easier and more secure for everyone. The smart card market is on the brink of realizing its full worldwide potential. This will be realized not only in GSM phones, but on a wider scale for cash-less transactions, loyalty schemes, access control systems, medical record cards, etc. This year we will see personal computers shipped with smart card readers as standard equipment. This combination will unlock widespread worldwide acceptance of multi-application smart cards. Handheld battery-powered smart card readers in taxis and buses will become common place.

Next-generation smart card manufacturers are looking to integrate fingerprint-sensors for market resurgence. The interest in fingerprint sensor technology is reemerging after years of being relegated to a niche application due to cost and technology limitations. Yet there is evidence that fingerprint ID devices may be breaking out of their niche. Some players incorporate the sensors into consumer-based systems like laptops, and sensor vendors tout improved ruggedness and accuracy.

Keyboards

A computer keyboard is an array of switches, each of which sends the PC a unique signal when it is pressed. Mechanical and rubber membranes are the two types of switches most commonly used. Mechanical switches are spring-loaded "push to make" types. They complete the circuit when pressed, and break it when released. These are the types used in click-type keyboards that provide tactile feedback.

Membranes are composed of three sheets. The first has conductive tracks printed on it, the second is a separator with holes in it, and the third is a conductive layer with bumps on it. A rubber mat covering provides the springy feel. When a key is pressed it pushes the two conductive layers together to complete the circuit. On top is a plastic housing that includes sliders to keep the keys aligned.

The force displacement curve is an important keyboard factor. It describes how much force is needed to depress a key, and how this force varies during the key's downward travel. Research shows that most people prefer 80 g to 100 g. Game consoles may go to 120 g or higher, while other keys could be as low as 50 g.

The keys are connected as a matrix and their row and column signals feed into the keyboard's microcontroller chip. The chip is mounted on a circuit board inside the keyboard and interprets the signals with its built-in firmware program. For example, a particular key press might signal row 3, column B which the controller decodes as an 'A' and sends the appropriate code to the PC. These scan codes are defined as standard in the PC's BIOS, though the row and column definitions are specific only to that particular keyboard.

Increasingly, keyboard firmware is becoming more complex as manufacturers make their keyboards more sophisticated. It is not uncommon for a programmable keyboard in which some keys have switchable multiple functions to need 8 KB of ROM to store its firmware. Most programmable functions are executed through a driver running on the PC.

A keyboard's microcontroller is also responsible for negotiating with the keyboard controller in the PC. It reports its presence and allows PC software to do things like toggle the status light on the keyboard. The two controllers communicate asynchronously over the keyboard cable.

Many ergonomic keyboards angle the two halves of the main keypad to allow the elbows to rest in a more natural position. Apple's Adjustable Keyboard has a wide, gently sloping wrist rest that splits down the middle. It enables the user to find the most comfortable typing angle. It has a detachable numeric keypad so the user can position the mouse closer to the alphabetic keys. Cherry Electrical sells a similar split keyboard for the PC. It is the Microsoft Natural Keyboard which sells in the largest volumes and is one of the cheapest. This keyboard also separates the keys into two halves. The manufacturer claims its undulating design accommodates the natural curves of the hand.

Mice

In the early 1980s the first PCs were equipped with only a keyboard. By the end of the decade a mouse device had become an essential for running the GUI-based Windows operating system. The most common mouse used today is optoelectronic. It has a steel ball for weight and it is rubber-coated for grip. As it rotates it drives two rollers, one each for x and y displacement. A third spring-loaded roller holds the ball in place against the other two. These rollers turn two disks that have radial slots cut in them. Each disk rotates between a photo-detector cell, and each cell contains two offset light-emitting diodes (LEDs) and light sensors. As the disk turns, the sensors see the light appear to flash, indicating movement. The offset between the two light sensors shows the direction of movement.

There is a switch inside the mouse for each button, and also a microcontroller that interprets the signals from the sensors and the switches. It uses its firmware to translate them into packets of data that are sent to the PC. Serial mice use 12V power. They also use an asynchronous protocol from Microsoft composed of three bytes per packet to report x and y movement and button presses. PS/2 mice use 5 volts and an IBM-developed communications protocol and interface.

1999 saw the introduction of the most radical mouse design advancement since its first appearance in 1968. This was Microsoft's revolutionary IntelliMouse. Gone were the mouse ball and the other moving parts inside the mouse used to track mechanical movement. They were replaced by a tiny CMOS optical sensor—the same chip used in digital cameras—and an on-board digital signal processor (DSP).

Called the IntelliEye, this infrared optical sensor emits a red glow beneath the mouse. It captures high-resolution digital snapshots at the rate of 1,500 images per second. These images are compared by the DSP and translated into on-screen pointer movements. The technique, called image correlation processing, executes 18 million instructions per second (MIPS). It results in smoother, more precise pointer movement. The absence of moving parts means that the mouse's traditional enemies (food crumbs, dust, grime, etc.) are all but avoided. The IntelliEye works on nearly any surface such as wood, paper, and cloth. It does have some difficulty, though, with reflective surfaces such as CD jewel cases, mirrors, and glass.

Newer optical mice are more accurate and less fussy than the roller-ball types. With optical sensors and a DSP chip, the latest Wireless IntelliMouse Explorer has a video sensor. It can take 6,000 video snapshots per second of the work surface, allowing smooth and precise cursor movement on the screen. And this can all be done wirelessly.

Summary

A peripheral device is an external device that attaches permanently or temporarily to a PC to provide a specific functionality. Devices such as printers, scanners, smart card readers, keyboards, and mice are examples of peripheral devices. Each device provides a service that enables consumers to input or output information.

Digital Displays

Introduction

Many recent advances have been made in flat-panel displays, also known as digital displays. Consumer interest runs high for the sharpest, largest screen money can buy. From high-definition television to scientific and engineering workstations to handheld personal organizers and cell phones, end users are asking for high resolution, great color, and small form factor, all at the lowest possible cost. Such expectations have display designers exploring innovative schemes for representing information in a flat-panel format. Significantly, the only non-flat-panel technology in use is the cathode ray tube (CRT), which can still offer a competitive performance/cost ratio but is becoming increasingly limited in terms of the wide variety of emerging applications. Three essential conditions driving the flat panel display market are its technological and commercial readiness (in view quality, manufacturing efficiency, and system integration), the substantial capacity investment, and the lower-end selling prices.

To provide such advances, manufacturers are looking at a various digital display technologies. Each technology, however, provides advantages that are ideal for only a few applications. Liquid crystal displays may be ideal for notebook PCs and desktop monitors, but Plasma Display Panels (PDPs) provide a good mid-range between high-definition TV and conventional television monitors. Meanwhile, LCoS (liquid-crystal-on-silicon) chips are looking to provide the essential low-cost driver for a new generation of rear-projection high-definition TV systems. In a similar vein, vendors are pursuing organic LED technology as an enabler for robust, small, and low-cost display elements in portable consumer appliances. Organic LEDs are easily processed onto a variety of substrates, and produce high-quality images when magnified by compact optical systems. Of course, all of these technologies will eventually run into more-exotic variants such as active-matrix liquid-crystal panel and thick dielectric electro-luminescent (TDEL) technologies. These technologies and their manufacturers are after a rich market with a production of 8 million units in 2003 and an estimate of 13 million units in 2007. Production location for digital displays will probably be the Asia Pacific region in order to gain maximum cost benefits.

Whatever the actual system used to display information, one continually increasing demand is the amount of information being fed to displays. Larger size, higher resolution, low power, and real-time video requirements all feed into this data bottleneck. Migration from the bulky CRT monitors to digital displays, however, is still occurring due to the reduced form factor, lower weight, and better pictures and sound of digital displays.

CRTs—Cathode Ray Tubes

In an industry in which development is so rapid, it is somewhat surprising that the technology behind monitors and televisions is over 100 years old. The CRT was developed by Ferdinand Braun, a German scientist, in 1897, but was not used in the first television sets until the late 1940s. Although the CRTs found in modern monitors have undergone modifications to improve picture quality, they still follow the same basic principles. Despite predictions of its impeding demise, it looks as if the CRT will maintain its dominance in the PC monitor and television market for the near future.

Anatomy of the CRT

A CRT is an oddly shaped, sealed glass bottle with no air inside. Its overall shape begins with a slim neck at one end, and gradually spreads outward (similar to the shape of a funnel) until it forms a large base at the other end. The surface of the CRT base, which is the monitor's screen, can be either almost flat or slightly convex, and is coated on the inside with a matrix of thousands of tiny phosphor dots. Phosphors are chemicals that emit light when excited by a stream of electrons, and different phosphors emit different colored light. Each dot consists of three blobs of colored phosphor, one red (R), one green (G), and one blue (B). These groups of three phosphors make up what is known as a single pixel.

The "bottleneck" of the CRT contains an electron gun, which is composed of a cathode, a heat source, and focusing elements. Color monitors have three separate guns, one for each phosphor color—R, G, and B. Combinations of different intensities of red, green, and blue phosphors can create the illusion of millions of colors. This effect is called additive color mixing, and is the basis for all color CRT displays.

Images are created when electrons are fired from the electron gun converge to strike their respective phosphor blobs (triads), and each is illuminated to a greater or lesser extent. When electrons strike the phosphor, light is emitted in the color of the individual phosphor blobs. The electron gun must be heated by a built-in heater before it will liberate electrons (negatively charged) from the cathode. The emitted electrons are then narrowed into a tiny beam by the focus elements. The electrons are drawn toward the phosphor dots by a powerful, positively charged anode, located near the screen.

The phosphors in a group are so close together that the human eye perceives the combination as a single colored pixel. Before the electron beam strikes the phosphor dots, it travels through a perforated sheet located directly in front of the phosphor layer known as the "shadow mask." Its purpose is to "mask" the electron beam, forming a smaller, more rounded point that can strike individual phosphor dots cleanly and minimize "over-spill," a condition in which the electron beam illuminates more than one dot.

The electron beam is moved around the screen by magnetic fields generated through a deflection yoke that encircles the narrow neck of the CRT. The beam always begins moving from the top left corner (as viewed from the front), and sweeps from left to right across the row of pixels. The beam is controlled so that it will flash on and off as it sweeps across the row, or "raster," allowing energized electrons to collide only with the phosphors that correlate to the pixels of the image that is to be created on the screen. These collisions convert the electron's energy into light. Once a pass has been completed, the electron beam moves down one raster and begins again from the left side. This process is repeated until an entire screen is drawn, at which point the beam returns to the top left to start again.

The most important aspect of any monitor is that it should give a stable display at the chosen resolution and color palette. A screen that shimmers or flickers, particularly when most of the picture is showing white, can cause itchy or painful eyes, headaches, and migraines. It is also important that the performance characteristics of a monitor be carefully matched with those of the graphics card driving it. It is of no consequence to have an extremely high performance graphics accelerator, capable of ultra high resolutions at high flicker-free refresh rates, if the monitor cannot lock onto the signal.

A monitor's three key specifications are the maximum resolution it will display, at what refresh rate, and whether this is in interlaced or non-interlaced mode.

Resolution and Refresh Rate

The resolution of a monitor is the number of pixels used by the graphics card to describe the desktop (the display area of the monitor), expressed as a horizontal-by-vertical figure. Standard VGA resolution is 640 x 480 pixels. The most common SVGA resolutions are 800 x 600 and 1024 x 768 pixels.

Refresh rate, or vertical frequency, is measured in hertz (Hz), and represents the number of frames displayed on the screen per second. Too few, and the eye will notice the intervals in between frames and perceive a flickering display. The worldwide accepted refresh rate for a flicker-free display is 70 Hz and above, although standards bodies such as the Video Electronics Standards Association (VESA) are pushing for higher rates of 75 Hz or 80 Hz.

A computer's graphics circuitry creates a signal based on the Windows desktop resolution and refresh rate. This signal is known as the horizontal scanning frequency, or HSF, and is measured in kHz. Raising the resolution and/or refresh rate increases the HSF signal. A multi-scanning or "autoscan" monitor is capable of locking on to any signal that lies between a minimum and maximum HSF for the monitor. If the signal falls out of the monitor's range, it will not be displayed.

CRT Mask Technologies

The different CRT mask technologies include interlacing, dot pitch, dot trio, aperture grill, slotted mask, and enhanced dot pitch.

Interlacing

An interlaced monitor is one in which the electron beam draws every other line, say one, three, and five until the screen is full, then returns to the top to fill in the even blanks (say lines two, four, six, and so on). An interlaced monitor offering a 100-Hz refresh rate only refreshes any given line 50 times a second, giving an obvious shimmer. A non-interlaced (NI) monitor draws every line before returning to the top for the next frame, resulting in a far steadier display. A non-interlaced monitor with a refresh rate of 70 Hz or greater is necessary to be sure of a stable display.

Masks and Dot Pitch

The maximum resolution of a monitor is dependent on more than just its highest scanning frequencies. Another factor is dot pitch, the physical distance between adjacent phosphor dots of the same color on the inner surface of the CRT. Typically, this is between 0.22 mm and 0.3 mm. The smaller the number, the finer and better resolved the detail. However, trying to supply too many pixels to a monitor without a sufficient dot pitch to cope causes the very fine details, such as the writing beneath icons, to appear blurred.

There is more than one way to group three blobs of colored phosphor—indeed, there is no reason why they should even be circular blobs. A number of different schemes are currently in use, and care needs to be taken in comparing the dot pitch specification of the different types. With standard dot masks, the dot pitch is the center-to-center distance between two nearest-neighbor phosphor dots of the same color, which is measured along a diagonal. The horizontal distance between the dots is 0.866 times the dot pitch. For masks that use stripes rather than dots, the pitch equals the horizontal distance. This means that the dot pitch on a standard dot-mask CRT should be multiplied by 0.866 before it is compared with the dot pitch of these other types of monitors.

The difficulty in directly comparing the dot pitch values of different displays means that other factors—such as convergence, video bandwidth, and focus—are often a better basis for comparing monitors than dot pitch.

Dot Trio

The vast majority of computer monitors use circular blobs of phosphor and arrange them in triangular formation. These groups are known as "triads" and the arrangement is a dot trio design. The shadow mask is located directly in front of the phosphor layer—each perforation corresponding with phosphor dot trios—and assists in masking unnecessary electrons, avoiding over-spill and resultant blurring of the final picture.

Because the distance between the source and the destination of the electron stream towards the middle of the screen is less than at the edges, the corresponding area of the shadow mask gets hotter. To prevent it from distorting and redirecting the electrons incorrectly, manufacturers typically construct it from Invar, an alloy with a very low coefficient of expansion.

This is all very well, except that the shadow mask used to avoid overspill occupies a large percentage of the screen area. Where there are portions of mask, there is no phosphor to glow and less light means a duller image. The brightness of an image matters most for full-motion video, and with multimedia becoming an increasingly important market consideration, a number of improvements make the dot-trio mask designs brighter. Most approaches to minimizing glare involve filters that also affect brightness. The new schemes filter out the glare without affecting brightness as much.

Toshiba's Microfilter CRT places a separate filter over each phosphor dot and makes it possible to use a different color filter for each color dot. Filters over the red dots, for example, let red light shine through, but they also absorb other colors from ambient light shining on screen – colors that would otherwise reflect as glare. The result is brighter, purer colors with less glare. Other companies are offering similar improvements. Panasonic's Crystal Vision CRTs use a technology called dye-encapsulated phosphor, which wraps each phosphor particle in its own filter. ViewSonic offers an equivalent capability as part of its new SuperClear screens.

Aperture Grill

In the 1960s, Sony developed an alternative tube technology known as Trinitron. It combined the three separate electron guns into one device; Sony refers to this as a Pan Focus gun. Most interesting of all, Trinitron tubes were made from sections of a cylinder, vertically flat and horizontally curved, as opposed to conventional tubes using sections of a sphere, which are curved in both axes. Rather than grouping dots of red, green, and blue phosphor in triads, Trinitron tubes lay their colored phosphors down in uninterrupted vertical stripes.

Consequently, rather than use a solid perforated sheet, Trinitron tubes use masks that separate the entire stripes instead of each dot; Sony calls this the "aperture grill." This replaces the shadow mask with a series of narrow alloy strips that run vertically across the inside of the tube. Rather than using conventional phosphor dot triplets, aperture grill-based tubes have phosphor lines with no horizontal breaks, and so rely on the accuracy of the electron beam to define the top and bottom edges of a pixel. Since less of the screen area is occupied by the mask and the phosphor is uninterrupted vertically, more of the screen can glow, resulting in a brighter, more vibrant display. The equivalent measure to dot pitch in aperture grill monitors is known as "stripe pitch."

Because aperture grill strips are very narrow, there is a possibility that they might move due to expansion or vibration. In an attempt to eliminate this, horizontal damper wires are fitted to increase stability. This reduces the chances of aperture grill misalignment, which can cause vertical streaking and blurring. The down-side is that damper wires obstruct the flow of electrons to the phosphors and are therefore just visible upon close inspection. Trinitron tubes below 17-inch or so manage with one wire, while the larger models require two. A further down-side is mechanical instability. A tap on the side of a Trinitron monitor can cause the image to wobble noticeably for a moment. This is understandable given that the aperture grill's fine vertical wires are held steady in only one or two places horizontally. Mitsubishi followed Sony's lead with the design of its similar Diamondtron tube.

Slotted Mask

Capitalizing on the advantages of both the shadow mask and aperture grill approaches, NEC has developed a hybrid mask type that uses a slot-mask design borrowed from a TV monitor technology that originated in the late 1970s by RCA and Thorn. All non-Trinitron TV sets use elliptically shaped phosphors grouped vertically and separated by a slotted mask.

In order to allow more electrons through the shadow mask, the standard round perforations are replaced with vertically aligned slots. The design of the trios is also different, and features rectilinear phosphors that are arranged to make best use of the increased electron throughput.

The slotted mask design is mechanically stable due to the crisscross of horizontal mask sections but exposes more phosphor than a conventional dot-trio design. The result is not quite as bright as an aperture grill but much more stable and still brighter than dot-trio. It is unique to NEC, and the company capitalized on the design's improved stability in early 1996 when it fit the first ChromaClear monitors to come to market with speakers and microphones and claimed them to be "the new multimedia standard."

Enhanced Dot Pitch

Developed by Hitachi, the largest designer and manufacturer of CRTs in the world, Enhanced Dot Pitch (EDP) mask technology came to market in late 1997. Enhanced Dot Pitch technology takes a slightly different approach, concentrating more on the phosphor implementation than the shadow mask or aperture grill.

On a typical shadow mask CRT, the phosphor trios are more or less arranged equilaterally, creating triangular groups that are distributed evenly across the inside surface of the tube. Hitachi has reduced the distance between the phosphor dots on the horizontal, creating a dot trio that's more akin to an isosceles triangle. To avoid leaving gaps between the trios, which might reduce the advantages of this arrangement, the dots themselves are elongated, and are therefore oval rather than round.

The main advantage of the EDP design is increased clarity, which is most noticeable in the representation of fine vertical lines. In conventional CRTs, a line drawn from the top of the screen to the bottom will sometimes "zigzag" from one dot trio to the next group below, and then back to the one below that. Bringing adjacent horizontal dots closer together reduces zigzag, and has an effect on the clarity of all images.

CRT Electron Beam

If the electron beam is not lined up correctly with the shadow mask or aperture grill holes, the beam is prevented from passing through to the phosphors, thereby causing a reduction in overall pixel illumination. As the beam scans, however, it may sometimes regain alignment and therefore pass through the mask/grill to the phosphors. The result of this varying realignment is that the brightness rises and falls, producing a wavelike pattern on the screen, referred to as moiré. Moiré patterns are often most visible when a screen background is set to a pattern of dots, as in a gray screen background consisting of alternate black and white dots, for example. The phenomenon is actually common in monitors with improved focus techniques, because monitors with poor focus will have a wider electron beam and therefore have more chance of hitting the target phosphors instead of the mask/grill. In the past the only way to eliminate moiré effects was to defocus the beam. Now, however, a number of monitor manufacturers have developed techniques to increase the beam size without degrading the focus.

A large part of the effort toward improving the CRT's image is aimed at creating a beam with less spread. This will allow the beam to more accurately address smaller individual dots on the screen without impinging on adjacent dots. This can be achieved by forcing the beam through smaller holes in the electron gun's grid assembly, but at the cost of decreasing the image brightness. Of course, this can be countered by driving the cathode with a higher current so as to liberate more electrons. Doing this, however, causes the barium that is the source of the electrons to be consumed more quickly and reduces the life of the cathode.

Sony's answer to this dilemma is SAGIC, or small aperture G1 with impregnated cathode. The SAGIC approach comprises a cathode impregnated with tungsten and barium material whose shape and quantity have been varied so as to avoid the high current required for a denser electron beam that consumes the cathode. This arrangement allows the first element in the grid, known as G1, to be made with a much smaller aperture, thus reducing the diameter of the beam that passes through the rest of the CRT. By early 1999 this technology had helped Sony reduce its aperture grill pitch to 0.22mm, down from the 0.25 mm of conventional Trinitron tubes. The tighter beam and narrower aperture grill worked together to provide a noticeably sharper image.

In addition to dot size, control over dot shape is also essential, and the electron gun must correct dot shape errors that occur naturally due to the geometry of the tube for optimal performance. The problem arises because the angle at which the electron beam strikes the screen must necessarily vary across the screen's width and height. For dots in the center of the screen, the beam comes straight through the electron gun and, undeflected by the yoke, strikes the phosphor at a perfect 90 degrees. However, as the beam scans closer to the edges of the screen, it strikes the phosphor at an increasing angle, with the result that the area illuminated becomes increasingly elliptical as the angle changes. The effect is most pronounced in the corners—especially with screens that are not perfectly flat—when the dot grows in both directions. It is essential that the monitor's electronics compensate for the problem in order to maintain image quality.

It is possible to alter the shape of the beam itself, in sync with the sweeping of the beam across the screen, by using additional components in the electron gun. In effect, the beam is made elliptical in the opposite direction so that the final dot shape on the screen remains circular.

Controls and Features of the CRT

Not so long ago, advanced controls were found only on high-end monitors. Now, even budget models boast a wealth of image-correction controls. This is just as well since the image fed through to the monitor by the graphics card can be subject to a number of distortions. An image can sometimes be too far to one side, or appear too high up on the screen, or need to be made wider, or taller. These adjustments can be made using the horizontal or vertical sizing and positioning controls. The most common of the "geometric controls" is barrel/pincushion, which corrects the image from dipping in or bowing out at the edges. Trapezium correction can straighten sides that slope in together, or out from each other. Parallelogram corrections will prevent the image from leaning to one side, while some models even allow the entire image to be rotated.

Making more common appearances too are on-screen controls. These are superimposed graphics that appear on the screen (obscuring parts of the main image), usually indicating what is about to be adjusted. This is similar to TV sets superimposing, say, a volume bar while the sound is being adjusted. There is no standard for on-screen graphics, and consequently there is a huge range of icons, bars, colors, and sizes in use, some better than others. The main point, however, is to render adjustments as intuitively, as quickly, and as easily as possible.

Sizes and Shapes

By the beginning of 1998, 15-inch monitors were gradually slipping to bargain-basement status, and the 17-inch size, an excellent choice for working at 1,024 x 768 resolution, was moving into the slot reserved for mainstream desktops. At the high end, a few 21-inch monitors were offering resolutions as high as 1800 x 1440.

In late 1997 a number of 19-inch monitors appeared on the market, with prices and physical sizes close to those of high-end 17-inch models, offering a cost-effective compromise for high resolution. A 19-inch CRT is a good choice for 1280 x 1024—the minimum resolution needed for serious graphics or DTP, and the power user's minimum for business applications. It's also a practical minimum size for displaying at 1600 x 1200, although bigger monitors are preferable for that resolution.

One of the main problems with CRTs is their bulk. The larger the viewable area gets, the more the CRT's depth increases. The long-standing rule-of-thumb is that a monitor's depth matches its diagonal CRT size. CRT makers have been trying to reduce the depth by moving from the current 90-degree deflection to 100 or 110 degrees. However, the more the beam is deflected, the harder it is to maintain focus. Radical measures to reduce CRT depth include putting the deflection coils inside the glass CRT; they normally sit around the CRT's neck.

The result of this development effort is the so-called "short-neck" CRT. In early 1998 17-inch short-neck monitors measuring around 15 inches deep reached the market. The new technology has taken over the 17-inch, 19-inch and 21-inch sizes, and a new rule-of-thumb was established where the monitor depth is about two inches shorter than its diagonal size.

The shape of a monitor's screen is another important factor. The three most common CRT shapes are spherical (a section of a sphere, used in the oldest and most inexpensive monitors), cylindrical (a section of a cylinder, used in aperture-grill CRTs), and flat square (a section of a sphere large enough to make the screen nearly flat). A flat square tube (FST) is standard for current monitor designs.

Flat Square Tubes (FSTs)

Flat Square Tubes improve on earlier designs by having a screen surface with only a gentle curve. They also have a larger display area, closer to the actual tube size, and nearly square corners. There is a design penalty for a flatter, more square screen, because the less of a spherical section the screen surface is, the harder it is to control the geometry and focus of the image on that screen. Modern monitors use microprocessors to apply techniques like dynamic focusing to compensate for the flatter screen.

These screens require the use of a special alloy, Invar, for the shadow mask. The flatter screen means that the shortest beam path is in the center of the screen. This is the point where the beam energy tends to concentrate, and consequently the shadow mask gets hotter here than at the corners and sides of the display. Uneven heating across the mask can make it expand and then warp and buckle. Any distortion in the mask means that its holes no longer register with the phosphor dot triplets on the screen, and image quality will be reduced. Invar alloy is used in the best monitors because it has a low coefficient of expansion.

By 2000, completely flat screens had become commonplace. One of the problems with flat screens is that they accentuate the problem of the electron beam shape being elliptical at the point at which it strikes the screen at its edges. Furthermore, the use of perfectly flat glass give rise to an optical illusion caused by the refraction of light, resulting in the image looking concave. Consequently, some tube manufacturers have introduced a curve to the inner surface of the screen to counter the concave appearance.

Multimedia Monitors

Sound facilities have become commonplace on many PCs, requiring additional loudspeakers and possibly a microphone as well. The "multimedia monitor" avoids having separate boxes and cables by building-in loudspeakers of some sort, maybe a microphone, and in some cases a camera for video conferencing. At the back of these monitors are connections for a sound card.

However, the quality of these additional components is often questionable, adding very little to the cost of manufacture. For high-quality sound, nothing is better than good external speakers, which can also be properly magnetically shielded.

Another development that has become increasingly available since the launch of Microsoft's Windows 98, which brought with it the necessary driver software, is USB-compliant CRTs. The Universal Serial Bus (USB) applies to monitors in two ways. First, the monitor itself can use a USB connection to allow screen settings to be controlled with software. Second, a USB hub can be added to a monitor (normally in its base) for use as a convenient place to plug in USB devices such as keyboards and mice. The hub provides the connection to the PC.

Digital CRTs

Nearly 99% of all video displays sold in 1998 were connected using an analog VGA interface, an aging technology that represents the minimum standard for a PC display. In fact, today VGA represents an impediment to the adoption of new flat panel display technologies, largely because of the added cost for the flat panel systems to support the analog interface. Another fundamental drawback is the degradation of image quality that occurs when a digital signal is first converted to analog and then back to digital before driving an analog-input liquid crystal display (LCD) panel.

The autumn of 1998 saw the formation of Digital Display Working Group (DDWG) that included computer industry leaders Intel, Compaq, Fujitsu, Hewlett-Packard, IBM, NEC, and Silicon Image. The objective of the DDWG was to deliver a robust, comprehensive, and extensible specification of the interface between digital displays and high-performance PCs. In the spring of 1999, the DDWG approved the first version of the Digital Visual Interface (DVI) specification, based on Silicon Image's PanelLink technology and using a Transition Minimized Differential Signaling (TMDS) digital signal protocol.

While primarily of benefit to flat panel displays (which can now operate in a standardized all-digital environment without the need to perform an analog-to-digital conversion on the signals from the graphics card driving the display device), the DVI specification potentially has ramifications for conventional CRT monitors too.

Most complaints about poor CRT image quality can be traced to incompatible graphics controllers on the motherboard or graphics card. In today's cost-driven market, marginal signal quality is not uncommon. The incorporation of DVI with a traditional analog CRT monitor will allow monitors to be designed to receive digital signals, with the necessary digital-to-analog conversion being carried out within the monitor itself. This will give manufacturers added control over final image quality, making consumer differentiation based on image quality much more of a factor than it has been hitherto. However, the application of DVI with CRT monitors is not all easy sailing.

Originally designed for use with digital flat panels, one of the drawbacks of DVI is that it has a comparatively low bandwidth of 165 MHz. This means that a working resolution of 1280 x 1024 could be supported at up to an 85-Hz refresh rate. Although this is not a problem for LCD monitors, it is a serious issue for CRT displays. The DVI specification supports a maximum resolution of 1600 x 1200 at a refresh rate of only 60 MHz—totally unrealistic in a world of ever-increasing graphics card performance and ever-bigger and cheaper CRT monitors.

The proposed solution is the provision of additional bandwidth overhead for horizontal and vertical retrace intervals, facilitated through the use of two TMDS links. Digital CRTs that are compliant with VESA's Generalized Timing Formula (GTF) would then be capable of easily supporting resolutions exceeding 2.75 million pixels at an 85-Hz refresh rate.

Another problem with DVI is that it is more expensive to digitally scale the refresh rate of a monitor than using a traditional analog multisync design. This could lead to digital CRTs being more costly than their analog counterparts. An alternative is for digital CRTs to have a fixed frequency and resolution like an LCD monitor, and thereby eliminate the need for multisync technology.

Digital Visual Interface technology anticipates that screen refresh functionality will become part of the display itself in the future. Using this methodology, new data need only be sent to the display when changes to that data must be displayed. With a selective refresh interface, DVI can maintain the high refresh rates required to keep a CRT display ergonomically pleasing, while avoiding an artificially high data rate between the graphics controller and the display. Of course, a monitor would have to employ frame buffer memory to enable this feature.

Safety Standards

In the late 1980s concern over possible health issues related to monitor use led Swedac, the Swedish testing authority, to make recommendations concerning monitor ergonomics and emissions. The resulting standard was called MPR1. This standard was amended and became the internationally adopted MPR2 standard in 1990, which called for the reduction of electrostatic emissions by infusing a conductive coating onto the monitor screen.

A further standard, called TCO, was introduced in 1992 by the Swedish Confederation of Professional Employees. The electrostatic emission levels in TCO92 were based on what monitor manufacturers thought was possible rather than on any particular safety level, while MPR2 had been based on what they could achieve without a significant cost increase. The TCO92 standard set stiffer emission requirements, and also required monitors to meet the international EN60950 standard for electrical and fire safety. Subsequent TCO standards were introduced in 1995 and again in 1999.

Apart from Sweden, the main impetus for safety standards has come from the U.S. In 1993, the Video Electronics Standards Association (VESA) initiated its DPMS, or Display Power Management Signaling standard. A DPMS compliant graphics card enables the monitor to achieve four states: on, standby, suspend, and off, at user-defined periods. Suspend mode must draw less than 8W so that the CRT, its heater, and its electron gun are likely to be shut off. Standby takes power consumption to less than about 25W, with the CRT heater usually left on for faster resuscitation.

The Video Electronics Standards Association has also produced several standards for plug-and-play monitors. Known under the banner of DDC (Display Data Channel), they should, in theory, allow a system to figure out and select the ideal settings. In practice, however, this very much depends on the combination of hardware.

The Environmental Protection Agency's (EPA's) Energy Star is a power-saving standard, mandatory in the U.S. and widely adopted in Europe, requiring a main power saving mode drawing less than 30W. Energy Star was initiated in 1993, but really took hold in 1995 when the U.S, government, the world's largest PC purchaser, adopted a policy to buy only Energy Star compliant products.

Other relevant standards include:

- ISO 9241 part 3 – The international standard for monitor ergonomics.
- EN60950 – The European standard for the electrical safety of IT equipment.
- The German TUV/EG mark – This means a monitor has been tested to both standards, in addition to the German standard for basic ergonomics (ZH/618) and MPR2 emission levels.

Ergonomics

The quality of the monitor and graphics card, and (in particular) the refresh rate at which the combination can operate, are of crucial importance in ensuring that users who spend long hours in front of a CRT monitor can do so in as much comfort as possible. These, however, are not the only factors that should be considered for comfort of use. Physical positioning of the monitor is also important, and expert advice has recently been revised in this area. It had been thought previously that the center of the monitor should be at eye level. It is now believed that to reduce fatigue as much as possible, the top of the screen should be at eye level, with the screen between 0 and 30 degrees below the horizontal and tilted slightly upwards. Achieving this arrangement with furniture designed in accordance with the previous rules, however, is not easily accomplished without causing other problems with respect to seating position and, for example, the comfortable positioning of keyboard and mouse. It is also important to sit directly in front of the monitor rather than to one side, and to locate the screen so as to avoid reflections and glare from external light sources.

The Future of CRTs

With a 100-year head start over competing screen technologies, the CRT is still a formidable technology. It is based on universally understood principles and employs commonly available materials. The result is cheap-to-make monitors capable of excellent performance, producing stable images in true color at high display resolutions.

Among the CRT's most obvious shortcomings are:

- Power consumption; it uses too much electricity.
- Mis-convergence and color variations across the screen.
- Its clunky high-voltage electrical circuits and strong magnetic fields create harmful electro-magnetic radiation.
- It is simply too big in size and weight.

There seems little doubt that the consumer of the future will demand a larger screen for home entertainment. Those who have experienced sporting events or PC games on larger screens recognize that what they get is a more compelling, more immersive experience—and they won't willingly go back. Sales of DVD players from manufacturers to retailers continues to grow aggressively, and in 2002 for the first time the sales of DVD players outpaced VCR sales. Many of those purchasers want to see their movies on the largest screen possible.

Vacuum-tube technology continues to be the conventional display technology of the home due to its low cost, simple manufacturing process, and good image quality. Yet, as consumers demand larger screen sizes, CRT displays are taxed for brightness while trying to maintain a sharp image. Their depth, weight, and sheer volume make them a bulky appliance to place in the home. The market-leading large screen (39 inches diagonal) direct-view CRT-based TV measures some 28 inches front to back, and weighs more than 250 pounds.

With even those who have the biggest vested interest in CRTs spending vast sums on research and development, it is inevitable that one of the several flat panel display (FPD) technologies will win in the long run. However, this race is taking longer than was once thought, and current estimates suggest that flat panels are unlikely to account for greater than 50% of the market before the year 2006. Flat panel displays are being developed to realize a thinner unit with low voltage requirements and low power consumption – objectives not easily achieved with CRTs.

Digital Displays/Flat Panel Displays

With many countries finalizing standards and launching broadcast services, digital TV (DTV) is becoming a force in the worldwide electronics industry. Manufacturers of DTVs and their components such as LCD panels, plasma display panels, and semiconductors are ramping-up production for this market. Digital TV set sales will increase steadily over the next several years as digital terrestrial, and in some cases satellite, broadcasts become more common and unit prices fall, due to technological advances and improved economies of scale.

By 2005, annual worldwide DTV set shipments will reach 26 million units, according to Cahners In-Stat Group, with most sales occurring in North America, Europe, Japan, and several Asia-Pacific nations. Manufacturers are planning to add several new features to the basic DTV set design:

- Digital cable ready – Very important for the U.S. market, where over 60 million households are cable subscribers.

- Digital connections – Either IEEE 1394 or DVI.
- Hard disk drives – Will be added to DTV sets to enable personal video recording and to cache data broadcasts.
- Internet access – Some manufacturers will integrate 56K dial-up, cable, or DSL modems.
- Electronic program guides – Already present in digital cable and satellite systems, some DTV manufacturers will build EPG software into their sets.

The projected numbers of units for PDPs, LCDs, and rear projection displays is shown in Figure 10.1, courtesy of Gartner Group.

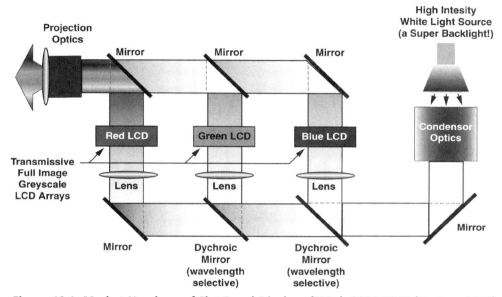

Figure 10.1: Market Numbers of Flat Panel Displays (FPDs), 2000-2007 (Gartner 2003)

While Japanese producers pioneered FPD manufacturing, they have had limited success in holding on to market share relative to the more aggressive Korean manufacturers, who have unquestionably benefited from a second mover advantage. Furthermore, other shifts are also occurring in the region, as can be seen with Philips, Samsung, and Sharp investing in display centers in Taiwan and moving design and manufacturing centers to China.

Manufacturers are also exploring new display technology to create brighter pictures by using alternatives to the CRT. Until very recently, most audio-video (AV) and data communications devices used CRT monitors as their main display device. These monitors, however, require high voltage for the emission and angle control of the electron beams, and it is difficult to slim down the size of CRT units. Flat panel display technology is the perfect answer to demands for more compact displays with greater energy efficiency and diversity. However, CRTs will continue to be used in the majority of direct view and rear projection sets due to their lower cost. In addition, CRTs still offer the best possible level of black reproduction.

Flat panel digital display technologies include:

- Liquid crystal displays (LCD) – A liquid crystal display consists of an array of tiny segments (called pixels) that can be manipulated to present information. This basic idea uses liquid crystal technology and is common to all displays, ranging from simple calculators and wristwatches to a full color LCD television and computer screen. Manufacturers are increasing the production of 20- to 30-inch panels to be used as TV displays. The resulting economies of scale will make these sets more affordable. The different LCD types include:

- **Active matrix LCDs (AMLCDs)** – AMLCDs give good resolution and, in the color version, vivid colors because each pixel is controlled by its own transistor. Thus, AMLCDs have become the standard in the hand-held market. However, AMLCDs are rather heavy and power-hungry because they are based on layers of glass and require a backlight. In addition, the contents of the screen can only be viewed if the user is facing the device head-on. There are two types of AMLCDs; TFT (thin-film transistor) and two terminal non-linear devices.

- **Liquid crystal on silicon (LCoS)** – Thomson developed the first LCoS DTV set in cooperation with Three-Five Systems, Corning, and ColorLink. Liquid Crystal on Silicon uses three microdisplay imagers, optics, and a prism to enable thinner sets with a brighter picture.

- **Passive matrix LCDs (PMLCDs)** – PMLCDs use fewer electrodes, which are arranged in strips across the screen, as opposed to AMLCDs, which control each pixel individually. While this makes them less expensive and easier to manufacture, the color and clarity are negatively affected. Passive Matrix LCDs are reasonably well suited to smaller devices with monochrome screens. They are also known as STN (super twisted nematic) LCDs.

- **Ferro-electric LCD**

- **Polymer-Dispersed LCD**

- **Cholestric LCD (ChLCD)** – ChLCDs are differentiated in that once an image is displayed on the screen, no additional power need be expended to keep it there, thus saving the battery. Cholestric LCD also have a wider viewing angle and can be made of plastic, improving ease-of-use and cutting weight. However, ChLCDs' ability to display color is limited, which makes the technology better suited for low-end applications.

- Plasma display panels (PDP) – The term plasma refers to a gas that consists of electrons, positively charged particles known as anions, and neutral particles. The PDP operates by passing electricity through neon gas, causing it to become temporarily charged; light is produced when the gas spontaneously discharges. The displays operate at high voltages, low currents, and low temperatures, resulting in long operating lifetimes. Plasma technology allows larger flat panel sets than LCD, ranging from 36 to 60 inches. The overwhelming majority of buyers are corporate, but consumer purchases will increase as prices decline. Plasma Display Panels types include DC operation and AC operation (include Color AC PDP).

- Organic light emitting diode (OLED) – OLED refers to the light-emitting diode devices made from organic materials. Types of OLED include monochromatic and full-color.

- Digital light processing (DLP) – Developed by Texas Instruments, DLP uses a digital micro-mirror device to modulate reflected light.

- Electro-luminescent (EL) displays – These displays directly convert the electrical energy into luminous energy. In EL displays, the electrons associated with the chemical elements in solid material acquire a high energy level from the electric field. Light is produced when the electrons return to their normal, or ground state, condition. Types of EL displays include AC thin-film electroluminescent (TFEL) display panels, color AC-TFEL displays, AM-TFEL displays, AC Hybrid EL displays, DC Powder EL displays, and transparent EL displays.

Organic EL is a self-emitting flat panel display that uses the color radiating character of organic material by electrical stimulation. Organic EL uses a silicon thin film transistor as the switching device for each picture element, and has a brighter display, lower power consumption, and faster response time than TFT-LCD. This type of display is not technologically stable at this time, and it is still under development for commercial production.

- Field emission displays (FED) – FEDs are flat, thin vacuum tubes that use arrays of electron emitters (cathodes) to produce electrons, which are accelerated towards a phosphorus-coated (anode) faceplate that emits visible light.
- Vacuum fluorescent display (VFD) – A VFD is a flat CRT that uses multiple cathodes and a matrix deflection system.
- Microdisplays – Microdisplays are very small displays (one inch or less diagonal measurement) that are viewed through the use of optics. These miniature displays are generally the size of the semiconductors from which they are made (the size of a quarter), and are viewed close to the eye—either in eyeglasses or an eyepiece—yielding an image that looks like a full-sized screen. To date, these devices have been limited by their cost, social acceptability, and time-to-market issues. We think, however, that this method for viewing data will gain wider acceptance over the next few years as costs come down and the unit's power efficiency is recognized. The different types of microdisplays include:
  - Transmissive microdisplays
  - Reflective microdisplays:
    - LCoS
    - Microelectromechanical Systems (MEMS)
    - Digital Micro-mirror Device (DMD)
    - Grating Light Valve (GLV)
    - Actuated Mirror Array (AMA)
  - Emissive microdisplays:
    - TFEL microdisplays
    - Vacuum Fluorescent on Silicon (VFoS) microdisplays
    - OLED microdisplays
  - Scanning microdisplays
- Digital light amplification (D-ILA) – The basis for D-ILA is electronic valve technology using liquid crystal on silicon. The technology was developed and is owned by JVC.
- Inorganic LED displays – Inorganic LEDs, or LEDs, were developed in 1960s as an outgrowth of semiconductor technology. The devices emit light when a forward bias voltage is applied to the PN junction in a single crystal of gallium arsenide (GaAs) or other III-V compounds. By appropriate doping and/or use of crystals containing III-V materials, it is possible to produce emissions of red, green, yellow, and blue light.
- Flat-thin CRTs – Because of the stiff competition from digital display CRTs, manufacturers are improving the manufacturing process for thinness, lightweight, flatness, and low power. Thin CRTs are emerging as a viable screen solution because they have wider viewing angles, they can smoothly display motion on the screen, and they have great color. However, at this point, they are expensive due to low availability. Long term, they could be a practical option because they are good for outdoor usage, easy to add to a CRT manufacturer's current production lines, and simple to produce.
- Light-emitting polymer (LEP) diode screens – Based on plastic, these screens can be flexible and based on any shape or size. As of 2003, LEP technologies were limited to prototypes and degrade after 10,000 hours of use.

- ■ Electromechanical displays
- ■ Rotatable dipole displays
- ■ Electro-phoretic imaging displays (EPID)
- ■ Electrochromic displays

The most popular display types used in consumer appliances, digital TVs, and PC monitors are LCDs, PDPs, DLPs, LCoS, and OLEDs, which are described in more detail below.

LCD—Liquid Crystal Displays

Liquid crystals were first discovered in the late 19th century by the Austrian botanist, Friedrich Reinitzer. The term "liquid crystal," itself, was coined shortly afterward by German physicist Otto Lehmann. Liquid crystals are almost transparent substances, exhibiting the properties of both solid and liquid matter. Light passing through liquid crystals follows the alignment of the molecules that make them up—a property of solid matter. In the 1960s it was discovered that charging liquid crystals with electricity changed their molecular alignment, and consequently the way light passed through them—a property of liquids.

LCDs consist of two glass plates with liquid crystal material (transparent organic polymers) between them. The LCD has an array of tiny segments (called pixels) that can be manipulated to present information. Many LCDs are reflective, meaning that they use only ambient light to illuminate the display. LCDs contain transparent organic polymers that respond to an applied voltage. To form the display, manufacturers deposit a polarizing film on the outer surfaces of the two ultra-flat glass (or quartz) substrates, with a matrix of transparent indium tin oxide (ITO) electrodes on the inner surfaces of these substrates. With micron-sized spacers holding the two substrates apart, the sandwich is joined together. The substrates are cut into one or more displays, depending on the original size of the substrates (from 12-inch to 22-inch square). The outer edges of each display are then sealed with a gasket, the interior air is evacuated, and the void is injected with liquid crystals.

The polarizers on the front and back of the display are oriented 90° from one another. With this orientation, no light can pass through unless the polarization of the light is altered. Liquid crystals are a means for changing the polarization. When no voltage is applied, liquid crystals can be aligned in twisted (90°) or super-twisted (270°) configurations. With these configurations the polarity of light is rotated, allowing the light to pass through the front polarizer, thus illuminating the viewing surface. When a voltage is applied, the liquid crystals align to the electric field created, the polarity of the incoming light does not change, and the viewing surface appears dark.

All LCDs must have a source of reflected or back lighting. This source is usually a metal halide, cold cathode, fluorescent, or halogen bulb placed behind the back plate. Since the light must pass through polarizers, glass, liquid crystals, filters, and electrodes, the light source must be of sufficient wattage to generate the desired brightness of the display. Typically, the internal complexity of the display blocks greater than 95% of the original light from exiting on the viewer's side. As a result, the generation of unseen light causes a major drain on a battery-operated LCD's power source.

Some LCD systems perform much better than others. The greater twist angle of super-twisted nematic (STN) liquid crystals allows a much higher contrast ratio (light to dark) and faster response than conventional twisted nematic (TN) crystals. For color displays, each visible pixel must consist of three adjoining cells, one with a red filter, one with a blue filter, and one with a green filter, in order to achieve the red-green-blue (RGB) color standard. While color decreases the resolution of the display, it also adds information to the display, particularly for desktop publishing and scientific applications.

Since its advent in 1971 as a display medium, liquid crystal displays have moved into a variety of fields, including miniature televisions, digital still and video cameras, and monitors. Today, many believe that the LCD is the most likely technology to replace the CRT monitor because there is no bulky picture tube and a lot less power is consumed than with its CRT counterparts. The technology involved has been developed considerably since its inception, to the point where today's products no longer resemble the clumsy, monochrome devices of old. It has a head start over other flat screen technologies and an apparently unassailable position in notebook and handheld PCs where it is available in two forms – low-cost, dual-scan twisted nematic (DSTN) and high image quality thin film transistor (TFT).

Principles

The LCD is a transmissive technology. The display works by letting varying amounts of a fixed-intensity white backlight through an active filter. The red, green, and blue elements of a pixel are achieved through simple filtering of the white light.

Most liquid crystals are organic compounds consisting of long rod-like molecules that, in their natural state, arrange themselves with their long axes roughly parallel. It is possible to precisely control the alignment of these molecules by flowing the liquid crystal along a finely grooved surface. The alignment of the molecules follows the grooves; if the grooves are exactly parallel, then the alignment of the molecules also becomes exactly parallel. In their natural state, LCD molecules are arranged in a loosely ordered fashion with their long axes parallel. However, when they come into contact with a grooved surface in a fixed direction, they line up parallel along the grooves.

The first principle of an LCD consists of sandwiching liquid crystals between two finely grooved surfaces, where the grooves on one surface are perpendicular (at 90 degrees) to the grooves on the other. If the molecules at one surface are aligned north to south, and the molecules on the other are aligned east to west, then those in between are forced into a twisted state of 90 degrees. Light follows the alignment of the molecules, and therefore is also twisted through 90 degrees as it passes through the liquid crystals. When a voltage is applied to the liquid crystal, the molecules rearrange themselves vertically, allowing light to pass through untwisted.

The second principle of a LCD relies on the properties of polarizing filters and light itself. Natural light waves are oriented at random angles. A polarizing filter is simply a set of incredibly fine parallel lines. These lines act like a net, blocking all light waves apart from those (coincidentally) oriented parallel to the lines. A second polarizing filter with lines arranged perpendicular (at 90 degrees) to the first would therefore totally block this already polarized light. Light would only pass through the second polarizer if its lines were exactly parallel with the first, or if the light itself had been twisted to match the second polarizer.

A typical twisted nematic (TN) liquid crystal display consists of two polarizing filters with their lines arranged perpendicular (at 90 degrees) to each other, which, as described above, would block all light trying to pass through. But the twisted liquid crystals are located in-between these polarizers. Therefore, light is polarized by the first filter, twisted through 90 degrees by the liquid crystals, and finally made to completely pass through the second polarizing filter. However, when an electrical voltage is applied across the liquid crystal, the molecules realign vertically, allowing the light to pass through untwisted but to be ultimately blocked by the second polarizer. Consequently, no voltage equals light passing through, while applied voltage equals no light emerging at the other end.

The crystals in an LCD could be alternatively arranged so that light passed when there was a voltage, and not passed when there was no voltage. However, since computer screens with graphical interfaces are almost always lit, arranging the crystals in the no-voltage-equals-light-passing configuration saves power.

Rules

Liquid crystal displays follow a different set of rules than CRT displays, offering advantages in terms of bulk, power consumption, and flicker, as well as "perfect" geometry. They have the disadvantage of a much higher price, a poorer viewing angle, and less accurate color performance.

While CRTs are capable of displaying a range of resolutions and scaling them to fit the screen, an LCD panel has a fixed number of liquid crystal cells, and can display only one resolution at full-screen size using one cell per pixel. Lower resolutions can be displayed by using only a portion of the screen. For example, a 1024 x 768 panel can display at a resolution of 640 x 480 by using only 66% of the screen. Most LCDs are capable of rescaling lower-resolution images to fill the screen through a process known as rathiomatic expansion. However, this works better for continuous-tone images like photographs than it does for text and images with fine detail, where it can result in badly aliased objects as jagged artifacts appear to fill in the extra pixels. The best results are achieved by LCDs that resample the screen when scaling it up, thereby anti-aliasing the image when filling in the extra pixels. Not all LCDs can do this, however.

Unlike CRT monitors, the diagonal measurement of an LCD is the same as its viewable area, so there is no loss of the traditional inch or so behind the monitor's faceplate or bezel. This makes any LCD a match for a CRT 2 to 3 inches larger. A number of leading manufacturers already have 18.1-inch TFT models on the market capable of a native resolution of 1280 x 1024.

A CRT has three electron guns whose streams must converge faultlessly in order to create a sharp image. There are no convergence problems with an LCD panel, because each cell is switched on and off individually. This is one reason why text looks so crisp on an LCD monitor. There's no need to worry about refresh rates and flicker with an LCD panel—the LCD cells are either on or off. An image displayed at a refresh rate as low as between 40-60 Hz should not produce any more flicker than one at a 75-Hz refresh rate.

Conversely, it's possible for one or more cells on the LCD panel to be flawed. On a 1024 x 768 monitor, there are three cells for each pixel—one each for red, green, and blue, which amounts to nearly 2.4 million cells (1024 x 768 x 3 = 2,359,296). There's only a slim chance that all of these will be perfect; more likely, some will be stuck on (creating a "bright" defect) or off (resulting in a "dark" defect). Some buyers may think that the premium cost of an LCD display entitles them to perfect screens—unfortunately, this is not the case.

LCD monitors also have other elements that you don't find in CRT displays. The panels are lit by fluorescent tubes that snake through the back of the unit, causing a display to sometimes exhibit brighter lines in some parts of the screen than in others. It may also be possible to see ghosting or streaking, where a particularly light or dark image can affect adjacent portions of the screen. And fine patterns such as dithered images may create moiré or interference patterns that jitter.

Viewing angle problems on LCDs occur because the technology is a transmissive system, which works by modulating the light that passes through the display, while CRTs are emissive. With emissive displays, there is a material that emits light at the front of the display, which is easily viewed from greater angles. In a LCD, as well as passing through the intended pixel obliquely, emitted light passes through adjacent pixels, causing color distortion.

Currently, most LCD monitors plug into a computer's familiar 15-pin analog VGA port and use an analog-to-digital converter to convert the signal into a form the panel can use. An industry standard specification for a digital video port has been created by VESA. Liquid crystal display monitors will incorporate both analog and digital inputs.

Creating color

In order to create the shades required for a full-color display, there have to be some intermediate levels of brightness between all light and no light passing through. The varying levels of brightness required to create a full-color display are achieved by changing the strength of the voltage applied to the crystals. The liquid crystals in fact untwist at a speed directly proportional to the strength of the applied voltage, thereby allowing the amount of light passing through to be controlled. In practice, though, the voltage variation of today's LCDs can only offer 64 different shades per element (6-bit) as opposed to full-color CRT displays, which can create 256 shades (8-bit). Color LCDs use three elements per pixel, which results in a maximum of 262,144 colors (18-bit), compared to true-color CRT monitors supplying 16,777,216 colors (24-bit).

As multimedia applications become more widespread, the lack of true 24-bit color on LCD panels is becoming an issue. While 18-bit is fine for most applications, it is insufficient for photographic or video work. Some LCD designs manage to extend the color depth to 24-bit by displaying alternate shades on successive frame refreshes, a technique known as Frame Rate Control (FRC). However, if the difference is too great, flicker is perceived.

Hitachi has developed a technique whereby the voltage applied to adjacent cells to create patterns changes very slightly across a sequence of three or four frames. With it, Hitachi can simulate not quite 256 grayscales, but still a highly respectable 253 grayscales, which translates into over 16 million –colors—virtually indistinguishable from 24-bit true color.

Figure 10.2 shows the anatomy of the LCD projector, and Figure 10.3 shows the block diagram of a high-resolution LCD monitor controller.

Types of LCDs

- TFT-LCD, or active matrix – A flat panel display that uses a thin film transistor (TFT) formed with silicon as the switching device of the picture element. A silicon layer forms a thin film transistor on glass substrate, and is able to have two physical properties, amorphous and polycrystal:
  - **Amorphous** – TFT-LCD in which amorphous silicon phase is amorphous TFT-LCD. Amorphous TFT-LCD has less electronics mobility than poly-type, and is normally used for ordinary product applications, such as portable/notebook PC and desktop LCD monitors, that do not absolutely require very fine pixel pitch.
  - **Polysilicon** – Poly crystal silicon phases are divided into low-temperature polysilicon (LTPS) and high-temperature polysilicon (HTPS) by the temperature at which the silicon layer forms a glass substrate. The LTPS phase has a better electron mobility that makes fine pixel pitch possible, and is usually adopted by applications employing small

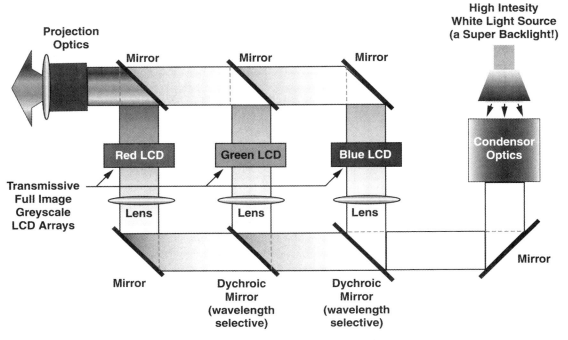

Figure 10.2: Anatomy of the LCD Projector

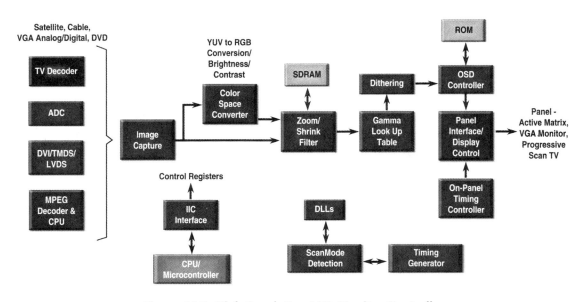

Figure 10.3: High-Resolution LCD Monitor Controller

and midsize, but high-resolution, handheld displays such as PDAs and DSCs. The HTPS phase is used in projector and viewfinder applications.

- The TFT is the most popular type of LCD in the market today.
- TFD-LCD – A flat panel display that uses thin film diode (TFD) formed with silicon as the switching device of the picture element. It has a simpler manufacturing process than TFT-LCD, but did not become the standard product in the market because most vendors focused on TFT-LCD due to its well-accumulated technology infrastructure.
- LCoS – Liquid-Crystal-on-Silicon is primarily used for projector applications.
- STN-LCD, or passive matrix – A flat panel display that uses super-twisted nematic liquid crystal technology, and adopts a passive switching circuit for the picture element. There are two forms of STN-LCD, Monochromatic and Color:
  - **Monochrome STN-LCD** – Monochrome STN-LCD does not have a color filter and only displays black-and-white images, with some grayscale output images. Monochrome STN-LCD is mainly used for applications that do not need color displays and are cost sensitive, such as cellular handsets.
  - **Color STN-LCD** – Color STN-LCD uses a color filter consisting of red, green, and blue unit cells for color display. It is applied to various applications from small cellular handsets to large portable PCs, owing to its relatively low price. However, it is gradually losing market share due to its limitations of slow response time and narrow viewing angle.
- Twisted Nematic (TN) LCDs – Used for watches and calculators.

Thin Film Transistor Displays

Many companies have adopted Thin Film Transistor (TFT) technology to improve color screens. In a TFT screen, also known as active matrix, an extra matrix of transistors is connected to the LCD panel—one transistor for each color (RGB) of each pixel. These transistors drive the pixels, eliminating the problems of ghosting and slow response speed that afflict non-TFT-LCDs. The result is screen response times of the order of 25 ms, contrast ratios in the region of 200:1 to 400:1, and brightness values between 200 and 250 cd/m^2 (candela per square meter).

The liquid crystal elements of each pixel are arranged so that in their normal state (with no voltage applied) the light coming through the passive filter is "incorrectly" polarized and thus blocked. But when a voltage is applied across the liquid crystal elements they twist up to ninety degrees in proportion to the voltage, changing their polarization and letting more light through. The transistors control the degree of twist and hence the intensity of the red, green, and blue elements of each pixel forming the image on the display.

Thin film transistor screens can be made much thinner than LCDs, making them lighter. They also have refresh rates now approaching those of CRTs because current runs about ten times faster in a TFT than in a DSTN screen. Standard VGA screens need 921,000 transistors (640 x 480 x 3), while a resolution of 1024 x 768 needs 2,359,296, and each transistor must be perfect. The complete matrix of transistors has to be produced on a single, expensive silicon wafer, and the presence of more than a couple of impurities means that the whole wafer must be discarded. This leads to a high wastage rate and is the main reason for the high price of TFT displays. It's also the reason why there are liable to be a couple of defective pixels where the transistors have failed in any TFT display.

There are two phenomena that define a defective LCD pixel: a "lit" pixel, which appears as one or several randomly placed red, blue and/or green pixel elements on an all-black background, or a "missing" or "dead" pixel, which appears as a black dot on all-white backgrounds. The former failure mode is the more common, and is the result of a transistor occasionally shorting in the "on" state and resulting in a permanently "turned-on" (red, green or blue) pixel. Unfortunately, fixing the transistor itself is not possible after assembly. It is possible to disable an offending transistor using a laser. However, this just creates black dots that would appear on a white background. Permanently turned-on pixels are a fairly common occurrence in LCD manufacturing, and LCD manufacturers set limits, based on user feedback and manufacturing cost data, as to how many defective pixels are acceptable for a given LCD panel. The goal in setting these limits is to maintain reasonable product pricing while minimizing the degree of user distraction from defective pixels. For example, a 1024 x 768 native resolution panel, containing a total of 2,359,296 (1024 x 768 x 3) pixels, that has 20 defective pixels would have a pixel defect rate of (20/2,359,296)*100 = 0.0008%. The TFT display has undergone significant evolution since the days of the early, twisted Nnematic (TN) technology based panels.

In-Plane Switching

Jointly developed by Hosiden and NEC, In-Plane Switching (IPS) was one of the first refinements to produce significant gains in the light-transmissive characteristics of TFT panels. In a standard TFT display, when one end of the crystal is fixed and a voltage applied, the crystal untwists, changing the angle of polarization of the transmitted light. A downside of basic TN technology is that the alignment of liquid crystal molecules alters the further away they are from the electrode. With IPS, the crystals are horizontal rather than vertical, and the electrical field is applied between each end of the crystal. This improves the viewing angles considerably, but means that two transistors are needed for every pixel instead of the one needed for a standard TFT display. Using two transistors means that more of the transparent area of the display is blocked from light transmission, so brighter back lights must be used, increasing power consumption and making the displays unsuitable for notebooks.

Polysilicon Panels

The thin-film transistors that drive individual cells in the overlying liquid crystal layer in traditional active-matrix displays are formed from amorphous silicon (a-Si) deposited on a glass substrate. The advantage of using amorphous silicon is that it doesn't require high temperatures, so fairly inexpensive glass can be used as a substrate. A disadvantage is that the non-crystalline structure is a barrier to rapid electron movement, necessitating powerful driver circuitry.

It was recognized early on in flat-panel display research that a crystalline or polycrystalline (an intermediate crystalline stage comprising many small interlocked crystals analogous to a layer of sugar) of silicon would be a much more desirable substance to use. Unfortunately, this could only be created at very high temperatures (over 1,000°C), requiring the use of quartz or special glass as a substrate. However, in the late 1990s manufacturing advances allowed the development of low-temperature polysilicon (p-Si) TFT displays, formed at temperatures around 450°C. Initially, these were used extensively in devices that required only small displays, such as projectors and digital cameras.

One of the largest cost elements in a standard TFT panel is the external driver circuitry, requiring a large number of external connections from the glass panel, because each pixel has its own connection to the driver circuitry. This requires discrete logic chips arranged on PCBs around the periphery of the display, limiting the size of the surrounding casing. A major attraction of p-Si technology is that the increased efficiency of the transistors allows the driver circuitry and peripheral electronics to be made as an integral part of the display. This considerably reduces the number of components for an individual display; Toshiba estimates 40% fewer components and only 5% as many interconnects as in a conventional panel. The technology will yield thinner, brighter panels with better contrast ratios, and allow larger panels to be fitted into existing casings. Since screens using p-Si are also reportedly tougher than a-Si panels, it's possible that the technology may allow the cheaper plastic casings used in the past—but superseded by much more expensive magnesium alloy casings—to stage a comeback.

By 1999, the technology was moving into the mainstream PC world, with Toshiba's announcement of the world's first commercial production of 8.4 in and 10.4 in low-temperature p-Si (LTPS) displays suitable for use in notebook PCs. The next major advance is expected is to see LTPS TFTs deposited on a flexible plastic substrate—offering the prospect of a roll-up notebook display.

CRT Feature Comparison

Table 10.1 provides a feature comparison between a 13.5-inch passive matrix LCD (PMLCD) and active matrix LCD (AMLCD) and a 15-inch CRT monitor.

Table 10.1: Feature Comparison Between LCDs and CRTs

Display Type	Viewing Angle (degrees)	Contrast Ratio	Response Speed	Brightness	Power Consumption (watts)	Life
PMLCD	49-100	40:1	300 ms	70 - 90	45	60K hrs
AMLCD	> 140	140:1	25 ms	70 - 90	50	60K hrs
CRT	> 190	300:1	N/A	220 - 270	180	Years

Contrast ratio is a measure of how much brighter a pure white output is compared to a pure black output. The higher the contrast the sharper the image and the more pure the white will be. When compared with LCDs, CRTs offer by far the greatest contrast ratio.

Response time is measured in milliseconds and refers to the time it takes each pixel to respond to the command it receives from the panel controller. Response time is used exclusively when discussing LCDs, because of the way they send their signal. An AMLCD has a much better response time than a PMLCD. Conversely, response time doesn't apply to CRTs, because of the way they handle the display of information (an electron beam exciting phosphors).

There are many different ways to measure brightness. The higher the level of brightness (represented in the table as a higher number), the brighter the white displays. When it comes to the life span of a LCD, the figure is referenced as the mean time between failures for the flat panel. This means that if it is runs continuously it will have an average life of 60,000 hours before the light burns out. This would be equal to about 6.8 years. On the face of it, CRTs can last much longer than that. However, while LCDs simply burn out, CRT's get dimmer as they age, and in practice don't have the ability to produce an ISO-compliant luminance after around 40,000 hours of use.

LCD Market

The total TFT-LCD revenue for PC-related applications, portable PCs, and LCD monitors is projected to grow to $22.2 billion by 2005. Non-PC applications, cellular handsets, PDAs, car navigation systems, digital still cameras, digital video cameras, Pachinko, and portable games revenue will increase to $7.7 billion in 2005. The revenue share of non-PC applications among the entire TFT-LCD market will be 24.5% through 2005.

In 2000, the total TFT-LCD demand reached 31.7 million units, while total supply was 32.2 million units. The demand for portable PCs in 2000 was 26.0 million units, with a 30.4% of year-to-year growth rate, and the demand for TFT portable PCs was 24.6 million units, with a 33.6% year-to-year growth. Meanwhile, LCD monitor demand (including LCD TVs) was recorded at 7.0 million units, with a high 60% growth rate. In 2001, the demand for TFTs was at 42.2 million units. However, price declines have continued from 2000 through 2004. Prices of 14.1-inch and 15.0-inch LCDs, the main products for portable PCs and LCD monitors, declined by 24% and 30%, respectively, in 2001. Key applications of TFT-LCDs are segments, such as portable PCs, LCD monitors, and LCD TVs, that require large LCDs.

The 15-inch grade monitor led the market in 2003, commanding about an 80% share. For the post-15-inch grade market, aggressive marketing by several LCD manufacturers and the development of low-end products that possess conventional viewing angle technology will offer a more easily accessible price level for the 17-inch grade monitor. Larger size models (i.e., greater than 20-inch) are not expected to see enlarged demand, and recorded a mere 2% share in 2003. That means bigger TFT-LCD models will have limited adoption in the monitor segment due to their high price.

Today, portable PCs account for about 80% of the consumption of LCDs. LCDs are practical for applications where size and weight are important. However, LCDs do have many problems with the viewing angle, contrast ratio, and response times. These issues need to be solved before LCDs can completely replace CRTs.

DisplaySearch reports that LCD TV sales accelerated in 2003 with 182% growth in 2003 after 113% growth in 2002. A total of 567,530 LCD TV units were shipped in first quarter of 2003, with Japan accounting for 69% of the market. Revenue generated for the same quarter for LCD TVs was $685 million. The expected LCD TV units are expected to reach 745,300. The value of LCD TV sets will increase from $3.4 billion in 2003 to $19.7 billion in 2007. Top vendors for LCD TVs are Sharp Electronics, LG Electronics, Samsung Electronics, Sony, and Panasonic (Matsushita Electrical). The optimism regarding the potential growth is based on a number of factors including:

- Larger substrate fabs, in which larger sized panels will be better optimized and more cost effective
- Lower panel prices
- Added competition
- Rapid emergence of HDTV

LCD Applications in PC Monitors

During 2000 through 2002, LCD monitor demand showed a 62% cumulative annual growth rate (CAGR), while CRT demand is expected to show a meager 7% CAGR for the coming few years. Consequently, the portion of TFTs in the total monitor market increased to 4% in 2002 and to 22% in 2004. Liquid crystal display monitor demand, excluding LCD TVs, was 7 million units in 2000, and reached 12.2 million units in 2002. As LCD module prices continuously decrease, the demand for

LCD monitors will act as a key LCD market driver rather than portable PCs. However, the consumer market that supports monitor demand seems quite price sensitive, and LCD monitor demand will maintain its vulnerable position, which can fluctuate easily as a function of pricing and economic status.

In terms of display size, the 15-inch grade will undoubtedly lead the market by 2004 with more than 75% share. For the post-15 inch grade market, the 17-inch grade will gain the second position in 2004 at 14% share, and the 18.1-inch grade will be the third position at 7.5% share due to relatively conservative marketing.

Total monitor sales were 120 million units in 2001, with CRTs claiming 108 million and TFT-LCDs claiming 12 million units, and with a respective market share of 90% and 10%. By 2004, the total monitor sales will exceed 157 million units with CRTs selling 122 million units and having a 78% market share and TFTs having 22% market share and 35 million units. In terms of the world-wide LCD monitor market demand by region, Japan's market share continues to decrease. Off the 12.2 million LCD monitor units shipped in 2001, the market share of respective geographies were: U.S. at 22.6%, Europe at 16.6%, Japan at 51.5%, Asia/Pacific at 4.5%, and ROW (Rest of the World) at 4.8%. Comparatively, of the predicted 34.8 million units of LCD monitors that will be shipped the geographical market share changes to: U.S. – 32.6%, Europe – 20.8%, Japan – 28.2%, Asia/Pacific – 11%, and ROW – 7.3%.

In 2001, the supplies of 14-inch and 15-inch based TFT-LCDs were 68.1 and 54.9 million units, respectively. In 2002, the market for 14-inch and 15-inch TFT-LCDs reached 90.6 and 75.3 million units, representing growth rates of 32.9% and 36.8%, respectively. In light of this business situation, many Japanese vendors accelerated their penetration into new and still profitable applications, such as cellular phones and PDAs. Due to an overall cellular market recession and the delay of third-generation (3G) mobile services, however, the adoption of TFTs in cellular handsets will not occur until 2004. During 2001 and 2002, Japanese, Korean and Taiwanese vendors focused on scaling up the LCD monitor market's size via reductions in production costs. If production yields catch up with current a-Si TFTs, LTPS will be a strong solution in the long term to overcome the price declines caused by a surplus market.

The continued oversupply market has pushed down the pricing of TFTs regardless of differences in TFT sizes. In 2000, however, among products aimed at portable PCs, the 13.3-inch model recorded the largest price decline (a 38.3% drop) due to its waning demand in the market. In the case of products for LCD monitors, the 17-inch model recorded the largest decline in 2000 (a 51.5% drop). However, the prices of large-sized TFTs (i.e., over 17-inch) were abnormally high due to low production yield. The large price declines were caused by both the oversupply market and yield improvement.

During 2001 and 2002, TFT prices for portable PCs and 15-inch LCD monitors dropped by about 30–40%, and the larger sizes dropped more by about 45%. The sharp price drops—almost collapse—prompted most vendors to sell their products for less than their total production cost. From 2Q01, price declines will break even with the variable cost for new entrants. Large-scale loss may act as a resistor to price declines. From 2001, however, as demand from the LCD monitor segment seems to mainly pull the TFT market, considering its cost-sensitive demand structure, it appears difficult to make the price trend level off as long as total supply exceeds total demand.

LCD demand was 42.8 million units in 2001 and 57.1 million units in 2002 with 33% growth. LCD monitor demand driver was the portable/notebook PC monitors.

In 2002, LCD monitor demand was 21 million units and continues to see a CAGR of 55.8%. The key driver for LCD monitor market growth will be the continuous price declines in TFT-LCDs and LCD monitor systems. Unlike the portable PC market, which mainly depends on corporate demand, LCD monitor demand will largely rely on consumer market and CRT replacement requirements. By 2005, TFT will have a 26.1% share of the total monitor market (including CRT and LCD). During the same period, LCD monitor demand will show a 32.8% CAGR while CRT monitor demand will show only a 3.6% CAGR. LCD monitor demand, excluding LCD TV, was 13.5 million units in 2001 and will reach 42 million units by 2005. As the portable PC demand will show a stable 15.6% CAGR for same period, booming LCD monitor demand will mainly tract the total TFT-LCD market from 2001. However, the consumer market that drives monitor demand seems quite price sensitive, and LCD monitor demand will maintain a vulnerable position that can be fluctuated easily by prices and economic status.

In terms of display size, in 2001, the 15-inch grade had an 80% share in the market and will keep its leading position through 2005, even though the portion will decrease to 65% in 2005. Owing to LCD price declines and increasing demand for multimedia environments, larger LCDs with higher resolution will gain more shares by 2005. LCDs larger than 15-inch had a 15% share in 2001, which will increase to 35% in 2005. However, due to continued difficulties in yield, the share of LCDs larger than 20-inch will not easily increase in market share, and this segment of the market will reach only 4% by 2005. In larger products, the 17-inch and 18-inch LCDs will be the main drivers of the market. The 17-inch grade will get more share than the 18-inch grade due to aggressive promotion and film technology adoption, but the 18-inch grade will continue to increase its share as production yield improves.

The different types of LCD monitors, based on inches of screen, are:

- 12.1-inch – Starting in the first quarter of 2001, the 12.1-inch TFT-LCD changed its staple resolution from SVGA to XGA in response to market demands, and it was able to maintain a relatively small price decline compared with other sizes. During 2000 and the first quarter of 2001, the 12.1-inch model maintained a stable demand increase due to strong market needs for real portability. However, following the 13.3-inch model's faster price decline in March 2001, the 12.1-inch and 13.3-inch TFT-LCDs have maintained an almost-identical price level. This may be one of the key inhibitors to demand growth in spite of the strong portability features of the 12.1-inch size. Additionally, less legibility caused by higher resolution may be a weak point in promotion.

- 13.3-inch – Sandwiched by the 12.1-inch and 14.1-inch sizes, the 13.3-inch model has been losing market share and recorded large and continuous price declines in the first quarter of 2001. However, starting in the second quarter of 2001, as the prices of the 13.3-inch and 12.1-inch models came close, it seemed to show an average level of price declines. Despite lots of negative outlooks in the market, many big portable PC companies such as Compaq, Hewlett-Packard, IBM, Toshiba, and NEC keep a certain number of 13.3-inch projects, and even have plans to add new ones due to the size's recovered competitiveness.

- 14.1-inch – The 14.1-inch TFT-LCD holds the top market share in the portable PC market, and this trend will be sustained for a long while. However, it may not avoid price decline trends despite high market demands. Basically, production facilities for the 14.1-inch model are also refitted for the production of 17-inch models, and the facilities for 13.3-inch TFT-LCDs are optimized for 15-inch or 18.1-inch TFT-LCDs. This means 14.1-inch-optimized facilities have difficulties in production control by size mix compared with 13.3-inch-, 15-inch-, and 18.1-inch-optimized facilities due to a premature 17-inch market. The 14.1-inch prices will decline for a comparatively long period until the 17-inch demand picks up.

- 15-inch – There have been lots of efforts to promote big screen-size portable PCs since the market changed to oversupply. However, the market still appears to be reluctant to use that wide, but quite heavy, product. In that sense, the prices of 15-inch displays for portable PCs could not differ from the overall market trend. The 15-inch displays for portable PC applications require a comparatively high technology level in development and production, and new industry entrants still have some difficulties in joining as strong players. Therefore, a certain level of price premium has been kept compared with 15-inch displays for monitor applications, and this trend will continue through the year. The toughest competition has been in the 15-inch-for-monitor applications market because it is almost the only playable market for Taiwanese new entrants, and giants in Korea also focus on this market. However, the market's booming demands seemed to ease the steep price declines in the second quarter of 2001. Unlike portable PCs, demand for LCD monitors mostly relies on the consumer market and is quite pricing sensitive. Consequently, the price movement of the 15-inch monitor applications market is restricted by a range, in which strong market demands can be kept regardless of the supply-versus-demand situation.

- 17-inch – During the first quarter of 2001, 17-inch prices declined relatively little. However, this was mainly because there was already an aggressive price reduction in the fourth quarter of 2000. Despite offensive promotion by a Korean vendor and several Taiwanese vendors, the proportion of 17-inch models in the LCD monitor market is still minor due to the still-high market price of monitor sets. As mentioned previously, the 17-inch and 14.1-inch sizes jointly own production facilities, and, consequently, 14.1-inch-focused vendors need an aggressive promotion for 17-inch to escape tough market competition. Based on that sense, 17-inch prices will decline continuously for some time regardless of the market situation.

- 18.1-inch – Even though the 18.1-inch model hit the road earlier than the 17-inch model, the 18.1-inch has smaller market share due to a passive promotion by vendors. Basically, the 18.1-inch TFT-LCD shares production facilities with the 15-inch TFT-LCD, and there was no strong need for aggressive promotion, ignoring the low production yield. However, the surplus market, steadily improved yield, and already decided next investment pulled 18.1-inch players into more aggressive promotion. During 2001, the price gap between 18.1-inch and 17-inch models rapidly narrowed.

Despite signals of a decrease in demand in the PC market, portable PC demand will show stable growth. Portable PC demand will not be deeply affected by slow demand in the consumer market due to its focus on the corporate market. In the portable PC market, demand for the dominant 14-inch TFTs will continue to be the mainstream offering. Meanwhile, 13-inch and 12.1-inch TFTs will continue their gradual descent. However, the application diversification of portable PCs and the introduction of new Web-based applications will increase the demand for less-than 10-inch products. The movement toward bigger display sizes in portable PC will converge at the 14 or 15-inch levels, and it will strongly drive the competition in display quality that meets multimedia requirements such as high resolution, brightness, portability, and power consumption rather than display size. This trend will be the penetration barrier to new entrants.

LCDs for Non-PC Applications

Currently, TFT-LCDs are the only flat panel displays that are mass produced and have a large infrastructure already established. These LCDs are mainly used for portable PC applications, but are now rapidly evolving into monitors and other more segmented applications. Despite the overall PC market contraction, the TFT-LCD market is showing a relatively high growth in volume, although prices have continued to collapse. The TFT-LCD is one of the typical equipment industries that has had a periodic discordance between demand and supply. This demand-supply imbalance has caused difficulties for vendors and customers doing production planning and purchasing.

Unlike the portable PC market, LCD monitor demand comes from the consumer market and is leveraged by low average selling prices. Similar to the memory industry, the competition in the TFT-LCD market is capital and cost intensive, and is driven by players in Korea, Japan, and Taiwan.

So far, TFT-LCD vendors have recorded huge financial loses because of the depressed state of the market. As the profit margins of TFT-LCD manufacturing continue to narrow, production cost reductions by process optimization and increasing capacity for economies of scale will become critical for the survival of many of the suppliers. As a result of this trend, many efforts have been made to extend TFT-LCD applications beyond PC products. The suppliers have been the major proponents of non-PC applications, mainly due to a strong captive market. Some of these applications are cellular handsets, PDAs, car navigation systems, digital still cameras (DSCs), digital video cameras, Pachinko, portable games, and LCD TVs.

Cellular Handsets

Looking at the large, fast growing mobile handset market, display manufacturers are targeting cellular handset displays. The TFT-LCD has been expected to form a genuine demand starting 2003, following the scheduled mobile service evolution to 2.5-generation (2.5G) and 3G levels, and the rapid growth of the handset market itself. However, despite the expectations, the following factors have been slowing the overall cellular handset demand:

- Expectation of the launch of next-generation service (2.5G and 3G) has delayed handset purchasing.
- Next-generation service technology is unfinished.
- European service providers spent too much on the 3G-service radio wave auction, weakening their financial structure and hindering full-scale investment for service infrastructure.

The market stagnancy caused by these factors has pushed most handset vendors to devote themselves to a survival competition in the low-cost, voice-focused 2G market. Consequently, the formation of a genuine market for TFT-LCD displays for cellular handsets has been delayed, and the displays still have comparatively expensive prices and immature operating features. Currently, several types of 2G (including technologies such as GSM, TDMA and CDMA) services exist by service regions and providers. This diversity will likely remain in 2.5G (including technologies such as GPRS/EDGE, TDMA/EDGE, cdma2000, W-CDMA) services. Despite its obscure service commencement schedule and demand growth potential, 3G services will take two standards; W-CDMA for European players, and cdma2000 for U.S. and Asian players.

In 2001, the overall cellular handset market was around 425 million units, and it will grow to 672 million in 2005, with a 13% CAGR (market research by IDC). In terms of switch-on schedule, 3G service began in Japan and Korea in 2002, because replacement demand is strong due to a short 1- to 1.5-year handset life cycle. The second wave for 3G service will take place in Europe and Asia around 2004, but the huge burden that radio wave auction places on European players may make the service schedule obscure. In the United States, where cellular service users' growth is slower than it is in other developed countries, 3G service will be launched during 2004–2005, a relatively late schedule. Also, 2.5G will become the mainstream solution, not the interim step, due to the huge investment requirement of 3G service.

3G service began in 2001 at the service testing level and have a minor demand level, but the initial service launching of 3G was in 2003. In 2002, 3G demand reached 8.8 million units and took a 2% share, and 2.5G demand was 66 million units with a 14.9% share, which is much more than 3G. During 2002–2005, 3G demand will show a high CAGR of 155% and will reach 146 million units in 2005, while 2.5G demand will show a 80.4% CAGR and will reach 388 million units in 2005. The overall cellular handset market will grow at a 14.7% CAGR during the same period, as 3G service will commence in Japan/Korea, Europe, and the United States. Despite the high growth rate and the spread of the 3G service area worldwide, 3G will only have a 21.7% share in 2005. The rest of the market will be taken by 2.5G and even 2G service due to new subscribers who are more focused on the cost of the handset than service quality and diversity. The market recession in cellular handsets will push many players into an unsound financial status and slowdown the construction of widespread 3G infrastructure. The recession will also lower the 3G share beneath its expected level. Basically, a small portion of TFT-LCD may be adopted in 2.5G handsets, but will mainly be used in 3G service, in which non-voice, visual content can be operated. Therefore, TFT-LCD demand from cellular handsets will be directly influenced by the 3G service schedule and its growth.

The STN-LCD is now commonly used for cellular handsets and had a 96% share, while other displays, including TFT and organic EL, reached only a 4% share in the 2.5G category in 2001. In 2005, the STN-LCD share will decline to 88%, while the other display types will share the remaining 12%. To smoothly operate video content in 3G services, displays must be in color with high response time features because amorphous TFT-LCD will capture the main market among active color displays. The LPS display, despite better features such as lower power consumption and high resolution, will still be relegated to a niche market because of the high production cost caused by low yields. Organic EL will be affected by large barriers such as premature technology and the industry being unprepared for mass production, and the display type will only take a 2% share in 2004. In 2005, the LPS portion of the market will be larger than conventional TFT-LCD, with a share of 7.8%, because of the shift in production and yield improvements. Conventional TFT-LCD will have a 7.5% share. Despite the small share of organic EL, only 4.4% in 2005, it will show 151% CAGR during 2001–2005 (higher than any other display), and will gain a meaningful share because of its excellent performance and high brightness during outdoor use. However, the high CAGR of non-STN active displays is caused by their small initial volume, and their demand will be limited in 2.5G and 3G service. In 2005, overall color displays, TFT, LPS, and organic EL will take a share of 19.7%, slightly under that of 3G service (21.7%), reflecting the 3G-dependent demand structure of color displays in cellular handsets.

PDAs

PDAs are grouped as part of smart handheld devices (SHDs). These devices are primarily grouped as handheld companions, including PC companions, personal companions, and vertical applications devices that include a pen tablet, pen notepad, and keypad. Most of the displays used in SHDs are larger than 8 inches, similar to PC products. Only personal companions such as PDAs use the small and midsize displays.

The PDA market continues to grow because:

- Flexible point of access – Rapid progresses in the mobile communication environment will enable a flexible point of access with versatile devices.
- Declining manufacturing costs – Continuous declines in the cost of key PDA parts, such as memories and display devices, will make PDAs affordable.

Until recently, the PDA market was led by dedicated companies such as Palm and Handspring. However, the success of this category of products has brought PC and consumer device manufacturing giants such as Compaq, HP, Sharp, Sony, and Casio to the market. Market researcher IDC reports that in 2001 the total PDA market exceeded 15 million units, and will reach 58 million in 2005, with a 39.2% CAGR. Palm OS has taken the top share in the built-in operating systems market, followed by Windows-CE. However, Palm OS will lose share gradually as many new market entrants will adopt PocketPC because of its compatibility with existing PC systems.

Monochrome STN has the dominant share of the PDA display market due to its unbeatably low cost. However, active driving displays, such as TFT-LCDs, will gradually gain share through 2005 for the following reasons:

- Price – Continuous TFT-LCD price declines due to excessive production capacity
- Features – Strong multimedia PDA display feature developments, required to operate various upcoming application software

Because PDAs are used outdoors and require full-time power-on to access mobile communications, they will need high brightness and an advanced level of power-saving features; and LPS will be the most adequate candidate among the currently available displays. However, higher production costs will make LPS concede its top share spot to the amorphous TFT-LCD color display market. Amorphous TFT-LCD's share will grow to 40%, and LPS will increase its share to about 25% in 2005. In display quality and specification, organic EL will exceed the TFT-LCDs, and it will be one of the potential PDA displays. However, due to its premature production infrastructure, the development of the organic EL market was retarded until 2003. In 2003, it garnered 7.4% of initial share, which will grow to 20.7% in 2005. In 2005, TFT-LCD, including LPS revenue for PDA, will exceed $2.3 billion, much higher growth than for overall PDA unit demand, 39% reflecting rapid growth of TFT-LCD adoption rate. During same period amorphous TFT-LCD and LPS market will grow with 66.8% and 66.7% unit CAGR respectively.

Car Navigation (Telematics) Systems

The car navigation system market has initially been growing without direct relation to the mobile communication environment, and its high system price has made the after-market installation category its mainstay. In addition, the major market for car navigation systems has been regionally limited to Europe and Japan, where road systems are quite complicated. The systems could not be successful in the U.S. market, which, although it has the largest automobile demand, has a relatively well-planned road system. However, the following factors will offer a driving force for the car navigation system market:

- Increasing demand – Demand for recreational vehicles and family traveling is gradually growing.

- Better displays – The entrance of a new service concept, "telematics," related to the fast-evolving mobile communications environment, is strongly pushing the introduction of a new display terminal for automobiles and car navigation systems.

Both mobile service providers that are suffering from a slow market and automobile manufacturers that have to cope with tough competition due to excessive production capacity are eager for the creation of a new business field. Besides the basic positioning feature, other functions including total service provisions for auto travelers (e.g., accommodations, food reservations, and an automatic rescue signal in emergencies) will be added to the car navigation system, and it will bring a new business opportunity to both automobile manufacturers and mobile service providers. These functions will accelerate the long-term demand growth of car navigation systems in both the installed base and after-market segments.

The total car navigation system market was just over 4 million units in 2001 and will reach 10.2 million in 2005, for a 25.2% CAGR. Because the car navigation system is basically used and powered inside the car environment, it does not strongly require a power saving feature; however, it needs a wide viewing angle due to its slanting location in relation to the driver. Therefore, amorphous TFT-LCD, which has a lower cost structure than LPS and wider viewable angles than STN, will be an adequate solution for car navigation system displays. Amorphous TFT-LCD will gradually increase to 94.2% by 2005. STN LCDs make up the remaining percentage for telematics systems.

The 5.8-inch and 7-inch wide format displays will be standard for car navigation systems. In 2001, 5.8-in. displays garnered a 52.8% share, which was larger than that of 7-inch displays due to the high TFT-LCD price. However, as TFT-LCD prices drop and the features of car navigation systems, such as DVD watching, gradually add-up, the 7-inch display will garner more market share. In 2004, the market share of the 7-inch display will exceed that of the 5.8-inch display, and the 7-inch display will maintain a larger share through 2005. Total TFT-LCD revenue from car navigation systems will grow to $891 million in 2005 (IDC research). Among the various non-PC applications, car navigation systems will use comparatively large TFT-LCDs and consume a relatively large input substrate capacity. However, the TFT-LCD revenue portion from car navigation systems will be smaller than that from small applications due to the low price-per-unit.

Digital Still Camera

As PCs add more upgraded features and the Web becomes the fundamental PC working environment, the demand for digital still cameras will be on track for sound growth. In 2001, the overall digital still camera market shipped 17.5 million units, and it will reach 38.9 million units in 2005, with a 22.1% CAGR. However, because digital still cameras are still behind conventional film cameras in image quality and feature-considered prices, the demand for full-scale replacement will be hindered. Therefore, digital still cameras will form a differentiated demand from that of conventional cameras, and will mainly be used under a PC-related working environment until the cost/performance ratio exceeds that of conventional cameras. In addition, the fact that a PC is necessary to achieve any output image will be a barrier to widespread demand.

Displays for digital still cameras require the following:
- High resolution for a clear output image
- Low power consumption for a portable user environment
- Comparatively high brightness to make images visible outside

In 2005, among all displays for digital still cameras, amorphous TFT-LCD demand will grow to 9.1 million units, with a 69% share. In 2005, 2.3 million LPS units will be shipped, and the demand will grow to 4.0 million in 2005, maintaining a similar market share. Currently, the display satisfying all of these needs with an affordable price is TFT-LCDs, including LPS. By 2005, TFT-LCDs share will increase to 77%. In 2001, amorphous TFT-LCD had a 27% share; a smaller share than that of LPS because of the better features and smaller display cost in digital still camera BOM. However, as more LCD vendors focus on low-cost amorphous TFT-LCD production and specification, amorphous TFT-LCD demand will show a slightly larger growth than LPS. During 2001–2005, the amorphous type will record a 26.4% CAGR, while LPS will show a 24.2% CAGR. In 2005, amorphous TFT-LCD will increase its share to 31%.

In 2001, the combined demand for TFT-LCD and LPS was 12.3 million units, and it will reach 29.9 million units in 2005, with a 25.0% CAGR. During the same period, the TFT-LCD price will drop by 31.4%, resulting in a lower revenue CAGR (13.7%) than volume growth. In 2005, total TFT-LCD revenue for digital still cameras will reach $512 million units in 2005. In comparison, TFDs will reduce its market share with the least CAGR over the five years, with only 8.9 million units shipping in 2005 compared with 29.9 million units for TFT-LCD and LPS.

Digital Video Camera

Based on basic demand for video cameras, a continuously evolving Web environment and PC upgrades will result in sound growth of digital video camera demand. In addition, replacement demand of analog products will accelerate the growth of digital still cameras, regardless of the overall slow demand of consumer electronic goods. In 2001, the total digital video camera demand was 7.2 million units, and it will grow to 13.1 million in 2005, with a 15.9% CAGR.

Displays for digital video cameras basically require the following:
- Fast response time – Needed for a smooth motion picture
- Additional brightness – Required for outdoor use
- Low power consumption – Required for a portable environment

Consequently, TFT-LCDs, including LPS, are the only possible solution for digital video cameras through 2005. This is in spite of the entrance of other active driving displays, such as organic EL, due to the TFT-LCD's low production cost. The TFT-LCD displays, including LPS, will record $353 million in 2005. During this period, digital video camera unit demand will record a 15.9% CAGR; however, because of continuing price cuts, TFT-LCD revenue will show a lower CAGR of 5.4%. The TFT-LCD market has been through a few supply shortage/supply surplus cycles because of its generic equipment industry character, and this market fluctuation will remain for a time in the PC-related applications market with portable PC and LCD monitors. However, despite the overall TFT-LCD situation, TFT-LCD prices will continue to decline during 2001–2005 because of the relatively small size of the market and low demand for customized product designs.

Pachinko

The Pachinko machine has quite a large market, but mostly in Japan. Pachinko is a mixture between a slot machine and pinball. Because it is a gambling machine, Pachinko requires comparatively large displays for various non-PC applications and consequently consumes a meaningful portion of input capacity. However, the specialized environment of the gambling machine industry, which has to be under official control and influence as well as regionally dependent demand, make it difficult to enter this market.

In 2001, the size of the Pachinko display market was 3.3 million units, which will decrease to 2.8 million in 2005, with –3.7% CAGR as the market continues to lose volume. The Pachinko market is quite saturated and does not have a large growth potential. Consequently, the upcoming display demand will be mainly for replacement needs. As prices have continued to decline, most Pachinko machines have already adopted TFT-LCD displays. For the forecast period, 2001–2005, the overall Pachinko machine TFT-LCD demand will gradually decrease. Therefore, the Pachinko machine display market will remain unattractive and inaccessible for non-Japanese LCD vendors.

Pachinko machine screens do not require high-level features such as power saving, but must be reliable due to an indoor and quite severe operating environment. Thus, mainly conventional amorphous TFT-LCD and even low-cost TFD-LCD screens will be used, along with a smaller number of STN displays. The continuous TFT-LCD price declines will expel the STN displays from this product category in 2005. In 2001, the overall TFT-LCD revenue for Pachinko machines was $353 million, which will decrease to $239 million, with a –9.3% CAGR in 2005. The revenue CAGR decline will be larger than the volume CAGR decline during the 2001–2005 period due to LCD price declines.

Portable Game

Like the Pachinko machine market, the portable game market also has unevenly distributed regional demand, with most of the machines sold in Japan. In other areas, console-box-type game machines without displays are quite popular. Game machines form a huge worldwide market, and many giant companies, including Nintendo, Sony, and Microsoft, focus on it with newly developed devices. As more complicated features are adopted by game machines, portable products with limited specifications and display quality will not grow enough to match the increasing demand in the game machine market overall.

Researcher IDC expects that the total demand for portable games will continue to decrease from 2003 to 2005. Amorphous TFT-LCD is the dominant portable game display and reported an over 90% share in 2003. The remainder of the market share goes to STN, at 4.7%. A low production cost and power-saving features are the main reasons that TFT-LCD is dominant. However, it is also because the specific supply structure has led to a sole vendor covering the market, which effectively bars LPS displays from portable games. In 2001, the overall portable game market shipped 24.5 million units, and a 95.3% share (23.4 million units) was provided by TFT-LCD. In 2005, the total market will decrease to 19.5 million units, with a negative 5.5% CAGR, but the market share for TFT will increase to 99.1% as the TFT-LCD prices gradually decline and replace STN displays in the market.

In 2001, the overall TFT-LCD revenue coming from portable game products was $815 million, but it will decrease to $462 million in 2005 as both the market volume and TFT-LCD prices will decrease. During this period, TFT-LCD revenue will show a negative 13.2% CAGR, which is a larger revenue decline in CAGR than that of the volume CAGR decline. However, the portable game display market is almost entirely covered by one LCD vendor, Sharp; thus, the decline in demand will not directly affect the overall industry.

Differences between CRTs and LCDs

An important difference between CRT monitors and LCD panels is that the former requires an analog signal to produce a picture, and the latter require a digital signal. This fact makes the setup of an

LCD panel's position, clock, and phase controls critical in order to obtain the best possible display quality. It also creates difficulties for panels that do not possess automatic setup features, and thus require these adjustments to be made manually.

The problem occurs because most panels are designed for use with current graphics cards, which have analog outputs. In this situation, the graphics signal is generated digitally inside the PC, converted by the graphics card to an analog signal, then fed to the LCD panel where it must be converted back to into a digital signal. For the complete process to work properly, the two converters must be adjusted so that their conversion clocks are running in the same frequency and phase. This usually requires that the clock and phase for the converter in the LCD panel be adjusted to match that of the graphics card. A simpler and more efficient way to drive a LCD panel would be to eliminate the two-step conversion process and drive the panel directly with a digital signal. The LCD panel market is growing from month-to-month, and with it the pressure on graphics adapter manufacturers to manufacture products that eliminate the two-step conversion process.

One of the most important aspects of a consumer device, whether a PC or handheld device, is the screen, since it is provides the ability to read the device. In the past few years, we have seen a gradual transition from monochrome gray-scale screens to color in PC companions, PDAs, and smart phones. Higher cost, power requirements, and the fact that it is hard to see what is on the screen in high light conditions, such as outdoors, have impeded the use of color screens. The majority of mobile consumer devices have heretofore been equipped with thin film transistor liquid crystal displays (TFT-LCDs), which are lighter and thinner, and consume less power than a standard CRT. As TFT-LCDs have become more prevalent, the average selling prices have also fallen. There have been intermittent shortages of LCDs because of pricing and product availability, and manufacturers from several countries ramping up production of new technologies such as PDPs, OLEDs, Thin CRT, and light-emitting polymer (LEP) diodes.

Key LCD advantages include:

- Sharpness – The image is said to be perfectly sharp at the native resolution of the panel. LCDs using an analog input require careful adjustment of pixel tracking/phase.
- Geometric distortion – Zero geometric distortion at the native resolution of the panel. Minor distortion is apparent for other resolutions because images must be re-scaled.
- Brightness – High peak intensity produces very bright images. This works best for brightly lit environments.
- Screen shape – Screens are perfectly flat.
- Physical dimension – Thin, with a small footprint. They consume little electricity and produce little heat.

Key LCD disadvantages include:

- Resolution – Each panel has a fixed resolution format that is determined at the time of manufacture, and that cannot be changed. All other image resolutions require re-scaling, which can result in significant image degradation, particularly for fine text and graphics, depending upon the quality of the image scaler. Therefore, most applications should only use the native resolution of the panel.
- Interference – LCDs using an analog input require careful adjustment of pixel tracking/ phase in order to reduce or eliminate digital noise in the image. Automatic pixel tracking/ phase controls seldom produce the optimum setting. Timing drift and jitter may require frequent readjustments during the day. For some displays and video boards, it may not be possible to entirely eliminate the digital noise.

- Viewing angle – Every panel has a limited viewing angle. Brightness, contrast, gamma and color mixtures vary with the viewing angle. This can lead to contrast and color reversal at large angles, and therefore, the panel needs to be viewed as close to straight ahead as possible.

- Black-level, contrast and color saturation – LCDs have difficulty producing black and very dark grays. As a result they generally have lower contrast than CRTs, and the color saturation for low intensity colors is also reduced. Therefore, they are less suitable for use in dimly lit and dark environments.

- White saturation – The bright-end of the LCD intensity scale is easily overloaded, which leads to saturation and compression. When this happens the maximum brightness occurs before reaching the peak of the gray-scale or the brightness increases slowly near the maximum. It requires careful adjustment of the contrast control.

- Color and gray-scale accuracy – The internal gamma and gray-scale of an LCD is very irregular. Special circuitry attempts to fix it, often with only limited success. LCDs typically produce fewer than 256 discrete intensity levels. For some LCDs, portions of the gray-scale may be dithered. Images are pleasing but not accurate because of problems with black-level, gray-level, and gamma, which affects the accuracy of the gray-scale and color mixtures. Therefore, they are generally not suitable for professional image color balancing.

- Bad pixels and screen uniformity – LCDs can have many weak or stuck pixels, which are permanently on or off. Some pixels may be improperly connected to adjoining pixels, rows or columns. Also, the panel may not be uniformly illuminated by the backlight, resulting in uneven intensity and shading over the screen.

- Motion artifacts – Slow response times and scan rate conversion can result in severe motion artifacts and image degradation for moving or rapidly changing images.

- Aspect ratio – LCDs have a fixed resolution and aspect ratio. For panels with a resolution of 1280 x 1024, the aspect ratio is 5:4, which is noticeably smaller than the 4:3 aspect ratio for almost all other standard display modes. Some applications may require switching to a letterboxed 1280 x 960, which has a 4:3 aspect ratio.

- Cost – Considerably more expensive than comparable CRTs.

Further requirements of a display include:

- Contrast ratio – The contrast ratio (the brightness of an image divided by the darkness of an image) contributes greatly to visual enjoyment, especially when watching a video or television. The LCD panel determines the contrast ratio by blocking out light from the backlight—the blackness of the black and the brightness of the white define a monitor's contrast ratio.

- Color depth and purity – The color filters of the LCD panel—the red, green, and blue sub-pixels—establish the colors shown on a flat panel display. The number of colors a panel can display is a function of how many bits of information make up each pixel on the screen.

- Response time – You need an LCD that has very fast response time when you are looking at full motion video if the scene is going to look sharp and crisp. Otherwise, you would see "comet tails" streaming behind moving images. Slow response times are unacceptable for video and fast animation applications.

The following lists the most common video processing algorithms that will be used for LCD-TVs:

- Aspect ratio conversion
- De-interlacing

- Color compensation – Compensates for variations in the color performance of a display, and allows any color to be addressed independently and adjusted without impacting other colors. For example, the color depth of an LCD panel may be only six bits per pixel, providing 262,144 different colors—but an incoming analog computer signal may be eight bits per pixel, providing 16.7 million colors.
- Noise reduction.
- Motion artifact reduction.
- Video sample rate conversion.

PDP—Plasma Display Panels

Plasma Display Panels (PDPs) consist of front and back substrates with phosphors deposited on the inside of the front plates. These displays have cells that operate similarly to a plasma or fluorescent lamp, where the discharging of an inert gas between the glass plates of each cell generates light. Depending upon the type of gas used, various colors can be generated. In a monochrome display, the light from the gas discharge is what is seen on the display. To obtain a multicolor display, phosphor is required. The plasma panel uses a gas discharge at each pixel to generate ultraviolet radiation that excites the particular phosphor, which is located at each pixel. The PDP requires a wide range of switching circuits and power supplies to control the on and off gas discharges, and to effect consistent quality images. It has been a complicated technology, in part because the Y-side driver circuit must handle both address scanning and image-sustaining functions.

The newest generations of PDPs feature phosphors that are continuously "primed"; that is, the panel remains illuminated at a low level so it stays active and responds to immediate luminescence. But the downside impact of continued illumination—until recently—was that it created an ever-present low-level glow, which reduced contrast. The vital design objectives for new generations of maturing PDP technology became constant illumination with high contrast and very high brightness, or luminosity.

Advantages of PDPs over cathode ray tube (CRT) TVs are:
- Accurate cell structures – PDPs have cell structures accurate enough to produce a geometrically perfect picture. Standard CRT TVs have a geometric distortion due to the inability of the electron beam to focus on all points.
- High brightness levels – PDPs are evenly illuminated without the typical dark/hot spots observed from CRTs.
- Perfect focus – PDPs can achieve perfect focus, whereas the CRT has regions that are less focused than others.
- No magnetic interference – PDPs are not susceptible to magnetic fields (while CRT's electron beam is influenced even by the earth's magnetic fields).
- Thin profile – A plasma monitor/TV is never more than 4 inches thick and is lightweight. This allows PDPs to be hung on the wall, making it an excellent solution when dealing with space restrictions.
- Large screen size.
- Full color with good color parity.
- Fast response time for video capability.
- Wide viewing angle of greater than 160 degrees in all directions.
- Insensitivity to ambient temperatures.

Overall, PDPs are believed to offer superior cost-to-view for TV applications in form factors greater than 30-inches in the near term. The TFT-LCD and LCoS alternatives have yet to match the combination of picture quality, price, form factor, or availability offered by PDPs, and OLED technology has yet to gain critical mass in the small format segments alone. While PDPs face serious competition from LCD systems in the sub-40-inch range and from rear- and front-projection in the 60-inch size range, they will find a sweet spot in the 40- to 60-inch range.

Disadvantages of PDPs are:

- Costly to manufacture – Plasma technology is complex and the plasma manufacturing process is time consuming. Hence, the yield when producing a line of plasma panels in a factory is very low. Currently, PDPs are incapable of taking significant volumes of sales away from CRT TV due to high consumer cost; approximately $10,000.

- Pixel burn-out – Older generations of color plasma TV/monitors have experienced gas pixels burning out on the panel, and creating a visible black spot on the screen. This leads to frustrated customers who need to pay for the display's repair. Also, due to phosphorus degradation (amount of time to 50% luminance), the panel has a shorter panel lifetime.

- Course resolution – Barrier ribs are one source of the tradeoffs inherent in PDP display technology. They are responsible, for example, for PDPs relatively coarse resolution, which appears to be limited to just over 30 pixels/inch. A PDP is required to be at least a 42- to 50-inch size. Some PDP makers are working to optimize their barrier rib manufacturing processes—sand blasting, screen printing, etc.—in order to achieve finer geometries and, thus, higher resolution. But shrinking the barrier ribs has two significant drawbacks. First, the smaller pixel cell area reduces the brightness of the display, and second, the smaller the ribs, the more fragile they become. The result of broken ribs is either low manufacturing yield (which raises cost), or the most disturbing of display defects, pixels that are always illuminated.

- No analog gray scale – Analog gray scale is impossible with a PDP, since its pixels are either on or off due to gas physics, with no gradations in between. Gray scale is generated, therefore, through digital frame-rate modulation, which introduces its own set of tradeoffs. To get full gray scale, a single PDP frame may contain up to twelve sub-frames, each consisting of combinations of so-called sustain pulses. Low gray levels typically require just a few pulses, while high gray levels may require up to several hundred pulses. The need for complex frame rate modulation to achieve gray scale means that PDPs require complex control circuitry - specifically a heavyweight DSP (digital signal processing) circuit to generate the data required, and a large high-frequency pulse circuit for the PDPs address and sustain voltages. The additional complexity introduces cost, of course, in both parts count and real estate.

- Poor image quality – Image quality of PDP displays has radically improved since they were first introduced a few years ago, and it is expected that image quality will continue to improve. Nevertheless, because of the PDPs inherent tradeoffs, they still lag the video quality and overall performance of the CRT. The more sub-frames, the less bright the display, for example. This can be compensated for, of course, by making cell sizes bigger, but this has the associated tradeoff of reduced resolution.

- Poor performance – Performance issues similar to those associated with LCDS including; luminance, contrast ratio, power efficiency (luminous efficiency), elimination of motion artifacts, increased resolution for advanced TV, driving voltage reduction, poor dynamic range and black levels, and high pixel count.

■ Low luminance (non-sunlight readable) – High luminance is required for outdoor applications.

■ Low contrast in high ambient light.

■ High power consumption and heat generation.

Among the flat-panel display candidates for large-screen TVs, the PDP has a clear head start, with 42-inch-diagonal and larger PDP TVs already available in the commercial market. Plasma Display Panels are capacitive devices that emit light through the excitation of phosphors. The PDP adds the complexity of a gas chamber, which necessitates a high vacuum, and a precise three-dimensional cell using an internal barrier-rib structure to define discrete pixel locations.

The cost of PDPs has also come down nicely over the past few years, driven in part by supply glut and in part by manufacturing efficiencies. This sharp price reduction will continue, but the ambitious cost reductions that the PDP industry touted early on have been repeatedly pushed back. Further improvements also continue in areas of picture quality, including contrast ratio and brightness levels. It is expected that PDP TVs will still be prohibitively expensive even by 2004, and will not make much of an impact on the consumer market. The rather complicated steps like barrier rib construction, phosphor deposition onto the side walls of the barrier ribs, sealing, and evacuation are major bottlenecks in the long and expensive PDP manufacturing processes. The PDP process requires about 70 tools and 70 process steps. A theoretical display fabrication facility with an annual capacity of 250,000 displays would require about a $250 million investment for PDP. Because the key components are fabricated using photolithographic or thick-film deposition technology, yield improvement holds the key to further cost reduction. Furthermore, a large number of driver ICs are required to create light emissions from each cell, and the plasma discharges consume a high voltage (150-200V) and a lot of power (300W). Finally, there continue to be technical challenges relating to picture quality; for example, contrast tends to deteriorate over time because of a continuous occurrence of slight plasma discharge.

PDP Market

Plunging prices, improving quality, and enhanced features have enabled PDP systems to make major inroads into the consumer market, thus, paving the way for rapid growth over the next several years, according to iSuppli/Stanford Resources. The year 2002 brought dramatic gains in PDP sales with revenue increasing by 62% and units sold increasing by 143%. The average factory price of a PDP system fell to $2,743, down 33% from $4,100 in 2001, with the drop in PDP system pricing a result of declines in panel pricing and material costs as well as a reduction in channel costs.

The worldwide PDP system factory revenue will grow to $11.2 billion in 2007, up from $2.2 billion in 2002. Unit shipments will rise to 8 million in 2007, up from 807,096 in 2002 and 1 million in 2003. The cost-per-diagonal-inch fell from in excess of $200 to $150 in 2001, to $100 by the end of 2002, and is predicted to fall to $75 by year-end 2003. It is predicted that by 2007 the consumer sector will account for 81% of systems.

The main reasons for the increasing popularity of PDP systems among consumers are:

■ Declining prices – A 50-inch PDP that cost $10,000 in 2001 is expected to cost only $1,200 in 2010.

- Performance benefits – Better viewing angles, improved picture quality, and thin form-factors compared to competing technologies like CRTs, LCDs, and projectors.
- Size availability – Plasma Displays are available in screen sizes ranging from 32-inch to 63-inches, covering the needs of different markets and geographies.
- Consumer outlet marketing – Availability of PDP TVs at mainstream consumer electronics outlets rather than only at specialty electronics stores or through audio-video integrators.
- Digital TV advancement – Worldwide migration toward Digital TV through cable and satellite networks and HDTV format. The progressive scanning format of PDPs works well with DTV content.

The global PDP TV industry is controlled (OEM) mainly by Sony, Panasonic (Matsushita), Pioneer, NEC, and Fujitsu. The top five plasma panel manufacturers are Fujitsu-Hitachi Plasma, Matsushita, Pioneer, NEC, and LG Electronics. The existing providers are, however, facing competition from Chinese, and Taiwanese companies are also set to join the market, using their ample production capacity for PC monitors.

A combination of factors is responsible for the high prices, including:

- Cost of driver ICs – Unless some way of significantly reducing the operating voltage can be found.
- Driver chip expense – Although column driver chips are less expensive, a large number are required. For example, 60 are needed to drive a 1920 x 1080 display, and, thus, represents a significant cost.
- Large power supply.
- Specialized glass substrates.
- Multiple interconnects.
- EMI filters.
- Metal electrode parts.
- Low manufacturing yield – Although it is now approaching 80%.
- Material costs – Plasma displays are driven by material costs such as low-cost ceramic and packaging materials.

A number of cost reduction possibilities exist for plasma displays, and are listed below:

- Cheaper substrate – Use ordinary soda-lime, float-process window glass. However, this suffers substantial dimensional change during high-temperature processing.
- Substrate layer replacement – Replace one of the glass substrates with a sheet of metal laminated to a ceramic layer.
- Different electrodes – Use all-metal electrodes instead of either transparent indium-tin-oxide or tin-oxide electrodes in parallel with high conductivity, opaque metal bus bars.
- Different scan driver scheme – Use single-scan addressing rather than dual-scan designs, which requires half the number of data address drivers. However, the increased number of horizontal lines requires more precise address time during the fixed frame time.

As for LCD TV, there are a number of commonly used video processing algorithms that are used with PDP TVs, and are listed as follows:

- Post-processing – Gamma correction, sharpness enhancement, color correction
- Aspect ratio conversion
- De-interlacing
- Noise reduction
- Motion artifact reduction
- Video sample rate conversion

Functionality

Plasma Display Panels are like CRTs in that they are emissive and use phosphor. They are also like LCDs in their use of an X and Y grid of electrodes separated by an MgO dielectric layer, and surrounded by a mixture of inert gases (such as argon, neon, or xenon) that are used to address individual picture elements. They work on the principle that passing a high voltage through a low-pressure gas generates light. Essentially, a PDP can be viewed as a matrix of tiny fluorescent tubes that are controlled in a sophisticated fashion. Each pixel, or cell, comprises a small capacitor with three electrodes. An electrical discharge across the electrodes causes the rare gases sealed in the cell to be converted to plasma form as it ionizes. Plasma is an electrically neutral, highly ionized substance consisting of electrons, positive ions, and neutral particles. Being electrically neutral, it contains equal quantities of electrons and ions and is, by definition, a good conductor. Once energized, the cells of plasma release ultraviolet (UV) light that then strikes and excites red, green, and blue phosphors along the face of each pixel, causing them to glow.

Within each cell, there are actually three subcells, one containing a red phosphor, another a blue phosphor, and the third a green phosphor. To generate color shades, the perceived intensity of each RGB color must be controlled independently. While this is done in CRTs by modulating the electron beam current, and therefore also the emitted light intensities, PDPs accomplish shading by pulse code modulation (PCM). Dividing one field into eight sub-fields, with each pulse-weighted according to the bits in an 8-bit word, makes it possible to adjust the widths of the addressing pulses in 256 steps. Since the eye is much slower than the PCM, it will integrate the intensity over time. Modulating the pulse widths in this way translates into 256 different intensities of each color—giving a total number of color combinations of 256 x 256 x 256 = 16,777,216.

The fact that PDPs are emissive and use phosphor means that they have an excellent viewing angle and color performance. Initially, PDPs had problems with disturbances caused by interference between the PCM and fast moving pictures. However, this problem has been eliminated by fine-tuning the PCM scheme. Conventional plasma screens have traditionally suffered from low contrast. This is caused by the need to "prime" the cells by applying a constant low voltage to each pixel. Without this priming, plasma cells would suffer the same poor response time of household fluorescent tubes, making them impractical. The knock-on effect, however, is that pixels that should be switched off still emit some light, reducing contrast. In the late 1990s, Fujitsu alleviated this problem with new driver technology that improved contrast ratios from 70:1 to 400:1. By the year 2000 some manufacturers claimed as much as 500:1 image contrast, albeit before the anti-glare glass is added to the raw panels.

The biggest obstacle that plasma panels have to overcome is their inability to achieve a smooth ramp from full-white to dark-black. Low shades of gray are particularly troublesome, and a noticeable posterized effect is often present during the display of movies or other video programming with dark scenes. In technical terms, this problem is due to insufficient quantization, or digital sampling of brightness levels. It's an indication that the display of black remains an issue with PDPs.

Manufacturing is simpler than for LCDs, and costs are similar to CRTs at the same volume. Compared to TFTs, which use photolithography and high-temperature processes in clean rooms, PDPs can be manufactured in less clean factories using low-temperature and inexpensive direct printing processes. However, with display lifetimes of around 10,000 hours, a factor not usually considered with PC displays—cost per hour—comes into play. For boardroom presentation use this is not a problem, but for hundreds of general-purpose desktop PCs in a large company it is a different matter.

However, the ultimate limitation of the plasma screen has proved to be pixel size. At present, manufacturers cannot see how to get pixels sizes below 0.3 mm, even in the long term. For these reasons PDPs are unlikely to play a part in the mainstream desktop PC market. For the medium term they are likely to remain best suited to TV and multi-viewer presentation applications employing large screens, from 25 to 70 inches.

Figure 10.4 shows the plasma display panel controller.

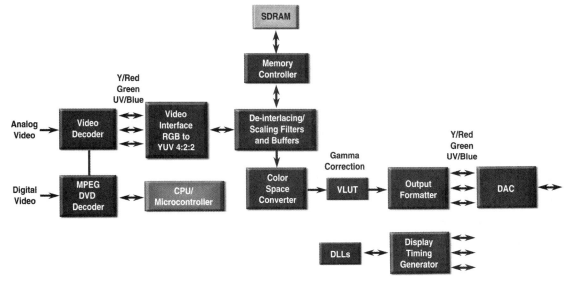

Figure 10.4: Plasma Display Panel Controller

Key Differences Between LCDs and PDPs

Key differences between LCDs and PDPs include form factor (display size), video quality, life span and power consumption, capital cost.

Form Factor

It is difficult for PDPs to compete effectively with TFT-LCDs in form factors below 30-inch because of the low manufacturing cost of TFT-LCDs. This is confirmed by the larger LCD shipments in unit terms for the smaller form factors. It is still unknown as to which solution will win for form factors exceeding 30-inch. However, TFT-LCD suppliers such as Sharp and Samsung are devoting substantial resources to overcoming technology and cost hurdles, which could propel TFT-LCD towards becoming the mainstream TV display technology. Furthermore, it is estimated that TFT-LCDs only have a 5% penetration within the TV market, and, therefore, the growth path appears to be extremely promising for the second-half of this decade. Currently, the maximum size available for PDPs is 70 inches, but is 54-inch (demos) and 40-inch (in reality) for LCDs. Also, cost-per-area PDP becomes cheaper at greater than 40-inches, while for LCD is much higher with larger screens.

The LCD is lagging behind plasma in the large-screen flat panel TV market as 50-inch plasma models are already in volume production. Non-Japanese vendors led by Samsung and LG Philips are moving to supply large-screen LCD TVs as they exhibit 46-inch thin-film (TFT) models at electronic shows. Samsung's 46-inch model has picture quality equivalent to HDTV with a Wide Extended Graphics Array (WXGA) standard resolution (1,280 x 720). It has high specs equivalent to those of the plasma model, that is, a wide field angle (170 degrees vertically and horizontally), a contrast ratio of 800:1, and a maximum brightness of 500 candles. Furthermore, Samsung has overcome the slow response time issue of LCD TV with a fast response time of 12 ms—the minimum response rate requirement to allow a viewer to watch an animated picture with comfort is 15 ms. Other companies are innovating in different ways, such as Sharp, which has developed leading edge technology with a 4-inch TFT color display using a plastic substrate. This substrate enables a thinner, lighter, and stronger display, which is suitable for mobile applications. It will be one-third the thickness of the glass-based substrate display, one-fourth the weight, and have more than 10 times greater shock resistance.

Video Quality

The PDP approaches CRT video quality. Liquid crystal displays, however, require improvement in transition speeds and black levels. Currently, TFT-LCDs account for an estimated 66% of total FPD sales because the high resolution, color, and refresh rate. Another reason is that the integration of driver circuits upon the transistor array continues to position the LCD as the best solution for mainstream applications.

Life Span and Power Consumption

Two of the main criticisms of PDPs include their relatively short lifetimes and high power consumption. However, they have superior picture quality and an attractive form factor, and therefore, may be entrenched as the premium technology platform within the ultra-large-area display market.

Capital Cost

Plasma display panel manufacturing facilities have lower amortized cost. However, LCD factories tend to be similar in cost to a wafer fab (~ $1.5 billion). It is a very risky proposition if the market is limited to a few thousand units. Anything 40 inches or higher is difficult to produce.

Other Considerations

Other considerations in the difference between LCDs and PDPs are viewing angle, light sources, and color technology, as follows:

- Viewing angle – PDPs allow 160 degrees plus horizontal viewing angle, and LCDs typically allow 90 degrees (up to 160 degrees possible) vertical viewing angle.
- Light sources – PDP has an emissive (internal) light source, while LCDs have a transmissive (external backlight).
- Color technology – PDPs use phosphor (natural TV colors), while LCDs use color filters (different color system than TV).

Key Differences Summary

To survive in the competitive flat panel market, vendors must continue to improve production technology. In addition to the need for yield improvement of the panel, other components and parts in the TV system must be integrated as they account for the major cost portions. For instance, in the case of LCD TV, no single vendor has a technical edge over the others, and reduction of component cost holds the key to competitiveness and profitability.

PALCD—Plasma Addressed Liquid Crystal Display

A peculiar hybrid of PDP and LCD is the plasma addressed liquid crystal display (PALCD). Sony is currently working, in conjunction with Tektronix, on making a viable PALCD product for consumer and professional markets.

Rather than use the ionization effect of the contained gas for the production of an image, PALCD replaces the active matrix design of TFT-LCDs with a grid of anodes and cathodes that use the plasma discharge to activate LCD screen elements. The rest of the panel then relies on exactly the same technology as a standard LCD to produce an image. Again, this won't be targeted at the desktop monitor market, but at 42-inch and larger presentation displays and televisions. The lack of semiconductor controls in the design allow this product to be constructed in low-grade clean rooms, reducing manufacturing costs. It is claimed to be brighter, and retains the "thin" aspect of a typical plasma or LCD panel.

FEDs—Field Emission Displays

Some believe FED (field emission display) technology will be the biggest threat to LCD's dominance in the panel display arena. The FEDs capitalize on the well-established cathode-anode-phosphor technology built into full-sized CRTs, and use that in combination with the dot matrix cellular construction of LCDs. However, Instead of using a single bulky tube, FEDs use tiny "mini tubes" for each pixel, and the display can be built in approximately the same size as a LCD screen.

Each red, green, and blue sub-pixel is effectively a miniature vacuum tube. Where the CRT uses a single gun for all pixels, a FED pixel cell has thousands of sharp cathode points, or nanocones, at its rear. These are made from material such as molybdenum, from which electrons can be pulled very easily by a voltage difference. The liberated electrons then strike red, green, and blue phosphors at the front of the cell. Color is displayed by using "field sequential color," in which the display will show all the green information first, then redraw the screen with red, followed by blue.

In a number of areas, FEDs appear to have LCDs beaten. Since FEDs produce light only from the "on" pixels, power consumption is dependent on the display content. This is an improvement over LCDs, where all light is created by a backlight that is always on, regardless of the actual image on the screen. The LCD's backlight itself is a problem that the FED does not have. The backlight of a LCD passes through to the front of the display, through the liquid crystal matrix. It is transmissive, and the distance of the backlight to the front contributes to the narrow viewing angle. In contrast, a FED generates light from the front of the pixel, so the viewing angle is excellent—160 degrees both vertically and horizontally.

Field Emission Displays also have redundancy built into their design, with most designs using thousands of electron emitters for each pixel. Whereas one failed transistor can cause a permanently on or off pixel on a LCD, FED manufacturers claim that FEDs suffer no loss of brightness even if 20% of the emitters fail.

These factors, coupled with faster than TFT-LCD response times and color reproduction equal to the CRT, make FEDs look a very promising option. The downside is that they may prove hard to mass produce. While a CRT has just one vacuum tube, a SVGA FED needs 480,000 of them. To withstand the differences between the vacuum and external air pressure, a FED must be mechanically strong and very well sealed. By the late 1990s, six-inch color FED panels had already been manufactured, and research and development on 10-inch FEDs was proceeding apace.

DLP—Digital Light Processor

Digital Light Processors (DLPs) enable TV to be completely digital and provide superior video performance required by home entertainment enthusiasts. Texas Instruments (TI) has developed and patented this new technology (the semiconductor chip and the engine).

The DLP mirror chip is one of the most exciting innovations in display technology, as it has been successfully exploited commercially. Fundamentally, the mirror chip is a standard static memory chip design. Memory bits are stored in silicon as electrical charges in cells. An insulating layer with a mirror finish above the cells is added, and is then etched out to form individual hinged flat squares. When a memory bit is set, the charge in the cell attracts one corner of the square. This changes the angle of the mirrored surface, and by bouncing light off this surface, pictures can be formed. The optical semiconductor chip has an array of 480,000 (SVGA), 786,000 (XGA) or 1,310,000 (SXGA) hinged, microscopic mirrors mounted on a standard logic device.

Color is also a complication, since the mirror chip is basically a monochromatic device. To solve this, three separate devices can be used, each illuminated with a primary color, and tiny mirrors operate as optical switches to create a high resolution, full color image. Alternatively, a single device can be placed behind a rotating color wheel with the chip displaying red, green, and blue components sequentially. The chip is fast enough to do this and the resulting picture looks fine on color stills, but has problems handling moving images. Complicated optics are needed with DLPs to convert a picture the size of a postage stamp into a projectable display. Heat is unavoidable because a lot of light is focused on the chip in order to make the final image bright enough, and a large amount of ventilation is required to cool the chip. This process is noisy, although the latest projectors have a chip encased within a soundproof enclosure.

The DLP can be designed into both front and rear projection TVs. While the mirror chip is currently only available in projectors, it is likely that they will appear in a back-projecting desktop display eventually. Close to 1 million projectors and displays based on DLP technology have been shipped in six years, and well over half those shipments have been made in the past two years. Using only a single panel, for example, it allows a very small, lightweight optical subsystem to be used, decreasing the size and weight of the projection engine, and allowing slim, lightweight, elegant products.

Recent DLP products feature contrast ratios of 1,300:1. This performance is due to the unique method that a DMD (Digital Micromirror Device) uses to digitally switch light on and off. Because the DMD steers light into or out of the projection path, optical designers can uniquely control and limit the off-state light, independent of on-state light, which controls overall brightness. Prototype DLP subsystems in the laboratory show the potential to deliver a contrast ratio in excess of 2,000:1 in the near future—a level of performance which rivals that of CRT and film.

Contrast ratio is not, of course, the only factor in image quality. The overall superiority of DLP technology in this area can be gauged, however, by the fact that a derivative—DLP Cinema technology—is currently the de facto standard projection technology as the movie industry begins the transition to a new, digital era. Only DLP Cinema technology reliably delivers digitally accurate images to a large screen, which, for most movie goers, is superior to images they see with traditional celluloid-based film.

Particularly encouraging is the cost to manufacture of DLP technology. Although proven to be highly manufacturable, yields continue to increase with significant scope for further improvement, allowing costs to continue to be driven down. It is estimated that by 2005 the bill of materials cost of a display product based on DLP technology will be comparable to that of a similar product based on CRT technology. Manufacturers have announced plans to bring to market 42-inch-diagonal screen tabletop TVs which, at $3,000, are priced comparably with CRT technology-based competitors. And they will offer equal or superior image quality at a fraction of the CRT weight and footprint.

Figure 10.5 (a) and 10.5 (b) shows the anatomy and system functionality of the DLP. Figure 10.6 shows the digital image processing board of the DLP or LCD projector.

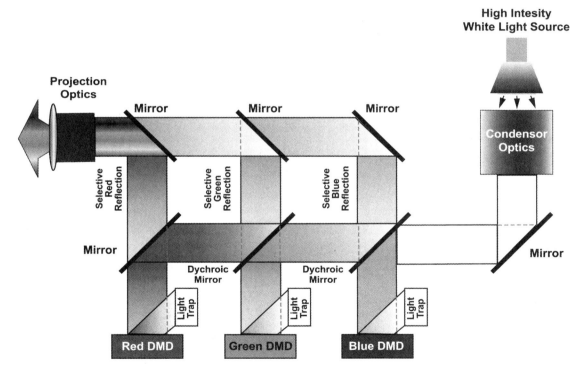

Figure 10.5 (a): Anatomy of the Digital Light Processor

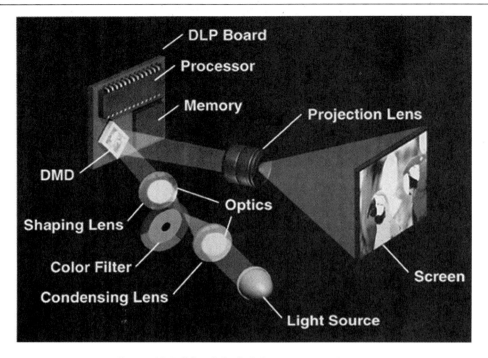

Figure 10.5 (b): Digital Light Processor System

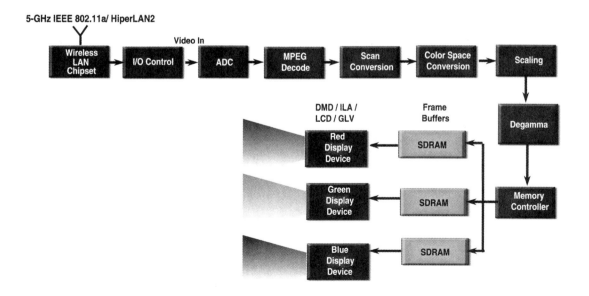

Figure 10.6: LCD/DLP Projector—Digital Image Processing Board

Organic LEDs

When magnified, high-resolution organic LED microdisplays less than one-inch diagonal in size enable large, easy to view virtual images similar to viewing a computer screen or large TV screen. The display-optics module can be adapted easily to many end products such as mobile phones and other handheld Internet and telecommunications appliances, enabling users to access full Web and fax pages, data lists, and maps in a pocket-sized device. The image can be superimposed on the external world in a see-through optics configuration; or, with independent displays for both eyes, a true 3-D image can be created.

Efficiency is always a concern for displays, but it is especially important for microdisplays that are to be used in handheld appliances or headsets for wearable products. The best match to the range of microdisplay requirements is achieved by the combination of OLEDs on silicon. The OLEDs are efficient Lambertian emitters that operate at voltage levels (3-to-10 V) accessible with relatively low-cost silicon. They are capable of extremely high luminance, a characteristic that is especially important for use in the helmets of military pilots, even though most of the time they would be operated at much lower levels. Luminance is directly linear with current, so gray scale is easily controlled by a current-control pixel circuit. Organic LEDs are very fast, with faster response than liquid crystals, an important feature for video displays.

Fabrication is relatively straightforward, consisting of vacuum evaporation of thin organic layers, followed by thin metal layers and a transparent conductor oxide layer. A whole wafer can be processed at one time, including a process for sealing, before the wafer is cut into individual displays. Up to 750 small displays can be produced from a single 8-inch wafer, including interface and driver electronics embedded in the silicon.

Organic LEDs were invented by C. W. Tang and S. A. Van Slyke of Kodak, who found that p-type and n-type organic semiconductors could be combined to form diodes, in complete analogy to the formation of p-n junctions in crystalline semiconductors. Moreover, as with gallium arsenide and related III-V diodes, the recombination of injected holes and electrons produced light efficiently. In contrast to the difficult fabrication of III-V LEDs, where crystalline perfection is essential, organic semiconductors can be evaporated as amorphous films, for which crystallization may be undesirable. The prototypical Kodak OLED, which is used in consumer products today by Pioneer, is a down-emitting stack, with light coming out through a transparent glass substrate.

For use on top of an opaque silicon chip, we modify the stack, starting with a metal anode with a high work function and ending with a transparent cathode, followed by a layer of transparent indium tin oxide. We also change the active layer to make it a white-light emitter. We do this by using a diphenylene -vinylene type blue-green emitter, which is co-doped with a red dye to yield a white spectrum.

Even though an OLED microdisplay on silicon may have millions of subpixels, the OLED formation can be rather simple because the complexity is all in the substrate. For each subpixel, corresponding to a red, green, or blue dot, there is a small electrode pad, possibly 3.5 x 13.5 microns, attached to an underlying circuit that provides current. The OLED layers can be deposited across the complete active area, using shadow masking in the evaporator. This includes the cathode, which is common for all pixels. This simple approach is made possible by the fact that the materials are not good lateral conductors, so the very thin organic films cannot shunt current from one subpixel to another. With such thin structures, light also does not leak into adjacent pixels, so contrast is maintained even between neighboring pixels.

Organic LED devices are very sensitive to moisture, which attacks the cathode materials; to a lesser degree, they are sensitive to oxygen, which can potentially degrade the organics. For this reason they must be sealed in an inert atmosphere after fabrication and before being exposed to ambient environmental conditions. Organic LEDs on silicon can be sealed with a thin film. This thin film seal has been critical to the development of full color OLED microdisplays, since microscopic color filters can be processed on top of a white OLED using photolithography after passivation with the thin film seal.

The current through the OLED in a basic active-matrix OLED pixel cell is controlled by an output transistor, and all emitting devices share a common cathode to which a negative voltage bias is applied. This bias is set to allow the full dynamic range. Since the output transistor has a limited voltage capability, depending upon the silicon process used, it is important to have a steep variation in the OLED of luminance versus voltage so that a large variation in luminance.

Organic LEDs will become a major competitor to the TFT-LCD market due to the promise of significantly lower power consumption, much lower manufacturing cost, thinner and lighter form factors, as well as potentially sharper and brighter images and greater viewing angles. The light emitting properties of AM-OLEDs eliminate the need for backlight units, backlight lamps, inverters, color filters, liquid crystal, alignment film, and the manufacturing steps associated with these in LCD production. A large number of companies are investing in research and development as well as manufacturing capital in this emerging market, and on last count there were approximately 85. Furthermore, there is a large willingness to cross-license technology in order to speed up commercialization. Organic LEDs are also highly appealing in the upcoming 2.5G and 3G wireless markets. It is unlikely that OLEDs will be able to compete with incumbent technologies for large display applications until the second half of the decade. Some of the companies involved in OLED development include Pioneer, Motorola, Samsung, Sanyo, Kodak, and TDK.

LED Video for Outdoors

Ever wonder what exactly was going on behind the scoreboard to keep these giant, super bright displays running in a large outdoor stadium display during a bitter rainstorm? Many large screen display systems are designed to operate much differently than your standard desktop monitor, as high power consumption, adverse weather conditions, and long operating hours make this market one of the most unique in the electronic display industry.

The popular LED video technology offers some of the best performance characteristics for the large-venue outdoor sports facility, especially brightness and lifespan. The market for large screen LED video displays is valued at nearly $1.5 billion by 2007, and is being targeted by all major suppliers such as Fuji, Matsushita, Mitsubishi, Sony, and Toshiba, among others.

LCoS—Liquid Crystal on Silicon

Liquid Crystal on Silicon (LcoS) technology has received a great deal of attention with its anticipated lower cost structure, and a number of companies have raced to ramp up volume production. However, LCoS development has been derailed by many of these companies due to more difficult manufacturing problems, and has reached the end-market consumer only in very small volumes.

One limitation of LCoS imagers is that they rely upon polarization to modulate the light, just as LCD panels do. This effectively creates an upper limit to the contrast ratio, negatively impacting, for example, the detail reproduction in dark scenes. Contrast ratio is a historic strength of CRT and film-based projectors. Poor contrast ratio also limits the overall perceived sharpness or clarity of an image. This is true even in relatively bright scenes, because the magnitude of the difference between bright and dark areas of a scene is a key factor in how we perceive an image. Liquid Crystal on Silicon, then, does not compete well with other display technologies in this key area of perceived image quality.

Considering the fundamental cost structure of LCoS, there is nothing unique in the wafer cost compared to the wafer cost of any other display technology, such as DLP. Yet, unlike products based on DLP technology that require only a single panel to deliver outstanding images, LCoS designers rely on three panels to achieve adequate picture quality. Single-panel LCoS designs have faltered because of manufacturing difficulties, or because of the problems inherent in complicated triple rotating-prism optical systems. Liquid Crystal on Silicon technology market penetration will lag until it can achieve a higher image performance from single-panel architectures.

There is potential for LCoS to build a sizeable niche within the flat panel TV market because of extremely high resolution, good picture quality, scalability above 30 inches, and competitive long-term pricing. This is likely to occur somewhere between PDPs at the premium high-end and TFT-LCDs, which are likely to become the mainstay technology. Furthermore, they believe that LCoS can gain momentum by 2005 due to the existing commercial success of two other microdisplay rear projection systems—Digital Light Processing (DLP), from TI and High Temperature Polysilicon (HTPS) TFT-LCD. However, this form of display is not as slim as a PDP, but is of equivalent weight, offers superior power consumption, and a substantially longer operating life. Compared with the CRT rear projection set, the LCoS form factor is far superior as it offers better picture quality, is far brighter, and provides wide aspect ratios at HDTV resolution.

Comparison of Different Display Technologies

Table 10.2 summarizes the pros and cons for the different display technologies.

Table 10.2: Summary of the Different Display Technologies

Technologies	Pros	Cons
Plasma (PDP)	Thin and Alluring. Prices, while stratospheric, are dropping. As think as 3 inches. Light enough to be hung from the wall.	Price. Plasma (gas) expands at high altitudes and makes hissing noises. Like CRTs, PDPs lose brightness as they age.
LCD Panel	Based on the same technologies as laptop PC screens. Offer best picture quality of the non-tubes. Brightness and color are superb. Viewing angle is 175 degrees. Even thinner than PDPs with no burn-in problems.	Price. Slow Response times – hence not ideal for fast video.
Rear Projection LCD	Lowest price per screen inch. Not subject to burn-in issues.	Narrower viewing angle. Issues with color consistency, contrast & display of true blacks.
CRT	Proven technology. Best image quality and response times. Screen sizes up to 45 inches.	Bigger the screen size – bulkier and heavier the set.
Rear Projection LCoS	Brighter and sharper than most rear projection LCDs. Slender as a DLP.	Slow response times. Cost
DLP	Performance. Form factor. Cost.	Not as thin as PDP or LCDs. Occasional flaws in video artifacts, flaws in source material.

Three-dimensional (3-D) Displays

With flat screens in big demand today, most of the familiar corporate names are working hard to develop increasingly flatter and thinner display technology. However, much research is also going into bringing the image to life using three-dimensional (3-D) volumetric techniques.

There are many types of volumetric displays, but they are most often portrayed as a large transparent sphere with imagery seeming to hover inside. Designers are required to grasp complex 3-D scenes. While physical prototypes provide insight, they are costly and take time to produce. A 3-D display, on the other hand, can be used as a "virtual prototyping station," in which multiple designs can be inspected, enlarged, rotated, and simulated in a natural viewing environment. The promise of volumetric displays is that they will encourage collaboration, enhance the design process, and make complex data much easier to understand.

Volumetric displays create images that occupy a true volume. In the still nascent vocabulary of 3-D display technology, it could also be argued that re-imaging displays, such as those using concave mirror optics, are volumetric, since they form a real image of an object or display system placed inside.

Unlike many stereoscopic displays and old "3-D movies," most volumetric displays are auto-stereoscopic; that is, they produce imagery that appears three-dimensional without the use of additional eyewear. The 3-D imagery appears to hover inside of a viewing zone, such as a transparent dome, which could contain any of a long list of potential display elements. Some common architectures use a rotating projection screen, a gas, or a series of liquid crystal panels. Another benefit of volumetric displays is that they often have large fields of view, such as 360 degrees around the display, viewable simultaneously by an almost unlimited number of people.

However, volumetric displays are difficult to design, requiring the interplay of several disciplines. Also, bleeding-edge optoelectronics are required for high-resolution imagery. In a volumetric display, fast 3-D rasterization algorithms are needed to create fluid animation. Once the algorithms 'rasterize' the scene by converting 3-D data into voxels, the data slices are piped into graphics memory and then into a projection engine. The term 'voxels' is short for "volume elements," which are the 3-D analog to pixels. Rasterization is the process in which specialized algorithms convert mathematical descriptions of a 3-D scene into the set of voxels that best visually represent that scene. These algorithms, which do the dirty work of mapping lines onto grids, are well-known in the world of graphics cards and 2-D CRTs.

A typical high-end volumetric display, as bounded by the performance of SLM technology, might have a resolution of roughly 100 million voxels. Because it's difficult to set up 100 million emitters in a volume, many 3-D displays project a swift series of 2-D imagery onto a moving projection screen. Persistence of vision blends the stack of 2-D images into a sharp, perceived 3-D picture.

There are many different types of volumetric displays, including:

- Bit mapped swept-screen displays – Those in which a high-frame-rate projector illuminates a rotating projection screen with a sequence of 5,000 images per second, corresponding to many slices through a 3-D data set.
- Vector-scanned swept-screen displays – Those in which a laser or electron gun fires upon a rotating projection screen.
- 'Solid-state' volumetric displays – Those that rely on a process known as two-step up-conversion. For example, 3-D imagery can be created by intersecting infrared laser beams within a material doped with rare-earth ions.

Three Dimensional LCDs

Three dimensional displays can be used to realistically depict the positional relationships of objects in space. They will find uses in medical applications requiring detailed viewing of body areas, as well as in computer-aided manufacturing and design, retail applications, electronic books and games, and other forms of entertainment. Therefore, several Japanese companies have announced that they will establish a consortium to promote products and applications for 3-D stereographic LCDs in an effort to bring the technology into the commercial mainstream. The consortium, which is spear-headed by Itochu, NTT Data, Sanyo, Sharp, and Sony, will aim to standardize hardware and software to produce stereographic 3-D displays requiring no additional viewing aids, and to build an industry infrastructure to develop and distribute 3-D content. However, analysts remain skeptical about how successful the technology can be, and therefore believe that the market will remain a niche for at least a few more years. The price premium for 3-D displays over standard displays will be a hurdle until the production volumes increase and technology advances cut cost. Many existing components such as glass and drivers can be adapted to 3-D displays; however, suppliers of GFX chipsets may have to re-design their products to support the viewing of stereo-graphic display signals.

The Touch-screen

A touch-screen is an intuitive computer input device that works by simply touching the display screen, either with a finger or a stylus, rather than typing on a keyboard or pointing with a mouse. Computers with touch-screens have a smaller footprint, and can be mounted in smaller spaces; they have fewer movable parts, and can be sealed. Touch-screens can be built in, or added on. Add-on touch-screens are external frames that have a clear see-through touch-screen, and mount over the outer surface of the monitor bezel. They also have a controller built into their external frame. Built-in touch-screens are heavy-duty touch-screens mounted directly onto the CRT tube.

The touch-screen interface—whereby users navigate a computer system by simply touching icons or links on the screen itself—is the most simple, most intuitive, and easiest to learn of all PC input devices, and is fast becoming the interface of choice for a wide variety of applications. Touch-screen applications include:

- Public information systems – Information kiosks, airline check-in counters, tourism displays, and other electronic displays that are used by many people who have little or no computing experience. The user-friendly touch-screen interface can be less intimidating and easier to use than other input devices, especially for novice users, making information accessible to the widest possible audience.
- Restaurant/point-of-sale (POS) systems – Time is money, especially in a fast paced restaurant or retail environment. Because touch-screen systems are easy to use, overall training time for new employees can be reduced. And work can get done faster, because employees can simply touch the screen to perform tasks, rather than entering complex key strokes or commands.
- Customer self-service – In today's fast pace world, waiting in line is one of the things that has yet to speed up. Self-service touch-screen terminals can be used to improve customer service at busy stores, fast service restaurants, transportation hubs, and more. Customers can quickly place their own orders or check themselves in or out, saving them time, and decreasing wait times for other customers.
- Control/automation systems – The touch-screen interface is useful in systems ranging from industrial process control to home automation. By integrating the input device with the

display, valuable workspace can be saved. And with a graphical interface, operators can monitor and control complex operations in real-time by simply touching the screen.

- Computer based training – Because the touch-screen interface is more user-friendly than other input devices, overall training time for computer novices, and therefore training expense, can be reduced. It can also help to make learning more fun and interactive, which can lead to a more beneficial training experience for both students and educators.

Basic Components of a Touch-screen

Any touch-screen system comprises the following three basic components:

- Touch-screen sensor panel – Mounts over the outer surface of the display, and generates appropriate voltages according to where, precisely, it is touched.
- Touch-screen controller – Processes the signals received from the sensor panel, and translates these into touch event data that is passed to the PC's processor, usually via a serial or USB interface.
- Software driver – Translates the touch event data into mouse events, essentially enabling the sensor panel to emulate a mouse, and provides an interface to the PC's operating system.

Touch-screen Technologies

The first touch-screen was created by adding a transparent surface to a touch-sensitive graphic digitizer, and sizing the digitizer to fit a computer monitor. The initial purpose was to increase the speed at which data could be entered into a computer. Subsequently, several types of touch-screen technologies have emerged, each with its own advantages and disadvantages that may, or may not, make it suitable for any given application.

Resistive Touch-screens

Resistive touch-screens respond to the pressure of a finger, a fingernail, or a stylus. They typically comprise a glass or acrylic base that is coated with electrically conductive and resistive layers. The thin layers are separated by invisible separator dots. When operating, an electrical current is constantly flowing through the conductive material. In the absence of a touch, the separator dots prevent the conductive layer from making contact with the resistive layer. When pressure is applied to the screen the layers are pressed together, causing a change in the electrical current. This is detected by the touch-screen controller, which interprets it as a vertical/horizontal coordinate on the screen (x- and y-axes) and registers the appropriate touch event. Resistive type touch-screens are generally the most affordable. Although clarity is less than with other touch-screen types, they're durable and able to withstand a variety of harsh environments. This makes them particularly suited for use in POS environments, restaurants, control/automation systems and medical applications.

Infrared Touch-screens

Infrared touch-screens are based on light-beam interruption technology. Instead of placing a layer on the display surface, a frame surrounds it. The frame assembly is comprised of printed wiring boards on which optoelectronics are mounted and concealed behind an IR-transparent bezel. The bezel shields the optoelectronics from the operating environment while allowing IR beams to pass through. The frame contains light sources (or light-emitting diodes) on one side, and light detectors (or photosensors) on the opposite side. The effect of this is to create an optical grid across the screen. When any object touches the screen, the invisible light beam is interrupted, causing a drop in the signal received by the photosensors. Based on which photosensors stop receiving the light signals, it is easy to isolate a screen coordinate. Infrared touch systems are solid state technology and have no moving mechanical parts. As such, they have no physical sensor that can be abraded or worn out with heavy use over time. Furthermore, since they do not require an overlay—which can be broken—they are less vulnerable to vandalism, and are also extremely tolerant of shock and vibration.

Surface Acoustic Wave Technology Touch-screens

Surface Acoustic Wave (SAW) technology is one of the most advanced touch-screen types. The SAW touch-screens work much like their infrared brethren except that sound waves, not light beams, are cast across the screen by transducers. Two sound waves, one emanating from the left of the screen and another from the top, move across the screen's surface. The waves continually bounce off reflectors located on all sides of the screen until they reach sensors located on the opposite side from where they originated. When a finger touches the screen, the waves are absorbed and their rate of travel thus slowed. Since the receivers know how quickly the waves should arrive relative to when they were sent, the resulting delay allows them to determine the x- and y-coordinates of the point of contact and the appropriate touch event to be registered. Unlike other touch-screen technologies, the z-axis (depth) of the touch event can also be calculated; if the screen is touched with more than usual force, the water in the finger absorbs more of the wave's energy, thereby delaying it even more. Because the panel is all glass and there are no layers that can be worn, Surface Acoustic Wave touch-screens are highly durable and exhibit excellent clarity characteristics. The technology is recommended for public information kiosks, computer based training, or other high-traffic indoor environments.

Capacitive Touch-screens

Capacitive touch-screens consist of a glass panel with a capacitive (charge storing) material coating on its surface. Unlike resistive touch-screens, where any object can create a touch, they require contact with a bare finger or conductive stylus. When the screen is touched by an appropriate conductive object, current from each corner of the touch-screen is drawn to the point of contact. This causes oscillator circuits located at corners of the screen to vary in frequency depending on where the screen was touched. The resultant frequency changes are measured to determine the x- and y- co-ordinates of the touch event. Capacitive type touch-screens are very durable, and have a high clarity. They are used in a wide range of applications, from restaurant and POS use, to industrial controls and information kiosks.

Table 10.3 summarizes the principal advantages and disadvantages of each of the described technologies.

Table 10.3: Summary of the Different Touch-screen Technologies

Technology Type	Touch Resolution	Clarity	Operation	Durability
Resistive	High	Average	Finger or Stylus	Can be damaged by sharp objects
Infrared	High	Good	Finger or Stylus	Highly durable
Surface Acoustic Wave	Average	Good	Finger or Soft-tipped Stylus	Susceptible to dirt and moisture
Capacitive	High	Good	Finger Only	Highly durable

Digital Display Interface Standards

Efforts to define and standardize a digital interface for video monitors, projectors, and display support systems were begun in earnest in 1996. One of the earliest widely used digital display interfaces is LVDS (low-voltage differential signaling), a low-speed, low-voltage protocol optimized for the ultra-short cable lengths and stingy power requirements of laptop PC systems. Efforts to transition LVDS to external desktop displays foundered when the rival chipmakers Texas Instruments and National Semiconductor chose to promote different, incompatible flavors of the technology; FPD-Link and Flat-Link, respectively. Other schemes such as Compaq's Digital Flat Panel (DFP), VESA Plug and Display, and National Semiconductor's OpenLDI also failed to achieve widespread acceptance.

Finally, the Digital Display Working Group (DDWG) came together at the Intel Developer Forum in September, 1998, with the intent to put the digital display interface standard effort back on the fast track. With an ambitious goal of cutting through the confusion of digital interface standards efforts to date, the DDWG—whose initial members included computer industry leaders Intel, Compaq, Fujitsu, Hewlett-Packard, IBM, NEC, and Silicon Image—set out to develop a universally acceptable specification. The DDWG is an open industry group with the objective to address the industry's requirements for a digital connectivity specification for high-performance PCs and digital displays. In April, 1999, the DDWG approved a draft Digital Visual Interface (DVI) specification, and in so doing, brought the prospect of an elegant, high-speed all-digital display solution—albeit at a fairly significant price premium—close to realization.

While DVI is based on TMDS (transition minimized differential signaling), a differential signaling technology, LVDS (low voltage differential signaling), a similar technology, is equally, if not more, relevant to the development of digital display interfaces. Both DVI and LVDS are discussed below.

DVI—Digital Visual Interface

Silicon Image's PanelLink technology—a high-speed serial interface that uses TMDS to send data to the monitor—provides the technical basis for the DVI signal protocol. Since the DFP and VESA Plug and Display interfaces also use PanelLink, DVI can work with these previous interfaces by using adapter cables.

The term "transition minimized" refers to a reduction in the number of high-to-low and low-to-high swings on a signal. "Differential" describes the method of transmitting a signal using a pair of complementary signals. The technique produces a transition-controlled, DC balanced series of characters from an input sequence of data bytes. It selectively inverts long strings of 1s or 0s in order to keep the DC voltage level of the signal centered around a threshold that determines whether the received data bit is a 1 voltage level or a 0 voltage level. The encoding uses logic to minimize the number of transitions, which helps avoid excessive electromagnetic interference (EMI) levels on the cable, thereby increasing the transfer rate and improving accuracy.

The TMDS link architecture consists of a TMDS transmitter that encodes and serially transmits a data stream over the TMDS link to a TMDS receiver. Each link is composed of three data channels for RGB information, each with an associated encoder. During the transmit operation, each encoder produces a single 10-bit TMDS-encoded character from either two bits of control data or eight bits of pixel data, to provide a continuous stream of serialized TMDS characters. The first eight bits are the encoded data; the ninth bit identifies the encoding method, the tenth bit is used for DC balancing. The clock signal provides a TMDS character-rate reference that allows the receiver to produce a bit-rate-sampling clock for the incoming serial data streams.

At the downstream end, the TMDS receiver synchronizes itself to character boundaries in each of the serial data streams, and then TMDS characters are recovered and decoded. All synchronization queues for the receivers are contained within the TMDS data stream.

A fundamental principle of physics known as the "Copper Barrier" limits the amount of data that can be squeezed through a single copper wire. The limit is about a bandwidth of 165 MHz, which equates to 165 million pixels-per-second. The bandwidth of a single-link DVI configuration is therefore capable of handling UXGA (1600 x 1200 pixels) images at 60 Hz. In fact, DVI allows for up-to two TMDS links—providing sufficient bandwidth to handle digital displays capable of HDTV (1920 x 1080), QXGA (2048 x 1536) resolutions, and even higher. The two links share the same clock so that bandwidth can be divided evenly between them. The system enables one or both links, depending on the capabilities of the monitor.

Digital Visual Interface also takes advantage of other features built into existing display standards. For example, provisions are made for both the VESA Display Data Channel (DDC) and Extended Display Identification Data (EDID) specifications, which enable the monitor, graphics adapter, and computer to communicate and automatically configure the system to support the different features available in the monitor.

A new digital interface for displays poses a classic "chicken and egg" problem. Graphics adapter card manufacturers can't cut the analog cord to the millions of CRTs already on the desktops, and LCD monitor makers can't commit to a digital interface unless the graphics cards support it. Digital Visual Interface addresses this via two types of connector; DVI-Digital (DVI-D), supporting digital displays only, and DVI-Integrated (DVI-I), supporting digital displays backwards compatibility with analog displays.

The connectors are cleverly designed so that a digital-only device cannot be plugged into an analog-only device, but both will fit into a connector that supports both types of interfaces. The digital connection uses 24 pins, sufficient for two complete TMDS channels, plus support for the VESA DDC and EDID services. In fact, single-link DVI plug connectors implement only 12 of the 24 pins; dual-link connectors implement all 24 pins. The DVI-D interface is designed for a 12- or 24-pin DVI plug connector from a digital flat panel. The DVI-I interface accommodates either a 12- or 24-pin DVI plug connector, or a new type of analog plug connector that uses four additional pins, plus a ground plane plug to maintain a constant impedance for the analog RGB signals. A DVI-I socket has a plus-shaped hole to accommodate the analog connection; a DVI-D socket does not. Instead of the standard cylindrical pins found on familiar connectors, the DVI pins are flattened and twisted to create a Low Force Helix (LFH) contact, designed to provide a more reliable and stable link between the cable and the connector.

Of course, the emergence of a widely adopted pure digital interface also raises concerns about copyright protection, since pirates could conceivably use the interface to make perfect copies of copyrighted material from DVD and HDTV signals. To address this, Intel has proposed the High-Bandwidth Digital Content Protection (HDCP) encryption specification. Using hardware on both the graphics adapter card and the monitor, HDCP will encrypt data on the PC before sending it to the display device, where it will be decrypted. The HDCP-equipped DVI cards will be able to determine whether or not the display device it is connected to also has HDCP features. If not, the card will still be able to protect the content being displayed by slightly lower the image quality.

High-Bandwidth Digital Content Protection is really for the day when PCs are expected to be used to output images to HDTV and other consumer electronic devices. It is even possible that first-run movies will someday be streamed directly to people's homes and displayed through a PC plugged into a HDTV. Without something akin to HDCP to protect its content, Hollywood would likely oppose the distribution of movies in this way.

LVDS—Low-Voltage Differential Signaling

Low-Voltage Differential Signaling (LVDS), as the name suggests, is a differential interconnectivity standard. It uses a low voltage swing of approximately 350 mV to communicate over a pair of traces on a PCB or cable. Digital TV, digital cameras, and camcorders are fueling the consumer demand of high-quality video that offers a realistic visual experience. These have become an integral part of our lifestyles. The other trend is to try connecting all these digital video equipment together so that they can communicate with each other. Moving high-bandwidth digital video data within these appliances and between them is a very challenging task.

There are very few interconnectivity standards that meet the challenge of handling high performance video data of 400 Mb/s. In addition to performance, invariably all of these applications require the interconnectivity solution to offer superior immunity to noise and low power, and be available at a low cost. No solution fits the bill better than LVDS.

Low-Voltage Differential Signaling was initially used with laptop PCs. It is now widely used across digital video applications, telecom, and networking, and for system interconnectivity in general across all applications. The integration of LVDS I/O within low-cost programmable logic devices now enables high-performance interconnectivity alongside real-time, high-resolution image processing.

LVDS Benefits

Low-Voltage Differential Signaling provides higher noise immunity than single-ended techniques, allowing for higher transmission speeds, smaller signal swings, lower power consumption, and less electro-magnetic interference than single-ended signaling. Differential data can be transmitted at these rates using inexpensive connectors and cables. Low-Voltage Differential Signaling also provides robust signaling for high-speed data transmission between chassis, boards, and peripherals using standard ribbon cables and IDC connectors with 100 mil header pins. Point-to-point LVDS signaling is possible at speeds of up-to 622 Mb/s and beyond.

Low-Voltage Differential Signaling also provides reliable signaling over cables, backplanes, and on boards at data rates up-to 622 Mb/s. Reliable data transmission is possible over electrical lengths exceeding 5 ns (30 inches), limited only by cable attenuation due to skin effect.

The high bandwidth offered by LVDS I/Os allows several lower data rate TTL signals to be multiplexed/de-multiplexed on a single LVDS channel. This translates to significant cost savings in reduced number of pins, number of PCB traces, number of layers of PCB, significantly reduced EMI, and lower component costs.

Mini-LVDS

Mini-LVDS is a high-speed serial interface between the timing controller and the display's column drivers. The basic information carrier is a pair of serial transmission lines bearing differential serialized video and control information. The number of such transmission-line pairs will depend on the particular display system design, leaving room for designers to get the best data transfer advantage. And the big advantage is a significant decrease in components—switches and power supplies—and a simpler design, one of the key parameters for product success in this highly competitive field.

Summary

There is no doubt that digital displays will penetrate the consumer market in products ranging from TV sets to cell phones. Consumers are attracted to brighter and crisper products based on alternative display technologies to replace the CRT in their homes. However, there are several contenders in this market. While today the flat panel market is dominated by LCDs, the future seems apparent that different technologies are ideal for different applications. The home entertainment screen technology of the future must address four key requirements: it must offer image quality comparable with direct-view CRT, it must come in an aesthetically pleasing package that will fit in any home, it must be reliable, and it must be affordable.

Of all these systems, the liquid crystal display is considered to be the most promising and the display technology of choice, with the world market for TFT-LCDs having reached 46 million units in 2001. In addition to being thin and lightweight, these displays run on voltages so low they can be driven directly by large-scale integration (LSI). And since they consume low power, they can run for long periods on batteries. Additional advantages over other types of flat panel display include adaptability to full color, low cost, and a large potential for technological development. Its main applications will be in laptop PCs, desktop monitors, multitask displays for PCs, information panels, presentation screens, monitor displays for conferences, wall-mount TVs, and more.

The plasma display technology, having evolved to a pivotal stage in high image quality and lower unit cost, are affordable and appropriate for use in consumer applications. For many people, plasma has become synonymous with their conception of what the TV technology of the future will look like; it is, after all, the closest thing we have to the sci-fi notion of a TV that hangs on the wall. Certainly, plasma fulfills the second requirement; plasma panels are, for the most part, attractive products. Unfortunately, this technology is too expensive in large screen sizes. The case for plasma is not helped by the fact that the more affordable products announced thus far are at substantially smaller screen sizes, presumably a function of low factory yields.

Liquid-crystal-on-silicon technology has received a great deal of attention with its anticipated lower cost structure, and a number of companies have raced to ramp up volume production. However, due to more difficult manufacturing problems, LCoS development has been derailed by many of these companies.

Digital Light Processing technology appears to offer the upside of plasma, with its ability to enable slim, elegant product designs, but without the downside of high price. It appears to offer the upside of LCoS technology in that it can be inherently inexpensive to manufacture while offering excellent image quality, but without the downside of uncertainty over the viability of a single-panel solution or the question marks over its manufacturability.

Portable products such as cell phones, PDAs, web tablets, and wristwatches are also vying for next-generation display technologies for sharper and color images. Output formats will range from the current low pixel count monochrome reflective graphic images in low priced cell phones, up to high definition full color video in high-end appliances. The high growth cell phone display market has already blossomed into a $4 billion business and it promises to be the next competitive battleground between passive and active matrix LCDs. But two new display technologies, microdisplays and OLEDs, are targeted at carving out a significant piece of the action.

However, even before they have any appreciation for the performance and features of flat-panel monitors, consumers will buy a digital display only if the price is right; that is, if it costs less than a CRT. There is some appreciation for the savings in desk space afforded by the thin flat-panel monitor, but few potential buyers place a premium on it, and few envision the wall- or boom-mount monitors that have been shown as prototypes. Also, until the consumer is educated on which technology is ideal for a relevant application the market confusion will continue.

Digital Imaging—
Cameras and Camcorders

Introduction

Camera technology is shifting from analog to digital. Most people have piles of photographs lying in the closet. While they bring back good memories, the storage and sharing of photos has always been an issue. With improvements in digital photography and the falling prices of hard disk drives and PCs, the digital camera is gaining in popularity. Digital cameras are replacing 35-mm cameras because of image quality, ease of use, compact size, and low cost. A digital camera is basically an extension of the PC. The photos can be viewed on the camera's built-in LCD screen or on a PC monitor if the camera is plugged into a computer. The photos can be edited (color adjustments, cropping, etc.) with a PC graphics application and sent to friends by e-mail, or printed.

Initially, digital cameras had difficulty capturing images of widely varying light patterns. They generated images that were fuzzy when enlarged or washed out when using a flash. And they carried a minimum price tag of $500. Although the concept was good and the promise of technology development was assured, digital cameras still took a back seat to film-based cameras. However, the market today for digital cameras has broadened significantly. Rapid technology changes in sensors, chipsets, and pixel resolution contributed to this progress. Advanced semiconductor products, such as CMOS sensors, helped improve digital imaging and overall camera performance. Meanwhile, the price of digital camera components—namely charged-coupled devices (CCDs)—continues to fall dramatically, reducing unit cost. Today, digital camera technology is transitioning from a niche market to mass consumer market. The unit price for some entry-level digital cameras has dropped below $50, giving the consumer greater choice in digital photography.

Just as digital cameras are taking over analog cameras, traditional video camcorders are migrating to digital format. Hence, this chapter covers two distinct types of camera devices—the digital still camera and the digital camcorder.

Digital Still Cameras

Digital still cameras are one of the more popular consumer devices at the forefront of today's digital convergence. They enable instantaneous image capture, support a variety of digital file formats, and can interoperate through an ever-growing variety of communications links. And, with the Internet providing the superhighway to disseminate information instantaneously, digital images are now at virtually everyone's fingertips. From e-mail to desktop publishing, captured digital images are becoming pervasive.

Capturing images of precious moments is made even easier with the use of digital still cameras. Unlike older film-based cameras, they have the ability to store images in digital formats such as JPEG, which enables instantaneous file exchange and dissemination. This also means that the information content is less susceptible to degradation as time goes by. Further, the Internet provides an easy, fast, and no-cost medium to share one's captured images with friends and family. There is also a growing trend towards recording video clips and appending audio notes in digital still cameras. This feature allows the user to have a more complete record of images and demonstrates the increasing value of digital convergence.

Market for Digital Still Cameras

The digital camera market is currently in the midst of unprecedented growth. The segments include entry-level models up through 5 MP (megapixel) point-and-shoot models. These are segmented by resolution level including VGA, XGA, 1 MP, 2 MP, 3 MP, and 4 MP. While digital cameras have eclipsed film-based cameras in the high-end professional camera market, high prices have limited their appeal to users of film-based cameras and personal computers. But these two groups will soon see the price of digital cameras significantly reduced, and large unit sales will result. The worldwide market for digital camera shipments is forecast by Dataquest to grow to over 16 million units by 2004. Semico forecasts digital camera sales growth from 5.2 million in 1999, to over 31 million by 2005. IC Insights reported that digital camera shipments in 2003 reached 29 million units and associated revenues reached $6.1 billion. iSupply predicts that in 2005 digital camera shipments will exceed 42 million units and factory revenues will reach $9.5 billion. This compares with the prediction of 35 million units of film cameras in 2005. Hence, 2005 will be the first year digital cameras will outsell film cameras.

Some of the drawbacks of traditional analog film cameras are:
- Cost of film
- Cost of film processing
- Cost of processing unwanted images
- Film development time
- Drop-off and pick-up time
- Duplication and delivery
- Time to retake poor quality photos
- Storage of the pictures

The main reasons for digital camera growth are declining average selling prices, an improved digital imaging infrastructure, and increased awareness of digital imaging. Not only have prices declined, but new entry-level products have been introduced that allow consumers to sample the technology without paying high prices. Entry-level models have peaked the curiosity of the larger mass-market segment. The entry-level segment has been a significant driver in terms of unit shipments for the overall market.

As opposed to just a year ago, digital camera users now have options when deciding what to do with their digital images. The digital imaging infrastructure has developed so that digital images may be used in a variety of ways, and not just viewed on a PC monitor. Such examples are:
- Personal inkjet printers are less expensive and of higher quality.
- Declining PC prices are leading to PC penetration into consumer homes.
- Internet sites are offering printing, storing, and sharing services.
- Software packages are offering more robust feature sets.

The awareness of such technology has increased dramatically and more consumers are giving digital imaging a try. Coupled with improved marketing campaigns by camera, software, and Internet vendors, this has pushed digital imaging into the spotlight as never before.

Some challenges for the digital photography technology include consumer awareness and the pervasiveness of digital infrastructure. Drawbacks facing the digital camera industry include:

- High initial cost
- Poor resolution
- The need for a TV or PC to see full-size pictures
- The need for a quality printer to develop the photo

The formula for success in the digital camera market is multifaceted. It is important to produce a picture with high-quality resolution and to view it as soon as the photo is taken. Decreasing prices of imaging sensors and inexpensive, removable picture storage (flash memory cards) are helping the market to grow. In addition, camera, PC, and IC vendors are working together to come up with combinations that lower prices and make their systems more attractive. This will produce affordable systems that provide seamless functionality with other equipment.

Some popular market trends include:

- **The PC and the Internet** – The Internet has become a very important component of the digital imaging infrastructure. Imaging sites are battling to become the premier imaging Website and a default site for those looking to print, store, manipulate, or share their images. One key issue on this front is how people are getting their images to the Internet or to a specific email address. At this point, the PC remains the epicenter of digital image processing. Images are downloaded to the PC from the digital camera and sent to their ultimate location.

- **Digital camera card readers** – In order to see the pictures, one traditionally had to hook the camera to a cable dangling from a computer port, and then run PC photo software to download pictures from a camera to the PC. This is doing it the hard way. It requires the camera to be turned on, which drains the camera battery while performing the download. Card readers solve this problem. They plug into the computer and accept the memory card to provide quick access to the photos. This allows the camera to be turned off and put away while viewing pictures. You simply slip the card into the slot and the photos appear on the PC screen. This extends the life of the camera's battery since it does not need to be re-charged as often. Depending on the model, readers that plug into a USB port can transfer photos at speeds up to 80-times faster than cards that plug into serial ports.

- **Bluetooth wireless technology** – Wireless technology has added a new dimension to the image-transfer process. The Internet could eclipse the PC as the dominant imaging medium, and the digital camera could truly become a mobile device. The adoption of Bluetooth wireless technology within the digital camera market will help the exchange of images between the digital camera and the PC. While digital camera usability issues are being worked out, Bluetooth will take a back seat. However, wireless technology—particularly Bluetooth—will become an important component of digital imaging. Initial incorporation of Bluetooth is expected via add-on devices because vendors will be slow to build the technol-ogy into the cameras. The add-on component could make use of a variety of camera ports such as the USB port. As the price of Bluetooth declines and digital cameras become more user friendly and widely used, Bluetooth will be integrated into a great many digital camera designs. The instant gratification of the digital camera will then be complemented with the instant distribution of Bluetooth. By 2004, IDC expects Bluetooth to be integrated into 20% of all U.S. digital camera shipments and 19% of worldwide shipments.

■ **Megapixe**l – Megapixel digital cameras are breaking through feature and price barriers. Megapixel is one-million pixels (picture elements) per image. However, the looser and more widely accepted definition says that at least one dimension in the file you upload must be at least 1,000 pixels or more. Thus, a 1024 by 768 camera, which technically delivers only 786,432 pixels, still counts as a megapixel unit. Megapixel cameras provide a marked improvement in resolution compared to early digital camera models. Improved resolution afforded by these cameras convinced some previously hesitant consumers to buy digital cameras. As a result, the market for megapixel cameras has grown quickly. Today it represents in excess of 98% of worldwide digital camera shipments.

■ **PDA and cell phone cameras** – As it continues to grow into a mass-market product, the digital camera is seeing convergence with other consumer devices such as PDAs and cell phones. PDA manufacturers are bringing out modules that allow users to snap images and make mini-movies which can then be exchanged with friends through Bluetooth or cellular technologies. Availability of such modules presents manufacturers with a unique value proposition.

■ **MP3 player and digital cameras convergence** – Since MP3 players and digital cameras require flash memory and similar processing power, -manufacturers are bringing out consumer devices that combine these functionalities.

■ **Business applications** – Business users have more to gain from digital photography than home users. The technology lets the user put a photo onto the computer monitor within minutes of shooting. This translates into a huge productivity enhancement and a valuable competitive edge. Digitally captured photos are going into presentations, business letters, newsletters, personnel ID badges, and Web- and print-based product catalogues. Moreover, niche business segments that have relied heavily on traditional photography—such as real-estate agents and insurance adjusters—now embrace digital cameras wholeheartedly. If the requirement is to capture images in electronic form in the shortest possible time, then a digital camera is the only choice. In fact, they are ideal for any on-screen publishing or presentation use where PC resolution is between 640 x 480 and 1024 x 768 pixels. A digital camera in this resolution range can quickly capture and output an image in a computer-friendly, bitmapped file format. This file can then be incorporated into a presentation or a desktop publishing layout, or published on the World Wide Web.

Digital Still Camera Market Segments

The digital still camera market can be subdivided into several segments:

■ **Entry-level** – This segment includes users whose image quality requirements are not critical. These cameras are usually low-cost and have limited picture resolution—generally 320 by 240 pixels, but as high as 640 by 480 pixels-per-exposure in VGA (video graphics array). They have limited memory and features. These entry-level digital cameras are also known as low-end or soft-display mobile cameras.

■ **Point-and-shoot** – These cameras provide more functionality, features, and control than the entry-level types. Point-and-shoot digital cameras also have better images (standard 640 by 480 pixel resolution) than the entry-level cameras. The common characteristics of point-and-shoot digital cameras are enhanced sensitivity, exposure control, photographic controls, high quality images, and ergonomics design. Some differentiated features include a zoom lens, color LCD screen, auto focus, auto flash, and removable memory cards. Closely equivalent to 35-mm film-image quality, photo-quality point-and-shoot digital cameras are specified with image resolution ranging from 800 to 3,000 pixels (0.8 to 3.0 megapixels). Typical 35-mm film has a resolution of about 4,000 by 4,000 pixels. This segment of the digital camera market, which is just getting started, is likely to be the fastest growing in the coming years.

- **Professional** – These digital cameras can produce images that equal or surpass film quality. They are the finest, most expensive, and sophisticated cameras available to photographers. These cameras have very high resolution, high memory capacity, and high-speed interface technologies such as IEEE 1394 and USB 2.0. They consistently produce professional quality images in 36-bit pixel color (12-bits per sub-pixel)—usually with resolutions greater than 2.0 megapixels. They have a superb array of interchangeable lens options and a great user interface complemented by numerous accessories. These cameras are designed to replace 35-mm film cameras in professional news gatherings and document applications. The professional-level market features expensive prepress, portrait, and studio cameras that produce extremely high-resolution images.

- **Gadget** – These cameras are built into toys and personal computers. They do not have as high an image quality, but they are very good for snapshots and spur-of-the-moment image captures. These cameras are derivatives of entry-level cameras and have stylized enclosures, customized user interfaces, powerful imaging capabilities, and enhanced features.

- **Security** – Security digital cameras are aimed at the surveillance market and provide image capture of the area in focus. They are more rugged than the typical digital camera cousins and can withstand outdoor weather. These cameras may be wired and/or wireless. They can have time lapse and continuous scan capabilities. These cameras are rugged, fit into small camera enclosures for covert placements, and are usually remote controlled.

- **Industrial digital cameras** – These cameras are designed for the industrial environment where normal non-rugged consumer type cameras would not last long. They function well in conditions such as high and low temperature extremes, humid or wet environments, corrosive chemical environments, vibration, shock, and vacuum. Hence, they are rated for harsh environments and have failsafe features. These cameras have automated and remote-controlled operations. They support a variety of industrial interface options such as Ethernet, CAN (Controller Area Network), I²C (Inter IC), and RS (Recommended Standard)-485.

- **Web cameras (webcams)** – These are entry-level cameras that are low cost and have poorer picture quality. They are commonly used for sending images across the Internet. They are also referred to as a PC peripheral scanner. These cameras enable the user to bring the picture image data directly into a PC application. They can also be used for creating new types of entertainment with the camera and PC. They serve as a handy personal communication tool for exchanging images, sound, and moving pictures. Higher resolution is necessary when printing, but such resolution is not necessary when enjoying images through TV or any PC display monitors. Webcams range from the silly to the serious. A webcam might point at anything from a coffeepot to a space-shuttle launch pad. There are business cams, personal cams, private cams, and traffic cams. At a personal level, webcams have lots of productive uses such as watching the house when one is out of town, checking on the babysitter to make sure everything is okay, checking on the dog in the backyard, and letting the grandparents see the new baby. If there is something that one would like to monitor remotely, a webcam makes it easy!

A simple webcam consists of a digital camera attached to a computer. Cameras like these have dropped well below $30. They are easy to connect through a USB port, whereas earlier cameras connected through a dedicated card or the parallel port. A software application connects to the camera and grabs a frame from it periodically. For example, the software might grab a still image from the camera once every 30 seconds. The software then turns that image into a normal JPEG file and uploads it to the Web server. The JPEG image can then be placed on any Web page.

Inside the Digital Still Camera

In principle, a digital camera is similar to a traditional film-based camera. There is a viewfinder to aim it, a lens to focus the image onto a light-sensitive device, and some means to store and remove images for later use. In a conventional camera, light-sensitive film captures images and stores them after chemical development. Digital photography uses a combination of advanced image sensor technology and memory storage. It allows images to be captured in a digital format that is available instantly—there is no need for a "development" process.

Although the principle may be the same as a film camera, the inner workings of a digital camera are quite different. Once a picture is snapped, an embedded processor reads the light level of each pixel and processes it to produce a 24-bit-per-pixel color image. Soon after the picture is taken, the JPEG image is projected onto a LCD display on the back of the camera, or it may be compressed in non-volatile flash memory storage via software. Digital cameras provide speed and the convenience of instant development.

In the process of creating the pixels, an image is focused through a lens and onto an image sensor which is an array of light-sensitive diodes. The image sensor is either a charge-coupled device (CCD) or CMOS (complementary metal-oxide semiconductor) sensors. Each sensor element (chip) converts light into a voltage proportional to the brightness which is passed into an analog-to-digital converter (ADC). The ADC then translates the voltage fluctuations of the CCD into discrete binary code. It does this through a series of photodiodes—each containing red, green, and blue filters— which respond to different ranges of the optical spectrum. The sensor chip is typically housed on a daughter card along with numerous ADCs. The digital reproduction of the image from the ADC is sent to a DSP which adjusts contrast and detail, and compresses the image before sending it to the storage medium. The brighter the light, the higher the voltage and the brighter the resulting computer pixel. The more elements, the higher the resolution, and the greater the detail that can be captured.

Apart from the image sensors, the semiconductor content of a digital camera's bill-of-materials includes embedded micro-logic, flash memory, DRAM, analog, ADC, other logic, and discrete chips. By design, digital cameras require a considerable amount of image processing power. Microprocessor vendors, ASIC vendors, and DSP vendors all view this requirement as a potential gold mine—a huge market poised to buy their wares. Semiconductor manufacturers are looking at integrating five key elements of this technology by incorporating them into the camera's silicon and software. They want to do this in order to deal with form factor and to gain maximum silicon revenue per digital camera. These five key elements of technology and design include:

- a sensor
- an image processing unit
- a microprocessor
- memory (DRAM)
- digital film storage (flash memory)

Most of these single-chip devices accept images from a CCD (charge-coupled devices) or CMOS (complementary metal-oxide semiconductor) sensor, process the in-coming image, filter it, compresses it, and pass it onto storage.

Image Sensors

An image sensor is a semiconductor device that converts photons to electrons for display or storage purposes. An image sensor is to a digital camera what film is to a 35-mm camera—the device that plays the central role in converting light into an image. The image sensor acts as the eye of the camera, capturing light and translating it to an analog signal.

Digital image-sensor devices such as CMOS sensors and CCDs are the key components for clear and bright resolution in digital cameras. The CCD or CMOS sensors are fixed in place and can continue to take photos for the life of the camera. There is no need to wind film between two spools, which helps minimize the number of moving parts. The more powerful the sensor, the better the picture resolution.

CCD Image Sensors

Charge-coupled devices are the technology at the heart of most digital cameras. They replace both the shutter and film found in conventional cameras. The origins of the CCD lie in the 1960s when the hunt was on for inexpensive, mass-producible memory solutions. Its eventual application as an image-capture device had not even occurred to the scientists working with the initial technology. Working at Bell Labs in 1969, Willard Boyle and George Smith came up with the CCD as a way to store data, and Fairchild Electronics created the first imaging CCD in 1974. It was the predominant technology used to convert light to electrical signals. Prior to use in digital cameras, CCDs were implemented in video cameras (for commercial broadcasts), telescopes, medical imaging systems, fax machines, copiers, and scanners. It was some time later before the CCD became part of the main-street technology that is now the digital camera.

The CCD works like an electronic version of a human eye. Each CCD consists of millions of cells known as photosites, or photodiodes. Photosites are essentially light collecting wells that convert optical information into an electric charge. When light particles, or photons, enter the silicon body of the photosite, they provide enough energy for negatively charged electrons to be released. The more light that enters the photosite, the more free electrons that are made available. Each photosite has an electrical contact attached to it that. When a voltage is applied to this contact (whenever photons enter the photosite), the silicon immediately below the photosite becomes receptive to the freed electrons and acts as a container, or collection point, for them. Thus, each photosite has a particular electrical charge associated with it. The greater the charge, the brighter the intensity of the associated pixel.

The next stage in the process passes this charge to what is known as a read-out register. As the charges enter and then exit the read-out register, they are deleted. Since the charge in each row is coupled to the next, this has the effect of dragging the next in behind it. The signals are then passed—as free of signal noise as possible—to an amplifier, and thence on to the ADC.

The photosites on a CCD actually respond to light of any hew—not to color. Color is added to the image by means of red, green, and blue filters placed over each pixel. As the CCD mimics the human eye, the ratio of green filters to that of red and blue is two-to-one. This is because the human eye is most sensitive to yellow-green light. As a pixel can only represent one color, the true color is made by averaging the light intensity of the pixels around it—a process known as color interpolation.

Today, CCDs are the superior product for capturing and reproducing images. They are excellent devices in terms of offering quality picture resolution, color saturation, and for having a low signal-to-noise ratio (i.e., very little granular distortion visible in the image). On the down side, CCDs are complex, power hungry, and expensive to produce. Depending on complexity, each CCD requires three-to-eight supporting circuits in addition to multiple voltage sources. For these reasons, some IC manufacturers believe CCDs have neared the end of their usefulness, especially in emerging applications that emphasize portability, low power consumption, and low cost.

Recognizing a glass ceiling in the conventional CCD design, Fujifilm (Fuji Photo Film Company) and Fujifilm Microdevices have developed a new, radically different CCD (Super CCD). It has larger, octagonal-shaped photosites situated on 45-degree angles in place of the standard square shape. This new arrangement is aimed at avoiding the signal noise that has previously placed limits on the densities of photosites on a CCD. It is also aimed at providing improved color reproduction, improvements in signal-to-noise ratio (SNR), image resolution, and power consumption. It also offers a wider dynamic range and increased light sensitivity, all attributes that result in sharper, more colorful digital images.

CMOS Image Sensors

CMOS sensor image capture technology first emerged in 1998 as an alternative to CCDs. The CMOS imaging sensors include both the sensor and the support circuitry (ADC, timing circuits, and other functions) on the same chip—circuitry needed to amplify and process the detected image. This technology has a manufacturing advantage over CCD because the processes for producing CMOS are the same as those currently used to produce processor, logic, and memory chips. CMOS chips are significantly less expensive to fabricate than specialty CCD since they have proven, high-yield production techniques with an existing infrastructure already in place. Another advantage is that they have significantly lower power requirements than CCDs, thereby helping to extend camera battery life. The also offer a smaller form factor and reduced weight as compared to CCDs. Furthermore, while CCDs have the single function of registering where light falls on each of the hundreds-of-thousands of sampling points, CMOS can also be loaded with a host of other supporting tasks such as analog-to-digital conversion, load signal processing, white balance, and camera control handling. It is also possible to increase CMOS density and bit depth without increasing the cost.

All of these factors will help CMOS sensors reduce the overall cost of the digital cameras. Some estimates conclude that the bill-of-materials for a CMOS image sensor digital camera design is 30% less as compared to CCD image sensor camera design. Other estimates note that CMOS sensors need 1% of the system power and only 10% of the physical space of CCDs, with comparable image quality. Also, as process migration continues to sub-micron levels, manufacturing CMOS sensors using these processes will allow greater integration, resulting in increasing performance and lower costs compared to CCDs. Increasing the on-chip processing of CMOS sensors (while also lowering costs) provides additional potential for CMOS to out perform CCDs in image quality.

For these and other reasons, many industry analysts believe that almost all entry-level digital cameras will eventually be CMOS-based. They believe that only mid-range and high-end units (for PC video conferencing, video cell phones, and professional digital cameras) will use CCDs. Current problems with CMOS, such as noisy images and an inability to capture motion correctly, remain to be solved. Currently, CMOS technology clearly has a way to go before reaching parity with CCD technology.

However, developments are ongoing to improve the resolution and image quality of CMOS image sensors. This, combined with the use of 0.15-and-smaller micron processing, enables more pixels to be packed into a given physical area, resulting in a higher resolution sensor. Transistors made with the 0.15-micron process are smaller and do not take up as much of the sensor space which can then be used for light detection instead. This space efficiency enables sensor designs that have smarter pixels that can provide new capabilities during image exposure without sacrificing light sensitivity. With the release of higher quality CMOS sensors, this sensor technology will begin penetrating digital camera design in high-quality professional markets. These markets include professional cameras, film scanners, medical imaging, document scanning, and museum archiving. In the longer term, it is anticipated that the sensor's underlying technology will migrate down to the larger consumer markets.

Differences between CCD and CMOS image sensors are described in Table 11.1:

Table 11.1: CCD vs. CMOS Sensors

CCD	CMOS
Small pixel size	
Low noise	Single power supply
Low dark current	Low power
High sensitivity	Single master clock
Multiple chips required	Easy integration of circuitry
Multiple supply voltages needed	Low system cost
Specialized manufacturing needs	

Memory

Two types of memory are available—removable storage and disk storage.

Removable Storage

Just like processing power is critical for high-quality images, flash memory is required to store digital images. Many first-generation digital cameras contained one or two megabytes of internal memory suitable for storing around 30 standard-quality images at a size of 640 x 480 pixels. Unfortunately, once the memory had been filled, no more pictures could be taken until they were transferred to a PC and deleted from the camera. To get around this, modern digital cameras use removable storage. This offers two main advantages. First, once a memory card is full it can simply be removed and replaced by another. Second, given the necessary PC hardware, memory cards can be inserted directly into a PC and the photos read as if from a hard disk. Each 35 mm-quality digital camera photo consumes a minimum of 800 KB of compressed data, or 3 MB of uncompressed data. In order to store 24 photos, a digital camera will typically incorporate a 40-MB or 80-MB memory card.

To store an equivalent number of images with quality similar to conventional film, a digital camera will use a flash memory card. Since 1999 two rival formats have been battling for domination of the digital camera arena—CompactFlash and SmartMedia. First introduced in 1994 by SanDisk Corporation, and based on flash memory technology, CompactFlash provides non-volatile storage that does not require a battery to retain data. It is essentially a PC Card flash card that has been reduced to about one quarter of its original size. It uses a 50-pin connection that fits into a standard 68-pin Type II PC Card adapter. This makes it easily compatible with devices designed to use PC Card flash RAM, with maximum capacities reaching 512 MB. Originally known by the awkward abbreviation SSFDC (Solid State Floppy Disk Card) when it first appeared in 1996, the Toshiba-developed SmartMedia cards are significantly smaller and lighter than CompactFlash cards. The SmartMedia card uses its own proprietary 22-pin connection. But like its rival format, it is also PCMCIA-ATA-compatible. It can therefore be adapted for use in notebook PC Card slots. The SmartMedia card is capable of storing 560 high-resolution (1200 x 1024) still photographs, with cost-per-megabyte being similar to that of CompactFlash.

Devices are available for both types of media to allow access via either a standard floppy disk drive or a PC's parallel port. The highest performance option is a SCSI device that allows PC Card slots to be added to a desktop PC. CompactFlash has a far sturdier construction than its rival, encapsulating the memory circuitry in a hard-wearing case. SmartMedia has its gold-colored contact surface exposed and prolonged use can cause scoring on the contact surface. Its memory circuitry is set into resin and sandwiched between the card and the contact. Storage capacity is becoming an increasingly important aspect of digital camera technology. It is not clear which format will emerge as winner in the standards battle. SmartMedia has gotten off to a good start, but CompactFlash is used in PDAs as well, and this extra versatility might prove an important advantage in the long run.

A third memory technology, Sony's Memory Stick, is also being adopted for digital still cameras. Smaller than a stick of chewing gum and initially available with a capacity of 32 MB, Memory Stick is designed for use in small AV electronics products such as digital cameras and camcorders. Its proprietary 10-pin connector ensures foolproof insertion, easy removal, and reliable connection. And a unique Erasure Prevention Switch helps protect stored data from accidental erasure. Capacities had risen to 128 MB by late 2001, with the technology roadmap for the product going all the way up to 1 GB. Infineon (formerly Siemens Semiconductor) and Hitachi announced their combined efforts in the new multimedia card development and production. It is essential for vendors to consider these memory card variations when planning future applications aimed at gaining a large share of the consumer market.

These cards will require at least 200 MB of flash memory, and will be able to write 12 MB of memory in a few seconds. At least one flash card is sold with the cameras, and it is used to store pictures until they can be printed or transferred to a PC. Consumers can purchase greater capacity flash memory cards as accessories. The highest capacity flash memory card on the market at the time of this writing is 256 MB. This is capable of storing up to eight hours of digital music, more than 80 minutes of MPEG-4 video, or more than 250 high-resolution digital images.

Disk Storage

With the resolution of still digital cameras increasing apace with the emergence of digital video cameras, the need for flexible, high-capacity image storage solutions has never been greater. Some higher-end professional cameras use PCMCIA hard disk drives as their storage medium. They consume no power once images are recorded and have much higher capacity than flash memory. For example, a 170-MB drive is capable of storing up to 3,200 "standard" 640 by 480 images. But the hard disk option has some disadvantages. An average PC card hard disk consumes around 2.5 watts of power when spinning idle, more when reading/writing, and even more when spinning up. This means it is impractical to spin up the drive, take a couple of shots, and shut it down again. All shots have to be taken and stored in one go, and even then the camera's battery will last a pitifully short length of time. Fragility and reliability are also a major concern. Moving parts and the extremely tight mechanical tolerances to which hard drives are built make them inherently less reliable than solid-state media.

One of the major advantages of a digital camera is that it is non-mechanical. Since everything is digital, there are no moving parts, and a lot less that can go wrong. However, this did not deter Sony from taking a step that can be viewed as being both imaginative and retrograde at the same time. They included an integrated 3.5-inch floppy disk drive in some of their digital cameras.

The floppy disk media is universally compatible, cheap, and readily available. It is also easy to use—no hassles with connecting wires or interfaces. However, the integrated drive obviously added both weight and bulk to a device that is usually designed to be as compact as possible. Sony has introduced a digital camera that can store images on an 8 cm/185 MB CD-R media. A mini-CD provides sufficient capacity to store around 300 640 x 480 resolution images using JPEG compression. They have also announced plans to provide support for mini re-writable media as well as write-once CD-R media. Performance implications that would have been unacceptable for digital camera applications made certain trade-offs (like write-time) unavoidable. Notwithstanding the constraints imposed, the primary benefit of CD-RW media—it's reusability—is fully realized. Users are able to delete images one at a time, starting with the most recent and working backward, and also have the option to erase an entire disk via a "format" function.

Connectivity

Despite the trend towards removable storage, digital cameras still allow connection to a PC for the purpose of image downloading. Transfer is usually via a conventional RS-232 serial cable at a maximum speed of 115 Kb/s. Some models offer a fast SCSI connection. The release of Windows 98 in mid-1998 brought with it the prospect of connection via the Universal Serial Bus (USB), and digital cameras are now often provided with both a serial cable and a USB cable. The latter is the preferable option, allowing images to be downloaded to a PC more than three times faster than using a serial connection. USB 2.0 offers 480 Mb/s support for transferring data and images. Bluetooth is the next-generation technology that will allow the wireless transfer of images from the digital camera to the PC.

Supplying a digital camera with drivers that allow the user to simply download images to a standard image editing application is also becoming increasingly common. Some digital cameras provide a video-out socket and S-Video cable to allow images to be displayed directly to a projector, TV, or VCR. Extending the "slide show" capability further, some allow images to be uploaded to the camera, enabling it to be used as a mobile presentation tool.

An increasing number of digital cameras have the ability to eliminate the computer and output images directly to a printer. But without established interface standards, each camera requires a dedicated printer from its own manufacturer. As well as the more established printer technologies, there are two distinct technologies used in this field: thermo autochrome and dye sublimation.

Picture quality

The picture quality of a digital camera depends on several factors, including the optical quality of the lens and image-capture chip, and the compression algorithms. However, the most important determinant to image quality is the resolution of the CCD. The more elements, the higher the resolution, and the greater the detail that can be captured.

Despite the massive strides made in digital camera technology in recent years, conventional wisdom remains that they continue to fall behind the conventional camera and film when it comes to picture quality, though they offer flexibility advantages. However, since this assertion involves the comparison of two radically different technologies, it is worth considering more closely.

Resolution is the first step to consider. While it is easy to state the resolution of a digital camera's CCD, expressing the resolution of traditional film in absolute terms is more difficult. Assuming a capture resolution of 1280 x 960 pixels, a typical digital camera is capable of producing a frame size of just over 2 or 3 million pixels. A modern top-of-the-line camera lens is capable of resolving at least 200 pixels per mm. Since a standard 100 ASA, 35 mm negative is 24 x 36 mm, this gives an effective resolution of 24 x 200 x 36 x 200, or 34,560,000 pixels. This resolution is rarely achieved in practice and rarely required. However, on the basis of resolution, it is clear that digital cameras still have some way to go before they reach the level of performance as their conventional film camera counterparts.

The next factor to consider is color. Here, digital cameras have an advantage. Typically, the CCDs in digital cameras capture color information in 24 bits-per-pixel. This equates to 16.7 million colors and is generally considered to be the maximum number the human eye can perceive. On its own this does not constitute a major advantage over film. However, unlike the silver halide crystals in a film, a CCD captures each of the three component colors (red, green, and blue) with no bias. Photographic film tends to have a specific color bias dependent on the type of film and the manufacturer. This can have an adverse effect on an image according to its color balance.

However, it is also the silver halide crystals that give photographic film its key advantage. While the cells on a CCD are laid out in rows and columns, the crystals on a film are randomly arranged with no discernible pattern. As the human eye is very sensitive to patterns, it tends to perceive the regimented arrangement of the pixels captured by a CCD very easily, particularly when adjacent pixels have markedly different tonal values. When you magnify photographic film there will be no apparent regularity, though the dots will be discernible. It is for this reason that modern inkjet printers use a technique known as "stochastic dithering" which adds a random element to the pattern of the ink dots in order to smooth the transition from one tone to the next. Photographic film does this naturally, so the eye perceives the results as less blocky when compared to digital stills.

There are two possible ways around this problem for digital cameras. Manufacturers can develop models that can capture a higher resolution than the eye can perceive. Or, they can build in dithering algorithms that alter an image after it has been captured by the CCD. Both options have downsides, however, such as increased file sizes and longer processing times.

Features

A color LCD panel is a feature that is present on all modern digital cameras. It acts as a mini GUI, allowing the user to adjust the full range of settings offered by the camera. It is also an invaluable aid to previewing and arranging photos without the need to connect to a PC. Typically, this can be used to simultaneously display a number of the stored images in thumbnail format. It can also be used to view a particular image full-screen, zoom in close, and, if required, delete it from memory.

Few digital cameras come with a true, single-lens reflex (SLR) viewfinder. (Here, what the user sees through the viewfinder is exactly what the camera's CCD "sees.") Most have the typical compact camera separate viewfinder which sees the picture being taken from a slightly different angle. Hence, they suffer the consequent problems of parallax. Most digital cameras allow the LCD to be used for composition instead of the optical viewfinder, thereby eliminating this problem. On some models this is hidden on the rear of a hinged flap that has to be folded out, rotated, and then folded back into place. On the face of it this is a little cumbersome, but it has a couple of advantages over a fixed screen. First, the screen is protected when not in use. Second, it can be flexibly positioned to allow the photographer to take a self-portrait, or to hold the camera overhead while retaining control over the framing of the shot. It also helps with one of the common problems in using a LCD viewfinder—difficulty viewing the screen in direct sunlight. The other downside is that prolonged use causes batteries to drain quickly.

To remedy this problem, some LCDs are provided with a power-saving skylight to allow them to be used without the backlight. In practice, this is rarely practical. Also, if there is sufficient ambient light to allow the skylight to work, the chances are that it will also render the LCD unusable.

Digital cameras are often described as having lenses with equivalent focal lengths to popular 35 mm camera lenses. In fact, most digital cameras feature autofocus lenses with focal lengths around 8 mm. These provide equivalent coverage to a standard film camera because the imaging CCDs are so much smaller than a frame of 35 mm film. Aperture and shutter speed control are also fully automated with some cameras, but also allow manual adjustment. Although optical resolution is not an aspect that figures greatly in the way digital cameras are marketed, it can have a very important role in image quality. Digital camera lenses typically have an effective range of up to 20 feet, an ISO equivalency of between 100 and 160, and support shutter speeds in the 1/4 of a second to 1/500th of a second range.

Zoom capability provides the camera with a motorized zoom lens that has an adjustable focal length range. It can be equivalent to anything between a 36-mm (moderate wide angle) and 114-mm (moderate telephoto) lens on a 35-mm format camera. But zoom doesn't always mean a close up; you can zoom out for a wide-angle view or you can zoom in for a closer view. Digital cameras may have an optical zoom, a digital zoom, or both. An optical zoom actually changes the focal length of your lens. As a result, the image is magnified by the lens (called the optics or optical zoom). With greater magnification, the light is spread across the entire CCD sensor and all of the pixels can be used. You can think of an optical zoom as a true zoom that will improve the quality of your pictures. Some cameras have a gradual zoom action across the complete focal range, while others provide two or three predefined settings. Digital zoom does not increase the image quality, but merely takes a portion of an image and uses the camera's software (and a process called interpolation) to automatically resize it to a full-screen image. Let's say that one is shooting a picture with a 2X digital zoom. The camera will use half of the pixels at the center of the CCD sensor and ignore all the other pixels. Then it will use software interpolation techniques to add detail to the photo. Although it may look like you are shooting a picture with twice the magnification, you can get the same results by shooting the photo without a zoom, and later enlarging the picture using your computer software. Some digital cameras provide a digital zoom feature as an alternative to a true optical zoom. Others provide it as an additional feature, effectively doubling the range of the camera's zoom capability.

A macro function is often provided for close-up work. This allows photos to be taken at a distance of as close as 3 cm. But it more typically supports a focal range of around 10–50 cm. Some digital cameras even have swiveling lens units capable of rotating through 270 degrees. They allow a view of the LCD viewfinder panel regardless of the angle of the lens itself.

Some cameras offer a number of image exposure options. One of the most popular is a burst mode that allows a number of exposures to be taken with a single press of the shutter—as many as 15 shots in a burst, at rates of between 1 and 3 shots-per-second. A time-lapse feature is also common which delays multi-picture capture over a pre-selected interval. Another option is the ability to take four consecutive images—each using only one-quarter of the available CCD array. This results in four separate images stored on a single frame. Yet another is the ability to take multiple exposures at a preset delay interval and tile the resulting images in a single frame.

Some cameras provide a manual exposure mode which allows the photographer a significant degree of artistic license. Typically, four parameters can be set in this mode—color balance, exposure compensation, flash power, and flash sync. Color balance can be set for the appropriate lighting condition—daylight, tungsten, or fluorescent. Exposure compensation alters the overall exposure of the shot relative to the metered "ideal" exposure. This feature allows a shot to be intentionally under- or over-exposed to achieve a particular effect. A flash power setting allows the strength of the flash to be incrementally altered. And a flash sync setting allows use of the flash to be forced, regardless of the camera's other settings.

Features allowing a number of different image effects are becoming increasingly common. This allows the selection of monochrome, negative, and sepia modes. Apart from their use for artistic effect, the monochrome mode is useful for capturing images of documents for subsequent optical character recognition (OCR). Some digital cameras also provide a "sports" mode that adds sharpness to the captured images of moving objects, and a "night shooting" mode that allows for long exposures.

Panoramic modes differ in their degree of complexity. At the simpler end of the spectrum is the option for a letterbox aspect image that simply trims off the top and the bottom edges of a standard image, taking up less storage space as a result. More esoteric is the ability to produce pseudo-panoramic shots by capturing a series of images and then using special-purpose software to combine them into a single panoramic landscape.

A self-timer is a common feature, typically providing a 10-second delay between the time the shutter is activated and when the picture is taken. All current digital cameras also have a built-in automatic flash with a manual override option. The best cameras have a flash working range of up to 12 feet and provide a number of different flash modes. These include auto low-light, backlight flash, fill flash for bright lighting shadow reduction, and force-off for indoor and mood photography, and for red-eye reduction. Red-eye is caused by light reflected back from the retina which is covered in blood vessels. One system works by shining an amber light at the subject for a second before the main burst of light, causing the pupil to shrink so that the amount of reflected red light is reduced.

Another feature now available on digital cameras is the ability to watermark a picture with a date and time, or with text. The recent innovation of built-in microphones provides for sound annotation in standard WAV format. After recording, this sound can be sent to an external device for playback, or played back on headphones using an ear socket.

There are additional features that demonstrate the digital camera's close coupling with PC technology. One such feature allows thumbnail images to be emailed directly by camera-resident software. Another is the ability to capture short video clips that can be stored in MPEG-1 format. Higher-end models also provide support for two memory cards and have features more commonly associated with SLR-format cameras. These features include detachable lenses and the ability to drive a flash unit from either the integrated hot shoe or an external mount.

Operation

It is important to note that shooting with a digital camera is not always like shooting with a film camera. Most units exhibit a 1- to 2-second lag time from when the shutter button is pressed to when the camera captures the image. Getting used to this problem can take some time, and it makes some cameras ill-suited for action shots. However, this is an area of rapid improvement and some of the most recent cameras to reach the market have almost no delay.

Most digital cameras also require recovery time between shots for post-capture processing. This includes converting the data from analog to digital, mapping, sharpening, and compressing the image, and saving the image as a file. This interval can take from a few seconds to half-a-minute, depending on the camera and the condition of the batteries.

In addition to regular alkaline batteries, most digital cameras use rechargeable nickel cadmium or nickel hydride batteries. Battery lifetimes vary greatly from camera-to-camera. As a general rule, the rechargeable batteries are typically good for between 45 minutes to 2 hours of shooting, depending on how much the LCD and flash are used. A set of four alkaline AA cells has a typical lifetime of 1 hour.

Figure 11.1 shows the block diagram of a digital still camera and camcorder.

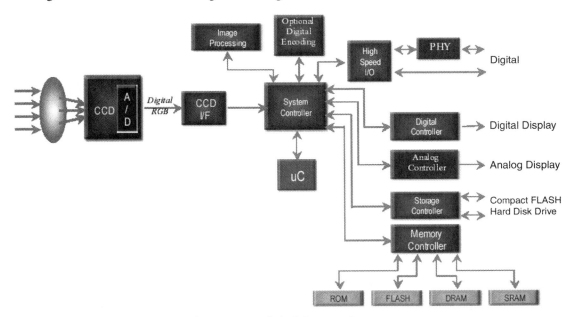

Figure 11.1: Digital Camera System

The basic components of a typical digital camera include:

- **Lens system** – Consists of a lens, sensor (CMOS or CCD), servo motor, and driver.
- **Analog front end** – Conditions the analog signal captured by the sensor, and converts it into digital format before passing the signal on to the image processor.
- **Image processor** – Major tasks of the image processor chip include initial image processing, gamma correction, color space conversion, compression, decompression, final image processing, and image management.
- **Storage** – Typically flash memory.
- **Display** – Typically a color LCD screen.

Digital Camcorders

Camcorders, or video camera-recorders, have been around for nearly 20 years. Camcorders have really taken hold in the United States, Japan, and around the world because they are an extremely useful piece of technology available for under $500.

Digital camcorders are one of the more popular consumer devices at the front lines of today's digital convergence. As such, they enable instantaneous image and/or video capture, typically support a variety of digital file formats, and can interoperate through an ever-growing variety of communications links. Additionally, with the Internet providing the medium to disseminate information instantaneously, digital video clips are now at virtually everyone's fingertips. From e-mail to desktop video editing, digital video clips are becoming pervasive.

Market Demand for Digital Camcorders

The market demand for video related equipment such as digital camcorders is rapidly growing. Gartner Dataquest reports that digital camcorders are expected to have a 28.7% CAGR (or compound annual growth rate) between 2001 and 2004. This compares to analog camcorders that expect to see a 7.5% CAGR for the same period. By 2004, the total market for digital camcorders is expected to grow to 17 million units and factory revenues of U.S. $10.5 billion.

The Camcorder History

The video cassette recorder (VCR) allowed consumers to watch recorded video media whenever they wanted. But the development of the VCR led naturally to the development of a combined video camera and recording technology—the camcorder. A new generation of video cameras that are entirely digital is emerging, and with it a trend towards totally digital video production. Previously, desktop video editing relied on "capturing" analog video from Super VHS or Hi-8, or from the professional level Betacam SP. Then, it would have to be converted into digital files on a PC's hard disk—a tedious and "lossy" process, with significant bandwidth problems. The digital videocassette is small, contains metal-oxide tape, and is about three-quarters the size of a DAT. This cassette format confers the significant advantage of allowing the cameras, capture cards, editing process, and final mastering to all remain within the digital domain, end-to-end, in the recording process.

The first glimmer of camcorder things to come arrived in September 1976 when, for the first time, JVC announced VHS in Japan. Along with its first VHS format VCR, JVC also showed two companion video cameras, each weighing about three pounds. They could be attached to a 16.5-pound shoulder-slung portable VCR. Two years later, Sony unveiled its first portable Beta format VCR/video camera combination. And RCA followed with its own two-piece VHS system in 1978.

These early portable video attempts were bulky and required a separate camera and portable VCR. Almost simultaneously in 1982, two companies announced CAMera/reCORDER, or camcorder, combinations. In 1982 JVC unveiled its new mini-VHS format, VHS-C. The same year Sony announced its Betamovie Beta camcorder which it advertised with the catch-phrase, "Inside This Camera Is a VCR." The first Betamovie camcorder hit stores in May 1983. In 1984 photo giant Kodak introduced a new camcorder format with its first 8-mm camcorder, the KodaVision 2000. Sony followed with its 8-mm camcorder the following January, and the first hi-band 8-mm, or Hi8, camcorder, in 1988.

Television producers took advantage of both our love of technology and our penchant for voyeurism by introducing a plethora of "reality-based" TV shows, using camcorder footage submitted by viewers. The first of these was ***America's Funniest Home Videos*** which premiered as a special in 1989. It was a top 20 show in its first three seasons. The show, spurred on by the popularity of the camcorder, stimulated even greater sales of the device. The camcorder reached its voyeuristic heights in 1991 when George Holliday caught a trio of Los Angeles policemen beating a motorist named Rodney King. The resulting furor prompted police departments across the country to install video cameras in their patrol cars and consumers to start using camcorders to record civil disturbances.

Many disasters and news events would subsequently be captured by amateur videophiles. As the quality of the footage produced by camcorders increased, many shoestring cable organizations started to employ camcorders instead of professional video equipment. At home, film records of family celebrations such as weddings and bar mitzvahs were replaced by video.

In 1992 Sharp became the first company to build in a color LCD screen to replace the conventional viewfinder. Nearly all camcorders today offer a swing-out LCD panel that keeps the user from having to squint through a tiny eyepiece. By then, however, a new and improved video format was also in development—the digital videocassette, or DVC (now simply DV or Mini DV). The DVC consisted of a quarter-inch digital videotape housed in a cassette about the same size of the standard digital audiotape (or DAT). This new format was created by a group of manufacturers in a cooperative effort called the HD Digital VCR Conference.

Panasonic, and later Sony, debuted the first Mini DV camcorders in September 1995, followed by Sharp and JVC two months later. The advantages of the new format were immediate. The cameras themselves were much smaller than either 8 mm or VHS-C. The resolution was also twice that of VHS, resulting in less generation loss when editing and making copies, and because of its digital nature and the IEEE-1394 interface on all DV camcorders, footage could be downloaded and edited on a PC.

The IEEE-1394 interface on the Mini DV camcorders then gave rise to personal computer-based home video editing, which has become as sophisticated as the editing suites found in many TV studios. It also enabled struggling filmmakers to use video to create their imaginative works with only an inexpensive camcorder. The best example of this camcorder-made movie trend is the wildly successful ***The Blair Witch Project*** in 1999.

With a DV camera, a traditional video capture card is not needed because the output from the camera is already in a compressed digital format. The resulting DV files are still large, however. The trend towards digital video includes digital video cameras now equipped with an IEEE 1394 interface as standard. Thus, all that is needed to transfer DV files directly from the camera to a PC-based editing system is an IEEE 1394 interface card in the PC. There is increasing emphasis on handling audio, video, and general data types. Hence, the PC industry has worked closely with consumer giants such as Sony to incorporate IEEE 1394 into PC systems to bring the communication, control, and interchange of digital, audio, and video data into the mainstream.

Camcorder Formats

One major distinction between different camcorder models is whether they are analog or digital, so they have been divided into those two distinct categories.

Analog

Analog camcorders record video and audio signals as an analog track on videotape. This means that every time a copy of a tape is made, it loses some image and audio quality. Analog formats lack a number of the impressive features that are available in digital camcorders. The main difference between the available analog formats focuses on the kind of videotapes the analog camcorder uses and the resolution.

Some analog formats include:

- **Standard VHS** – Standard VHS cameras use the same type of videotapes as a regular VCR. One obvious advantage of this is that, after having recorded something, the tape can be played on most VCRs. Because of their widespread use, VHS tapes are a lot less expensive than the tapes used in other formats. Another advantage is that they give a longer recording time than the tapes used in other formats. The chief disadvantage of standard VHS format is that the size of the tapes necessitates a larger, more cumbersome camcorder design. Also, they have a resolution of about 230 to 250 horizontal lines, which is the low end of what is now available.

- **VHS-C** – The VHS-C camcorders record on standard VHS tape that is housed in a more compact cassette. And VHS-C cassettes can also be played in a standard VCR when housed in an adapter device that runs the VHS-C tape through a full-size cassette. Basically, though, VHS-C format offers the same compatibility as standard VHS format. The smaller tape size allows for more compact designs, making VHS-C camcorders more portable, but the reduced tape size also means VHS-C tapes have a shorter running time than standard VHS cameras. In short play mode, the tapes can hold 30-to-45 minutes of video. They can hold 60 to 90 minutes of material if recorded in extended play mode, but this sacrifices image and sound quality considerably.

- **Super VHS** – Super VHS camcorders are about the same size as standard VHS cameras because they use the same size tape cassettes. The only difference between the two formats is that super VHS tapes record an image with 380 to 400 horizontal lines, a much higher resolution image than standard VHS tape. You cannot play super VHS tapes on a standard VCR, but, as with all formats, the camcorder itself is a VCR and can be connected directly to the TV or the VCR in order to dub standard VHS copies.

- **Super VHS-C** – Basically, super VHS-C is a more compact version of super VHS, using a smaller size cassette, with 30-to-90 minutes of recording time and 380 to 400 resolution.

- **8 mm** – These camcorders use small 8-millimeter tapes (about the size of an audiocassette). The chief advantage of this format is that manufacturers can produce camcorders that are more compact, sometimes small enough to fit in a coat pocket. The format offers about the same resolution as standard VHS, with slightly better sound quality. Like standard VHS tapes, 8-mm tapes hold about two hours of footage, but they are more expensive. To watch 8-mm tapes on the television, the camcorder needs to be connected to the VCR.

- **Hi-8** – Hi-8 camcorders are very similar to 8-mm camcorders, but they have a much higher resolution (about 400 lines)—nearly twice the resolution of VHS or Video8, although the color quality is not necessarily any better. This format can also record very good sound quality. Hi-8 tapes are more expensive than ordinary 8-mm tapes.

Digital

Digital camcorders differ from analog camcorders in a few very important ways. They record information digitally as bytes, meaning that the image can be reproduced without losing any image or audio quality. Digital video can also be downloaded to a computer, where it can be edited or posted on the Web. Another distinction is that digital video has a much better resolution than analog video—typically 500 lines.

There are two consumer digital formats in widespread use:

- **Digital Video (DV)** – DV camcorders record on compact mini-DV cassettes, which are fairly expensive and only hold 60 to 90 minutes of footage. The video has an impressive 500 lines of resolution, however, and can be easily transferred to a personal computer. Digital video camcorders can be extremely lightweight and compact—many are about the size of a paperback novel. Another interesting feature is the ability to capture still pictures, just as a digital camera does.
- **Digital-8** – Digital-8 camcorders (produced exclusively by Sony) are very similar to regular DV camcorders, but they use standard Hi-8 (8 mm) tapes which are less expensive. These tapes hold up to 60 minutes of footage which can be copied without any loss in quality. Just as with DV camcorders, Digital-8 camcorders can be connected to a computer to download movies for editing or Internet use. Digital-8 cameras are generally a bit larger than DV camcorders—about the size of standard 8-mm models. Digital 8-mm technology is the digital extension of the Hi8/8 mm camcorder format which was invented by Sony and is now available in several models. Digital 8 offers up to 500 lines of resolution. This is superior to both Hi8 and 8-mm which have 400 and 300 lines, respectively. Digital-8 also offers playback and recording for both Hi8 and 8-mm cassettes, and provides CD-like sound and PC connectivity. Newly designed record/playback heads in Digital-8 camcorders can detect and play analog 8-mm or Hi8 recordings.
- **Mini DV** – Mini DV images are DVD-quality, and the sound quality is also superior to other formats. Captured footage can be viewed through a TV (using the camcorder's standard AV outputs), a PC (using the camcorder's IEEE 1394 digital output), or by capturing digital still frames, then printing them out on a color printer. Also, video clips can be edited and enhanced on a PC.

Inside the Digital Camcorder

To look inside the digital camcorder, let us explore traditional analog camcorders. A basic analog camcorder has two main components—a video camera and a VCR. The camera component's function is to receive visual information and interpret it as an electronic video signal. The VCR component is exactly like the VCR connected to your television. It receives an electronic video signal and records it on videotape as magnetic patterns. A third component, the viewfinder, receives the video image as well so you can see the image being recorded. Viewfinders are actually tiny black-and-white or color televisions, but many modern camcorders also have larger full-color LCD screens that can be extended for viewing. There are many formats for analog camcorders, and many extra features, but this is the basic design of most all of them. The main variable is the kind of storage tape they use.

Digital camcorders have all the elements of analog camcorders, but have an added component that takes the analog information the camera initially gathers and translates it to bytes of data. Instead of storing the video signal as a continuous track of magnetic patterns, the digital camcorder records both picture and sound as ones and zeros. Digital camcorders are so very popular because ones and zeros can be copied very easily without losing any of the recorded information. Analog information, on the other hand, "fades" with each copy—the copying process does not reproduce the original signal exactly. Video information in digital form can also be loaded onto computers where you can edit, copy, e-mail, and manipulate it.

Camcorder Components

The CCD

The image sensor technology behind both camcorders and digital still cameras are CCDs. But since camcorders produce moving images, their CCDs have some additional pieces that are not in digital camera CCDs. To create a video signal, a camcorder CCD must take many pictures every second, which the camera then combines to give the impression of movement.

Like a film camera, a camcorder "sees" the world through lenses. In a film camera, the lenses serve to focus the light from a scene onto film treated with chemicals that have a controlled reaction to light. In this way, camera film records the scene in front of it. It picks up greater amounts of light from brighter parts of the scene and lower amounts of light from darker parts of the scene. The lens in a camcorder also serves to focus light, but instead of focusing it onto film, it shines the light onto a small semiconductor image sensor. This sensor, a CCD, measures light with a half-inch panel of 300,000-to-500,000 tiny light-sensitive diodes called photosites.

Each photosite measures the amount of light (photons) that hits a particular point and translates this information into electrons (electrical charges). A brighter image is represented by a higher electrical charge, and a darker image is represented by a lower electrical charge. Just as an artist sketches a scene by contrasting dark areas with light areas, a CCD creates a video picture by recording light intensity. During playback, this information directs the intensity of a television's electron beam as it passes over the screen. Of course, measuring light intensity only gives us a black-and-white image. To create a color image, a camcorder has to detect not only the total light levels, but also the levels of each color of light. Since the full spectrum of colors can be produced by combining red, green and blue, a camcorder actually only needs to measure the levels of these three colors to reproduce a full-color picture.

To determine color in some high-end camcorders, a beam splitter separates a signal into three different versions of the same image. One shows the level of red light, one shows the level of green light, and one shows the level of blue light. Each of these images is captured by its own chip. The chips operate as described above, but each measures the intensity of only one color of light. The camera then overlays these three images and the intensities of the different primary color blends to produce a full-color image. A camcorder that uses this method is often referred to as a three-chip camcorder.

This simple method of using three chips produces a rich, high-resolution picture. But CCDs are expensive and eat lots of power, so using three of them adds considerably to the manufacturing costs of a camcorder. Most camcorders get by with only one CCD by fitting permanent color filters to individual photosites. A certain percentage of photosites measure levels of red light, another percentage measures green light, and the rest measure blue light. The color designations are spread out in a grid (see the Bayer filter below) so that the video camera computer can get a sense of the color levels in all parts of the screen. This method requires the computer to interpolate the true color of light arriving at each photosite by analyzing the information received by the other nearby photosites.

A television "paints" images in horizontal lines across a screen, starting at the top and working down. However, TVs actually paint only every other line in one pass (this is called a field), and then paint the alternate lines in the next pass. To create a video signal, a camcorder captures a frame of video from the CCD and records it as the two fields. The CCD actually has another sensor layer behind the image sensor. For every field of video, the CCD transfers all the photosite charges to this second layer which then transmits the electric charges at each photosite, one-by-one. In an analog camcorder, this signal goes to the VCR which records the electric charges (along with color information) as a magnetic pattern on videotape. While the second layer is transmitting the video signal, the first layer has refreshed itself and is capturing another image.

A digital camcorder works in basically the same way, except that at this last stage an analog-to-digital converter samples the analog signal and turns the information into bytes of data (ones and zeros). The camcorder records these bytes on a storage medium which could be, among other things, a tape, a hard disk, or a DVD. Most of the digital camcorders on the market today actually use tapes (because they are less expensive), so they have a VCR component much like an analog camcorder's VCR. However, instead of recording analog magnetic patterns, the tape head records binary code. Interlaced digital camcorders record each frame as two fields, just as analog camcorders do. Progressive digital camcorders record video as an entire still frame which they then break up into two fields when the video is output as an analog signal.

The Lens

The first step in recording a video image is to focus light onto the CCD by using a lens. To get a camera to record a clear picture of an object in front of it, the focus of the lens needs to be adjusted so it aims the light beams coming from that object precisely onto the CCD. Just like film cameras, camcorders let the lens be moved in and out to focus light. Of course, most people need to move around with their camcorders, shooting many different things at different distances, and constantly refocusing is extremely difficult.

A constantly changing focal length is the reason that all camcorders come with an autofocus device. This device normally uses an infrared beam that bounces off objects in the center of the frame and comes back to a sensor on the camcorder. To find the distance to the object, the processor calculates how long it took the beam to bounce and return, multiplies this time by the speed of light, and divides the product by two (because it traveled the distance twice—to and from the object). The camcorder has a small motor that moves the lens, focusing it on objects at this distance. This works pretty well most of the time, but sometimes it needs to be overridden. You may want to focus on something in the side of the frame, for example, but the autofocus will pick up what's right in front of the camcorder.

Camcorders are also equipped with a zoom lens. In any sort of camera, you can magnify a scene by increasing the focal length of the lens (the distance between the lens and the film or CCD). An optical zoom lens is a single lens unit that lets you change this focal length, so you can move from one magnification to a closer magnification. A zoom range tells you the maximum and minimum magnification. To make the zoom function easier to use, most camcorders have an attached motor that adjusts the zoom lens in response to a simple toggle control on the grip. One advantage of this is that you can operate the zoom easily, without using your free hand. The other advantage is that the motor adjusts the lens at a steady speed, making zooms more fluid. The disadvantage of using the grip control is that the motor drains battery power.

Some camcorders also have something called a digital zoom. This does not involve the camera's lenses at all. It simply zooms-in on part of the total picture captured by the CCD, magnifying the pixels. Digital zooms stabilize magnified pictures a little better than optical zooms, but you sacrifice resolution quality because you end up using only a portion of the available photosites on the CCD. The loss of resolution makes the image fuzzy.

The camcorder can also adjust automatically for different levels of light. It is very obvious to the CCD when an image is over- or under-exposed because there won't be much variation in the electrical charges collected on each photosite. The camcorder monitors the photosite charges and adjusts the camera's iris to let more or less light through the lenses. The camcorder computer always works to maintain a good contrast between dark and light, so that images do not appear too dark or too washed out, respectively.

Standard Camcorder Features

The basic camcorder components include:

- **Eyepiece** – This provides a very small black-and-white screen, viewed through a portal, where the user can see exactly what is being taped.
- **Color viewfinder** – The only difference between this and the black-and-white eyepiece is the color which provides a more accurate depiction of what you are recording.
- **LCD screen** – With this feature, the user can watch the color LCD screen instead of viewing through the eyepiece (some models provide both). The screen can usually swivel in order to adjust your viewing angle. But LCDs do use more power than a simple eyepiece, so it could decrease your battery life.
- **Simple record button** – When you hold the camcorder by the grip, your thumb will rest on the record button. All you have to do to switch the record mode on-and-off is press the button. This acts as a sort of pause on recording, so that you continue recording at the exact spot on the tape that you last stopped recording.
- **Zoom function** – This lets the user magnify images that are farther away. Zooms are operated with a simple toggle control on the camcorder grip. Optical zoom changes the size of the object in view without moving the camcorder. Usually limited to 24x, the larger the optical zoom number the more flexibility when shooting. A high number is especially important if the cam's lens is not detachable and some control is needed over the shooting. Digital zoom uses the camcorder's processor to expand an image beyond optical zoom capabilities (up to 300x). But using this feature can degrade image quality, and this feature should be turned off for clear images (the higher the optical zoom number, the clearer the digital zoom image).
- **Auto focus** – The camcorder senses where objects are in front of it and adjusts the focus accordingly.
- **VCR controls** – These controls let the camcorder operate as a standard VCR.
- **Battery and AC adapter** – Nickel-cadmium batteries are normally standard, and allow for one-to-two hours of recording time before they need to be recharged. Smaller or higher-end

cameras may use proprietary batteries specific to the manufacturer, and this can make extra or replacement batteries expensive. Camcorders come with a rechargeable battery and a power cord that attaches to a standard 115V outlet.

- **Audio dub** – On most camcorders, the user can record new audio over video that has already been recorded.
- **Fade-in and fade-out** – This function often works by simply underexposing or overexposing the image to the point that the entire screen is black or white.
- **Clock** – With the correct time and date programmed, the camcorder can display it on your recorded video.
- **Headphone jack** – This lets the user monitor sound quality as the footage is being shot or reviewed, using the onboard VCR.
- **Microphone** – Most camcorders come with a built-in microphone to record sound, which is ideal for personal use. For professional sound quality, look for a camera that attaches an external microphone.
- **Light** – Most camcorders do quite well in low-light conditions without help from additional lights. Some cameras come with a built-in light source, and many can be programmed to turn on automatically when conditions require additional light. Some cameras allow the user to connect an external light source to the unit for greater control over lighting conditions.
- **Image stabilization** – This feature electronically stabilizes the image being filmed, but can decrease its clarity. Digital stabilization performs the same function by using digital technology, and can also correct for tilting and panning movement—again, this feature can decrease image quality. Optical stabilization uses a series a lenses to decreases the effects of camera movement and vibration on the image being filmed—very important with handheld cameras and when filming a moving object.
- **Exposure modes** – Most cameras will set the proper exposure mode for the conditions in which you are filming. Some cameras allow the user to adjust the exposure for certain conditions, such as low light, backlight, or motion.
- **Camera control** – Some camcorders automatically adjust everything—for amateur moviemakers, these settings are more than satisfactory. Professional users, or those who want more control, look for features such as manual exposure control, manual focus, manual zoom, manual white balance, and so on.
- **Still image capability** – Digital camcorders let the user pick still images out of the video. Camcorders with a built-in or removable memory device allow still pictures to be taken as with a digital still camera.
- **Detachable lens adapter** – Some camcorders have a detachable lens adapter. For example, a wide-angle lens attachment is a common accessory.
- **Low-light responsiveness** – Camcorders come with specifications regarding the minimum recommended level of light during recording.
- **Progressive scan** – Progressive scan is only available in digital formats, and records an image with a single scan pass instead of as odd and even fields. This technology increases image quality, and is especially important in using a camcorder to take still pictures.
- **Analog video input** – A digital camcorder with an analog input allows the user to convert existing VHS format tapes to digital format for editing or viewing purposes.
- **16x9 recording mode** – This is "wide-screen" recording mode.
- **Audio recording formats** – Most digital camcorders can support both 32 kHz, 12 bit and 48 kHz, 16 bit audio formats. The 48 kHz, 16 bit audio is better than CD quality.

- **IEEE 1394 (FireWire, iLink) compatibility** – Most of the newer digital camcorders come with IEEE 1394 compatibility which allows extremely fast downloading to a computer.
- **Playback features** – Almost all camcorders come with VCR-type features like rewind, play, and pause.
- **Special effects** – Some higher-end camcorders come with special effects features, such as fade-in/fade-out, and special recording modes like sepia or negative.
- **Motion/audio sensing** – Some camcorders have special sensors that turn the camera on in the presence of movement or sound. This is useful for security purposes.

Summary

Both camcorders and still cameras moved toward digital technology at the same time. The concept of digital photography grew out of television and space technologies. In the 1960s NASA developed digital imaging as a way of getting clearer pictures from space in preparation for the moon landing. During the 1970s digital imaging was further developed for use with spy satellites. Digital imaging finally reached the consumer in 1995 when Kodak and Apple both unveiled the first consumer digital cameras. It took two years for the first million-pixel model to reach stores. Each succeeding year has seen an additional million-pixel increase in resolution.

It is easy to understand the booming business that digital camera manufacturers are experiencing these days. The host of easy-to-use personal and business publishing applications, the dramatic expansion of the Web and its insatiable appetite for visual subject matter, and the proliferation of inexpensive printers capable of photo-realistic output make a digital camera an enticing add-on. Those factors, combined with improving image quality and falling prices, put the digital camera on the cusp of becoming a standard peripheral for a home or business PC. Digital cameras are the 21st century's solution to the old buy/process film and wait routine. Budget permitting, one can buy a digital camera that can produce photos equal to the quality of 35 mm. For a frequent vacationer and traveler, carrying camera bags jammed full of film, lenses, lens covers, extra batteries, and fully-equipped 35-mm cameras has been a constant source of inconvenience. A digital camera is easy to carry, and photographs can be e-mailed to friends and relatives over the Internet. Valuable digital photographs can also be retained on a PC hard drive, while others can be deleted. The camera's memory can simply be erased and reused. Hence, digital cameras provide conveniences in the development, storage, and distribution of photographs.

The formula for success in the digital camera market is multifaceted. It is important to produce a picture with high-quality resolution and to view it as soon as the photo is snapped. Low-cost imaging sensors and inexpensive, removable picture storage (flash memory cards) will help the market to grow. In addition, camera, PC, and IC vendors are working together to come up with different combinations that lower prices and make their systems attractive in terms of providing seamless functionality with other equipment.

A camcorder is really two separate devices in one box—a video camera and a video tape recorder. Creating videos of precious moments is made even easier with the use of digital camcorders. Unlike older analog camcorders, digital camcorders have the ability to store videos in various digital formats. One such is MPEG which enables instantaneous file exchange and dissemination. This also means that the information content is less susceptible to degradation over time. Additionally, the Internet provides an easy, fast, no-cost medium over which digital videos can be shared with friends, families, and associates.

The image quality available on camcorder playback depends on both the camera and the recording device. The weak link for analog formats is the tape recorder, but the camera section (CCD chips, lens, and video processing circuits) is usually the limiting factor for D8 and MiniDV camcorders where the digital recording is very good. Most MiniDV cameras give a picture quality that is better than most Hi8 cameras, but this may not always be the case. In particular, the cheapest MiniDV cameras are in some ways inferior to good Hi8 cameras. Image quality tends to closely follow the price of the camera. There is also a difference between single-chip and 3-chip cameras. The less expensive MiniDV cameras use a single CCD chip to capture the image through the lens which yields less attractive color images than a 3-chip CCD camera. One advantage of digital video is that the signal can be copied without loss of quality, as inevitably happens with analog formats.

CHAPTER **12**

Web Terminals and Web Pads

Introduction

Web terminals (also known as Web pads) are stand-alone devices used primarily for Web browsing and e-mail. These appliances are typically based on embedded operating systems, are packaged in an all-in-one form factor that includes a monitor, and are tethered to a Web connection. Other names for these devices include Web appliance, countertop appliance, Net appliance, or desktop appliance. They are marketed as an instant-on, easy-to-use device for browsing the World Wide Web (WWW) or Web-like services such as e-mail. They do not typically allow a user to install off-the-shelf applications that run outside a browser.

The all-in-one form factor can include a built-in monitor or can connect to a dedicated external monitor. It cannot, however, use a TV set for a monitor. While telephony features can be included, devices with traditional telephone-like form factor and a monitor are called screen phones, and will be discussed in further detail in Chapter 14.

Market

Market accelerators include:

- Web and e-mail access
- Quick access and no long boot-up like PCs
- Low cost and refined utility
- Varied deployment configurations
- Vertical markets such as hotels and schools

Major market inhibitors include:

- Questions about support from service providers
- Lower PC prices and new subsidized distribution models
- High price of built-in displays

Additional market inhibitors are poor promotion, lack of consumer awareness, and consumer confusion. These have come about because:

- **The message is not effective** – Consumers do not comprehend what a Web terminal can truly offer. They understand it is a stripped-down version of the PC simple enough for a "non-techie," but do not understand exactly what it can actually do or not do.
- **Changing naming conventions** – They have been called Web terminals, Internet terminals, and Web appliances, which does not accurately depict the Web terminal.
- **Constant comparison with the PC** – The thought that this stripped-down PC will one day eliminate all PCs.

- **Using low cost as a draw** – Reducing costs through elimination of components has led to elimination of key functions of the product. For example, not having a hard disk prevents storage of MP3 music files.
- **Over-expectations have backfired** – The press has now taken the approach that the Web terminal will never succeed. This is because of the tremendous amount of press that these products originally received and their failure on the first try. Market analysts predicted that Web terminals would have shipment volumes similar to the PC, but this has not materialized.

The primary market for Web terminals is the consumer, but they are sometimes used in vertical applications such as hotels and schools. Plenty of doubt surrounds the Web terminal market as companies drop out and reported volumes remain low. First generation Web terminal products meet the expectations of instant-on, easy-to-use appliances, but the designs need further improvement. Consumers are used to having functionality such as full support of Shockwave, flash, real audio/video, Windows Media and audio/video applications. They have been disappointed when they try to link to a flash-based greeting card or media stream. Many Web terminals use embedded operating systems, and many of the third party applications cannot be ported to Web terminals. Web terminals cannot, therefore, provide a true Web experience for consumers. Most Web terminals only support up to 56 Kb/s data rates and are not enabled for broadband access and home networking. The Web terminal also cannot exchange downloaded MP3 files with a portable MP3 player.

The primary focus of the Web terminal is to provide easy Web access for the consumer. But half the households in the U.S. have at least one PC, and Web access is easier to perform on the PC rather than paying over $1,000 for a Web terminal. The selling points for these devices are instant-on, and quick and convenient access to the Web. However, they are not suitable for prolonged "surfing" sessions. The Web terminal's entire value proposition falls through because of its technical short-comings and the fact that it is overpriced for the mainstream market population. However, recent Web terminal products have added applications such as calendars and scheduling tools that can be synchronized with PDAs. But other highly desirable applications, such as Sony's AirBoard (a wireless video screen for viewing DVD movies), need to be incorporated in order for these products to succeed. Web terminals could conceptually serve as PC replacements and PC adjuncts, although they appear functionally as stripped-down PCs. They are therefore targeted at non-PC users who have little technical sophistication and require only limited Web usage.

Distribution has been a problem because Web terminals are still emerging and are unproven. Retailers have been hesitant to put Web terminals on shelves with competing devices (such as set-top boxes and gaming consoles) that have better business models and higher return on investment. Telephone companies and cable companies have also been hesitant to promote these devices because of the requirement to carry inventory (which they never prefer). Web terminals must be seen as a well-executed product and be sold through both direct sales and retail channels in order to have maximum success in the market place.

Key variables that will affect the Web terminal market include:
- PCs with near instant-on Web access. The latest generation of Windows operating systems have shorter boot-up times in PCs. The PC will become as easy to access as the Web terminal if boot time continues to shorten.
- Broadband access to residences is becoming cheaper and more necessary as Internet traffic expands to include audio, video and images.

■ Home networking and residential gateways will allow Internet access to multiple consumer devices in the home. The PC adjunct consumer devices will also require Internet access and connection to the other appliances.

Web terminal shipments worldwide are predicted to exceed 2.7 million units and $540 million in value by 2004.

Components

Web terminals generally use passive matrix color LCD screens around the 10-inch size. Display variations include LCD panels in sizes of 8-to-12 inches, touch-screen capabilities, and the much cheaper (and bulkier) CRT monitors. Some Web terminals do not offer screens and allow the user to choose a PC display. Cost of the LCD screens is the major contributor to the high bill of materials when manufacturing Web terminals.

Web Pads/Tablets

Introduction

The Web pad is a wireless, portable, low-cost, easy-to-use, tablet-shaped, consumer-targeted digital device that usually has a touch-screen display/control. It has a browser-based interface to simplify and enhance the Internet experience. Web tablets are similar to Web terminals, but with the key difference that Web tablets are portable. While that makes the Web tablet very appealing, it does add to the overall cost.

The two separate components of a Web tablet/pad are:
1. A portable LCD built into a tablet shaped device, and
2. A base station with an analog or broadband hard-wired connection to the Internet. The base station also sends and receives wireless (RF) transmissions to and from the pad/tablet.

Transmissions between the Web tablet and the base station use RF or wireless home networking technologies such as HomeRF, Bluetooth, or IEEE 802.11b. These technologies are not based on cellular or mobile telephony. This limits the wireless range of the Web tablet to approximately 150 feet. Multiple tablets can be used with the same base station, and sometimes the residential gateway can be used in place of a base station. The base station connects to the network through DSL or a cable modem. The tablet's portability brings more value than a Web terminal and makes an attractive product. But it is less usable when configured with a soft keyboard that consumes half of the Web surfing screen. Handwriting recognition is also being introduced in some products, but is still emerging as a technology. Long battery life allows consumers to use the Web pad for many hours without recharging.

Applications

Primary applications for Web tablets include:
■ Internet and Web browsing and surfing
■ E-mail
■ Multimedia
■ E-Book reading
■ Combining Internet access with home environment and security controls.

Some Web tablets also have hard disk drives, and can be used in more sophisticated vertical applications that have intensive computing requirements such as:

- Inventory control
- Order entry systems—merchandising
- Healthcare and hospitals
- Education
- Field and sales force automation
- Shipping
- Government agencies.

Web pads have succeeded in the less price-sensitive vertical markets. Tablet PC vendors are trying to penetrate the same vertical markets as Web tablets due to large volumes and less price sensitivity.

Market Data and Dynamics

Market researcher IDC predicts that the worldwide market for Web pads is expected to exceed 1.5 million units with a value of $1.35 billion by 2005. Allied Business Intelligence (ABI) predicts that more than 23 million Web pad units will be sold annually by 2006.

Some of the issues preventing Web tablets from becoming a mass-market product are:

- **High bill of materials (BOM)** – The high cost of wireless home networking components, battery integration, and the LCD display causes significant consumer price issues. A $500 BOM results in discouraging end user pricing of over $800.
- **Limited value proposition** – The Web tablet's unique value proposition is its wireless mobile connection to the Web. This is not enough to drive mass-market success of the device.
- **PC competition** – The PC's and notebook PC's broad value position overshadows the single, unique value proposition of the Web tablet.
- **Lack of vendor support** – Top consumer electronics and PC manufacturers have not embraced this technology.
- **Tablet PCs** – These are fully functional PCs that have a touch-screen LCD display.
- **Other competition** – PDAs and PC companions offer some of the functionality offered by Web tablets and (with wireless connectivity) can be taken along on the road.

Failure of Web Pads

This market has seen the failure of several high profile products such as 3Com's Audrey and Netpliance. The Web pad provides several nice features such as one-touch information access (to weather, e-mail, news, and so forth) and the ability to synchronize with PDA products. It is impossible, however, to make reasonable profits by selling a $500 device if one cannot also charge recurring monthly fees for its use.

Some of the reasons Web pad companies have left the market include:

- **Crowded market** – Multiple vendors announced products that had similar value propositions—primarily access to the Web. However, reduced PC prices and an excessive number of companies introducing Web pads caused Web pad price reductions that eliminated any profit margin.
- **Economy** – With the economy faltering after the Internet boom, venture capitalists and the public market could not sustain business models that showed no profits for years to come.
- **Failure of strategic partnerships.**

Hybrids

Web pad derivatives are finding applications in refrigerators, home automation, and TV remotes. One such hybrid is the Nokia MediaScreen, which is a tablet prototype that combines digital TV and the Internet. It uses mobile phone technology with a 12-inch TFT screen, and its main application is in cars and trains (not residences). Web tablets are ideal devices for providing e-book services—portable, dedicated reading devices with flat screens and Internet access (hence requiring connectivity components).

Integrating Web pad devices (small portable tablets) into traditional home appliances such as refrigerators and kitchen stoves has been a common application discussion. The Web pad device would attach to, or be integrated with, the refrigerator door. Several companies are entering this market due to the high unit volumes of white goods sold globally. They hope to make money through the sale of services and white goods content provisioning. Appliance manufacturers such as General Electric, Sunbeam, and Whirlpool are also interested in this market. They believe Web pad integration could be used to reduce the number of visits by technicians to service appliances. For example, a malfunctioning refrigerator could provide detailed failure information via an Internet connection. A service technician could then schedule a single visit (or no visit), already knowing the challenge that lay ahead. Other companies such as Cisco Systems and Sun Microsystems are also joining this market to provide equipment and services for connectivity.

A future capability of the "fridge pad" could be the ability to monitor expiration bar codes of perishable and regularly consumed goods. These could then be automatically re-ordered when required through a favorite online grocery store. Such conveniences make the Web pad seem like an ideal consumer device with a strong future. One can also download recipes to a screen with Web access in the kitchen. Recipes could then be e-mailed to friends or saved on the base station hard disk drive.

Companies such as Honeywell and Qubit are also introducing products based on the Web tablet for home automation. Implementations include tethered, untethered, and wall integrated solutions that usually have a LCD touch screen. These devices will act as a Web browser and e-mail client, in addition to providing home control functions.

Another interesting and promising effort is the super remote control. This application integrates iTV (integrated TV) platforms into Web pads. A constraint of iTV, however, is the physical limitations of the TV set. Since the iTV normally displays the menu of the set-top box, manufacturers are looking to coordinate the set-top box operation with the Web tablet in order to display an interactive interface on the tablet screen. This would allow the iTV screen to be unblocked and pristine. The Web tablet can also download interactive content from the set-top box, and control other entertainment appliances.

Hence the integrated Web pad has other applications in addition to providing access to the Web. An integrated Web pad could control several appliances around the house, control entertainment, monitor and manage white goods, control home temperature and more.

Phased Success

Several companies have already been unsuccessful in a Web pad venture. But this market is still in its infancy and is experiencing growing pains. There are three phases in the evolution of every product (such as PCs, hand-held devices, and mobile phones) as that product is introduced into the consumer market. The Web pad has experienced phase one, and will be pushing through phase two and three. These phases are described as follows:

- **Phase one** – Phase one saw the introduction of several Web pads and business models. The market began to grow, and there was rampant experimentation. The Web pad market segment attracted a lot of venture capital. There were targeted acquisitions and new partnerships created. Web pad products have gained some ground at the hands of the consumer, but acceptance has been shaky. Several companies will cease to exist following their failure in this market.

- **Phase two** – Better-positioned companies remain intact into phase two and benefit from the painful but valuable lessons learned during the shakedown in phase one. Those with staying power take these lessons, refine their business plans, and foster formidable relationships. It is much easier for a General Electric to introduce a Web pad-based refrigerator to try out the market, for example, than for a start up. Bigger companies can always hedge the losses from one product with the profits made from another. Start-up companies may have only one product, which is their only chance to survive. Drastic product improvements are sometimes needed during phase two in order to meet the acceptable usability threshold for more potential users. Market volume and awareness increase slowly but steadily in this phase. As the Web pad market grows out of phase one and into phase two, more appealing product designs, better applications, and sustainable business models will push the Web pad from experimentation (gadget geeks and early adopters) toward the mainstream market.

- **Phase three** – In phase three, the market finally matures and mass-market consumers begin to accept the product and the business model. Newer and more elegant products, and better and more innovative applications have more appeal to consumers. Increased volumes bring the price down and awareness is spread more widely. Some vendors begin consolidating into larger entities as the market matures further, possibly pushing the market into a final phase of saturation.

Consumer products such as PCs, handheld devices, and mobile phones first appealed to, and succeeded in, the large business community. They then made their way into the consumer's hands. The Web pad, in contrast, is directly targeted at the consumer and will not see volume gains through the business environment. This approach may cause a delay in product market success. Vertical products based on Web pads (such as health care devices) may also have a future, however.

The average selling price of the Web tablet is expected to fall, which may help its market growth. This price reduction will be achieved through a drop in the cost of its LCD display and by unbundling the tablet from the base station before marketing. The price of the tablet as a stand-alone device (no base station) will appear more attractive to the consumer. Consumers will no longer need to purchase the Web tablet base station for wireless connection as the residential gateway market penetration increases. This market is already showing signs of saturation with a very high number of vendors.

The Tablet PC Threat

The recently announced tablet PC is a formidable threat to the Web tablet. The tablet PC is an 8-by-11-inch device with high-powered computing capability for the office and the home, and provides much more functionality than Web pads. The tablet PC will also support all applications supported by a standard PC. It can be used to create and edit converted versions of Word and Excel documents and to view PowerPoint and PDF formatted files. You can browse the Web with Internet Explorer and perform instant messaging. Multimedia applications include Microsoft Media Player for viewing video clips, MP3 Player, and Picture Viewer. In fact, the tablet PC is a convenient alternative to a full-size notebook PC. It is a portable tool for taking notes, editing documents, and managing PIM data and it delivers more screen area than a PDA. The tablet PC is also a device for vertical-market deployments, such as health care and insurance.

Some tablet PC products also have handwriting recognition applications. An alternative and convenient on-screen input mode (virtual keyboard) is more user-friendly than virtual keyboards on smaller PDAs. The tablet PC also includes both a Type II PC card slot and a Type II CompactFlash card for expansion and wireless communications accessories. Its accessory line includes an external keyboard and a sturdy docking station with USB upstream and synchronization ports, P/S2 keyboard and mouse ports, and an RJ-45 Fast Ethernet port. This product is far different from Web tablets that have embedded operating systems that are produced by third party developers.

However, the pricing and the value proposition of both products (Web tablet and tablet PC) will determine the success of each. But Microsoft's formidable presence will give tremendous clout and brand recognition to the emerging tablet PC market.

The tablet PC comes with the latest generation Microsoft OS and runs nearly any Windows application. It has a big bright touch-sensitive screen, a processor, RAM, and has an internal multi-gigabyte hard drive. Desktop and notebook PCs have reached the end of their evolutionary advancement in the corporate market, but they are not going away—they will get faster, smaller, and cheaper. A notebook PC will still be better for text-entry-focused applications than a Web pad.

Recent introductions of tablet PC variants from a number of manufacturers show significant advantages of using tablet PCs in the business environment. Tablet PCs are available in two primary designs: One features an attached keyboard and can be configured in the traditional laptop "clamshell" mode, and the other uses a variety of detachable keyboard designs in a so-called "slate" form factor. All tablet PCs are designed to be a user's primary business PC, recognizing input from a keyboard, mouse or pen. Tablet PCs are powered by chips optimized for low-power consumption and longer life from Intel Corp., Transmeta Corp. and Via Technologies Inc. Tablet PCs are available at a number of retailers throughout the United States.

The tablet PC operating system enables Windows-based applications to take advantage of various input modes, including keyboard, mouse, pen and voice. With software developed and optimized by Microsoft for the new platform, the tablet PC can function as a sheet of paper. Handwriting is captured as rich digital ink for immediate or later manipulation, including reformatting and editing. The link between the pen input process and a wide range of Windows-based applications will give users new ways in which to collaborate, communicate, and bring their PCs to bear on new tasks. Its high-resolution display makes the tablet PC ideal for immersive reading and rich multimedia applications.

The tablet PC's full Windows XP capability enables it to be a primary computer. Utilizing a high-performance x86-compatible chip architecture, the Tablet PC takes advantage of key technology improvements in high-resolution low-power LCDs, efficient batteries, wireless connectivity, and data storage to deliver a rich set of functions, with the added dimension of pen-based input. A number of third-party software vendors are now engaged in developing software specifically for the tablet PC.

Components of a Web Pad

Some of the key components of the Web pad include:

- **Microprocessor** – Its comparatively low-power consumption and low-heat producing characteristics are ideal for Web pad applications. Some of these processors include National Geode, Transmeta Crusoe, Intel XScale, Intel StrongARM, Intel Celeron, and ARM.

- **Embedded operating systems** – Web terminals and Web tablets are limited in their appeal because they lack large amounts of storage memory and dedicated functionality. Consumers will not adopt these devices unless manufacturers select operating systems that boot-up rapidly and are available instantaneously when the screen is turned on. Some of the popular operating systems for Web tablets and terminals include BeOS (by BeIA), Linux (open-source), QNX's RTOS, Windows PocketPC (by Microsoft), VxWorks (by Wind River), and Jscream (by saveJe). Price and performance are key factors when choosing an operating system for these products. But the most critical factor in choosing an operating system is the software applications that the operating system supports. One of the main reasons for the success of the PC is the number of applications that it can support. The Web pad needs to provide multimedia capabilities as well. Of key importance, the Web pad requires a Web browser to provide Web support—the most popular include Microsoft's Internet Explorer and AOL Time Warner's Netscape Communicator. Embedded Web browsers are also available from Espial, Opera, QNX, and others.

- **Display** – A color touch-screen LCD is the main user interface. This is a simple-to-operate consumer friendly solution that varies in size between 8 and 10 inches. Display options include TFT (active matrix) screens for improved resolution. This, however, makes the finished product even more costly.

- **Battery** – The Web pad can typically operate for 2 to 5 hours on batteries.

- **Flash memory and RAM**

- **Power management unit**

- **Network Interfaces** – This enables communication between the tablet and the base station. The network interface is provided by wireless technologies such as wireless LANs (like IEEE 802.11b), HomeRF, Bluetooth, DECT, and others:

  - **LAN standards** – The IEEE 802.11b wireless LAN, or Wi-Fi, is a common method of connecting portable PCs to the network. The IEEE 802.11b is also gaining ground as the dominant home networking technology because the same wireless LAN card used in industry can also be used in the household. But the 802.11b offers 11 Mb/s data rates without quality of service. Two other LAN standards, the IEEE 802.11a and HiperLAN2, offer both quality of service and high data rates. They are also well-suited for home environments that require voice, video and data access. Future Web tablets will be based on wireless LAN technologies such as IEEE 802.11a and HiperLAN2.

  - **HomeRF** – HomeRF is a popular wireless home networking technology that offers 1.6 Mb/s data rates at a lower price than 802.11b. But reduced HomeRF membership and the dropping price of IEEE 802.11b make the future of HomeRF look rather bleak. Several Web tablet vendors have also chosen 802.11b instead of HomeRF as their wireless technology.

  - **Bluetooth** – Bluetooth is gaining ground as a network interface in the home. It has the support of over 2000 companies and the promise of a $5 component price, which makes it attractive. Its personal area networking features also make it an ideal technology to replace Infrared networking technology. Bluetooth should be ideal for synchronizing calendars and address books between different consumer devices such as PDAs, PCs, and Web tablets. However, Bluetooth is not an ideal technology for cordless communication because of its limited range and data rates.

- **Digital Enhanced Cordless Telecommunications (DECT)** – DECT is a cordless phone standard that has limited data rate support and a voice-oriented specification. Few Web pads will make use of DECT technology because of these limitations.
- **Keypad interface**
- **Built-in smart card reader** – A smart card reader has been included in anticipation of applications like secured online transactions, personalized authentication services, access to personal data, and e-Cash storage options.

The base station comes with a built-in 56K V.90 modem and an integrated 10/100 Base-TX Ethernet port as standard. There is also a wireless module that is designed for communication with the Web tablet using 802.11b or other wireless technologies.

Figure 12.1 shows a block diagram of a Web pad/tablet.

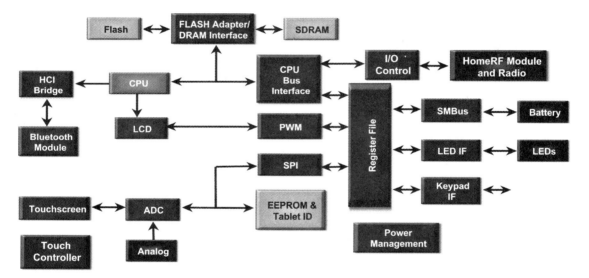

Figure 12.1: Web Pad Block Diagram

Role of the Service Provider

Some Web pads and Web terminals fail primarily because of poor business models. A consumer is not motivated to pay up to $1,500 for a Web tablet when their PC supports far more applications. The Web pad will succeed in coming years only if service providers play a key role in their deployment. These appliances will become customer acquisition and retention tools for ISPs, telecommunication companies, and cable companies.

Cable companies have generally stayed out of the Web pad and Web terminal markets so far. They have been focused on set-top boxes as the appliance that provides cable services and Internet access to the consumer. The Web pad base station, however, could very easily connect through a cable modem. The telephone companies could also provide DSL service to the base station for high-speed Internet access. Meanwhile, the ISPs are currently driving the Web pad and Web terminal market.

The long-term goal of all of these companies is to provide high-speed data, voice, and video access to multiple devices (PCs, TVs, gaming consoles, and Web pads) within the connected home. The ISPs, telephone companies, and cable companies are all trying to add new users and often try schemes such as offering free Web cameras to attract customers.

The ISPs try to increase their revenues by offering free online devices like Web pads to new customers. This translates into additional users who are willing to pay monthly fees to use their portals. New users then become customers of online retailers and order merchandise through those portals. The ISPs sell additional portals that are then used to provide advertising real estate on the consumer screen. Microsoft and AOL Time Warner are the major ISPs with online portals. They would like to sell easy-to-use, instant-on Web terminals and Web pads in order to attract the PC-shy user. These ISPs could offer rebates on the Web pads and Web terminals in exchange for a year's Internet service at $22 a month—similar to what they charge to connect a PC through a regular phone line. Some ISPs have offered rebates for the purchase of PCs in order to lock-in recurring revenues for years of subscription. The ISPs are always trying to extend their reach from the traditional PC to consumer devices like TVs, handheld devices, and screen phones. Web terminals and tablets are simple extensions of this strategy.

The strategy of service providers such as EarthLink and DirecTV that do not have online properties like portals to sell is to simply earn revenues from subscriptions. They also provide e-mail services, which brings consumers back to the ISP. They prefer selling the cable or DSL modems through several direct or indirect channels. Other companies may also offer rebates on appliances such as Web pads with Internet service through one of these providers. This strategy allows the ISP to be more focused on technology when providing service. These subscription providers have offered broadband service far quicker than rival ISPs with portals.

The attraction of online portals that have no ISP services is the quality and quantity of content provided by their portal. Some ISPs with portals are becoming the de facto portal of choice for the consumer by partnering with ISPs such as Yahoo that do not have portals of their own.

Summary

The rise of the World Wide Web has stimulated an onslaught of Web-based applications and services. Everything from calendars, shopping, address books, customized news services, and stock trading can be done over the Web. This has made the standalone PC based on the Windows operating system less relevant for those who want only Web-based information and services. New machines that are easier and simpler to use than a PC are on the way, and promise to liberate non-techies from the headaches of dealing with a PC. Two such attempts are Web terminals and Web pads.

While the initial outlook for Web terminals has been gloomy, they will have their place in the consumer market. They are currently in the evolution phase of market acceptance. This is where business models need to evolve and the product value proposition needs to be further strengthened. Unit volumes will not be enough for the next few years to justify the current number of vendors building Web terminals. These vendors are all fighting over a very small pie, and several vendors will exit the market.

The Web pad provides an easy-to-use, portable tool for wireless, high-speed connection to the Internet and centralized home control applications. Most Web pads/tablets weigh less than three pounds, measure less than an inch thick, and are about the size of a standard sheet of paper. A user can surf the Internet, send and receive e-mail, and enter information via a wireless touch screen display—all this at higher speed and with more features than the Web terminal. The Web pad's transmission and reception signals link to an 802.11b wireless access point connected to broadband cable, DSL, or a dial-up phone connection. It allows the consumer to enjoy the benefits of a high-bandwidth wireless connection from anywhere inside or outside the home from within 200 feet of an access point or base station. The tablet is currently fashionable, or "in," which also helps to sell the product. The Web pad will become more appealing as word of its usefulness spreads. This will certainly help sales and accelerate the market for this product.

The basic math still holds: Consumers currently will pay $500 for a PC that gathers dust, but not for a less powerful, less usable Net appliance. The Web tablet is not to be confused with the Microsoft Tablet PC. Though they look similar, the Tablet PC is a full-fledged PC with a fast processor and a Windows operating system that's marketed toward businesses. The Web tablet is a wireless Internet device that shares a household's PC Web connection and other core hardware, and better yet, one person can surf the Web on the tablet while another connects via the PC.

Web pads and Web terminals have seen discouraging results thus far in the consumer market, but it is important to note that these products are not completely out of the running. History has shown that failure of one product configuration does not kill an entire category. Discouraging results so far could be due to improper product execution and poorly thought-out business models. For example, the Apple Newton and other handheld devices failed in the PDA market before the Palm Pilot was finally a success due to its correct product design. Web pads and Web terminals are still waiting for the perfectly executed product strategy to emerge.

Internet Smart Handheld Devices

Introduction

Internet smart handheld devices (SHDs) provide direct Internet access using an add-on or integrated modem. Vendors that have announced Internet SHDs include Palm Computing, Dell Computer, Nokia, Sony, Samsung, and Hewlett-Packard, with numerous others following suit. Over the last few years, SHDs have been the hottest and largest growth segment in the consumer market.

Smart handheld devices include handheld companions, smart handheld phones, and vertical application devices (VADs). The SHD market has grown significantly over recent years because of market accelerators such as their acceptance by corporations and the presence of content (e-mail and Internet services such as stocks, news, and weather). Key market inhibitors, however, have been the slow acceptance of the Windows CE operating system, high cost (and the inability to eliminate the PC), and the lack of applications for these devices. Market researcher IDC predicts that annual SHD shipments will exceed 35 million units and revenues of over $16.5 billion by 2004.

Vertical Application Devices

Vertical application devices (VADs) are pen- and keypad-based devices that are used in specific vertical applications in a variety of industries. Key applications for VADs include:

- Routing, collecting, and delivering data for a vendor in the transportation industry
- Providing physicians access to patient's records in hospitals

These devices include pen tablets, pen notepads, and keypad handheld segments. The pen tablet is used to gather data from the field or in a mobile situation, such as entering orders in a restaurant or collecting data in a warehouse or rental car return lot. Pen tablet products typically weigh two pounds or more and have a display that measures five inches or more diagonally.

Pen notepads are small pen-based handheld devices that feature a five-inch or smaller display (measured diagonally). This product is typically used in business data collection applications (e.g., route accounting).

Keypad handheld products are used in vertical business data-collection applications. Typically, these devices are built to withstand harsh environments, such as rain or dust, and they can continue to work after being dropped three feet onto a hard surface. The worldwide market for vertical application devices, including pen tablets, pen notepads and keypads, will exceed 3.4 million units and $5 billion in revenue by 2005.

Smart Handheld Phones (or Smart Phones)

Smart phones are generally cellular voice handsets that also have the ability to run light applications, store data within the device, and, in some cases, synchronize with other devices such as a PC. The ability to run these applications is what differentiates smart phones from handsets that are merely Wireless Application Protocol (WAP)-enabled (which will be virtually all handsets over the next four years). The expanded capabilities of the smart phone can include the functions of a personal companion such as PIM (personal information management), as well as the ability to access data through a WAP-enabled micro browser such as Openwave's UP browser. Smart phones will initially be able to take advantage of multiple standards of data connectivity, including:

- Short messaging systems (SMS, which will most likely be concentrated in "dumb" handsets)
- WAP (Wireless Application Protocol)
- Services that pull data from existing web-based content providers and reformat them for consumption on a small screen.

Smart handheld phones include the emerging enhanced, super-portable cellular phones that enable both voice and data communications. Some of the applications for smart handheld phones include cellular voice communications, Internet access, calendar, and Rolodex[a] data such as names, addresses, and phone numbers.

The smart phone market is in constant flux as new players (e.g., Hewlett-Packard) enter the market to capitalize on its explosive growth, and old ones (e.g., Psion) abandon it due to competitive issues. The constant change in the market dynamics is beneficial to the growth of the market, however, since the result is an environment that fosters innovation and the adoption of new technologies. Smart phones generally have a larger screen and perhaps smaller buttons than the simple WAP-enabled handset, and some will include a pen for touch-screen input. Ericsson, Neopoint, Qualcomm, Kyocera, Samsung, and Nokia have introduced smart phones. The leading operating systems for smart phones include Palm OS, Symbian's EPOC, and Microsoft's Stinger. The worldwide market for smart phones will reach about 18 million units and $7.7 billion by 2004

Handheld Companions

Handheld companions include personal and PC companions, and personal digital/data assistants (PDAs). Applications for handheld companions include:

- Personal information management (PIM)
- Data collection
- Light data creation capabilities (such as word processing for memos)

Personal and PC Companions

Personal computer companions normally feature a keyboard, a relatively large screen, a Type I and/or II PC card expansion slot, the Windows PocketPC operating system, data synchronization with a PC, and, in some instances, a modem for wire-line Internet access, and a pen input for use with a touch screen. PC companions are generally used for PIM and data creation activities such as e-mail, word processing, and spreadsheets. This form factor will increase in popularity (but will not come near the explosive growth of PDAs) as prices fall and as technology currently associated with notebook computers is added to the product segment. Examples of current products in this category include the Hewlett-Packard Jornada and the Psion Revo.

The most popular class of devices among handheld companions is the PDA, which is discussed in further detail below.

Personal Digital (Data) Assistants—The PDA

In the 1980s the Franklin planner or Filofax organizer was the visible sign that you were a busy and important person. The end of the 1990s replaced that badge of distinction with a digital equivalent— the Personal Digital Assistant (PDA). A PDA is effectively a handheld PC, capable of handling all the normal tasks of its leather-bound ancestor—address book, notepad, appointments diary, and phone list. However, most PDAs offer many more applications such as spreadsheet, word processor, database, financial management software, clock, calculator, and games.

What made PDAs so attractive to many PC users was the ability to transfer data between the handheld device and a desktop PC, and to painlessly synchronize data between the mobile and desktop environments. Early PDAs were connected to the PC by a serial cable. Modern PDAs connect to the PC via an infrared port or a special docking station.

The allure of the PDA in the realm of business is not hard to understand. Small, portable, and powerful technologies have always held great general appeal. For style-conscious users looking for the latest and greatest in gadgetry, the PDA is a natural accompaniment to that other essential business item of the twenty-first century—the mobile phone. The increasing power of PDAs has led to a growing interest in the corporate arena. If simple data manipulation and basic Internet connectivity are the only applications required, the PDA is an attractive option—likely to be a much lighter burden to bear than a notebook PC for both the mobile worker and the company bank account.

Because of the PDA's size, either a tiny keyboard or some form of handwriting recognition system is needed for manually entering data into a PDA. The problem with the keyboard is that they are too small for touch-typing. The problem with the handwriting recognition system is the difficulty in making it work effectively. However, the Graffiti handwriting system has proven to be the solution to the handwriting recognition problem. This system relies on a touch-screen display and a simplified alphabet (which takes about 20 minutes to learn) for data entry. Typically, PDAs with the Graffiti system provide the option to write directly onto the display, which translates the input into text, or to open a dedicated writing space, which also provides on-line examples and help.

The PDA market has become segmented between users of the two major form factors—devices that have a keyboard and others that are stylus-based palm size devices. The choice depends on personal preference and the level of functionality required. Following this trend, Microsoft has evolved its CE operating system into the PocketPC. However, PDAs will still require a universally sought-after application to make them truly ubiquitous. The advent of multifunction universal communications tools that combine the capabilities of mobile phones and PDAs may be set to deliver that. The ability to conveniently and inexpensively access the Internet with a single device is the holy grail of mobile computing. In fact, Palm Computing brought this ability a significant step closer with the launch of their wireless Palm VII in the autumn of 1999. One possible outcome is for the market for the wireless PDA class of device to split into two approaches. One approach would be for those whose desire is basically for an upscale mobile phone, and who only require modest computing power such as a built-in web browser. The other would be for those who require a portable computer and who want to be in touch while in transit.

The PDA has been one of the poster children for putting digital computing into the hands of the consumer. In general, they are palm-sized devices—about the size of a package of cigarettes. Users input data by tapping the keys of a mini keyboard pictured on-screen or, more commonly, by writing with a stylus on a note-taking screen. The notes are then "read" by handwriting recognition programs that translate them into text files. The PDAs are designed to link with desktop or laptop PCs so that users can easily transfer dates, notes, and other information via a docking unit or wireless system. Personal digital assistants come in many shapes and sizes, and are synonymous with names like handheld computer, PC companion, connected organizer, information appliance, smart phone, etc. A PDA or handheld computer is primarily a productivity and communications tool that is lightweight, compact, durable, reliable, easy to use, and integrates into existing operations. Typically, it can be held in one hand leaving the other to input data with a pen type stylus or a reduced size keyboard. It is undoubtedly one of the most successful appliances to have arrived in the palms of the consumer.

Personal digital assistants fall into one of several general categories—tablet PDAs, handheld PCs (HPCs), palm-size PCs (PPCs), smart phones, or handheld instruments. Ease of use and affordability are two essential factors driving consumer demand. Handheld computers and PDAs are not a new concept, but only recently have they begun to find broad appeal. Palm Computing, Inc.'s Palm-brand "connected organizers" and Microsoft's PocketPC were introduced in 1996, and have revived consumer demand for PDAs. The PocketPC operates the new generation Windows CE operating system, which is a stripped-down version of its Windows operating system, and is tailored for consumer electronics products.

History of the PDA

The idea of making a small handheld computer for storing addresses and phone numbers, taking notes, and keeping track of daily appointments originated in the 1990s, although small computer organizers were available in the 1980s. In the late 1990s, a combination of primarily chip, power, and screen technologies resulted in an avalanche of newer and increasingly sophisticated information and communication gadgets for both home and mobile use. The growth of what would become the handheld computing market was driven by the transformation of the corporate environment into an extended, virtual enterprise. A mobile, geographically dispersed workforce requiring fast and easy remote access to networked resources and electronic communications supported this transformation. The emergence of corporate data infrastructures that support remote data access further encourages the growth of the handheld computing market.

The easy acceptance of mobile computing may well have been influenced by the TV show *Star Trek*, where producer Gene Roddenberry forbade pen and paper on the 23rd century starship U.S.S. Enterprise. This requirement gave rise to the "Tricorder" and the concept of mobile information devices. In the 1970s, Xerox's PARC research center explored the Dynabook notebook computer concept. The first mobile information device (in the real world) was the Osborne 1 portable computer in April 1981. The first Compaq appear in July 1982, and was followed by the Kaypro 2 in October of that same year. All three "luggable" computers were the size of small suitcases, and each weighed about 25 pounds. The Kaypro was mockingly nicknamed "Darth Vader's Lunchbox."

These early portable computers and their successors, laptop and then notebook computers, merely served as replacements for their full-sized counterparts. Consumers were seeking a new type of device—one that would supplement the computer and replace paper-based appointment calendars and address books.

Psion and Sharp

It is generally accepted that UK-based technology firm Psion defined the PDA genre with the launch of its first organizer in 1984. The Psion 1 weighed 225 grams and measured 142mm x 78mm x 29.3mm—narrower than a large pack of cigarettes, and slightly longer and thicker. It was based on 8-bit technology, and came with 10K of nonvolatile character storage in cartridges. It had two cartridge slots, a database with a search function, a utility pack with math functions, a 16-character LCD display, and a clock/ calendar. The optional Science Pack turned the Psion into a genuine computer, capable of running resident scientific programs and of being programmed in its own BASIC-like language, OPL. The Psion I was superseded by the Psion II—500,000 Psion IIs were produced between the mid-1980s and the early 1990s. Many of these were commercial POS (point of sale) versions that ran specialized applications and did not include the standard built-in organizer functions.

In 1988, Sharp introduced the Sharp Wizard, which featured a small LCD screen and a tiny QWERTY keyboard. This, however, saw very little success. The Psion Series 3a, launched in 1993, and based on 16-bit microprocessor technology, represented the second generation in Psion's evolution. The Series 3a was housed in a case that looks remarkably like a spectacle case and opened in a similar way to reveal a 40-character x 8-line mono LCD and 58-key keyboard in the base. The Series 3a broke new ground with its ability to link to a desktop PC and transfer, convert, and synchronize data between the two environments. Psion's domination of the PDA market was assured for a couple of years. The more powerful Series 3c and the third generation 32-bit Series 5 were launched in 1997, built upon the success of the 3a. The Series 5 boasted the largest keyboard and screen—a 640x240 pixel, 16 gray-scale—of any PDA to date. But these features did not prevent Psion from losing its PDA market leadership position to 3COM's groundbreaking PalmPilot devices.

Apple Computer

Psion's success prompted other companies to start looking at the PDA market. Apple Computer made a notable attempt to enter this market in mid-1993, when the launch of its first Newton Message Pad was heralded as a major milestone of the information age. Faster and more functional chips led to the development of the electronic organizer into the personal digital assistant (PDA), a term coined by then Apple CEO John Sculley. Several other established electronics manufacturers such as Hewlett-Packard, Motorola, Sharp, and Sony soon announced similar portable computing and communication devices.

An ambitious attempt to support data entry via touch-sensitive LCD screens and highly complex handwriting recognition software differentiated Apple's Newton technology from its competitors. In 1997 Apple launched the eMate, a new PDA that continued the Newton technology. But Newton's handwriting recognition technology never became fast or reliable enough, even though it had advanced by leaps and bounds in the years since its first appearance. The Newton was also too large, too expensive, and too complicated. In 1998 Apple announced its decision to discontinue development of the Newton operating system.

Palm Computing

In 1995, a small company called Palm Computing took the idea of the Newton, shrunk it, made it more functional, improved the handwriting recognition capability, halved Newton's price, and produced the first modern PDA, the Palm Pilot. U.S. Robotics acquired Palm Computing in 1995, and transformed the PDA market one year later by introducing the company's keyboard-less Pilot products. Data was entered into these devices with a stylus and touch-sensitive screen, using the company's proprietary Graffiti handwriting system. This process relies on a touch-screen display and a simplified alphabet—which takes about 20 minutes to learn—for data entry. Typically, PDAs with the Graffiti system provide the option to write directly onto the display, which translates the input into text. The user can also open a dedicated writing space that provides on-line examples and help.

Palm products became formidable players in the handheld computing arena following a further change in ownership in mid-1997, when U.S. Robotics was purchased by 3Com. This led to a burgeoning PDA market. Their success has also led to a segmentation of the market into users of the two major form factors: devices that have a keyboard, and stylus-based palm size devices that don't. The keyboard-entry devices are increasingly viewed as companion devices for desktop PCs, and often run cut-down versions of desktop applications. The palm-size form factor retained the emphasis on the traditional PIM application set, and generally has less functionality. The choice depends on personal preference and the level of functionality required.

In 1996 Palm Computing, Inc.—then a part of U.S. Robotics—led the resurgence of handheld computing with the introduction of its Pilot 1000 and Pilot 5000 devices. Designed as companion products to personal computers, Palm PDAs enable mobile users to manage their schedules, contacts, and other critical personal and business information on their desktops and remotely. They automatically synchronize their information with a personal computer locally or over a local or wide area network at the touch of a button. Their most distinguishing features include their shirt-pocket size, an elegant graphical user interface, and an innovative desktop-docking cradle that facilitates two-way synchronization between the PC and organizer.

The Pilot devices introduced the "palm-sized" form factor. Early devices were about the size of a deck of playing cards and weighing around 155g. By 1999 sizes had become smaller still and the design was much sleeker; the Palm V weighing in at 115g at a size of 115mm x 77mm x 10mm. At that time devices were equipped with a 160 x 160 pixel backlit screen and came complete with a comprehensive suite of PIM software. The software includes a date book, address book, to-do list, expense management software, calculator, note-taking applications, and games. The software bundle also included an enhanced version of the award-winning Graffiti power writing software by Palm Computing, which enables users to enter data at up to 30 words a minute with 100% accuracy. Functionality has made this easy to use device the de facto standard in the handheld computing market.

By the end of 1999 Palm Computing had once again become an independent company from 3Com and had consolidated its market leadership position. It has since led the market with the launch of its much-anticipated Palm VII device, which added wireless Internet access to the familiar suite of PIM applications. Several web content providers collaborated with Palm to offer "web-clipped" versions of their sites—designed specifically for the Palm—for easy download. Palm Computing is undoubtedly set to dominate the palm-size segment for some time to come as the overall PDA market continues to grow. The Palm Pilot was a hit with consumers because it was small and light enough to fit in a shirt pocket, ran for weeks on AAA batteries, was easy to use, and could store thousands of contacts, appointments, and notes.

The Present and the Future

Following Palm's success in the PDA market several companies have licensed the Palm operating system, which in many ways has been the reason for Palm's success. This licensing trend has also been stimulated by the existence of a few thousand Palm-based application software developers. A second camp, based on Microsoft's PocketPC, has emerged and continues to gain significant ground as the market looks for PDA applications that are compatible with notebook and desktop PCs. Microsoft has been part of this market since September, 1996 with the introduction of its first PDA operating system called Windows CE. This PocketPC operating system has been licensed to PDA makers including Casio, Sharp, and Hewlett-Packard.

Today, you can buy Palm-like devices from major PC hardware and consumer electronics manufacturers. Though originally intended to be simple digital calendars, PDAs have evolved into machines for crunching numbers, playing games or music, and downloading information from the Internet. The initial touch-screen PDAs used monochrome displays. By 2000, both Palm and PocketPC PDAs had color screens. All have one thing in common—they are designed to complement a desktop or laptop computer, not to replace it.

Meanwhile the PDA market is seeing a growing threat from the mobile phone market. This market continues to provide similar functions such as address books, calculators, reminders, games, and Internet access, not to mention voice.

PDA Applications

Early PDAs were designed as organizers and came with little variation on the set of familiar PIM applications—perhaps with a couple of games. They could store addresses and phone numbers, keep track of appointments, and carry lists and memos. The traditional PIM functionality has matured and grown in sophistication over the last few years but continues to be present today. Personal digital assistants today are more versatile, and can perform the following functions:

- Manage personal information
- Store contact information (names, addresses, phone numbers, e-mail addresses)
- Make task or to-do lists
- Take notes and write memos on the notepad
- Scheduling—keep track of appointments (date book and calendar), and remind the user of appointments (clock and alarm functions)
- Plan projects
- World time zones
- Perform calculations
- Keep track of expenses
- Send or receive e-mail
- Internet access—limited to news, entertainment, and stock quotes
- Word processing
- Play MP3 music files
- Play MPEG movie files
- Play video games
- Digital camera
- GPS receiver

Applications in Vertical Markets

Personal digital assistants can help technicians making service calls, sales representatives calling on clients, or parents tracking children's schedules, doctor's appointments, or after-school activities. There are thousands of specialty software programs available in addition to the PDAs basic functions. These include maps, sports statistics, decision-making, and more. Some applications that show the versatility of PDAs include uses by health professionals, amateur astronomers, truck drivers, and service technicians.

Health Professionals

Many health professionals (physicians, nurses, pharmacists) need to regularly keep track of patient information for medications and treatment. Computer terminals are habitually limited in number and seldom available in a clinic, and especially so at the patient's bedside. In addition, many health professionals need access to information about pharmaceuticals (pharmacopoeias, Physician's Desk Reference, Clinician's Pocket Reference). They also need access to emergency room procedures and other medical or nursing procedures. Doctors and nurses can put all of this information on a PDA instead of carrying manuals with procedures, or references, or index cards with patient information in their pockets. They can note patient monitor readings and other patient information at the bedside on the PDA for later upload into a PC. They can download drug and procedural reference materials onto a PDA for consulting at bedside. They can have programmed drug dosage calculators in their PDAs.

Another promising application for mobile technology is the management of home health services. Hospitals trying to save money and maintain a reasonable staff workload can benefit greatly by sending patients home as soon as possible. But to ensure proper follow-on care, the traveling medical staff caring for those patients must be able to schedule appointments on the fly, and communicate with hospital-bound professionals. They also must have remote access to any given patient's dossier of current and historical medical information. The cost-reduction benefits of using the PDAs mobile technology extend beyond the patient's early departure from the hospital. The constant communication afforded by a portable device can also reduce the cost of rescheduling nurses and other professionals in the field. And when a home visit is covered by a healthcare professional previously unfamiliar with the patient's case, the ability to instantly access the patient's complete medical history via the portable device allows the visiting case worker to get up to speed quickly, potentially shortening the visit and certainly providing better-informed care.

Amateur Astronomers

The equipment that an amateur astronomer takes out in the field when observing is sometimes daunting. Not only must they take the telescope, telescope mount, eyepieces, and cold weather gear, but also include star charts, field guides, and notebooks. Some astronomers have to carry a laptop computer to drive their computer-guided telescopes. Consequently, many of these loose items get scattered across an observer's table and are easily misplaced in the dark. If an amateur astronomer had a PDA, he or she could download a planetarium program into the PDA, which could then serve as a star chart and field guide. They could then take observation notes on the PDA and later upload them into a PC. The PDA could also replace a laptop and drive the computer-guided telescope to the desired celestial coordinates. All of these functions could be done with one PDA instead of several other items.

Truck Drivers

Truck drivers on the road must frequently communicate with their companies and their homes. They consult e-mail and keep track of expenses, shipping records, maps, and schedules. Many drivers use laptop computers in their trucks to take care of these tasks. However, laptops have relatively short battery life and are bulky. Modern PDA models and software can now do many of these functions, including personal information management to mapping and wireless e-mail.

Service Technicians

Utility companies (telephone, cable, etc.) issue PDAs to their customer service technicians to allow the technicians to receive dispatching information, mapping, and diagnostics. This lets the technicians spend more time in the field with customers. In general, this translates into hours saved, happier customers, and minimized operating costs.

Vertical and Horizontal Market Considerations

Some firms developed businesses selling PDA automated systems to companies that needed handheld computers to solve various business problems. Applications included delivery-route accounting, field-sales automation, and inventory database management in manufacturing, warehouse, health care, and retail settings. The horizontal-market PDA companies have now recognized there is a very real need for PDAs within vertical markets.

Unlike horizontal-market PDAs, vertical-market PDA buyers are not overly concerned with cost. The payback period for the PDA can be calculated. In many vertical-market application cases, a $2,000 cost/PDA makes sense because of the vast cost savings achieved with the implementation of a PDA-based system. In another well-known example, Avis introduced handheld computers to handle car rental return transactions. The handheld computers allowed Avis to provide a much higher level of customer service regardless of possible cost savings. The car-return process was much faster, and that made renting from Avis attractive to customers. The tangible cost savings of having handheld devices like PDAs either provides a company with a competitive advantage (such as fast check-in for harried business travelers) or saves a company a significant amount of money (more time available to spend with customers).

One universal avenue that PDA marketers are seeking to let them translate the device's benefits from vertical applications to the mass market may be literally at hand in the ubiquitous World Wide Web (WWW). With its enormous, built-in word-of-mouth notoriety and enough plausible remote-access scenarios to support a coherent large-scale marketing push, a PDA permitting low-cost wide-area web connectivity could be the key to attracting home users. The identification of user needs will no doubt be crucial to the PDA's future. Vendors are facing a marketplace where the basic selling proposition will be to convince a consumer who has already spent $2,500 for a full-service desktop PC to spend an additional $500 to $1,500 for a wireless device to communicate with the PC.

With virtually everyone a potential customer, the market seems almost limitless. Someday, if the conventional wisdom holds, grandma will produce digital coupons at the checkout counter, courtesy of the supermarket's memory-card advertising flyer. The kids will order interactive music videos via their handheld link to the set-top TV box. Mom and dad will effortlessly transfer the day's e-mail from their traveling PIM to the home desktop for automatic identification, prioritization, and file synchronization.

PDA Market

Personal digital assistant shipments are expected to display impressive growth. From 4 million units in 1998, annual shipments of PDAs increased to more than 12 million units in the year 2003, and are expected to reach 13.6 million units in 2004. The worldwide end-user revenue reached $3.5 billion in 2003 and is expected to top $3.75 billion in 2004. While the PDA ASP is expected to continue to decline, the PDA market is still forecast to expand at a solid rate. Consequently, demand for ICs in PDA applications will also grow significantly. The average IC content in a PDA, in terms of dollars, is about 40%, correspondingly, the PDA IC content market is expected to exceed $1.6 billion by 2004. The market for PDAs overall is expected to grow to $4.3 billion by 2004. Global PDA retail revenues are expected to increase five-times-over by 2006, with total unit shipments exceeding 33 million. Low cost PDAs from Dell and Palm should boost the market for PDAs. The PDA is expected to take the lead in the device race as the most suitable device for using these services.

The leading manufacturers of PDAs are Casio, Dell Computer, Hewlett-Packard, LG Electronics, Palm Computing, Nokia, Philips, Psion, Samsung, Sharp, and Sony. Apple Computer, the pioneer in PDAs and the first to market without a keyboard, discontinued its Newton product line in early 1998. Apple is also expected to announce their line of new PDA products. Personal computer manufacturers such as HP, Toshiba, NEC and Acer are also targeting the PDA market. As the annual growth rate for the PC industry has started to slow, these manufacturers are looking to boost revenue growth.

PDA Market Trends

There are a number of popular trends affecting the PDA market. These include everything from form factor, to the efforts of individual vendors, individual features of the PDA device, wireless connectivity, competing products, and more.

Form Factor

The PDA provides access to important personal and business data via a small, portable, and lightweight handheld device without an expensive and relatively heavier notebook or laptop computer. It has become very popular in the consumer market, with much potential in vertical markets. Popularity looks to continue its healthy growth in the consumer market, with anticipated further growth in a widening vertical market.

Palm Computing

The PDA products from Palm Computing are the most popular PDAs today. The success of Palm has been tied to its utter simplicity. It was designed to provide a few personal and business "house keeping" functions and as a companion to PCs. Palm PDAs have been termed as a 'connected organizer' by Palm Computing, since it is simple to exchange contact and calendar information between a PC and the handheld unit. The latest Palm generations can wirelessly connect to the Internet or to corporate networks through third-party attachments. The success of Palm-based PDAs has been due to the real-time operating system (RTOS), called Palm OS. Palm OS has garnered wide support—there are several thousand registered Palm developers. The Palm OS is also established as a standard operating environment in the industry. As such, it has been licensed to other manufacturers (such as Sony, IBM, TRG and Symbol) for use in products such as smart phones, pagers, PDAs, and data collection terminals. Consumers continue to prefer the Palm OS and its clones to the various handheld and palm-sized computers running Microsoft's "mini-me" version of its Windows operating system, Windows CE. Palm OS continues to enjoy a larger market share than Microsoft's Windows CE operating system. While Microsoft will gain larger market share in the coming years, Palm OS will remain the PDA operating system market leader for the next few years.

Handspring

Four members of the team that developed the first Palm PDA later formed a company called Handspring. While the Handspring PDA looks very similar to the Palm, its most notable difference is a special slot for hardware and software upgrades. This slot, called the Springboard, allows the easy addition of extras such as pagers, modems, MP3 players, cell phones, digital cameras, Internet access programs, GPS receivers, bar code scanners, video recorders, games, portable keyboards, and other applications and utilities. Palm Computing acquired Handspring in 2003.

Microsoft

Microsoft has provided its Windows CE RTOS (real-time operating system) to various PDA clone and handheld computers manufacturers. This provided the consumer with an interface and applications they are familiar with on the PC. Windows CE was similar to the popular Windows 95/98 operating system, but had more limitations than features. The Windows CE 2.0 generation had several enhancements and was more widely accepted than its predecessor. It was considered as a potential threat to nonstandard platforms such as Palm, Psion, and Avigo (by Texas Instruments). However, Microsoft was not able to monopolize the PDA industry with Windows CE as it did the desktop and notebook PC market with its Windows operating system. However, Microsoft has become a more formidable player in the PDA market since the introduction (in April 2001) of its PocketPC operating system.

The PocketPC platform runs a much improved and simplified version of Microsoft's Windows CE operating system. Previous versions of this OS had been derided as nothing more than a slimmed down version of its desktop OS and not conducive to the handheld environment. Some of the key adopters of PocketPC are Dell Computer, Hewlett-Packard, and Casio. Windows-based PDAs have advantages over Palm PDAs that include color screens, audio playback, several application support, and better expansion/connectivity capabilities via card slots. The Windows-based PDAs are intended for users who need most, but not all, of the functionality of larger notebook PCs. Windows PocketPC is available in larger platforms from companies such as HP, LG Electronics, Sharp, and so forth. These platforms offer more memory, a keyboard, a track pointer or track-pad, a color screen, weighed about two pounds, and cost around $800.

Microsoft is expected to continue to attack as wireless phones, handheld devices, and the Internet continue to converge into multipurpose devices that are expected to soar in popularity over the next few years, and gain in this potentially lucrative market. Windows PocketPC is a small-scale version of the Windows operating system that has been adapted for use with a variety of diverse portable devices, ranging from AutoPC systems in cars to cable TV set-top boxes to handheld and palm-sized computers. Windows PocketPC gets by on relatively little memory, while still offering a familiar interface to Windows users, complete with a Start button and "pocket" versions of programs such as Microsoft Outlook, Word, Excel, and PowerPoint. Microsoft also introduced PocketPC 2003 in June 2003. While this is not a major upgrade, some improved features provide a better wireless connection. Microsoft licensees are launching new models based on this OS and Intel's XScale PXA255 processor, which will provide improved speed and battery life.

EPOC and Symbian

The EPOC takes its name from the core of Psion's Series 3 OS. It was called EPOC to mark Psion's belief that the new epoch for personal convenience had begun. For the Series 5, EPOC had become the name of the OS itself and had evolved into a 32-bit open system. It originally ran only on RISC-based processors using ARM architecture. Subsequently, EPOC32 became portable to any hardware architecture. Psion did not begin to license its EPOC32 operating system until 1997. Support was disappointing though, with Philips as the only major manufacturer to show any interest. However, in mid-1998 Psion joined forces with Ericsson, Nokia, and Motorola to form a new joint venture called Symbian. Symbian had the aim of establishing EPOC as the de facto operating system for mobile wireless information devices. It was also seeking to drive the convergence of mobile computing and wireless technology—enabling Internet access, messaging, and information access all within a device that fits in a shirt pocket. These three handset manufacturers (Ericsson, Nokia, and Motorola) would share their development resources with Psion, which would continue to develop the EPOC32 operating system in conjunction with its new partners. Symbian believes that there will either be smart phones combining communication and PIM functionality, or wireless information devices with more features that will combine today's notebook, mobile phone and PDA in a single unit. Symbian claims that EPOC32 has a number of characteristics that make it ideal for these devices, such as modularity, scalability, low power consumption, and compatibility with RISC chips. As such, Symbian plans to evolve its EPOC technology into two reference designs. One will be a continuation of the current form factor, supporting fully featured PDAs and digital handsets. The other will be an entirely new design providing support for a tablet-like form factor with stylus operation, handwriting recognition, and powerful integrated wireless communications—a device that sounds remarkably like the Palm VII. Color support by EPOC became a reality in late 1999 with the launch of two sub-notebook models, the Series 7 and netBook.

Internet Connectivity

Wireless connectivity to corporate and faster Internet data via PDA devices provides users with the ability to obtain information at anytime, from anywhere. New PDAs will communicate more with the Internet over a wireless connection. Instead of downloading entire web pages to your PDA, Palm devices use a process called 'web clipping' to slice out bits of text information and send the text through the airwaves to your PDA. For example, say that you want to get a stock quote from an online broker such as E-Trade. You tap the E-Trade icon, fill out a form on your PDA listing the ticker symbol and tap the Send button. Your text query is sent via a data packet-paging network to an Internet server. Software on the servers searches the E-Trade site and then transmits the answer back to your PDA. News headlines, phone numbers, e-mail, and other information can be transmitted in the same way. Eventually, PDAs will merge with cell phones and use a cellular network to communicate via voice as well as text. It is also likely that PDAs will become faster, have more memory, and consume less power as computer technology advances.

Wireless LANs

The value proposition of the wireless-enabled PDA becomes even more compelling with the advent of wireless LANs (WLANs) on PDAs. Wireless LANs (particularly WiFi—Wireless Fidelity or IEEE 802.11b) allows lower cost-per-minute charges and higher bandwidth to end users when compared to a wide area connectivity solution such as 3G technology. This also creates revenue opportunities for carriers and service providers. In usage scenarios where spontaneity is important, the wireless LAN-enabled PDA presents an attractive form factor for accessing the mobile Internet, when compared to the notebook PC. This is compelling in public WLAN applications such as shopping malls, where retailers can provide services such as online coupons to consumers. This creates revenue opportunities for WLAN carriers, retailers, service providers, device manufacturers, and software application developers. Secondly, WLAN has the potential to augment the bandwidth of cellular networks. There is an opportunity for cellular carriers to partner with WLAN carriers to offload the cellular network in scenarios where calls are made within range of a WLAN network. This would provide the consumer with a lower cost of making voice and data calls over the LAN. These scenario models have the potential to accelerate the adoption of WLAN technology and drive the sales of PDAs. The proliferation of WiFi "hot spots" entices more people to carry WiFi-capable computing devices with them more of the time.

Wireless Synchronization Using Infrared

There's little benefit in having a word processor or similar feature on a PDA without the capability to transfer and synchronize data back to a desktop system—particularly as relatively few devices support printing via a parallel printer port. It is no surprise then that data transfer and synchronization is a feature that has improved significantly in recent years. This has been due to the efforts of third parties who have developed both hardware accessories for use with PDA docking cradles and software applications designed to make the synchronization task as comprehensive and as simple to execute as possible. Most PDAs employ a similar docking design, which enables the device to be slotted into a small cradle that is connected to the desktop PC via a serial cable. Many cradles also provide a source of power, as well as facilitating connection to the desktop device, and recharge the PDA's battery while the device is docked. The synching feature has proved one of the most popular features of the PDA. It allows a user to update and organize the Palm Rolodex function and calendar as frequently as they might like. Today, the synchronization is done through a cable that connects the PC and the PDA (either directly or in a cradle). This, however, is quite inconvenient.

Hence, two wireless technologies—Infrared Data Association (IrDA) and Bluetooth—are likely to play an increasing role in the synchronization task in the future. Since its formation in June 1993, the IrDA has been working to establish an open standard for short-range infrared data communications. The IrDA chose to base its initial standards on a 115 Kb/s UART-based physical layer that had been developed by Hewlett-Packard, and an HDLC-based Link Access Protocol (IrLAP) originally proposed by IBM. It is a point-to-point, narrow angle (30° cone) data transmission standard designed to operate over distances up to 1 meter at speeds between 9.6 Kb/s and 16 Mb/s. The IrDA is commonly used in the mobile computing arena for establishing a dial-up connection to the Internet between a portable computer and a mobile phone. This standard also specifies the IrLAN protocol for connecting an IrDA-enabled device to a wired network. Although the worldwide installed base reached more than 50 million units, many consider IrDA to have been a failure. The manner in which many manufacturers implemented the standard resulted in numerous incompatible "flavors" of IrDA. In addition, software support was poor. The result was that IrDA is difficult to use and has never worked as well as it was intended—which is why so many are hoping that the Bluetooth initiative, started in mid-1998, will fare better.

Chapter 13

Bluetooth

Bluetooth personal area networks allow multiple devices to exchange information between each other via a wireless connection. For example, a single electronic phonebook, calendar, and to-do list can be synchronized and easily maintained between the PC, cell phone, and the PDA using Bluetooth. Ultimately, this synchronization will have to extend to documents other than data. Messages exchanged between cellular users can be displayed on a PDA through Bluetooth technology. Bluetooth allows wireless exchange between devices in close proximity (10 meters), while cellular technology allows the exchange of information between the cellular and Internet networks. Bluetooth also allows the use of a wireless headset for voice communications and commands to the PDA (like the cell phone). Manufacturers are also interested in using the Bluetooth short-range radio system to connect PDAs to other devices, such as printers.

High-Capacity Storage

Personal digital assistants support multiple memory types such as ROM, RAM, flash, and micro hard disk drive.

Cellular/Mobile Phones

Cellular phones with PDA-like features first surfaced in 1996 when Nokia began selling GSM phones—particularly the Communicator. This device has a display and keyboard similar to that of a handheld PC. Apart from sending and receiving phone calls, it can send faxes and email, browse the Web, maintain the calendar, and connect to a digital camera to transmit pictures. Other companies that have announced integrated cellular phones with PDA functions include Qualcomm, Kyocera, and others. Common functions like electronic phonebook, messaging, and web access are the primary reason for the convergence of cellular phones and the PDA. One problem with this convergence, however, is that usability is compromised at both ends when a wireless phone and a handheld computer are integrated—a situation that has limited the adoption of these smart phones. They provide several functions, but hardly perform any one function well. For example, there is no way one can get a useful data screen mounted into a product that is as small as a cellular phone. This category is similar to the smart phones, as described in Smart Handheld Phones section. In the longer term, some of the functionality provided by PDAs will be widely integrated into cellular phones and other consumer devices as processors get smarter and memory gets cheaper. But the PDA's value proposition is slowly eroding as these cellular phone features become wireless and Internet enabled—the integrated voice and data capability combined with the larger screen size of the PDA device can provide voice capability as well as ease in viewing content.

Additional Functionality – Add-ons and Multifunction Devices

Personal digital assistants are using expansion slots to integrate other functionality. These functions include:

- Multimedia
- Color screens
- Audio
- Voice recorder
- MP3 players
- Voice and data GSM/GPRS (Global System for Mobile Communications/General Packet Radio Service) cell phone unit

- Global positioning system (GPS) unit
- Optical character recognition scanner
- Zoom-lens digital cameras.

Personal digital assistants and GPS receivers will be combined into one handheld device. Some PDAs will be able to capture and store images like a digital camera. This trend can already be seen in production through add-on devices for the Handspring Visor.

Each interface accommodates a specific type of card inserted into the interface:

- **Springboard** – A proprietary interface designed by Handspring for their Visor family (Palm platform) of devices. This has the largest physical volume of the peripheral cards. It also provides the highest data transfer rate because it allows attached I/O devices to connect directly to the processor bus. It can accommodate a separate battery to support higher power functions.
- **CompactFlash** – An industry-standard expansion interface for portable devices. It is widely supported, and although the CompactFlash card size is relatively small, it provides enough room to implement many functions. A popular interface for adding memory to digital cameras and portable MP3 players. In addition, serial ports, modems, Ethernet cards, cellular network attachments, and GPS devices are implemented on CompactFlash cards. This interface is expected be supported by many more devices in the future because of its wide exiting support.
- **SmartMedia** – The proprietary design currently manufactured by Toshiba and Samsung. Toshiba has trademarked the name, but has made the specification available without cost to the industry in the hope of creating a de facto standard. Not all SmartMedia cards are interchangeable. Some require 5.0 volts from the connecting PDA, some use 3.3 volts, and others can adapt to either voltage. The voltage of the card must be matched to the voltage of the specific portable device.
- **Memory Stick** – Developed by Sony and Fujitsu, the Memory Stick media are relatively small, and thus, accommodate relatively few functions. Its most common use is as a memory platform. There is a Memory Stick format supported by the music industry called Magic Gate. Magic Gate supports the Secure Digital Music Initiative (SDMI) specification designed to eliminate illegal copying and distribution of copyrighted content over the Internet. Sony is driving this interface to become an industry standard, and it appears in various consumer devices such as cameras and portable MP3 players.
- **MultiMediaCard** – Created by joint agreement between Siemens, Hitachi, and SanDisk. The MultiMediaCard is the size of a postage stamp and is designed to provide a small memory storage medium for portable MP3 music players. Its small size limits its ability to support a variety of other I/O attachments.
- **Secure Digital** – Based on the MultiMediaCard interface, this interface also supports the SDMI specification. Similar in size to the MultiMediaCard, the Secure Digital card is used mainly for simple memory expansion. Although the specification enables development of I/O devices using cables or other extensions, this approach can be cumbersome because of its small size. An advantage of the Secure Digital interface is that it can also accept a MultiMediaCard.
- **PC Card (also referred to as PCMCIA)** – The oldest industry-standard expansion interface for portable devices. It supports a wide range of peripherals. The physical hardware interface is not currently integrated into PDAs because of the relatively large size of the connector and the battery power required. Instead, it is implemented in an optional

"sled" device that attaches to the PDA. These interface cards are less likely to appear on devices that use small batteries because they can be high-power consumers. Because PC Card slots are standard on portable computers, PC Cards have been developed that act as adapters for other smaller expansion cards. For example, a PC Card-to-CompactFlash adapter card allows a CompactFlash card to be inserted into the PC Card. Similar adapters have been developed for Memory Stick, MultiMediaCard, Secure Digital, and SmartMedia cards. These adapters can be inserted into a PC Card slot on a portable computer and used to transfer data between the handheld and portable PC. The card sizes for these interfaces vary—smaller cards typically being used for memory expansion. The larger cards' greater physical volume allowing the accommodation of additional functions such as digital camera and GPS.

Data Entry

Typing can be slow and frustrating because of limited space with a PDA keyboard. Hand writing data into the PDA would obviously be a much more sensible idea, but the complexity involved in making handwriting recognition work is immense. The problem is that vector, line-drawn shapes do not make words. The human eye recognizes the difference between a drawing and a word. There are two approaches a computer can take to resolving the matter.

The easy approach is the one taken by the Graffiti recognition system. It understands each letter shape as a unique pattern for that letter, and converts that shape to its corresponding letter. There is no attempt at understanding words or context. To make it even easier, Graffiti uses some special character shapes, which the user of the system must learn. The approach taken by Newton was much more ambitious. It attempted to read the user's writing and convert it into words. It sought to "learn" the way the user wrote, based on some standard test scripts and some interactive tutorials. This approach becomes increasing effective as the system recognizes more and more of the user's hand-writing style. With Newton, it was possible to set up multiple user writing personalities, and was therefore important to ensure that Newton always used the correct personality for a given user to avoid its knowledge base from becoming "polluted." The Apple computer was the key to Newton, but sadly the Newton was a costly mistake for the company. It will be remembered for the over-hype and for its bold, but ultimately unsuccessful, attempt to solve the problem of handwriting recognition. Other developers, on the other hand, emulated the success of the Graffiti approach. Communication Intelligence Corporation's (CIC) "Jot" and ParaGraph's CalliGrapher are two such systems often used by Windows PocketPC and EPOC-based PDAs.

The Palm's data entry function is difficult to use, although it is very useful and allows you to take notes. The user must first learn the Pilot's "Graffiti" pen strokes (done with the Palm's "stylus"), but the feature can prove invaluable for taking meeting or conference notes. Initially, it takes some time to learn to perfect the note-taking feature, although most of the Graffiti strokes are similar to the English alphabet. The recognition capability of the Graffiti is nearly flawless, and stands in marked comparison to "proto-PDAs" like the late-lamented Apple Newton. The Palm also boasts a small QWERTY keyboard that is part of the screen, and is handy for inputting address information and other material. You can also use your PC keyboard for large-scale data input to the Pilot via a synchronization link (a plug in "cradle").

Touch-screen

Most PDAs, PC companions, and smart phones have a touch-screen, which allows navigational and input ease of use. The pen input allows for quick finger clicking through a variety of fields, and allows for signature capture, drawings, and other note-taking means. However, touch-screens come at a cost. They are slightly heavier, offer less clarity, and are less durable, and are more prone to long term wear and tear. They are also priced higher than devices without a touch-screen (usually between $5 and $10). In the long term, the goal is to integrate touch screens on the LCD manufacturer level as they become more pervasive. This will allow for tighter integration of touch, resulting in decreased weight of the device and improved durability.

Battery

Battery technology continues to be the Achilles' heel of the computing industry. Memory storage, CPUs, and other components are developing at breakneck speeds; nevertheless, the chemical battery processes that supply the juice for all PDA devices continue to plod along their linear development curve. However, the basic difference between computer parts and battery processes is a fundamental one. Computer components can develop at exponential rates, according to Moore's Law, because their underlying processes are built on physics-based equations. This is in opposition to the chemical reaction-based science of battery power. The distinction is an important one, because, while components develop along their exponential curves, battery power continues to disappoint as it develops at tediously slow rates.

This problem has been confronted in three ways: chemically, physically, or both. Physically, the developments are happening on the chipset side. Low power consumption chips can increase the overall time a single charge will last within the machine by carefully and consistently monitoring the low power chip's own use of power. Similarly, the power from one charge can be made to last longer if displays and backlights can control power consumption by automatically shutting off backlights or through multiple-level settings. Battery technology relies on changing the chemical makeup to enable better, longer-lasting batteries and charges. This happens in two categories: disposable (alkaline) and rechargeable batteries. The popular battery technology for PDAs will continue to migrate from alkaline to rechargeables. Common rechargeables include Nickel cadmium (NiCd), Nickel metal hydride (NiMH), Lithium ion (LiION), and Lithium polymer (Li-polymer), with each technology having unique pros and cons. Lithium ion will continue to be the most prevailing battery standard. Power management and newer battery technologies, such as lithium polymer, will become available in next generation PDAs.

Tablet PCs and Handheld PCs

As the capabilities of handheld computers expand and tablet PCs continue to proliferate in the market, the two categories are headed toward a collision course. Tablet PCs are essentially full-fledged PCs with touch screens, wireless Internet connections, and speech and handwriting input. While larger than handheld devices today, tablet PCs will likely shrink and grow lighter over time. Microsoft and a handful of PC makers are pushing tablet PCs as the future of the PC industry. Comparatively, handheld computers such as PDAs are further along in terms of available products—they have interesting features but are viewed as an emerging product category. While neither type of device is close to challenging the dominance of the PC in the technology landscape, the two may be battling for the title of second banana. Tablet PCs are more powerful than current handheld devices. But with continuing advancements in processors and improved multimedia capabilities, handheld devices are slowly closing the gap. Tablet PCs use components (such as Celeron or Pentium III processors, hard disk drives, and screens) that are closer in size and performance to the PC than handheld devices. While tablet PCs initially cost about $2,000, they have come down in price as manufacturers are seeing increased volumes. Tablet PCs have the width and length of a legal-size pad of paper, are between 1.5 inches and 2 inches thick, and come with color touch-screens. Their touch-screen resolutions are sharp enough to allow people to read electronic books (eBooks). Handheld devices are more portable than tablet PCs, and are in a different price range—from $100 to $600. Handheld devices are meant to come closer to fitting inside consumers' shirt pockets and to provide contact and scheduling information. Manufacturers have been touting higher-end features, such as wireless communications, in next-generation devices. Despite the increasing popularity of handheld devices, some major manufacturers will remain on the fringes of the market because of price competition and the lack (for now) of a standard platform.

The tablet PC market should see accelerated growth as Microsoft and other independent software vendors develop applications that take advantage of the tablet-specific extensions to Windows XP. It is likely that some vendors will bring a lightweight (less than 1 kilogram), landscape-oriented, wireless tablet PC to the market that could be an attractive alternative to a Pocket PC. Tablet PCs and landscape-oriented PDAs will be primarily adopted for vertical applications and niche markets in the next year or two. Users will make trade-off decisions between Pocket PCs and tablets.

PDAs for Telematics

Manufacturers targeting vertical markets are planning to use the PDA for business professionals and automotive telematics users. These PDAs are particularly strong in the areas of browser capabilities and wireless connectivity. They will also provide proximity and business applications, as well as entertainment (including films), video, animation, and games. Automotive vendors and PDA makers have come to believe that PDAs could offer a better form of control for dashboard-based telematics systems. Palm and Delphi Automotive formed MobileAria Inc., which will develop a docking station to link the information stored in the personal Palm device to a Delphi voice-recognition-based telematics system. Delphi has already introduced its Mobile Productivity Center, which uses a low-power RF transmitter to enable a Palm V computer to communicate with an unused frequency on a car radio. This would allow it to read text messages, addresses, and other information aloud. By employing a cell phone, the system can also download e-mail, read Internet text, access news, stock quotes, and sports scores. The introduction of the PDA-based telematics program signals a step up in the competition between cell phone manufacturers and PDA makers. Cell phone manufacturers think that their products should control telematics systems, while PDA makers believe that PDAs should do it. The key to winning this contest will be delivering location-based wireless Internet applications to automobile drivers. The PDA has the potential to become the leader in controlling telematics systems in the car. With a PDA, one can get information at home or in the office, take it to the car, and tap into it at the dashboard. But with cell phone-based systems, all information remains remote from the user—essentially left behind. Both Palm and Microsoft are planning to penetrate the automobile telematics dashboard market with their respective operating systems. Palm OS delivers the simplicity, while Microsoft's Windows CE is a key telematics enabler because it provides strong browsers and wireless capabilities. The Internet and the PDA will allow future automobiles to tightly integrate dissimilar electronics subsystems. A driver will likely plug a PDA into the dashboard and let it configure itself for the task at hand. The PDA will communicate wirelessly, and transfer diagnostics to and from a service shop or download entertainment content (MP3 audio/music files and satellite radio). The same PDA acts as the window to the car's navigation system by integrating GPS capabilities. The PDA-cum-console also operates as a web browser, speech recognizer, digital camera, bankcard, and smart phone.

PDA Market accelerators

Market accelerators for PDAs include:

- Enterprise adoption
- High numbers of vendors have entered this market, which has had the effect of taking the average selling price of PDAs much lower and bringing them to mainstream acceptance
- Recent form factors include slimmer and sleeker designs
- The simplicity and mass adoption of Palm OS
- High number of shareware, freeware, enterprise applications, custom development, development tools, and other forms of platform enhancement.
- Adoption of Microsoft Windows and applications that can be ported to desktop and notebook PCs
- Enhanced functionality such as digital cameras, MP3 players and Bluetooth
- Software

Market inhibitors for PDAs include:

- Other devices (e.g., cell phones and two-way pagers) providing some functions that are the value proposition for PDAs.
- Slow adoption of Windows CE

PDA Form Factors

- **Palm-Size PCs** – These devices are also known as tablet PDAs, and have vertically-oriented touch-screen (monochrome or color) displays. These usually have stylus interface and operation software, with a keyboard for alternate data input. Some models have handwriting recognition. They have a relatively smaller amount of memory and synchronize with the PC and have the capability to send and receive messages. They typically weigh less than one pound.
- **Handheld PCs** – These devices are also known as clamshell PDAs, and are general-purpose computers designed to fit comfortably in a user's hand. Clamshell PDAs look and operate similar to notebook PCs, with similar style keyboards. Input can be keyboard-based, pen-based, or both, with a potential for voice- and mouse-based inputs. They are based on similar operating systems, microprocessors, and platforms, and are distinguished from a notebook/laptop PC by size, weight, and physical usage. They have a horizontally-oriented monochrome or color touch-screen display, and have relatively large amounts of memory. Their primary power source is batteries. They also have PC card slots for modems, Ethernet LAN, etc. They adhere to hardware and software compatibility standards. Some key applications include PIM, e-mail, web browsing, as well as subset versions of command desktop applications such as word processors, spreadsheets, and presentation applications. These can be considered, to some degree, as the ultra-portable computer.

- **Smart Phones** – These are Digital cellular phones combined with PDA-like features. Smart phones provides access to e-mail and the Internet, and have the capability to exchange faxes and data. They have a keyboard, keypad, or stylus interface.

- **Handheld Instruments** – These are PDA-like instruments used for data acquisition, logic analysis, test, measurement, and have PC-like capabilities. They have an alphanumeric keypad and graphical capabilities. They also have ports to connect to PCs and printers.

- **Wireless Messaging Handheld Devices** – These are e-mail machines that provide business users with access to their corporate e-mail when on the road. They are still emerging and are very popular in the finance business community. Wireless messaging (e-mail) devices support e-mail, very light applications, and PIM. These devices are best suited for deployment by enterprises seeking to better connect their mobile professionals. They are very popular in metropolitan areas, and include devices by manufacturers such as RIM (Research in Motion), Motorola, Sharp, and Franklin. They operate something like two-way pagers and related messaging devices that offer the ability to send and receive text messages. They also offer the promise of limited WAP-compliant Internet access. Bandwidth constraints typically limit the amount of information that can be sent over these two-way messaging networks. However, they do offer benefits as opposed to cellular, PCS, and other networks because their coverage is typically greater and they offer an always-on ability of receiving and storing messages. They can be used over paging, narrowband PCS, private packet radio, and wireless WAN packet data networks such as CDPD. These devices have evolved from being simple beepers to devices that allow full mobile computing over the last few years. They have full keypads, offer PIM functionality, and have manufacturers who are creating applications for the devices.

Components of a PDA

There are three differences between the two major PDA categories (hand-held computers and palm-sized computers)—size, display, and mode of data entry. Specifications of PDA categories vary according to the market and average selling price they are targeting. Hand-held computers tend to be larger and heavier than palm-sized computers. They have larger liquid crystal displays (LCD) and use a miniature keyboard, usually in combination with touch-screen technology, for data entry. Palm-sized computers are smaller and lighter. They have smaller LCDs than hand-held computers, and rely on stylus/touch-screen technology and handwriting recognition programs for data entry.

Regardless of the type of PDA, they all share the same major features and circuitry—microprocessor, operating system (OS), solid-state memory (RAM, ROM, flash), batteries, and a digital LCD display. They also have an input device (buttons in combination with a touch-screen or keyboard), communications functions, input/output ports (chip interfaces), and desktop PC software. These functions can be combined on a single highly integrated IC or several ICs can be used for individual functions. Microprocessor functions such as LCD control, PC card control, modem, graphics acceleration, and interface to headphones, speakers, microphones are normally provided by an application specific integrated circuit (ASIC).

Figure 13.1 is a block diagram of a typical handheld PDA. The key components are described in greater detail.

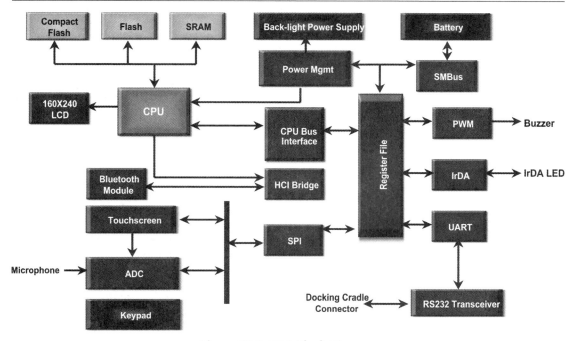

Figure 13.1: PDA Block Diagram

Microprocessor

The microprocessor is the heart of the PDA, just as in all standard desktop and laptop computers. This processor coordinates all PDA functions according to programmed instructions. Unlike desk and laptop PCs, PDAs use smaller, cheaper microprocessors. Although these microprocessors tend to be slower than their PC counterparts (25 to 100 MHz, compared with over 2 GHz in PCs), they are adequate for the tasks that PDAs perform. The benefits of small size and low price outweigh the cost of slow speeds.

Some have high-performance RISC processors, while others have embedded versions of processors that were used in earlier mainstream desktop PCs (such as the Intel 386 or Motorola 68K). Popular processors are produced by Hitachi (SH-3), NEC, Motorola (68K, DragonBall), Intel (386, XScale), Toshiba, Philips, MIPS, and ARM, among others. The processor combines several functions such as power-saving features, UART with IrDA or Bluetooth support, and a monochrome or color LCD controller. Next-generation processors from manufacturers such as Intel, called the Xscale, will enable wireless communications and multimedia capabilities.

Memory

The PDA usually also has a small amount of flash memory or ROM to hold the operating system and basic built-in applications (address book, calendar, memo pad), but does not have a hard disk drive. It stores basic programs that remain intact even when the machine shuts down. Data and any programs added after purchase are stored in the device's RAM. This approach has several advantages over standard PCs. When the PDA is turned on, all the programs are instantly available and one does not have to wait for applications to load. When changes are made to a file, they are stored automatically without the need of a Save command. And when you turn the device off, the data is still safe because the PDA continues to draw a small amount of power from the batteries.

All PDAs use solid-state memory; some use static RAM and others use flash memory. The amount of RAM in a system varies from 2 MB to 64 MB, depending on functionality. All PDAs normally come with a minimum of 2 MB of memory—1 MB of memory can store up to 4,000 addresses and 100 e-mail messages. However, application programs occupy memory space, so more advanced models usually have more memory (5 to 32 MB). The PocketPCs usually have 16 or 32 MB because the operating system occupies more memory space. Memory can be upgraded (increased) in some PDA models.

Some PDAs will even incorporate removable forms of memory. In the next few years the amount of flash memory in PDAs will grow. A Type II PC-Card credit card-sized hard drive (HDD) and the IBM MicroDrive have been unveiled to give PDA users an added 2 GB of storage capacity. This will allow storage of a large number of streaming video, digital images, and MP3 files.

Operating Systems

The PDA's operating system contains preprogrammed instructions that tell the microprocessor what to do. These operating systems are not as complex as those used by PCs. They generally have fewer instructions and take up less memory. For example, the Palm operating system fits in less than 100K of memory, which is less than 1% of the memory size required for Windows 98 or the Mac OS.

Popular operating systems used in PDAs include Microsoft's PocketPC (formerly called Windows CE), Palm OS, Geoworks GEOS, and EPOC. Palm OS requires less memory, runs faster, and easier to use. PocketPC easily supports color displays, graphics, miniaturized Windows packages (Word, Excel), and other devices (such as built-in MP3 players or MPEG movie players). However, PocketPC takes up more memory, is slower, and is more complicated to use. But if it is important to be able to exchange files with Windows packages, then PocketPC might be a better choice. Some PDAs have used GEOS and EPOC, but the Palm and Microsoft operating system primarily dominate the market. The Palm OS dominates the market because its operating system is specifically tailored to the basic uses of a PDA, but PocketPC is challenging Palm OS. Third-party software developers exist for both operating systems.

Software Applications

The palm-size, stylus-based class of PDA comes with some form of handwriting recognition software in addition to a number of other useful applications. These software applications can include diary/scheduler, to-do list, phone/address book, notepad, drawing application, finance software, calculator, alarms, world time, file manager, voice recorder, data synchronization, and printer connection. Data used by these various applications can be synchronized between the PDA and a PC via a synchronization utility—HotSync for Palm OS, ActiveSync for PocketPC. The synchronization utility is installed on the PC's hard drive and communicates with the PDA by either cable, IR, wireless, or modem. Versions of the handheld device's address book, calendar, and other important applications must also be installed on the desk or laptop. It is also possible to use a personal information manager (PIM) like Lotus Organizer or Microsoft Outlook that supports syncing. The PDA assigns each data record a unique ID number and notes the date it was created; a record being one appointment, one contact, one memo, etc. You simply press a button on the PDA or its cradle in order to synchronize data between the PDA and PC. This causes the synchronizing software to compare a data record on the PDA to the duplicate ID numbered stored data record stored on your PC, and accepts the most recent one for both locations. All identical data record ID numbers on both the PDA and PC are synchronized ("updated") in this manner. The beauty of synchronization is that you always have a copy of your data on your PC, which can be a lifesaver if your PDA is broken, or stolen, or simply runs out of power.

Web applications have become standard offerings in addition to the PIM software and data synchronization application (which is included with the PDA software). This is the result of recent developments in communications and because PDAs now provide wireless connections to the Internet for access to e-mail and web browsing. Keyboard-based designs may also come with scaled-down versions of desktop applications such as a word processor, spell checker, custom dictionary, and spreadsheet. These scaled-down software applications are needed for the ability to exchange data and work with the full-scale Windows applications found on a PC, but are currently available only on PocketPC PDAs.

There are thousands of other specialty programs for PDAs that can be used with both major operating systems. These include maps, medical software, decision-making, and astronomy programs, for example, which are available as freeware or shareware. The explosion in the PDA market has resulted in the development of several third-party applications aimed at both the consumer and corporate sectors. One of the most interesting new areas of development—both because of its inherent applicability to the mobile user and because it is an excellent example of the exploitation of converging technologies—are the various route planning and journey tracking systems now available. Combining the power and convenience of PDA devices with the intelligence and precision of a GPS, these packages allow users to plan their trip, customize maps and directions, and track their progress on land and sea, or in the air, whatever their mode of transport. The user indicates the required journey, either by pointing and clicking on a map, or by entering the names of the start and destination points. Particular places of interest—or places to be avoided—can be entered before the route is calculated. Any section of the road map can be displayed at any scale and at any level of detail. Location names, routes, legends, locators, and scales can be displayed in a variety of ways. Similarly, driving instructions can be displayed in several "natural language" formats and at different levels of detail. When used in conjunction with a compatible GPS satellite receiver, the traveler's position on the map will be updated as they move. A "split view" allows progress along the route to be viewed at the same time as displaying the driving instructions.

Display

All PDAs have some type of LCD screen. Unlike the LCD screens for desktop or laptop computers, which are used solely as output devices, PDAs use their screens for output and input. The LCD screens of PDAs are smaller than laptop screens, but vary in size. Hand-held computers generally have larger screens than palm-sized computers. Most PDAs shipping today are either thin-film transistor (TFT) LCD—active matrix displays or color super-twist nematic (CSTN) types, which will not change in the next few years. While PDAs are mostly monochrome today, they will most likely to migrate to video graphics array (VGA) color and super VGA (SVGA). Alternate display technologies for PDAs include ThinCRT and light-emitting polymer (LEP) diodes.

These displays have pixel resolutions that range from 160 x 160 and 240 x 320. Display options include monochrome/black-and-white (16 grayscale) or color (65,536 colors). They also offer passive or active matrix—active matrix displays have sharper images and are easier to read. They are either reflective or backlit. Backlit screens are good for reading in low light. Hand-held PDAs tend to have larger screens. Most palm-sized PDAs have four-inch (10 cm) square screens. Some PDAs only allow you to write in special areas of the screen, while others allow you to write anywhere on the screen surface.

Batteries

Batteries power all PDAs. Some models use alkaline (AAA) batteries, while others use rechargeable batteries (lithium, nickel-cadmium or nickel-metal hydride). Battery life depends on the kind of PDA one has and what it is used for. Some of the things that can increase the drain on batteries include the type of operating system (PocketPC takes more memory), additional memory, color screens, and applications such as MP3 and video files.

Battery life can vary from two hours to two months, depending upon the PDA model and its features. Most PDAs have power management systems in place to extend the battery life. Even if the batteries are so low that you can no longer turn the machine on (the PDA will give you plenty of warning before this happens), there is usually enough power to keep the RAM refreshed. If the batteries do run completely out of juice or if you take them out of the machine, you will have about one minute to replace the batteries before the transistors inside the device lose their charge. When that happens, most PDAs lose all their data, which makes backing up a PDA on a desktop or a laptop extremely important. Personal digital assistants also come with AC adapters to run off household electric current.

Input Device

Personal digital assistants vary in how you can input data and commands. Hand-held computers typically use a miniature keyboard in combination with a touch screen. Palm-sized computers use a stylus and touch screen exclusively in combination with a handwriting recognition program. Each model also has a few buttons to initiate screens or applications.

The tiny, four-inch screen on a palm-sized computer serves as an output and an input device. It displays information with an LCD, similar to that on a laptop. But a touch screen sits on top of the LCD that lets you launch programs by tapping on the screen with a pen-like stylus. The stylus is also used to enter your data by writing on the screen. Think of the Palm's screen as a multilayer sandwich. On top is a thin plastic or glass sheet with a resistive coating on its underside. The plastic or glass floats on a thin layer of nonconductive oil, which rests on a layer of glass coated with a similar resistive finish. Thin bars of silver ink line the horizontal and vertical edges of the glass. Direct current is applied alternately to each pair of bars, creating a voltage field between them.

When you touch the stylus to the screen, the plastic pushes down through the gel to meet the glass (called a "touchdown"). This causes a change in the voltage field, which is recorded by the touch screen's driver software. By sending current first through the vertical bars and then the horizontal bars, the touch screen obtains the X and Y coordinates of the touchdown point. The driver scans the touch screen thousands of times each second and send resulting data to any application that needs it. In this way, the PDA knows when you're tapping an on-screen icon to launch a program or gliding it across the screen to enter data.

Now let's look at how the handwriting recognition works. Using a plastic stylus, you draw characters on the device's touch screen. Software inside the PDA converts the characters to letters and numbers. However, these machines don't really recognize handwriting. Instead, you must print letters and numbers one at a time. On Palm devices, the software that recognizes these letters is called Graffiti. Graffiti requires that each letter to be recorded in one uninterrupted motion, and you must use a specialized alphabet. For example, to write the letter "A," you draw an upside-down V. The letter "F" looks like an inverted L. To make punctuation marks, you tap once on the screen and draw a dot (for a period), a vertical line (for an exclamation point), and so on. To help Graffiti make more accurate guesses, you must draw letters on one part of the screen and numbers in another part. The disadvantage of handwriting recognition software is that you have to learn a new way to write, it is slower than normal handwriting and the device's character recognition is rarely letter-perfect. On the other hand, it's surprisingly easy to learn and it works. Some PDAs let you enter data anywhere on screen and employ different recognition software that does not require a special alphabet (but the process still works better if you draw your letters a particular way).

If you cannot get the hang of PDA handwriting, you can use an onscreen keyboard. It looks just like a standard keyboard, except you tap on the letters with the stylus. An accessory to some palm-sized computers is a collapsible keyboard that plugs into your PDA; this device is more practical than handwriting if you use the device to send e-mail. Eventually, most PDAs will incorporate voice recognition technology, where you speak into a built-in microphone while software converts your voice waves into data.

Communicating With PCs or other PDAs

Because PDAs are designed to work in tandem with your desktop or laptop, any data entered or changed on one device (PDA or PC) needs to be updated to the same state on the other device (PDA or PC). If you make an appointment on your desktop computer, you need to transfer it to your PDA; if you jot down a phone number on your PDA, you should upload it later to your PC. You also need to be able to save everything on the PDA to a desktop computer in case the batteries go dead in the PDA. So, any PDA must be able to communicate with a PC. The communication between PDA and PC is referred to as data synchronization or syncing. This is typically done through a serial connector or Universal Serial Bus (USB) port on the PDA. Some PDAs have a cradle that they sit in while connected to the PC.

In addition to communicating through a cable, many PDAs have an infrared communications port that uses infrared (IR) light to beam information to a PC or another PDA. This requires an infrared transceiver. Some PDAs also offer wireless methods to transfer data to and from a PC/PC network through a wireless e-mail/Internet service provider like those available on new models of cell phones. Finally, some PDAs offer built-in telephone modem accessories to transfer files to and from a PC/PC network. Some PDAs also have Bluetooth functionality, and hence have a Bluetooth radio and baseband processor that interfaces to the host processor.

Expansion Modules

Personal digital assistants can increasingly also be extended by attaching a cell phone, printer, additional memory module, modem, or other peripheral. Most devices have infrared ports that implement the Infrared Data Association (IrDA) standard. This port can be used to connect to a desktop or portable PC, exchange contact information with other mobiles, or connect to cell phones, printers, and other infrared devices. As Bluetooth personal area network (PAN) technology becomes more prevalent, it is expected to replace IrDA ports on handheld devices. Many PDAs also have either a USB or serial port. A USB or serial cable can be attached to the port, or the entire device can be inserted into a cradle that provides a USB or serial connection to a desktop or portable PC, or to USB or serial devices. They may also support expansion interfaces that allow connection of peripherals, addition of memory, or provision of network connectivity. Some generations of PDAs have expansion modules that can add functions such as digital cameras, video recorders, and so forth. This capability emerged from Handspring's Springboard platform.

Summary

The SHD market is seeing continuing acceptance by corporations and content providers. While this group of devices broadly consists of several subcategories, the popular SHDs among consumers and corporate users include smart phones, PDAs, and PC companions.

The PDA is a lightweight, robust mobile device characterized by its ability to do PIM tasks. It can also run applications enabling a considerable number of productivity tasks—address book, daily planner, to-do lists, memo pads, calendar, project lists, expense reports, and data collection and manipulation; just about everything that you need to keep your life organized. It also synchronizes data with the desktop, and today accepts input through a pen-based input on a touch-screen. Over the next few years, PDAs will change significantly in terms of form factor, and will become much smaller and lighter. However, they will most likely continue to be transformed by the addition of incremental functions. These will include voice communication (e.g., the VisorPhone™), multimedia applications such as streaming audio and video (e.g., Macromedia's Flash Player), and geographic positioning. The market will also see networking with surrounding devices (e.g., Bluetooth, IEEE 802.11b), digital camera and video functionality, the addition of expansion slots (e.g., Springboard, Secure Digital), and voice recognition for navigation and data input being added.

The PDA will also provide consumers with Internet access—the rapidly gaining popularity of the WWW is opening doors for PDAs. The WWW is beginning to change the rules of computing. Servers now are maintaining information, storage, and applications, thereby freeing web access units (terminals) from the significant requirements of memory, storage, and processing capability. The key element is communications. Communications must be fast, inexpensive, and seamless. Assuming this type of communication will be available in the next few years, PDAs will grow to become web terminal devices able to access company databases, stock quotes, and a customer's corporate pages. To date, PDAs have been successful in vertical markets because there is a business need for wireless computers within those markets. A system ties together database management and data collection. However, in horizontal markets, a PDA is simply a productivity tool. For the most part, it is not as good as what it replaces (the traditional paper planner). But horizontal-market acceptance will occur because of two changes. First, the technology and application of the technology will improve so that future PDAs will be better than paper planners. Second, PDAs will become more integrated into corporate computer networks. When that happens, corporate information will be accessed readily via the PDA using the WWW network concept. In addition, the PDA's primary use will evolve from a simple organizer to a mission-critical corporate tool.

Personal digital assistants are also increasingly being used in mainstream corporate, consumer, government, education, and other vertical markets. The consumer market is split between the low-priced PDA such as Palm Zire and media rich devices such as the Sony Clie or HP-Compaq iPAQ. Therefore, vendors must focus their development and marketing efforts to meet the needs of the specific market segments, and dominate a specific market niche as opposed to a horizontal product targeted at all end markets. In addition, handheld vendors and wireless solution providers must demonstrate that their solutions increase employee productivity and have a positive impact on the company's financials because corporate end users are concerned with cost controls.

Personal digital assistant vendors are competing not only with other PDA device manufacturers but also with Internet and data-capable cellular handsets, known as smart phones. Cellular handsets have the advantage of a greater penetration in the market, as well as the primary functionality of providing voice services.

Screen and Video Phones

Introduction

The telephone is the single most pervasive consumer device to penetrate the home, and although it has migrated from being a black tethered monster to a sleek wireless device with added other features such as call forwarding, conferencing, and answering machines, it still performs its original basic function of transmitting and receiving voice calls. A new and emerging category of telephones will help leapfrog the "boring" phone of the past into an Internet screenphone that helps bring family and friends closer. The Internet screenphones (also popularly known as video phones or display phones), have an LCD screen that allows the exchange of audio, video, and data between people during a telephone call. The Internet screenphone also provides instant access to the benefits of the Internet. Exchange of video between calling parties is in many ways the obvious next step in simple voice telecommunications. It is important to note that this chapter does not include much discussion about smart phones, since these have been covered in detail in Chapters 5 and 13.

Mobile phone makers are stepping up efforts to convince customers that it is worth buying new handsets and services. They are introducing phones with built-in cameras that make it easier for users to exchange images and sounds. These will be multimedia replacements for popular text messaging services. In the future, even cellular phones, known as smart phones, will include video exchange between users.

Screenphones are the corded cousins of smart phones. Screenphones offer many of the same options as smart phones, including Internet access, shopping, and email. However, screenphones do have one important advantage over smart phones—the viewable screen area can be significantly larger than that of a smart phone, thus making Web browsing easier on the eyes.

Definition

The Internet screenphone is a high-end desktop telephone with an LCD screen. It includes a base module, voice communications module (corded and cordless handset and speakerphone), keypad, and screen display. Screenphone commands and functions are activated on the virtual keyboard by touching desired keys and icons with a finger or specially designed pen. The screenphone communicates cordlessly via Bluetooth with the compact, color screen that can be carried from room-to-room. It provides Internet access for applications such as email, messaging, informational services, and Web browsing. The screenphone delivers features driven by concept familiarity, product suitability, and existing network/service support.

History

It has been decades since AT&T wowed the crowds at the 1964 World's Fair with the promise of a new telephone concept, the picturephone, that would let people see one another when they made a call. The picturephone never succeeded, but the fascination has endured. Much like the parade of ill-fated contraptions that preceded the first airplane, new renditions of the picturephone have littered the years, ever inspired by the presumption that people have the same instinctive desire to make eye contact on the phone as they do in person. These days, the concept is being reborn with a wireless twist, while gaining new life in the wired world through high-speed Internet connections and desktop computers equipped with video cameras.

These latest incarnations may take time to reach widespread use for much the same reason that past success has been limited to video conferencing for businesses. The main hurdle remains the fact that the nation's telephone networks, both regular and wireless, were designed to carry voices, not pictures—especially the moving kind. However, these phones can only transmit video images at the still-photo speed of four frames per second, jumping from freeze-frame to freeze-frame every few seconds.

A new brand of phones using the IP (Internet Protocol) network are also making their way into consumer homes. These phones have broadband access connections, similar to the DSL and cable modems used with PCs. Broadband access allows them to gain Internet access and quality of service algorithms to provide high-speed voice, video, and data. At the same time, modern interpretations of the picturephone and ever-cheaper computer cameras still face an obstacle at both ends of every phone call, caused by the ubiquitous copper wire connection. While the major arteries of today's communications networks have been upgraded with fiber-optic cables, the wiring that reaches into most buildings is still made of copper that was never meant to carry a heavy-duty video signal.

Faster Access Still Is Not Enough

Today, thanks to explosive demand for faster Internet access, billions of dollars are being spent to upgrade telephone lines for DSL service and cable TV systems for high-speed connections. It will take years, however, to deliver those services nationwide, leaving dial-up phone service as the only option for most people. But even for those who can get DSL and high-speed cable, cost and capacity remain obstacles.

Equipment and installation for these services can cost more than $300, and the most affordable versions, costing $40 or $50 per month, are designed for downloading from the Internet and may not provide enough bandwidth to send a live video signal. From the beginning, AT&T and other would-be purveyors of videophone service also needed to install special lines to reach a home or business, an expense they tried to share with would-be customers.

Lower Price, Higher Quality Needed

When the picturephone debuted in 1964, prices ranged from $16 to $27 for a three-minute call between special booths that AT&T had set up in New York, Washington, and Chicago. When AT&T launched residential picturephone service in Pittsburgh and Chicago in 1970, customers had to pay about $160 a month—a fee that included the equipment rental and just 30 minutes of calling time. Extra calling time was an additional 25 cents a minute.

But the picturephone also doubled as a non-picture standard voice telephone, since there weren't many people with whom to share the visual calling experience. To obtain the visual experience, it was first necessary to hook up one picturephone at home and another at work or a relative's house, doubling the sticker shock. In the end, fewer than 500 people signed up for picturephone service by the time the plug was pulled in 1974. The next quarter-century brought a stream of similar non-starters. One of these non-starters was the Videophone, attempted by both Panasonic and AT&T in 1992. In all cases, either the price was too high, the quality too poor, or both.

Vendors and service providers will need to foster strategic partnerships and alliances in future years in order to come up with creative business plans that allow the proliferation of this product. An offer of a free Videophone with every commitment for local and long distance service may yet help the screenphone become a highly successful consumer device. Many instruments have been invented, promoted, and died. People are willing to pay very little more for both a picture and a voice than they are willing to pay for a voice alone. Combined picture and voice communication will only become common when the service is extremely inexpensive.

The long-elusive breakthrough, meanwhile, is occurring through computers rather than a specially designed telephone. With a PC, all you need is a camera that costs as little as $20. Most consumers already have a high quality monitor and a machine to process the bits.

Screenphone Applications

Useful applications provided by the screenphone include:

- **Complete communications center** – A phone, browser, address book, and message center all-in-one for telephony, Internet access, and email services.
- **Broadband transmission and reception** – Full-motion color video and audio transmitted over standard analog telephone lines.
- **Email** – Some screenphones provide automated periodic retrieval of email without the need for a PC.
- **Internet connectivity** – Screenphones provide the easiest and fastest method (one-button) to surf the Web for information, entertainment, and e-commerce. One does not have to wait for the system to boot-up, and a simple touch on the screenphone's screen allows you to immediately surf the Internet. Faster connections are possible using both home and office Internet screenphones. With built-in fast cable or ADSL modem, the Internet screenphone provides non-interrupted Internet connection. Along with Internet access, the built-in magnetic strip reader can be used for home shopping, home banking, and bill payment. The on-line banking service allows you to transfer funds, make account inquiries, and find out loan and deposit rates. The 24-hour bill payment services eliminate the need for writing checks and can be done at any time. This makes life simpler and more convenient for many. Internet access could also be used to get the real-time sports updates, horoscopes, lottery results, FedEx tracking, etc. For professional use you could get the live stock quotes, business credit information, investment and stock reports, 24-hour news coverage, etc.
- **Internet phone calls allow call and surf at the same time** – Screenphones allow the user to make phone calls, see each other, and perform Internet activities at the same time. This does, however, require two telephone lines—one for telephony and another for Internet connection (or a single phone line and a broadband access connection). This allows the consumer to never miss any important call.

- **Safe and secure transactions** – Advanced encryption (SSL) allows you to safeguard sensitive information transmitted through the public Internet. You can enjoy a world of possible applications via SSL that includes home banking, Internet shopping, stock trading, and much more—all at the simple touch of a fingertip.
- **PC-like features** – The screenphone includes software like DOS, Windows 3.1, Netscape Navigator, games, word processor, and spreadsheet that allow you to do many basic things that PCs can do.
- **All normal phone features** – Screenphones retain all telephone features, such as call waiting, three-way calling, call blocking, call log, caller ID, hands-free speakerphone, and voice mail.
- **Snapshots** (for high resolution still-picture transmission)
- High-quality built-in video camera
- Preview mode (enables you to see the image the caller will see)
- Privacy mode (video off/audio on)
- Calendar
- Address Book
- Fax
- Send SMS messages as you would on cell phone

The Screenphone Market

Worldwide Internet videophone shipments are expected to reach 3.5 million units by the year 2004, with revenues exceeding $600 million. However, Internet screenphones and videophones are expected to ship 9.5 million units, with $2.2 billion in revenue by the same year.

The screenphone business is primarily focused on the average person—especially those who cannot afford, or do not need, all the power of a PC. It also focuses on small to medium-sized companies. This market is not focused on the high-tech user. Screenphones are priced at approximately $240, with a basic monthly service fee of $60 to $70. One of the methods used to sell the screenphone is direct selling (compensation plan), which many people think is a flourishing and beneficial business. Of course, many big telecommunication companies also provide the screenphone. Other businesses involved in this field include device manufacturers, software vendors, network operators, and service providers.

Market Growth Accelerators

There are numerous factors that will impact the growth of the screenphone market.

Convenience

The Internet screenphone is specifically designed for consumers. It is user-friendly and intuitive to operate, and will provide a completely new experience no matter whether you surf the Web, write/receive e-mails, or arrange your personal information. It comes with everything you will ever need—7- to 8-inch high-quality VGA touch screen, digital answering machine, hands-free telephone, caller ID, call log, address book, notepad, and more. Various integrated Web functions such as JavaScript, image viewing, proxy, and more provide an easy and convenient access to the Internet. In fact, it is designed to meet all your communication and information needs in daily life.

Vendors

There will be more vendors with more products that incorporate the Internet screenphone. This will lead to competition and lower prices. Big vendors with large marketing departments will help promote and validate the screenphone concept.

The World Wide Web

The World Wide Web has become a pervasive channel for disseminating information and distributing products and services. Consumer electronics companies are racing to create a new class of low-cost consumer devices designed to make getting online even simpler, less expensive, and more routine. Telephony devices that are designed to support Internet connectivity and act as databases for storing personal management information include Internet screenphones and wireless smart phones. It is the simplicity and familiarity of the telephone platform that makes the screenphone or smart phone easier for consumers to understand and use than other Internet appliances. Digital handsets accounted for 85% of all mobile phone sales in 1998, and analysts predict that data traffic (data and applications) will represent one third (roughly 33%) of all traffic by the end of 2007, versus a mere 3% in 1998. Screenphones will provide access to the PSTN network as well as the IP network, thereby enabling voice, data, and video communication through a single product. Also, there will be increasing demand for Internet connections at multiple points in the home.

Revenue Enhancement

The screenphone is a revenue enhancement device for telephone companies, Internet Service Providers (ISPs), and cable companies. Telephone companies can be expected to bundle screenphone services in order to enhance their revenue stream as they did for advanced calling features such as caller ID or voicemail and Internet access. Bundling will also be used to acquire and retain new telephone company customers in scenarios in which a screenphone is provided at reduced cost in exchange for a customer's term commitment. Bundling services with screenphones can even help to tie-in contracts to other services, such as requiring a commitment to long distance services. Potential advertising and e-Commerce revenue can also be realized. The inherent advantage of a screenphone is its screen, which can be used to display advertisements (or links to merchants) while a consumer is using the phone. Furthermore, telephone companies have the advantage of seamless billing. By integrating Internet access and value-added services into a telephone bill, telephone company customers will not have to open another envelope with another bill, thus precluding the possibility of the consumer contemplating the value of this additional fee on a separate billing. Seamless billing as part of an existing telephone bill will make this scenario much less likely. As of late, ISPs have been very focused on customer acquisition and retention. Account attrition rates are relatively high, and ISPs are spending a good amount of money giving away free disks and months of service to attract customers.

Screenphones can also provide an alternative approach to getting and keeping new customers. The promotion and distribution of the screenphone as an Internet access device can help ISPs in their customer retention efforts by requiring customers to commit for a period of time in exchange for a subsidized phone, much the same way the cellular phone industry works today. The strong appeal of a phone with an integrated touch-screen for Internet access could also help attract customers. Internet Service Providers are also in an ideal position to promote screenphones by leveraging their existing analog infrastructure. Furthermore, the private-label branding of these devices with an ISP's logo can help promote a brand name.

Another compelling reason for the distribution of screenphones by ISPs is the increase in advertising opportunities. Advertisements are displayed on the PC only when the computer is powered-up and the user logged in. Additional devices sitting in the kitchen or bedroom that display simple advertisements will increase viewer exposure to ads, thereby increasing advertising revenue. Cable companies can also take advantage of screenphones as one more access device for their data networking services. Granted, few cable companies have jumped into telephony services with both feet as of yet, but it is coming. The ultimate goal of providing an integrated platform (or a fully networked home using services provided by cable) will require a device to access these services. The screenphone is well suited to that purpose. The customer acquisition and retention methods used by telephone companies and ISPs, described in the preceding sections, are also applicable for cable companies.

Service Provider Distribution and Promotion

Telephone service providers ("enterprises") and RBOCs or ISPs can pursue one or more equipment manufacturer to produce solutions that meet the needs of their specific service. The service-driven approach is the dream of CPE vendors. They are always waiting for a well-financed service provider to step up with an OEM order for hundreds-of-thousands of units. Traditionally, the key targets for this type of approach would be the RBOCs in the United States and their telecommunications equivalent in other countries. More recently, there is the hope that an ISP such as AOL Time Warner would be the ideal candidate. Unfortunately, RBOC-driven efforts thus far have had little impact on the market beyond phones offering basic Analog Display Services Interface (ADSI) Caller ID-type features. The involvement of ISPs, although likely, will tend to be focused on the NetTV form factor and unlikely to drive significant volumes in the consumer area. With the exception of one or two potential major opportunities, most opportunities in this area will be for smaller unit volumes. These types of sales will need to be driven more by the CPE side across multiple service providers rather than by one big leader.

CPE-Driven Spanning Service

Equipment manufacturers can individually pursue service providers ("enterprises") to develop services for their equipment and gain their support to subsidize the equipment cost. The CPE-driven approach is currently well under way by many vendors. In this approach, CPE vendors will need to contract multiple deals with multiple enterprises to use their equipment for the delivery of services on a non-traditional platform. The key opportunity is to use screenphones as a customer-retention and loyalty device while simultaneously using them to encourage higher business volume to the enterprise. For example, a bank may order thousands or tens-of-thousands of units to serve as home banking terminals for its customers, a brokerage may do the same for its top-dollar clients, or a retailer may do the same for its regular customers. This process will need to be driven by the CPE manufacturers, and it will require numerous smaller agreements to drive big volumes. There is moderate yet solid volume potential with the CPE-driven approach.

Manipulation by Manufacturers

Equipment manufacturers can develop low-cost devices and arrange their own low-cost service solutions, or leverage existing service solutions (such as existing ISP contracts). This is similar to the approach taken with the Microsoft Xbox and Palm VII PDA.

Concept Familiarity

Consumers are familiar with the concept of telephones and of billing for telephony services. They understand that there is a worldwide network of connections working behind the scenes and that a service fee is associated with the use of the product. They are also accustomed to having a handset on the kitchen countertop or on a bedroom nightstand for communication and information purposes. Given the extremely high penetration rate and virtual ubiquity of telephone devices, it is no doubt relatively easy for consumers to grasp the concept of an integrated screen for Internet access as a natural extension of telephony.

Product Suitability

Telephones fit the demands of the consumer: communication and information access, simply and pervasively. In addition to basic telephony services, screenphones provide access to e-mail and quick Web-browsing capabilities without the hassle of PCs. Screenphones also have applications in vertical markets, from hospitality to banking. Furthermore, by combining the physical space required by separate phone and PC components, the screenphone can satisfy consumers' needs with a smaller physical footprint.

Available Infrastructure

Screenphones can easily take advantage of the existing network and account management infrastructure (billing, marketing, etc.) used by telephone companies for caller ID and other services. Internet service providers and cable TV operators can also take advantage of their existing support structures.

Inexpensive Calls Using VoIP

Screenphones work on both IP and traditional telephony networks, providing built-in savings on the telephone bill. By complying with H.323, screenphones can interoperate with PC software such as Microsoft NetMeeting and multiple vendors' products.

Web Touch and Call

When the user connects to a designated corporate home page, they can read product information and simultaneously contact a salesperson with a PC at the company. It is also designed to make a call simply by touching an icon on the company website. Meanwhile, the company can leverage the cost of their toll-free number with increased sales opportunities.

PC-Hatred

The PC is too complex, over-powered, and over-priced. Especially for those who are only interested in connecting to the Internet, using e-mail, and browsing the World Wide Web, and are afraid of the complex computer commands and database management requirements. The PC also becomes obsolete quickly, causing the consumer to wonder if it is worth it to buy a PC. These are the main factors that encouraged the invention of the Internet screenphone and smart phones—devices that are designed specifically to connect online and access the Information Super Highway to allow home-based businesses or small- to medium-sized companies to get the information they really need.

Gadget Fascination

Gadget fans cannot help being wowed by the responsive touch-screen, speed, simplicity, and integrated phone functions. Names and numbers logged in the caller ID directory can be automatically stored in the phone's directory with the tap of a Rolodex-like icon. Once names and addresses are stored in the directory, another tap of an icon takes you to e-mail or speed dial.

Consortium Formation

The world's leading computer, consumer electronics, and telecom companies need to jointly work towards releasing a standard for connecting screenphones to the Internet. This will accelerate worldwide access to the Internet and encourage the development and adoption of screenphones. A basic device and service standard would also moderate equipment cost. This standard would then assure service providers of an ample installed base that crosses several service efforts. Device manufacturers could also be assured of reasonable device volumes. This issue can be solved by the consortium approach. With a minimum device standard, CPE manufacturers could manufacture high volume units and drive down costs. At the same time, service providers could be assured of an ample device base for their services. In total, each device manufacturer and service provider would be able to leverage the entire development community for a common solution platform and shared installed base.

A consortium called the Internet Screenphone Reference Forum (ISRF), initially supported by leading device manufacturers, software vendors, network operators and service providers, has closed its doors. Supporters of this initial consortium included Acer, Alcatel, Belgacom, Deutsche Telekom, Ericsson, France Telecom, IBM, KPN Research, Lucent, Nortel, Philips, Samsung, Siemens, Sun Microsystems, Swisscom, Telenor, Telia, and U.S. West. The consortium disbanded because the group was based upon fundamental support for Java technologies, which is overkill for this market. The ISRF specification was intended to standardize screenphone access to the Internet, giving consumers an easy-to-use, low-cost alternative for getting online. The consortium believed that the creation of a single standard would encourage the development of low-cost Internet appliances for consumers and small businesses, further speeding the adoption of e-business. A new consortium must be created that focuses on the lowest common denominator (e.g., HTML support), and then aggressively promotes a vendor- and technology-neutral screenphone concept to enterprises and consumers alike. This will enable the growth and introduction of services and devices while protecting the current investments made by customers, content developers, and systems, service and applications providers. It will provide an open standard for accessing a vast array of services that interoperate with any compliant Internet screenphone. Consumers, businesses, and industry alike will benefit when all parties involved in the design and manufacture of screenphones build their solutions using a common, open standard for Internet access. Adoption of this proposed standard would drive lower costs and a broad choice of service and appliances. Worldwide access to the Internet will then accelerate, especially in areas where personal computers are not a practical or affordable solution.

Market Growth Inhibitors

Growth inhibitors for the screenphone market include:

- High cost (due to the high cost of components such as LCD screens)
- The inability to completely replace the PC. A $400 device that does not replace the PC and only provides some Internet capability will not have great success.
- Uncertain business models
- Shift to IP resulting in a higher manufacturing bill of materials
- Minimal consumer awareness
- Lack of a 'killer' application
- Competing digital consumer devices (such as NetTVs) and the PC

- **Issues with form factors** – Small screens, small keyboards, and large physical sizes. The mini keyboard reduces everyone to hunt-and-peck typing, you can't store Net information, and the phone has a small screen that cannot display a full frame of Internet information.
- **Are people ready to watch and talk?** – Even as all the pieces come together, video calling faces one challenge that may prove insurmountable: The telephone has lacked a visual component for so long that it has evolved as a unique form of communication with its own distinct nuances. Most people drift both mentally and physically as they talk on the phone. Whether it's pacing, doodling, rest, or work, it cannot be very easy to multitask while staring at a camera. In fact, people may even relish the visual solitude of a standard phone call.

Screenphone Market Trends

Some of the trends in the screenphone market include:

- Touch-screen control
- Bio-metric security features
- Speech recognition
- Directory assistance telephone
- E-mail telephone
- Personal message center
- Evolving standards in security and data processing
- Advanced LCD technology
- **Cordless screenphones** – Cordless screenphones provide a complete cordless communication center for the home. These are a combination of cordless telephone, answering machine, address book, message center, and Internet appliance—all in one compact and portable unit. Basically, the cordless screenphone consists of a wireless web pad with color LCD touch screen and built-in cordless speakerphone, plus a small base station. This consumer device is easy to install and use, just as a normal cordless telephone. Installation is as simple as connecting the base station to power and a phone line, and then touching the screen to begin operating the device. Once enabled, you can use it to make phone calls, surf the Web, check your e-mail, and send voice clips—wirelessly. One unusual aspect is that, unlike ordinary cordless phones, it communicates with its base station via Bluetooth or IEEE 802.11b—both are 2.4 GHz short-range spread spectrum digital wireless technology championed by makers of cell phones, PDAs, and other portable devices.

Categories and Types

Screenphones are consumer-oriented devices combining a telephone with a basic Web browser, letting people easily connect to the Internet. They can be divided into three broad product categories:

- **Low-end screenphone products (current price range $100–$150)** – These use a very small LCD display and provide only the most basic access to limited Internet content such as text e-mail or text news. They usually allow only three to four lines of information on the screen.
- **Midrange screenphones (current price range $199–$399)** – Midrange devices have a larger screen with a miniature keyboard. These products provide robust interactivity with limited Web browsing capabilities, through the use of ADSI (Analog Display Services Interface) scripts. Vertical applications such as home banking and information-based service such as stock quotes, news feeds, weather, e-mail, sports scores, and more can be accessed from the Internet. In some cases, a user can create e-mail and browse the Web on a minia-

ture keyboard. The variation in cost is due to features such as digital answering machines, cordless handsets, etc.

- **High-end products (current price range $400+)** – These have a 7- to 8-inch color screen with graphics and touch-screen capabilities that allow e-mail, and Web browser (IP) functionality. Other, smaller applications include address books and notepads, as well as support for advanced network services such as caller ID and voicemail. They have a touch-screen, retractable keyboard, and smart card reader. They are intended for quick Web-based transactions such as ordering pizza or downloading stock quotes. They usually include speakerphones and feature two-line capability, allowing for simultaneous Web browsing and voice communications over standard telephone lines. Some of these also include digital answering machines. The first generation of this phone will connect with a 56-Kb/s modem that shares one phone line for Internet and regular phone calls. Future models will be able to handle two lines, as well as ADSL (using phone lines) high-speed connection techniques. The Internet screenphone bears some similarity to other consumer Internet appliances, such as TiVo, that are intended to shield users from some of the complexities of the Internet. TiVo boxes automatically dial into the Internet over a phone line, but, in the future, TiVo plans to achieve high-speed Internet access using cable TV infrastructure.

The Public Switched Telephone Network (PSTN) vs. the Internet Protocol (IP) Network

Because screenphones access the two different networks, the Public Switched Telephone Network (PSTN) and the Internet Protocol (IP) Network, it is important to discuss how calls are made and packets are routed in the two networks. While all of us are familiar with the steps involved in making a telephone call, it is important that we also understand the significance of the underlying infrastructure that is used to make a call today.

Making Calls Today using the PSTN

Current voice telephony is based on a circuit-switched infrastructure. This means that when a call is placed, the PSTN dedicates a specific 64 Kb/s connection. This reserved end-to-end bandwidth is allocated for the duration of the call on a fixed path or channel. This mode of operation is referred to as 'connection oriented,' and there is a set of mechanisms that are required to establish the required connections. While we are not aware of it, picking up the phone and making a call involves a significant amount of activity within the telephone system.

These activities map to user actions in the following way:

- **User picks up the phone** – The local central office (CO) detects the off-hook condition, verifies that sufficient resources are available within the subscriber switch, and then gives the user a dial tone.
- **User dials number** – After dialing is complete, the subscriber switch creates a connection request that is forwarded through the packet network used for managing connections. This process uses the Common Channel Signaling System No. 7 (SS7) protocol, and is referred to as signaling. Resources are then allocated as the connection request propagates through the network.
- **User waits for party to answer** – If end-to-end resources are available, the user will hear a ring indication. If there is congestion in the network, the user will hear a fast busy signal and an "all circuits are busy" announcement or some other indicator.

The user's voice is transmitted as analog signals over 2-wire connection from the handset to the CO. The CO performs 2-wire to 4-wire conversion and may require line echo cancellation for certain calls, especially long distance. At the CO, voice is digitized at 8 KHz rate using 8-bit companded format (to reduce signal noise). The digitized voice signal reaches the destination CO via bridges, switches, and the backbone, where it is then converted back to analog signals and sent to the receiving handset. The process is the same from the remote handset. The PSTN assigns a fixed path or channel for voice transfer and uses SS7 with dual tone multi-frequency (DTMF) tones to set up and tear down the calls.

A voice call generally does not use the full channel bandwidth. While PSTN supports full duplex transfer, phone calls involve one person talking and the other listening and vice versa. There are also many periods of silence where the network transmits no information. Hence, there is a pending issue about how we could better use the network bandwidth—with certain features, voice calls would use less bandwidth.

Voice and Video Exchange over the IP Network

The IP network is a complementary network that is seeing huge growth and expansion. While primarily meant for data traffic, carriers are looking for efficiency to maintain a single network and are therefore looking for voice and video to be exchanged over the same network. Consumers are also demanding Internet, voice, data, video, and images over multiple devices over a single network. This has given rise to the term convergence, which, when used in the context of networking, refers to the ability to transfer data and voice (and/or video) traffic on a single network. Voice over Internet Protocol (VoIP) is the transmission of voice traffic in packets. It is also known as Internet telephony, IP telephony, packet-voice, and packetized voice.

Voice over Internet Protocol networks packetize voice in real-time. Packetized voice also uses significantly less bandwidth, since multiple packets can be transmitted simultaneously. The SS7 and TCP/IP networks are used together to set up and tear down the calls. Address Resolution Protocol (ARP) is also used in this process.

The process of creating IP packets (voice, data, video, etc.) includes:

- **Step 1** – An analog voice signal is converted to a linear pulse code modulation (PCM) digital stream (16 bits every 125 msec).
- **Step 2** – The line echo is removed from the PCM stream. The stream is further analyzed for silence suppression and tone detection.
- **Step 3** – The resulting PCM samples are converted to voice frames and a vocoder compresses the frames. The vocorder, following communications standard G.729a, creates a 10-ms long frame with 10 bytes of speech, and compresses the 128 Kb/s linear PCM stream to 8 Kb/s.
- **Step 4** – The voice frames are then integrated into voice packets. First, an RTP packet with a 12-byte header is created. Then an 8-byte UDP packet containing source and destination address is added. Finally, a 20-byte IP header containing source and destination gateway IP addresses is added.
- **Step 5** – The packet is sent through the Internet where routers and switches examine the destination address, and route and deliver the packet to the appropriate destination. Internet Protocol routing may require jumping from network-to-network and may pass through several nodes.

- **Step 6** – When the destination receives the packet, the packet goes through the reverse process for playback.

The IP packets are numbered as they are created and sent to the destination address. The receiving end must reassemble the packets in their correct order (when they arrive out of order) to able to create voice. All IP addresses and telephone numbers must be properly mapped for the process to work.

Table 14.1 shows the differences between the PSTN (circuit-switched) and IP (packet-switched) networks.

Table 14.1: PSTN (Circuit-switched) vs. IP (Packet-switched) Networks

Description	PSTN	Internet
Designed for	Voice Only	Packetized data, voice & video
Bandwidth Assignment	64Kbps (dedicated)	Full-line bandwidth over a period of time
Delivery	Guaranteed	Not guaranteed
Delay	5-40ms (distance dependent)	Not predictable (usually more than PSTN)
Cost for the Service	Per minute for charges for long distance, monthly flat rate for local access	Monthly flat rate for access
Voice quality	Toll quality	Depends on customer equipment
Connection Type	Telephone, PBX, switches with frame relay and ATM backbone	Modem, ISDN, T1/E1, Gateway, Switches, Routers, bridges, Backbone
Quality of Service	Real-time delivery	Not real-time delivery
Network Management	Homogeneous and interoperable at network and user level	Variuos styles with interoperability established at network layer only
Network Characteristics (Hardware)	Switching systems for assigned bandwidth	Routers & bridges for layer 3 and 2 switching
Network Characteristics (Software)	Homogeneous	Various interoperable software systems
Access Points	Telephones, PBX, PABX, switches, ISDN, high-speed trunks	Modem, ISDN, T1/E1 Gateway, high-speed DSL and cable modems

Challenges in Designing IP-Based Voice Systems

There is a rapid shift in moving to standards-based architectures. While corporate telephony (PBX) has been based on proprietary designs, IP-telephony products are all based on Internet Protocol—an open standards-based technology. Designers must adhere to these standards, placing a tougher load on product-validation and testing. Also, IP standards are constantly evolving and enhancements are being added.

Voice quality for VoIP products must match the voice quality of circuit-switched voice systems. Factors effecting voice quality are line noise, echo, the voice coder used, and network delay. Any additional features and functions provided by a packet-switched network need to have been already in use for several years in a circuit-switched network. Features such as 911-support, call waiting, toll-free numbers, credit card billing, caller ID, and three-way calling will have to be supported by the network infrastructure.

Quality of Service

An important feature of converged IP networks is that they support quality of service (QoS) features for the transfer of voice and video streams. Quality of service refers to a network's ability to deliver a guaranteed level of service to a user. The service level typically includes parameters such as minimum bandwidth, maximum delay, and jitter (delay variation).

An essential concept with QoS mechanisms is that they must be negotiated up front, before the data transfer begins—a process referred to as signaling. The purpose of this negotiation process is to give the network equipment that is responsible for providing QoS an opportunity to determine if the required network resources are available. In most cases, the network will reserve the required resources before granting a QoS guarantee to the client. The overall QoS process consists of the following steps:

1. Client requests resource
2. Network determines if request can be fulfilled
3. Network signals yes or no to request
4. If yes, it reserves resources to meet need
5. Client begins transferring data

Another contentious issue in the quest for converged networks is in determining the appropriate layer of the protocol stack in which to merge traffic. The old method was to combine traffic at Layer 1 of the seven-layer ISO protocol by using separate TDM circuits for voice and data traffic. However, this approach is cumbersome to configure and made inefficient use of bandwidth since there was no statistical multiplexing between separate circuits. Until recently, the vision for voice data convergence was through the use of ATM (asynchronous transfer mode) at Layer 2. The convergence of voice and data was the reason that ATM was developed over a decade ago. The quality of service features built into ATM were defined specifically for this application. However, ATM has a fixed cell length and that leads to added overhead. Also, one must manage ATM and IP networks. The most recent trend is to merge voice and data traffic at Layer 3 over IP networks. This approach takes advantage of new IP QoS features such as the Resource Reservation Protocol (RSVP) and Differentiated Services (DiffServ) technology. These technologies can also take advantage of Layer 2 QoS features, if available.

The Internet Engineering Task Force (IETF) has developed several technologies that are being deployed to add quality of service features to IP networks, and include the following:

- **Resource reSerVation Protocol (RSVP)** – RSVP is defined in RFC 2205, and is used by a host to request specific qualities of service from the network for particular application data streams or flows. Routers also use this protocol to communicate QoS requests to all nodes along the flow path, and to establish and maintain state. Quality of service requests through RSBP usually results in reserved resources at each node along the data path.
- **Resource Allocation Protocol (RAP)** – RAP is a protocol defined by the IETF, and is used by RSVP-capable routers to communicate with policy servers within the network. Policy servers are used to determine who will be granted network resources and which requests will have priority in cases where there are insufficient network resources to satisfy all requests.
- **Common Open Policy Service (COPS)** – COPS is the base protocol used for communicating policy information within the RAP framework. It is defined in RFC 2748.
- **Differentiated Services (DiffServ)** – DiffServ is defined in RFCs 2474, 2475, 2597, and 2598, and uses the Type of Service (TOS) field within the IP header to prioritize traffic. DiffServ defines a common understanding about the use and interpretation of this field.

- The **Real Time Protocol (RTP)** – RTP is used for the transport of real-time data, including audio and video. It is used in both media-on-demand and Internet telephony applications, and uses the User Datagram Protocol (UDP) for transport. This protocol consists of a separate data and control part; the latter is called Real Time Control Protocol (RTCP). The data part of RTP is a thin protocol that provides timing reconstruction, loss detection, security, and content identification.
- The **Real Time Streaming Protocol (RTSP)** – RTSP is defined in RFC 2326, and is a control extension to RTP. It adds VCR like functions such as rewind, fast forward, and pause to streaming media.

The H.323 Standard

International standard H.323 is the IP Telephony market leader. It is defined as an umbrella recommendation for video conferencing over LANs that do not provide a guaranteed QoS (Ethernet, Token Ring, FDDI, and so on). It is also the ITU-T standard for video conferencing and multimedia communications data transfer between user terminals, network equipment, and assorted services over packet-based local, metropolitan, and wide area IP networks, such as TCP/IP. It includes the Internet standards concerned with high-quality video over LAN, as well as low-quality video over the 28.8 Kb/s circuit.

Networks using H.323 today carry hundreds of millions (perhaps billions) of minutes per month. This standard has proven to be an extremely scalable solution that meets the needs of both service providers and enterprises. Products that use H.323 range from stacks and chips to wireless phones and video conferencing hardware. Companies such as Cisco, Clarent, Genuity, iBasis, ITXC, Lucent, Microsoft, Pagoo, PictureTel, Polycom, RADVision, Siemens, Sonus Networks, VocalTec, and many others have embraced H.323.

The following H.323 standards are used in screenphone products and voice over IP (VoIP) gateways:

- **H.225** – Call Signaling Protocols that perform signaling for establishment and termination of call connections based on Q.931.
- **H.245** – A Control Protocol that provides capability negotiation between the two end-points (such as a voice compression algorithm for conferencing requests, etc.).
- **Registration, Admission, and Status (RAS) Protocol** – RAS is used to convey the registration, admissions, bandwidth change, and status messages between IP Telephone devices and servers called Gatekeepers (that provide address translation and access control to devices).
- **Real-time Transport Control Protocol (RTCP)** – RTCP provides statistics information for monitoring the quality of service of voice calls.

Figure 14.1 describes the H.323 protocol stack in detail.

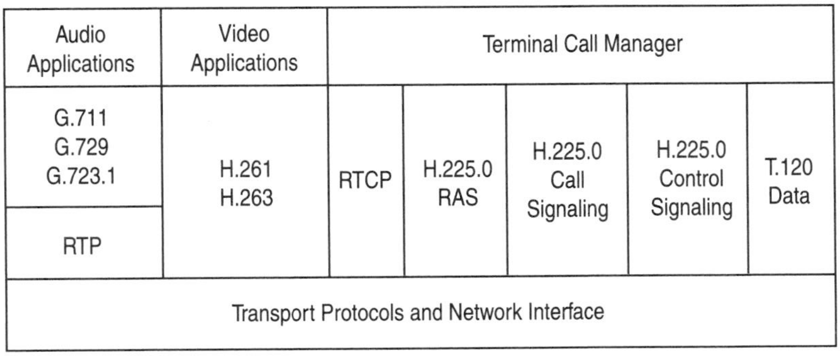

Audio Applications	Video Applications	Terminal Call Manager				
G.711 G.729 G.723.1	H.261 H.263	RTCP	H.225.0 RAS	H.225.0 Call Signaling	H.225.0 Control Signaling	T.120 Data
RTP						
Transport Protocols and Network Interface						

Figure 14.1: H.323 Protocol Stack

The H.324 Standard

The H.324 standard is new. It provides a foundation for interoperability and high quality video, voice, and data-based phone calls, and was recently adopted by the ITU. The H.324 standard specifies a common method for simultaneously sharing video, voice, and data over high-speed modem connections. It is the first standard to specify interoperability over a single analog line. This means that the next generation of videophone products will be able to talk to one another and provide a foundation for market growth.

The H.324 standard uses a normal 28,800 b/s modem connection between callers. This is the same type of modem connection that is used today to connect PC-users to the Internet and other on-line services. Once a modem connection has been established, the H.324 standard specifies how digital video and voice compression technologies are used to convert sounds and facial expressions into a digital signal. The H.324 standard defines how these signals are compressed to fit within the data rate allowed by analog phone lines and modem connections. The data rate allowed by modems is up to a maximum of 28,800 b/s. Voice is compressed down to a rate of around 6,000 b/s. The video picture is also compressed, and uses the rest of the bandwidth allowed by the modem connection.

The H.324 standard and its capability will be added to new phones in a variety of shapes and sizes. The following are examples of videophone products consumers will see over the next couple of years as a result of this standard:

- **Stand-alone videophone** – This device will appear like a wall telephone, but with a camera mounted on top and an LCD display.
- **TV-based videophone** – This videophone will look like a cable set-top box with a built-in camera and it will sit on top of a television. The TV will be used for displaying a videophone image.
- **PC-Based Videophone** – There will also be H.324 videophone software applications for PCs. Software-based videophones will use color monitors to display images, and 28,800 b/s modems to connect to other videophones. Software-based videophones will take advantage of the processor's performance in compressing and decompressing audio and video signals. There will also be PC-based videophone products available that will run on slower PC's. These products will be more expensive than software-based videophones. They will use add-in boards with advanced digital signal processors (DSPs) to compress audio and video and to perform modem functions.

Although video quality is much better than in earlier versions of analog videophone products, H.324 products do not offer "TV" quality. The audio (voice) will sound like a normal phone call, but the video picture will vary depending on screen size. For example, if an image is displayed inside a 176 x 132 pixel window on both ends of a videophone connection, the video picture can be delivered up to 15 frames per second. This rate is roughly half that of the frame rate used for television. But, while the H.324 standard states that videophone products can deliver up to 15 frames per second, the actual frame rate for a videophone will vary depending on the size of the video window selected and the movement of participants during a video phone call.

Screenphone Components

The technology behind screenphones is very similar to that of an analog or digital modem used with a PC. Because the technology behind screenphones uses the PSTN, first generation and lower-end screenphones use analog modems. Next generation and higher-end screenphones use digital modem technologies, which interface the CO to the PSTN network.

Key components of an Internet screenphone include the following:

- System processor
- Video processor/CODEC
- Video encoder
- Memory controller
- Memory (flash and RAM)
- DSP (for telephony and quality of service algorithms)
- User interface logic (LCD controller, keyboard controller)
- Color display screen
- Built-in magnetic strip reader
- Built-in data/fax modem.

The screenphone also has a full computer keyboard, and ports for a mouse, printer, and other communications and video peripherals. Figure 14.2 shows the block diagram of a typical screenphone.

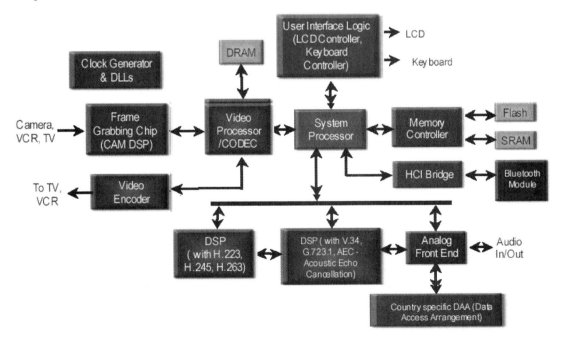

Figure 14.2: Screenphone Block Diagram

Display

The most obvious characteristic that makes screenphones different from ordinary telephones is the video display. This provides either text or graphical information on the accessible screen. Unfortunately, the display is usually the most expensive component in the screenphone, and the success of these products is largely dependent on lowering future costs of LCDs—the LCD is the most common type of display used in screenphones. Most LCDs used in screenphones are black and white (with a few in color). Some of the LCDs also offer touch-screen capability.

Low-end screenphones only have the ability to display basic text for e-mail and/or news, although small applications (such as an address book) could be possible on these limited screens as well. These requirements are easily satisfied by small monochrome LCDs with a width of 18-to-20 characters, and a height of three-or-four lines. Limited screens such as these are readily available and affordable. Mid-range products also require only basic text capability on a monochrome screen (although limited graphics are possible). Screen sizes in this segment are approximately 4-to-5 inches diagonally. However, functional keys designed into the hardware could allow full use of the screen space. High-end screenphones, on the other hand, require larger displays (current models range from 7.4-to-8.2 inches) with touch-screen capabilities. Most vendors' models use passive matrix color screens supporting VGA. A grayscale screen would be less expensive and would require less processing power and memory, and could therefore reduce the price of the high-end unit. Grayscale displays can easily and efficiently satisfy the requirements of these screenphones, although the flashiness of a color screen will no doubt be more appealing to impulse-buying consumers.

Active-matrix screens could be integrated into the high-end screenphones, but the higher cost is not justifiable at this stage. Most high-end phones today do not require high-speed motion (no video, no cursor, etc.), and active-matrix screens would drive the purchase price far beyond the already-high price points they now have.

Processors

Screenphone manufacturers use processors that can provide limited functionality and multiple functions, depending on the need. These include DSPs as well as embedded processors from manufacturers such as MIPS, ARM, XScale, and so forth.

Communications and Interfacing

Bandwidth requirements for low-end screenphones are minimal, because most rely on downloading limited amounts of data (basic text e-mail without attachments, or information services only). This is especially true for devices that conduct unattended, automated data downloads or uploads. Because no user will actually be waiting for the transaction to take place, the difference between a 15-second transmission sequence and a 30-second transmission sequence will be undetected. Analog modem technology is the correct solution in this case, with speeds at or below 9,600 b/s. This situation could eventually move forward as higher speeds become cheaper.

Midrange screenphones require higher communications speeds as a result of their increased functionality (e.g., sending e-mail). Analog modem speeds from 9,600 b/s to 28.8 Kb/s can be used; anything greater adds unnecessary cost. Software-based modems can also be considered. With their graphical Web browsing capability, high-end screenphones demand much more bandwidth than the other segments. The current solution is a V.90 56 Kb/s chipset over plain old telephone service (POTS) lines, although some vendors have adapted their phones for use on Integrated Services Digital Network (ISDN).

High-speed access through technologies such as DSL (digital subscriber line) and cable will certainly become more prevalent in the future, as the technologies become more widespread. As a result, many vendors have designed a flexible "communications unit" into their devices that can be easily modified for upcoming units as broadband becomes more common.

Telephony

A primary difference between the screenphone form factor and other digital consumer appliances (such as NetTV and Web terminals) is the integrated handset and core telephony functionality. This functionality also includes advanced features such as caller ID and voicemail (or even digital answering machines).

Verbal Communications

All screenphones on the market today integrate basic voice-calling functionality to conduct a voice telephone call through a corded handset. Efforts at cordless handsets have been minimal thus far, although it would not be unreasonable to think that they could become more readily available in the future. Most screenphones also feature speakerphones, many of which are full-duplex to prevent clipped conversations.

Basic Call Management

Nearly every screenphone offers basic telephony management functions such as hold, mute, line status indicator, auto redial, last number redial, flash, and speed-dial memory. Speed-dial functionality is also offered. Traditional phones activated this function by pressing speed-dial hard buttons. Mid-range and high-end screenphones, however, can provide entire personal information management (PIM) applications with on-screen lookup by name, and have the ability to store addresses or other information. Ideally, the screenphone would be able to connect with a PIM application (such as Microsoft Outlook or Lotus Organizer) to synchronize data (forgoing the need to re-enter each contact). Few screenphones have yet explicitly shown this capability.

Screenphones can also take advantage of their basic voice operation by implementing voice recognition for Web browsing and the like. Again, there have been no implementations of this function yet, but hopefully there will be developments of this sort in the future.

Dual-Line Functionality

Most screenphones on the market today offer two-line functionality, allowing users to browse the Web and talk on the phone at the same time. This allows several usage patterns, including looking up Internet-based information in support of a telephone conversation (e.g., "What movies are playing tonight?") and multitasking during conversations. In addition, dual-line phones can enable conference calling—a key feature for small office/home office (SOHO) users.

Of course, this capability requires a second phone line in the house. Nonetheless, most screenphone vendors have modified their designs to include the second line, regardless of whether the second line is actually used or not.

Answering Machines and Memo Recording

The inclusion of a digital answering machine (possibly with memo and/or conversation recording) is also prevalent in today's screenphones. Built with "voicemail waiting" and "e-mail waiting" indicator lights, the answering machine makes the screenphone a central communications center. Multiple mailboxes would make the devices even more convenient. Voice processing functions include the following:

- **PCM Interface** – This Conditions PCM data, and includes functions such as companding and re-sampling. This block also includes the Tone Generator, which generates DTMF tones and call progress tones.
- **Echo Cancellation Unit** – This unit performs echo cancellation on sampled, full-duplex voice port signals in accordance with the ITU (International Telecommunications Union) G.165 or G.168 standard.
- **Voice Activity Detector** – Suppresses packet transmission when voice signals are not present. If activity is not detected for a period of time, the voice encoder output will not be transported across the network. Idle noise levels are also measured and reported to the destination so that "comfort noise" can be inserted into the call.
- **Tone Detector** – The Tone Detector detects the reception of DTMF tones and discriminates between voice and facsimile signals.
- **Voice Coding Unit** – The Voice Coding Unit compresses voice data streams for transmission. There are several different codecs used to compress voice streams in VoIP applications. Each has been targeted at a different point in the tradeoff between data rate (compression ratio), its processing requirements, the processing delay, and audio quality. Table 14.2 compares the various ITU codecs with respect to these parameters. The MOS (mean opinion score) parameter is a measure of audio quality that is rated by a subjective testing process, where a score of 5 is excellent and a score of 1 is bad.

Table 14.2: Voice Coding Standards

Standard	Description	Data Rate (Kb/s)	Delay (mS)	MOS
G.711	Pulse Code Modularion (PCM).	64	0.125	4.8
G.721, G.723, G.726	Adaptive Differential PCM (ADPCM).	16, 24, 32 & 40	0.125	4.2
G.728	Low-Delay Codebook Excited Linear Predictors (LD-CELP).	16	2.5	4.2
G.729	Conjugate-Structure Algebraic CELP (CS-ACELP).	8	10	4.2
G.723.1	Another CS-ACELP codec: Multi Pulse – Maximum Likelyhood Quantizer (MP-MLQ)	5.3 & 6.3	30	3.5 & 3.98

- **Voice Play-out** – Voice Play-out buffers received packets and forwards them to the voice codec for play-out. This module provides an adaptive jitter buffer and a measurement mechanism that allows buffer sizes to be adapted to network performance.
- **Packet Voice Protocol** – This encapsulates the compressed voice and fax data for transmission over the data network. Each packet includes a sequence number that allows the received packets to be delivered in correct order. This also allows silence intervals to be reproduced properly and lost packets to be detected.

Advanced Network Services Support

Telephones today can be designed to support a broad range of telephone company-offered services. Given the broad marketing support available from the telephone companies, screenphone designs should try to support as many of these services as possible. These services include:

- **Distinctive ringing** – Permits users to know which person is being called via different-sounding tones. This feature is typically priced at an additional $2 to $4 a month.

- **Caller ID** – Permits display of the caller's telephone number. There are different variations of this service, including caller ID II (the ability to display the caller's name or directory information) and call waiting/caller ID (allows information to display while the line is in use). These features cost an additional $3 to $15 a month and are rather profitable for telephone companies.
- **Voicemail** – An answering service provided over the telephone company's network, precluding the need for a mechanical cassette-based machine. A message waiting is indicated by the dial tone wavering into a "stutter" pattern. Screenphones can integrate a voicemail-waiting light that automatically senses this wavering tone.

The Analog Display Services Interface (ADSI)

Analog Display Services Interface (ADSI) is a protocol developed by Telcordia (formerly Bellcore) in 1993. This protocol can either take the form of real-time interactive applications or offline ADSI scripts, which are miniature programs downloaded into the phone (similar to the concept of Java applets). Applications of ADSI include telephone directory white pages and home banking. Another application is a script that provides caller ID "with disposition," giving the user the choice of accepting an incoming call, putting it on hold, sending it to voicemail, or bringing it into an existing conversation.

Although the interactive side of ADSI has had only limited success, many early screenphones have seen a good number of sales due to the use of offline ADSI scripts. Some vendors have had success in distributing screenphones to regional Bell operating companies (RBOCs) and Canadian telephone companies. Nonetheless, ADSI functionality faces a challenge with the rise of Internet and Web browsers. Most screenphone vendors are incorporating Web browsing capability (i.e., IP instead of ADSI) to appeal to consumers. Many applications such as home banking can no doubt be obtained on the Web in lieu of an ADSI script. Of course, ADSI phones (or midrange phones) generally cost much less than high-end, Web browsing phones.

Operating Systems

Operating systems used in consumer devices are a hot topic, especially given the threat they pose to Microsoft's dominance as consumers move to non-PC devices to access the Web. The real-time operating system is an operating system that guarantees a certain capability within a specified time constraint. For example, an operating system might be designed to ensure that a certain object was available for a robot on an assembly line. This operation uses what is usually called a "hard" real-time operating system, and would terminate with a failure if the calculation could not be performed for making the object available at the designated time. In a "soft" real-time operating system, the assembly line would continue to function but the production output might be lowered as objects failed to appear at their designated time, causing the robot to be temporarily unproductive. Some real-time operating systems are created for a special application; others are for more general purpose. Some existing general-purpose operating systems claim to be real-time operating systems. To some extent, almost any general-purpose operating system (such as Microsoft's Windows NT or IBM's OS/390) can be evaluated for its real-time operating system qualities. That is, even if an operating system does not qualify, it may have characteristics that enable it to be considered as a solution to a particular real-time application problem. In general, real-time operating systems are said to require multitasking, process threads that can be prioritized, and a sufficient number of interrupt levels.

Real-time operating systems are often required in small, embedded operating systems that are packaged as part of micro-devices. Some kernels can be considered to meet the requirements of a real-time operating system. However, since other components, such as device drivers, are also usually needed for a particular solution, a real-time operating system is usually larger than just the kernel.

Real-time operating systems have been prevalent in this area, with vendors and products such as QNX Software Systems' QNX, Wind Rivers' VxWorks, and Integrated Systems' pSOS. Leveraging its acquisitions of Diba and RTOS developer Chorus, Sun Microsystems has been promoting Java as a screenphone platform with JavaPhone, an extension of Sun's PersonalJava environment. Although PersonalJava is more of a development environment than an actual operating system, Sun also makes available the JavaOS, based on the Chorus kernel and embracing PersonalJava. Several vendors, including Nortel Networks, Samsung, and Alcatel have adopted Java. Motorola, together with systems integrator AudeSi, also developed Touchstone, a Java-based reference for digital consumer devices. Unfortunately, vendors using the Java platform have met with only varying degrees of success. Nortel was said to have been developing a Java-based screenphone but never came through with the device. Alcatel only recently released its Java-based phone, and even then the device has been limited to France. Samsung has also been developing a screenphone based on the Java platform but has yet to fully deploy it.

In addition, Lucent Technologies has stopped development of its Inferno operating system, which ran devices such as screenphones and supported PersonalJava. Vita Nuova of the United Kingdom is tentatively scheduled to take over support for the product. Rest assured, Microsoft has not been standing idly by. In 1997, the software giant made an investment in Navitel Communications, a company that was originally developing screenphone hardware but has since focused on screenphone systems development instead. Navitel has since been acquired by Spyglass, and, together with Microsoft, the firms continue their developments for screenphones in the Windows PocketPC area. Microsoft also released the Windows PocketPC Web Telephone Edition, which targets the Windows PocketPC operating system for screenphones and uses Internet Explorer Web browser technology.

Video Phone (Webcam) Using the PC

While dedicated devices have the potential of growing into Internet screenphones, another popular category of Internet videophones has hit the market. These phones incorporate an Internet videophone, the PC, digital camera, microphone, speakers, and a modem. These videophones use the Internet/IP backbone to transmit voice, video, and data between users based miles apart.

These phones are gaining market acceptance with consumers because of the low (and sometimes even total lack of) cost, as they allow consumers to connect to friends and family around the world without paying a dime. However, the quality needs to improve through use of higher bandwidth connections and quality of service algorithms. Also, while this technology may finally work one day, it is hard to conceive a business model for person-to-person communications. Multiuser applications and peer-to-peer videoconferencing also seem to be gaining ground. But the quality will never improve and users will remain unsatisfied when someone else is paying the bill until the business models have been well thought through.

e-Learning

The one business that seems like a natural for videophones is micro-seminars or e-seminars/e-Learning. It uses low-end Internet multimedia technology to change the economics of the booming seminar and training business. A complete industry known as e-Learning is gaining popularity as businessmen look for venues to avoid travel due to cost and security concerns. This removes the high cost of making product brochures on hot new technologies for customers. It also saves money by reducing employee travel expenses. In addition, one does not need to spend millions of dollars setting up big conferences, and pay five speakers to present to the same audience. Also, the economics of this approach are not very attractive to lecturers—those supplying the actual knowledge. The conference organizers keep most of the money—as well they should, since they bear the cost of promotion, take the risk of booking the hotel space, feed everybody lunch, and collect and assemble the conference material.

Using desktop videoconferencing to unbundle the package cuts out the middleman, drops the price of attending by a factor of ten, and allows the messenger to take the message to a global market simultaneously. This requires lecturers to be equipped with only a high-end PC, a digital camera, and a broadband access connection. Presentations would consist of multicast audio, with one video talking head and controls to let the lecturer drive everyone's browser through the material. Payment can be collected online using a credit card or smart card. Even at a reasonable price of $100 per person, groups as small as 20 would be quite profitable.

These sessions can also be scheduled in real time. Videotapes just don't have the performance value of a live presentation, nor do they allow the presenter to respond to questions. Best of all, anyone that has some expertise to share, knows how to do stand-up, and can draw a crowd of a few dozen people can get in the game without even quitting their day job. Numerous start-ups are working on this. Also, Microsoft has joined the bandwagon with its NetMeeting software. The market should soon start seeing integrated packages for providing such services.

Summary

Since Alexander Graham Bell invented the telephone in 1876, telecommunication devices have developed and evolved beyond our imagination. Now we can communicate with people who live on the opposite side of the world on the corded/cordless telephone, through the Internet, and by other means.

There has been a great deal of fanfare in the press recently regarding an emerging set of consumer devices designed to provide easier access to the Internet. One such device is the screenphone, an intelligent telephone with a screen and a small optional keyboard that offers Internet access and custom services through a phone line at the touch of a button. This device is designed so that users can quickly access the Internet and transact business online. The device is targeted at the business and home consumer who needs to access information on the Web, as well as store, retrieve, or send specific data using e-mail or a phone number to connect to an address.

The screenphone cannot truly be defined as a new technology, since it is actually a combination of many communications media—a good example of technology convergence. The Internet screenphone integrates a personal computer, telephone/speakerphone, fax modem, digital answering machine, video conferencing camera, and debit/credit card swipe. In a word, the Internet screenphone is a multifunctional device designed specifically to connect online and access the information super highway—thus enabling voice, video, and data communications.

Several manufacturers have announced easy-to-use, multifunctional screenphones that have an integrated high-quality LCD at a moderate-cost. These screenphones combine advanced video and audio compression technology to transmit full-motion color video and audio over standard analog telephone lines. Additionally, these products are capable of Internet access. These phones are capable of functioning as an Internet phone, peer-to-peer with another screenphone, or through a gateway provider between a screenphone and a standard phone. Both connection options allow users to make low-cost long distance calls to nearly anywhere in the world.

The era of communications with simultaneous exchange of data, voice, and video is upon us. Whether it uses the telephone network or IP network, communications convergence will occur over the next few years. Screenphones will succeed as consumer products if manufacturers can offer next-generation video and audio performance over standard phone lines and the ability to trade-off the video transmission rate for image quality. Manufacturers and service providers will need to provide subscribers with audio, video, text, and Internet capabilities that will establish new levels of consumer interaction and ease-of-use. For screenphones to be successful, service providers and utility companies will need to increase revenues by offering bundled services to the same consumer to whom they have been selling phone line services. Having a destination site portal will also provide after-sale consumer benefits such as those offered by advertisers and merchant partners. Audio-video ads and promotions will also be conveniently enabled through this Internet appliance.

The screenphone saves customers money by connecting calls through the Internet. The Web screenphone combines the Internet features of a PC (such as e-commerce, Web surfing, and e-mail) with IP telephony and the simplicity of a regular telephone. Screenphones provide an out-of-the-box plug-and-play solution for full e-mail send/receive functionality, WWW browsing, and other high-end features. Furthermore, these products provide the non-PC user with the opportunity to browse the Internet. Advantages of most screenphones include ease-of-use, telephone functions, answering machine, and a two-line telephone that allows simultaneous voice conversations and Web access. However, screenphones cannot download applications or allow users to listen to streaming audio tracks. Users must also sign up for a separate ISP account for the phone, even if they already have an account with another provider.

Many people in the communications and media industries see this product as the third mass media for consumer markets, in addition to TV and the computer. From this point of view, this lightweight, compact, portable, and multifunctional screenphone should have a bright future. The human desire for convenience and accessibility is seemingly unlimited, and the future will see even more functions integrated into this device.

Automotive Entertainment Devices

The hit 1980s TV show *Knight Rider* featured the crime fighter, Michael Knight, whose partner, Kitt, was a futuristic talking car. Kitt constantly updated Knight with information about the quickest route to their destination or the whereabouts of criminals. Cars of the future may not be able to drive themselves like Kitt (at least in the near future), but will mimic the technology by using telematics.

Mobile multimedia has emerged as the hottest growth area in transportation. Consumer electronics, wireless communications, and the Internet are converging within the automobile, and creating a new type of node on the Internet. Mobile multimedia brings a host of new features, applications, and functions into the cockpit, providing advanced forms of communication, entertainment and productivity. Mobile multimedia is also sparking new business collaborations among manufacturers of consumer-electronics, software, vehicle-systems, and automakers, thereby providing opportunity for new and increased revenue streams.

Introduction

"Slow down, Lola!" Charlie warns Alice in the driver's seat. But Charlie isn't her boyfriend or her husband. The voice comes from the car's audio system. Charlie is her Personal Agent, and he has been built into her car's computer system. Alice and her friend Cindy are running late for their biology class. "You don't have to rush," Charlie advises. "Professor Wilson has just cancelled today's class." Cindy touches the LCD screen on the dashboard and accesses the college web page. "Hey, that's right! Class has been cancelled," she grins. "Which means we've got a few hours to kill. Feel like a cup of Cappuccino?"—Alice asks Charlie for the directions to the nearest coffee shop. "Would you like to try something new?" he suggests. Charlie recommends Timeless Tea, a drive-through Chinese teahouse that recently opened in the neighborhood. "You've got an electronic discount coupon for one pot of Jasmine tea and almond cookies," he reminds her. At the Chinese drive-through, Alice places her order and pays for it by inserting her IC card into the onboard card reader. The payment is wirelessly transmitted to the restaurant. As they wait for their snack, the two friends watch some movie previews that are being streamed to the car. "I love that song! What is it?" asks Alice as she turns up the volume. "It's the theme song from *Toy Story 2*," Charlie replies after searching the Internet. "I can download the MP3 to your car stereo if you like. The price is $2.00 plus tax." "Charge it to my credit card," Alice instructs him.

Another example: a man leaves his home at 8:30 a.m. He downloads his e-mail so he can listen to it on the way to work. His car door opens automatically when he touches the door handle, which recognizes his fingerprints. His seat and mirror are also automatically reset to suit his preferences. He places his PDA in its cradle. When he starts the car, a "low oil" warning message notifies him to make an appointment to add oil, which he does. He drives to work listening to the news on satellite radio.

While these scenarios may sound futuristic, automobiles like these are not all that far off. That is because cars and trucks have rapidly become more than just simple modes of transportation. They have evolved into mobile offices for people on the move, and entertainment centers for kids and family to use while traveling. The telematics landscape is growing and changing at a rapid rate, and a lot of research and development dollars are pouring into it. In the not-so-distant future, drivers can expect traditional radio, tape, and CD audio players to be augmented with interactive computer games, MP3 players, and satellite-broadcasting services. In-car services will range from news headlines to personal investment portfolios, from stock price updates to weather, sports scores, and movie or theatre show times. Personal productivity applications such as calendar, event planning, and portable versions of business applications will also be provided. Mobile phones and PDAs, which are already evolving with wireless Internet capabilities, will have docking facilities in automobiles.

These devices will also provide conventional or Internet phone services, and will be enhanced with voice synthesis to allow drivers and passengers to hear e-mail or navigation instructions. E-commerce is possible with wireless connectivity, with services that order convenience items such as fast food or movie tickets. In addition, today's GPS will be enhanced with dynamic routing instructions that are updated with real-time traffic reports. Other geographic services like roadside assistance and stolen car recovery can be easily incorporated. A combination of sensor technology with onboard advanced diagnostics and communication with automotive service networks will automatically diagnose potential problems and proactively alert drivers and service personnel to the need for preventive repairs and maintenance. And finally, different levels of access can be granted to drivers, passengers, auto mechanics, and valets.

The cars of today are nothing like what they were just a few years ago. They are much safer and smarter machines. Transitioning from all mechanical systems to electrical systems by embedding microcontrollers and other electronics has provided consumers with not only better reliability but also features such as smart hydraulic and braking systems, seat warmers, parking distance control, rain-detecting windshield wipers, and advanced entertainment devices.

The automobile is the highest-volume mobile consumer product. And it is fast becoming the highest-volume mobile computing platform as its electronics content increases. In the area of multimedia electronics, cars are evolving from having a simple radio with perhaps a cassette or CD player to having a variety of information systems that need to communicate and interact with one another. The trend toward having networked information and entertainment systems is accelerating as consumers become accustomed to the conveniences afforded by those systems. These systems must also provide increased occupant safety by allowing the driver to concentrate on controlling the car rather than on the intricacies of the individual components.

Cars today can come with an optional GPS and security system to help locate the car when stolen, but high-end systems being deployed today are going to be commonplace within the next few years. But vehicle safety requires interaction between the car telephone and stereo system in order to automatically mute the stereo when a call is requested or received. Voice control and hands-free speakerphones require a microphone to digitize voice. Display systems are needed for navigation information and DVD playback.

To be effective, all these subsystems must interface with the driver. Audio and visual information must be safely presented in many formats to inform the driver and/or to entertain the passengers. The most efficient and cost-effective way to continue innovations in those areas is to allow needed devices to be developed independently. They can then be networked together using standard hardware and software (bus) interfaces. Such an arrangement will require digital interoperability. Options will be easy to add, since the network provides the infrastructure to transfer information from one device to another. Cars will be customized to each buyer's preferences right at the dealership, and will not depend on a pre-selected list. Safety will be enhanced by integrated user interfaces that require minimal driver distraction in order to control installed interoperable components.

Consumers will also require the ability to connect the portable devices they use to the automotive 'infotainment' network. The most widespread interfaces are still analog, whether it be the audio output of an MP3 player or the audio and video signals from portable DVD players and game consoles. All portable products have analog audio outputs for headphones. If the device has a video output it is usually analog, even when a digital option is present. The standard Universal Serial Bus is the most common interface among digital devices, whether MP3 players, digital cameras, or laptop PCs. However, the IEEE standard 1394 (FireWire™ or iLink™) has gained some favor in high-end video cameras, gaming consoles, and digital video recorders. Personal digital assistants and handheld GPS units usually have an RS-232 interface and/or wireless connectivity. Hence, a common interface must be available to allow devices to be interconnected and to interact as part of the network.

Everyone is at risk when it comes to driver distractions. The National Highway Traffic Safety Administration estimates that activities that distract drivers—everything from changing the radio station to eating while behind the wheel—are a contributing factor in one-in-four motor vehicle collisions. This represents more than 1.5 million crashes a year, or more than 4,300 crashes a day. Reducing driver distraction is paramount for every telematics solution provider.

These devices provide entertainment, security, communication, and information in automobiles, and are commonly known as automotive entertainment devices. They also go by popular terms such as auto PCs, telematics, Web on Wheels, Automotive Net Devices, Intelligent Transmission Systems (ITS), and car infotainment devices.

Automotive electronics industry products are segmented into groups:

- In-car entertainment:
  - Auto stereo
  - AM/FM radio and CD/tape combinations
  - Digital audio broadcasting (DAB) radios
  - Multi-disc CD changers
  - Graphic equalizers
  - Power amplifiers
- Vehicle body control:
  - Multiplex systems
  - Driver's door console
  - Keyless entry systems
  - Memory seats and mirrors
  - Lighting controls
  - Memory seats/steering wheel position
  - RFID/Remote Keyless Entry
- Driver information:
  - GPS navigation systems
  - Location systems/vehicle tracking devices

- Dashboard instrument clusters
- Electronic compasses
- Electronic thermometers
- Head-up displays
- Digital (cellular) telephones
■ Power train:
  - Alternators
  - Engine control units
  - Cruise control
  - Ignition control
  - Engine sensors
  - Transmission control
  - Collision warning systems
  - Two-wheel-drive/four-wheel-drive control
■ Safety and convenience:
  - Air bag control units
  - Climate control systems
  - ABS (antilock braking systems)
  - Traction control systems

The Evolving Automobile

It is easy to understand the evolution of mobile multimedia with a brief look at the historical innovation roadmap of the automobile. This explains how changing lifestyles have affected vehicle innovation, and helps put the current era of mobile multimedia integration into perspective. There have been three eras of vehicle evolution—power/body, convenience/comfort, and intelligent/information.

The power/body era roared in following World War II, with such technology advances as high-compression engines, wider tires, and larger chassis. New innovation concentrated on performance. Lifestyles strongly reflected military and industrial strength during this period. This was also the era for the growth of middle-income families. Then, in the early 1960s, came the convenience/comfort era. With more disposable income, Americans craved technologies such as power steering, electric windows, air conditioning, power brakes, and automatic windshield washers.

The third era focused on the development of intelligent/information vehicles, and was prompted by energy shortages and social issues. Armed with more education and passion for social issues, maturing baby boomers demanded vehicles more closely aligned with their values. In the late 1970s and early 1980s electronics integration was introduced on vehicles to improve fuel economy and reduce emissions to better protect the environment. Technological innovations such as electronic ignition, fuel injection, and electronic engine controls were added. Electronics integration enhanced the safety and security of vehicles with antilock brakes, airbags, and traction control.

Rapidly Rising Intelligence

The intelligent/information era expanded rapidly. In 1974, the total electronic content of a car was only about 2%, and this consisted of a radio and the solid-state voltage regulator in the alternator. A recent study by Roland Berger & Partner LLC, a strategy consulting firm, puts current vehicle electronic content at 22% and expects it to grow to as much as 40% by 2010. The report states that

80% to 90% of all new innovations in high-end vehicles are related to electronics, with these vehicles already close to the 40% content level.

Anywhere, Anytime

The proliferation of information on the Internet and the growth in personal use of consumer multimedia products at home and work have increased demand for these same products in vehicles. An "anywhere, anytime" society exists today. People expect to be able to receive and send information from wherever they happen to be, including the vehicle. The phenomenal growth of mobile multimedia is made possible by advancements in microprocessors, human/machine interfaces, and wireless communications. Automakers are moving at lightning-like speeds to add electronic media to their vehicles to differentiate their products, increase revenues, and satisfy consumer demand.

Telematics combines audio and a hands-free cell-phone module with a GPS module to provide automated communications between a vehicle and a service center. The system provides such varied services as emergency assistance, voice navigation, vehicle tracking, unlocking doors, and concierge help. Telematics is creating new business models and new businesses. Vehicle manufacturers can even extend telematic income streams for the entire life of the vehicle. Rather than collect income from the purchase and financing, the vehicle can provide a lifetime income stream much like telephone or cable TV services. New businesses such as service centers, location-based merchandising, and advertising will also play an important role. For instance, a system that has a customer profile and location could advertise discount coupons on hotels or meals. The system could also allow a driver or passenger to access sports or stocks, or make purchases. A recent global study by Roland Berger forecasts that (in addition to mobile telephone services) banking and shopping from the car will account for the greatest share of telematics applications.

Customized Channels

Advanced audio systems such as the Satellite Digital Audio Receiver Service are already gaining ground. Satellite radio will revolutionize traditional AM/FM radio by adding a satellite band that allows drivers to listen to approximately 100 customized channels of clear digital music and entertainment from coast-to-coast. Rear-seat entertainment systems demand has already outpaced forecasts. The systems are especially popular among van or sport utility vehicle (SUV) owners with rear-seat passengers. Most rear-seat entertainment is in the form of DVD movies or video games.

Experts also expect to see mobile multimedia enter the vehicle through smart radios. These systems add a PC card and high-resolution display panels to provide the driver with added information such as navigation, e-mail, news, integrated telephone, and Internet information. The driver can execute all of those functions via voice recognition, text-to-speech, and drive-by-wire steering-wheel controls without taking his or her eyes off the road.

What is Telematics?

The word "Telematics" is derived from *Tele*communication and Infor*matics* (Information Technology). Telematics stands for combining the power of information technology and the Internet with advanced telecommunication technologies. It is defined as the convergence of telecommunications, networking, and computing, and is the single largest change agent in the automotive electrical arena. Telematics is an emerging industry and offers location-based voice and data communication tools. In other words, telematics provides "smart" information, tailored to where customers are and to what they are doing—providing enhanced security, navigation, and convenience to mobile consumers.

Telematics systems offer a never-before possible level of safety and security, traffic congestion and route guidance information, advanced diagnostics capabilities, productivity, personalized information, and entertainment in the automobile. Computing capabilities in the vehicle are connected to the network using wireless technologies, and allowing the user to access and act upon content, applications, and services residing on the network.

Telematics brings together the fusion of computers and wireless communications technology inside cars to enable, for example, motor vehicles to "speak" instructions to drivers about such things as traffic conditions and weather conditions. The cars will not only be giving you advice as to the correct freeway to take, but they will also tell you which freeway has the most traffic on it. In-Vehicle Information Systems (IVIS) use either voice recognition and interactive audio technology or a simple touch-screen to communicate with the driver or passenger—all enabled by telematics technology. This provides hands-free phone dialing, Web browsing, and the ability to listen to or compose e-mail. It also provides access to a wider variety of personalized entertainment, dynamic route guidance with real-time traffic and weather information, remote vehicle diagnostics, safety, and emergency help.

As an example, telematics can come to the immediate rescue of a person who has locked his keys in the car. To active the rescue, the forgetful driver deftly retrieves a phone from his pocket and calls an operator. The operator then remotely commands the vehicle to unlock. This happy scenario enables drivers not only to have their cars unlocked via wireless two-way communication, but also to get roadside assistance, directions, or location of the nearest gas station by merely pressing a dashboard button and making a request.

Telematics (or location-based services) functionality can include:
- Vehicle location determination through GPS capabilities
- Emergency response upon collision or breakdown
- Theft avoidance, prevention, and detection
- Traffic and congestion alerts integrated with route guidance information
- Advanced diagnostics capabilities for automobile performance, updates, and maintenance
- Access to e-mail and other productivity applications
- Personalized news, sports, and weather information, all based on personal preferences
- New levels of entertainment, such as music, games, and even more (one day, even movies on demand)

The very words "automotive safety" conjure up images of air bags and seat belts. There is, however, more to the safety movement than this. Today's automakers see a mountain of potential safety hazards on the horizon. And some of those problems can be traced to the very features that are designed to keep us safe.

Telematics, for example, can be an automotive blessing or a curse. Onboard cell phones and navigation systems can save stranded drivers or they can distract overloaded ones. Similarly, by-wire systems could one day enable vehicles to tie into intelligent transportation infrastructures. However, if improperly implemented, they could render brakes and steering systems virtually useless at the wrong times. Safety and security are currently the driving forces behind the decision to put telematics in the car. However, one of the problems today is that many cell phones are completely independent, and there is no way for the phone to 'know' if you are able to accept a call.

Telematics is not the only area in which electronics will play a critical safety role. Research is also taking place on the development of time-triggered, fault-tolerant architectures that could serve as the electronic backbone for brake-by-wire, steer-by-wire, and throttle-by-wire systems, similar to the fly-by-wire controls of modern military and commercial aircraft.

The Automotive Electronics Market

The registration of over 100 million new vehicles every year in the U.S. alone is a powerful motivation in itself for invading the car with infotainment electronics. In addition, the average American daily commute is over 82 minutes. These two factors provide huge opportunities for improving productivity while driving. Potential product applications include communication, music-on-demand, real-time traffic information, and remote vehicle maintenance.

Several opportunities for automotive manufacturers exist, including:

- Product differentiation
- Recurring revenues from monthly services
- Mobile commerce enabled by GPS: Advertisements/promotions based on location
- Multimedia specialized-content transmission: Stocks, news, weather, sports

Leading research firms and various market vendors have provided the following Telematics market data and projections:

- McKinsey and Company – According to a recent McKinsey and Company report, the market for in-vehicle applications will reach $100 billion by the year 2014.
- General Motors – As of year-end 2000, some 1.5 million autos were telematics-equipped. General Motors' OnStar Division reported that 70% of its customers opted for a continued $30 monthly subscription service at the end of the complimentary first year.
- UBS Warburg – Investment bank UBS Warburg projects that telematics (which includes the suite of services powered by wireless communications, GPS, and onboard electronics) will be worth $9 billion a year in the U.S. by 2005, and $13 billion by 2010. They envision the worldwide telematics market to reach $47.2 billion in revenues by 2010.
- International Data Corporation (IDC) – Industry research firm IDC has estimated that the telematics market will grow from $1 billion in 1998, to $42 billion by 2010.
- Dataquest – The worldwide market for automotive electronics equipment is a multi-billion dollar industry, which is expected to exceed $81 billion by 2004 and reach $92 billion by 2006, according to analyst firm Dataquest. The automotive electronics equipment included in this category is airbags, ABS, auto stereo, climate control unit, dashboard instrument cluster, engine control units, GPS navigation systems and remote/keyless entry systems. The semiconductor content for this equipment is expected to exceed $17 billion by 2004 and $21.5 billion by 2006.
- Strategy Analytics – Strategy Analytics predicts that over 50% of all cars produced in North America will have built-in telematics-capable terminals by 2006. They further predict that 85%-to-90% of new large/luxury cars will be telematics-enabled. Strategy Analytics also predicts the world market for in-car telematics will grow from $7.7 billion at the end of 2000, to $24.3 billion by 2006. A full $9.4 billion of that market in North America alone.
- Cahners In-Stat Group – A recent study by the Cahners In-Stat Group estimated that the sale of speech-recognition systems for telematics, handheld devices, appliances and other applications is expected to reach $2.7 billion by 2005. Researchers who conducted the study said that automotive applications for voice recognition could explode, especially in one key area. The study suggests that only limited money can be made with a speech product that

adjusts your radio or your air conditioner, but that there is a considerable voice-recognition market for connecting the car to the Internet and to the outside world.

■ Dataquest – A study by Dataquest estimates that every new car assembled in Europe will be fitted with daylight driving lamps and an antilock braking system (ABS) as standard by 2004. In addition, every car manufactured will be a "virtual WAN node," and able to communicate with other cars on the road, by 2010.

While these numbers vary depending on the market segments included, research from these leading firms indicates huge future growth in telematics revenues. Safety features will be the initial driving force to get telematics into the car, but ultimate overall market success will depend on other telematics applications that drive up monthly subscription charges. Telematics providers currently offer route assistance, vehicle diagnostics, e-mail, and concierge services. However, they are developing entertainment options such as backseat pay-per-view movies and personal music collections beamed to the car stereo. In the near term, features such as high-pressure direct injection (both gasoline and diesel), tire pressure monitors, adaptive and pedestrian protection airbags, night-vision enhancement systems, and radar cruise control will gain ground.

Current navigation and traffic applications must rely on data on a CD-ROM mounted in the car trunk because of the lack of online bandwidth into the car. Essentially, these applications are electronic road maps. However, navigation on a two-way network, with traffic and route information accessible in real time, would truly be innovative and far more useful. Yet live telematics in the U.S. would rely on a fractured wireless infrastructure to communicate with the network. This concept works well for voice traffic, and is the reason why call-center based telematics services such as GM's OnStar are successful. But the infrastructure is totally incapable of carrying the amount of data necessary for live navigation or real-time traffic information. That would require a third-generation (3G) cellular system, which will be deployed in three-to-five years, and provide data rates of 384 Kb/s when in vehicular motion and up to 2 Mb/s when stationary.

If telematics is to grow, GPS technology must develop, too. The standards for real-time traffic and navigation systems depend on the ability of telematics to deliver data from cars to a centralized collection system. Instead of using people or cameras to generate traffic reports, cars equipped with locator devices and designed to send information could contribute as well as receive information.

Apart from automotive manufacturers and telematics providers, the development of automotive entertainment devices presents an opportunity for increasing revenues beyond traditional schemes for telecommunications and wireless service providers. It covers many different fields and has enormous untapped potential. Wireless telecommunications carriers will naturally play a leading role in the advancement of telematics within next-generation mobile phones. Their telematics services will keep users seamlessly connected to the networks wherever they might be—behind the wheel of a car or walking on the street.

Some key considerations of the telematics market include:

■ Channel – The "channel" describes the vendor route of how automotive consumer appliances make their way into the automobile. Unlike personal or home solutions, automotive appliances are not normally bought off the shelf. Automotive-based solutions either come preinstalled at the factory or installed at the dealership. They are also installed by aftermarket specialists that may be stand-alone vendors, possibly attached to the local auto body shop or (as is increasingly prevalent) as a part of a consumer electronics chain.

- Partnerships – Partnerships are critical to bringing a viable solution to market for all consumer devices. The traditional digital companies do not have the knowledge of the car market, and those that specialize in putting devices into cars (radios and alarms) do not have experience with the Internet and computing. We think there will be opportunities for technology enablers to supply the overall solutions providers (the factory and the well-known car equipment vendors) with pieces of the puzzle. Promising areas include auto-based services and content.

- Interface – While the technology in auto consumer appliances is basically the same as that in mobile and home solutions, the user interface for in-car systems is very different. This is because these products (in general) are used when the user is actually driving. In addition, major opportunities exist for companies that can make the user interface process easier due to the long product life cycle of automobiles (typically five-to-six years for a model to be phased out).

- Potential User Base – The key age bracket for after-market car stereos is 16 to 24. We think, however, that auto consumer devices such as navigation systems will generally appeal to more affluent high-end mobile professionals for whom greater access or information velocity outweighs the significant associated cost.

The Controversy

The firestorm of controversy over the use of cell phones in cars while driving is being welcomed as a potential windfall by automakers and their suppliers. As engineers struggle to defeat the problem of driver distraction, the industry is charging forward with plans to build advanced telematics systems. The car phone controversy spiked after New York State passed a law banning the use of handheld phones while driving cars. The controversy has hit a nerve with consumers, who appear torn over whether to condemn car phones or embrace them. The New York law and proposed legislation in virtually every other state is aimed at handheld cellular phones. This leaves telephone-addicted automobile drivers little choice but to move toward built-in car phones, and especially toward hands-free, voice-operated units. Hands-free units are said to offer the best solution now available to deal with the driver distraction problems that car phones pose. And if car phone legislation proliferates around the country, as most experts expect, the factory-installed systems may emerge as the best choice for consumers. When consumers purchase a new car, telematics options look a lot more attractive because they offer the hands-free capability built right into the vehicle. Many experts believe this legislation could fuel broad consumer demand for the quick fix of a hands-free phone, which may be the biggest factor yet in igniting sales of factory-installed telematics units. However, there are several challenges such as speech recognition and text-to-speech, which will need to be supported.

Hands-free does not necessarily translate to risk-free. Further research and a continued search for alternative solutions are required because it is really not enough to say that a phone is hands-free. It also requires the addition of contextual intelligence to voice recognition systems.

Ironically, many of the same individuals who have pressed for car phone legislation are them-selves admitted car phone users. In a recent survey conducted by Microsoft Corporation's MSN Carpoint.com, a Web-based car ownership service, 47% of respondents said they thought operating a handheld cell phone should be illegal while driving. At the same time, in a survey conducted by Forrester Research, 63% of handheld cell phone owners said they had used their cell phones in their vehicles. A similar survey quoted by General Motors put the figure even higher, with as many as 75% of handheld cell phone owners' admitting they had operated their phones while driving.

Love the law or hate it, more legislation that prohibits drivers from using handheld cell phones is inevitable, and will start a cascade of new laws nationwide. Much of the debate about such laws centers on the issue of driver distraction. Automakers and telematics vendors believe the problem has already been largely resolved through the development of voice recognition systems. Voice recognition lets drivers dial the phone, answer it, converse, and hang up—all without ever taking their eyes off the road or their hands off the wheel. Voice recognition therefore eliminates two of the most obvious safety hazards of using the phone while driving.

Voice recognition, however, is not a total answer. Studies have shown that a significant component of driver distraction is cognitive, not physical. Drivers may have their eyes on the road and their hands on the wheel when using a hands-free cell phone, but their mind may not be on driving. In a study conducted while subject were operating a driving simulator at Oak Ridge, researchers placed cognitive loads on drivers by asking them to solve a simple math problem while driving. At the same time, the drivers were given driving instructions. The vast majority of drivers failed to follow the instructions while mulling the math problem. Studies such as this show that hands-off alone is not going to solve the entire problem. The best solution today is to require that drivers pull over and stop while using a cell phone. However, the "pull-over-and-stop" legislation will not be accepted in a society that is already reliant on mobile phones.

One point on which the experts agree is that car phones should be less invasive. Consequently, software makers and telematics companies are spending much of their time developing improvements to today's hands-free systems. Much of their effort is centered on reducing cognitive load. To do that, automakers and telematics manufacturers are creating systems that know when, and when not, to communicate with the driver. Such systems monitor the driver's environmental context in order to understand what is happening around the vehicle. They then use that information to decide whether or not to bother the driver. If you are stopping hard or avoiding an accident, it is probably not a good time to be listening to gas prices over the Internet. But it seems perfectly acceptable to do those kinds of tasks while in normal driving conditions. Automakers and telematics vendors expect hands-free phone operation to trickle down to all vehicle models within a few years. The underlying theme for vendors is a conviction that car phones are here to stay. Cell phones are not just being used by the affluent anymore.

The newer systems will alter that model by enabling consumers to switch brands and update phones as often as they like. It would also eliminate the need for extra phone numbers, contracts, and billing procedures. The rise of such advanced systems is seen in part as a response to passage of a law in New York State and other cities banning the use of handheld cell phones while driving. Automakers and consumers alike are anxious to find new techniques that will let them legally keep their car phones even if more states pass similar legislation. By some estimates, approximately 80% of new-car buyers own portable phones, and three-quarters of phone owners admit to having used their handheld units while driving. Without a doubt, the handheld ban is going to provide a big boost for telematics.

The new technology is seen as an important step forward, not only for automakers but also for portable cell phone manufacturers. Up to now, cell phone makers have maintained a law-abiding public image, recommending that users pull over and stop their cars before dialing. Privately, however, many are surely seriously concerned over the proliferation of cell phone laws that are under consideration in almost every state. Such laws could slow sales of hand-held cell phones because factory-installed units (which are the most logical solution to the problem) would remain in place for the life of a vehicle. In contrast, phone makers hope customers will update their phones every year-or-two. Cell phone market leaders such as Nokia, Motorola, and Ericsson need to ensure that their market is not hurt by this legislation. Ironically, phone suppliers initially resisted the idea of a universal docking station, but are now extremely interested in cooperating with the cell phone cradle manufacturers.

Factory Installed and After-market Cell Phone Cradles

Car manufacturers are scrambling to meet the dictates of pending cell phone laws by means of advanced cradles that offer hands-free dialing, sophisticated voice recognition, and the ability for consumers to dock any brand of cell phone. But some automotive engineers are already wondering aloud whether factory-installed phones are a safer and more secure alternative. The very reason that they cannot be removed from the vehicle makes them unappealing to some users. That, however, can be a benefit during an emergency such as an airbag deployment, because users can be assured the phone will be there. There would also be times when a customer leaves their cell phone on the desk when they leave the office or home, and might not have it when they really need it in the car.

Ford Motor Company was the first automaker to position itself for universal cell phone capabilities when it announced a deal with Cellport Systems Inc., a maker of wireless connectivity hardware for automobiles. The deal, reportedly valued at $48 million, would enable Ford to sell Cellport's universal docking station as an aftermarket device or as a factory-installed option. Several other carmakers have similar programs in the works. Car manufacturers are recognizing that people are bringing their own phones into their cars because consumers want a product that gives them a choice—and that preference is not going to change. Industry experts say the availability of such universal docking systems could trigger a change in the car phone business. In the past, automakers have typically plumbed-in phones for the life of the vehicle, usually locking users into a single cellular carrier. Such methods are currently the norm even for the most advanced telematics service providers, such as GM's OnStar Division.

These products could open up the market for handheld phone makers because it enables users to change both brands and cellular carriers without tearing up their car's electrical system or ripping out their console-mounted phones. A sub-$60 pocket adapter connects the cell phone of choice to the docking station. The popular systems will be compatible with TDMA, CDMA, and GSM communication networks, which are used by the world's biggest cellular carriers. Several automakers are said to be integrating the universal docking station into their factory-installed lineup, as well as offering it as an aftermarket device. This is the only possible way to service a plethora of different personal phones with a hands-free, single-platform system.

The advent of such systems is likely to stir debate among automotive engineers about which technique is better suited to handle the complexities of hands-free operation and voice recognition. Many engineers believe that cell phones can be more effectively developed for in-car use if they are specifically engineered for each car's acoustic characteristics and permanently installed. Because console-mounted phones are not held close to a user's head, FCC regulations allow them to operate at 3 watts, instead of 0.6 watts like handheld phones. Factory-installed units also typically have more processing power and memory, access to external antennas, and the advantage of more efficient microphone placement. It is hard to optimize voice recognition in a portable cell phone for the acoustical cavity of the car. If the phone gets embedded in the car, one can do a better job with the voice-recognition engine. General Motors' OnStar was founded on the idea of using embedded phones, and now has over 1.3 million cars using its telematics services.

Many of the deficiencies of handheld phones may soon be alleviated, however, especially if automakers undertake programs to connect them to factory-installed "plumbing." Such factory-installed units could be pre-engineered for better voice-recognition performance, and could use onboard amplifiers to boost transmitting power. Indeed, several automotive IC component manufacturers already offer a power compensator for their hands-free cell phone kits, thus enabling them to operate at 3 watts instead of 0.6 watts. For those reasons, vendors and automakers say they plan to watch the market and the impending legislation before deciding on built-in vs. mobile. The market will drive what is most appropriate, and manufacturers need to be prepared to address either side of that equation.

The Human Touch

Tomorrow's smart car is likely to give drivers considerable control over their interaction with the telematics interface. The driver will most likely use a voice recognition interface when driving, and a touchscreen interface when stationary. After all, driving by itself makes many exacting demands of the human in the driver's seat. Telematics must aim to provide safety, comfort, and efficiency for humans. However, it is important to answer the following questions:

- What additional sensory detail is required when telematics come into vehicles?
- Will the system turn out to be more hindrance than help?
- Will drivers have enough control over the systems to protect themselves from distracting input?
- If drivers don't have control, who or what will make those decisions and on what basis?

These and other similar questions must be answered before the full potential of telematics can be realized. But to get the answers, manufacturers must first ground themselves in the findings of human factors research, and especially research into how the brain deals with input.

Most such research related to driver and passenger behavior relies on indirect measures to quantify the effects of potential distractions. Most of these are related to the eyes, given the importance of vision to driving. The duration, frequency, and scanning patterns of drivers' glances are fairly common measures. Driver-vehicle performance measures are also useful, such as lane changing frequency, speed maintenance, car-following performance, and driver reaction times. Another category of measures focuses on driver control actions, such as steering wheel inputs, accelerator modulations, gear shifting, brake pedal applications, and hands-off-the-wheel time. These are used to infer the extent to which the driver is distracted during an experiment. Finally, measures of task completion time have been used (or are proposed for use) as an index of the distraction potential of a device. Subjective assessments of driver workload and device design are also studied.

Driver Distraction

Recent research in which IBM participated at the University of Michigan Transportation Research Institute has started to explore the differences in the level of driver distraction based on the type of speech (recorded or synthetic), as well as type of message (navigational messages, short e-mail messages, or longer news stories). Current research into human factors leads us to think of the role of technology in terms of three broad categories: safety, comfort, and efficiency.

Safety is always the most important consideration. The ideal telematics system will implement diagnostic routines to let the consumer know when the car is in a dangerous state and to pull off the road, such as when the engine overheats. When there is an accident, an automatic collision-notification system component should try to analyze accident severity and report the location as part of an automatic distress call. The car's diagnostic information will automatically report the location to a dispatcher when the vehicle breaks down.

Telematics systems will expand the role already played by adding embedded comfort-control systems in many vehicles. The car's interior temperature, seat position, steering wheel position, and mirrors will all be adjusted through the system. The consumer will be able to obtain traffic updates and accident reports, combined with a navigation system that will identify a route around problem areas. The system will also provide the driver with information regarding food and rest information in the area—adding another useful element to the future of telematics.

To successfully address the many roles technology can play as a driver assistant, a framework for the human-computer interactions must first be developed. The potential for information overload from incoming messages or notifications is a major concern facing designers. A driver notification framework must be developed that specifies an organized, systematic way for how the device will communicate with the driver. Such a framework for an IVIS (In-Vehicle Information System) must take into account two distinct types of communication possibilities between the driver and the IVIS: synchronous and asynchronous. In synchronous communication the driver first acts and the IVIS responds with visual or audible feedback. Because the driver initiates synchronous communication, the response from the IVIS is expected. By contrast, the IVIS initiates communication in asynchronous communication and notifies the driver of an event, such as an emergency, a phone call, or new e-mail. The driver does not anticipate this IVIS communication event, and therefore information must be disseminated carefully, maybe even sparingly, to reduce the likelihood of unduly distracting the driver.

Sound Cues

Driver notification in today's conventional cars is limited to sound cues and a few status lights on the dashboard. The driver is notified that something requires immediate attention (such as overheating), but few details are available until the driver takes the car to a service center. In an automobile with a speech-based IVIS, it will be tempting to have the IVIS begin speaking as soon as the system becomes aware of a problem or concern. But how many of these can the driver handle and still drive safely? To answer this question, researchers need to consider how drivers manage their tasks and, specifically, how they respond to speech-based vs. visual-based input.

One approach to studying how the brain processes input focuses on the ways drivers manage their tasks. Initial data suggests that drivers may not prioritize and manage speech-based tasks effectively. For example, studies have shown that a concurrent verbal task may increase a driver's propensity to take risks, and they may not compensate appropriately when verbal interactions slow reaction times. In short, drivers may not fully recognize that speech-based interaction with an IVIS is a distraction to driving, and may therefore fail to compensate. Unfortunately, research on speech communication via cell telephones and on standard concurrent verbal tasks is not helpful because these tasks are fundamentally different from talking to a computer.

Special Demands

The special demands of navigating a complex menu structure may introduce a cognitive load that would compete with the spatial demands of driving in a way that a conversation would not. In addition, current in-vehicle computers cannot yet modulate their interaction with the driver as a function of the immediate driving situation. In fact, interaction with a speech-based system may prove to be less distracting than conversations with a passenger, because it is easier to end a conversation with a machine. Further research is needed to clarify these matters before telematics systems are implemented.

Extensive research in the area of driver task management makes it clear that telematics designs must take into account how the brain processes speech-based vs. visual-based input. But research on the suitability of one method over the other is mixed. One set of answers is based in what is called the "multiple resource" capability of the brain. According to this theory, speech-based interaction will distract a driver from the primary task of manually controlling the car less than a visual display. Speech-based interaction demands resources associated with auditory perception, verbal working memory, and vocal response—the physical act of driving demands resources associated with visual perception, spatial working memory, and manual response. Because these resources are independent, time-sharing should be quite efficient.

However, other points of view are in ready supply. Some researchers, for example, are concerned that speech-based interaction might place attention demands on common central processing resources could undermine time-sharing and compromise driving safety. We will be able to better understand how speech-based interaction might undermine safety if we set aside theories about time sharing, and focus instead on theories that see the brain as a single-channel entity with limited capacity.

A long-term goal of telematics human factors engineers is to establish a workload-management framework. Much effort in this regard is in progress at the University of Michigan. A workload-management component would decide what information to give the driver and when to give it. Decisions would be based on a wide range of variables, including weather, road conditions, driving speed, driver's heart rate, the time of day, and importance of the message. The workload manager would then decide whether to allow a distraction to reach the driver. While ideas about workload management are brought to maturity, IBM is working on an IVIS that alerts the driver to pending information with a subtle tone or by turning on a status light (driver's choice). This lets the driver make a decision about when to listen to the message.

As the trend toward automotive multimedia grows, pressure is slowly building for automakers to apply a more human touch to the design of user interfaces on infotainment, navigation, and telematics systems. Many automakers will revert to tried-and-tested design techniques as they strive to bring those products to market faster. The problem is, however, that such techniques are less likely to be successful as electronic features grow in complexity. Most products available today have features that are buried, difficult to use, and unclear. Often, the people who are designing these systems do not understand the bigger picture. As a result, many radios and navigation systems present a confusing array of buttons that are difficult for consumers to manipulate without frequently consulting the user manual.

Too many companies bring in human-factors experts as an afterthought, leaving the bulk of product design work to software developers and electronics engineers. The engineers and programmers are often the same people who write code for engine and transmission controllers. Too often, the device gets designed and assessed, and then they call in the human-factors department to bless the end product. But human factors should be involved in the conceptual stage of the design, not after the fact. When human-factors experts are called in late in the development process, substantial changes are often impossible. Manufacturing engineers may have already begun tooling up, and deadlines have drawn too close.

Further, engineers and programmers are not always well suited to the task of designing an interface. Engineers and software developers have a very different mindset and a far higher comfort level with complex systems than does the average consumer. By necessity, they are people who are able to keep a great deal of detail in their heads. They are capable of dealing with a lot of complexity, and that makes them fundamentally unsuited to judge whether an interface will work well for the average consumer. Concepts that seem perfectly reasonable to them are not going to work well for consumers. Failing to consider the user interface from the onset could result in systems that compromise the safety of drivers and passengers. While the driver can cope safely with a lot of distractions, all it takes is a split second when they cannot cope and the driver ends up in a major accident.

Changing the Process

A growing number of automakers and vendors are adopting design techniques intended to create simpler, more intuitive user interfaces in order to deal with that potential problem. By employing rapid-prototyping software that automatically generates code for graphical user interfaces (GUIs), some engineering teams are building and debugging prototypes more quickly. As a result, the engineering teams are also bringing human-factors experts into the design process earlier. Hence, software engineers, human-factors engineers, and product designers all work together to devise simulation models. These models allow customers to interact with the system and to see how it works before the design is committed to manufacture. They also allow usability and ergonomics studies early in the development process.

Product design engineers say the solution to the problem is to put the design in the hands of a graphics designer from the outset. Graphics designer typically have experience in art and layout. The next step would be to consult with human-factors experts to find a simple, intuitive design for average users. This is the approach several companies are employing for next generation systems. Cross-functional teams include mechanical and electrical engineers, industrial and graphical designers, and human-factors experts. The team employs rapid-prototyping software to do initial bench-top simulations on PCs. Using a so-called fail-fast approach, designers debug the system before taking it to a full-fledged driving simulator and testing it again in a limited hardware setting.

A key to such efforts is the availability of rapid-prototyping software, which allows graphical designers and human factors experts to get involved earlier in the process. Until recently, programmers were needed to hand-write the code for the GUI. Typically, they created the display on the basis of hand sketches. Human-factors experts then stepped in after the GUI was finished.

Today, software application programs provide an environment in which a non-programmer can build a user interface. The GUI design application will also automatically generate deployable C code for the graphical interface, which can then be used in logic and control software. Such programs may provide the impetus to pry engineering teams away from the time-honored technique of handwriting the graphics code for user interfaces and the logic associated with them. It is estimated that approximately 80% of user interfaces still use handwritten code. However, this is generally because the requirements of embedded systems are so tight (with small memories and low-performance microprocessors) that they require detailed hand-massage to get the code to work. For that reason, some users prefer to separate graphical code-generating capability and logic generation. However, some experts believe that the trend will swing toward using code-generating programs, especially for graphical displays.

Automotive companies say they are using such programs to develop complex user interfaces for radios, audio systems, climate controls, and telematics units. These software simulators let the automakers build behavioral models that they can test in "caves" (computer-automated virtual environments) and in vehicle simulators costing millions of dollars.

Automotive engineers say these development tools enable them not only to improve the quality of their displays but to also cut weeks from development time. The same techniques are being applied in aerospace for cockpits and heads-up displays. They are also used in developing products such as washing machines and refrigerators, which are also starting to incorporate complex user interfaces. In the process, such techniques will enable human factors engineers to make automotive products safer. Human factor experts should be involved in the design from the beginning if a product is going to affect safety. That, however, almost never happens today.

Pushbutton Controls Versus Voice Recognition

Pushbutton Controls

Automakers are uncertain about the near-term future of voice recognition technology, and are therefore also developing alternative interfaces for their in-car PCs. New "infotainment" systems will employ mouse-like knobs and buttons, rather than voice interfaces, as a means of accessing electronic features. But others remain committed to voice recognition as the interface that best supports the conventional "eyes on the road, hands on the wheel" notion of safety, and this alternative solution can only slow the implementation of the in-car PC.

The BMW company is pursuing iDrive, an infotainment system for its high-end 7-Series cars. It is heavy on tactile feel and light on voice technology. The system uses a large knob on the vehicle's center console to enable drivers to gain access to the navigation system, climate controls, telephone, and e-mail. Similarly, Volvo displayed a concept car PC that employs steering-wheel-mounted buttons for input. The Volvo California Monitoring and Concept Center is pursuing the tact of using voice recognition to operate the phone only. Johnson Controls, a major automotive supplier, demonstrated the Harmony Infotainment Generation 2, that company engineers said uses voice recognition only as an "overlay." They believe that voice recognition should be a complement to methods that use pushbutton controls.

This philosophy is now part of a growing debate among automotive engineers, many of whom were staunchly committed to voice recognition only three years ago. Some car manufacturers, such as General Motors and DaimlerChrysler, are still betting heavily on voice recognition; others are inching away from it. Such beliefs fly in the face of the commonly held notion that the only safe auto infotainment systems are those that let users keep "eyes on the road, hands on the wheel." The problem with techniques other than voice recognition, according to this school of thought, is that they require users to divert their attention momentarily from the road. Visteon Corporation and others believe that in-car PCs distract drivers because the radically different cognitive processes of driving cars and reading screens do not go hand-in-hand. They believe that speech recognition is the ultimate solution. Automakers and consumers will support development of in-car computers that use only tiny displays that fit within the opening used by a conventional car radio. Such displays accommodate only a few words of text and no graphics.

In contrast, the new tactile systems from several companies use large liquid crystal color displays. Such systems are far larger than conventional radio displays: BMW's, for example, measures 3.5 inches high and 9 inches wide, and Johnson Controls' is 7 inches across.

The wide screens work in unison with large console-mounted knobs or steering-wheel-mounted buttons. In essence, the mechanical interfaces use operational techniques similar to those of a PC mouse. On BMW's iDrive, for example, users can change functions—from communications to climate control, navigation, or entertainment—by pushing the console knob forward or back, or side-to-side. The user can scroll through menus by twisting the knob, and select functions by clicking a button located in the middle of the knob. Similarly, Volvo engineers support an interface that allows users to change functions by pressing a "three-button mouse" on the steering wheel. The system displays menus on a high-mounted, 5-inch by 8-inch screen.

Designers of the new systems emphasize that their products are created from scratch as automotive infotainment systems, rather than fitted, after market-style, into an existing dashboard opening. More important, the engineers say their systems can be employed intuitively and with a minimum of distraction. They claim that, in most cases, drivers can still concentrate on the road while using the systems' knobs and buttons. Force feedback in the knobs and tactile marks in the handles give each function a distinct feel, so users seldom need to look at the screen.

Makers of the new tactile systems say their designs are based on extensive user research. They claim that the designs represent an effort to reduce the cognitive load on the driver, which looms as an important issue as various states consider legislation to prohibit the use of cell phones by drivers while vehicles are under way. In an effort to prevent consumers from being intimidated by the tactile interface, many OEM and auto manufacturers are using human factors experts and industrial design teams to determine the easiest way to enable these features.

Voice Recognition—A Must for Telematics

The term automotive infotainment conjures up images of buttons, knobs and display panels on the car dashboard. However, if automotive infotainment systems are to reach their full billion-dollar market potential during the next few years, the key to success will be voice-recognition software and not a glitzy instrument panel. Voice recognition will filter out the hum of a tire, the whistle of the wind, and the din of a radio while deftly picking out the subtleties of human speech in order to provide functions such as e-mail retrieval and Internet access. Speech-to-text and text-to-speech will be used as soon as acceptable quality is available.

According to experts, voice recognition is a non-negotiable component of automotive infotainment systems for a simple reason of helping drivers avoid accidents. Automotive experts and legislators fear that a profusion of electronic gadgets in automobiles will distract drivers. Humans can cope with a lot of distractions while driving, but all it takes is one split second when they cannot cope and one has a major accident.

Given such concerns, it is no wonder that industry analysts are predicting a sharp upswing in the use of voice-recognition software in automobiles. They expect voice recognition to serve as an interface for car radios, CD players, navigation systems, and cell phones. Experts also see it as a means to send and retrieve e-mail when used in conjunction with speech-to-text and text-to-speech programs, as well as providing access to the Internet. Onboard speech recognition can also be used to access radios, CD players, and navigation systems.

Much Work Needs to Be Done

The drive to use voice recognition in automobiles is hardly new. General Motors' OnStar division already employs voice recognition in its Virtual Advisor, which is expected to reach an installed base of more than a million vehicles. Clarion's AutoPC also uses it, and a speech-recognition system from Visteon Corporation made a big media splash in Jaguar's S-Type vehicle.

The biggest technical challenge facing voice recognition reliability is noise. Under ideal conditions, makers of such systems say they can be extremely accurate. While 97% accuracy can be constantly delivered under certain dead quiet conditions, a different accuracy level is obtained at 30 miles per hour with the windows open. Next-generation software programs will need to deal with a wide variety of noise sources in order to achieve speech recognition's full potential. Noise from the road, wind, engine, defroster, fan, radio, windshield wipers, and back-seat occupants, in addition to other cabin sounds like doors, windows, and moon roofs opening and closing, all conspire to confound speech recognition systems. Designers will also need to solve less obvious problems, such as line echo, electrical interference, and signal strength. Those problems have slowed implementation and acceptance of some in-car PCs for several manufacturers. The few successful voice-recognition systems employ off-board servers.

Such problems reportedly are far more manageable at 30 mph than they are at 70 mph. In particular, engineers say that road noise, wind noise, and even back-seat occupant (kids) noises become magnified at higher speeds. Some engineers believe that accuracies at high speeds drop to as low as 70%. The fan is a problem year-round in terms of vehicle noise—the real killer seems to be the defroster in winter and the air conditioner in summer.

Engineers from several companies are dealing with noise problems by developing specialized routines for noise and acoustic echo cancellation. They design the speech engines by collecting a great deal of information for noise models, language models, and word models. Data is collected in vehicles in varying conditions such as with the windows up, windows down, on cobblestone roads, on dry roads, on muddy roads, with the fan on, with the fan off, with the radio on, and with the radio off. The results help improve the voice recognition software.

Some software makers, such as Clarity LLC and Conversational Computing Company, say that the solution to the noise dilemma lies in the use of specialized software. Clarity, for example, offers a technology that extracts the voice signal-of-interest. This technology provides an improvement over noise suppression systems, which have difficulty with noise signals that have components overlapping voice signals. Some of these products use time-domain signal-processing techniques to extract the audio signal of interest from a mixture of sounds, including background noise and music.

Some companies have also tried dual microphones and so-called "array" microphones. However, the cost of array microphones is currently outside the acceptable cost limit for such automotive applications.

Divergent Paths

Ultimately, voice recognition's automotive success is not so much a matter of if, but when. As 'speech engines' improve, most believe that voice recognition will take on a bigger role in automobiles. The consumer is becoming more aware that pushing a button, twisting a knob, or flipping a switch is less required as speech technology becomes more natural-language-oriented and consumers grow more comfortable with it.

Many proponents of voice recognition believe that speech technology will ultimately win out. Accuracy is not as big a problem as voice recognition's detractors have made it out to be. Most voice recognition makers claim that their existing systems achieve an accuracy of higher than 90%. They claim that voice recognition systems balk primarily because consumers are intimidated by the concept, not because the systems are inaccurate. Proponents believe that the automotive industry will move away from the transition phase of knobs and buttons once consumers become comfortable with using speech. Developers of tactile (knobs and buttons) systems say, however, that if automakers do lean toward tactile techniques, pushbutton systems will not go away anytime soon. This is because the systems must be integrated into the dashboard very early in the design process. Unlike voice-based systems, they cannot be placed in existing console or dashboard radio slots. As a result, some tactile systems being explored now will debut four or five model years down the road. And since not every car model is fully redesigned every year, the tactile systems are likely to be in use for some time after that. Hence, tactile techniques will not go away completely in the near future, even though speech recognition is, without a doubt, in the future of telematics.

Standardization of Electronics in Automobiles

Innovations in automotive electronics have become increasingly sophisticated. As a result, high-end cars now contain upwards of 80 microprocessors for engine control, cruise control, traction control, intelligent wipers, anti-lock brakes, collision avoidance systems, car navigation, and emergency service calls, among other functions. Traditional automotive applications of automotive electronics have been focused on under-the-hood applications, including the power train, body, safety, and chassis. However, the automobile sector is now focusing on telematics in order to cover a growing number of in-car devices that will be linked to, and interoperable with, the external world. These applications are primarily focused on bringing information and entertainment services to drivers and passengers. They are made possible by advances in embedded microprocessor technology, both in hardware and software, and by standards adopted by the automotive industry. Creating such devices, however, can present tremendous challenges for embedded developers. Some of the challenged they face include ensuring that devices can be connected through multiple networks, coping with software complexity, and designing systems that can be upgraded over time.

Typical vehicle functions can be separated into individual networks such as power train, body electronics, and entertainment. While the first two are considered safety-critical applications, non-safety applications are communicated on the entertainment network. The first two parts of the network include sensors, controls, displays, door locks, engine data, and drive train data. Manufacturers are scrambling to incorporate the latest in consumer electronic devices—from cellular phones to mobile DVD. The challenge in the multimedia area is to combine consumer technology with automotive knowledge in order to make truly automotive parts for these systems. This also includes incorporating audio systems such as tuner, satellite radio, preamplifier/amplifier, in-dash CD player, speakers, anti-theft equipment, emergency call system, etc.

The communication and multimedia bus system is one of the hottest topics in car development. It would provide a common bus for connecting and controlling all environmental control and telematic/infomatic devices. The array of devices that would be connected in a typical car environment include a cell phone, radio, sound system, climate control, navigation equipment, and, at a later date, a digital satellite radio receiver. Putting all these into separate boxes that have point-to-point connections is a waste of resources. In the perfect environment, the heart of the system is a master controller where most of the intelligence is concentrated. This is similar to the PC, where you connect the different peripherals to the host. Migration from consumer devices to the onboard network must be carefully controlled. It is one thing for a consumer connection to fail at home, but quite another if the rogue device brings down the automotive network while the vehicle is traveling at highway speeds. The scenario where Junior buys a device at a discount store, drops it a few times on the way to the car, and finally manages to plug it into the onboard network just as Mom pulls onto the freeway is not far-fetched. If this device disrupts the network, causing, for example, the stereo system to suddenly screech loudly, the consequences could be disastrous. Many carmakers are reluctant to allow retail products to be connected to their onboard systems. They prefer giving access only to those devices they certify as safe, and that meet the stringent environmental requirements of the automotive world.

Another issue is the design-time and life span differences between consumer products and the automobile of today. The typical new consumer product design-time might be nine-to-twelve months, with a life span of six-to-nine months. However, the product design-time for a car could be over three years, with a product life span of another three years. The long design-in time for cars results in buses that tend to be used for many years. Car systems will evolve on their own cycles without being tied to consumer product cycles. They will undergo the stringent testing periods required to ensure safety and long life. Carmakers will continue to control the standards they implement.

Regardless of the challenges, consumer convenience is important. Therefore, gateways will become more common over time in order to provide a wide variety of consumer interfaces and to insulate the critical onboard systems from external interference. Gateways will provided a method to accommodate the current consumer need. Analog audio and video signals will be digitized and reproduced on integrated audio and video equipment. Other digital interfaces will be accommodated as they are developed. Gateways will give carmakers the opportunity to adapt their products to specific customer segments, while allowing them to build a robust, mission-critical infrastructure across their product lines. Manufacturers are similarly challenged to introduce OEM versions of these devices in a timely fashion. Keeping up with the marketplace, however, has traditionally been difficult since new electronics products are introduced at a faster pace than automotive. Automakers have typically integrated devices that were developed especially for their proprietary OEM networks, thus slowing the process and increasing costs. Several approaches have emerged to address the OEMs' need to incorporate entertainment devices in vehicles, but they differ in both technology and assumption.

For telematics to succeed, proponents know they must first promote openness and encourage standardization. Without it, the technologies face an almost-impossible uphill battle toward acceptance. In contrast, standardization simplifies the task of automakers and vendors alike. By agreeing on a standardized design, automakers can drop-in new electronic components at the eleventh hour of the design process, and vendors need not redesign their products for each new implementation. As the auto industry defines its customer requirements, technology providers can start to fulfill its needs. By relying on open industry standards, all key players—from manufacturers to service centers and retailers—can focus on delivering core expertise to the end user. This would certainly be better than expending the time and effort it would take to develop separate, incompatible designs for specific vehicles or proprietary computing platforms. Future systems will be highly integrated, open, and configurable.

The next two sections of this chapter detail some of the popular standards for power train, body electronics, and multimedia buses.

Standards for In-vehicle Power Train and Body Electronics

Like most designers of electronics-based devices, those engineering today's automobiles search for higher speeds. But they must also ensure the highest reliability while trying to remain at the cutting edge. One of the interesting trends of late is the use of "portable" technology in the automobile. One obvious reason for this is the number of microcontrollers that now reside within the car. With so many devices on the bus, it is imperative that they consume as little power as possible. Hence, the need for low-power (portable) techniques has increased.

In North American autos, there are mainly two types of networks in a car. First is a power-train network that connects the various pieces of the power train, and sometimes the anti-locking braking system (ABS). Here, the high-speed CAN bus is most commonly employed. The electronics for the power-train systems are in large part controlled by the IC vendors, and there is limited differentiation among systems. In the body electronics, however, it is a different story. This is where the car manufacturers can differentiate themselves. Cars made by different manufacturers such as DaimlerChrysler or BMW, for example, show their own particular style for the climate control, telephone interface, or instrument cluster. This is one method of branding for automakers. Unlike the power train network, automakers are very much in control of the body buses and electronics.

The automotive industry has been working to develop industry standards for in-vehicle networks that will ultimately simplify incorporating electronics. Networks allow easy connection of electronic devices, while reducing cost, weight, and complexity, as compared with point-to-point wiring. Networked devices simplify information sharing between the devices.

Following are some of the different power train and body electronic standards used in the worldwide automotive market.

The Society of Automotive Engineers (SAE) J-1850 Bus

In the search for a less costly bus for simple on-off applications, automakers in recent years have turned to the SAE J-1850 bus. This bus has been commonly adopted in the USA as a mid-speed bus standard aimed at non-real-time communications and diagnostics, such as body electronics. This includes components like power seats and dashboard electronics. The J-1850 bus has been around for about 15 years, and has been implemented in General Motors, Ford, and Chrysler vehicles. The J-1850, however, has failed to become an industry-wide standard, and most automakers have created their own proprietary versions of it. Toyota, for example, uses the Body Electronic Area Network bus. Ford, Chrysler, and GM are said to each have their own flavors of J-1850 as well. The result is that suppliers must rewrite drivers and redesign interfaces for each proprietary version of J-1850.

All J1850 networks share a common message format, which was developed as a standard by the Society of Automotive Engineers. But that is where the commonality stops, as the manufacturers employ different development tools, test tools, and protocols—it is even possible for different models from the same manufacturer to exhibit differences in protocol and message lists.

The Philips Lite Automotive NETwork (PLANET)

Philips Semiconductors is a supplier of dedicated automotive systems and components focused on the core areas of IVN (in-vehicle networks), vehicle access and immobilization, infotainment, and telematics. The company has developed an adaptive passenger restraint system called PLANET. It ensures that the car will react in a collision to protect the driver and passengers by causing 'smart' deployment of airbags. This serial communications network is highly fault tolerant, meaning that electronic messages will still get through even in the event of severe damage to the network.

This system combines reliable performance and robustness while remaining extremely cost-effective. By making such systems more affordable, airbag system manufacturers can meet public demands for increased car safety in all car models. The United States Council for Automotive Research (USCAR) is still evaluating PLANET as a possible worldwide standard for all safety-critical systems.

The X-by-Wire Project

The X-by-Wire effort is a European project that defines a framework for introducing vehicle safety related, fault-tolerant electronic systems that do not have mechanical backup. The "X" in "X-by-Wire" is a 'wild card' that represents any safety-related application such as steering, braking, power train, suspension control, or multi-airbag systems. These applications will greatly increase overall vehicle safety by liberating the driver from routine tasks and assisting the driver to find solutions in critical situations.

Traditional steering and braking system components, such as the steering column, intermediate shaft, pump, hoses, fluids, belts, and brake power booster/master cylinder, are completely eliminated. The X-by-Wire systems are fully programmable and provide the maximum opportunity for optimizing ride and handling. They also provide active safety performance, including vehicle stability control, rollover prevention assistance, and collision avoidance.

The European consortium developing the X-by-Wire telematic standard includes DaimlerChrysler AG (the project coordinator), Centro Ricerche Fiat, Ford, Magneti Marelli, Mecel, Robert Bosch GmbH, Vienna University of Technology, Chalmers University of Technology, and Volvo.

The successful development of a common European X-by-Wire framework, accompanied by broad and fast dissemination, would translate into a significant strategic advantage for European automotive, supplier, and semiconductor industries. It would also have the potential to become a European or even a worldwide standard. The success of this project will put the European Industry in a pole position in an important emerging high technology market, with the vehicle customer gaining a direct benefit. Safe, intelligent driver assistance systems based on X-by-Wire solutions will make affordable safety available to everybody. Gaining the technological leadership will also bring a number of benefits to other industry sectors.

Car (or Controller) Area Networks (CAN)

The Car (or Controller) Area Networks (CAN) bus is a simple two-wire differential serial bus system that operates in noisy electrical environments with a high level of data integrity. It has an open architecture and user-definable transmission medium that make it extremely flexible. The multi-master CAN bus is capable of high-speed (1 Mb/s) data transmission over short distances (40m), and low-speed (5 Kb/s) transmissions at distances of up to 10,000 meters. It is highly fault tolerant, with powerful error detection and handling designed-in. It is typically operated at half speed (500 Kb/s) in order to control electromagnetic radio emissions. In some instances, where speed is not necessary, it can run at even slower speeds of around 125 Kb/s.

The German company Bosch developed the CAN specifically for the automotive market in the mid-1980s, which remains its primary application area today. In fact, the CAN is currently the most widely used vehicle network with more than 100 million CAN nodes installed. Many automakers now use CAN buses to enable safety-critical systems, such as antilock brakes, engine controllers, road-sensing suspensions and traction controllers, to talk to one another. A typical vehicle can contain two or three separate CANs operating at different transmission rates. A low speed CAN, running at less than 125 Kb/s, can be used for managing body control electronics such as seat and window movement controls and other simple user interfaces. Low speed CANs have an energy saving sleep mode in which nodes stop their oscillators until a CAN message wakes them. Sleep mode prevents the battery from running down when the ignition is turned off. A higher speed (up to 1 Mb/s) CAN controls the more critical real-time functions such as engine management, antilock brakes, and cruise control. Controller Area Network protocols are becoming standard for under-the-hood connectivity in cars, trucks, and off-road vehicles. One of the outstanding features of the CAN protocol is its high transmission reliability, which makes it well suited for this type of application. The CAN physical layer is required for connections to under-the-hood devices and in-car devices.

The CAN bus system has also found its way into general industrial control applications such as building management systems and lift controllers. Surprisingly, it is now also being considered for use in telecommunication applications such as Base Transceiver Stations (BTS) and Mobile Switching Centers (MSC) equipment. This is due to CANs protocol error management, fault isolation, and fault tolerance capabilities.

The CAN specification gives full technical details for both CAN 2.0A and CAN 2.0B protocols. Part A describes the message format as identified in CAN specification 1.2; part B describes both the standard (11-bit) and extended (29-bit) message formats. All implementations must fully meet the specifications in either part A or B in order to be compatible with this revision of the CAN specification. High-speed CAN is already the standard in Europe for automotive power train applications, and is now going global. It has also gained popularity in the United States. The requirement for reliable, fault-tolerant networks increases as the number of powered systems in cars increases. A fault-tolerant version of CAN is also used in the body area.

The CAN system, fault-tolerant by definition, is a two-wire bus. With fault tolerance built-in, the network switches to a one-wire mode and continues to operate if one of those two wires is shorted to ground or is open. The specification requires that the transition from two wires to one happen without losing an information bit. For this reason, the physical layer chip for the body area is more complicated than the physical chip power-train device, even though it runs at a much lower speed.

But when entering the fault condition, timing becomes unpredictable. Using CAN, if a message transmission does not go through, the initial sender repeats the message. The sender knows that a fault has occurred because it has not received an acknowledgement of receipt from the receiver. This results in unpredictable timing, but only under the fault condition. A node that is broken and continually transmits messages, thereby monopolizing the bus, is called a "babbling idiot."

Even though some parts of the car have a "fail-safe" condition, others do not. For example, the air bags continually run diagnostics. If a fault is detected, a warning lamp illuminates, asking the consumer to get it fixed. Under this condition the air bag will not deploy, but it will also not stop the car from being operated. The same case holds true for the ABS. The ABS will not function if a fault is detected, but the fault will not stop the vehicle's brakes from operating properly in their normal (non-ABS) mode. However, steering is an example of a component that requires true fault tolerance—some components have no fail-safe condition. Steering would be a time-triggered system that would stay predictable even under fault conditions. For example, a separate bus-guardian chip is used to prevent the babbling-idiot problem. The chip opens the window for a node to put anything onto the bus only during the time slots allocated. This network is called byte-flight, and is an optical network for safety systems. The protocol handles connections between essential modules, such as the air bags. Under normal conditions, the bus operates in a low-speed mode, around 10 Kb/s, to minimize emissions. But during a crash, it switches to a high-speed mode, because one is no longer concerned about emissions. This is called the 'distributed systems' interface.

The Vehicle Area Network (VAN) Bus

Another architecture available to automotive designers is the VAN bus, developed in France. It is a modern communication protocol similar to CAN, as both are multi-master architectures. However, the VAN has a different bit-coding scheme than CAN. The VAN is especially recommended for very low cost applications. It became an ISO standard in 1994, and is now implemented in mass production on vehicles. The VAN protocol was mainly employed by French automakers. Today, it is used in some body applications. However, even the French have converged around the CAN bus for the power train.

The NEC CAN

The Nippon Electric Company (NEC) developed the NEC CAN as an alternative to the European CAN. The NEC CAN system incorporates a proprietary hardware-based CAN bridge in its latest 32-bit microcontrollers that lets the microcontroller operate as a gateway for communication between various networks within the car. This function allows the microcontrollers to manage multiple tasks simultaneously for up to five CAN networks, such as the power train, the navigation system, or the climate control system. By implementing the bridge in hardware, the controller can relieve the host CPU of some of its routine tasks. Unlike traditional CAN-based products, NEC devices offer "time-triggered" functionality that allows signals to be delivered based on time slots rather than the arbitration of external events. For added safety, controllers can perform "time-out monitoring" to anticipate the receipt of critical messages within a defined time frame. The bridges contain three components; a programmable state machine, a memory area to store commands to be executed on the state machine, and temporary buffers. The state machine lets automakers determine the number of memory resources assigned to each CAN interface, based on their own requirements.

The Local Interconnect Network (LIN)

Another bus that is increasingly gaining popularity is the Local Interconnect Network, or LIN. The LIN is a new low cost serial communication system designed for distributed electronic systems in vehicles. It also complements the existing portfolio of automotive multiplex networks. This network was developed to supplement CAN in applications where cost is critical and data transfer requirements are low.

Developed in 1998, the LIN consortium defines the LIN bus specification and provides strong support to the proliferation of the standard. The founding partners of the LIN consortium include car manufacturers Audi AG, BMW AG, DaimlerChrysler AG, Volkswagen AG, and Volvo Car Corporation AB. Tool vendor Volcano Communications Technologies and semiconductor manufacturer Motorola are also founding partners. The consortium was developed to standardize a serial low-cost communication concept in conjunction with a development environment that enables car manufacturers and their suppliers to create, implement, and handle complex hierarchical multiplex systems in a very cost competitive way.

From a performance standpoint, the LIN it is slower than the CAN bus and is intended for sub-buses in local areas, such as the doors. In actual application, the CAN bus would go to the door, where there would be one CAN node. From that node, there would be LIN sub-buses going to windows, mirrors, and door locks. The deployment of air bags is handled in a similar manner. The CAN node provides the main interface, from which the LIN sub-buses then go directly to the sensors. Segmenting out to something like the LIN is mostly driven by cost. A LIN node does not cost as much as a CAN node—each LIN node costs approximately $1 less per node than a CAN bus. Also, the CAN buses are nearing their limit in terms of the number of nodes they can support. The LIN is a holistic communication concept for local-interconnect networks in vehicles. Its specification defines interfaces for development tools and application software, in addition to protocol and the physical layer.

The LIN enables a cost-effective communication for smart sensors and actuators where the bandwidth and versatility of CAN is not required. Communication is based on the SCI (UART) data format, a single-master/multiple-slave concept, a single-wire 12V bus, and clock synchronization for nodes without stabilized a time base.

An automotive low-end, multiplex communication standard had not been established until today. The LIN consortium was conceived with the goal to standardize a low cost serial communication concept within the development environment. The goal of this communication concept would then be to provide a tool that enables car manufacturers and their suppliers to create, implement, and handle complex, hierarchical, multiplex systems in a very cost competitive way. The LIN standard will reduce the manifold of existing low-end SCI-based multiplex solutions, and will cut the cost of development, production, service, and logistics in vehicle electronics.

The LIN specification covers transmission protocol, transmission medium, interfaces for development tools, and application software. It guarantees the interoperability of network nodes from the viewpoint of hardware and software, and assures a predictable EMC (electromagnetic compatibility) behavior. This concept allows the implementation of a seamless development and design tool chain, and enhances the speed of development and reliability of the network.

Key Features of The LIN Standard

The LIN is a low-cost, single-wire serial communications protocol based on the common SCI (UART) byte-word interface. Universal Asynchronous Receiver/Transmitter interfaces are available as a low cost silicon module on almost all microcontrollers. They can also be implemented in a software equivalent or pure state machine for ASICs. Each LIN node costs approximately $1 less per node than a CAN bus. Communication access is controlled in a LIN network by a master node, eliminating the need for arbitration or collision management in slave nodes. This provides a guarantee of worst-case latency times for signal transmission and predictable systems.

Each LIN bus can consist of a maximum of 16 nodes—one master and 15 slaves. A crystal/quartz or ceramics resonator clock source is not needed because the bus incorporates a self-synchronization mechanism that allows clock recovery in the slave nodes, allowing the microcontroller circuit board to be both less complex and less costly. Power for the single-wire bus comes from the car's 12-volt battery.

There is a significant cost reduction of hardware the platform. The specification of the line driver and receiver is following the ISO 9141 (VBAT-based) single-wire standard with some enhancements. The maximum transmission speed is 20 Kb/s in order to meet electromagnetic compatibility (EMC) requirements and maintain clock synchronization. This is, however, an acceptable speed for many applications.

A LIN networks node does not use any system configuration information except for the denomination of the master node. Nodes can also be added to the LIN network without requiring hardware or software changes in other slave nodes. The size of a LIN network is typically under 12 nodes (though not restricted to this), resulting from the small number of 64 identifier and the relatively low transmission speed. Major factors affecting the cost efficiency of LIN networks include clock synchronization, the simplicity of UART communication, and the single-wire medium.

A LIN network comprises one master node and one or more slave nodes. All nodes include a slave communication task that is split into a transmit and receive task, while the master node includes an additional master transmit task. Communication in an active LIN network is always initiated by the master task the master transmits a message header that contains a synchronization break, synchronization byte, and message identifier.

Exactly one slave task is activated upon reception and filtering of the identifier. This then triggers transmission of the message response. The response comprises two, four, or eight data bytes and one checksum byte. The header and the response data bytes form one message frame.

The message identifier denotes the content of a message but not the destination. This communication concept enables the exchange of data in various ways: from the master node (using its slave task) to one or more slave nodes, and from one slave node to the master node and/or other slave nodes. It is possible to communicate signals directly from slave-to-slave without the need for routing through the master node, or to broadcast messages from the master to all nodes in a network. The sequence of message frames is controlled by the master and may form cycles including branches.

Target Applications for LIN

The LIN is used for distributed body control electronic systems in automobiles. It enables effective communication for smart sensors and actuators where the bandwidth and versatility of CAN is not required. Typical applications for the LIN bus are assembly units and simple on-off devices. These devices include door control systems (window lift, door locks, and mirror control), steering wheel, car seats, climate regulation, window controls, lighting, sun roofs, rain sensors, alternator, HVAC flaps, cruise control, windshield wipers, mirrors, as well as controls for a host of other simple applications.

The cost sensitive nature of LIN enables the introduction of mechatronic elements in these, units such as smart sensors, actuators, or illumination. They can be easily connected to the car network and become accessible to all types of diagnostics and services. Commonly used analog signal coding will be replaced by digital signals, leading to an optimized wiring harness. It is expected that LIN will create a strong market momentum in Europe, the U.S.A., and Japan.

Proponents of the new LIN bus said that it could accomplish many of the same tasks as J-1850 standard, but at lower cost. Operating at speeds of up to 20 Kb/s, LIN offers sufficient performance to send signals back and forth between, for example, the steering wheel and cruise control system or radio. It does not, however, offer the kind of bandwidth needed for multimedia applications such as cell phones, navigation systems, and CD players. Most automakers now plan to use a CAN-based bus known as IDB-C for high-bandwidth applications, since the IDB-C bus operates at speeds of up to 250 Kb/s.

The LIN's first applications appeared in the 2001 model year. Most of LIN-based applications incorporate from three to 10 LIN nodes, probably in door modules. Ultimately, however, LIN consortium members expect the number of nodes to grow to about 20 per vehicle within five to ten years, reaching a worldwide volume of approximately 1.2 billion LIN nodes per year by the end of the decade.

The LIN is also used as a sub-bus for CAN for machine control outside the automotive sector. In this case, the LIN is a UART-based, single-master, multiple-slave networking architecture originally developed for automotive sensor and actuator networking applications. The LIN master node connects the LIN network with higher-level networks, like CAN, extending the benefits of networking all the way to the individual sensors and actuators.

Fault-Tolerant Systems

Over the next five-to-ten years, traditionally mechanical vehicle systems, such as braking and steering, will become electrical systems connected to a high-performance CPU via a high-speed, fault-tolerant communications bus. Features such as brake-by-wire, steer-by-wire, and electronic valve control will configure the ultimate driving experience through an integrated driver-assisted vehicle. A high-speed, fault-tolerant, time-triggered protocol is required to provide the communication between these systems, especially where safety is concerned.

The same factors that drove fly-by-wire in the aerospace industry are now powering drive-by-wire technology—the substitution of mechanical and hydraulic links by purely electronic systems—in the automotive industry. Electrical systems allow reduction of overall cost during the complete life cycle of a car. The use of software solutions significantly reduces development and production costs. Meanwhile, an appropriate electronic design helps to solve the expensive differentiation of product variants. Less space and weight lead to less fuel consumption. And electronic systems provide new concepts for remote diagnosis and maintenance. Finally, electronic systems enable new technologies like fuel cells and intelligent power-train control systems. It is estimated that electronic parts will comprise at least 25% of the total cost of a passenger car by 2006. Electronic parts will comprise more than 35% of the total cost for high-end models.

There are several safety arguments for in-car networks and electronics. Mechanical components, such as the steering column, represent a remarkable risk to the driver. The replacement of such parts reduces injuries in case of an accident. Electronic control systems enable intelligent driver assistance in dangerous situations, such as collision avoidance and, if inevitable, controlled collision with minimal harm to the passengers. Integrated vehicle control systems can integrate different subsystems into a single, harmonized driver support system and promote active safety.

By-wire systems enable many new driver interface and performance enhancements such as stability enhancement and corrections for cross wind. Overall, by-wire systems offer wide flexibility in the tuning of vehicle handling via software. In a by-wire car, highly reliable and fault-tolerant communication networks control all components for braking and steering. Moreover, steer-by-wire provides the opportunity for significant passive safety benefits; the lack of steering column makes it possible to design better energy-absorbing structures. Finally, an important potential benefit of by-wire subsystems is active safety; a capability only fully realized when integrated into systems. Integrated by-wire systems referred to as drive-by-wire or X-by-wire, permit the implementation of a full range of automated driving aids, from adaptive cruise control to collision avoidance. While by-wire technologies promise many benefits, they must be carefully analyzed and verified for safety because they are new and complex.

As the size and complexity of these new networks increase, ease of integration has become a major challenge for design engineers. Additionally, new in-system testing, software upgrade service, and diagnosis capabilities offer opportunities for aftermarket business. Safety is intimately connected to the notion of risk, and is popularly conceived as a relatively high degree of freedom from harm. Risk is a combination of the likelihood and the severity of an unplanned, undesirable incident. A system is generally considered to be safe if the level of risk is reasonable. This must be evaluated according to societal, legal, and corporate concerns.

This paradigm shift in automotive electronics to drive-by-wire has led to fault-tolerant communication architectures, which must fulfill hard real-time requirements. Future control systems will need bandwidth of up to 10 Mb/s. They require a deterministic data transmission with guaranteed latency and minimal jitter. In addition, they must support scalable redundancy, error detection and reporting, fault tolerance, a global time base, and support of such different physical layers as fiber optics and wire. Time Triggered Architecture (TTA) and FlexRay are the main standards for automotive fault tolerance.

Time-Triggered Protocol (TTP)

In the field of real-time control systems there are two fundamentally different principles of how to control the activity of a computer system—time-triggered control and event-triggered control. In a time-triggered system, all activities are carried out at specific times that are known *a priori*. Accordingly, all nodes in time-triggered systems have a common notion of time based on approximately synchronized clocks. In contrast, all activities are carried out in response to relevant events external to the system in event-triggered systems.

Time-triggered technology has been developed over the past fifteen years at Vienna University of Technology. It was refined in co-operation with leading industrial pacesetters in European Research Projects, such as X-by-Wire and TTA (Time-triggered Architectures). The centerpiece of this technology is the innovative Time-triggered Protocol (TTP) that integrates all services needed to design composable fault-tolerant real-time systems with minimal effort.

Some of the important advantages of a time-triggered architecture are:

- Composability – The various components of a software system can be developed independently and integrated at a late stage of software development. The smooth and painless integration of components helps to manage the ever-increasing complexity of embedded real-time systems.

- Predictable temporal behavior – System behavior is based on a periodic pattern and is controlled by the progression of time; this makes it possible to predict behavior of a temporal system. This approach simplifies system validation and verification.

- Diagnostic ability and testing – The interface between components is precisely specified in both the value and temporal domain, thus permitting each component to be tested in isolation. Moreover, testing can be carried out without the probe effect.

- Reusability of components – The "glue logic" (or glue ware) needed to interconnect existing components into a new system is physically separate from the component application software in a time-triggered distributed architecture. Existing components may therefore be reused in a new context without any modification of the tested and proven application software.

- Fault-tolerance – The replica deterministic behavior of time-triggered architectures supports the implementation of fault-tolerant systems by active redundancy. This facilitates the construction of highly dependable systems.

Increasing demand for safer and more efficient cars has led to a digital revolution in automotive designs. Electric brakes will replace hydraulics, thereby increasing safety, lowering operating cost, and eliminating the use of environmentally hazardous fluids. Steering columns will disappear and will be replaced by electric steering, improving driver safety and making it easy to provide either left-hand and right-hand drive models. Meanwhile, power-train control systems will become more tightly coupled with vehicle stability and ride comfort systems. Integrated vehicle control systems will combine such elements as active braking, steering, and suspension control. The "by-wire" systems raise new challenges for car manufacturers and suppliers—they mandate real-time and fault-tolerant communication systems as defined by TTP.

This technology supports the partitioning of a large system into a set of nearly autonomous subsystems that have small and testable interfaces. It is therefore possible to build large and complex systems by integrating independently developed components with minimal integration effort.

Recent advances in dependable embedded-system technology, as well as continuing demand for improved handling and passive and active safety improvements, have led vehicle manufacturers and suppliers to work to develop computer-controlled, by-wire subsystems with no mechanical link to the driver. These include steer-by-wire and brake-by-wire, and are composed of mechanically decoupled sets of actuators and controllers connected through multiplexed in-vehicle computer networks. Time-triggered architectures are being used as the basis for by-wire systems. And, although there has been considerable reinterpretation of the protocol requirements, the industry believes that the overall strategy is correct. Also, various fault-tolerance strategies are possible, leading to many possible configurations even with the "standard" architecture. Additional fault-tolerant strategies can also be layered on top of the basic strategy.

Basic time-triggered architecture includes multiple controller modules or nodes and communication via a TTP over multiple channels (for fault tolerance). The TTP approach is backed by German carmakers Audi AG and Volkswagen and other key manufacturers such as NEC, Honeywell, Delphi

Automotive Systems, PSA, Renault, TTChip, Visteon ARM Holdings, and OKI Semiconductor. The push for TTP focuses on safety-critical systems, and is led by the TTA Group (a governing body for time-triggered protocol). Time-triggered Protocol is designed for real-time distributed systems that are hard and fault-tolerant, and ensures that there is no single point of failure. Aerospace companies have also adopted this protocol due to its rigorous and plain safety approach. Messages are primarily state-oriented (as opposed to event-oriented)—each persists until it is changed, and can only change at defined intervals. The network appears as a global memory to nodes. All nodes are internally self-testing and drop out of participation in the network when value or timing errors are detected.

Communication errors (timing and value) are detected by the network, rather than by application programs. The primary fault-tolerance strategy is replication of fail-silent components: When one task or node fails, another is already running and the system employs the alternative. Many variations on this strategy are possible, and the goal typically is to take advantage of natural redundancy when possible. Signal definition, timing, and scheduling are done off line, deemed correct before the system is assembled. This makes it possible to easily compose overall systems from independently designed and tested subsystems.

These TTP characteristics translate into five key features:

- Predictability – Computation latency is predictable because of deterministic time-triggered scheduling.
- Testability – Time-triggered protocol and arrival time scheduling and computations are automatically checked.
- Integration – Systems can easily be composed from independently designed and tested components or subsystems.
- Replica determinism – Replicated component behavior is consistent other components of the same type; each is doing the same thing at the same time, or at some predetermined offset time.
- Membership – Fault status is automatically broadcast in a time-triggered architecture by virtue of "membership" on the communications network.

The first four features primarily support design correctness and complexity concerns; the fifth supports fault-tolerant operation, a continuous safe state, and multi-layer diagnostics. The goal of TTP is to provide fault tolerance without excessive complexity. Currently, TTP is the main protocol supporting time-triggered architectures, but others are possible and the industry must reach a consensus to get beyond the development stage.

Time-triggered architecture includes fault-tolerant units with redundant network nodes, fail-silent nodes that cease communicating when an error is detected, and dual communication paths for network fault tolerance. An essential aspect of time-triggered architecture design is fail-silent components. Fail-silence depends on the self-testing quality or coverage within the components, which applies to all aspects of the system: Software, controller hardware, sensors, and actuators. It is assumed that additional testing and diagnosis at the subsystem and vehicle levels may result in actions other than fail-silent. Software components can be made self-testing by a variety of techniques. These include acceptance or "sanity" checks on computed results, diverse, replicated components, and redundant computation with diverse data.

Manufacturers have developed a bootstrap approach for controller hardware, beginning with complete self-testing of the core (the CPU), memory, and main data channels into and out of the CPU. Other on-chip functions are self-testing or can be easily tested by the CPU since it can be trusted. Data memory has error-detecting codes. Special data monitors test the program memory and buses. This self-testing strategy offers some additional features and benefits. First, it is software transparent, which means that special-purpose or CPU-specific functions do not have to be added to test the core. Therefore, software can be easily maintained and ported to other CPUs. Second, this approach generally has low overhead. The CPU accounts for only a small part of the chip area for most microcontrollers as compared to memory and other on-chip circuits. The memory space required for additional tests is typically more expensive. Third, the technique does not require complex self-testing strategies at the controller module or board level. Manufacturers have generally had good experience with this approach. Several million units are currently used in electric power steering and antilock brake system controllers. It does catch real errors and does not appear to create new problems such as timing glitches.

Finally, anomalies must be detected and actions taken at the subsystem and vehicle behavior levels. In other words, it is necessary to detect inappropriate behavior regardless of the source. Anomalies may arise from complex interactions of components or incomplete specifications rather than faults. Promising research techniques may lead to systematic solutions to developing diagnostics, a strategy that is commonly called model-based diagnostics. Abstract models of the control system behavior are executed on line and compared to actual system behavior. However, specialized models with appropriate features and real-time performance must still be developed. A variant called TTP/C refers to the SAE's Class-C classification for fault tolerant, high-speed networks; TTP/A is a low-cost Class A derivative for smart sensor and actuator networks.

The basic functional principle of TTP is the time-division multiple access bus. An *a priori* defined time schedule controls all activities of this communication system, and all members of the communication system know their assigned sending slots. A distributed algorithm establishes the global time base with steady clock synchronization. To prevent the so-called "babbling idiot" problem and slightly-out-of-specification failures, TTP/C provides an independent bus guardian that guarantees exclusive access to the bus. Precisely defined interfaces in the value and time domains ensure the integration ease of TTP-based systems. Time-triggered Protocol supports soft real-time or event-triggered data traffic such as remote access and advanced diagnosis strategies.

Time-triggered Protocol is the mature network solution that is low cost and can handle safety-critical applications. Silicon support for TTP has been available since 1998. Second generation silicon, supporting communication speeds of up to 25 Mb/s, is available today.

The Time-triggered CAN (TTCAN)

Communication in the classic CAN network is event triggered; peak loads may occur when the transmission of several messages is requested at the same time. The CAN's non-destructive arbitration mechanism guarantees the sequential transmission of all messages according to their identifier priority. Real-time systems, however, require that a scheduling analysis of the whole system to be done to ensure that all transmission deadlines are met even at peak bus loads. The TTCAN was conceived to overcome this potential issue.

The TTCAN is an extension of the CAN protocol and has a session layer on top of the existing data link and physical layers. The protocol implements a hybrid, time triggered, TDMA schedule, which also accommodates event-triggered communications. The TTCAN is intended for use with engine management systems, transmission controls, and chassis controls with scope for x-by-wire applications.

The FlexRay

The FlexRay is a new network communication system designed specifically for the next generation of automotive applications or "by-wire" applications that demand a high-speed, deterministic, fault-tolerant bus capable of supporting distributed control systems. The FlexRay consortium was established in October of 2000 to rally the industry around one safety-critical bus standard for by-wire systems. The biggest names in the automotive industry—BMW, DaimlerChrysler, Motorola, Philips Semiconductors, and most recently Bosch—have united to develop and establish the FlexRay standard. This advanced automotive communications system supports a time-triggered network architecture, considered key to reliable operation of such safety-critical applications. Standards such as FlexRay promote consistency, ease of development and use, reliability, competition, efficiency, and lower costs. Figure 15.1 shows the FlexRay network in detail.

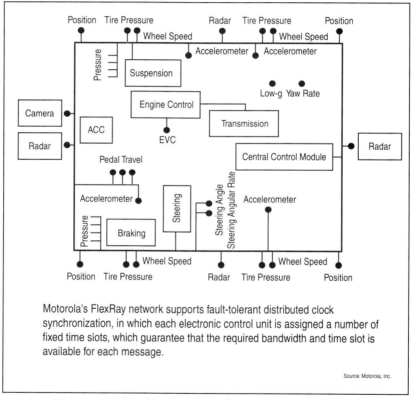

Figure 15.1: FlexRay Network (Source: Motorola, Inc.)

The FlexRay enables vehicles to advance to 100% by-wire systems, eliminating dependency on mechanical backups. By-wire applications demand high-speed bus systems that are deterministic, fault-tolerant, and capable of supporting distributed control systems. In addition, the technology is designed to meet key automotive requirements like dependability, availability, flexibility, and a high data rate, all to complement major in-vehicle networking standards such as CAN, LIN and Media-Oriented Systems Transfer (MOST).

The FlexRay communication system serves as more than a communications protocol. It also encompasses a specifically designed high-speed transceiver and the definition of hardware and software interfaces between various components of a FlexRay node. The FlexRay protocol defines the format and function of the communication process within a networked automotive system. Unlike current automotive protocols such as MOST, CAN, LIN, and J1850, the FlexRay protocol meets the requirements for dependable systems and plays an important role in the communication between the vehicle's various electronic systems.

With the increasing amount of data communication between electronic control units, it is important to achieve a high data rate. The FlexRay is initially targeted for a data rate of approximately 10 Mb/s; however, the protocol design allows much higher rates. It is a scalable communication system that allows both synchronous and asynchronous data transmission. Synchronous data transmission allows time-triggered communication to meet the requirement of dependable systems. Asynchronous FlexRay transmission is based on the fundamentals of the byteflight protocol, and allows each node to use the full bandwidth for event-driven communications. Byteflight is a high-speed data bus protocol for automotive applications.

The FlexRay's synchronous data transmission is deterministic, with a minimum message latency and message jitter guaranteed. This is an advanced feature as compared with CAN and its priority arbitration scheme. The CAN arbitration scheme delays lower priority messages in favor of high-priority ones. Therefore, the delay for any message, except the highest-priority message, is not defined beforehand. Predefined transmission times are guaranteed for each message in the FlexRay synchronous mode, without interference from other messages.

FlexRay supports redundancy and fault-tolerant distributed clock synchronization for a global time base, thus keeping the schedule of all network nodes within a tight, predefined precision window. In the synchronous mode, each electronic control unit is assigned a number of fixed time slots. These fixed time slots guarantee that the required bandwidth and time slot is available for each message. This assignment ensures that all messages get through without competing for bandwidth or an arbitration scheme.

An independent bus guardian prevents data collisions even in the event of a fault in the communication controller. This contributes towards error containment on the physical layer. By-wire distributed systems require a communications protocol link that ensures that each application function within the system will be carried out in a known time frame.

The FlexRay system has been designed to support both optical and electrical physical layers. This allows the user to implement a wiring scheme that best suits his or her needs. Optical data transmission has the advantage of not being sensitive to electromagnetic interference coupling that can cause significant communication disturbance.

Standards for In-vehicle Multimedia Electronics

Automobiles have evolved from having a simple radio with perhaps a cassette or CD player to having a variety of sophisticated information systems that need to communicate and interact with each other and with a human user. Cars today include GPS navigation systems that can work in conjunction with a security system to locate a stolen car. Car occupant safety requires the driver to concentrate on controlling the car rather than on the intricacies of the individual components. The car telephone needs to interact with the stereo system to mute it when a call is requested or received. Voice control and hands-free speakerphones require a microphone to digitize the voice. Display systems are needed for navigation information and DVD playback. Figure 15.2 shows the multimedia network, which consists of the different consumer devices found in the automobile.

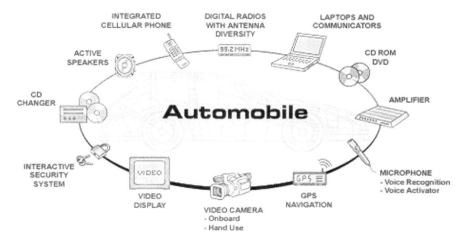

Figure 15.2: The Multimedia Automobile Network

All of these subsystems must interface with the car driver in order to be effective. They must also present audio and visual information in a wide variety of formats to inform the driver and/or to entertain the passengers. The multimedia network should also be able to manage the information in order to safely present it to the user as it comes from the various components. The most efficient and cost effective way to continue innovations in all these areas is to allow individual devices to be developed independently, and then be networked together using standard hardware and software interfaces. Digital interoperability will also be required.

Options will be easy to add since the network provides the infrastructure to transfer information from one device to another. Cars will be customized to each buyer's preferences right at the dealership and will not depend on a pre-selected list. Safety will be enhanced, as components are easily controlled from their user interfaces and have well defined interfaces that allow them to interoperate.

Areas that need higher speed, such as the entertainment system, are usually not mission-critical applications. For example, if there is a fault in the line that controls the car's CD player, it is not as critical as a fault in the ABS or air bags. Electromagnetic emissions are the biggest problem with high-speed buses. The faster a bus operates, the more emissions become an issue. One candidate for the high-speed multimedia bus is IEEE 1394 (FireWire). Unfortunately, FireWire cannot be used in the car because IEEE 1394 specifies connectors that radiate electromagnetic emissions. The emergence of optical (known as fiber optic or light wave) technology eliminates emissions as the biggest potential problem. Essentially, PC technology is used here, only in an improved version that meets automotive grade standards. This means taking temperature constrains, mechanical issues, vibration, acceleration, and other issues into consideration—all details that do not appear in the normal PC arena. Automobile audio and video require a bandwidth on the order of 50-to-60 Mb/s. The bandwidth requirement is much lower if the transmission is audio only. But since the introduction of DVD and video games, it is much more desirable to send video signals over the bus. Those signals could be sent to the rear seats for viewing.

Technologies available to meet the requirements of in-car video and multimedia distribution include IDB-1394, MML, MOST and AMI-C, among others. Some of the multimedia standards gaining popularity in automobiles are listed below.

The Intelligent Transportation Systems Data Bus (IDB) Family

The IDB Forum manages the IDB-C and IDB-1394 buses, and standard IDB interfaces for OEMs who develop after-market and portable devices. The IDB family of specifications provides an open-architecture approach that is spearheaded by a collaborative group of automotive OEMs and electronics device manufacturers. This family includes an entry-level network based on CAN 2.0B, called IDB-C (intelligent transportation systems data bus – CAN) running at data rates of 250 Kb/s. Initial applications of IDB enable developers to gain access to the body electronic network (with the automotive OEMs permission). For example, this will allow devices such as navigation systems to access the vehicle odometer and sensor information to improve navigation accuracy and performance. This will also allow fleet applications like rental cars to use odometer and vehicle fuel level to automate the vehicle rental process. This will be done with simple plug-in devices that can easily be added to a vehicle when it enters the fleet, and moved to subsequent vehicles just as easily. Other applications for IDB-C are expected to include connectivity through consumer devices such as digital phones, PDAs, and audio systems.

Implementation of IDB-C is further simplified and low cost since most onboard vehicle networks are already implemented on the CAN bus or some variant thereof. While relatively slow, IDB-C still provides vehicle sensor information fast enough for the auto market. Vehicle OEMs consider IDB-C to be the entry point of open-architecture networking that will pave the way for introduction of value-added services to telematics products such as the GM OnStar. Commercial introduction of IDB-C equipped vehicles is expected in 2004.

IDB-1394 (FireWire)

The IEEE 1394 (FireWire) bus is moving in the direction of the automobile, as seen by a draft specification jointly introduced by the 1394 Trade Association (1394ta) and the Intelligent Transportation System Data Bus (IDB) Forum. This specification defines the enabling requirements for FireWire-equipped vehicles, and, therefore, addressing high-speed multimedia applications that require large amounts of information to be moved quickly. It establishes the basic architecture, plastic fiber, and copper connector specifications for 1394-equipped embedded devices such as CD or DVD players, displays, games, and/or computers. The IDB-1394 specification defines the automotive-grade physical layers, including cables and connectors, power modes, and higher layer protocols required so that all 1394 devices can interoperate with embedded automotive IDB-1394 devices. This specification also allows 1394 portable consumer electronic devices to connect and interoperate with an in-vehicle network. True high-speed open-architecture networking is expected to arrive in 2004, when vehicle manufacturers will introduce the IDB-1394 to select vehicle platforms.

The IDB-1394 auto architecture is divided into a fiber-optic embedded network and a Customer Convenience Port (CCP), and the spec is supplemental to the IEEE 1394-1995, 1394a-2000, and 1394b standard. The next step is completion of the 1394 automotive power management specification, which should occur fairly soon. The CCP, which consists of a 1394b physical layer and connector, lets users connect their consumer-electronics devices in a hot-pluggable fashion. The embedded network will support devices running at 100, 200, or 400 Mb/s, with architectural support to speeds over 3.2 Gb/s.

In comparison, IDB-C is geared toward devices with data rates of 250 Kb/s (such as digital phones, PDAs, and audio systems). The IDB-1394 specification was designed to take advantage of the large number of consumer IEEE 1394 devices already in the marketplace. It maintains compatibility with legacy 1394 consumer devices, but is designed specifically for vehicle applications that are implemented on plastic optical fiber cable with an automotive grade connector and system components. The specification includes all seven layers of the OSI model to ensure compatibility among devices already available in the marketplace, as well as new devices designed specifically for vehicle use.

The IDB-1394 specification details a power budget of 14.5 dB and low-cost in-line connector technology assembly to ensure vehicle integrity. A minimum bend radius of 15 mm has been specified to assure ease of implementation in even the smallest vehicle compartments. The IDB-1394 specification builds on the strength of applications already introduced in consumer electronics and computer applications, is ideally suited for Internet Protocol (IP), video, and HDTV.

A draft specification for an automotive version of the IEEE 1394 bus has been released, and is moving forward as a high-speed serial interface for future vehicles. As a potential competitor to the fiber-optic MOST (Media-Oriented Systems Transfer) bus, it is backed by several European carmakers. Support from the powerful Automotive Multimedia Interface Collaboration (AMI-C), another automotive multimedia industry group with support from some the world's biggest automakers, will be the key to success for the 1394 bus.

The Mobile Media Link (MML)

The MML is an optical bus that can transmit or receive at rates up to 110 Mb/s, and was developed by Delphi Automotive Systems. It is built around a hub concept, meaning that there is one central node that connects all the branches (unlike ring architecture).

MOST (Media-oriented Systems Transfer) Cooperation

The Media-oriented Systems Transfer (MOST) Cooperation is an alternative to open-architecture IDB technology, and is being considered by several German automakers and suppliers as the standard multimedia bus for connecting multiple devices in the car. The MOST Cooperation is based on a partnership of carmakers, set makers, system architects, and key component suppliers. Together they define and adopt a common multimedia network protocol and application object model. The MOST Cooperation a venue for all members to combine their expertise and ambitious efforts in an attempt to establish and refine a common standard for today and tomorrow's automotive multimedia networking needs. The concept for the MOST Cooperation began as an informal cooperative effort in 1997. Since the cooperation was founded in 1998, 17 international carmakers and more than 50 key component suppliers are working with the MOST technology and contributing towards its innovation.

The automotive industry has settled on using plastic optical fiber as the underlying physical infrastructure for these "infotainment" networks. Both the MOST Cooperation (with 14 carmakers and close to 60 of their suppliers) and the AMI-C (with eight members) specify plastic optical fiber as their physical medium. Companies that support MOST or that have plans to use the approach in future car platforms include; Audi, BMW, DaimlerChrysler, Fiat, Ford, Jaguar, Porsche, PSA/

Peugeot/Citroen, Volvo, Volkswagen, Harman/Becker, and OASIS Silicon Systems. A recent notable example of MOST implementation was its use by Harman Becker in the latest BMW 7-series. The MOST Cooperation's membership is continuously growing due to its philosophy of openness to any company. The Cooperation invites and welcomes other companies to become associated partners in order to contribute their collective efforts to the development and enhancement of the MOST Technology.

Vehicle manufacturers considering MOST technology envision controlling all devices added to the vehicle system, resulting in a closely controlled network. Likewise, devices designed for implementation on manufacturers' platforms will be unique to those platforms to assure manufacturer oversight of all in-vehicle devices. The MOST Cooperation is currently implemented at 24.8 Mb/s on plastic optical fiber cable with an automotive grade connector and interface. It is highly reliable, is scalable at the device level, and offers full support for real-time audio and compressed video.

MOST Technology Overview

The MOST Cooperation is a multimedia, high-speed fiber-optic network developed by the automotive industry, and optimized for automotive applications. Its design provides a low-overhead and low-cost interface for the simplest of devices, such as microphones and speakers. At the same time, more intelligent devices can automatically determine the features and functions provided by all other devices on the network. They can then interact with the other devices on the network and establish sophisticated control mechanisms to take away distractions as the different subsystems try to communicate information to driver. The MOST Cooperation network make it suitable for any application inside, or outside, the car that needs to network multimedia information, data, and control functions.

The MOST Cooperation is a synchronous network—a timing master supplies the clock and all other devices synchronize their operation to this clock. This technology eliminates the need for buffering and sample rate conversion, and can connect very simple and inexpensive devices. For an automotive technology standard to be widely used across all vehicle lines it must first be cost-effective—the MOST network employs low cost ring structures to answer this need. Each device requires only one fiber-optic transmitter and one receiver. Other, more costly topologies are also allowed, including star and tree configurations.

The technology is similar to that of the public switched telephone network. Data channels and control channels are defined. Control channels are used to set up and define the data channels for the sender and receiver. Once the connection is established, data can flow continuously and no further processing of packet information is required. The MOST technology also supports isochronous data transfer. This feature is important because, in most cases, the order and time at which data arrives is more important than ensuring that every bit gets there. Isochronous data transfer is the optimum mechanism for delivering streaming data (information that flows continuously). The most obvious example is music, where the bits must be played in the right sequence, even if a bit is lost. This technology uses a ring-based architecture. When the ring is broken, causing a fault, the bus stops operating. For this reason the bus is only used in non-mission-critical applications, like the entertainment system.

Computer based data, such as Internet traffic or information from a navigation system, is typically sent in short bursts and (often) to many different destinations. The MOST uses circuit-switched architecture, similar to the public telephone system, and eliminates the need for routing information required by packet switching architecture. This results in very efficient bandwidth utilization (close to 95% efficiency for synchronous streaming data) and low-cost silicon implementations. The MOST Cooperation has defined efficient mechanisms for sending asynchronous, packet based data.

The control channel in this technology ensures that all devices to can cleanly start up and shut down the data they are using. This is possible because the control channel permits devices to send control messages while the data channels are in use. Figure 15.3 provides an illustration of the MOST network.

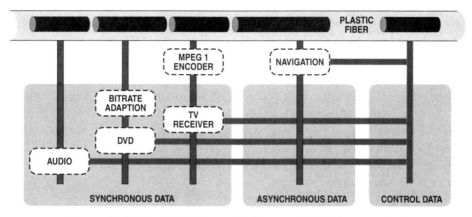

Figure 15.3: The MOST Network (Source: MOST Cooperation)

System software and application programming interfaces (APIs) are just as important as the hardware, and are crucial to insure that devices from different manufacturers can interact with each other. The API's need to be object oriented so applications can concentrate on the functions they provide rather than on the details of the underlying hardware. The software and APIs need to be able to control all the features that devices provide on the network, whether from AV equipment, GPS navigation systems, telephones, or telematics systems. The MOST specification encompasses both the hardware and software required to implement a multimedia network. This specification also defines all seven layers of the OSI data network reference model. This allows designers developing applications to concentrate on the functions that affect the end user rather than the complexities of the underlying network.

The MOST Cooperation has tapped its partners' expertise to design APIs that can work with all of the functional units that need to be networked together in an automobile. Compatibility is assured with this approach because these APIs will be published as standard function catalogs, and all MOST devices have been designed using these APIs.

The key features of the MOST network are:
- Ease-of-use because of:
  - Simple connectors
  - No hum loops, no radiation
  - Plug-n-Play; self identifying devices with auto initialization
  - Dynamically attachable and re-configurable devices

- Virtual network management, including channel allocation, system monitoring addressing and power management
- Wide range of applications:
  - Applications from a few Kb/s up to 24.8 Mb/s
  - High degree of data integrity with low jitter
  - Support of asynchronous and synchronous data transfer
  - Support of multiple masters
  - Support of up to 64 devices
  - Simultaneous transmission of multiple data streams such as control, packet, and real-time information
  - Devices can be constructed out of multiple functions
  - Low overhead due to embedded network management
- Synchronous bandwidth – Synchronous channels provide guaranteed bandwidth with no buffering required and up to 24 Mb/s of synchronous data throughput.
- Asynchronous bandwidth:
  - Variable asynchronous data throughput of up to 14.4 Mb/s
  - Asynchronous data dedicated control channel with more than 700 Kb/s
- Flexibility:
  - Wide range of real-time channel sizes and packet sizes
  - Remote operation and flow control
  - Variable arbitration mechanisms
  - Protocol independent
- Synergy with consumer and PC industry:
  - Operates with or without PC
  - Consistent with PC streaming and the Plug-n-Play standards
- Low implementation cost:
  - Low cost sub-channel at 700 Kb/s
  - Optimized for implementation in consumer devices
  - Suitable for implementation in low cost peripherals
  - Low cost cable and connectors
  - Low cost integrated circuits

MOST networks are used for connecting multiple devices in the car, including car navigation, digital radios, displays, cellular phones, and DVD players. Simple MOST devices are the basic building blocks that can be used by more complex devices (such as radios, navigation systems, video displays, and amplifiers) to implement sophisticated information and entertainment systems. Consumer devices always have analog interfaces in addition to a changing variety of digital interfaces. These applications also provide the most flexible method of connecting external devices to an onboard network. The MOST technology is optimized for use with plastic optical fiber. It supports data rates of up to 24.8 Mb/s, is highly reliable and scalable at the device level, and offers full support for real-time audio and compressed video.

The Automotive Multimedia Interface Collaboration, Inc. (AMI-C)

The AMI-C represents the first attempt by some of the world's biggest automobile manufacturers aimed at standardizing in-vehicle electronics. It is a non-profit organization of automakers that was created over three years ago to facilitate the development, promotion, and standardization of electronic gateways to connect automotive multimedia, telematics, and other electronic devices within the car. The AMI-C establishes a common interface for adding cellular phones, PDAs, navigation systems, CD players, video screens, digital radios, in-car PCs, and a host of other products to vehicles. This interface commonality allows all of the devices to work interchangeably. Automakers participating in the AMI-C include Ford, Fiat Auto, General Motors, Honda, Nissan, PSA Peugeot-Citroen, Renault, and Toyota.

The AMI-C is a worldwide organization of motor vehicle manufacturers created to facilitate the development, promotion, and standardization of automotive multimedia interfaces to motor vehicle communication networks. This organization was formed in 1999, and is organized on a membership basis.

The AMI-C sponsors a "proof-of-concept" demonstration to help demonstrate the potential of today's latest technology. This demonstration features a variety of vehicles from AMI-C Members, equipped with multimedia devices that highlight the flexibility and adaptability of AMI-C specification. Demonstration vehicles showcase integrated cellular telephones, PDAs, emergency call systems, and compact disc changers. Each device plugs into the network, and works by sharing information with other devices and vehicle systems. Devices are also exchanged between vehicles equipped with both low-speed and high-speed networks in order to demonstrate compatibility and interoperability. The AMI-C specifications help expand product markets by moving single-vehicle, proprietary interface to a cross-vehicle, open interface. Electronic devices that support this specification will work in any vehicle with an AMI-C-compliant interface, regardless of the manufacturer.

The AMI-C mission is to develop a set of common specifications for a motor vehicle electronic systems multimedia interface that will accommodate a wide variety of computer-based electronic devices in the vehicle. The AMI-C's objectives are to:

- Provide a method that makes it convenient for consumers to use a wide variety of emerging media, computing, and communications devices in their vehicles
- Foster innovation of new features and functions in automotive related media, computing, and communications electronics by creating a stable, uniform hardware and software interface in the vehicle
- Reduce time to market and facilitate upgrades of evolving vehicle electronics
- Support deployment of telematics by defining specifications for both telematics and information interfaces between the vehicle and the outside world
- Reduce relative costs of vehicle electronic components
- Improve the quality of vehicle electronic components through reduction in variations (thus improving first time capability)

Component variation has caused a multitude of problems for automakers and vendors. Since components do not follow a common design, electronic products are not interoperable from vehicle-to-vehicle. Vendors, therefore, must reengineer products for each OEM. As a result, significant development time is lost and products are often two generations behind state-of-the-art systems by the time they reach automotive production. The AMI-C representatives hope to eliminate this exact problem by creating a foundation that would enable plug-and-play interoperability of products from vehicle-to-vehicle.

Three of AMI-C's key members, BMW, DaimlerChrysler, and Volkswagen, have discontinued their memberships at AMI-C, but continue to support the MOST cooperation however. While some companies have moved on from the AMI-C, the consortium continues to see strong support from other automotive heavyweights. The AMI-C has announced that they will support the MOST bus in their new specification. The AMI-C specification can accommodate either IEEE 1394 or MOST as its high-speed bus.

The IDB-1394, AMI-C, and MOST

The AMI-C has met with representatives from the 1394 Trade Association and the IDB Forum in an effort to introduce a specification that would put IEEE 1394 serial buses on board everyday vehicles. If they can reach agreement, the bus would provide a high-speed connection for the efficient transmission of data in vehicles. The agreement would also enable consumers to plug handheld electronic devices, such as camcorders and DVD players, into vehicle networks.

The availability of such bus technology is considered critical because of the increasing amount of digital data that now passes through vehicles. In particular, automotive audio and video systems will require high-bandwidth buses to handle huge data streams in the future. The 1394 serial interconnection is a strong candidate for such applications because of its high speed and compatibility with consumer electronics. The IEEE Standard 1394 is already used in more than 10 million camcorders and 8 million PCs, as well as in millions of printers, scanners, video games, and mass-storage devices. It offers speeds up to 800 Mb/s, does very well with video, and is well suited for rear-seat entertainment systems in minivans and SUVs.

The meeting between AMI-C and the standards groups raises questions about the future of automotive network buses. The AMI-C is currently drafting the first release of a specification that will include a low-speed CAN bus, and possibly a high-speed fiber optic bus. Until recently, the fiber-optic MOST bus was the leading high-speed candidate for endorsement by AMI-C. But OEM engineers and representatives from the 1394 Trade Association and the IDB Forum have now introduced an 80-page draft specification of the so-called IDB-1394 bus. This specification could be thrust into a prominent position with the AMI-C.

The AMI-C group of manufacturers has stated that it wants to standardize electrical architectures because these architectures could lead to simpler implementation of devices in vehicles. These devices would include navigation systems, CD players, video screens, digital radios, cell phones, and a host of others. Car manufacturers now complain that they must often re-engineer vendor products because they are not designed to any specific standard. As a result, manufacturers lose valuable development time and risk falling behind the rate of innovation in the electronics industry.

The IDB-1394 could help by offering carmakers a high-speed fiber optic bus that takes advantage of the economies of scale of widely used 1394 technology when 1394 is incorporated into automotive applications. The new IDB-1394 describes the basic architecture, plastic fiber, and copper connector specifications for the 1394 multimedia backbone. It provides consumers with a foundation for 1394-equipped embedded devices, and provides a way to connect portable computer products such as CD players, DVD players, games, or laptop computers in a vehicle.

The specification also includes a so-called "Customer Convenience Port" to make a connection to those outside devices. The convenience port, which uses copper wire, provides a common port type that consumer devices can plug in to. Embedded 1394 dashboard devices, however, would connect directly to the bus.

Whether the IDB-1394 will be endorsed in the AMI-C specification still needs to be determined. The AMI-C board members said that technical issues must first be resolved before IDB-1394 can be endorsed. In particular, engineers question whether IDB-1394 meets the electromagnetic interference (EMI) and temperature specifications needed for an automotive environment.

The AMI-C's members know, however, that they will need a high-speed bus specification at some point in the near future. The only bus set for definite endorsement by AMI-C right now is the IDB-C, which offers a speed of 250 Kb/s. In contrast, the MOST system features a speed of 24.8 Mb/s, while the IDB-1394 operates at 100 Mb/s or greater.

Neither the MOST nor the IDB-1394, however, appear likely to make the first release of the AMI-C spec. The MOST (co-developed by Oasis Silicon Systems AG, Becker Automotive Systems, BMW, and DaimlerChrysler) has been stalled in the AMI-C proceedings by intellectual property issues. The AMI-C's members say those issues that involve "lack of openness" must be resolved before they can endorse it. For that reason (along with IDB-1394's technical problems) they question whether either high-speed bus can be endorsed in their first release specification. But AMI-C member companies said they have no preconceived notions about how many high-speed buses AMI-C can endorse. They could endorse a single high-speed bus or even multiple standards.

The Digital Data Bus (D2B) Communications Network

The Digital Data Bus (D2B) is a networking protocol for multimedia data communication when integrating digital audio, video, and other high data rate synchronous or asynchronous signals. The D2B can run up to 11.2 Mb/s, and can be built around an unshielded twisted pair (UTP), named "SMARTwire," or a single optical fiber. This communication network is being driven by C&C Electronics in the UK, and has seen industry acceptance from Jaguar and Mercedes Benz. The Integrated Multimedia Communication System deployed in the Jaguar X-Type, S-Type, and new XJ saloon is an example of its use.

The D2B optical communications network is self-configuring on start up, and will adapt to all devices present on the network at the time. This means that new devices can be easily fitted to the network at any time during its life. Automakers that use the D2B optical multimedia system will find that the standard evolves in line with new technologies as they are introduced. The standard will be backwards compatible, ensuring that new products can be added to a cars system during its lifetime. The D2B optical is based on an open architecture. This simplifies expansion because changes are not required to the cable harness when a new device or function is added to the optical ring. The bus uses just one cable, either a polymer optical fiber or copper, to handle the in-car multimedia data and control information. This gives better reliability, fewer external components and connectors, and a significant reduction in overall system weight.

Wireless Standards—Cellular, Bluetooth, and Wireless LANs

Telematics refers to the communications to and from a car. Its goal is to accomplish these communications in a safety-conscious manner as well as providing other services, which could be entertaining and/or informative. As a result, one of the telematics control channels could also possibly be wireless enabled, and allow transmission of the API via wireless standards. This includes a uniform, consistent set of commands to answer a phone, place a call, and perform manual or dual-tone multi-frequency (DTMF) dialing, and other data transports. Wireless connectivity adds complexity because there are so many competing wireless standards to accommodate.

Analog cellular connectivity is the vehicle standard today due to its ubiquity, and despite a peak data throughput of only 9600 b/s. Since safety and security applications are currently the most common services, coverage and reliability are paramount. Digital cellular standards provide better data rates, voice clarity, and cost. However, the digital infrastructure doesn't yet cover enough geographical area to support these applications. There is, however, an impetus to add digital service for cost-sensitive voice users.

A couple of non-cellular wireless standards could also play a role in the telematics picture. Bluetooth, the short-range, low-implementation-cost wireless technology, connects user-carried handheld devices (including cell phones, notebook PCs, and PDAs) to some form of the gateway device and/or access point. Bluetooth is a computing and telecommunications industry specification that allows users to easily interconnect with each other, and to effortlessly synchronize data (such as a phone list) between their handheld and vehicle databases. Hence, Bluetooth enables a high-value, information-centric solution. A driver will be able to use a Bluetooth cordless headset to communicate with a cellular phone in his or her pocket, for example. As a result, driver distraction can be reduced and safety increased. Bluetooth is supported by over 2500 companies worldwide, and will penetrate PC, consumer electronics, and mobile phone markets. The automotive industry has created a special-interest group (SIG) for the definition of Bluetooth car profiles. This SIG includes members such as AMI-C, BMW, DaimlerChrysler, Ford, GM, Toyota, and VW. One example of Bluetooth deployment in cars is a hands-free system called BlueConnect™ by Johnson Controls that allows drivers to keep their hands on the wheel while staying connected through a Bluetooth-enabled cellular phone. There has, however, been some concern voiced over long-term support of Bluetooth devices, and also about how a noisy in-car environment will effect its operation. The lifecycle of cars and other vehicles is very much longer than that of consumer products or mobile phones—this mismatch between support and service timescales must be addressed by silicon manufacturers. However, the Chrysler Group used Bluetooth connectivity in its vehicles at Convergence 2002. Bluetooth support will provide both OEM and after-market installations with an even-greater level of electronics integration in vehicles. Clearly, the industry is moving toward openness in architecture to allow simplified plug-and-play technologies to penetrate the marketplace in response to increased consumer demand.

One challenge is to create Bluetooth applications that do not interfere with the frequency band of other services such as satellite digital audio radio—Bluetooth uses 2.4 to 2.483 GHz and satellite radio 2.32 to 2.345 GHz. Bluetooth offers several advantages for mobile phone applications, such as low power requirements, and can be manufactured at low cost. Future Bluetooth automotive applications will have the potential to play a big role in shaping the emerging telematics market.

Vehicle applications that require greater bandwidth than that provided by today's cellular links, and greater transmission distance than that provided by Bluetooth, may rely on the IEEE 802.11b and IEEE 802.11a wireless LAN technologies. These technologies support data rates of 11 Mb/s and 54 Mb/s, respectively. Wireless LANs enable relatively quick transfer of multimedia files within a 100-meter range—important feature for content providers and resellers. However, the vehicle wireless LAN infrastructure is nascent, and its full potential will emerge only when content providers recognize its value.

Using cellular technologies and a combination of the other short-range technologies such as Bluetooth to provide some of the same services as telematics providers, enables wireless telecommunication providers to give birth to new business models and enable additional revenues. The Chrysler Group, backed by a partner team that includes IBM, Intel, and AT&T Wireless, has decided on a hands-free telematics approach. This lets consumers continue to use portable cell phones in their cars and alleviates their concerns about driver distraction. Chrysler's approach, which relies heavily on Bluetooth radio technology, departs dramatically from General Motors' OnStar model, which has dominated the market until now. Instead of employing factory-installed, console-mounted phones, Chrysler's platform will allow consumers to bring their cell phones into their cars, synchronize them with in-car voice-recognition systems, and then lay the handsets on the front seat or even on the floor while they talk.

Automotive industry analysts said that the technology's simplicity could help fuel the growth of telematics, expanding it beyond its current clientele (composed mostly of early adopters) to include a broader swath of automotive customers. The communications platform allows consumers to bring any kind of cell phone into a vehicle as long as it is Bluetooth-enabled.

Chrysler's offering also differs from the status quo in how it plans to make money from its telematics system. Chrysler will not attempt to profit from the service revenue in this case, but will leave that to one of its partners—AT&T Wireless, which will provide cellular services. Instead, Chrysler plans to profit by using its telematics platform as an enticement for consumers to buy its cars, and by also providing the hardware for the system. The company has worked with several major suppliers, including Intel, IBM, Johnson Controls, and Gentex, to develop the electrical architecture and software backbone for its system.

Open Standards—Enabling Telematics

It is clear that the adoption of open standards for vehicle networks will wave a green flag for the telematics market. Automakers will have to implement standards-based networks alongside their proprietary networks. The automotive world has, so far, remained insulated from the standardization drive that has helped accelerate the development of other markets. The Internet, low-cost networking technology, and advances in embedded computing have driven consumer-electronics manufacturers to develop standards for network systems. Automotive manufacturers, however, have preferred to develop internal standards that have remained proprietary from all but the closest partners.

Fortunately, the automotive industry is beginning to change. And telematics is an environment where open standards make sense for automotive manufacturers. Electronic wide range of electronic services and applications, regardless of manufacturer, brand, or model, will not be deliverable to vehicles until there is a standard method to communicate with them. The average high-end vehicle has 25 to 50 embedded microprocessors collecting data from anti-lock brake, navigation, engine, and other vehicle systems. Electronics will soon comprise 35% of the average vehicle's cost. Add to this scenario the new breeds of mobile computing devices networked into the vehicle, and you have the potential for a convergence that produces not just a computer on wheels, but a powerful distributed computing platform.

Standardization of the original IBM PC platform engendered astonishing growth in applications and services for the platform. With the advent of the Internet, the standardized platform moved to a network model. Telematics will advance this network model by making dynamic, real-time, in-vehicle networks accessible. Standardization is rare within the automotive world. Only the recent promise of supplemental revenue streams, deeper customer relationships, and the potential for operation efficiencies from advanced telematics has prompted automakers to consider implementing standards for vehicle communication technologies.

Whatever open standards automakers ultimately adopt for interfacing consumer electronics with the vehicle, there is great benefit in adopting network standards that follow those already established for the consumer-electronics industry. That benefit would be to enable the greatest degree of interoperability. By adopting, or adapting, a standard interface from the outside world, automotive manufacturers will allow customers to enjoy more services, applications, and devices in the vehicle environment. This approach will also enhance opportunities for non-traditional revenue sources.

Consumer-electronics manufacturers will also see benefits (such as additional product usage time) from standardized vehicle network interfaces. Americans spend 500 million hours in their cars every week, and 80% to 90% of cell phone users state that they use their cell phones in their vehicles. The potential for extra usage and, by extension, extra product value, is enormous.

Network standardization also provides consumer-electronics manufacturers with the ability to create new product categories. For example, consumer-electronics manufacturers can develop location-aware gaming and education products by interfacing with vehicle assets. Mobile-commerce providers will benefit by developing applications that have value in the vehicle space. Products like location-aware travel narratives and advertising represent new categories of location-based services, which will also feed consumer demand for vehicles with integrated electronics and Internet connections.

Automotive electronics suppliers, telematics-system providers, and third-party developers will also benefit from standardized interfaces. Specifically, standards will produce greater markets for telematics technology and applications that have a lower cost of entry. This will encourage the proliferation of more advanced telematics services. In this scenario, the ultimate beneficiary of vehicle networking standards will be the consumer.

Unfortunately, no single communications network will suffice to meet all the cost and performance requirements in an automobile of the future. Instead, leading automotive OEMs will support a reasonable number of network protocols (possibly including CAN, LIN, MOST, and others) as the standard.

Summary of In-car Buses

Telematics systems exhibit characteristics more like those of consumer products—short time to market, short time in market, and changing standards and protocols. These characteristics are the complete opposite to those of traditional in-car electronics. This disparity could potentially cause issues from design right through to manufacture and servicing. These issues also have an impact on the way engineers are approaching designs and choosing in-car bus systems. Traditional in-car bus systems based on a serial, event-triggered protocol (such as the CAN and J1850) have been used successfully in the body control area for many years. Bandwidth and speed restrictions, however, will make it difficult for these to be used in the newer real-time applications. A range of new bus standards has been emerging, such as time-triggered protocols and optical data busses, in order to meet these new data throughput challenges.

In-car buss networks can be divided into four main categories:

- Body control – Covering data and control signals between car seat controls, dashboard/instrument panel clusters, mirrors, seat belts, door locks, and airbags (passive safety)
- Entertainment and driver information systems – Communication and control between radio, Web browser, CD/DVD player, telematics, and infotainment systems

- Under the hood – Networking between ABS brakes, emission control, power train, and transmission
- Advanced safety systems – Data transfer between brake-by-wire, steer-by-wire, and driver assistance systems (active safety)

Figure 15.4 summarizes the buses used for in-vehicle networking applications, Figure 15.5 summarizes the safety critical buses, and Figure 15.6 summarizes the speed of the in-car networks.

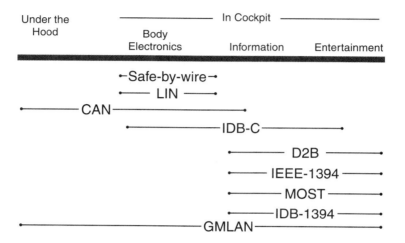

Figure 15.4: Summary of Buses for In-vehicle Networking Applications

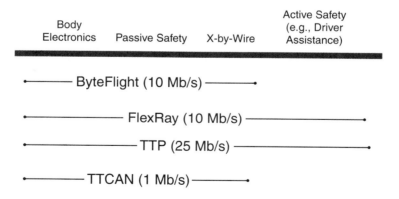

Figure 15.5: Summary of Safety Critical Buses

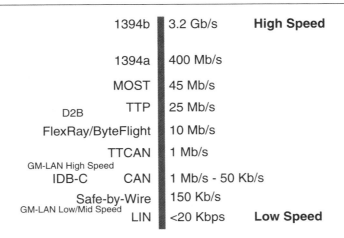

Figure 15.6: Summary of In-Car Network Speeds

Components of a Telematics System

Figure 15.7 shows a block diagram of a typical telematics system.

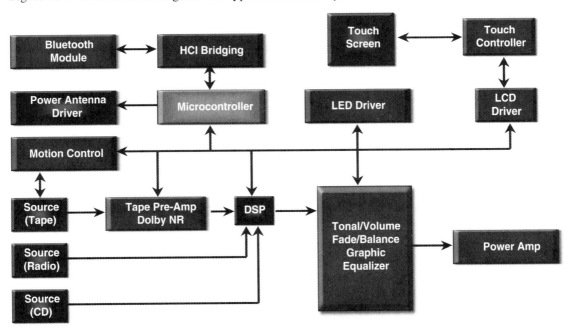

Figure 15.7: Block Diagram of a Telematics System

379

32-Bit Microprocessors/Microcontrollers

With the advent of intelligent cars, 32-bit microcontrollers are on the fast track to winning many applications within the smart vehicle. Last year, according to a research study by Strategy Analytics, 32-bit processors powered 10% of the vehicle. By 2005, 32-bit solutions are expected to take over 25%of the processing power in the vehicle. Today's automotive design engineers are seeking solutions for complex applications, such as communicating via driver information systems, adding enhanced safety features, maximizing engine performance, and ensuring the reliability and performance of power trains.

High-performance 32-bit microcontrollers can provide a solution to some of the most pressing requirements for successful telematics systems design. These design requirements include in-car networking and higher levels of on-chip system integration (to reduce the weight, materials cost, and design time of vehicles), and robust, high-speed, cost-effective embedded solutions.

Strategy Analytics predicts that 41% of all cars will be sold with a telematics system on board by 2005. Automotive manufacturers will increasingly need more computing power to handle consumers' desires for in-car information services, wireless connectivity, messaging, and productivity. Entertainment, OEM/dealer functions of the future, and driver-related services such as graphics, large-vocabulary speech and emergency services will also require increased performance. The chip solution that handles the new features will also need to power enhanced speech capabilities, Java applications, graphics, Bluetooth, GFX, the IEEE 1394 interface, and audio codecs. The 32-bit processors, such as the PowerPC 823e, can enable the multifaceted aspects of these driver information systems.

Global positioning systems (GPSs) are very popular in Japan, and are beginning to become standard equipment on upscale U.S. vehicles. Americans spend, on average, far more time in their vehicles than people in other countries due to their commuting habits and the wide-open spaces in North America. As a result, GPSs will inevitably become standard on U.S. cars. In addition, digital audio (from both satellite and terrestrial sources) will become very popular, and high-quality "better-than-CD" digital sound will be available in the automotive environment. Meanwhile, several consumer entertainment manufacturers have demonstrated rear-passenger video, movies, games, and other entertainment. These entertainment devices have high-resolution color LCDs and fast, high-resolution graphics-features that make them desirable to the user. These desirable capabilities, however, also require the computational power of a 32-bit processor to bring them alive. For obvious safety reasons, these features will not be available for front-seat passengers.

In the front, the 32-bit processor provides the computational power needed by screen-less driver-information systems. Screen-less units will have an excellent speech-recognition and generation system. While a speech-enabled system is less distracting than a screen, it needs to be friendly, accurate, and tolerant to avoid causing driver "cognitive distraction."

Over the next five years, the primary market drivers for 32-bit power trains applications will be spark-ignition engine management, diesel control, and electronic transmission. Power-train solutions based on a 32-bit architecture deliver enhanced performance and improved fuel consumption, while simultaneously minimizing exhaust emissions. Already a number of global automakers have specified this controller architecture because of its aptness for model-based design. It is also noted for executing the high-performance floating-point algorithms needed to provide robust, accurate control of vehicle power-train systems.

The trend to provide on-chip digital signal processing (DSP) capability further increases the attractiveness of a 32-bit design. It allows, for example, the integration of misfire and knock detection—a function currently performed by expensive external components. The rich set of intelligent I/O on many 32-bit controllers is tuned specifically for applications such as conventional, direct-injection, and hybrid engines; combined alternator-starter systems; electronic valve control; and continuously variable transmissions. Auto manufacturers' demands for cost-saving mechatronic solutions, such as oil-immersed transmission controllers, has added the new challenge of harsh environmental conditions for new controllers. Semiconductor manufacturers must ensure high-performance operation—up to 200 MHz and beyond—in operating temperatures of 150 degrees C. At the same time, they must ensure flash data retention of 10 years.

The next 10 years will produce a proliferation of direct-injection and advanced transmission-control hybrid engines, and combined alternator starters. In addition, many automotive manufacturers are finding 32-bit solutions to be effective for such applications as instrumentation, suspension control, and dashboard controls. One advantage of using standard processor architecture across applications is a reduced learning curve for the engineers, thus enabling them to develop new applications quickly and take the product to market faster. The potential reuse of code, such as lookup tables or algorithms, across a wide range of applications is another benefit of standard architecture. Libraries of code could be reused directly in different applications. And there is the obvious benefit of being able to reuse tools.

Digital Signal Processors (DSPs)

The DSP provides functions such as acoustics and speech processing. In addition to hands-free operation, the DSP-based chip set enables drivers using cellular phones to be heard more clearly without having to speak loudly into their phones. It also lets them turn up the volume on their phones to hear more clearly. The DSP, which operates at 100 MIPS, can simultaneously enable voice recognition; hands-free, full-duplex speakerphone functions (including adaptive acoustic echo cancellation), line echo cancellation, and noise suppression; simultaneous voice and data; and voice memo-recording functions.

The DSP enables automakers to incorporate sophisticated voice-recognition packages. This feature may turn out to be the most important aspect in meeting "hands-free" laws in many states. It enables hands-free operation by allowing drivers to make a call by using a voice command and speaking out a person's name (voice dialing). The DSP chip also suppresses background noise in order to help the voice software distinguish between voice commands and background sounds. And it also offers acoustic and line echo cancellation software. Such software improves the quality of phone conversations because it enables full-duplex operation, allowing people at both ends of the line to talk at the same time. Today's systems cannot yet respond directly to a dictated phone number, only names and commands. These features will be included in the next-generation products, and will be largely possible due to falling DSP costs. In two years, engineers will be able to add substantially more processing power and memory to the DSP for the same price as a DSP of today.

High-end Features

The adoption of high-end processors directly impacts the market and features that automotive manufacturers can offer. With consistent processor types across different applications, economies of scale will eventually drive down the price of a part by creating a larger market base. High-end processors are normally first seen in state-of-the-art, high-end vehicles. With improvement in technology and the economies of scale that will be realized with widespread adoption, the same technology can be used in lower-cost, lower-scale applications and eventually in low-end vehicles, bringing more to the average consumer.

One of the fastest-growing areas in automotive electronics is advanced chassis systems. Chassis systems include electronic braking, power-assisted steering, active suspension, dynamic stability-control systems, and collision warning and avoidance systems. These systems monitor the vehicle's behavior sensors in real-time, and take appropriate action by executing complex control algorithms that require significant processing power. Typically, 32-bit RISC microcontrollers running at high frequencies are used in these applications.

Automotive Software Technologies

These powerful 32-bit embedded processors are characterized by low cost, minimum power consumption and low heat generation, compact packaging, high speed, and advanced software support. As 8-, 16- and 32-bit microcontrollers and microprocessors become less expensive, more sensors with more sophisticated control algorithms are becoming increasingly common. Mechanical components are being replaced by electromechanical systems that contain software components. On top of all of this, car manufacturers are expected to continually add functionality to maintain their competitive edge.

In order to manage this growth, car manufacturers need to drastically decrease the high recurring expenses in the development, management and non-application-related aspects of control-unit software. They also need to make control units made by different suppliers compatible by standardizing the interfaces and communication protocols. One solution has been to use a standard real-time operating system (RTOS), built around a basic real-time kernel and a common application programming interface (API). Car infotainment and telematics devices, especially car navigation systems, require highly functional operating systems and connectivity. Connectivity is required between telematics systems and buses in order to open the door of a vehicle remotely or to diagnose under-the-hood devices. This type of functionality can be implemented on top of a highly functional RTOS used for the telematics device.

A very broad range of services will be delivered through telematics devices and interfaces. These services will be based upon activities that interact with information derived directly from the vehicle and its related subsystems. Interfaces or this information will be from the engine compartment, fuel monitors, entertainment devices, proximity sensors, GPS receivers, and touch-panel user-interface panels. Speech interpretation and text-to-speech interfaces, in-cab passenger monitors, safety devices (such as crash sensors and air bags), and even driver activity monitors will also be part of such a system.

The telematics processor must access those devices through vehicle buses and networks like CAN, MOST, and J1850. It will also connect with a variety of off-vehicle communications interfaces through cell phone or next-generation digital-radio-based mobile connections. There will be a broad range of vendors offering to provide wireless interconnectivity, safety, navigation, security, insurance, entertainment, maintenance, fuel, and voice or text/image communications services. The key to successful telematics deployment is the creation of an integrated, uniform, and flexible program execution environment on the vehicle that connects to powerful services that reside away from the vehicle. It is in this way, that companies who want to sell and deliver services to all drivers and passengers can efficiently build the necessary components and expect them to run everywhere.

Drivers and passengers will demand an integrated approach to interacting with those services. They will require it to either be available on demand or completely avoid bothering them until necessary. The issues of driver distraction we are seeing now are rooted in the interaction with things that are not aware of each other or of the driver's cognitive environment.

A key part of telematics-based computing will determine how and when to interact with drivers and passengers. Requirements in this area are already surfacing as a result of driver distraction issues relating to cell phones and in-car navigation interfaces. New research is now being launched on the cognitive issues relating to driver activities and focus.

Traditional communication protocols are also required for telematics systems to enable communication between the vehicle and the external infrastructure (cellular phones, electronic tolls, broadband access, roadside transponders, satellite, and radio broadcast). Standards, protocols, and specifications such as DHCP, TCP, UDP, IP (IPv4), IPv6, IPSec, PPP, IKE and IEEE 802.11 are some of the additional run-time components that are required for the development of telematics devices.

Consumers around the world are awaiting the imminent launch of a range of low cost cars that include a plethora of telematic services. Meanwhile automotive manufacturers across the globe are busy putting their software platforms in place to run these advanced services. The telematics platform enabling technologies include hardware, software, and IT services. The hardware platform primarily consists of CPU, memory (RAM, flash), local devices, and vehicle bus interfaces. The software platform includes the embedded RTOS and device drivers. Device drivers provide Ethernet, CAN, MOST, IEEE-J1850, TCP/IP, serial, PCMICA, audio, touch, video, etc. Automotive applications such as cell phone, GPS receiver, vehicle bus access, user interface management, and entertainment, among others, require subsystem software or virtual machines.

A number of companies have developed subsystem software products for this emerging industry including Microsoft Corporation, with its Windows based AutoPC software, and Sun Microsystems with its Java based environment.

For several years, the Microsoft AutoPC has been marketed as an open solution. But it is open only if software is written for a particular OS platform and back-end service solution. It does not encompass multiple embedded operating systems, underlying hardware, user interfaces, and back-end solutions.

Software specifically designed for automotive applications, such as the newly upgraded Microsoft Windows CE for Automotive 3.5 in-car platform, provides a foundation for building a safer telematics system. Key components in Windows CE for Automotive 3.5 that help customers address driver distraction include patented driver distraction controls, advanced speech control, personalized communications, and a flexible system enabling developers to create tailored solutions easily. Windows CE for Automotive provides customers with a standard interface that understands whether or not a car is operating. Microsoft's driver distraction controls enable manufacturers to build a device that meets their individual safety requirements.

A car can have more than one owner, and may be shared among roommates or family members. The OS must therefore adjust settings such as the seat, mirrors, and radio stations every time a new driver slips behind the wheel. With the advent of in-car information systems, drivers will be able to easily change such personalized settings as preferred radio presets, contact lists, personal credentials, and other favorites.

Microsoft Windows CE for Automotive gives developers the building blocks, such as a rich speech interface, for solutions that are easy to implement. This is because Windows CE for Automotive is based on Windows CE 3.0, an OS developed from the ground up as a small-footprint, highly customizable, and modular operating system for embedded applications. By creating the flexible building blocks to develop a system that addresses driver distraction on many different levels, Microsoft strives to address customer requests for an in-car software platform that is not only reliable and robust, but that also provides the foundation for safe devices.

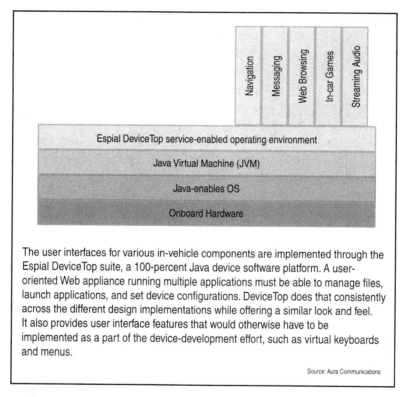

Figure 15.8: User Interface Implemented in a Java-Enabled Device
(Source: Aura Communications)

Sun Microsystems has been working closely with the AMI-C. Among the standards that have been incorporated into the AMI-C specifications are Java technology, Bluetooth, Open Services Gateway Initiative (OSGi), Wireless Application Protocol (WAP), CAN, and MOST.

Manufacturers retain control over the choice of CPU (hardware) and RTOS components that power the execution platform through software-based Java virtual machine technology that delivers a uniform application execution environment and hot-code replacement features. Java technology is available for every level of computing, from embedded devices through mainframe servers. It executes within developer workstations as well as the target environment in which it will eventually be used. Many of the elements that are key to an integrated driving experience can be automated through Java applications. It is these applications, often in the form of applets and servlets, that interact with sensors and controls within the vehicle. They also communicate with devices on the vehicle bus network, intercept the onboard diagnostic codes, and calculate driver distraction and focus. Emerging technologies such as Java-based Smart Cards could become intelligent replacements for ignition keys. Once communication is established between a device and an application in the telematics platform, its interaction can be used and controlled directly, both on-platform and off-platform, through applications using normal socket-based TCP/IP connections.

A Java based device manages files, sets device configurations, and runs multiple applications (navigation, messaging, web browsing, in-car gaming, streaming audio, etc.)—all while offering a consistent look and feel. It also provides user interface features that would otherwise have to be implemented as part of a device-development effort, such as virtual keyboards and menus.

Manufacturers and Products

As the vehicles of the future present a multi-billion dollar revenue opportunity, several automotive and consumer electronics manufacturers are targeting this arena. Some of these companies building automotive entertainment products include; Clarion, General Motors/OnStar, Delphi Automotive Systems, Ford Motors, Visteon, InfoMove, Intel, IBM, BMW, Motorola, Ericsson, Nokia, LoJack, and DaimlerChrysler's Mercedes-Benz. A recent study by Strategy Analytics predicts that partnering, collaboration, and product positioning will be the key to addressing the mobile multimedia opportunity. Common open standard solutions are key. Standard software and serial communication bus interfaces will enable significant reuse of building blocks and reduce product development cost and cycle time.

Telematics products provide navigation, entertainment, security/emergency, PIM, e-mail/messaging, Web access, news/information, and e-Commerce capabilities. However, the following issues must be addressed before these products are mainstream:

- First-generation products need improvements
- High cost of ownership: Up front costs and service fees
- Safety issues with the vehicle in motion
- Underdeveloped wireless infrastructure: Slow data transmission speeds and spotty cellular coverage
- Several substitute products exist, including cellular phones that are seeing an increase in subscribers and in services offered. Handheld PDAs and pagers with wireless access (Palm VII) are also alternatives.

The Microsoft Corporation, with its Windows based Auto PC software, has made an aggressive move in this market. Some of the features provided by this software include speech recognition, synthesis for hands-free operation, and access to information such as address book, e-mail, and GPS. It is also inter-operable with portable PCs and handheld devices. Alpine, Daewoo, Delphi Automotive Systems, Harmon Kardon, Hyundai, Infinity, JBL, Nissan, Peugeot, Citroen, Samsung, and Volkswagen support Microsoft's Auto PC. This push into automotive entertainment will provide the software giant a huge untapped market opportunity.

The technologies that will enable these new products are being developed in partnerships among diverse companies with a vision to connect every area of a person's life. For instance, Visteon is working with partners such as Samsung, Intel, Microsoft, Agere Systems (formerly Lucent Micro-electronics), IBM, and Nintendo to provide a telematics environment that is easy to use and robust enough to endure the harsh vehicle environment. Commercial fleets such as trucks already use portable navigation devices from Visteon (Portable NavMate) to increase predictability and profitability. Visteon recently announced that the next generation of its Visteon speech recognition technology had already been used on the Jaguar X-Type.

The market opportunity presented by telematics is so big that Delphi Automotive Systems' Mobile MultiMedia Business unit booked $2.9 billion worth of business before it was even 18 months old. The Delphi business is expected to grow at approximately 25% to 30% annually. Delphi classifies mobile multimedia products into four broad categories of technologies: telematics advanced digital audio, rear-seat entertainment, and smart radio systems. Delphi, which got into mobile media as early as 1936 with the first in-dash radio, more recently helped develop and manufacture the first telematics units for North America. The technology has emerged as the single highest growth sector in vehicles made in North America.

Several companies are developing a software program that enables telematics makers to build dialog systems that examine the context of data within the car and then make intelligent decisions. One such company is I/Net, which is developing the Conversational Interface for Vehicles. This system is an offshoot of robotics work the company's principal officers conducted while working for NASA. Such systems look at context and understand what's happening around the vehicle, then use that information to either remain quiet or speak to the driver. I/Net's Conversational Interface also offers the advantage of understanding natural language, thus eliminating the need for users to memorize word-for-word commands. Using such schemes, software makers believe they can alleviate one of the most maddening problems facing voice recognition: the need for users to repeat commands until they recall, verbatim, the command that the system understands. Such repetition can cause huge cognitive burdens. I/Net software engineers say their system can already understand such driver commands as "turn it down," and know whether the driver is referring to the radio volume or the heater controls. The trick is to keep recent commands in memory and use them as context when understanding language. The company claims that its program (which is less than 1 MByte in size) can fit within the memory confines of any telematics system that incorporates voice recognition.

In addition to I/Net, Lernout and Hauspie (L&H) is among the companies working on that dilemma. Lernout and Hauspie engineers are working on systems similar to those of I/Net, and see them as part of a long-range plan. The first milestone involves breaking away from rote command menus, thus enabling drivers to use more natural language. The second is the ability to put such language in context in order to understand commands that might be obvious to a person, but not to a machine. Lernout and Hauspie also made a major move to bring voice recognition down from luxury vehicles to low-end cars. Its recently introduced Distributed Speech Recognition technology effectively cuts the costs of telematics hardware by eliminating the need for large onboard memories in the vehicle. By moving much of the processing to an off-board server, L&H says its speech recognition systems can operate on just 128 KBytes of memory. Typical speech recognition programs usually have a footprint of 1.2 to 2 Mbytes, and are expected to get larger as capabilities are added.

IBM, along with tier-one vendors and automakers, is looking for solutions. In their contribution, researchers from the IBM Telematics Solutions Group have examined the human-factors issues posed by the proliferation of multimedia electronics, and offer some solutions. IBM is developing a "workload manager," that could one day consider such factors as vehicle speed, distance between cars, weather, time of day, and even the driver's heart rate before allowing a phone call to go through to the driver. This research would enable us to come up with a driver workload number that could tell the system whether or not to bother the driver, making the auto industry the real usability expert.

Currently, several after-market products exist. Also, companies are forming partnerships to develop standards, and are working out the feature mix for Auto PC products. Market researchers believe that the automotive entertainment market will be a low-volume, high-and niche for the next several years.

Telematics Providers—MobileAria, OnStar, and Wincast

Palm Incorporated and Delphi Automotive Systems Corporation have created a new company, MobileAria, to begin selling voice-activated services such as e-mail and news. It plans to sell wireless Internet services for the new hands-free communications system they designed for automobiles. The system is called the Mobile Productivity Center (MPC), and works by linking a Palm organizer and mobile phone to a car's audio system.

The MPC is a cradle-like device the companies have developed to hold a Palm computer and cell phone. It is designed for easy installation on a dashboard or a cup holder, is activated through a microphone, and responds through a car audio system over an unused FM radio frequency. The first version of the device could only hold a Palm V computer and certain mobile phones made by Ericsson. Hands-free and voice-activated technologies are gaining popularity due to growing worries about the road hazard posed by people who use their wireless devices while driving.

Delphi, the world's largest auto supplier and a former division of General Motors, does not see MobileAria as a direct competitor to GM's OnStar vehicle communications system. In particular, there are no plans to offer safety and security services through live operators like OnStar. However, OnStar is launching two new services called Virtual Advisor and Personal Calling that would likely duplicate many of MobileAria's offerings.

General Motors' OnStar Division has made inroads with 32 of GM's 54 North American models, and the company has said it expects to cover the entire line soon. At the same time, OnStar will use the SpeechWorks speechify text-to-speech engine on its Virtual Advisor. Still, because there is no immediate plan to retrofit cars with the current OnStar system for the new services, MobileAria will have an advantage with drivers who are not planning to buy a new car.

MobileAria's first services will focus on providing voice-activated access to the personal information stored on a Palm device and wireless applications (such as having e-mail or news read by a computerized voice over the car's speakers). The primary source of revenue for MobileAria will be through customer subscriptions, though the company might try to generate revenues from subscriber transactions or technology licensing down later on.

Oracle Corporation recently signed a partnership with Wingcast, a joint venture between automaker Ford Motor Company and wireless software developer Qualcomm Incorporated, to help bring talking cars to the masses. Now Oracle, Ford, and Qualcomm plan to mimic that technology in modern cars using telematics. Cars using this service would be equipped with GPS technology, which beams the location and speed of the car to a central computer at Wingcast's headquarters. The computer then uses that information to send relevant data back to the driver's location. Using speech recognition technology, the data is spoken back to the driver minutes later. In addition to traffic information, the service also enables drivers to connect wirelessly to the Internet, and check their corporate e-mail and have the messages read back to them in the car. A salesperson will not only be able to pick up their mail while watching the road with their hands on the wheel, but they will also be able to receive new leads and directly link up with a perspective customer. The idea is to enable all of Oracle's software so that they can be delivered wirelessly to automobiles via Wingcast's service.

San Diego-based Wingcast was founded in October 2000. And, although Ford and Qualcomm have a majority holding in the venture, Nissan Motor Company Limited has agreed to include Wingcast's telematics technology in its vehicles next year. Wingcast has mentioned that it will also sign deals with telecommunications service providers. The deal with Oracle was not exclusive, and Wingcast would also offer connections to corporate data like e-mail, stored in other software firms' applications, as well as information on the Internet.

Having originally said the service would be available in 2001, Wingcast delayed the launch of its in-vehicle communications service until mid-2002. The delay sets Wingcast further behind rival General Motors Corporation's OnStar telecommunications service, which, since its launch in 1996, has grown to over one million subscribers, and is available on about half of GM's fleet.

The following services and companies provide telematics services to consumers:

- Global positioning system (GPS) – GPS provides an automobile's position, which is essential for emergency and navigation services.
- Internet services – Web-based service providers deliver the personalized information that drivers want in their automobiles. Some of the Internet service providers include Yahoo!, AOL Time Warner, Reuters Group, and ComROAD.
- Telematics equipment providers – A number of manufacturers build onboard telematics equipment, such as antennas, transmitters and interfaces. Some of these providers include Delphi Automotive Systems, Motorola, Visteon, and Siemens Automotive.
- Wireless networks – Telematics service providers have partnered with existing wireless companies to allow seamless nationwide access. Some wireless service providers include Sprint PCS, SBC Communications, AT&T Wireless Group, and Verizon Communications.
- Telematics services – Telematics centers coordinate all the information and services delivered to the automobile, using both the Internet and their own databases. These telematics providers include BMW Assist, Mercedes-Benz's Tele Aid, General Motors' OnStar, Ford/Qualcomm's Wingcast, and Palm/Dephi's MobileAria.

A brief look into the product manufacturers shows that partnerships are integral to the success of telematics. The need for companies from varying industries to work together to bring solutions to market faster is promoted because these devices are deployed in cars, often require use of the telecommunications infrastructure, and require development of components that integrate into the automobile. This market opportunity can be turned into a big success for a few players if they partner at the right points, and realize effective ways for converting their investments into revenue generating strategies.

Satellite Radio

Satellite radio systems beam CD-quality signals from space to special radios, most of which are installed in cars. Customers can also listen to these systems at home. Satellite radio offers 100 channels of commercial-free music, news, sports, and entertainment. All channels broadcast directly to your vehicle, anywhere in the continental United States. This is a revolution that will change the way we listen to radio. Current research indicates that millions of Americans are dissatisfied with conventional radio. Considering that nearly half of the conventional radio stations broadcast in one of only three formats—country music, contemporary music and news—it is understandable. Even assuming one finds a station they like, they must endure up to 18 minutes of commercials each hour. The listener must also be prepared for additional interruptions—a common one is that most radio signals begin to fade at a distance of 30 miles from their source.

Broadcast, Satellite, and the Car

The satellite radio system comprises three parts—broadcast, satellite, and the car. National broadcast studios and transmission facilities encompass production facilities capable of broadcasting 100 radio stations. These include extensive music libraries that are continually updated with the latest new recordings, and more obscure selections that are no longer available commercially.

Line-of-sight (the ability to "see" the satellite) is required in order for an antenna to receive a satellite transmission. Companies either maximize this line of sight by placing their satellites in orbits directly above the United States, or positioning satellites in geostationary orbits over the equator. By placing satellites above the U.S. the customers receive improved signal strength and coast-to-coast coverage. Content is also fed to a number of transmitters located in major urban areas. These transmitters, called ground repeaters, supplement the satellite coverage in dense urban areas where tall buildings might block the satellite signal.

After the satellites and ground repeaters send their signals, the vehicle's radio picks them up and converts them to music, talk, and data. The car receiver is made up of two parts: the antenna module, and the receiver module. The antenna module is an active system with several elements that "look" along the horizon for the terrestrial signal, and higher for the satellite signal. This module picks up available signals simultaneously, amplifies them, filters out the noise and interference, and passes them on to the receiver module. The chipset inside the receiver module down-converts the signals from 2.3 GHz to a lower intermediate frequency, and then to the digital base band. The signals are converted from analog to digital as part of the down-conversion process. Within the digital realm, available signals are inspected for quality and combined in an optimal way to use the best information from each source. Instead of merely using the best signal, manufacturers use the best input from all the signals. The combined digital signal is then processed for a number of things. First, it is digitally filtered. The forward error correction is reversed and the music is decrypted. The stereo digital audio is converted back to analog for delivery through the speakers.

The Players

Two broadcasters Sirius Satellite Radio (based in New York) and XM Satellite Radio (based in Washington, D.C.) offer nationwide coverage. They are aggressively marketing satellite radio to consumers desiring more than the limited programming currently available. Consumers will have access to more than 100 channels of eclectic digital sounds, from hip-hop to jazz and opera, for $9.99 (for XM) or $12.95 (for Sirius) a month. The car systems are available on many new models for about $300, and are expected to proliferate to other cars, as carmakers are eager for a stake in this new, fast-growing business. General Motors funded XM (they invested $100 million in 1999) and DaimlerChrysler is backing Sirius. Other car manufacturers are picking up on these two technologies, which are not compatible. For home and portable systems, one can buy mobile, palm-size receivers from Sirius and XM at retailers such as Best Buy, Circuit City, and Wal-Mart for about $200.

Both Sirius and XM say advertising will make up a small portion of their revenue, a strategy that could backfire. Demand for satellite radio reached 1.5 million subscribers by end of 2003 (1.2 million for XM and 300,000 for Sirius). Analysts believe that satellite radio will reach 3 million subscribers by 2004. Both companies continue to see low revenues, high losses, and continuing debt burden. That is bad news considering that Sirius and XM each spent $80 million for FCC licenses in 1997. They have since spent a combined $3 billion building satellites and developing programs, and have been forced to yield equity to auto manufacturers in order to get their radios into cars. The enormous cost of building a system that relies on satellites, repeaters, and a vast programming network might knock these companies right out of the radio orbit. Analysts predict that satellite radio needs to reach 7.5 million subscribers and about $1 billion in revenues to cover the cost of capital and interest.

To reach this target, these companies are getting some help from the auto manufacturers. General Motors is offering XM in 44 of their 57 car lines, up from 25 in 2002. Sirius, which had a five-year head start on XM, has already launched satellites and hit the market with radios available in Ford and DaimlerChrysler's 2003 models. Toyota and Honda are also signing on and customers can buy XM radio for their homes through major retailers. Based on some of the early success of XM and convinced that its potential was enormous, GM led a group of investors in a $450 million refinancing for XM on Dec 23, 2002. GM deferred $250 million in payments that XM owed it for loans, bonds, and revenue sharing plans until 2006. A group that included Haywood, Barry, Honda Motor Company and the Hearst Corporation put in $200 million in new cash. In return, XM agreed to develop data services, such as weather and traffic, for Honda cars.

XM, meanwhile, made several smart moves that allowed it to pull away from rival Sirius. Five years ago, XM decided to develop its radio chipset in-house while Sirius had outsourced its design to Agere Systems. Sirius' chips were delayed and this gave XM a year's head start in the market. For this technology to be successful, both vendors must focus on increasing their subscriber base.

The Vision for Telematics

Just as the Internet has reached high levels of ubiquity in our daily lives, telematics will similarly provide delivery of location-sensitive information seamlessly. Unlike ever before, information and entertainment will be selected, modified, presented, and used relative to location. The impact on how we do business, on how we market products and services, and on how we live will be tremendous. Telematics is in a state of dramatic growth in multiple dimensions. This interest in telematics is from automotive OEMs, wireless carriers, and prospective and existing consumers' customers. While the industry took four years to reach 100,000 customers, there were nearly 1.5 million customers by the end of 2001. Telematics will not only provide customers with added safety and security, but also create a relationship between OEMs and their customers.

Telematics in the vehicle today are a lot like the Internet ten years ago. Back then, few Internet-based applications existed and access was painfully slow. Today, as wireless technologies proliferate and data rates increase, expect to see some truly innovative products for the Internet. Consumer expectations about telematics have shifted as the perceived need for connectivity increases. The car radio is no longer a simple device for tuning in radio broadcasts, or even for playing cassettes or

CDs. Instead, it is becoming a communications center, providing navigation, Internet access, in-vehicle entertainment, automatic emergency assistance calls, and vehicle tracking.

Several companies are focused on developing speech technologies to enhance the user-vehicle interface. In-vehicle computing requires a completely new paradigm, the main impetus for which is to minimize driver distraction. Forget the mouse and keyboard—voice recognition will be the primary input device for vehicle-based computing. Text-to-speech technology will supplant the video monitors of today's desktop systems.

Voice technology is already in production in several high-end, luxury automobiles. It allows drivers to control their phone, audio system, and climate-control system without using their hands. The system can be configured in one of six languages. As a speaker-independent system, it does not require a lengthy training process as with most of today's PC-based voice-recognition software packages.

Two key technologies make telematics products possible. One is GPS technology that can help pinpoint a vehicle's location or provide destination location information to the driver. The second key enabler involves the various wireless technologies—cellular, satellite, wireless LANs, Bluetooth, and others—that provide a means to move data in, around, and out of the vehicle. By combining GPS and wireless technologies, designers can cater to a whole new set of consumer needs. Telematics suppliers have the advantage of historical precedence—namely, the Internet. Much can be learned from the evolution of the Internet and other service-based business models. The vehicle could be another node on the Internet.

Safety is at the core of telematics design and development and will likely follow the cell phone's evolutionary path. Early on, cell phones were primarily used to aid motorists in the event of a breakdown or accident. Now cell phones are essential to modern life. Telematics will likely follow the same evolutionary path: safety/security, communications, convenience, and entertainment.

The second wave of telematics involves hands-free phone dialing. The third wave will provide drivers real-time data such as traffic information and weather updates. Real-time traffic information, combined with a navigation system, will enable dynamic updating of turn-by-turn navigation instructions. This will be a much-anticipated convenience that will help justify the cost of such systems, according to suppliers. Wireless delivery of all forms of data (e-mail, compressed audio/video, and financial information) is integral to the telematics vision, and promises to make the vehicle another access point in cyberspace.

Telematics will reach further into the fabric of society, on both personal and commercial levels. Nearly everyone needs to squeeze more time out of the day and wants to improve their quality of life. Telematics-enabled vehicles will lead to instant billing to the credit card as one pump's gas. Or, if one wants to reserve a parking space, they will be able to communicate and alert a gate that recognizes the car's electronic signature to open and let the person park. Then they will be billed for the service that was used.

Commercial uses, too, are restricted only by imagination. Truck fleets will gather data from engine modules to determine when an individual truck needs maintenance, when a truck's load will reach its destination, and if a load has been tampered with. Commercial fleets already use portable navigation devices to increase predictability and profitability.

Telematics and navigation systems are now almost standard in high-end models and rental cars. Drivers are demanding simple guidance help when traveling in unfamiliar locations, safety and convenience support, and assistance. However, current systems are still rather expensive, and, for the cost-conscious owners of lower-end models, telematics is still more of a tempting option rather than a standard feature.

It is not just the availability of advanced silicon solutions that is driving developments forward. The number of service providers is also increasing rapidly, with most automotive manufacturers and rental companies around the globe offering various telematics and navigation packages. In addition, the infrastructure on which these systems are based is growing. Global positioning systems and cellular communications networks are already well advanced, while digital satellite broadcast services and traffic-monitoring systems are expanding daily.

Automotive navigation systems have evolved rapidly since the introduction of simple GPS systems, with the availability of embedded high-performance microprocessor cores and systems-on-silicon approaches. This delivers open-platform flexibility together with the processing performance for a range of solutions from basic telematics features to fully featured high-end navigation and route-guidance systems.

With a dedicated telematics core, it is possible to add any number of key peripherals to the overall design, such as on-chip GPS correlator, UARTs for direct connection of cellular phones, or a CAN controller to interface with a car's power train network. Other software-based features, such as "dead reckoning," can also be implemented either by drawing data from dedicated sensors or from vehicle-management systems (e.g., wheel sensors and odometer) via the in-vehicle network.

Providing a fully integrated car infotainment/telematics system requires a wide selection of peripherals. Such a system should deliver a much greater range of features, such as dynamic navigation, route calculation, and map matching/rotation as well as car radio, Internet access, and e-mail. The basic car radio has been the standard bearer for in-car entertainment for many years. Recently, however, this has changed as the digital era has taken hold. The introduction of satellite and terrestrial digital radio services changed the audio entertainment picture, and digital technology has vastly improved the listening quality of analog radio. Even today we are seeing digitally enhanced, high-quality car radios with integrated CD players appearing as standard in most vehicles, with an increasing range of other audio systems, such as MP3 players, available as after market sales.

The all-important user interface is evolving into a true multimedia interface. Dedicated media processors offer the performance to handle all required audio, video, and graphics processing. Speech-recognition systems are set to deliver true hands-free control over a wide range of features within the car, from simply turning accessories such as the air conditioning on or off, to automatically muting the various audio sources when answering or making a phone call.

Online Integration

Mobile communications technologies are allowing the introduction of Internet-based services from basic e-mail retrieval to online booking, along with emergency calling in case of an accident. In-car television, DVD players, and gaming consoles are all soon to be available as entertainment for passengers, heightening the enjoyment factor on long journeys. Monumental changes are occurring within the car infotainment market, all of which are possible only as consumer-based automotive technologies come online.

What is required, however, is a lot more than just a selection of individual components and systems. An integrated, comprehensive, and detailed design and development approach is necessary.

In fact, it is the functionality of such systems that dictates how the system is built, with individual blocks being fully interoperable. For example, data exchange between telematics and multimedia processors is a necessity in order to incorporate traffic-routing information into route calculation. This requires both systems to be connected via a bus bridge, in a way that guarantees that bus saturation and interrupts from one system do not adversely affect the performance of the other.

To achieve this interoperability while maintaining time-to-market constraints requires a modular design approach. This not only guarantees hardware compatibility but also ensures that pre-developed software for the telematics processor can be directly reused on a mobile multimedia platform. Additional peripherals such as a graphics accelerator, SDRAM and PCI interfaces, and standard peripheral interfaces can also be integrated. Dedicated systems, from a basic telematics processor right up to a fully-featured, high-end infotainment system can be quickly developed by initially developing a range of core infotainment hardware and software blocks that are fully interoperable and reusable.

Summary

The era of telematics is upon us. Providing consumers with safety, information, and entertainment has a potential of billions of dollars of revenues for automakers, wireless service providers, and consumer electronics manufacturers. However, several challenges exist for this industry. Telematics technology is happening now and has already made a broad and all-encompassing change to the way we think of the facilities available to us while we're traveling. Powerful 32-bit embedded processors, characterized by low cost, minimum power consumption and low heat generation, compact packaging, high speed and advanced software support, and smart connected systems are becoming available to telematics developers worldwide.

Telematics are looking to erase the safety concerns of using cell phones while driving. A study in the New England Journal of Medicine in 1997 found that talking on a cell phone while driving increased the chances of being involved in an accident by four times. Cities such as Brooklyn, Ohio, and New York City have banned the use of cell phones in vehicles. As other cities look to follow suit, the backlash against cell phones could follow a similar backlash against in-car e-mail, which will be delivered as speech. In fact, many believe that the future of telematics will rely heavily on voice recognition. But the technology is not very advanced, and the car is hardly a quiet space.

Another challenge for telematics providers is the lack of open standards. Startups may have to build different versions of their products for each automaker without the existence of uniformity for hardware or operating systems. That would raise costs immensely for bootstrap operations with good ideas. And it could turn into a disappointment for consumers. Clearly, adopting open standards for vehicle networks will wave the green flag for the telematics market. Automakers will have to implement standards-based networks alongside their proprietary networks. The automotive world has, so far, remained insulated from the standardization drive that has helped accelerate the development of other markets. While the Internet, low-cost networking technology, and advances in embedded computing have driven consumer-electronics manufacturers to develop standards for network systems, automotive manufacturers have preferred to develop internal standards that they have kept proprietary from all but the closest partners. Fortunately, the auto industry is beginning to change, and their needs to be cooperation between standards such as MOST, AMI-C, IDB-1394, and others.

eBooks

Introduction

Reading a book or magazine is taking on a new meaning with the emergence of eBooks, a revolutionary technology and product that is changing the way we read.

The hardbound book has been around since the age of Gutenberg, giving readers a highly functional, easy-to-use technology. Paperback books, introduced in the last 60 years, have combined this technology with low cost and convenience to become well established in the publishing world. The time and effort required to photocopy or re-type the contents of a printed book provide a simple but effective form of copyright protection.

The current publishing process for books printed on paper utilizes a complex infrastructure for printing, distribution, and sales. Authors and publishers have time-tested methods for selecting, editing, and marketing books. And they have well-established contractual traditions governing the division of profits.

The creation and distribution of paper books raises additional issues:

- Harvesting trees to create books reduces the ecosystem's capacity to generate oxygen and remove carbon dioxide from the atmosphere.
- Only 60% of an average book's total print run is sold. Additional resources are expended when excess books are recycled back into paper pulp.
- Transportation of books to retail outlets consumes additional resources.
- Purchasing a book usually involves driving a motor vehicle. Either the reader must drive to a bookstore, or someone must deliver the book to the reader.
- Size and weight limitations make it difficult to carry or store large numbers of books.
- A shortage of bookstores and poor distribution channels in many parts of the world make books expensive and difficult to find.

What is an eBook?

The eBook book is an electronic version of content normally contained in newspapers, magazines or books. The content is created and stored in a computer file format that can be accessed by a variety of computer hardware and software applications. The content can be as unique as the electronic medium itself and contain audio, video, or live hyperlinks. Or it can be as familiar as its print counterparts. eBook content can be downloaded from the Web or received as an email file attachment. eBooks on diskette or CD-ROM are sent through the mail and sold in bookstores.

An eBook is similar to a paper book and contains:

- Cover art
- Title page

- ISBN number
- Copyright notice
- Editor
- Publisher.

eBooks allow users to:

- Electronically search the text for specific words and phrases
- Change the font size and style
- Type notes electronically and organize them within the text
- Create highlights and bookmarks
- Hyperlink to specific parts of the text.

In the future, eBooks will be able to link to other websites with related topics and access a dictionary that pronounces the words aloud.

Benefits of Using eBooks

- Once a book has been converted to an electronic form (such as Microsoft Word or Adobe Acrobat document), it can be stored and transmitted at minimal cost and impact to the environment.
- Storing eBooks on computer drives, diskettes, and CD-ROMs take less shelf space and weighs less than printed books. Dedicated handheld reading devices weigh approximately 15 ounces and store up to twenty books. Adding memory allows the user to store additional books.
- Updated documents can be downloaded and accessed immediately.
- Storing books in electronic format reduces warehousing and shipping costs.
- Writers and publishers have the freedom to explore small niche markets bringing readers original works unlimited by genre lines, market size, or print capabilities.
- eBooks can be enhanced with live hyperlinks, sound, animation, and simulation capabilities.
- Using the self-contained software users can bookmark, annotate, and search the entire eBook.
- eBooks files can be carried and accessed anywhere with a portable PC, PDA, or eBook reading device
- Business and recreational travelers can download eBooks to their notebook PCs without adding weight or taking up space in their luggage.

Professionals such as doctors, lawyers, and pharmacists are already utilizing eBooks. Businesses, government personnel, colleges, universities, and schools are attracted to eBooks for a number of reasons. EBooks offer convenient storage capacities and the ability to update and download the most current documents. Schools, colleges, universities, and libraries are piloting content delivery via portable and desktop PCs and eBook reading devices. These provide students with the latest e-textbooks and digital libraries without adding weight to backpacks.

People with special reading needs now have more books and magazines available to them with eBooks. A simple font size change turns an eBook into a large print edition and some file formats are compatible with screen reading technology.

Overall, eBooks provide improved onscreen reading quality, portability, storage capabilities, current content, and quick delivery to readers. Consumers have access to full-length novels and texts, annotated editions, short fiction and nonfiction, magazines, articles, news and current events.

Reading eBooks

- eBooks can be read using computers, and handheld eBook reading devices such as the REB1100™ or REB1200™ (licensed by Gemstar-TV Guide International and manufactured by RCA). Some file formats allow you to print pages and read them on paper.

Desktop PC users can read eBooks anytime they use a computer that has software compatible with their eBook file format. eBooks can be transported anywhere the user brings their portable PC, PDA or eBook reading device.

eBooks come in standardized file formats such as:

- HTML
- PDF
- RTF
- Palm OS
- Windows CE
- eBook Formats

eBook reading software is often available as a free download and offers the following advantages over using standardized applications on your PC or PDA:

- Better onscreen reading quality
- Library-style book management
- Book-marking
- Annotation
- Adjustable font sizes
- Content search capabilities
- Access to library of reading material from software vendor's website
- Ability to create your own eBook files using publishing software

An eBook can be defined as an electronic version of a book that requires both hardware and software for onscreen viewing. The content can be contained in a downloadable file or placed on a CD. If viewed from this perspective, eBooks could be considered software.

Software is intellectual property (IP) that is traditionally licensed to a specific user. After registering, the user is generally entitled to technical support, bug fixes, and minor updates. Print book owners have traditionally been able to resell or trade their book purchases. However, reselling licensed software is often prohibited because it is too simple and inexpensive to replicate and distribute unauthorized copies.

eBooks have now joined today's publishing mix along with hard covers, trade paperbacks, mass-market paperbacks, and audio books. In its simplest form, eBooks contain an electronic version of the text from a paper book. But in its advanced form an eBook's content can contain film, animation, and audio. eBooks can link to a myriad of websites where readers can find additional information, be entertained, learn more about the author, and more. While printed research may become outdated a year or two after it is printed, eBook research can be updated, rewritten, corrected, and remain fresh.

Market

In its first three years the eBook industry has attracted major hardware, software, and publishing companies. According to Forrester Research sales of downloaded books totaled $12 million in 1999. The industry marked a milestone in 2000 with the distribution of 400,000 copies of the first exclusive eBook release, Stephen King's *Riding the Bullet*. Forrester Research estimates that eBook downloads could constitute a $426 million market by 2004. Global consulting firm Accenture projects the overall eBook industry could be worth several billion dollars by 2006-2007.

To accelerate eBook acceptance, the Open eBook Forum (OeBF) was established by the leading international trade and standards organizations in the electronic publishing industry. The members of the OeBF consist of hardware and software companies, publishers, accessibility advocates, authors, readers of electronic books, and related organizations. Members of the OeBF share common goals for establishing specifications and standards to advance the electronic publishing industry. The forum's work is intended to foster the development of applications and products that will benefit content creators, manufacturers of reading systems, and consumers.

Copy Protection

Technology providers, publishers, and device makers are collaborating to develop standards and copy protection mechanisms to avoid the pitfalls that have tripped up other trailblazers in digital content. Issues that need to be standardized include content platforms, file formats, and systems for copy protection, and the secure exchange of content. Government officials and industry executives are advising the eBook industry to act swiftly to avoid the same types of problems that have plagued the music and video industries—piracy of MP3 audio and disabling the content scrambling protection for DVDs.

To be profitable the eBook industry must protect their most valuable asset, the content, by focusing on digital rights management and copy protection. Early eBooks were protected from piracy by requirements for specialized electronic devices made available through exclusive deals between the device manufacturer and the book publisher. Today, eBook content can be read on any Windows PC using standardized software such as Acrobat Reader, from Adobe Systems Inc., or Microsoft's Microsoft Reader software.

eBooks can also be read with products such as:

- A Palm PDA running Peanut Press reader software from Peanut Press Inc.
- Aportis Doc software from Aportis Technologies
- TealDoc from TealPoint Software.

eBooks support on Windows CE or Pocket PC devices is provided via Microsoft Reader or the Peanut Press reader. Support for specialty eBook hardware such as Gemstar's Nuvomedia or Softbook's eBook platform is available via reader software based on the Open eBook (OEB) File Format.

Consumers expect that when they purchase eBook content they can view it on any of the devices currently available. This is not always possible because of the use of different platforms and file formats. If the read-anywhere eBook is to become a reality then standards must be established for content and file formats, digital rights management, distribution, and book product information. The recently released publication structure and file format standards from the OEB Forum are still in development and only address electronic content. Standards on device classes and how content is displayed remain unresolved.

Portability must be addressed in the next 18 to 24 months if the industry hopes to see convergence devices accommodate the eBook technology with telephony, other PDA content, and DVD storage.

The lack of a standard file format is a major barrier to market growth. Publishers converting printed texts undergo time-consuming and costly processes to accommodating divergent file formats for existing reader systems. Microsoft and Adobe are locked in a battle for platform dominance. Until standards are established publishers need to support all eBook platforms. Adobe's Portable Document Format (PDF) is the market leader for existing digital content. But PDF's conventional-sized page display is not optimized for small screens. eBook readers that support the OEB File Format, such as Microsoft Reader, the Peanut Press reader, Aportis Doc, and TealDoc are growing in popularity. However, translating files from Adobe Systems' PDF to the OEB file format is a significant task for publishers.

The largest obstacle to a dramatic increase of eBook titles is that publishers are reluctant to release additional titles until there is a standard and secure method for exchanging and protecting eBook content. One industry effort that is working towards a standard for secure content exchange is the Electronic Book Exchange System (EBX). Spearheaded by Glassbook Inc. and the EBX Working Group the EBX system currently under development would secure content transfers via public/private key encryption.

The industry is also considering other technologies for content exchange and digital rights management (DRM). Extensible Rights Markup Language (XrML), a secure and royalty-free language developed by Xerox Corp. and ContentGuard Inc., is under consideration as a standard for all DRM systems.

Time is running out for the eBook industry to stay a step ahead of the piracy problem. Some eBook titles have already been pirated. Several illegal postings of the latest in the Harry Potter series appeared for download shortly after the volume's print release. In addition, some users are suspected of having downloaded copies of King's *Ride the Bullet* without payment.

Making digital content widely available to consumers will help control piracy if publishers move quickly to provide digital content through a variety of channels. Consumers must be educated that selling, obtaining, or using unlicensed electronic content it is a criminal offense.

Technology Basics

eBook content is converted to an electronic file format similar to a Microsoft Word or Adobe Acrobat document. The largest cost in creating eBooks is related to the display technology. Our ability to inexpensively convert information into electronic format has outpaced our ability to display them clearly. eBook companies are challenged to utilize current technology such as silicon-based integrated circuits, lithium ion batteries, and liquid crystal displays to mimic the look and feel of a paper book.

A book is not merely a collection of words, it is also the container for those words. If eBooks are to take their place alongside and perhaps replace paper books, hardware and software improvements are needed to make reading an eBook as comfortable and convenient as reading a paper book. Currently none of the devices on the market meet this standard. This is not an indictment of the companies that make eBook devices or of the people who buy them. It is simply an acknowledgement of the opportunities for improvement. The current crop of eBook dedicated reading devices and multi-purpose PDA devices offer attractive features for the early-adopter and average tech-friendly reader.

eBook Technology Options

There are three main ways to read an eBook:

- On a dedicated reading device such as the REB1100™ or REB1200™ from RCA
- On a PDA or other multi-purpose device such as a Palm handheld, the Franklin eBookman, or a PocketPC device
- On a desktop or laptop PC using software from Microsoft, Adobe, or a variety of smaller vendors

Most current eBook devices and programs rely on competing and mutually incompatible file formats including:

- .rb (Gemstar)
- .lit (Microsoft)
- .pdf (Adobe).

Lack of a single standardized format requires ePublishers to convert their content into multiple formats. Consumers are inconvenienced when content purchased for one device or program is incompatible with others. And there is no indication that the industry is close to selecting a standard format. A temporary solution under development is a cross-platform standard called OEB. This standard would enable publishers to digitize content then convert it to any OEB-compliant format for final delivery.

Manufacturers and Products

The technology, key manufacturers, and their products for reading eBooks are described below.

Dedicated Reading Devices

A dedicated reading device is the best choice for purists who insist on a reading experience as close as possible to reading a paper book. These devices consist of a large, touch-sensitive LCD screen inside a book or tablet-sized plastic casing.

Dedicated reading device controls include simple buttons or a rocker switch for moving forward and backward in the text and onscreen menus for advanced functions like bookmarks, highlighting, annotation, key word search, and a dictionary. Dedicated reading device screens hold less text than a hardcover or paperback page. Unlike smaller devices, they hold enough text to avoid having to flip pages every few seconds. Device owners must purchase specially formatted content either with their PCs at online bookstores such as Barnesandnoble.com or directly from the device maker via a built-in modem.

One of the best selling dedicated reading devices is the RCA REB1100. It weighs 1.1 lbs. and sells for less than $300. Text is displayed on a 3 x 4.5 inch, backlit, black-and-white LCD screen. Gemstar-TV Guide International licenses REB1100 eBook technology to RCA. Gemstar also owns NuvoMedia, manufacturer of the Rocket eBook, and SoftBook Press, the manufacturer of the SoftBook Reader.

Owners of the Rocket eBook purchase and download titles from the Internet and transfer them to their devices via a serial cable. After acquiring NuvoMedia, Gemstar modified the second generation NuvoMedia and added a modem. This allowed Genstar to sell content through interactive catalogs downloaded via a phone connection. This telephone-based catalog ordering system is based on the original SoftBook Reader, which has now been replaced by the RCA REB1200. The REB1200 weighs 2.1 lbs. and sells for less than $700. The REB1200 has a 4.9 in. x 6.5 in. color LCD screen.

Gemstar is working with major publishers to convert their latest titles from established authors to the proprietary REB1100 and REB1200 formats.

RCA/Gemstar products are the only dedicated reading devices currently being sold in the U.S. Several other companies are developing devices for the North American market or selling them in Europe. In 2000, the French firm Cytale introduced the Cybook. Cybook has a 6.3 in. x 8.3 in. color LCD screen. The goReader device, under development by a Chicago company, weighs about 5 lbs., has a 7.3 in. x 9.5 in non-glare color LCD touch screen. The goReader's primary market is college-level electronic textbooks. Microsoft's Myfriend is optimized for the Microsoft Reader eBook software and is being built by Italian manufacturer IPM-NET. Myfriend weights 1.75 lbs. and will cost about $1200. It features a high-resolution 640 x 960 pixels on a 6.2-in. x 4.2-in. area color LCD screen.

Only dedicated eBook devices offer full security features. Unique ID's for each device allows retailers to encrypt content being downloaded. With publishers worried about content piracy, many of the latest best-selling trade books are only available for these platforms.

Multi-purpose Devices (including PDAs)

While many bestsellers are not yet available for nondedicated devices, there is plenty of content to read. PDA's are probably the most common way people are introduced to electronic books. Palm alone has sold 13 million of its PalmOS PDAs. Reading an eBook on a Palm device can be an enjoyable pastime. Palms excel in tight quarters or traveling. The Palm devices' screen displays the same width as a column of newspaper text. Some of the popular PDA product manufacturers are:

- Casio
- HP
- Handspring
- Sony.

For several years PalmOS software such as Peanut Reader, AportisDoc, Qvadis Express, and Mobipocket has allowed owners of PalmOS devices to read eBooks on PDAs. Palm, Inc. acquired Peanutpress the developer of Peanut Reader and pre-installs a version of the software called Palm Reader on its Palm m500 line (A PocketPC version of Palm Reader is also available.). Readers download content from the online Peanutpress bookstore and transfer it to their devices with the HotSync procedure used to install new software. The Palm's screen size of about 2.25 inches and resolution of 160 x 160 pixels can only display 12–14 lines of text fit on each page. The small size results in frequent page flipping.

The Franklin eBookMan is another option in the multi-purpose eBook device market. The eBookMan dedicates most of its front surface to a large (200 x 240 pixel) grayscale LCD screen. The eBookMan includes organizer-type software such as a calendar, address book, contact list, and entertainment software. Franklin's eBookMan 900, 901, and 911 models are available for $150 to $250.

Handheld computers running Microsoft's PocketPC operating system are Palm Inc.'s major competition in the eBook market. Handheld computers manufactured by Casio and Hewlett-Packard sell between $359 and $599. Handheld computers are heavier and bulkier than the PalmOS devices, have larger color LCD screens (240 x 320 pixel), and utilize Microsoft's ClearType system. ClearType is a feature of PocketPC market leader Microsoft Reader. ClearType's "sub-pixel render-ing" smoothes the edge of onscreen fonts by adjusting the red, green, and blue elements in each pixel. ClearType and color screens allow handheld computers to optimize graphics and text. PocketPC devices display eBook content better than the Franklin eBookMan and PalmOS devices. However, small screens remain a major limitation for eBook consumers.

PC-based eBook Software

Computers are the third option for reading an eBook. PC based eBook software is focused on reading rather than developing content. Security measures are also included to protect the content from duplication. eBook software strives to simulate the experience of reading a paper book.

The two major products in this market are Microsoft Reader and the Adobe Acrobat eBook Reader. Both products are available for free download. Microsoft Reader includes ClearType (only effective on LCD screens) to create the look and feel of paper books. Readers see generous white space and line leading, running headers and page numbers, and an easily accessible table of contents. Acrobat eBook Reader provides greater graphical and multimedia capabilities and better text interaction (in the form of highlighting and annotation). A growing list of original and converted eBook titles is available for both platforms.

Challenges

Current eBook reading devices offer the advantages of capacity, connectivity, and convenience. But, no device is as readable, portable, or reliable as a paper book. The larger screens for dedicated eBook devices increase their weight and the high manufacturing cost of LCD screens dictate higher prices. PDAs are more portable than dedicated devices but have small hard-to-read screens. Despite the advances in hardware and software PC screens are not comfortable for extended reading. Numerous studies show that users tend to print any document greater than three screens in length.

In order to appeal to a mass market audience an eBook device would have the following features:

- Thin (significantly less than 1 inch thick)
- Light (weighs less than 1 pound)
- Large screen approximating the size of a hardcover book.
- High-resolution screen (at least 200 dots per inch). The display should be full-color, readable in sunlight, and with a contrast ratio approaching that of paper.
- Multi-gigabyte memory storage
- Low power consumption and very long battery life (measured in days or weeks).
- Cost less than $100

One way to increase the acceptance of eBooks is for publishers to price them significantly below printed content. This would allow consumers to offset hardware costs after a number of eBook purchases. However, hardware designers must still cut costs.

Current display, storage, and battery technologies are barriers to further eBook growth. But certain innovations under development provide hope:

- Electronic paper (by E-Ink and Gyricon Media)
- Bi-stable color displays (by Advanced Display Systems)
- Roll-to-roll electronics manufacturing using flexible substrates such as plastic (by Lucent and Rolltronics Corp.).

These emerging technologies may lead to displays that do not resemble "devices" at all but merely sheets of self-printing rewritable paper. And that might be something Gutenberg could relate to.

Summary

Technology is emerging to convert all readable documents into digitized content. Digital content is changing the way information is bought, stored, and read. Exciting new hardware, software, and technologies continue to emerge. The publishing market is developing improved strategies to encrypt, format, and price eBooks. Media, consumer, and technology giants are vying for a piece of the multi-billion dollar printable media industry. Interestingly, a U.S. District Court judge in New York has passed a ruling that gives authors rather than their publishers the electronic rights to previously printed books.

At the same time, the growth of digital content is being inhibited by undeveloped standards for content platforms, file formats, copy protection, and the secure exchange of content.

eBooks will never totally replace printed books, but they are an experiment in reading form and function with capabilities to take us light years from Gutenberg's press.

Other Emerging and Traditional Consumer Electronic Devices

Introduction

While we have discussed a number of consumer devices and PCs, there are other appliances emerging with a good chance to penetrate the consumer market in the near future. Depending on their acceptance by consumers, cost, and business models, some of these products may fail and their functionality integrated within other appliances. Many of these appliances will complement the PC. Some of these appliances are fixtures in consumer homes. Their value and acceptance will be affected by the introduction of other consumer products. For example, the introduction of digital answering machines and voice mail servers reduced the market for stand-alone analog answering machines. The following products will affect the consumer market over the next few years. Also discussed are traditional appliances such as VCRs and pagers.

NetTV

NetTVs are TV-centered consumer appliances that provide Internet access using the TV as their primary display. These standalone products are set on top of the TV ("set-top"). They also include TVs with built-in Internet connectivity hardware. A typical NetTV includes:

- Communications module (modem)
- Core processor
- Operating system
- Display driver
- Applications such as a Web browser and e-mail client

NetTV manufacturers offer two options. Basic service includes limited interactive electronic programming guides or customized information tickers. Advanced services provide full graphical Web browsing, e-mail, and streaming video. Analysts predict that worldwide shipments for NetTV products will exceed 19 million units in 2004.

Key market accelerators include:

- Infrastructure upgrade: back end to clients
- Consumer interest in the Internet and new services
- Market need to generate new revenue streams

Some market inhibitors limiting the NetTVs market are:

- Lack of consumer interest in interactive services
- High costs
- Regulatory issues

In the future NetTV's functionality may be integrated in set-top boxes, digital TVs, enhanced traditional cable boxes, and direct satellite receiver devices.

E-mail Terminals

Most people think of a PC as the tool of choice for sending e-mail. However, a new breed of devices focused only on e-mail applications are springing up. Now users can send and receive e-mail using e-mail terminals. These standalone devices provide access to e-mail without Web browsing capabilities. E-mail terminals appeal to users who don't have access to a PC or want something easier to use. A basic e-mail terminal contains:

- Keyboard with soft-function keys
- Small built-in LCD display
- Modem
- Printer port

Upgraded models can include:

- Ability to change font sizes
- E-mail filters
- Increased e-mail storage capacity
- Address book
- Memory expansion slots

Some larger models offer full-sized keyboards, while smaller models come with small keyboards designed to fit their diminutive sizes.

E-mail terminals do not include Web browsers or allow the users to add applications. E-mail terminals use hard wire technology to provide Internet connectivity and are intended for use in residences. Devices such as Palm PDAs and Blackberry pagers provide e-mail access but are not considered e-mail terminals.

Recently released e-mail terminals offer users more than just simple text. Newer models allow you to open attachments such as HTML documents, digital photos, and pictures. Users can view, save, and print photo attachments in JPEG and GIF formats. These e-mail terminals include a photo album for storing a limited number of photos. Local storage allows e-mail storage, spell checker, calculator, and an address book.

These products typically use analog modems (56 Kb/s). Powered by 2–4 AA or AAA batteries, these devices can be used in any room with a phone jack. Newer models with a 900 MHz or 2.4 GHz RF interface for access to a cordless base station can be used throughout the house. Most models have a monochrome text-only screen and parallel printer port. High-end e-mail terminals capable of supporting graphics utilize a 6″ x 2.25″ 16-level grayscale display. Onboard memory provides storage for about 800 e-mail messages. The e-mail terminal weighs less than two pounds.

Low cost and simplicity have fueled the acceptance of e-mail terminals. The simplified design translates to a street price of under $100. Worldwide e-mail terminal shipments for 2005 are predicted to exceed 1.7 million units, generating revenue in excess of $200 million.

E-mail terminals are being sold in a wide variety of retail chains, including electronics stores such as Best Buy and Circuit City and mass-market stores such as Target and Wal-Mart. This strategy exposes e-mail terminals to a large segment of the consumer market. Service providers such as EarthLink bundle a charge for the e-mail terminal with a monthly service fee based on the types of services the consumer requires. E-Mail terminals can also provide one-touch access to daily weather, TV listings, and more. E-mail service providers offer nationwide toll-free dial-up numbers.

Market leaders for these services include:

- EarthLink (through their CIDCO acquisition)
- Landel Telecom
- ATLINKS (a joint venture between Alcatel and Thomson Multimedia)
- Askey Computer Company
- Sharp
- Vtech

The restricted functions of an e-mail terminal limit its overall appeal within the consumer market. E-Mail terminals could see more appeal and higher adoption rates in foreign markets. Higher long-distance charges in international markets make e-mail a cost effective alternative. Distribution channels in international markets are poorly defined with only telephone companies selling e-mail terminals.

Power users will continue to use PCs and laptops for e-mail. E-mail devices primarily appeal to less technical e-mail users. Lack of access to Internet services is not a barrier to acceptance for this market segment.

Wireless E-mail Devices

Wireless e-mail devices provide users with access to e-mail while traveling. The business financial community is embracing this technology. Wireless messaging (e-mail) devices support e-mail, light applications, and Personal Information Management (PIM). These devices provide improved connectivity for mobile professionals. The most popular device is the BlackBerry product produced by Research In Motion. Other wireless e-mail devices include BellSouth's MailBug, Motorola's Talkabout T900, and Motorola's Timeport P935 with Arch Wireless service.

Productivity savings are used to offset the cost of a wireless e-mail device. Users can utilize one hour per day of otherwise unproductive time and cut wireless phone use by 15%. This product is designed as a serious business tool, not a handheld game complete with address book, and calendar. The device uses a black-and-white screen and tiny keyboard to access a desktop e-mail account. Users check and respond to their desktop e-mail messages without dialing into their corporate network. Manufacturers of wireless e-mail devices have resisted adding multimedia features or color screens arguing it would reduce battery life and increase cost.

These devices make use of otherwise idle time—such as time spent commuting or waiting for meetings—to read and respond to e-mail. Users say that on average 53 minutes per day of previously unproductive time was converted into productive time. BlackBerry users also report they spend less time using laptop computers and wireless phones. Being able to immediately respond to e-mail reduced network dial-in and phone conversation times. The annual ownership cost of a BlackBerry device is virtually offset by direct savings from reducing other communications costs. This product is being used by 300,000 subscribers in 13,000 organizations. The typical user is away from their office 39% of the time and sends and receives about 8,500 time-sensitive e-mails per year. Such examples highlight the impact wireless e-mail devices are having on businesses. In the near future, cell phones and PDA's with wireless e-mail functionality will compete with traditional wireless e-mail devices.

Pagers

Pagers can alert a person to call back, store a message, or display additional information. Each pager has a unique identification number. The caller dials this number and then can leave a phone number or message. The pager service carrier transmits the information to the pager's display panel.

Early pagers only displayed phone numbers. Newer pagers display alphanumeric messages and offer scrolling for long messages. Some offer access to full-page text documents and e-mail. Others can receive voice mail, news, sports, stock quotations, lottery numbers, traffic, and weather reports. Information and service suppliers transmit this information to the pager. Many pagers allow business users and consumers to respond via two-way networks. These pagers often provide preset responses and tiny keyboards enable users to compose personalized responses.

Cell phones with integrated pager functionality are attracting an increased share of the overall market for pagers.

Internet-Enabled Digital Picture Frames

Just a few years ago, the idea of displaying a photo in a picture frame around the world in less than an hour seemed like a dream. We all have pictures we want to share with at least one technology-challenged friend or relative. The perfect device for sharing photos with these people is the digital picture frame.

This picture frame allows people without a computer or computer skills to enjoy digital photos. The person who wants to share purchases the digital picture frame, creates an account ($50 to $100 a year), and sends the frame to a friend or relative. Once the account is established the computer user uploads photos to the service provider's web site. Each night the picture frame uses a phone line to connect to the service provider and download any new pictures. The new pictures are then automatically added to the picture frame's slideshow.

The Web site provides frame setting controls to:
- Adjust the time the frame turns on and off
- Slideshow interval (how long photos display)
- Dial-up phone numbers

Using a digital picture frame requires a regular phone line, power outlet, and a subscription to the provider's network. Several start-up companies are using innovative ways of exchanging digital images without needing a computer or Internet access.

The digital picture frame is the size and shape of an ordinary picture frame. The frame contains an LCD screen with multiple photos in a slideshow format. The frame connects to the Internet via a phone line to download new pictures to display on the screen. The digital picture frame style or color can be modified to fit with almost any decor. A large (approximately 8 in. x 10 in.) high-resolution LCD screen provides easy viewing. Smaller and less expensive picture frames are also available.

Options for viewing photos from the digital picture frame are:
- Pausing the frame on a single image
- Using the slide show feature for viewing multiple pictures
- Clicking through the pictures at your own pace

The frame can display more than just photos. Users can add personal messages and turn photos into one-of-a-kind post cards. Famous works of art or other images can be displayed from the service provider's library. The frame can also display customized information including sports scores, news headlines, lottery numbers, and even local weather reports.

Each night between midnight and dawn the picture frame connects via a local access number to the service provider's network. Using a dual phone jack allows your phone and picture frame to plug into a single wall jack.

A subscription to the service provider's network is required to receive photos and the other information. The typical \$5–\$10 monthly fee is less than purchasing and developing a single roll of film.

Other frames contain memory slots that allow the consumer to display images from their digital camera without a computer or subscription fees. Users simply insert a CompactFlash™ or SmartMedia™ memory card from the digital camera into the digital frame.

Polaroid, Kodak, Kensington, Digi-Frame, and Ceiva are leading digital picture frame manufactures.

Inside the Digital Picture Frame

The frame is actually a very simple computer. It has most of the same components as the computer on your desktop in a simpler form designed to perform a single task. Some of the key components inside the digital picture frame include:

- **Processor** – The central processing unit (CPU) is similar to those used in electronic handheld games. The most processor-intensive task is downloading pictures from the Web site.
- **Memory** – The frame's operating systems is stored in ROM memory. Flash memory stores the pictures, settings and a portion of the operating software. Persistent ROM and Flash memory prevent data loss when the unit is unplugged.
- **Modem** – The frame uses a 56 Kb/s modem connect to the Internet and download new photos.
- **Display** – Typically a 640 x 480-pixel, passive-matrix LCD with a viewing area of either 8″ x 10″ or 5″ x 7″. Using a thin display means the digital frame is not much thicker than an ordinary picture frame. Approximately 4,100 different colors can be presented on the 12-bit color screen.
- **Controls** – Buttons adjust the brightness of the display, turn the frame on when the user first plugs it in, and initiate manual dialing.
- **Operating System** – The embedded operating system is similar to those designed for PDAs and electrical-testing equipment.

Creating a Digital Gallery

Once plugged into a phone line and power outlet, the frame displays the originally stored pictures. Pressing and holding a button on the back of the frame initiates a dial-in session to the Internet. The Internet connection is only used to download new photos and settings to the frame. The device is designed to behave like a picture frame and not a computer, so there is no interactivity, web access, or e-mail.

The frame connects to the Internet and logs into the provider's server using a unique serial number. The frame then compares the pictures in memory to those stored on the server and downloads any new ones. The frame also downloads any new settings. When this is completed, the frame hangs up the phone and starts displaying the new pictures. The frame turns itself on in the morning and off in the evenings based on user preference.

The Web Side

Once purchased, the frame needs to be registered at the service provider's Web site. Payment arrangements are also made for the ongoing service. This process creates an account for uploading pictures and adjusting the settings on the frame.

Now anyone can use the frame. Plugging the frame into a phone line and a power outlet and hitting the control button on the back starts the activation process—simple tasks for even the most technologically challenged people. The first time, the frame dials a toll-free number and downloads the settings you created on the Web site. One of the settings is a local dial-up number for the frame. The frame makes a second connection to the Internet using the local dial-up number and downloads the pictures you posted on the provider's Web site. By registering the frame on the Web site, you establish a personalized area to control the frames on your account.

At this personalized section the user can:

- Set up channels – Various content providers have agreements with photo service providers for daily content in the form of images that can be downloaded by the frame. For example, the Weather Channel provides an updated local forecast each time the frame dials in.
- Send pictures – Designate pictures to be sent to the frames you have registered. Frames can be set to randomly select images from your album. Other users can grant you permission to send pictures to their frames. Online storage at the website is approximately 1,000 pictures. Users can also e-mail photos from their album to anyone with a computer and e-mail account.
- See what's on the frame – See a list of pictures currently displaying on each frame on your account. A list of pictures waiting to be downloaded and ones that have been deleted are also displayed.
- Change the frame's settings – Allows control of the frame name, slideshow interval, time when the frame turns on and off, and local dial-up numbers.

Tomorrow's Digital Frames

The next-generation digital picture frame will allow users to:

- Print pictures from a local printer
- Print and mail pictures using third-party photo service providers
- Play recorded sounds with each picture
- Display pictures directly from a digital camera using a flash memory slot
- Adjust features such as print settings and slideshow intervals via small remote control

Even with these added features, the frame can be controlled from the Web. This functionality maintains the hassle-free interaction that has attracted many frame owners.

Pen Computing or Digital Notepad

Most of us are familiar with PDAs (Personal Digital Assistant) and notepads. PDAs excel at storing information like phone numbers, addresses, and to-do lists. Despite the popularity of portable computers, most students, lawyers, and business executives prefer to take notes, scribble down ideas, draw pictures, write letters, or take down meeting minutes, using pen and paper. Many business professionals have a need to organize, archive and distribute their handwritten notes. They also see a need for a product that can save Web addresses found in articles or ads. Notes taken with conventional notepads are hard to organize and retain, and paper notes have a nasty habit of disappearing just when you need them. Recently, pen-computing (also known as digital notepad) products have been introduced. The products utilize a paper pad set on a handwriting recognition-enabled tablet. A special pen translates the users notes into digital format for downloading to a PC.

These products are designed to bridge the gap between paper and computer. A special pen is used to write on a pad of 8-inch by 11-inch paper secured to the digital notepad. The 1MB of flash

ROM can store up to 100 pages of notes, sketches, and diagrams. Handwritten notes can be uploaded in seconds with a serial cable connection between the digital notepad and the PC. Special PC software displays the handwritten notes that can be filed, reorganized, faxed, e-mailed, or printed.

Companies developing pen-computing products include the Cross-Pen Computing division of A.T. Cross, Digital Ink, and E-Ink. The most famous of these companies, who also has a big name in the regular pen business, is A.T. Cross. Cross-Pen Computing, in partnership with IBM, develops pen-based tools for mobile data input and manipulation.

Inside the Digital Notepad

CrossPad by Cross (Pen Computing Group) is described as a Portable Digital Notepad (PDN). The product contains a writing board like tablet called the Pad. Users insert a paper notepad into the holder and take notes with the CrossWriter Digital Pen. The pen looks and handles like its ink-based counterpart. The pen contains a miniature low-power radio-frequency transmitter, a AAAA battery, and a redesigned compact ink source. As the user's hand moves the pen across the page, the pen transmits the movement to an x-y grid inside the the digital notepad. The pad's microprocessor decodes each pen position into x-y coordinates and stores them in memory. Ink Manager software converts those coordinates into handwriting. Users number each page and press the "page-forward" or "page-back" buttons on the dashboard. This allows CrossPad to put their notations on the correct page. Recent increases in processor speed and memory make this kind of calculation-intensive application possible.

The CrossPad is powered by four AAA batteries (average life 3-4 months). The digital pen is powered by a single AAA battery (average life 6-12 months). Its light weight and small size (only slightly bigger than the notepad), make the CrossPad very portable. Since you are writing on paper with a Cross pen the experience mimics using a regular notepad. However, these notes are digitally stored. Once you have finished your notes, you upload them to your PC with CrossPad's data transfer cable and the IBM Ink software.

Uploaded notes are sorted using the dashboard. Users can mark a keyword or phrase for later reference, for example, a supplier's name or the title of a project. The software then provides the capability to search for documents based on keyword notations or by creation and upload dates.

Once the notes are uploaded to your PC the IBM Ink Manager software allows you to:
- Organize notes based on keywords
- Save notes on your hard disk or other media
- Cut and paste sketches, diagrams, and signatures to other applications
- Search the notes using user identified keywords or bookmarks
- Save notes in a variety of formats like .exe, .ps, .jpg, .tif,
- E-mail handwritten notes and a self-executing viewer software

IBM also has released a software developer's kit to enable users to create CrossPad forms and applications for the legal, medical, insurance, finance, and transportation industries.

Ink Manager can transcribe certain handwritten notes into ASCII text. Success depends largely upon the neatness and regularity of the user's handwriting. Recognition can reach 90% or higher if the user "trains" the software to recognize their handwriting.

Digital Ink, Inc. has developed a wireless pen technology that allows users to communicate with handwritten text or drawings. The n-scribe system looks and writes like a regular pen. N-scribe allows users to transmit handwritten information over cellular phones or other Internet-ready appliances to the Web. This information can then be distributed as an instant message, e-mail, or fax, or stored for future reference. The n-scribe pen, with its proprietary patent-pending technology can be packaged for use as an accessory to today's most popular electronic products and services. This list includes cell phones, handheld computers, desktops, and laptop computers. Digital Ink also plans to develop, license, and sell its technology to a variety of strategic partners to create specialized products for the health care, education, consumer, and entertainment markets.

Another venture, E-Ink has developed a display called Immedia using technology developed at the MIT Media Labs. Immedia technology uses a liquid ink containing millions of microcapsules filled with light-colored particles suspended in dark-colored dye. This ink is applied to sheets of plastic and laminated to a layer of circuitry. When an electric charge is applied to one or more microcapsules, the light-colored particles move either toward or away from the charge. The dark dye or the light-colored particles are then exposed to the outside. Common display drivers control which microcapsules receive a charge. The ink can be printed on almost any surface, requires little power, and can be used on large surface areas.

This revolutionary product is targeted at students, executives and other people who take notes and later use these notes to write letters, memos, research papers, and books.

Robot Animals—Robot Dogs

Advancements in robotics and artificial intelligence are fueling the convergence of electrical and mechanical equipment technologies. One such result is the introduction of robot animals. Sony, Toshiba, Matsushita, and Tiger Electronics have all introduced robotic animals. Robotics animals include dogs, cats, snakes, and fish. The most popular of these animals are robotic dogs.

Robo-pups are designed for entertainment and as replacements for their animal counterparts. The latest generation of robo-pups cost $200–$1,500, the equivalent of a weekly food bill or mortgage payment. Still vendors believe that there is a market for these luxury robotic pets. Apart from their entertainment value, robotic dogs can also benefit the visually impaired.

Robotic dogs combine 16 different motors, sophisticated sensors, and a remote control. The robotic dog walks, responds to voice and clap commands, and performs tricks like headstands and push-ups. When it's time to tinkle, the robo-dog lifts a back leg and makes a musical sound, but leaves no mess. Robotic dogs such as the second-generation Aibo from Sony, with a head-mounted digital camera, carry a hefty $1,500 price tag.

The target market for these products are tech-savvy men and women ages 18–24 and baby boomers in their 40s. Retailers who cater to a more affluent market segment such as The Sharper Image and Neiman Marcus provide the primary distribution channels for this product. Robotic animals are available with messenger software that alerts owners they have e-mail. They can also fetch the digital news from text-based Web sites via a home PC and wireless local area network. You can also program the dog for more entertaining tasks. Using navigation software you can view streaming video from his camera on your PC. Robotic animals are also great for tasks like calling the kids to dinner. You record your voice on a PC using an inexpensive microphone and a .WAV file, and Aibo will go to their room and speak in your voice. Aibo is programmed to perform random movements, creating the illusion that it has a mind of its own. Upgrades allow it to respond to voice commands such as sit, dance, and to raise a metallic paw when its name is called.

But don't expect this robotic dog to fetch your slippers. The robot, a fad in gadget-crazy Japan, is meant to soothe its owner rather than perform menial tasks. Turning robots into home appliances will require more sophisticated technology.

Sony Memory Stick software gives the Aibo series the ability to recognize additional voice commands. The software also allows the head-mounted digital camera to take. JPEG format pictures when it's in "surveillance mode." In the future, expanded wireless LAN capability through a personal computer connection will allow owners to control the dog from up to 300 feet.

Japanese manufacturer Omron has introduced a feline version of the robotic dog called Tama. Tama interacts with her owner, needs love and attention and will develop her own personality. Just like a real cat, the robotic cat has emotions, purrs when stroked, and sleeps whenever she wants. Microphones embedded in the cat's head enable her to react to her own name by turning her head and blinking coyly. The goal was to create a cat that responds to the tone of your voice instead of just commands. Tama determines where you are and how happy or annoyed you feel by measuring the volume of your words. The robotic cat's fur mimics a central nervous system. The more she's stroked the happier she is, but mistreating her with a smack will bring an angry hiss. Using real cat sounds, Tama displays the six basic animal emotions of satisfaction, anger, uneasiness, dislike, fear, and surprise. Tama was originally designed as an entertainment device. But her interactive personality traits provides possibilities for use in pet therapy.

Robotic animals are a high-end niche market. Japanese consumers account for 85% of 100,000 fancy pedigree units sold.

White Goods

Digital consumer appliances that utilize the power of the Internet are not limited to products like PDAs or cell phones. New versions of white goods products such as refrigerators, stoves/ovens, washing machines, dryers, dishwashers, microwaves, and toasters will tap into the Web for additional features, service updates, and improved performance. In the near future, "smart appliances" will have technology embedded that allows them talk to each other, the Internet, and you. Vendors are also developing refrigerators that utilize wireless Internet access and the ability to control household appliances via the Web.

Consumer demand for kitchen-related devices has resulted in smart stoves and intelligent refrigerators. According to the U.S. Energy Information Administration, only 42% of households cook a daily hot meal. Women entering the workforce and busy family schedules have combined to reduce home cooking creating a demand for connected kitchens.

The future of white goods includes networking these products to provide operation, service, support, and information between customers and manufacturers. Companies such as Sunbeam, General Electric, Whirlpool, Sharp, Electrolux, Maytag, and Merloni Eletrodomestici are participating in partnerships, announcements, and rapid product development. These established companies are partnering with Microsoft and Sun Microsystems to provide network services for the kitchen. Microsoft, in partnership with HP, Intel, Matsushita, and Sony is pushing its UPnP (Universal Plug and Play) as the standard for appliance communication. Sun, in conjunction with Bosch, Cisco, Echelon, GTE, Motorola, Nokia, Oracle, Sears, Siemens, Sony, and Whirlpool is promoting its own Java and Jini code in a standard called OSGi (Open Services Gateway Initiative). These alliances are still in development and only a few products using these technologies are available today. These capabilities will exist in niche and high-end white goods. A large support infrastructure will be required for the success of these products.

An Internet-enabled oven brings a unique experience to everyday cooking. The ability to view recipes online, order ingredients over the Internet, and make quick meals from the material already on hand makes the networked oven an integral part of the future home. The home network brings you the capability of being notified on your digital TV when your laundry is done and also provides energy management capabilities.

In Chapter 12, we showed how white good products will integrate a web pad/tablet on the door of the refrigerator. However, it is important to understand the importance of the underlying technology. The tablet can be used for viewing entertainment, recipes, and serving ideas. To access the Web the appliance will need access to dial-up or digital modem technology. This will give the user the ability to control devices and distribute energy consumption via the Internet. Using this technology a microwave could be turned on when you leave work and have your meal ready when you enter the house.

Products such as the screen-fridge or fridge pad incorporate a built-in screen on the front of the refrigerator. While you're waiting for your Internet-ordered dinner, you can send and receive e-mail or access recipes for the food stored in your refrigerator. The fridge pad's "kitchen manager" will alert you to the yogurt's expiration date. A virtual keyboard on the touch-screen allows families to enter data, send e-mail, record video mail, browse the Web, shop online, and retrieve food recipes. One possibility for lowering the price of the fridge pad is the display of ad banners on the refrigerator's touch-screen.

Next generation smart microwaves provide sensors that calculate the right cooking time and level based on the ingredients. Future interactive microwaves will be able to record a recipe from the Internet or a TV cooking show. The screen touch-pad will allow you to enter a list of what's in your refrigerator and then display a recipe list for those ingredients.

The home of the future will contain the ability to control appliances from almost anywhere. For example, say you're relaxing in the family room. You click a Web site on your home intranet and a grill cheese sandwich starts cooking. Simultaneously, your wired refrigerator confirms the expiration dates for cheese, bread, and margarine.

CMI Worldwide has introduced iCEBOX, combining e-mail, Internet, TV, audio video CD, and home monitoring into a kitchen device that fits on your counter. Grocery lists stored on your handheld computer can be synced to your iCEBOX. Ordering the items is handled by the HomeAccess Grocery Shopping Web Phone appliance. This appliance includes a laser "bar code" scanner to make grocery shopping even faster. This category, together with home controls, is referred to as "smart homes." Sales of smart home equipment in the United States will reach $1.7 billion in 2005.

Internet connectivity will allow vendors to provide remote software upgrades. Service technicians could be automatically dispatched to handle device failures or scheduled maintenance without consumers having to place service calls.

Unlike sitting in the recliner watching TV or surfing the Internet, interacting with devices in the kitchen while your hands are full of pots and pans is difficult. Refrigerators, microwaves, stoves, and toasters are not where one will watch TV or surf the net. Appliances will be interactive doing new things in new ways. Also, business models still need to be defined. For example, appliance manufacturers want to offer services for free while the network companies are looking to charge for these services and capabilities.

Lighting Control

An automated home utilizes the Internet and wireless technology to remotely control the home lighting system. Controls adjust the lighting for different activities or times of day from different locations in the home. A forgotten lamp at home can be switched off from your office. At the touch of a button the illumination in your entertainment room can be dimmed for television viewing or raised to check the schedule or find a snack. A remote light control is also convenient in the bathroom. Remote controls easily adjust the lights when you're in the tub or dim them to serve as night-lights. Lighting controls provide added independence for individuals with reduced or limited mobility.

Lighting control systems combine a hand-held infrared wireless remote control and a dimmer with a built-in infrared receiver. The dimmer uses existing wiring and easily replaces a standard light switch. To control the lights, you aim the remote at the dimmer on the wall. The infrared signal turns the lights on or off or adjust the brightness. This device can also recall a preset light level and includes a fade-to-off feature so the lights dim gradually as you leave the room.

Some systems come with a group of modules that can be simultaneously set to predetermined levels with a single command.

For example, assume you have four different lighting scenes:

- **Evening** – Main family room lights at full brightness. Accent lights throughout the house are dimmed to 70%. All other lights are left unchanged.
- **Theater** – A few lights in the TV area are dimmed to 20% and remainder off. Other lights in the house unchanged.
- **Dinner** – Dining room lights at full brightness. Kitchen dimmed to 30%. Family room dimmed to 20%. Other lights in the house unchanged
- **Off** – All lights are turned off.

Once configured, a single command will initiate each scene and all the lights will change to their appropriate levels simultaneously. Inexpensive lamp modules have capabilities previously found only in high-end devices. Inexpensive complex lighting scenes now duplicate the functions of more expensive dedicated lighting systems.

Home Control

Modern home control devices allow the user to remotely identify someone pressing the doorbell and then unlock the door. The door phone is linked to your mobile or office phone. This allows a two-way conversation with the person at the door. The lock release can be activated from your telephone keypad. This principle can also enable people with mobility problems to automate the basic home appliance operations.

Other applications for home control include:

- Inactivity monitors for the elderly
- Medication reminders
- Child at home alert (when the child enters their personal code the dialer notifies a parent that they are home safely)
- Controlling lights and appliances to conserve energy

The web tablet is often used as the basis for home control. Getting ready to start your day, you reach for the web tablet that has replaced your normal bathroom light switch. Touching the 7-inch full-color display turns on the lighting for the house, turns off the security system, selects your favorite morning news program, and starts your coffee. In the kitchen, your spouse uses the touch screen to check the baby via closed-circuit camera. Seeing that the baby is awake, your spouse uses the touch screen to adjust the upstairs temperature and open the drapes. Another touch and Mozart starts playing softly in the nursery. In your home office, a touch screen in a desktop "valet" provides up-to-the-minute stock reports and allows you to view a visitor at the front door.

This full-color LCD touch panel web tablet can be used to control numerous audio/video components and whole-house systems such as lighting, temperature, and security. The tablet's high-quality color monitor allows you to view video sources such as cable and satellite TV, DVD, and closed circuit TV cameras. The panels can be flush-mounted in a wall or installed in attractive wood tabletop "valets." With seven different trim plates and selectable screen motifs, each panel can be designed to match the décor of any room. The large 7-inch screen permits user-friendly graphics and eliminates the need to scroll through numerous pages to control audiovisual components or house-wide systems. Most touch panels are proprietary to their specific system's controllers. The controller translates commands into a "language" the audio/video components in the system understand. Commands are translated via infrared, Bluetooth, wireless LANs, or HomeRF. The cost of a system depends on its size and complexity.

Home Security

Consumers can now adjust the temperature in their living room, verify that their kids have arrived home from school, or unlock the house for an unexpected guest—all over the Internet. With the click of a mouse, users can adjust settings for lighting, heating and air, security, and appliances.

The security industry is being challenged to incorporate this rapidly developing technology into their products. Companies are also being led to design systems and products that focus more on ease of use, value (not just price), and the Internet. This is a change to their existing business model of providing the least expensive systems and ongoing monitoring contracts. In order to take advantage of this market shift, many previous security-only installers are moving to product lines that provide more than just security system features. Installation companies are looking to sell tightly integrated products and systems. Until recently, systems that combined security, fire protection, comprehensive temperature control, extensive lighting control, flexible programming, and open communication capabilities were too expensive for the average consumer.

Home security can be as simple as door locks or as complex as high-tech network systems controlling sensors, locks, and surveillance cameras. Prices for residential security systems range from tens to thousands of dollars. A complete and practical home security system contains video cameras, alarm systems, network access, and intelligent circuitry. Depending on its complexity, the security system can be connected to home networks using power lines, Ethernet, USB, phone lines, or wireless technologies.

Security cameras at the doors or in your yard protect the home from strangers and allow the consumer to monitor visitors at the door. These devices use the Internet to provide home monitoring while at work or on vacation.

Several companies have announced integrated security and automation systems designed to meet the needs and budgets of people in smaller homes, condos, townhouses, smart apartments, and small businesses. These systems accept standard sensors for intrusion, fire, temperature control, lights, and appliances. These systems are accessed via telephone, PC, and the Internet.

Code Grabbers

Merging home automation and home security increases convenience. A single remote on your key chain can control different devices without entering a code on the keypad. This device combines ease of use, elimination of false alarms from entering the wrong code, and a panic alarm. The three-channel receiver and remote system turns your home alarm on/off, opens and closes your garage door, and includes a panic button for emergencies.

When looking for an all-in-one wireless system, choose one with code encryption or code hopping technology. Code encryption and code hopping technology defeat a thief's ability to use a Code Grabber. Code Grabbers are small digital recording devices used to duplicate and playback RF signals. The signals are then used to defeat remote-based vehicles and home security systems or for opening electric garage doors to gain access to your home.

Today, many manufacturers offer products with "anti-code grabbing" technology. This technology combines convenience and high-tech security. Each time the user presses a button on the remote control, the system delivers a new, nonrepeating code defeating Code Grabbers and Code Scanners. Homes protected with this technology will virtually eliminate the threat of these high-tech burglary tools while providing unparalleled convenience.

Figure 17.1 shows the block diagram of a typical home security system.

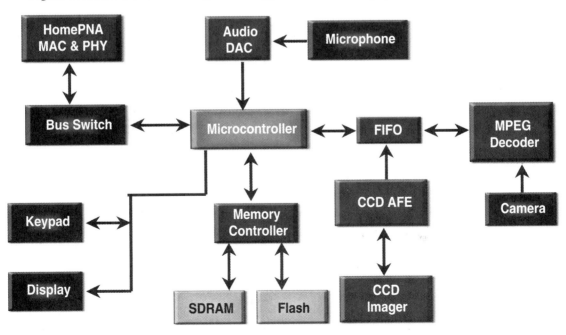

Figure 17.1: Home Security System Block Diagram

Energy Management Systems

In a home network, RF metering devices can perform energy management and automated meter-reading (AMR) functions. This technology allows metered devices around the home including gas meters, electric meters, and water meters to be read remotely. This allows consumers to monitor and manage their monthly utility bill. Energy management systems can also be used for controlling appliances, fire detection, and temperature surveillance, and control. Commercial applications include cable theft detectors, remote usage meter reading, and security systems.

Today, most consumer energy usage is collected manually by reading the meter outside the home. Using vehicles equipped with RF meter reading and wireless technology to drive by and collect meter reading saves time and increases productivity. The Internet will allow utility companies to gather energy consumption information over the Web. The Internet also allows users to pay their bills, saving millions of dollars in expenses for meter reading and bill paying.

Home Theater and Entertainment Systems

A networked entertainment system connects the set-top box/residential gateway, DVD player, digital VCR, and digital TV to the Web. This enhanced online experience gives you the ability to download MP3 files, video-on-demand (VoD), movies, and more.

Magnetic Recording

Magnetic recording is a backbone technology of the electronic age. Tape recorders and CD players helped define the current music industry. In the audio realm, magnetic tape (in the form of compact cassettes) is a popular way to distribute music. People either buy pre-recorded tapes or create their own from CD's. In the video realm, videotape is widely used in the broadcast industry and at home to store programming for later viewing. In the computer realm, magnetic recording is used for data storage on floppy disks, hard disks, and magnetic tape. The same technology is used in both cassette tapes and video tape recorders. Magnetic recording provides an easy and inexpensive technology with 10 to 20 year storage capabilities.

The magnetic recording system consists of two components: the recorder itself (which also acts as the playback device) and the tape it uses as the storage medium. The tape is simply a coating of ferric oxide powder bonded to a plastic base material. A dry lubricant is also incorporated to minimize wear on the recorder.

Iron oxide (FeO) is common red rust. Ferric oxide (Fe_2O_3), commonly called maghemite or gamma ferric oxide, is another iron oxide. This oxide is a ferromagnetic material and is permanently magnetized when exposed to a magnetic field. This gives magnetic tape two of its most popular features; the ability to record sound for playback at a later time, and the ability to erase the tape and record something else.

The original model for audiotapes was a thin steel wire used in a wire recorder invented in 1900 by Valdemar Poulsen. German engineers perfected the first tape recorders using oxide tapes in the 1930s. Reel-to-reel tapes were common until the compact cassette or "cassette tapes" were introduced. Patented in 1964, cassette tapes eventually became the dominant tape format in the audio industry.

A compact cassette is a fairly simple device. It consists of two spools, two rollers, a plastic outer shell and a 443 feet (135 meters) of tape.

The Tape Recorder

Every tape recorder from a Walkman to a high-end audiophile deck utilizes the same simple fundamental principles. An electromagnet applies a magnetic flux to the oxide on the tape. The oxide permanently "remembers" the flux. A tiny pea-sized electromagnet consists of an iron core wrapped with wire. During recording, passing the audio signal through the coil of wire creates a magnetic field in the core. The magnetic flux forms a fringe pattern to bridge the gap (shown in red) and magnetizes the oxide on the tape.

A varying magnetic field is pulled across the gap during playback. This creates a varying magnetic field in the core and a signal in the coil. This signal is amplified to drive the speakers. In a normal cassette player there are actually two small electromagnets one-half as wide as the tape. The two heads record the two channels of a stereo program. When you turn the tape over the two electromagnets align with the other half of the tape.

There are two sprockets that engage the spools inside the cassette. These sprockets spin one of the two spools to take up the tape during recording, playback, fast forward, and reverse. Below the two sprockets are two heads. The head on the left is a bulk erase head to wipe the tape clean of signals before recording. The head in the center is the record and playback head containing the two tiny electromagnets. On the right are the capstan and the pinch roller. The capstan revolves at a very precise rate to pull the tape across the head at exactly the right speed. The standard speed is 1.875 inches per second (4.76 cm per second). The roller simply applies pressure so that the tape is held tight against the capstan.

Tape Types and Bias

Higher-end tape decks have controls for different tape formulations and bias. There are four types of tape in common use today:

- **Type 0** – The original ferric oxide tape. No longer in general use.
- **Type 1** – Standard ferric oxide tape. Also referred to as "Normal Bias."
- **Type 2** – "Chrome" or CrO2 tape. The ferric oxide particles are mixed with chromium dioxide.
- **Type 4** – "Metal" tape. Metallic particles rather than metal oxide particles are used in the tape.

Sound quality improves as you go from one type to the next. Metal tapes provide the best sound quality. Only tape decks with special settings can record onto a metal tape, However, any tape player can play a metal tape.

The controls on the tape deck let you match the recording bias and signal strength to the type of tape you are using. "Bias" is a special signal applied during recording. Early tape recorders applied the raw audio signal to the electromagnet in the head producing distortion on low-frequency sounds. Bias adds a 100 KHz signal to the audio signal. Using bias moves the recorded signal into the "linear portion" of the tape's magnetization curve. This movement provides a more faithful reproduction.

VCRs—Video Cassette Recorders

The history of television is marked by important milestones beginning with the invention of the black and white TV set and the first broadcasts of television signals in 1939 and 1940. Another milestone was the advent and popularity of color in the 1950s. In the 1970s, cable television and cable channels like HBO and CNN started competing with the three big networks. And just as important was the development and popularity of the VCR in the 1970s and 80s. The VCR is important to the history of TV because it gave people control of what they could watch on their TVs. Outside of automobiles, VCRs are the most complex mechanical systems most people own. They contain motorized tape loading and ejection systems, complex motorized tape paths, and drum-mounted rotating read/record heads and can cost under $100.

The Tape

With a price under $2, a VCR tape is a classic miracle of mass production. The exterior of a video-tape consists of a two piece outer shell and a spring-loaded door to protect the tape. Inside there are two spools to hold the tape. The recording medium is an 800-foot long, 1/2-inch wide piece of oxide-coated Mylar tape. Several low-friction rollers guide the tape across the front of the shell. Two spring-loaded locks prevent the tape from unrolling inside the cassette. When the tape is inserted a lever in the VCR releases and opens the door exposing the tape. It also inserts a pin to disengage the two locks on the spools. At that point the drive can extract the tape and play it.

The VCR

Ampex employees Charles Ginsburg and Ray Dolby created the first commercial reel-to-reel videotape recorder in 1956. This new device was a major development for television broadcasters. It marked the first time that shows could be recorded and broadcast at a later time. Prior to 1956, all shows on television were live. Sony created the first inexpensive VCR in 1969. In 1972 the VHS tape format appeared and began its domination of the market. Video stores soon followed. Blockbuster Video has grown to over 4,000 outlets since opening it first video store in Dallas, Texas in 1985.

The VCR itself has two functions. It must deal with an extremely thin, somewhat fragile, and incredibly long piece of plastic tape. Also, it must read and convert the tape's signals for display on a television screen. Both of these are formidable tasks. Displaying video is a big technological challenge. In sound recording, the sound information is stored in a linear manner. As the tape moves past the recording head (at a speed of 2-3 inches per second) the sound information is laid down as a long line along the length of the tape. The same approach does not work for a video signal containing 500 times more information. To solve this problem, two recording heads are mounted on a tilted rotating drum. A television image is divided into a series of 525 horizontal scan lines. Half of these lines are displayed in a 60th of a second. This allows each pass of the VCR's rotating head to read or write the data for one field (262.5 scan lines) of the television image.

The recording head of the rotating drum lays down bands of individual fields. The drum contains two opposing heads (180 degrees apart). The two heads alternate reading or writing every other band. There are two tracks that represent the audio and control tracks. The control track is especially important. It provides key information such as the recorded speed, how fast to pull the tape past the drum (since the tape may stretch or shrink over time), and lining up the heads with the bands during playback.

A linear control track helps synchronize the rotating heads with the actual recording bands. Adjusting the "tracking" control on the VCR adjusts the skew between the control track and the actual head to get a closer match to the bands on the tape. Tracking adjustments can be used if a tape is badly worn or stretched. The head rotates at 1,800 RPM, or 30 revolutions per second. In SP mode the tape is moving past the head at 1.31 linear inches (33.35 mm) per second, in LP mode at 0.66

linear inches (16.7 mm) per second, and in EP mode it is 0.44 linear inches (11.12 mm) per second. Because of head rotation, the head moves over the tape at 228.5 inches (5804 mm) per second, or about 25 miles per hour (41 kph)! If the video information were being stored linearly, a two-hour movie would require a 50-mile long tape.

In order to record or play, the videotape must wrap around the rotating head. The VCR must also read the audio and control tracks to keep the tape moving at exactly the right speed and detect the end of the tape. The drive mechanism in the VCR must extract a long piece of tape and wind it through a tortuous path around a variety of rollers, drums, and heads in order to play the tape.

The Future of the VCR

The mechanism for playing and recording tapes is prone to errors. Consumers experience problems when the tape sticks and ruins the movie. The arrival of digital formats such as video CD (VCD), laser disk, and DVD marks the end of the analog videotapes. A large installed base ensures the short-term popularity of videotapes and VCRs. However, the budding DVD format provides a formidable challenge. Digital VCRs and digital video recorders (DVRs) will compete to provide key recording functions of traditional VCRs.

Integrated Audio Systems

Manufacturers are packaging speakers and electronics into all-in-one systems and other audio "solutions" to expand the appeal of home audio products. The integrated audio market continues to grow and is expected to reach $3.2 billion in 2004. These systems account for almost 60% of all factory-level home audio sales excluding small home radios and clock radios. The remainder are component audio products such as AV receivers and separate CD players. Integrated systems appeal to consumers who are too busy to shop, don't want to choose separate components, or have limited space.

Shelf Entertainment Systems

Shelf entertainment systems are compact audio systems small enough to sit on desks or shelves. Most shelf systems contain a CD player or changer, AM/FM tuner, amplifier, and cassette deck connected to a pair of speakers. In late 1999, the first integrated CD-recorder shelf systems with appeared. Since that time, additional components are being offered including integrated CD-recorders, DVD/CD players, and MP3 decoders. As consumers opt for smaller and more versatile systems, sales of shelf systems have risen strongly and sales for larger floor-standing rack entertainment systems have declined.

Sales growth in this market is also being driven by the popularity of small, high-quality shelf systems known as micro systems. These systems are smaller than shelf systems (most are only about 9.5- to 12.5-inches wide). They typically incorporate a tuner, amplifier, and CD player in a single chassis.

Home-theater-in-a-box (HTiB) systems package a full-size audio component, such as an AV receiver, with five or more speakers to re-create surround sound. Another type of HTiB consists of five speakers and a separate subwoofer containing amplifiers and other system electronics. Some HTiB packages include a component-size receiver with built-in DVD player. HtiB systems deliver all of the components for a home theater system and have proved popular with consumers.

Custom-Installed Audio Systems

Shelf and HTiB systems appeal to the convenience-oriented customer. Another market exists for custom-installed distributed-audio systems. These systems are built around a central audio system that distributes music to multiple rooms in a house through wall, ceiling, and outdoor speakers. Control panels in each room allow you to control the system remotely.

High-end home theater systems use custom cabinetry to conceal the video display, associated electronics and front speakers. In some sophisticated systems, a single button on a remote control activates the components, lowers a projection screen, and tilts the speakers to direct sound to the listeners.

The custom-installed system market continues to grow 30% annually. But it accounts for 10% of total U.S. home audio and home video sales. Sales of video products are projected to reach $2.5 billion in 2004, with sales of audio products expected to reach $400 million. In-wall speaker sales are expected to rise to 20% in 2004 to $60 million.

Building custom systems kept many independent AV retailers in business as high-volume electronics chains increased their share of the electronics market. Unable to out-price or out-promote the chains, independent dealers deliver a level of service that chains have been slow to master. Some custom AV installers install computer Ethernet and fiber optic networks and distribute broadband Internet access throughout the home. Others have diversified into lighting and phone-system installation as part of their effort to become one-stop solutions for homebuilders and homeowners. Many AV installers are integrating multiple home systems, such as security, lighting, climate-control, and distributed-audio systems. This allows the coordination of multiple systems from a single control panel.

Receivers/Amplifiers

The amplifier is the central component of any system using speakers. The amplifier takes the signal from an output device such as a CD player then amplifies and transmits it to a set of speakers.

Amplifiers are rated by the power produced in watts per channel. Each channel of the amplifier provides power amplification for one speaker. Most amplifiers power multiple channels. A two-channel amplifier powers a pair of speakers for stereo sound. A four-or six-channel amplifier can power two or three pairs of speakers. Bridging a multi-channel amplifier creates a single channel with twice the power.

Amplifiers are available producing 15-200 watts per channel. Most moderately priced amplifiers produce from 30 to 100 watts-per-channel, adequate power for an average room. A preamplifier tells the amplifier which signal source to amplify as output to the speakers. A receiver obtains radio signals via antennas. The tuner allows the selection of different radio signals.

Amplifiers commonly come with integrated preamplifiers, radio receivers, and tuners. These integrated packages are referred to as amplifiers and receivers.

Almost all amplifiers today are based on solid-state transistor technology. However, reaching back to early days of radio, a few audio companies have reintroduced vacuum tube amplifiers, eschewing transistors in favor of power-hungry sounding tubes. Only small quantities of these expensive high-end amplifiers are being manufactured.

Home Speakers

Speakers are one of the most critical pieces of stereo equipment that consumers purchase. Choosing the wrong speakers will lead to disappointing sound from any system.

Speaker drivers are designed to handle a particular range of frequencies. The tweeter is the soprano of speaker drivers handling the upper-end treble scale. Starting as small as in inch in size, this is the smallest speaker driver. The largest driver, the woofer, reproduces the bass and lower frequencies adding thump to the music. Woofers range from five inches to more than a foot in diameter.

Most speaker enclosures are two-way, with both a tweeter and a woofer. Three-way speakers include a tweeter, woofer, and a mid-range driver that reproduces a broad range of sound between highs and lows. Vocals are often reproduced in these middle frequencies. An enclosed electronic crossover network directs the sound frequencies to the appropriate driver

In recent years, speaker manufacturers have prompted consumers to replace older models by improving visual appeal, reduced size, and improved performance. Factory-level speaker sales will grow to $600 million in 2004. Sales for installing or upgrading home theatre systems has helped fuel this growth. Narrow floor-standing profile speakers improve visual appeal by replacing a single large woofer with multiple vertically stacked small woofers. Designs incorporating curves are being built from aluminum and molded vinyl.

Subwoofers

Speakers are now available with built-in subwoofers and dedicated amplifiers to reproduce the thundering realism of movie soundtracks. Stand-alone amplified subwoofers have also grown in popularity. Because their low-frequency sounds aren't localized, these low-profile speakers can be tucked away in an inconspicuous location. The sound they generate blends with the mid-bass, midrange, and high frequency sounds produced by other home theater speakers.

Combining stand-alone subwoofers with pint-size mini speakers can deliver the remainder of the audible listening range. Packaging these speakers with a powered subwoofer simplifies the purchasing process. Using identical drivers and crossovers helps sonically match the speakers. Sonic matching is critical to reproducing realistic sound effects. The rush of wind through the trees will sound the same from a set of properly matched speakers.

On-Wall, In-Wall Options

Wall-hanging speakers are designed for small rooms and flat wall-hanging plasma TVs. By using stands, these speakers can be placed closer to walls than traditional speakers. Some models are only 1.2 inches thick. Many thin speaker models use traditional vibrating cones to generate sound. But their woofer's magnet structure and basket have been redesigned to minimize depth. Other thin speakers feature a single vibrating flat panel or combine cone speakers with distributed-mode panels.

In-wall speakers with grilles that match the wall's paint or wall covering can blend into a wall and almost disappear. Consumers are turning to sophisticated in-wall and in-ceiling speakers as their primary stereo or home theater speakers. However, the primary market remains custom-installed distributed-audio systems that pipe music into multiple rooms from a single set of audio components.

PC Speakers

Home theater isn't the only factor influencing speaker-purchasing patterns. Growing sales of multimedia PCs and CD-ROM drives have fueled the development of amplified two-channel and multi-channel speaker systems for PCs. These systems improve the quality of sound effects, music, and voices on game and multimedia software. The speakers also allow users to listen to music through their PC.

Vehicle Security

The vehicle security market is one of the fastest-growing segments in the mobile electronics market with projected sales of $300 million in 2004. Vehicle security systems are adding features such as vehicle tracking, remote starting, and wireless communications. Another option being marketed is a GPS powered stolen car recovery system. The system can place a page, e-mail or phone call to an owner if a vehicle travels outside of a prescribed area.

Vehicle Radar Detectors

Radar detector sales are projected to climb to $200 million in 2004. Early radar detectors alerted drivers to the presence of speed monitoring devices. Studies show that drivers tend to slow down and be more attentive when alerted by radar detectors prompting some to label them as safety tools. Municipalities and states are using unmanned transmitters or "drones" at accident-prone locations to set off radar detectors. Microwave technology takes the drone concept a step further by providing drivers with specific information about high-speed emergency vehicles, stationary road hazards, or trains equipped with safety and radar transmitters. The transmitters cause older radar detectors to emit the same signal as when a speed-monitoring device is detected. New generation "smart" radar detectors gives drivers both an audible and visual warning describing the approaching hazard

Summary

Most, but not all, devices that affect consumers are covered here. Key applications are discussed, but many devices serve niche markets and appeal to only specific consumers. Note that in the future, devices such as VCRs, pagers, and audio cassette players will be replaced by their digital counterparts, or their functionality will be incorporated into other devices.

The Digital "Dream" Home

In the last few chapters we covered most of the consumer devices currently in use or that we expect to see in the next few years. However, there are other key trends that will be affecting the future home. In this chapter we will discuss the dynamics leading to the growth of an ideal digital home. The digital home is the result of trends leading to increased communications, entertainment, home control, and convenience. The design of the digital home is affected by five key trends. These are digital consumer devices, broadband access, home networking, residential gateways, and middleware.

Emergence of Digital Homes

The dawning of the digital home era has been a slow process. For the past few years most homes used a single PC to access the Internet, share files, and access a printer. With the expansion of Internet applications such as web browsing and web messaging, more homes own multiple PCs and other types of convergence appliances. The integration of the Internet into our lives is often taken for granted. Email has made communication easier. The Internet has enabled communities to be connected. And e-commerce has changed how business is transacted. More importantly, the Internet has enabled the construction of a robust network infrastructure. It has also consolidated and concentrated information and fueled the development of digital services. This opened an era marked by easy access to entertainment, communications, and services. In the digital home era services will be natural and intuitive to use.

Service providers will be able to bundle plain vanilla and value-added services such as interactive TV, TV commerce, interactive advertising, video-on-demand (VoD), smart home services, voice over broadband, home networking, and connected and intelligent appliances. Instead of driving to the video store a few clicks will order Internet-delivered movies. The digital home will bring service providers additional revenue streams through subscription-based models (such as cable TV) and usage-based models such as video-on-demand. Business such as phone companies boost profits by offering value-added services such as caller ID and voicemail. Digital homes will allow them to offer a whole new set value-added services.

The bandwidth required for these services is bringing connectivity to the digital home. Today houses have a disparate electronic ecosystem, including:

- **PC centric ecosystems** – Modems, scanners, digital cameras, and printers connected via a localized network.
- **Multimedia centric ecosystems** – Set-top boxes, digital televisions, digital video recorders, speakers, stereos, and DVD players.
- **Wireless centric ecosystems** – PDAs and mobile phones.

Broadband connectivity will enable digital homes to send and receive information. Connecting a multitude od digital devices will enable Internet access throughout the home. The evolution of a fully-fledged digital home will develop in three separate stages (Figure 18.1).

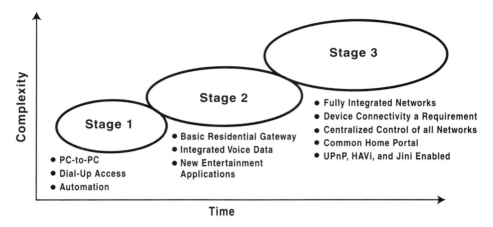

Figure 18.1: Migration to Digital Homes

Stage one is already prevalent in millions of households around the world. This stage is typically marked by a simple computer-to-computer connection, dial-up connectivity to the Internet, and a simple home automation system. Worldwide the digital household's evolution to stage two is beginning. The installation of a basic residential digital set-top box or a broadband centric modem gateway and the use of new entertainment applications mark this next step. Stage three, where consumers have almost unlimited connectivity and communication capabilities, remains a vision of the future. Around the world organizations and companies are developing technological standards and products to facilitate a fully integrated digital home. A tighter integration of data, video, and voice services will mark the final stage in this evolution. Home networking software and middleware will connect different makes and models of digital appliances. Stage three encompasses a complete provisioning of entertainment, conveniences, and information sharing.

The Digital Home Framework

The emergence of the digital home is driving demand for high-speed Internet access and new smart networked appliances. The scope of the digital home can be defined by the following five-part equation:

- **Broadband – High-Speed Internet Access**
- **Residential (or Media or Home) gateways** – A device for distributing voice, video, data, and Internet-enabled services. The gateway connects an intelligent device/router via broadband connectivity to a network provider's server infrastructure and the consumers' home network. This allows the use of smart network routing devices throughout the home.
- **Home networking technologies** – Home networking is the connection and distribution of audio, video, and data between different household consumer devices.
- **Middleware** – This software layer provides an interface between basic hardware/software systems and supported applications and devices. Middleware is a critical component for the deployment of value added services on a residential gateway.

■ **Digital Consumer Devices** – PCs, consumer electronics, and home control information appliances (IAs)

Figure 18.2 shows the architecture of the digital home as defined by the five parts.

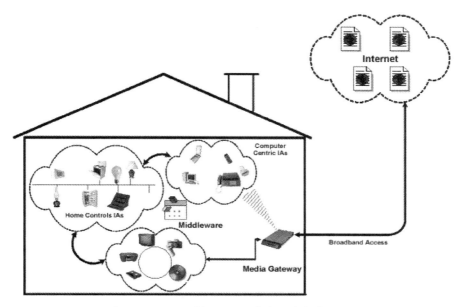

Figure 18.2: Architecture of the Digital Home

Many companies and organizations are involved in managing this change and evolution of the digital home, including:

■ Broadband access providers
■ ISPs
■ Consumer electronic manufacturers
■ Gateway manufacturers
■ PC manufacturers
■ Networking companies
■ Semiconductor manufacturers
■ Network operators
■ Application developers
■ Middleware and software vendors
■ International Standards organizations

Productivity Increases in the Connected World

The increasing penetration of the Internet has played a major role in enhancing our productivity. In the 19th century the transatlantic cable reduced the cost of communication. The mass-market introduction of the telephone was a further cost and quality improvement. However, it still cost $15 a minute and took 2 seconds to connect. Today the Internet allows us to communicate in less than 30 milliseconds without incremental costs. The Internet uses data, voice, and video to bring people closer and help make businesses more efficient. The 1980s saw productivity of corporations improved by PC's. The late 1990s were the beginning of the Internet/communications era. This life-changing revolution has affected the way we communicate, share information, and perform business.

The Internet has taken productivity to unprecedented levels. Internet use has translated into time reductions, lower costs, lower investments in labor, and consumer convenience. FedEx estimates it would need 20,000 customer service employees to replace its online service. E-mails have become an established and non-intrusive way of communicating. In the past year we exchanged 6.9 trillion e-mails. This level of communication is unprecedented in the history of mankind.

The productivity acceleration we see today through connecting people can be sustained only by increasing networking infrastructure bandwidth. John Sidgmore, COO of Worldcom, said that "Bandwidth demand increases by 1000% every year." Networking solution providers are scrambling to satisfy an insatiable demand for bandwidth. This demand is leading to tremendous advances in connectivity and productivity.

Increased demand for bandwidth is leading and forcing the evolution of all aspects of networking—the backbone, metro access, enterprise access, transoceanic cables, and more. Networks of all sizes are being upgraded to achieve better performance. The demand for increased bandwidth is forcing a rapid and fundamental change in the networking infrastructure. The only thing constant about this phenomenon is change. Today's networks are demanding higher performance, scalability, upgradability, increased security, product differentiation, and lower cost of ownership. Leading networking solution providers and Internet service providers (ISPs) are scrambling to meet the needs of this constantly changing marketplace.

Consumers want to use a single access point to satisfy the needs for all their appliances that access the Internet. Dial-up connections are no longer sufficient to satisfy the burgeoning demand. Multiple PCs and appliances need network connectivity to share digital content and increase productivity. Applications such as online shopping, MP3 files, digitized photographs, and video-on-demand require high-speed Internet access. Broadband access technology options such as DSL, cable, ISDN, fixed wireless, and T1 are becoming popular choices for today's consumer. Broadband is the best solution for providing multiple consumer devices in today's homes with high-speed access.

Broadband Access

The Internet has become an integral part of our lives. It has been around in one form or another since the late 1960's. The Internet has caught the imagination of millions of people around the world. Every six months the number of people connecting to the Internet doubles. The World Wide Web (the Web) is the Internet's best-known feature and provides resources such as text, graphics, video, audio, and animation. However, many users that access the Web with an analog modem experience the frustration of waiting for information to download to their computers. Earlier users accessed the Web with a 14.4 Kb/s modem and then upgraded to 28.8 Kb/s device. Even improved 56 Kb/s modems are not fast enough. For example, downloading a 3 Mbyte MP3 music file takes 7.5 minutes using a 56 Kb/s dial-up modem. The same file downloads in 3 minutes using ISDN (144 Kb/s) and 15 seconds using DSL or cable modem (1.5 Mb/s).

But for many consumers the wait is about to end. Companies such as AT&T, Microsoft, America Online, Yahoo, and Cisco are spending billions of dollars to bring broadband access to our homes. Digital broadband access to the home has provided a very fast two-way communication channel providing simultaneous up-link and downlink capabilities. This allows users to access and enjoy rich multimedia Web content. The ability to transmit real-time video anywhere at anytime is a revolutionary concept helping drive broadband acceptance. Rapid advances in technology have extended the list of possible broadband platforms that can access the Internet. High-speed communications links (higher than 128 Kb/s) provide the conduit needed to access advanced Internet services such as rich multimedia content, video-on-demand, and private networking services. Broadband connections eliminate the need to login and logoff to the Internet by staying continually connected.

Demand for high-speed Internet access has fueled the proliferation of rival technologies. Alternative solutions include xDSL, cable, fixed wireless (IEEE 802.16), power line, satellite, T1, and ISDN. Other broadband access technologies coming to market include Long Reach Ethernet (LRE), FTTH (Fiber to the Home), and PONs (passive optical networks).

Broadband Access Market Data

Demand for bandwidth is exploding as businesses and individuals seek faster access to increasingly complex Internet content. Worldwide Internet use is expected to climb from 142 million subscribers in 1998 to more than 500 million by 2004. The adoption of broadband Internet access is a powerful motivation to install home networks in multiple-PC households. According to IDC's residential broadband and telecommunications group, broadband access was installed in nearly 2.1 million U.S. households in 1999. This number is expected to reach 28.4 million in 2004 and 52.4 million in 2007. Deutsche Bank's Alex Brown reports that broadband subscribers exceeded 15 million users worldwide in 2001. He expects this number to exceed 75 million users by 2004. The acceleration of broadband access accounts in the U.S. and worldwide is displayed in Figure 18.3.

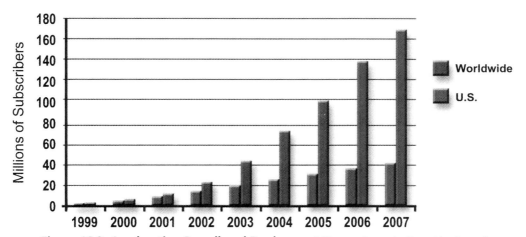

Figure 18.3: Accelerating Broadband Deployment (Source: Deutsche Banc Alex Brown)

Broadband Market Drivers

Some of the key market drivers for broadband access penetration are:

- **Increasing popularity of the Internet** – The Internet has grown ubiquitous to a large number of consumers. Internet applications use voice, video, and streaming video. Web browsing, e-mail, MP3 files, VoIP, digitized photographs, and online gaming fuel the demand consumer high-speed Internet access.

- **Increasing demand for high-speed access to the Internet** – Increasing numbers of Internet users are seeking high-speed Internet access providers to avoid the problems of dial-up services. People are signing up for broadband services to avoid the "world wide wait" associated with dial-up access.

- **Lower prices for Broadband services** – The demand for broadband services continues to increase as prices decline. The average monthly cost for a broadband service like DSL is approximately $40. Bundling services allows broadband providers to compete on prices with other types of Internet access services.

- **Growth in telecommuters and day extenders** – The growth of telecommuting is expanding the number of people who need high-speed Internet access to work from home. Telecommuting uses information and communications technologies to allow people to work outside the traditional office. The employee's residence is the most common alternate worksite. Others include satellite offices, hotel rooms, client offices, airplanes, trains, and automobiles. Many telecommuters use broadband to access the Internet at speeds much faster than those obtained through dial-up access. VPN Virtual Private Network (VPN) technologies provide telecommuters with secure access to corporate networks.

- **Home businesses** – A large number of small offices and home offices (SOHOs) are looking for high-speed Internet access.

- **Home networking for multiple PCs and digital consumer devices** – The demand for faster Internet access is accelerating as more households install multiple PC 's and consumer appliances. Networking homes appliances requires a continuous high-speed Internet connection. High-speed allows more Internet data to be shared among multiple access points within the home. Streaming media and other Internet delivered content provide consumers with more reasons to invest in-home networks.

- **Increasing availability of multimedia and interactive applications** – The demand for high-bandwidth broadband services grows as more multimedia and interactive applications are introduced. The number of video-based Internet applications will grow substantially in the next several years. These applications require a fast access to the Internet. Greater bandwidth enables higher quality high-definition (HD) signals streaming across the Web.

- **CD-quality Internet Radio** – Internet radio broadcasts are subject to poor audio quality marked by skips and pauses. Online radio places a heavy burden on a network or direct connection This makes it difficult to use the music as background noise while trying to use other application.

- **Telemedicine** – Telemedicine provides virtual medical services from the hospital to physicians' offices and homes. This Internet interaction with specialists allows physicians to provide expert care to the individual. It also allows individuals to spend less time in the hospital. A webcam, keypad, and an Internet connection can provide images and other patient data to the hospital.

- **eLearning** – eLearning provides the ability to learn cutting-edge information from your desk. It provides users with unrestricted access to education-oriented content, programming, and services. Individuals, consultants, and employees in SOHOs and large businesses have access to continuous learning in their homes, workplaces, and on the road. eLearning Web sites enhance the educational experience by providing people with access to educational material, mentors, professors, scientists, and fellow students. This kind of access to learning is critical as companies compete in fast changing markets. It allows them to turn out products in record time by providing employees learning-on-demand training. This market, estimated at $15 billion by 2004, requires high-speed Internet access to support interactivity and multimedia applications.

- **Education** – Computing and networking technologies are spilling over into the world of education. Recent technological advancements in personal computers, handhelds, and network communications take advantage of many systems and uncover new strategies for effectively using computing and networking in the classroom.

- **Online gaming** – The graphic content of games continues to move towards realism with more information displayed on a single screen. This complexity increases as players in rapidly shifting views control several independently moving characters. Using the full capability of online game-play means an ever-increasing demand for high-bandwidth connections. More bandwidth will provide gamers with access to new online universes.

- **File sharing** – At its peak usage Napster demonstrated the mass-market appeal of file-sharing technologies, as well as the impact massive file sharing has on Internet traffic. Apart from the legal questions of sharing copyrighted material, a host of less controversial uses for file sharing technologies exists. Sharing large photo files and home videos across the Web require high-bandwidth.

- **Web-based movies and software delivery** – Many software titles are too large to send across the Internet. Downloading gigabit-sized movie files is not feasible. More bandwidth makes sending larger files more manageable and opens new avenues of e-commerce.

- **Telewebbing/Interactive TV** – Millions of people enjoy surfing the web and watching TV at the same time. Broadband technology allows broadcasters to synchronize information from the web with TV broadcasts. And it can all be viewed on the TV screen. This allows Internet sports enthusiasts to research relevant data about a player during the game. At the same time fans can zoom in and out of plays, or watch the game through the eyes of their favorite player and view replays in slow motion from any angle.

Broadband Cable

The cable companies have taken the lead in supplying broadband connections to consumers. Broadband services typically include telephony, interactive multimedia, high-speed Internet access, video-on-demand, and distance learning. The service offerings vary between cable companies. A multimedia cable service is very much like an ISP in the telecommunications industry with similar issues and problems. The key difference is the greater bandwidth available for supporting more applications. The cable broadband network architecture is based on a network configuration of fiber-optic and coaxial cable known as hybrid fiber-coax (HFC).

Networks built using HFC technology have characteristics that make it ideal for handling the next generation of communication services. These networks simultaneously transmit broadband analog and digital services and meet the expandable capacity and reliability requirements of new digital data services. The HFC's expandable "pay as you go" architecture allows cable companies to add services incrementally without major changes to the overall network infrastructure. This allows new revenue streams, operational savings, and reliability enhancements to be matched with infrastructure investment. The HFC network architecture consists of fiber transmitters, optical nodes, coaxial cables, and distribution hubs. The signal is transmitted using fiber-optic feeders from the head-end in a star-like fashion to the fiber nodes. (The head-end is an industry term used to describe a service provider's main operations center.)

The fiber node distributes the signal over coaxial cable, amplifiers, and taps throughout their customer base. Customer base area sizes can range from 500–2000 home networks. HFC networks are capable of two-way communication at speeds of 40 Mb/s and above. Cable companies expect to generate new revenue streams with the combination of HFC and home networking technologies.

The digital revolution and the Internet are driving demand for broadband cable access to the home. Industry standards are crucial for the continued adoption of this type of broadband access technology. Some of the best-known International organizations are contributing to standardizing cable based broadband connections. They include EuroCableLabs (ECCA or EuroDOCSIS), EuroModem and DAVIC in Europe, CableLabs (formerly known as DOCSIS™—Data over Cable Service Interface Specification) in the United States, and DVB (Digital Video Broadcast).

The growth of the cable modem market remains restrained because:

- **Limited availability of two-way HFC networks to residential customers** – Cable television services are now available to the majority of households. But interested customers still experience delays in receiving broadband cable services. Many cable service providers must upgrade their networks to provide two-way communication. The number of subscribers for broadband cable services will be limited until older one-way communications networks are upgraded.

- **Shared nature of broadband cable services decreases speeds** – Cable broadband services operate in a similar manner to an Ethernet based Local Area Network (LAN). In-home cable network connections share available bandwidth with other broadband subscribers.

- **Cost of Service** – The costs of service vary among different broadband cable service providers based upon other services bundled with the broadband connection.

- **Cost of Equipment Purchase and Installation** – Purchasing and installing a cable-based modem is expensive compared to dial-up Internet access service.

- **Competition from ADSL Services** – Residential customers are monitoring the deployment and availability of ADSL services before deciding to purchase broadband cable services.

Cable Modem Operation

A standard cable modem has two connections. One end is connected to the wall-mounted TV outlet and the other to the in-home network. The cable modem communicates over the network to a CMTS (Cable Modem Termination System). The CMTS is the central device for connecting the cable TV network to a data network.

The speed of the cable connection depends on traffic levels and the overall network architecture. Theoretically cable modems are capable of receiving and processing multimedia content at 30 Mb/s, hundreds of times faster than a normal dial-up connection to the Internet. However, since bandwidth is shared by a number of home networks throughout the neighborhood subscribers typically download information between 1 and 1.5 Mb/s.

Cable operators define a portion of the frequency spectrum to carry the data. Typically the downstream path (head end to home network) will lie between 50 MHz and 750 MHz. The frequency range 5 to 42 MHz is used when transmitting information from the in-house network to the head end.

A major benefit for consumers who use this type of residential gateway is the fact that the connection is always open, eliminating the slow process of establishing a dial-up connection. The architecture of a typical cable modem is shown in Figure 18.4.

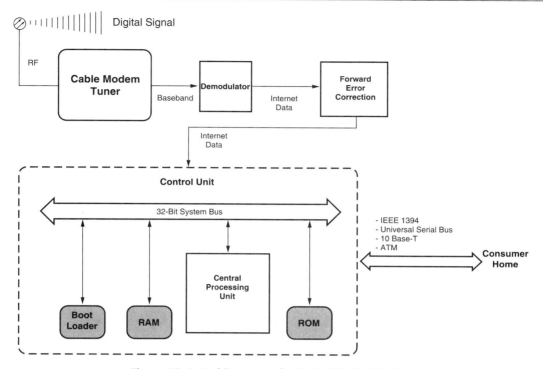

Figure 18.4: Architecture of a Typical Cable Modem

From Figure 18.4, the tuner in the cable modem receives a digital signal from the network and isolates a particular channel that contains Internet data. It then converts the signal from RF levels back to baseband. The baseband output signal from the tuner is then forwarded to a demodulator. The demodulator samples the signal and converts it into a digital bit-stream containing video, audio, and IP data. Once the bit-stream has been recovered it is sent to a forward error correction unit and checked for problems. The signal is then passed into the control unit. From the control unit the data is passed on to the home network using one of the following high-speed data port interfaces.

Broadband DSL (Digital Subscriber Line)

Until recently the plain old telephone system (POTS) and a standard analog modem offered the most convenient means to access information on the Internet. However, the existing telephone infrastructure was designed for voice signals. Its sensitivity to external noise sources and attenuation losses over long distances make it a poor choice for carrying digital information.

Voice-optimized networks typically handle less than 10% of their potential users at one time. Established network usage patterns allow resources to be allocated according to probabilistic models. The usage pattern for a typical data network is very different. Users typically remain connected to the Internet for longer periods of time. An average telephone call lasts 3 minutes. Internet connections average over 3 hours. During this time the network resources for that circuit are unavailable to handle voice calls. The increased level of Internet access combined with the very nature of this data traffic has prompted telecommunication companies and large carriers to deploy a new class of technologies known as DSL. It carries voice, video, and data at multi-megabit speeds over standard telephone wires. DSL services achieve much higher bit rates than traditional analog modems. Bit rates from 1.5 Mb/s to 51 Mb/s can be achieved with DSL service. The family of DSL services is collectively referred to as xDSL, where x refers to the different types of DSL technologies—ADSL, HDSL, SDSL, and VDSL.

In the late 1980s Bellcore Laboratories in the United States developed the first xDSL services. Telecommunications companies needed a cost-effective way of making high-capacity connections to the homes of subscribers to compete directly with cable television providers for the video-on-demand market. Video-on-demand (VoD—the ability to download movies to the player using the Internet) is expected to generate large revenues for its providers. DSL was developed to provide video-on-demand over existing copper wiring. However, this application was never fully developed or implemented properly. There are several types of DSL. The primary DSL technology used for the home networking and consumer market is Asymmetric DSL (ADSL). It was designed for the local loop, or last mile of copper wire from the Telephone Company's central office to a customer's home. ADSL allows users to download data to a home network at speeds that are comparable with Ethernet standards.

DSL connections offer speeds up to 100 times faster than existing dial-up connections and 25 times faster than ISDN. The technology allows simultaneous use of a single telephone line for both data transmission and voice calls.

Unlike a cabled TV environment users do not experience performance degradations or slow-downs from network sharing. The ADSL equipment is relatively easy to install. Once connected the home network has the convenience of having "always-on" Internet access. Typical ADSL broadband access links show downstream data transmission rates of between 6 and 8 Mb/s and upstream rates of between 640 and 800 Kb/s.

Factors affecting transmission rates include line length and the capabilities of the modem that is connected to the in-home network. Typically any home networks within a three-mile range of the DSL service providers' central office will receive a good a reliable broadband service. Data rates start to degrade when the distance between the home and the central office is greater than three miles.

Apportioning the line bandwidth asymmetrically helps ADSL achieve high data rates. An ADSL circuit has three different information channels:

- A plain old telephone service (POTS) channel
- An upstream channel with a capacity of up to 1 Mb/s
- A downstream channel with a capacity of up to 8 Mb/s

The POTS channel frequency is used for normal voice communications over the broadband connection. The ADSL upstream channel connection is considerably smaller than the downstream channel. Transmission bit rates range from 16 Kb/s to 1 Mb/s. Upstream rates can be set to be compatible with standard U.S. and European digital rate hierarchies.

The upstream channel is a duplex channel configured to handle upstream transmissions only. This is why ADSL is called an *asymmetric* xDSL service. ADSL's asymmetric configuration it is not well suited for video conferencing purposes. However, ADSL technology is ideal for a wide range of interactive services, delivering Internet content, and video-on-demand services for digital appliances connected.

ADSL Standards

Making ADSL a reality requires the cooperation of a variety of industries and companies, along with the development of many new standards.

A wide variety of International organizations and companies contributed to the standardization of ADSL. The International Telecommunications Union (ITU) is one of the most prominent organizations promoting DSL technology standards. In 1999 ITU approved two specific ADSL standards, G.dmt and G.Lite.

G.dmt ADSL technology uses existing phone lines between the central office and the customer to provide always-on, high-speed Internet access at rates up to 8 Mb/s downstream and up to 2 Mb/s upstream.

While the G.dmt standard is quite popular with businesses, the G.lite ADSL standard is focused on the consumer market. It provides data over existing phone lines at rates of up to 1.5 Mb/s downstream and 512 Kb/s upstream. G.lite offers high-speed, always-on Internet access combining convenient installation and a low cost. It is the mass-market vehicle for unlocking the broadband consumer market for DSL services. G.lite was specifically designed to eliminate the need for a DSL provider to send an engineer to the customer's premises. G.lite only requires the installation of new modem for a home network to access the DSL broadband connection. An industry group called the Universal ADSL Working is spearheading the effort to introduce G.Lite by supporting and expediting the development of a worldwide standard.

The consumer DSL market is currently in a transitional phase common to emerging markets. Manufacturers are actively developing DSL products that comply with the G.lite and G.dmt standards to accelerate this transitional period and enable a mass market for DSL broadband access connections.

An ADSL circuit consists of two ADSL broadband modems connected by a copper twisted-pair telephone line. A passive filter called a 'POTS splitter' maintains backward compatibility with the standard telephone system. It also avoids disruption of service due to equipment failure by separating the voice portion of the frequency spectrum from the digital modem circuitry. If the ADSL modem fails the POTS service is still available. This configuration allows users to make voice calls and transmit Internet data over the same broadband DSL connection. When ADSL transmission signals are received at the central office a more advanced POTS splitter is used to send the voice traffic to the public telephone network and data to the Internet.

Starting in 1999 using a DSL modem became significantly easier for consumers when manufacturers such as Dell and Compaq began shipping high-speed DSL modems with computers.

Fixed Wireless Broadband (IEEE 802.16)

With the limited availability of bandwidth from traditional telephone service providers and the worldwide deregulation of the telecommunications industry, a large market exists for fixed wireless broadband. Fixed wireless broadband is used by telecommunication companies to carry IP data from central locations on their networks to small low cost antennas mounted on a subscriber's roof. High-speed Internet access is among the primary advantages of broadband fixed wireless technology. It is capable of delivering speeds more than 100 times faster than those of traditional dial-up connections.

Wireless broadband is seen as an alternative to cable and DSL in suburban and rural markets. It is rapidly deployable, scalable, and has lower implementation costs. Like DSL and cable, broadband fixed wireless is an always-on, Internet-access technology designed to meet the growing demand for rich multimedia and voice applications.

Current deployments of fixed wireless broadband systems are based on a technology called MMDS (Multi-channel multi-point distribution system). The basic components of a MMDS based fixed wireless broadband system are shown in Figure 18.5.

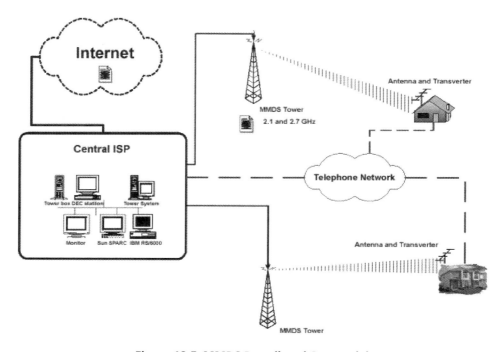

Figure 18.5: MMDS Broadband Connectivity

A wireless modem at the customer's location converts the request for Internet data from the personal computer or connected network to a signal suitable for transmission over the MMDS network. This modem is connected to a transverter, which converts intermediate frequencies to radio frequencies and passes them to the antenna. The transverter is typically mounted on a customer's roof or on a mast to gain line-of-site visibility of the service provider's transmitter. The transmitter is often located on top of a mountain or a tall building. Another antenna at the service provider's transmitter communicates with the wireless modem located at the customer's home. The transmission site then relays the request to the ISP facility. The ISP receives the request and retrieves data either from its servers or from the Internet over its own high-speed backbone connection. The ISP than returns the data via the MMDS network to the computer or network located at the customers location. Fixed wireless broadband solutions are attractive to people who live in rural areas where it is cost prohibitive for cable companies to build systems or telephone companies to supply DSL service.

Broadband Satellite

Communication satellites are orbiting microwave relay stations used to link two or more earth based microwave stations. Communication satellite providers typically lease some or all of a satellite's channels to large corporations. They are then used for long distance telephone traffic, private data networks, and distribution of television signals. The high costs of leasing these huge communication devices make them unsuitable for the residential market. Consequently a new suite of consumer services was developed called direct broadcast satellite (DBS) systems with a range of high-speed Internet access services. A DBS system includes of a mini dish that connects an in-house network to satellites located about 35,000 km above the Earth. These satellites have the ability to download multimedia data to home networks at speeds in excess of 45 Mb/s. Uploading information to the Internet requires the use of a slower telephone connection. A digital satellite modem is required to access this type of broadband platform.

A satellite modem includes a modem chipset that receives a digital signal from the satellite network. It then converts the analog signal to a digital format, checks for errors, and isolates a particular Internet data channel. The host interface provides communication between the satellite modem and the home network. The two most popular host interfaces are USB (Universal Serial Bus) and PCI (Peripheral Component Interconnect).

Power Line Broadband

The first patent for sending signals over power lines was in 1899. But the dream of using this technology for providing high-speed, two-way communications by utility companies is only now becoming a reality. The idea is simple. The utilities' existing infrastructure of stout copper lines, long distance cables, and in-house wiring has the potential to become a communications platform. Every electrical outlet in every building can be a port to the ultimate communications network. However, electrical grids were not originally intended to transmit data. Low impedance, no specific topology, multitudes of fuses and circuit breakers, and transformers that remove encoded signals from the voltage wave confound efforts to use this technology. Despite these problems the energy industry continues to seek a way to use the power grid as a broadband communications platform.

Power line communications technology enjoys several important strengths. The electric power grid provides an ideal communications platform because it is the most extensive network in the world. Power lines are already installed in most homes. The power lines are extremely robust and modern. Power lines carry signals for long distances without regeneration. A power line's near light speed propagation makes them ideal for the fast delivery of video and audio data. Power lines have a large capacity. And there is no topology limitation for the power lines. The main drawbacks of power line communications are the lack of established standards and that only a portion of industry is supporting its developmental.

ISDN (Integrated Services Digital Network) Broadband

ISDN was originally developed in the mid-80s as a means of delivering integrated voice, data, and video services to consumers. ISDN uses of a set of standard communication protocols for digitizing the telephone network.

Analysts indicate there are approximately 2 million ISDN lines deployed in the U.S. Though plagued by sluggish growth in the past few years, ISDN may be a good solution where cable modem or DSL service is not available. ISDN is a circuit switching technology. In circuit switching a dedicated communications path is established between two stations. The process of establishing these dedicated paths is referred to as signaling. Signaling is carried out over special channels referred to as D (Delta) channels. A channel is a conduit through which a digital or analog signal comprising user data or network information flows.

Signaling establishes one or more 64 kilobit-per-second (Kb/s) B (Bearer) channels between locations. These channels can be used for voice, data, or video. There are two main ISDN alternatives. Basic Rate ISDN (BRI) and Primary Rate ISDN (PRI). BRI services are targeted at home networks and small business users. BRI service is delivered over a single twisted pair wiring that is used to deliver POTS. It provides 2 B channels and one 16 Kb/s D channel. PRI service is targeted at larger corporate customers. A PRI line requires two sets of twisted pair telephone lines. In North American PRI service consists of 23 B channels and a single D-channel running at 64 Kb/s. This yields a total bit rate of 1.544 Mb/s. In Europe, Australia, and other parts of the world telecommunication companies provides 30 B and one 64 Kb/s D-channels. The total bit rate for this system is 2.048 Mb/s.

The limited availability of DSL services has prompted service providers to look for ways to use ISDN technology to provide DSL-like services. While ISDN cannot compete with the bandwidth available from the newer DSL services, they have developed an always-on ISDN. This is accomplished by using the D channel for signaling and forwarding IP traffic using the X.25 protocol. The D channel's permanent connection provides the subscriber with up to 16 Kb/s of continuously available bandwidth. When user traffic exceeds the bandwidth of the D channel one or both of the B channels are connected. The user's ISP, Phone Company, and the ISDN bridge or router must all provide always-on support for this to work properly.

The International Consultative Committee for Telegraph and Telephone (CCITT) developed the initial ISDN standards and specifications. The CCITT standards were replaced by the ITU-T (International Telecommunication Union-Telecommunication) standard. The ITU-T is a supplementary body of the United Nations. The ITU-T's definition of ISDN is "a network evolved from the telephone network, that provides end-to-end digital connectivity to support a wide range of services". These services provide for the transmission of both voice and other types of communication.

Long-Reach Ethernet (LRE) or Ethernet in the First Mile (EFM)

The IEEE 802.3ah subcommittee is developing a standard for Ethernet over voice-grade copper wires. Dubbed the Ethernet in the First Mile subcommittee, it has made great advances toward its goal of securing a standard for 10 Mb/s service over at least 2,500 feet of copper wire. The group expects to publish a final standard in 2004.

Long-Reach Ethernet (LRE) is an early example of this technology. LRE or Ethernet in the First Mile (EFM) broadband networking uses Ethernet technology to deliver 5-15 Mb/s performance over existing telephone-grade (Category 1/2/3) wiring. Delivering "last-mile" metropolitan-area network (MAN) services is made feasible with LRE. LRE reaches up to 5,000 feet and enables simultaneous voice, video, and data applications. It provides a cost-effective and easy-to-deploy technology for service providers to use in equipping multiunit buildings (MxU) and enterprise campus environments with broadband access. MxU buildings include hotels, multi-dwelling unit (MDU) housing, and multi-tenant unit (MTU) office buildings. Building owners can make their properties more attractive and create additional revenue by offering high-speed connectivity. Hotels use their existing copper telephone lines to offer high-speed Internet access. Cahners In-Stat Group predicts the LRE market to hit $3.4 billion by 2004. Residents, telecommuters and business travelers want secure, high-speed broadband connections. These connections provide them with VPN, high-speed Internet access, video-on-demand, IP telephony, multimedia entertainment, and office productivity applications. Manufacturing, education, and medical facilities use LRE to provide high-speed connectivity throughout their buildings for monitoring test equipment, providing online training, connecting classrooms, or monitoring the equipment in a patient's room.

LRE technology provides far more bandwidth in the "last mile" than has previously been available. This added bandwidth combined with the increased deployment of fiber cable solves the "last mile" bottleneck. The LRE solution delivers high-speed broadband access without the time, expense, and inconvenience of having to rewire the entire facility with Ethernet-grade cabling. It is based on Ethernet (Category 5) and utilizes a building's existing telephone-grade wiring. LRE also overcomes the traditional 100-meter Ethernet distance limitations. It provides Ethernet-performance of up to 15 Mb/s at distances up to 5,000 feet. LRE enables service providers to deliver an unprecedented number and variety of new revenue-generating broadband services to users.

LRE provides full-duplex transmission at tremendous speeds and distances:

- 15 Mb/s symmetric rate (up to 3,200 feet)
- 10 Mb/s symmetric rate (up to 4,000 feet)
- 5 Mb/s symmetric rate (up to 5,000 feet)

The hospitality industry, including hotels, hospitals, convention centers, and airports, already offers a broad variety of business and entertainment services to their guests. Value-added services generate substantial revenue in this industry. Broadband connectivity will dramatically increase the number and variety of services offered to guests and provides additional revenue to the provider. There are currently 51,000 hotels in the United States with 39 million rooms. According to Jupiter Communications, only 15% of these rooms featured high-speed access by the end of 2000. However, Jupiter reported that by 2003, hotels planned to extend broadband to over 50% of all their rooms. Full-service high-end hotels are among the first to offer this service.

MDUs including apartment complexes, condominiums, university dormitories, townhouse complexes, and other buildings represent a largely untapped market for the delivery of broadband services. According to market research firm Yankee Group fewer than 5% of an estimated 21 million MDU households in the United States had high-speed Internet access in 1999. But 95% want broadband service. This trend will turn 210,000 high-speed users into 3 million users by the end of 2004. The Yankee Group estimates that 14 million households are interested and willing to pay $40 per month for broadband services.

MTU commercial properties such as office buildings and campuses and industrial campuses represent a third major market for LRE. According to the U.S. Department of Energy there are 705,000 commercial office buildings in the United States, and 150,000 contain more than five tenants. These tenants represent a significant amount of pent-up demand for greater bandwidth. The high cost and disruption of wiring existing buildings with Category 5 cabling has previously discouraged building owners from providing broadband access. MTU buildings present a great need for a cost-effective, high-speed solution that does not require threading additional wire.

The three devices required to effectively deliver LRE to users are an LRE switch, LRE customer premises equipment (CPE), and a POTS splitter. LRE switches condition incoming and outgoing

Ethernet packets to run over standard copper wiring. Otherwise, they act as traditional Ethernet switches—directing, storing, and forwarding packets as well as matching destination and source addresses together into virtual channels. LRE switching units simply need to be installed with the telephone equipment. They then use 10/100 Mb/s or 1000Base-T to uplink to the Internet router and local servers. Multiple LRE switches and 10/100 Mb/s Ethernet switch ports can then be daisy chained together.

Each LRE port is terminated in with a CPE device. The device splits LRE and POTS traffic, converting the LRE traffic into Ethernet traffic and vice versa. This provides Ethernet access to the LRE signals without interfering with the service already in place on the telephone wire. A POTS splitter lets LRE and POTS co-exist on the same telephone line. This configuration is necessary for a deployment when there is an on-site PBX system and the POTS traffic and LRE traffic must coexist over the same copper wiring.

Components of robust LRE broadband solution are:

- **Low-cost** – Building owners do not need to rewire their buildings to achieve remarkable increases in available bandwidth
- **High-speed** – LRE delivers bandwidth of up to 15M bit/sec
- **Extremely functional** – LRE supports a large range of broadband capabilities

LRE provides a cost-effective solution that preserves the existing communications infrastructure of multiunit buildings and enterprise campus customers.

T1

Plesiochronous Digital Hierarchy (PDH) refers to the three worldwide digital communication standards used for point-to-point 2-wire-pair links based on the fundamental concepts of Time Division Multiplexing (TDM). These include the T-n carrier system in North America (NADH), the E-n carrier system in Europe (EDH), and the J-n carrier system in Japan (JDH). All PDH systems begin with the same base rate of 64 Kb/s. Their difference is the rate hierarchies and overhead/bit-stuffing schemes. Line rates range from 1.544 Mb/s (NADH T1) all the way to 274.176 Mb/s (NADH T4). With PSTN, the PDH carriers are used for trunking and transporting voice channels. Service Providers (SP) use higher rate PDH carriers to create Wide Area Network backbones. Businesses lease lower rate PDH lines such as T1 and E1 from an SP to gain access to the SP's WAN allowing digital information (voice or data) to be exchanged between geographically dispersed office locations and providing access to the World Wide Web.

T1 is a high-speed digital network (1.544 Mb/s) developed by AT&T in 1957. It was implemented in the early 1960's to support long distance pulse-code modulation (PCM) voice transmission. The primary innovation of T1 was the introduction of "digitized" voice signals. It also created a digital network representing an analog telephone system. It is described as a "two-point, dedicated high capacity digital service, provided on terrestrial digital facilities capable of transmitting 1.544 Mb/s". The interface to the customer can either be a T1 carrier or higher order multiplexed facility such as those used to provide access from fiber optic and radio systems.

In the basic definition there is a "higher order" or hierarchy for T1. T1 has a speed of 1.544 Mb/s. It was designed for voice circuits or "channels" (24 per T1 line or "trunk"). In addition there is T1-C that operates at 3.152 Mb/s. The T-2, which was implemented in the early 1970's operates at 6.312 Mb/s and can carry one Picturephone channel or 96 voice channels. The T-3 operates at 44.736 Mb/s and T-4 at 274.176 Mb/s. These are known as "super groups" and their operating speeds are generally referred to as 45 Mb/s and 274 Mb/s, respectively.

The early success of T1 is being phased out by other technologies offering higher data rates and lower costs.

Fiber-to-the-Home (FTTH) and Fiber-to-the-Curb (FTTC)

Fiber-to-the-home (FTTH), fiber-to-the-curb (FTTC), and fiber-to-the-business (FTTB) technology provides scalable voice, video, and high-speed data services over a single fiber optic network. This enables the delivery of bundled services to homes and businesses. This technology requires an optical fiber cable from the telephone switch to the subscriber's premises. For service providers this can be an extremely cost-effective alternative to DSL and HFC.

Synchronous' FTTH/FTTC Optical Node Receivers deliver two-way analog or digital video (CATV) and scalable high-speed data. This system offers a wide optical window supporting 1310nm and 1550nm transmissions. It supports a bandwidth of 40 MHz to 870 MHz and a full complement of analog and digital signals and return path capacity.

For years telecom companies have been working diligently to provide pseudo-broadband Internet connections using copper wire (DSL) and cable (cable modem). The term "pseudo-broadband" is used because the existing telecom infrastructure can only provide speeds of up to 1.5 Mb/s. In theory cable modems can provide up to 2.5 Mb/s. But in reality the shared aspects of their architecture result in lower speeds. Over time improvements will continue to be made to copper and cable systems. But they are going to be challenged by the introduction of fiber to the home. And fiber is coming much faster than most people predict. FTTH download speeds of up to 155 Mb/s are 100 times faster than pseudo-broadband DSL and cable modem connections.

Fiber to home has always been an attractive option. By avoiding electromagnetic interference (EMI) problems it features the highest reliability available today. It does not need electrical power and is immune to lightning. The properties of the fiber provide the lowest costs for maintenance, provisioning, and facilities planning.

Fiber to the home technology has seen significant and continuous progress. In the eighties FTTH was considered impractical. It was seen as too expensive, incompatible with analog TVs, and the bandwidth needed for digital TV was too high. Issues of lifeline covering and lack of product standardization also hindered its introduction. Recent advances in Passive Optical Networks (PONs), loop lasers, and chips for compressing digital video, and fiber have reduced system costs dramatically.

Challenges still exist for making FTTH cost effective. It will take years before most residences in America have true broadband Internet access. But FTTH is inevitable, for a number of reasons:

- **Speed** – FTTH is much faster than copper, cable, and wireless.
- **Demand** – Once FTTH starts being installed in residential infrastructure, homebuyers will seek it out. When the cable television infrastructure was first being deployed the availability of cable in a neighborhood became an important purchase consideration for homebuyers. The same thing will happen with FTTH. In the next few years FTTH will become the standard for all new construction. This will lay the groundwork for FTTH's rapid acceptance.
- **The Principle of Competitive Disadvantage** – Imagine a country where all buyers of new residential construction have Internet connections of 155 Mb/s. People with DSL and cable modem connections to the Internet will be at an enormous competitive disadvantage.

Imagine two people working from home, one with a 1.5 Mb/s Internet connection and another with a 155 Mb/s Internet connection. Which person do you think will be more productive? Which person has the fundamental competitive advantage in an Internet based society? The principle of competitive disadvantage will drive prices to the point where it becomes cost effective to deploy fiber to existing homes. The intrinsic competitive nature of our society and economy will not tolerate a situation where the majority of homes are at a huge technology-based competitive disadvantage. Once FTTH has become the standard for new construction the principle of competitive disadvantage will cause it to spread rapidly to existing construction. A majority of homes will have FTTH Internet connections within ten years with the principle of competitive disadvantage as a driving force behind such a rapid deployment.

Fiber in the home has all the makings of a 21st-century technology. FTTH is the best short-term and long-term option available for last-mile connectivity. International operators and suppliers, with or without U.S. involvement, are forging ahead with the FSAN (Full Service Access Network) initiatives to create internationally agreed upon standards. It has all-fiber-plant, future-proof bandwidth potential, low powering and operations cost, leading to cost parity in rural areas now. Based on life cycle costs, we will see parity in suburban and urban installations as costs continue to fall.

Passive Optical Networks (PONs)

A passive optical network (PON) system brings optical fiber cabling and signals up to, or close to, the end user. Depending on where the PON terminates, the system can be described as full form fiber-to-home, business, and curb. The passive component indicates that the optical transmission does not have power requirements or active electronic parts to move the signal through the network. The PONs work is based on network basics. Carriers want to connect each customer site with a wavelength of light without dedicating a fiber to every wavelength. To address this issue PONs bundle multiple wavelengths (up to 32 at present). By doing so a single access line can carry them from the carrier's central office (CO) to a manhole or controlled environmental vault close to a cluster of customer sites. At that point each wavelength is broken out and redirected into a different short length of fiber for an individual site. A different scheme is used for collecting traffic traveling from user sites to the CO in the opposite direction. In this case each site is given a specific time slot to transmit using a polling scheme similar to the one used in older networks.

Several PONs customers share the fiber and much of the service provider's equipment costs. Since there is no equipment installed between the service provider and the user, the optical path is "transparent" to bit rate, modulation format (digital or analog), and protocol (SONET/SDH, IP, Ethernet). In the future this transparency will allow easier upgrades and service bundling. New services and customers are added without affecting existing customers by changing service-specific equipment at the ends of the network. Most of today's other access network architectures do not offer this flexibility.

Despite their advantages PONs face significant obstacles on the road to success. The amount of upstream bandwidth transmitted over a PON is divided among the number of users at the customer site. On a 155-Mbit/s PON link with four splits each subscriber receives 38.75 Mb/s. The available bandwidth is also lowered by the addition of splitters on links that have already been split . Since PONs do not regenerate or convert optical signals mid-network they are less expensive to operate. But this also limits their reach. Without regeneration light signals lose power quickly limiting transmission capabilities. These disadvantages combined with the availability of other broadband access alternatives the market will limit the acceptance of PONs.

Analog (Dial-Up)

While hardly qualifying as broadband access, analog modems are the most popular type for providing Internet access. The modem is designed to operate with any dial-up phone worldwide. These modems support high-speed analog data, voice, and fax operations. Being host controlled allows integrated modems to reduce the overall chip count and the need for a separate microcontroller.

Despite market enthusiasm for the many flavors of DSL, cable modems, and other residential access approaches, their deployment continues to be gradual, sporadic, and regionalized. This means that analog modems still have an important role in enabling Internet connectivity. However, despite an increasing volume of sales, eroding margins, and significant declines in shipments plague the analog modem market. At the retail level modems can be purchased for less than $20.

The overall analog modem market can be categorized in three distinct segments:

- **Branded modems** – In this category the customer purchases a specific product. Branded modems may be sold individually or in multi packs through a variety of channels. The majority are sold through retail outlets.
- **PC OEM modems** – These modems are shipped directly to manufacturers and are then installed in new PC's.
- **PC Card modems** – This modem is designed to fit into PCMCIA standard laptop. This category also includes the emerging mini-PCI form factor.

Some popular analog modem market trends include:

- On a worldwide basis Internet and PC growth continue to sustain the analog modem market sales.
- Currently the worldwide availability, ease of use, and low cost of analog modems are unmatched by any other technology. PCs will continue to incorporate analog modems even for consumers using an alternative access technology.
- Volume sales will continue to increase even as margins continue to decline.
- Gradual market erosion will continue as new technologies such as xDSL, cable modems, and ISDN achieve broader achieve market acceptance.

IDC predicts that the worldwide analog modem shipments will increase from 76 million units in 2000 to over 120 million units by 2004. While faster access options continue to expand at a much faster rate, analog modems will continue to account for over two-thirds of total modem shipments.

Broadband Access Summary

The worldwide modem market will continue to be buoyed by the expansion of the Internet. Modems will continue to be the primary solution for residential Internet access. Cable and xDSL modems will continue their strong movement into the residential market.

The development and implementation of broadband technologies are being driven by the worldwide use of computers and new appliances that allow people Internet access from the comfort of their homes. A key factor encouraging the proliferation of home networking technologies is the spread of broadband connections that provide lighting-fast Internet services. Internet access and high-speed remote access to storage media require more data capacity than traditional telecommunications services can provide. Replacing copper wiring with fiber optic cabling is still an expensive option for delivering this capacity to your home. This high cost has focused development on deploying new technologies for existing network infrastructures.

Cable, wireless, DSL, and satellite companies have marketed their networks as an infrastructure that can provide home networking users with a variety of broadband Internet services. The new era of digital TV presents network service providers with an opportunity to deliver profitable Internet centric services. Cable companies can offer an Internet access service for home networks at a much higher rate than traditional and specialized ISPs. Wireless operators are exploring MMDS technology to boost data rates between our homes and the Internet. Satellite companies have joined the mix with cable, wireless cable, and telecommunication service providers in providing home networking users with a powerful broadband platform for accessing the Internet. Power companies are also exploring the possibility of using their electrical power grids for carrying near light-speed data for telephone, radio, video, Internet, and satellites. ISDN still remains the most widely deployed digital broadband service in the world. In the future new budding technologies such as fixed wireless broadband, LRE and FTTH will be competing in this market space.

Home Networking

The growth of home networking is being fueled by the emergence of digital consumer devices such as PCs, PC peripherals, control devices, telecommunication products, and digital consumer devices. The home network enables the distribution of data, voice, and video between consumer devices. Building islands of connected appliances is not a new market. A USB connected PC island with multiple PCs and PC peripherals such as scanners, printers and a fax machine exist today. The current multimedia island includes the TV, amplifier, speakers, DVD player, and VCR networked using IEEE 1394 and proprietary video cables. Pressure from consumers to expand the scope from networking islands to networking the entire homes is leading the emergence of the home networking market.

What is Home Networking?

From a purely technical perspective a home network connects a number of devices or appliances within a small geographical area such as a home and SOHO. A home network improves communication and allows the sharing of expensive resources. A home network consists of two basic categories: hardware components and software components. Home networking has evolved from its roots in automation and security to include the distribution of audio, video, and data content in the home. Home networking is also the interconnection and interoperation of different home electronic appliances, entertainment devices, PC hardware, and telecommunication devices.

Motivations for Networking Consumer Devices in the Household

Home networking is an ideal fit for technology early adopters. However, for many people keeping up with their nearest neighbors is not enough to motivate the installation of new home networking technologies. Motivating factors that do encourage people to network different classes of household appliances and computers together include:

- **Leverage existing investments** – Sharing hardware resources is the number one motivating factor for consumers investing in new home networking interconnection technologies. People want to leverage expensive appliances investments such as computers, set-top boxes, personal video recorders, digital cameras, and cable modems.
- **Shared Internet access** – Shared Internet access is the second most popular reason for the deployment of home networks. Combining home networks and residential gateways allow different members of a family to simultaneously use a single Internet access point.

- **Interconnecting subsystems** – The ability of a home networking infrastructure to interconnect different types of subsystems together is another motivating factor. For example, home security systems connect different types of sensors with a central controller. Integrating this type of network into an existing Pc-based home network helps to expand the functionality of the security system.
- **Rise of multi-PC households** – The rapid growth of multi-PC homes increases the need for multiple node PC networks.
- **In-home application evolution** – Each generation of standard software applications increases required file sizes. For example, over the last few years the file size of a PowerPoint demonstration with the same content has doubled. Graphics and digital photography are commonly sent via e-mail.

Figure 18.6 below shows sharing resources and digital content are big drivers for networking the different digital consumer devices and PCs in homes.

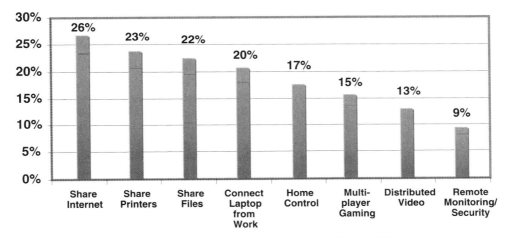

Figure 18.6: Applications Driving Home Networking

The primary functions enabled by home networks fall into five basic categories: entertainment, communications, computing/information, safety/security and home control/automation. The specific functions within each of these categories refer to the type of content or service provided by the network. Key home networking applications include:

Entertainment:
Listening to music in any part of the house is a popular option. Families can hear a single music source from several locations in the house. Or they can select a different local source by linking the separate stereos in the house. Speakers can be installed in ceilings and walls to fill a home with music during parties or to listen to the morning news in the shower. Existing stereo equipment can be used to play music from the Internet. The PC or a dedicated device can serve as your Internet radio providing access to music, live broadcasts, and news stories from around the world. Consumers can download the latest MP3 from their favorite band or highlights from today's game for playback in their living room. Music from MP3 files on the PC can be heard in the study and or while jogging. Users can create their own mixes by combining tracks on different CD's. Audio media types include

digital prerecorded physical format (such as CD, DVD Audio, SACD, MiniDisc, and Digital Tape), Internet music/MP3s, AM/FM radio broadcasts, Digital Radio, and prerecorded analog such as an audiocassette.

Home entertainment networks let family members watch the same videotape, DVD, or digital satellite program from any room in the house. With a DVD player in the family room and a television attached to a cable box in the bedroom, homeowners can use the remote control to turn on the DVD and go directly to their favorite scene. If their PC is equipped with a DVD drive they can watch a movie playing on the computer from the living room's home theater system. Parents can use a picture-in-picture television to monitor what the kids are watching or see the kids in another part of the house while watching the evening news. Video media types include external sources such as analog cable, digital cable, digital satellite (DBS), analog satellite (C-Band), analog terrestrial broadcast, digital off-air broadcast (DTV), and MMDS. Internal sources include video cassette, DVD, personal video recorder (TiVo), digital video camera, analog video camera, PC, and digital still camera.

With a networked gaming system kids can play the latest interactive video games against an opponent in another room or in another country as though they were sitting on the same couch. Digital picture frames can display the entire family digital photo album and digitized fine art.

Communications:

Local voice intercoms including public address (through mounted speakers and microphones), phone to phone (PBX), and wall-mounted intercoms form another need. Stand-alone intercom systems have been available for many years. Large, simulated wood-grain wall-mounted boxes have been replaced by modern, integrated intercom and paging systems featuring small flush-mounted keypads. Some systems use the telephone as both a paging and intercom system allowing you to press a key on the phone and speak through the home's AV speaker system. Pages are answered by picking up any phone in the house.

A call handling system (PBX) provides call routing, intelligent call forwarding, call priority, call blocking, message notification, call logging, call transfer, call attendant, voice mail, music and more. Multi-line phone systems can support several extensions and reduce the number of phone lines. Systems can support up to 16 extensions on just four leased lines. Adding features to a phone system like intercom, paging, auto attendant, voice mail, wireless and cordless phone access, and caller ID increase power and flexibility. The caller ID capability of some systems allows you to provide individual callers with separate outgoing messages and handling options.

With video conferencing users can see and hear the people they are talking to by using a PC and Web cam or a dedicated video conferencing device. They can also participate in distance learning courses, check in on kids in college, or their parents in another state. Video clips captured from a camera above the crib can be e-mailed to grandparents.

Computing/Information

Networks can receive external data via Internet, WAN and VPN access. They can receive local file access from another PC, file server, or PDA.

Home computer networks allow entire families to simultaneously share documents, printers, and high-speed Internet connections. Parents can check e-mail from the computer in the den while a child researches their homework assignment on the Web. Users can collect a list of kitchen ingredients from the Web site of a cooking show, add additional items to the shopping list, print it out in the den, and e-mail a copy to their spouse at work. Networking personal computers in a home can add mobility to a worker's list of home-office benefits. Network connections in other areas of the house allow you to back up your laptop, receive e-mail, and print from multiple locations. Wireless networking solutions can even let you bring your work outside.

Safety/Security

Home networks can provide a coordinated electronic emergency response. Action triggers such as smoke blower shutdown, flood-water shutoffs, intrusion lights, and freeze service calls can be controlled by a home network. Integrated sensors can provide a complete picture of what is going on inside and outside a home by detecting smoke, carbon monoxide, moisture, daylight, wind, power outages, and even the sound-signature of a tornado. In the event of an emergency, home systems can be configured to respond by turning on all the lights for emergency responders. They could also turn off all appliances to avoid the risk of fire if the injured person was cooking at the time. Some systems call many different numbers until they reach a neighbor or family member who can then listen and talk to the injured person via speakerphone or in-wall speakers and microphones. The simple act of arming your security system can provide significant energy savings by resetting your thermostat when your home is not occupied. It could also turn on exterior lights, turn off the interior lights, and dimly light a path to the bathroom. A fire alarm can trigger a signal to light a path to the exits and flash exterior lights to help firefighters find your home. Driveway, walkway, and front door lights can be programmed to turn on just before you arrive home or be triggered by driveway sensors when you drive up.

Local video monitoring includes closed-circuit television (CCTV) such as pool cameras, crib cameras, door cameras, driveway, and walkway cameras. For added safety and security the view from small cameras mounted inside and outside the house can be distributed to all the televisions in a home. Parents of young children can monitor the nursery while working in the study. Older children watching television can see who is at the front door before answering. For parents or people working at home, small cameras and microphones placed in an infant's room are an example of things that can help keep children safe. The sound can be channeled through any or all of the stereos in the house and the video viewed on a computer screen or on any TV.

Another option is remote monitoring. Traditional monitored security systems report emergencies to a central monitoring station. An operator is then available to take the appropriate action. Home-owner remote monitored security systems notify users directly if there is a problem. Auto-dialers dial phone numbers and play a prerecorded alert message. Many systems call several locations until the homeowner is found. By integrating home systems such as fire and burglar alarms, appliances, speakers and microphones, closed circuit TV, and the telephone parents can maintain a virtual home presence while at work. For example, the house can be programmed to call if the children are late getting home from school. The TV can be set to stay off until after homework is finished. And various appliances can be rendered inoperable without adult supervision. You can even be notified at home if your aging parent does not get out of bed at the usual time. Using a link to fire and burglar alarms the home automation system can notify users and their closest neighbors if there is an emergency. Remote monitored cameras, microphones, and sensors offer homeowners the ability to hear and see what is happening inside the house. Computers or PDAs can view strategically placed cameras within the home via the Internet. Remote users can even control networked devices through a Web-based interface

Home Control/Automation

Local control provides multi-room and scene lighting control for a room or an area of a room from any location in the home. Networking can enhance the performance of home theater systems. Lighting controls can be configured to respond to a home theater's remote control. Pressing a key on the remote can turn on the AV components, lower the projection screen, and dim the lights. Windows, blinds, and skylights can even be programmed to respond to a "movie" command. When the phone rings in the middle of a movie a simple connection to the telephone system can turn down the sound and display a caller's name and number on the screen. Modern integrated lighting systems incorporating many different appliances allow users to set moods or activities. For example, by pressing an "entertain" button you can dim the dining room and bring the lights in the living room to a comfortable level.

Home network control systems are available that let you perform functions remotely using a telephone keypad, voice, or via the Web. Logging in to the home's network from a remote computer or PDA allows users to see a live picture from cameras mounted in and around the house. The home control interface provides the ability to lock doors, turn on lights, or start the sprinkler. Telephone home control (touch-tone or voice) systems with remote monitoring and control capabilities let users monitor and control home systems. Forgetful users can turn off the iron or the coffeepot. People who will be away for a few days can use their phone to start the pet feeder, set the furnace to 60 degrees, and arm the security system. The system will also record a number to be called in case of an emergency. While homeowners are away the system can give the house a lived-in look by turning lights off and on periodically. The system can even call your phone and let the homeowners speak with the person who has just rung your doorbell.

Automated Control

Lighting control systems can automatically turn on or off from several locations in the home. These systems include multi-room and outdoor lighting that can automatically adjust lights based on the time of day or ambient light levels.

Room-by-room (or area-by-area) energy controls can be operated based on lifestyle factors (wake-up, go-to-work, arrive-home, and bedtime inputs). Integrating lighting systems allows homeowners to create scripts across categories of products and systems based on lifestyle preferences. For example, a script for "good morning" might include turning the heat up, putting the music on in living room, and the lights on in hall and kitchen. An evening house script might turn on the security system, turn off all lights except in a hall, set back the thermostat, and turn on the washing machine to take advantage of time-of-use billing. A "bedtime" script could arm the alarm system, turn down the water heater, adjust the humidity level, turn off all the lights, and close all but the bedroom vents.

More precise and comfortable temperatures within the home are created by using automatic temperature controls and knowledge about outdoor temperature and wind. Most security systems can turn down the heat when armed. However, a modern security system often knows more than just whether or not someone is home. Information from motion sensors, pressure-sensitive doormats, window sensors, and door sensors gives the security system a complete picture of what is going on in the house. The HVAC system can use this same information to improve the comfort level, energy efficiency, and air quality of a home. A home control unit can be programmed for several energy-saving routines using inexpensive motorized dampers. For example, a lack of motion in one room for a given period of time can close all that room's vents to avoid heating or cooling unused rooms.

Water management is another area that can be integrated with home networks. Water heaters can be integrated with home control systems to provide hot water only when it is needed. Spa and pool pumps can be set to run on schedules or to start when someone gets home in the evening. During freezing weather temperature sensors outside the house can turn off the water to external faucets. In the summer months, sprinkler systems can take temperature and even forecast information into account to avoid watering the grass if it is unnecessary.

User Requirements for Consumers and Corporate Users

End users requirements for a home network are different than those of a traditional enterprise network environment. These requirements include:

- **Low complexity and easy to use** – Without a network system administrator or MIS department home networking must be easy to use and simple to install. A home network must be "invisible," providing seamless operation with little or no user intervention or maintenance.
- **Reliability** – Similar to enterprise networks, home networks must be reliable. The network needs to resolve interference from home networking devices, microwave ovens, and cordless phones.
- **Scalability** – Scalability provides lower lifetime costs as consumers buy home networking products. Buying a network device establishes the foundation for an entire home network. Informed consumers avoid purchasing obsolete technology. A home network needs to protect the consumer's initial investment by maintaining interoperability and accommodating future applications.
- **Standards compliance** – As proved in corporate networks, industry standards are crucial for enabling mainstream consumer adoption of home networking products.
- **Support for high bandwidth multimedia content** – Most applications and data types that traverse a corporate network have relatively low bandwidth requirements. In contrast, home networks support all types of bandwidth hungry digital content. Existing and emerging digital devices such as televisions, DVD players, Digital Video Recorders, digital audio/MP3 players, DBS systems, flat-panel displays, digital set-top boxes, and PCs create the need to support multimedia content in the home. The mass adoption of home networks hinges on the ability to support multimedia.

Profiles of People Adopting Home Networking Technologies

The following profiles describe the millions of new home networking users in the next five years:

- **Home office workers** – A study by the IDC indicates nearly 50 million households in the United States include rooms for a dedicated home office.
- **Telecommuters** – Gartner Group market technology consultants are predicting that over one third of the workforce in the U.S. will engage in part-time telecommuting. Worldwide the trend is mirrored as workers remotely connect to corporate networks using dial-up and broadband connections.
- **Self employed professionals** – This group of software engineers, Web designers, freelancer workers and graphic artists enjoy the flexibility of working from a home network.

- **Students** – Information technology, the Internet, and education are inextricably linked. Students are very interested and knowledgeable about in-home networking technologies.
- **Families** – Home networking offers many enticing family benefits. Children use home networking technologies to access different Internet learning resources. Parents use the same network for keeping a closer eye on their offspring.

Home Networking Market Data

Consumers are spending more time than ever in their homes enjoying digital audio, video and data, high definition television (HDTV), computers, Internet access, and rich entertainment systems. This is fueling a demand for home networking technologies combining convenience and flexibility. A growth rate of up to 500% in home networking is projected in the next four years. Market researcher Cahners In-Stat predicts that worldwide the home networking and residential gateway market will exceed $6 billion by 2004 and $9 billion by 2006. Market numbers could exceed projections as the definitions and scope of home networks extend beyond basic technologies and products. Next-generation devices such as set-top boxes, digital TVs, and gaming consoles will integrate home networking functionality within existing appliances and provide a key market differentiator.

Home Networking Technologies

Home networking consists of the high-speed data networking technologies for distributing Internet access to multiple access points and appliances. The networking technologies provide interconnectivity for home-networked appliances. A wide variety of technologies exist for interconnecting devices within the home. No current single technology meets all of the requirements for the diversity of applications that will be created. While traditional Ethernet systems offer a robust and proven solution, most consumers do not have the time, interest, or knowledge to rewire their homes. Fortunately, the emergence of "no new wiring" technologies offers prospects for offering mass-market home networking solutions. The new technologies include wireless, phone line, and power line solutions. Each solution presents distinct benefits and drawbacks. Many organizations are beginning to suggest that all of these technologies will exist in a multi-layered home network architecture. Types of home networking technologies include:

- **No New Wires** – Utilizes pPhone lines and power lines
- **New Wires** – Ethernet, Optic Fiber, USB/USB 2.0, IEEE 1394, IEEE 1355
- **Wireless** – IrDA, HomeRF, Bluetooth, Wireless LANs (IEEE 802.11b, IEEE 802.11a, HiperLAN2)

Most home networking technologies define standards and specifications for the physical and data-link layers of the OSI network model. The physical layer provides the electrical, mechanical, and procedural specifications for the transmission of bits through a communication link, medium, or channel. The data link layer, which consists of the medium access sub-layer and the logical link control sub-layer, ensures error control and synchronization between two entities.

Home Networking Using "No New Wires"

Commercial networks are designed to carry data between computers. They typically use fiber optic, twisted pair, or coaxial cables to minimize noise and interference. Most homes do not have dedicated high-speed network cabling installed, and the labor cost of cable installation is cost prohibitive to most homeowners. Successful home networking solutions must utilize existing wiring infrastructures.

Companies creating home networking technologies need to focus on the following criteria:
- The technology needs to leverage existing wiring infrastructure
- It needs to be easy to install and maintain
- The technology needs to use existing standards and software platforms to reduce complexity
- Must include a quality of service (QoS) mechanism to provides low latency for telephony and other voice applications
- Consumers need data rates in excess of 10 Mb/s to distribute live video around their homes
- Needs to be inexpensive
- Technology must provide a level of security

Phone Lines

Until recently home networks depended on professionally installed special cables to link PCs, audio-video equipment, and peripheral devices. With hardware components in different rooms of the house this solution was often expensive and problematic. Now consumers can use their previously installed telephone wiring systems to link multiple computers and digital appliances around the house. Phone line technology promises to deliver home networks without installing hundreds of meters of new data cables inside the walls of households.

Home networking uses phone lines and phone jacks to connect consumer devices such as PCs, TVs, DVD, and MP3 players to each other and the Internet. Several phone line-based home networking system issues must be addressed for it to succeed. These include:

- **Random Wiring Topologies** – Business networks are built around a hub structure. A home phone line wiring system is a random "tree" system. Something as simple as plugging in a telephone or disconnecting a fax machine changes the tree structure.
- **Signal Attenuation** – The random tree network topology of phone line wiring system can cause signal attenuation. Attenuation is the reduction of signal strength during transmission of data across the home network. Open plugs and improperly terminated appliances are the primary causes of attenuation on a phone line network.
- **Signal Noise** – Consumer appliances, heaters, air conditioners, and telephones can cause unwanted signal noise on phone lines.
- **Consistency in Service Levels** – The network must be able to function reliably and deliver consistent service levels despite changes that result from someone picking up the phone, accessing a Web site, or an answering machine recording a message.
- **Telephone Jacks** – Phone jacks are not found everywhere in the home. U.S. households tend to have multiple phone jacks. Households in other countries, particularly Europe, are often limited to one or two phone jacks. Another problem can be the physical location of existing jacks with respect to the devices that need to be networked.

HomePNA (Home Phone line Networking Alliance)

The HomePNA is a consortium of more than 130 companies seeking to develop specifications for interoperable home-networked devices using existing phone wiring. It was established to overcome technical issues by defining standards and technologies. The group created an industry standard when it published an easy-to-use, cost-effective, and proven 1Mb/s home phone line-networking technology as its 1.0 specification. The technology allows PCs, peripherals, and other consumer devices to connect to the Internet and each other without interrupting standard telephone service. Using existing telephone wiring avoids costly or disruptive rewiring of the home. HomePNA has since introduced the second-generation home phone line networking technology (HPNA 2.0). And has embarked on the development of a third generation. The HPNA 2.0 specification brings a faster 10Mb/s technology to phone line networking. At the same time it maintains backward compatibility with HPNA 1.0. The new technology uses selective portions of the 2-30 MHz frequency band to achieve these increased data rates. HomePNA is working to incorporate their technologies into a range of electronic appliances including PC's, ADSL modems, cable modems, digital televisions, set-top boxes, and IP-based Web phones.

HomePNA Technical Architecture

HPNA has become the industry standard for telephone based home networking. It is a robust technology that can achieve data rates up to 32Mb/s in approximately the same bandwidth as the HPNA 1.0 system. It will be forward compatible with future appliances operating at speeds up to 100Mb/s and support up to 500 feet of phone wire between devices connected to RJ-11 jacks.

The following key components make up a phone line based home network:

Network Transport Technologies

- **Leveraging Ethernet Technology** – Home phone line networking uses standard Ethernet technology and adapts it where necessary to overcome the challenges presented by the home phone line environment. The access method for sharing the base-band signal on the home network bus is standard IEEE 802.3 compliant Media Access Control (MAC) and CSMA/CD (Carrier Sense Multiple Access/Collision Detect). Figure 18.7 depicts the home phone line networking data frame. Under the Ethernet standard information is bundled into a package called a frame. Data originating from home network connected applications is formed into standard 802.3 Ethernet data frames and passed to the phone line physical layer (PHY). The PHY circuitry then strips off the first 8 octets of the Ethernet frame (the preamble and delimiter fields). It replaces them with a PHY header designed specifically for the rigors of phone line networking. The reverse process is executed at the receiver. This allows home phone line networking to meet the needs of the home environment and leverage the tremendous amount of today's Ethernet-compatible software.

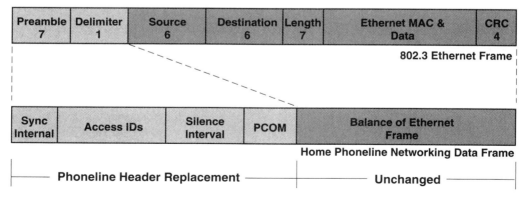

Figure 18.7: HomePNA Data Frame—Leveraging Ethernet Technology
(Source: HomePNA)

- **Spectral Compatibility** – Home phone line networking requires multiple services to coexist on a single piece of telephone wire. Members of the household need to make telephone calls while other members are using the home network for data transfer purposes. Multiplexing is a common method for simultaneously operating multiple data and voice services over a single pair of wires. Multiplexing is the technical term for describing the combination of multiple analog or digital signals for transmission over a single line or media. There are a number of multiplexing techniques used to combine different signal types. HomePNA uses a technique called Frequency Division Multiplexing (FDM). This technique assigns each communications service a different frequency spectrum. Services can exchange information using frequency-selective filters devices without interference from services communicating on a different frequency band. A pattern within the devices allows matching data to pass through the filter. The home network operates between the 5.5MHz and 9.5MHz. Voice communications operate in the 20Hz-3.4KHz and UADSL (universal asynchronous DSL) services occupy the 25KHz-1.1MHz ranges. Figure 18.8 depicts the spectral usage of three services that can share home phone wiring. POTS, UADSL Internet connectivity, and home phone line networking share the same line by operating at different frequencies.

Home Phoneline Networking is Compatible with Regular Telephone Usage and UADSL Internet Connectivity

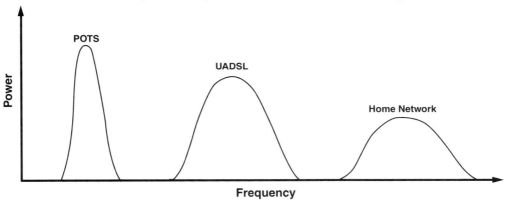

Figure 18.8: HomePNA Spectral Usage—Spectral Coexistence

- **High Performance Encoding** – Home phone line networking technology uses a time modulation line coding method. This increases the data throughput and allows reliable data transmission over home phone wiring. Its adaptive circuit dynamically adjusts to meet the varying environmental conditions of residential phone lines. Both transmit and receive circuits continually monitor line conditions and adjust settings accordingly. The receiver circuit of the PHY layer adapts to the varying noise levels on the wire. The transmitting circuit adapts the output signal strength to match requirements of other receivers. The 'squelch' algorithm sets the minimum and maximum signal levels. The receiver filters out extraneous noise that might compromise transmission and reception data.
- **Wiring** – The Ethernet technology found in corporate office environments was originally designed to support four types of wiring systems; thick coaxial cable, thin coaxial cable, unshielded twisted pair, and fiber-optic cable. Since most homes cannot use these expensive

cabling systems HPNA leveraged existing infrastructure provided by phone wire inside the home. The use of the phone wiring system means that every standard phone line connector (RJ-11 modular jack) in the house becomes both a port on the home network and a phone extension.

■ **Network Interface Cards** – All the appliances on a HomePNA based home network need an adapter to control data transmissions. The network interface card (NIC) is the physical interface between the appliance and the telephone cable. Digital appliances would be unable to connect to the network or each other without this card. Network cards are typically connected to each computer or appliance by an interface slot. After the card has been installed in the device the telephone cable is attached to the card's port. Once connected the computer is physically linked to the home network. All network cards are equipped with onboard microprocessors. The microprocessor is the central point from which the card's various functions are coordinated. The roles of the network card are to:
  • Prepare data for transmission
  • Store data prior to transmission
  • Send data across the in-house network
  • Control the flow of data between the digital appliance and the transmission medium

The NIC also acts as a translator. When it receives data the NIC translates the electrical signals into electrical pulses that the telephone cable can carry and the digital appliance can understand. HomePNA cards contain the necessary hardware and software routines that are stored in read-only memory allowing a home network to use existing in-home phone wiring system. Some HomePNA certified adapters come with connectors known as RJ-45. These interfaces are slightly wider than RJ-11 connectors and can be used to connect into a sophisticated data wiring system.

■ **Software** – Operating Systems (OS) – Each device on a home network requires a network capable operating system. An installed NIC requires a properly configured driver to communicate with other appliances on the network. To operate efficiently the driver must communicate commands to the network card quickly and clearly and be configured correctly to optimize the card's performance . Failure to do so will slow up network performance. HomePNA uses the NDIS (Network Driver Interface Specification) driver model. This driver is integrated with most Microsoft Windows operating systems. NDIS provides the consumer with a simplified plug-in driver architecture. At the lowest boundary

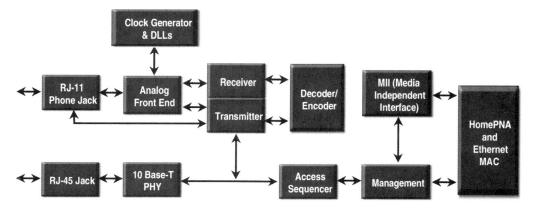

Figure 18.9: HomePNA PHY and MAC

layer NDIS contains a driver that is specific to the telephone wiring transmission medium. The layer above this contains a platform independent driver called a mini-port. This layer interfaces through a standard Application Programming Interface (API) to the NDIS layer. That layer in turn communicates with the home network's transport protocols. A major advantage of the NDIS software model is that network cards can be installed in a telephone based home network without a technician visit from your local service provider.

Figure 18.9 shows the HomePNA PHY and MAC components.

Key Features of HomePNA are:

- **Leverages existing standards** – HPNA has chosen a standard that uses 802.3 framing and Ethernet behavior to leverage existing IEEE-802.3 Layer 2 networking standards.
- **Quality of Service (QoS)** – Applications that will dominate home networks will utilize the transport of digital audio, digital video, and digital voice (IP telephony). To maintain voice quality latency in voice connections must be controlled below 10 to 20 ms. Streaming video and audio connections must receive an application-determined minimum bandwidth from the network. The aggregate throughput rate of 10 Mb/s for HPNA 2.0 is more than adequate for most application scenarios. Without a QoS mechanism burst loads presented by TCP transfers between PCs would jeopardize network latency and guaranteed bandwidth service requirements. Bandwidth allocation within a given class of service must be fair. Packet capture on a traditional Ethernet MAC 2 layer results in long access latency distributions. The HPNA 2.0 MAC layer eliminates packet capture by introducing eight priority levels and an improved collision resolution technique.
- **Robustness** – The quality of the communications channel is the primary difference between twisted-pair Ethernet and other technologies. Running Ethernet over Category-5 cable has many advantages. They include point-to-point communication, proper termination, a well-characterized channel response, and very low cross talk. In contrast, all of the no-new-wires media available for home networking suffer from problems leading to a severely impaired communications channel. HomePNA has developed a robust suite of technologies capable of overcoming the challenges associated with networking appliances on a typical in-home phone wiring system.
- **Performance** – Several external influences indicate home networking environments need a minimum 10 Mb/s throughput to operate efficiently. Technologies such as ADSL and the DOCSIS cable modems require data rates of 6 Mb/s or more to share the access bandwidth. Applications such as multiple DVD streams or high-definition digital video may even exceed the 10 Mb/s threshold.
- **Future Proof** – Once installed home networks are likely to remain in place for many years. In the future replacing home network interfaces embedded in appliances will be difficult. A good home-networking technology has built-in interoperability with future generations. For example, HomePNA is designed to be forward compatible with future devices and applications operating at speeds up to 100 Mb/s.
- **Security** – HomePNA combined with each home's unique phone circuit (phone number), provides excellent security .
- **Cost** – A successful home-networking technology must be inexpensive with the cost of a typical HomePNA card approximately $100.

The ratified HomePNA 2.0 specification defines a multipoint CSMA/CD packet network that supports uni-cast, multi-cast, and broadcast transmissions. While it has the look and feel of Ethernet, HomePNA 2.0 differs from 10Base-2 and 10Base-T in a number of aspects. First it has no wiring type, wiring topology, or termination restrictions. And HPNA 2.0 does not require a switch or hub to use a shared physical medium. Some of the products that use HPNA 2.0 technology include PCs and PC peripherals, digital modems, network hubs, IP telephones, digital TVs and set-top boxes, home security and automation, and Internet network appliances. According to IDC, phone line home networking projections are positive for both short-term and long-term growth. The analyst group predicts that by 2004 phone line technologies will account for 72% of the total installed base for home networking.

The HomePNA organization has finalized HomePNA 3.0. This specification boasts unprecedented data rates of 128 Mb/s with optional extensions up to 240 Mb/s. HomePNA is the only home networking industry specification combining speeds above 100 Mb/s and inherent deterministic Quality of Service (QoS). HomePNA technology complement wireless networking technologies and provides the ideal high-speed backbone for a home multimedia network. It provides a fast and reliable channel to distribute multiple feature-rich digital audio and video applications throughout a home without interfering with standard telephone services. HomePNA 3.0 enables consumers with multiple PCs to take advantage of a single high-speed Internet connection allowing users to simultaneously check e-mail, browse the Web, and share peripherals as well as to stream audio and video content to consumer electronics devices and PCs.

Deterministic Quality of Service makes HomePNA 3.0 unique among "no new wires" home networking specifications. It has the ability to deliver multiple high-speed real-time audio and video data streams without disrupting other services. HomePNA 2.0's QoS enabled equipment manufacturers to prioritize telephone voice data above computer data. But multimedia home networking requires much stronger QoS. HomePNA 3.0's greatly enhanced QoS provides support for real-time data. The technology permits users to assign specific time slots for each data stream. This guarantees that real-time data is delivered at the appropriate time without interruption and with predetermined latency. Providers can increase their revenue per customer and increases retention by offering "triple play" services of POTS, high-speed Internet access, and broadcast and on-demand video over home networks. It also allows HomePNA V3 to transport data with inherent QoS requirements such as IEEE1394.

The International Telecommunication Union (ITU) has already adopted global phone line networking standards G.989.1, G989.2, and G989.3. All are based on the HomePNA 2.0 specification. HomePNA member companies are working together to present version 3.0 recommendations to the ITU-T. Based on version 2.0 physical-layer technology, the HomePNA 3.0 physical interface is backward compatible as well as fully interoperable with HomePNA version 2.0 network components. This technology has been tested in hundreds of homes. Starting in 2001, leading manufacturers have included it in their products.

Silicon solutions incorporating the HomePNA 3.0 specification are being developed to power a variety of devices including pre-configured PCs, network interface cards and adapters, residential gateways, broadband modems, printers, multimedia devices, Internet appliances, set top boxes and consumer electronics.

Power Lines

Home networking technology allows consumers to use their existing electrical wiring system to connect home appliances to the Internet and each other. Home networks using high-speed power line technology can control anything that plugs into an outlet including lights, televisions, thermostats, and alarms.

The power line connects the home to the electric utility company in order to supply power. Power line communications are categorized as either access or in-home (see Figure 18.10).

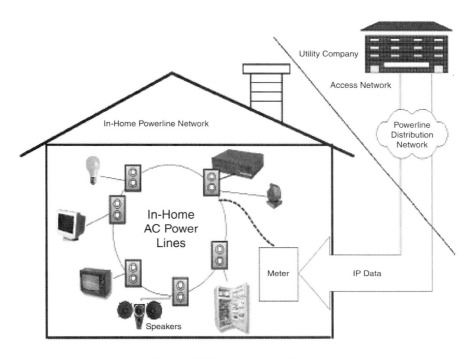

Figure 18.10: Power line Networks

Access power line technologies send data over the low-voltage electric networks that connect the consumer's home to the electric utility provider. The power line access technologies enable a "last mile" local loop solution that provides individual homes with broadband connectivity to the Internet. In-home power line technology communicates data exclusively within the consumer's home utilizing the electrical outlets. The same electric outlets that provide power serve as access points for the network devices. Although access and in-home solutions both send data signals over the power lines, they are fundamentally different technologies. Access technology competes with xDSL and broadband cable technologies to deliver a long-distance solution. The in-home power line technologies focus on delivering a short-distance and high-bandwidth (10 Mb/s) solution. They compete against other interconnection technologies such as phone line and wireless. (For a discussion on power line access technologies see Chapter 2.)

Advantages of in-home power line technologies:

- **Availability of electrical outlets** – One of the main advantages of using power lines for home networking is the availability of multiple power outlets in many rooms. This can eliminate the need for additional home rewiring.

- **Capable of transmitting data** – Power line networking takes advantage of the unused capacity of the power cable to transmit data over the existing home power cabling.
- **Distribution of audio** – Connecting your stereo to the in-home power line wiring network will allow you to distribute music throughout the house.
- **Speed** – Advancements in technology allow power-lines to distribute data at speeds up to 10 Mb/s.
- **Cost effective** – Using existing power line and power source for communications infrastructure reduces costs
- **Worldwide availability** – Available in most developed countries
- **Ease of use** – Easy customer installation

Disadvantages of in-home power line technologies:

- **Noise** – The amount of electrical noise on the line limits practical transmission speeds. Vacuum cleaners, light dimmers, kitchen appliances, and drills are examples of noise sources that affect the performance of a power line based home network.
- **Minimum-security levels** – Power lines do not provide a secure media.
- **Data attenuation** – Numerous elements on power line networks cause data attenuation.
- **High costs of residential appliances** – Power line network modems cost more than phone line network modems.
- **Lack of global standards** – Regularity issues in some International markets are preventing the development of a global data distribution standard for existing in-home power line systems.

Elements of an in-home power line network include:

- Inside house wiring
- Appliance wiring (power cords)
- Appliances (load devices)
- Circuit breakers
- Power line networking modems

Technical Obstacles for In-Home Power line Networks

Typical data and communications networks (including corporate LANs) use dedicated wiring to interconnect devices. But power line networks, from their inception, were never intended for transmitting data. Instead these networks were optimized to efficiently distribute power to all the electrical outlets in a building at frequencies typically between 50-60 Hz. The original designs of electrical networks never considered using the power line medium for communicating data signals at other frequencies. The physical topology of the electrical network, the physical properties of the electrical cabling, the appliances connected, and the behavioral characteristics of the electric current itself all combine to create technical obstacles. This makes the power line more difficult to use as a communications medium than other types of isolated wiring like the Category 5 cabling used in Ethernet data networks.

Typical Applications of Power Line Technologies

Power line technologies have two primary uses within the context of a home network—home control and data networking. Data networking is used to distribute IP data around the house.

Power lines have been used for home control and automation for many years. The most important types of home automation applications include controlling lights, ventilators, security systems, sprinklers, and temperature levels within the house. The home control networking systems market is undergoing a significant transition from closed-loop solutions to open IP-aware solutions. According to research by Allied Business Intelligence's (ABI) the U.S. home automation and controls equipment market is expected to grow from $1.1 billion in 1999 to $3 billion in 2005. Home control and automation systems are normally based on one of the three major power line technologies—CEBus, LonWorks, or X-10.

CEBus

CEBus is a standard proposed by the Electronic Industries Association. It defines a set of rules for consumer products to communicate with each other. The CEBus based products consist of two fundamental components, a transceiver and a micro controller. Data packets are transmitted by the transceiver at about 10 Kb/s. The CEBus protocol uses a peer-to-peer communications model with each node on the network having access to the media at any time. The CEBus standard includes commands such as volume up, fast forward, rewind, pause, skip, and temperature up or down one degree. These commands are based on a common application language (CAL). CEBus uses spread spectrum technology to overcome communication impediments found within the home's electrical power line. Rather than using a single frequency, spread spectrum signaling works by spreading a transmitted signal over a range of frequencies. The CEBus power line carrier spreads its signal over a range from 100 Hz to 400 Hz during each bit in the packet.

CEBus uses a Carrier Sense Multiple Access/Collision Detection and Resolution (CSMA/CDCR) protocol to avoid data collisions. Similar to HomePNA, this media access control protocol requires a consumer appliance to wait until the line is clear.

A CEBus based home network is comprised of a control channel and multiple data channels for each CEBus media. CEBus control channel communication is standardized across all media with a consistent packet format and signaling rate. This is used exclusively to control devices and resources of the network including data channel allocations. Data channels typically provide selectable bandwidths that can support high data rates. They are used to send data such as audio, video, or computer files over the network. The characteristics of a data channel can vary depending on the medium and connected device requirements. All data channel assignments and functions are managed by CEBus control messages via the control channel.

The advantages of CEBus are that every CEBus HomePnP™ device is capable of communicating with every other CEBus HomePnP device over the power line without the need for new wires. Many CEBus HomePnP devices can be installed without the need for a central controller. This enables devices to be used to provide solutions for many simple automation problems. And CEBus HomePnP devices can be networked with a central controller for larger and more extensive automation projects.

Without a requirement for additional wiring CEBus products are ideal solutions for retrofitting a house or building using the existing power lines. CEBus products also provide an ideal solution for new construction. CEBus enables the automation designer to work with the builder's preferred electrician without the need for extensive training. Because the CEBus HomePnP standard is a nonproprietary protocol based upon an open standard (EIA 600) manufacturers are not tied to an obscure and proprietary standard.

The CEBus Industry Council continues its support and development of the Microsoft led Simple Control Protocol (SCP) standard for the Universal Plug and Play (UPnP) standard.

The CEBus Industry Council's (CIC) mission is to provide information for the design and development community information about CEBus and CEBus Home Plug and Play. The Council involves all applicable industries and organizations in the development of interoperable products. The goal is to offer the homeowner multiple products to choose from that can communicate with each other and work as a system. These products can ask each other questions, answer questions, and provide unsolicited status reports based on what they know about the home's environment. These messages are passed back and forth through the home's power lines, telephone wires, television cable, infrared signals, and radio signals.

Home Plug and Play (HPnP)

HPnP is an industry specification describing the way consumer products cooperate with other products. The specification allows consumers and retailers to have systems that offer a diverse array of interoperable components and systems. Prior to HPnP, Home Automation Systems lacked the technical standards needed for uniformity and cost efficiencies.

The HPnP specifications provide the tools needed to transform stand-alone products into interactive network products allowing the home networks industry to flourish.

CIC's objective is to establish a thriving Home Networks Industry for its members by accelerating the evolution from stand-alone products to in home-networked products . Members participating in CIC's various committees and sub-committees direct CIC programs and activities. CIC has an open invitation to those wanting a prosperous Home Networks industry.

Network Products are created when stand-alone products are given network features allowing unrelated products the ability to communicate. Early standardization efforts specified how to send these messages between products and how to structure messages. The CEBus Standard was developed around Common Application Language (CAL). CAL is a universal Communications Language for Home Network Products. CAL was designed to be understood by most home electronic products. Originally CAL was only available as an integral component in the CEBus Standard. Industry sectors wanting to offer whole house network features recognized the need for a common application language that can be transported by multiple carriers. Many industry sectors found the best option was CAL if it was available outside the CEBus standard. In April 1997, EIA/CEMA agreed to publish CAL as a separate EIA Standard known as 721. This made EIA's CAL truly the CAL for all Network products in home.

Having a CAL with a robust set of contexts creates the basics for interactive communication between in-home network products. However, as with any language, grammatical usage rules are needed. For Home network products the HPnP specification provides the rules that yield truly understandable communication. CIC's HPnP specifications are not a language. The HPnP specification provides uniform implementation or rules for those using EIA's CAL as their communication language for home network products. HPnP is the definitive reference for those using CAL because it contains CIC's Approved Industry Specific Contexts. By the nature of its charter CIC is responsible for the development, maintenance, and publication of all Industry Specific CAL contexts.

Understandable communication between unrelated products requires common 'identifiers' for the products. The needed identifiers are not proprietary names. They are commonly acceptable names much like a dictionary of terms for network products in home. Those identifiers are essential for a successful home networks industry. Public meetings held by the CEBus Technical Steering

Committee and CIC's Context Development Steering Groups created a consensus method for establishing the needed identifiers. The method looks at the context in which a message will be sent (i.e,. from whom and to whom). Thus the identifiers used by the EIA's CAL are known as 'contexts'.

CAL provides the common application language needed for Home Networks. The HPnP specification provides uniform implementation rules for CAL. The CEBus standard is an EIA Standard that was developed at the same time as EIA's CAL Standard. CAL published (as EIA-721) and CIC's HPnP specifications are both structured for transport by multiple carriers. Together CAL and HPnP allow industry sectors with specialized transports the ability to meet their individual industry needs. They also interoperate with home network products using their industry specific communication transport protocols.

One such case currently receiving attention is the IEE-1394 transport protocol. This high-speed digital bus was developed to meet PC and PC peripheral needs. With HPnP, PCs and related peripherals can employ 1394 to meet their industry specific needs. And they still retain the ability to transmit understandable messages to other home network products using alternative transport protocols such as the CEBus standard.

LONWorks (Local Operation Networks)

LONWorks technology is an important new solution for control networks developed by Echelon Corporation. Worldwide there are over 6 million installed LonWorks based appliances. A LONWorks system includes all the necessary hardware and software components for implementing complete end-to-end control[1] systems—from silicon to software. A LONWorks network does not require central control or master-slave architecture. Intelligent control devices called nodes communicate with one another using a common protocol. Each node in the network contains embedded intelligence that implements the protocol and performs control functions. Each node includes a physical interface (transceiver) that couples the node's microcontroller with the communications medium.

LONWorks is an "open" technology and accessible to anyone. A typical node in a LONWorks control network performs a simple task. Nodes on a home network include devices such as proximity sensors, switches, motion detectors, and sprinkler systems .

The following elements make up a LONWorks home control system:

- **Neuron chip** – A microcontroller specifically designed to offer the most cost-effective solution available for network enabling and embedding of intelligence into home control devices.
- **Appliances** – Appliances on a LONWork enabled home network use a protocol to communicate with each other. This protocol is known as LonTalk and has been approved as an open industry standard by the American National Standards Institute (ANSI) – EIA 709.1.
- **The LONWorks Network Services (LNS)** – Architecture that provides a range of network services to appliances that are connected to a control system.

LonWorks technology provides a solution to the many problems of designing, building, installing, and maintaining device networks. Networks can range in size from two to 32,000 devices. They can be used in everything from supermarkets to petroleum plants; from aircraft to railway cars; from fusion lasers to slot machines; and from single-family homes to skyscrapers. Most industries have moved away from proprietary control schemes and centralized systems. Manufacturers are using open off-the-shelf chips, operating systems, and parts for products that feature improved reliability, flexibility, system cost, and performance. LonWorks technology is accelerating the trend away from proprietary control schemes and centralized systems. It provides interoperability, robust technology, faster development, and economies of scale.

Devices in a LonWorks network communicate using the standardized language LonTalk. LonTalk consists of a series of underlying protocols that allow intelligent communication among various devices on a network. The protocol provides a set of services that allow a device's application program to send and receive messages from other devices over the network. It can do this without needing to know the topology of the network or the names, addresses, or functions of other devices. The LonWorks protocol can provide end-to-end acknowledgement of messages, authentication of messages, and priority delivery to provide bounded transaction times. Support for network management services allow the remote network management tools to interact with devices. This interaction can include the reconfiguration of network addresses and parameters, downloading application programs, reporting network problems, and start/stop/reset of device application programs. LonTalk and LonWorks networks can be implemented over many mediums including power lines, twisted pair, radio frequency (RF), infrared (IR), coaxial cable, and fiber optics.

LonWorks control networks can be easily integrated with the Internet. This built-in capability allows for seamless networking between IP-based devices and control devices. LONWorks power line based systems also support remote monitoring of home appliances through standard Web browsers.

The LonMark Interoperability Association was created in May, 1994 by 36 companies to help develop interoperable multi-vendor systems based on LonWorks networks. The LonMark association provides an open forum for member companies to work together on marketing and technical programs promoting the availability of open interoperable control devices. Today thousands of companies are using LonWorks control networks to provide systems and solutions for building, home, industrial, telecommunications, and transportation industries. Worldwide installations of LonWorks based devices number in the millions.

There are many ways to design and control networks for automated controls. Flat peer-to-peer (P2P) architectures are the best. P2P architectures lack single points of failures inherent in any hierarchical architecture where messages from one device must first go to a controlling master or gateway before the signal can get to the target device. Every communication between two non-master devices includes an extra step or fault possibility. Designs using P2P allow direct communication between two devices. This eliminates the fault possibility of the master controller and removes a potential performance bottleneck. Device failures in a P2P design are much more likely to affect just the one device.

X-10

X-10 is another communications protocol that allows compatible home networking products to talk to each other using the existing electrical wiring in the home. Basic X-10 power line technology is almost 20 years old. It was initially developed to integrate low cost lighting and appliance control devices. X-10 originally was unidirectional, but the capability for bi-directional communication has been added. However, the vast majority of X-10 communication remains unidirectional. Controllers send signals over existing AC wiring to receiver modules. The X-10 modules are adapters that connect to outlets and control simple devices. With X-10's transmission rate limited to only 60 b/s it is unsuitable for handling Internet traffic. It does allow users to control lights and other electrical device from anywhere in the house with no additional wiring. Some control applications include telephones, swimming pools, temperature controls, ventilation controls, lighting, doors and gates, sprinklers, security, and kitchen appliances.

The X-10 Technology and Resource Forum designs, develops, manufacturers, and markets products based on this standard. Scores of manufacturers make X-10-compatible products. According to X-10 group more than 100 million such products have been sold. These home automation products are called "power line carrier" (PLC) devices. Builders who offer home automation often install them as an additional selling feature. The home automation line consists of "controllers" that automatically send signals over existing electric power wiring to receiver "modules" which in turn control devices such as lights, appliances, heating, and air conditioning units.. The local electronic store stocks all the necessary equipment required to automate a home with the X-10 standard. The main disadvantage for legacy X-10 technology is its limited capability in terms of speed and intelligence. Due to its low data rate and rudimentary functionality the technology is relegated to control applications. However, the ultimate goal of the X-10 technology is be a higher-speed protocol that is capable of facilitates communication between home PCs and controlled home appliances.

HomePlug

A not-for-profit organization called the HomePlug power line alliance was established in 2000 to create an open specification for home power line networking products and services. The goal is to allow power line home networks to compete effectively with phone line and wireless technologies. The 13 founding HomePlug members include 3Com, AMD, Cisco Systems, Compaq, Conexant, Enikia, Intel, Intellon, Motorola, Panasonic, Radio Shack, SONICblue, and Texas Instruments. Past challenges of utilizing power lines include a lack of industry specifications and multiple sources of electric noise. HomePlug is overcoming these challenges through the Alliance's efforts to create specifications and advanced optimized algorithms in semiconductor technologies. HomePlug has chosen Intellon's high-speed power line networking technology as the baseline upon which to build the Alliance's first-generation specification. Since 2001 extensive field trials have been used to validate the v1.0 specification.

The power line medium is a harsh environment for communication. The channel between any two outlets in a home has the transfer function of an extremely complicated transmission line network. It also has many stubs with terminating loads of various impedances. Such a network has an amplitude and phase response that varies widely with frequency. At some frequencies the transmitted signal may arrive at the receiver with relatively little loss and other frequencies may be driven below the noise floor. And the transfer function can change with time such as when the homeowner plugs a new device into the power line. Or if some of the devices plugged into the network have time-varying impedances such as switching power supplies or motors. As a result the nature of the channel between outlet pairs may have a wide range of variance. In some cases a broad swath of bandwidth may be suitable for high quality transmissions while in other cases the channel may have a limited capacity to carry data. Due to these frequency variations efficient use of the medium requires an adaptive approach that compensates for the channel transfer function. HomePlug technology includes an effective and reliable method of performing adaptation. This allows it to achieve high rates on typical channels and adjusts the bit rate to fight through really harsh channels.

Interference on the power line must also be considered. The most severe interference sources are difficult to analyze. The interference can be both impulsive or frequency selective in nature. Typical sources of noise are brush motors, fluorescent and halogen lamps, switching power supplies, and dimmer switches. In addition, a significant source can be amateur band radio transmitters. The net impact of these different interference sources is that raw received data bits tend to have significant numbers of bit errors that must be corrected. The HomePlug technology contains a combination of sophisticated forward error correction (FEC), interleaving, error detection, and automatic repeat request (ARQ) to ensure that the channel appears reliable to the network layer protocols.

The topology of power distribution to the home is another factor that must be considered. In a typical U.S. neighborhood a distribution transformer provides power to a relatively small number of homes. The distribution transformer effectively blocks the power line networking signals from crossing into the main power grid. But it does little to stop signals in one of the homes it powers from propagating to another home. Thus networking signals generated in one home may show up on the power line in another home. This creates urgent concerns about privacy similar to those encountered in wireless systems.

Manufacturers of low-speed power line networking equipment in the band below 1 MHz have struggled for years with the impact of circuit breakers and two-phase power distribution. Circuit breakers have substantial attenuation in the band used by these devices. And paths from one circuit to another may experience substantially more attenuation than same-circuit paths. Most homes have two-phase wiring and the only physical connection between some circuits is the connection at the distribution transformer. Power line communication between such circuits relies on coupling. This can easily create 20 dB of loss at frequencies below 1 MHz. Fortunately the losses from circuit breakers and cross phase coupling are less severe in the band occupied by the HomePlug signal. Typically they cause only a few dB of additional loss.

HomePlug Technology Overview

Any workable solution to reliable communication on the power line medium must include both a robust physical layer (PHY) and an efficient media access control (MAC) protocol. The MAC protocol controls medium sharing among multiple clients. The PHY specifies the modulation, coding, and basic packet formats.

The HomePlug PHY uses OFDM as the basic transmission technique. OFDM is widely used in DSL technology and in the terrestrial wireless distribution of television signals. HomePlug uses OFDM in a burst mode rather than in the continuous mode employed by those other technologies. HomePlug technology also uses concatenated Viterbi and Reed Solomon FEC with interleaving for payload data. Sensitive control data fields use turbo product coding (TPC).

The MAC protocol in the HomePlug technology is a variant of the well-known carrier sense multiple access with collision avoidance (CSMA/CA) protocol. Several features have been added to support priority classes, provide fairness, and allow the control of latency. The use of CSMA/CA means the PHY must support burst transmission and reception. Each client enables its transmitter only when it has data to send and when finished turns off its transmitter and returns to receive mode.

The PHY Layer

OFDM divides the high-speed data stream to be transmitted into multiple parallel bit streams at a relatively low bit rate. Each bit stream then modulates one of a series of closely spaced carriers. The property of orthogonality is a result of choosing the carrier spacing equal to the inverse of the bit rate on each carrier. The practical consequence of orthogonality is that if a fast Fourier transform (FFT) is performed on the received waveform over a time span equal to the bit rate on an individual carrier, the value of each point in the FFT output is a function only of the bit (or bits) that modulated the corresponding carrier and not impacted by the data modulating any other carrier. Channel equalization becomes simplified when the carrier spacing is low enough that the channel response is relatively constant across the band occupied by the carrier. If implemented in the frequency domain, equalization can be achieved by a simple weighting of the symbol recovered from each carrier by a complex valued constant. Many different types of modulation can be used on the individual carriers.

The need for equalization in HomePlug is completely eliminated by using differential quadrature phase shift keying (DQPSK) modulation. With this technology the data is encoded as the difference in phase between the present and previous symbol in time on the same subcarrier (see Figure 18.11). Differential modulation improves performance in environments where rapid changes in phase are possible.

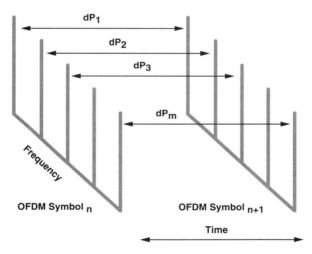

Figure 18.11: Differential Phase Encoding Across Symbols
(Courtesy: HomePlug and Communications Systems Design)

Unlike DSL, HomePlug does not use higher order quadrature amplitude modulation (QAM). With relatively short packets the overhead required for channel assessment and estimation of gain and carrier phase create a capacity penalty that offsets any potential gain from the modulation efficiency.

OFDM waveforms are typically generated using an inverse FFT (IFFT) where the frequency domain points consist of the set of complex symbols that modulate each carrier. The result of the IFFT is called an OFDM symbol. Each symbol has duration equal to the reciprocal of the subcarrier spacing. This is typically much longer than the data rate. At the receiver, the data can be recovered via a forward FFT and converted back to the frequency domain. Figure 18.12 shows the process of conversion between the frequency domain and time domains.

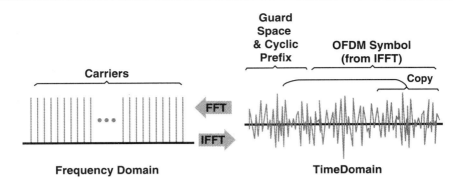

Figure 18.12: Transformation from the Frequency to the Time Domain and Adding the Cyclic Prefix
(Courtesy: HomePlug and Communications Systems Design)

Note that the time domain waveform also includes a cyclic prefix, essentially a replication of the last few microseconds of the OFDM symbol. The purpose of the cyclic prefix is to absorb the intersymbol interference that results from the delay presented by the irregular frequency of the channel.

Without the cyclic prefix some of the samples used in the FFT would contain energy from either the previous or the following OFDM symbol. If the cyclic prefix is as long as the worst-case delay variation across the frequency band, the FFT is not degraded by the neighboring symbols by waiting until the end of the prefix to start taking samples. The HomePlug data-bearing packet is formed from a series of OFDM symbols consisting of a start-of-frame delimiter, a payload, and an end-of-frame delimiter (see Figure 18.12). The destination station responds to uni-cast transmissions by transmitting a response delimiter indicating the status of the reception (ACK, NACK, or FAIL).

The delimiter consists of a preamble sequence followed by a TPC encoded frame control field. The preamble sequence is chosen to provide good correlation properties for each receiver. With it they can reliably detect the delimiter even when substantial interference and a lack of knowledge of the transfer function exists between the receiver and the transmitter interference.

The frame control contains MAC layer management information such as packet lengths and response status. The low rate TPC and interleaving used on the frame control provide good immunity to frequency selective impairments and broadband interference. All three delimiter types have a similar structure. But the data carried varies depending on the delimiter function.

Unlike the delimiters the payload portion of the packet is only intended for the destination receiver. Payload data only travels on a set of carriers that have been previously agreed on by the transmitter and intended receiver during a channel adaptation procedure. Because only carriers in the "good" part of the channel transfer function are used heavy error correcting code is not required. This combination of channel adaptation and simplification of the coding for uni-cast payloads allows HomePlug to achieve high data rates over power line.

The adaptation has three degrees of freedom:
- De-selection of carriers for badly impaired frequencies
- Selection of modulation on individual carriers (DB/SK or DQPSK)
- Selection of convolutional code rate (1*2 or 3*4).

The payload can also be sent using the highly robust ROBO mode that uses all carriers with DB/SK modulation and heavy error correcting code with bit repetition and interleaving. ROBO mode does not use carrier de-selection and can generally be accepted by any receiver. This mode is used for initial communication between devices that have not performed channel adaptation. It is also used for multi-cast transmission or uni-cast transmission in cases where the channel is so poor that ROBO mode provides greater throughput than de-selection of carriers with lighter coding.

The HomePlug PHY occupies the band from about 4.5 to 21 MHz. The PHY includes reduced transmitter power spectral density in the amateur radio bands to minimize the risk of radiated energy from the power line interfering with these systems. The raw bit rate using DQPSK modulation with all carriers active is 20 Mb/s. The bit rate delivered to the MAC by the PHY layer is about 14 Mb/s.

The MAC layer

The HomePlug transmission format is shown in Figure 18.13. The MAC uses a virtual carrier sense (VCS) mechanism and contention resolution to minimize the number of collisions. The receiver attempts to recover the frame control upon receipt of a preamble.

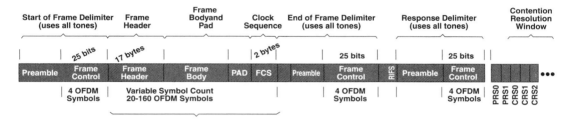

Start of Frame Indicates	Payload	End of Frame Indicates	Response Indicates	PRS0 & PRS1
Start of frame Contention control Length of frame Tone map index	Up to 13.75 Mb/s (PHY) rate Adapted modulation and tones Decoded based on tone map Extensible to higher rates	End of frame Contention control Channel access priority	ACK -- good packet NACK -- errors detected FAIL -- receiver busy	11 - Highest priority (3) 10 - Priority 2 01 - Priority 1 00 - Lowest priority (0)

Figure 18.13: HomePlug Transmission Format
(Courtesy: HomePlug and Communications Systems Design)

The frame control indicates whether the delimiter is the start of frame, end of frame, or response delimiter. Start of frame delimiters specify the duration of the payload to follow. The other delimiters define where the end of the transmission lies. If a receiver can decode the frame control in the delimiter it can determine the duration the channel will be occupied by this transmission and sets its VCS until this time ends.

If it cannot decode the frame control the receiver must assume a maximum-length packet is being transmitted and set the VCS accordingly. It may subsequently receive an end-of-frame delimiter and be able to correct its VCS.

The destination always acknowledges uni-cast packets at the MAC layer by transmitting the response delimiter. If the source fails to receive an acknowledgment it assumes that a collision has caused the failure. The destination may also choose to signal FAIL if it has insufficient resources to process the frame. Or it can signal NACK to indicate that the packet was received with errors that could not be corrected by the FEC.

The contention resolution protocol includes a random back-off algorithm to disperse the transmission times of frames queued, or being retransmitted due to collisions, while the channel was busy. It also provides a way to ensure that clients obtain access to the channel in priority order.

When one node completes a transmission other nodes with packets queued to transmit signal their priority in a priority resolution interval (indicated by PRS0 and PRS1 in Figure 3). The signals use on/off keying and are designed so easily extracted the priority of the highest priority even when multiple users signal different priorities at the same time.

Slot Choices

Nodes with queued priorities frames equal to the highest priority signaled choose a slot in a contention resolution window. They can then initiate transmission if no other node begins transmission in an earlier slot. Each node chooses a slot at random. The interval grows with increasing numbers of unsuccessful attempts to access the channel. Nodes that were preempted in a previous contention resolution window continue counting slots from where they left off rather than choosing a new random value. This approach improves the fairness of the access scheme.

Collision can occur if a node wishing to transmit fails to recognize a preamble from another node or if the earliest chosen slot in the contention resolution window is selected by more than one node. The preamble design is robust enough to ensure that the missed preamble rate is so low that this source of collisions has only minor impact.

Segmentation and reassembly are provided to improve fairness, reliability, and to reduce latency. In cases where there is no higher priority frames queued with other nodes the MAC includes features that allow the transmission of multiple segments with minimal delay. It also provides a capability for contention less access where channel access may be passed from node to node.

A common misconception is that contention-based access schemes have potentially unbounded latency. In the HomePlug MAC the latency is bounded by the method of discarding packets that cannot be delivered in the time required by the application.

It has been shown that the percentage of HomePlug packets discarded through this approach is low enough to be encompassed by the tolerated missed packet rate for low latency applications such as VoIP or streaming media. The combination of this feature and priority classes makes HomePlug well suited to applications requiring QoS.

Channel adaptation typically occurs when clients first join a logical network based on either a timeout or detected variation in the channel transfer function. This can signal either an improving or degrading condition. Any node can initiate a channel adaptation session with any other node in its logical network. This adaptation is a bi-directional process that causes either node to specify to the other the set of tones, modulation, and FEC coding to use in subsequent payload transmissions.

Privacy is provided through the use of a 56-bit data encryption standard (DES) applied at the MAC layer. All nodes on a given logical network share a common encryption key. The key management system includes features that enable the distribution of keys to nodes that lack an I/O capability.

Home Networking Using "New Wires"

The computer industry has been promising users the ability to easily connect electronics devices such as digital TVs, cameras, cable set-top boxes, and stereo equipment to each other and PCs for years. USB and IEEE 1394 interconnection technologies are the two solutions that have been developed specifically to meet these commitments. USB and 1394 are complementary technologies that differ in their application focus. USB is the preferred connection for most PCs and PC peripherals such as keyboards, digital cameras, and scanners. The primary target of 1394 is consumer electronic devices such as digital camcorders, digital PVRs, DVD players, and digital televisions. Ethernet was not originally designed to operate in a home entertainment environment. However, some of its variants are beginning to provide consumers with the ability to distribute high-speed data and video around the house. IEEE 1394, USB, and Fast Ethernet technologies are ideal for advanced entertainment networks. New wire technologies are known for the high data rates they support. However, they all require additional special wiring around your house often at an additional cost.

IEEE 1394 (FireWire or iLink)

The expansion of digital technology has allowed more and more people to share video, still images, and audio. Consumers are constantly searching for faster and easier ways of transferring and sharing this information. This phenomenon is driving the convergence of computers, consumer equipment, and communications. Communication is the force that draws these separate market segments together.

Convergence will happen when seamless high-speed communication becomes readily available. There are interconnection technologies that enable us to connect these devices across home. The IEEE 1394 protocol appears to be a strong contender for the communications channel that will make this happen.

Also known as the FireWire™ or iLink™, IEEE 1394 is a versatile, high-speed, and inexpensive method of interconnecting a variety of consumer electronics devices. Current applications include home theatre equipment and personal computer peripherals. Using IEEE 1394 requires new optical fiber or high-grade copper wiring throughout the home.

Origins of IEEE 1394

Apple Computer originally created the FireWire bus. It was born out of the need for a low-cost and consumer-oriented connection between digital-video recorders and personal computers. It grew into a standard called the IEEE-1394.

The 1394 Trade Organization was formed in 1994 to support and promote the adoption of the 1394 standard. Since 1994 over 300 specifications have been written and developed. The 1394 specification was formally introduced in 1995. Specifications for 1394a were introduced in 1998. The following year they released 1394b that is fully backwards compatible with the current 1394 and 1394a specifications. Each revision of 1394 has added new features, performance, and capabilities. These specifications have been "standardized" by industry standards organizations such as the IEC. Recognized in 2003, the 1394 Trade Association is the official standards organization. It is an international consortium of 170 companies devoted to the advancement of the IEEE 1394 standards.

IEEE 1394 Architecture

To understand how 1394 operates we need to now briefly examine the elements of a 1394-based home network. The components that form an IEEE 1394 based home network may be classified as the actual protocol, the cabling system, and the architectural design of the network itself.

Similar to other high-speed networking systems, IEEE 1394 adopts a layered approach to transmitting data across a physical medium. The four layers used by the IEEE 1394 include the physical, link, transaction, and serial soft API layers. These layers are graphically depicted in Figure 18.14.

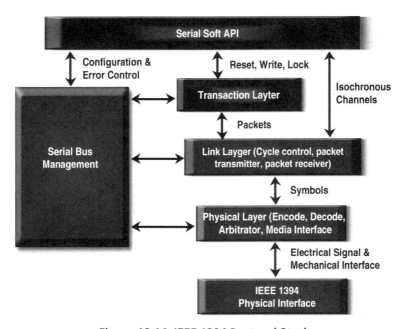

Figure 18.14: IEEE 1394 Protocol Stack

The physical layer provides the encoding-decoding, electrical signaling, and mechanical interface/connection between the 1394 appliance and the cable itself. In addition to the actual data transmission and reception tasks the physical layer also provides arbitration to insure all devices have fair access to the bus. The link layer provides cycle control, packet transmit, packet receive, CRC, and the host and application interface. The link layer takes the raw data from the physical layer and formats it into two types of recognizable 1394 packets—isochronous and asynchronous. Isochronous data transfer puts the emphasis on the guaranteed timing of the data and less emphasis on delivery. Isochronous transfers are always broadcast in a one-to-one or one-to-many fashion. No error correction or retransmission is available for isochronous transfers. Asynchronous data transfer puts the emphasis on guaranteed delivery of data with less emphasis on guaranteed timing. The third layer used by the IEEE 1394 protocol is called the transaction layer. It is responsible for managing the commands that are executed across the home network. The fourth and final logical function grouping is responsible for the overall configuration control of the serial bus.

Users can connect up to 63 devices with 1394. When 1394.1 bus bridges become available they will support connecting over 60,000 devices. Digital camcorders, surround sound processors, scanners, printers, hard disk audio recorders, videoconferencing cameras, and disk drives all share a common bus connection. They can share to an optional host computer and to each other as well. This makes 1394 a prime candidate for the "Home Network" standard initiated by VESA and other industry associations.

The 1394 standard defines three signaling rates. They are: 98.304, 196.608 and 393.216 Mb/s (megabits per second). These rates are referred to in the 1394 standard as S100, S200, and S400. In early 2002, the finalized 1394b specification expanded the standard to include 800 and 1,200 Mb/s speeds. Users can mix and match different speeds devices on the same bus. By using "isochronous" data transmission even the S100 implementation supports two simultaneous channels of 30 fps (frames per second) broadcast-quality video along with stereo audio.

Benefits of IEEE 1394:

- **Broad support** – The home networking industry has a very positive attitude towards the 1394 technology. Additionally, 1394 has the advantage of being adopted by consumer electronics manufacturers such as Sony, Panasonic, Philips, and Grundig.
- **Low cost** – 1394 is a low-cost digital interface available for audio and video applications.
- **Endorsed by international standards bodies** – The European Digital Video Broadcasters (DVB) have endorsed IEEE 1394 as their digital television interface.
- **Speed** – Multimedia entertainment is the most frequently used application in today's homes. A high quality distribution of video for entertainment applications requires larger bandwidth than audio and data. IEEE 1394 is capable of transporting data at 100, 200, 400, or 800 Mb/s. The next version of the standard will be capable of transporting data at 3.2 Gb/s.
- **Plug and play** – Consumers can add or remove 1394 devices without resetting their home network
- **Non-proprietary** – IEEE 1394 is an open and royalty-free standard.
- **Different applications** – There are many different applications for IEEE 1394. Almost all of the consumer electronic, office automation, industrial, biomedical, and networking devices can benefit from 1394 features and capabilities.

IEEE 1394 is the technology for connecting multimedia devices such as digital camcorders and VCRs, Direct-to-Home (DTH) satellite, cable TV set-top boxes, DVD Players, gaming consoles and home theater systems. IDC forecasts that by the year 2004 over fifty million different types of informational and entertainment appliances will come with a digital IEEE 1394 interface. This advanced technology will have a profound affect on the proliferation of home networks.

USB—Universal Serial Bus

Traditionally interfaces between PC components and peripheral devices have been analog rather than digital. Dramatic improvements in content quality are pushing the demand for computers and informational appliances with digital interfaces. The most common interface for high-speed data is the USB serial bus. USB stands for Universal Serial Bus. The specifications for this interface standard were released in 1995. The major goal of USB was to define an external expansion bus. It made adding peripherals to a PC as easy as hooking a telephone to a wall-jack. Virtually all-new PCs come with one or more USB ports. USB has become a key enabler of the Easy PC initiative. This initiative led by Intel and Microsoft is designed to make PCs easier to use. This effort sprung from the recognition that users need simpler and easier to use PCs that don't sacrifice connectivity or expandability. USB is one of the key technologies used to provide this. Most peripheral vendors around the globe are developing products to this specification.

Users planning the physical layout of a USB network need to make sure that the distances between devices are less than five meters. Connecting two or three computers and printers together requires a device called a USB bridge.

USB 1.1

USB version 1.1 has gained tremendous success in the marketplace. Most PC and peripheral vendors worldwide develop products based on this specification. USB supports isochronous and asynchronous high-speed data transfer protocols. Isochronous connections from the PC USB port to peripherals such as scanners, video devices, digital cameras, and printers support data transfers at a guaranteed fixed rate of 12 Mb/s. The slower asynchronous protocol is used to communicate with peripherals such as keyboards and mice at 1.5 Mb/s.

Data rates of up to 12 Mb/s are sufficient for low-to-medium speed peripherals. Many different serial and parallel connectors at the back of the PC are replaced one standardized USB plug-and-play connection. Home networking devices with a USB port can connect up to 127 different USB peripherals. USB's data rate also accommodates a whole new generation of peripherals including MPEG-2 video-based products, data gloves, and digitizers. Computer-telephony integration is expected to be a big growth area for PCs and USB provides an interface for ISDN and digital PBXs.

USB 2.0

USB 2.0 is the next generation USB version. From a home networking user's perspective USB 2.0 is just like USB 1.1. However, its much a higher 480 Mb/s bandwidth supports higher performance devices. All older version USB peripherals will work in a USB 2.0-capable system. Consumers primarily use USB 2.0 with high-bandwidth electronic products such as high-resolution video conferencing cameras and next generation scanners and printers.

USB On-The-Go

Due to its widespread acceptance USB is becoming the industry standard for connecting peripherals to PC's and laptops. Many of the new peripherals now using USB are also portable devices. As portable devices increase in popularity there is a growing need for them to communicate directly with each other when a PC is not available. The On-The-Go Supplement addresses this need for mobile interconnectivity by allowing a USB peripheral to have the following enhancements:

- Limited host capability to communicate with selected other USB peripherals
- A small USB connector to fit the mobile form factor
- Low power features to preserve battery life

In December 2001 the USB Implementers Forum released Revision 1.0 of the USB On-The-Go Supplement.

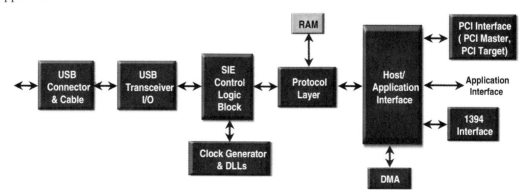

Figure 18.15: USB 1.1 Core

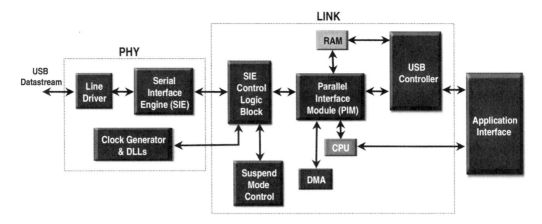

Figure 18.16: USB 2.0 Core

Figure 18.15 and 18.16 detail the USB 1.1 and USB 2.0 cores.

Benefits of USB include:

- **Plug and play** – USB fully supports plug-and-play technology. With plug-and-play hardware devices such as digital speakers, joysticks, and video cameras can be automatically configured as soon as they are physically attached to a home network.
- **Hot swapping capabilities** – USB also supports hot swapping of devices on a home network eliminating the need to shut down and restart devices.
- **Single port connection** – USB can replace all the different kinds of serial and parallel connectors on the back of a PC with one standardized plug and play combination.
- **Multiple connections** – Home networking devices that have a USB port can connect 127 different peripheral devices at one time.
- **Support for high-speed protocols** – USB supports two high-speed data transfer protocols: isochronous and asynchronous. USB 2.0 will support data rates up to 480 Mb/s.
- **No need for dedicated power supplies** – USB distributes the power to all connected devices eliminating the need for separate power supply boxes.
- **New home control mechanisms** – USB allows the transfer of data to flow in both directions between PC and consumer electronics devices. This makes it ideal for a desktop PC to control home appliances in new and creative environments.
- **Operating system support** – Microsoft operating systems provide complete support for USB technologies.

USB vs. IEEE 1394

Many people confuse IEEE 1394 and USB. Both are emerging technologies that offer a new method of connecting multiple peripherals to a computer. Both permit peripherals to be added or disconnected from a computer without the need to reboot. Both use thin flexible cables that employ simple durable connectors. But that is where the similarities end. Although 1394 and USB cables may look the same the amount of data flowing through them is quite different.

Today, 1394 offers a data transfer rate that is over 16 times faster than USB. In addition 1394 has a well-defined bandwidth roadmap with projected speed increases to 1Gb/s+ (125 MB/sec) and beyond in the next few years. Such dramatic improvements in data transfer capacity will be required to keep pace with bandwidth demanding devices, such as HDTV, digital set-top boxes, and home automation systems. A more complex protocol and signaling rate allows 1394 to move more data in a given amount of time. But 1394 is considerably more expensive than USB. So does this mean that IEEE 1394 will become the new wires technology of choice for most home networking users? Most industry analysts expect 1394 and USB to coexist peacefully with both replacing the myriad of connectors found on the back of today's PCs and consumer devices. USB will be reserved for low-bandwidth peripherals such as mice, keyboards, and modems, and 1394 will be used to connect to the new generation of high-bandwidth computer and consumer electronics products.

Fast Ethernet

Ethernet is a popular and internationally standardized networking technology for both hardware and software enabling computers to communicate with each other. Ethernet was developed by DEC, Intel, and Xerox. The Institute of Electrical and Electronics Engineers (IEEE) later standardized it as IEEE 802.3. People tend to use the terms Ethernet and IEEE 802.3 interchangeably. Ethernet supports data transfer rates of 10 Mb/s. A newer version of Ethernet called Fast Ethernet supports data transfer rates of 100 Mb/s. Fast Ethernet is the wired standard for enterprise networks and forms the basis of many home networks. The technology is fast, secure, reliable, and inexpensive. The newest version is called Gigabit Ethernet and supports rates of 1,000 Mb/s (1 Gb/s). This variation of Ethernet is mostly used for enterprise networking and is rarely deployed within the home.

To understand how Fast Ethernet operates on a home network we need to understand the elements of an Ethernet based home network. The components that form an Ethernet based home network may be classified as the actual protocol, the cabling system, and the interconnection devices.

Similar to 1394 and USB, Ethernet adopts a layered approach to transmitting data across a home network. Ethernet defines the lower two layers of the OSI Reference Model as the physical and data link layers. The physical (PHY) layer transmits the unstructured raw bit stream over a physical medium and describes the electrical, mechanical, and functional interface to the network. The physical layer can support a wide range of media specifications. Each of these specifications provides different data rates, media, and topology configurations. The data link layer is responsible for getting data packets on and off the home network. The data link layer is subdivided further into the LLC (Logical Link Control) and MAC (Medium Access Control) sub-layers. The LLC on the upper half of the layer does the error checking while the MAC on the lower half is solely responsible for getting the data on and off the home network. Like Ethernet, Fast Ethernet uses the CSMA/CD access method to handle simultaneous demands on a home network.

CSMA/CD stands for carrier sense multiple access with collision detection. Home networks that use the CSMA/CD access method allow a device to send data at any time. When an electronics or PC device has data to send, it listens to the phone line to see if it is busy. The device is sensitive to any carrier on the line. This access method is said to have carrier sense. If there is traffic on the line the device enters a waiting mode. If the line is free the station transmits its data immediately. If another device in the home decides to send data at the same time a collision may occur. Collision detection allows the two devices to detect this event and perform the required recovery. The devices back off for a period before re-transmitting. Of course it is essential that the two devices do not back off for the same length of time. If all appliances on the home network were set to back off and retry after half a second the same two data broadcasts would collide again. To prevent continual collisions each appliance on the network backs off for a random amount of time.

Benefits of Fast Ethernet based Home Networks:

- **Proven technology** – There are several million 10 Mb/s Ethernet users in the world today. By keeping the essential characteristics of the Ethernet technology unchanged in the 100 Mb/s world, home-networking users can benefit from years of Ethernet expertise.
- **Reliability** – Ethernet technology can reliably and efficiently network most types of consumer appliances including PCs, high definition televisions, set-top boxes, security cameras, and modems.
- **Support for high bandwidth in-home applications** – Multimedia entertainment is the most important application in today's homes. The bandwidth requirement of high definition television and streaming video cannot entirely be satisfied by the 'no new wires' technologies that are available in today's homes. Ethernet technology can reliably and efficiently deliver this level of functionality to home networking users.
- **Wide Industry Support** – Fast Ethernet is widely supported by many different companies.

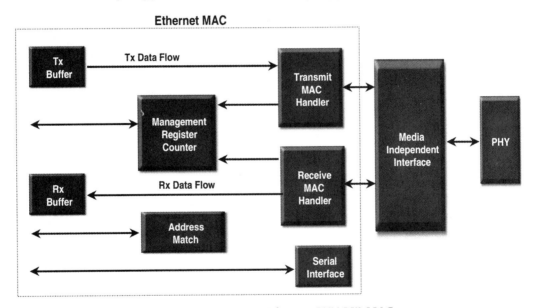

Figure 18.17: 10/100 Ethernet PHY-MII-MAC

Figure 18.17 shows a 10/100 Ethernet core including the PHY, MII (media independent interface), and the MAC (medium access control).

Users are required to use either Category 3 or Category 5 (CAT-5) unshielded twisted-pair (UTP) copper wire cabling between devices. This can be a problem if it requires physical modifications within the home such as drilling holes through walls or floors.

One of the most significant advantages that Ethernet has over other technologies is that it is incredibly affordable. The vast majority of new PCs ship with Ethernet cards or Ethernet LAN on

Motherboard (LOM). In addition, a number of consumer electronic (CE) vendors have begun to embed Ethernet into CE products such as set-top boxes, PVRs, DVD players, videogame consoles, and digital audio receivers. PCI Ethernet cards are available for as low as $9.99. PCMCIA cards were slightly higher at $24.99 and home routers start at $69.99.

IEEE 1355

A low-cost serial switched packet technology, IEEE 1355, is capable of delivering 2 Gb/s. Other key benefits include fault tolerance, scalability, hot plugging, and extremely low overhead. It can also be connected in a variety of topologies and this technology is can simultaneously transport multiple protocols on the same network.

Home Networking Using "Wireless" Technologies

Using a wiring system, either pre-wiring the house or using an existing wiring infrastructure, will suit many home networking users. However, there are people who want to have access to their home network as they wander around their homes. The new demand for a greater degree of mobility and flexibility has an obvious solution—wireless interconnection technologies. Market research analysts are predicting that wireless network technologies will eventually become more widespread than the various wired solutions. Technology consulting company Strategy Analytics forecasts that 19% of households in the U.S. and 15% of European households are expected to have wireless home networks by 2005. Wireless communications present the ideal solution for the home network but provide a variety of technical and deployment obstacles. Wireless technologies such as DECT, HomeRF, Bluetooth, and wireless LAN technologies are described below.

Wireless home networking solutions originate from four simple consumer demands:

- No new wiring infrastructures.
- The solutions must also be simple to install and easy to use.
- Interoperability with other networks such as phone line based home networks.
- Solutions need to be economical and home security cannot be compromised.

Wireless networking offers solutions to meet most of these requirements. Deploying a wireless network in a household brings advantages and disadvantages. On the plus side wireless home networking technology solutions provide consumers with the ability to access data from anywhere at anytime. The main disadvantage of deploying a wireless home networking solution is cost. While they may turn out cheaper in the long-term, wireless networks have additional costs. The distances that they operate can also be a problem for consumers who own large homes. Security is also a concern for wireless networks. Because there is no physical connection, eavesdropping on a wireless network is much easier.

Most home networks are linked over a physical medium. However, physical media can be expensive, inflexible, and error or failure prone. Inflexibility is often a significant problem that can be easily overcome by wireless communication. Wireless solutions are ideal platforms for extending the concept of home networking into the area of mobile devices. Consequently wireless technology is portrayed as a new system that complements phone line and power line networking solutions. It is unlikely that wireless technology will emerge as a home network backbone solution. But it will instead serve to interconnect the class of mobile communications devices into a sub-network. These mobility sub-networks will interface with other sub-networks and the Internet by connecting to the home network backbone. Wireless home networks transmit and receive data over the air minimizing the need for expensive wiring systems. With a wireless based home network you can access and share expensive entertainment devices without pulling new cables through walls and ceilings.

A transmitter and a receiver are the core of wireless communications. The user interacts with the transmitter. For example, one of the children inputs a Web address into their PC. This input is then converted by the transmitter to Electromagnetic (EM) waves and sent to the receiver. The receiver then processes these electromagnetic waves. For two-way communication each user requires a transmitter and a receiver. Many home networking device manufacturers build the transmitter and receiver into a single unit called a transceiver. The operation and functionality of wireless and wired home networks remain the same. However, there are distinctions in the technologies used to achieve the same objectives.

There are two main core technologies used in the design of wireless home networking products: infrared (IR) and radio frequency (RF).

IR

Most of us are familiar with everyday devices that use IR technology such as remote controls for TVs, VCRs, and CD players. IR transmission is categorized as a line-of-sight wireless technology. This means that the workstations and digital appliances must be in a direct line to the transmitter in order to operate. An infrared-based network suits environments where all the digital appliances that require network connectivity are in the one room. Now new IR technologies have been developed that can work outside of the line-of-sight. Expect to see these products in the very near future. IR home networks have the benefits of quick implementation but people walking across the room and moisture in the air can weaken their signals.

RF

The other main category of wireless technology is radio frequency. RF technology is more flexible technology that allows consumers to link appliances that are distributed throughout the house. RF can be categorized as narrow band or spread spectrum. Narrow band technology requires a clear channel uninterrupted by other digital appliances. Since each transmitter/receiver appliance transmits using its own frequency it is unlikely to interfere with other RF appliances connected to the home network. However if the wireless appliance gets moved to another part of the house then it is possible that interference may occur. This limitation makes the use of this technology unsuitable for a number of home networking applications.

Spread spectrum technology (SST) is one of the most widely used technologies in wireless home networks. SST was developed during World War II to provide greater security for military communications. As it entails spreading the signal over a number of frequencies, spread spectrum technology makes it harder to intercept. There are a couple of techniques used to deploy SST. A system called frequency hopping spread spectrum (FHSS) is the most popular technique for operating wireless home networks. FHSS transmissions constantly hop over entire bands of frequencies in a particular sequence. To a remote receiver not synchronized with the hopping sequence these signals appear as random noise. A receiver can only process electromagnetic waves by tuning to the relevant transmission frequency. The FHSS receiver hops from one frequency to another in tandem with the transmitter. At any given time there may be a number of transceivers hopping along the same band of frequencies. Each transceiver uses a different hopping sequence that is carefully chosen to minimize interference on the home network. Because wireless technology has roots in military applications, security has long been a design criterion for wireless devices. Security provisions are normally integrated with wireless home networking devices making them more secure than most wire-line based in-house networks. Complex encryption techniques make it near impossible for hackers to gain access to traffic on home wireless network.

A number of industry initiatives are underway to develop interoperable wireless in-home appliances. The following sections describe the most popular of these technologies.

IrDA

Anyone who owns a handheld device or a notebook PC is already IrDA-enabled. However, most people probably have never used the connectivity it allows because they did not own any devices to which they could connect. IrDA is a communications standard based on infrared in which data can be transmitted up to three feet in both directions on a point-to-point basis at speeds of 9-16 Mb/s. IrDA is well suited for user-initiated data exchanges where the user can point his or her device's IrDA port at the intended target like a remote control and start the transfer. However, the standard has limited functionality since the user must line up the sending and receiving devices, be within three feet, and be communicating with a device that understands its IrDA standard. An example of how people could use IrDA is the business card exchange function. Two people can point their Palm devices at each other and exchange phone numbers and addresses. Palm also uses IrDA to provide wireless connectivity between Palm devices and cellular phones to connect to the Internet.

Technical Summary of "IrDA DATA" and "IrDA CONTROL"

IrDA Data is recommended for high-speed, short range, line of sight, and point-to-point cordless data transfers. This makes it suitable for HPCs, digital cameras, and handheld data collection devices

IrDA Control is recommended for in-room cordless peripherals using a host PC and a lower speed, full cross range, point-to-point or to-multipoint cordless controller. It is suitable for keyboards (one way) and joysticks (two-way and low latency). IrDA Data and IrDA Control require designer's attention to ensure spatial or time sharing techniques to avoid interference.

IrDA Data

Developed in 1994, IrDA DATA defines a standard for an interoperable universal two-way cordless infrared light transmission data port. IrDA technology is already in over 300 million electronic devices including desktop, notebook, palm PCs, printers, digital cameras, public phones/kiosks, cellular phones, pagers, PDAs, electronic books, electronic wallets, toys, watches, and other mobile devices.

IrDA Data Protocols consist of both a mandatory and optional set of protocols. The mandatory protocols are listed below.

- PHY (Physical Signaling Layer)
- IrLAP (Link Access Protocol)
- IrLMP (Link Management Protocol and Information Access Service (IAS))

Characteristics of Physical IrDA Data Signaling:

- Continuous operation from contact to at least 1 meter, and typically 2 meters, can be reached. A low power version relaxes the range objective for operation from contact through at least 20 cm between low power devices and 30 cm between low power and standard power devices and consumes 10 times less power. These parameters are termed the required maximum ranges by certain classes of IrDA featured devices and set the end user expectations for discovery, recognition, and performance.

- Bi-directional communication is the basis of all specifications
- Data transmission from 9600 b/s with primary speed/cost steps of 115 Kb/s and maximum speed up to 4 Mb/s
- Data packets are protected using a CRC (CRC-16 for speeds up to 1.152 Mb/s and CRC-32 at 4 Mb/s).

Characteristics of IrDA Link Access Protocol (IrLAP):
- Provides a device-to-device connection for a reliable and ordered transfer of data.
- Device discover procedures.
- Handles hidden nodes.

Characteristics of IrDA Link Management Protocol (IrLMP):
- Provides multiplexing of the IrLAP layer.
- Handles multiple channels above an IrLAP connection.
- Provides protocol and service discovery via the Information Access Service (IAS).

Optional IrDA Data Protocols
- **Tiny TP** – Provides flow control on IrLMP connections with an optional Segmentation and Reassembly service.
- **IrCOMM** – Provides COM (serial and parallel) port emulation for legacy COM applications, printing, and modem devices.
- **OBEX**™ – Provides object exchange services similar to HTTP.
- **IrDA Lite** – Provides methods of reducing the size of IrDA code while maintaining compatibility with full implementations.
- **IrTran-P** – Provides image exchange protocol used in Digital Image capture devices and cameras.
- **IrMC** – Specifications on how mobile telephony and communication devices can exchange information. This includes phonebook, calendar, message data, and how call control and real-time voice are handled (RTCON).
- **IrLAN** – Describes a protocol used to support IR wireless access to local area networks.

IrDA Control

IrDA Control is an infrared communication standard that allows cordless peripherals such as keyboards, mice, game pads, joysticks, and pointing devices to interact with many types of intelligent host devices. Host devices include PC's, home appliances, game machines, and television/web set top boxes. IrDA Control is well suited to deal with devices that leverage the USB HID class of device controls and home appliances.

IrDA Control Protocols consist of a mandatory set of protocols:
- PHY (Physical layer)
- MAC (Media Access Control)
- LLC (Logical Link Control)

Characteristics of IrDA Control Physical Signaling:
- Distance and range equivalent current uni-directional infrared remote control units (minimum 5 meter range).
- Bi-directional communication is the basis of all specs.
- Maximum data transmission of 75 Kb/s.

- The data is coded using a 16-Pulse Sequence multiplied by a 1.5 MHz subcarrier allocated for high-speed remote control in IEC 1603-1. However, this base band scheme has harmonics that can intrude upon other IEC bands.
- Data packets are protected with a CRC (CRC-8 for short packets and CRC-16 for long packets). The physical layer is optimized for low power usage and can be implemented with low-cost hardware.

Characteristics of IrDA Control MAC:

- Enables a host device to communicate with multiple peripheral devices (1:n) and up to 8 peripherals simultaneously.
- Ensures fast response time (13.8 ms basic polling rate) and low latency.

Asymmetric MAC:

- Provides for dynamic assignment and re-use of peripheral addresses.
- Scheduling of media access is actually buried in the HID LLC.

Characteristics of the IrDA Control LLC:

- Provides reliability features that provide data sequencing and retransmission when errors are detected.
- Works with an HID-IrDA Control Bridge to enable the link control functions of USB-HID.

DECT (Digital Enhanced Cordless Telecommunications)

DECT is a flexible digital radio access standard for cordless communications in residential, corporate, and public environments. The first commercial DECT-based cordless phone systems were introduced in 1993. By the end of 1996 accumulated worldwide sales of DECT systems amounted to over 5 million lines.

By 2000 worldwide sales of DECT equipment were more than 30 million lines a year. The DECT standard was initially conceived in the mid-1980s as a pan-European standard for domestic cordless phones. DECT is a digital radio access standard for single and multiple cell cordless communications. It is based on multi-carrier TDMA (time division multiple access) technology. This is the same technology used in the major digital cellular standards. The central difference is that cellular systems were developed for wide-area coverage. The DECT standard was optimized for local coverage with high user densities. It is supported and promoted by the DECT forum with representatives in all the major geographical regions around the world.

Three applications for the DECT standard have reached widespread commercial deployment. They are home cordless phones, business cordless systems, and a radio alternative to wired subscriber accesses in public fixed telecom networks known as Wireless Local Loop (WLL). In a home networking environment a typical DECT system consists of a phone handset and a base unit that contains the radio base station. The DECT standard permits cordless data communication so that a home network could be used to simultaneously support both cordless data and cordless voice services.

DECT Technology overview

DECT provides for voice and multimedia traffic and contains many forward-looking technical features. This allows DECT-based cordless systems to play a central role in important new communications developments such as Internet access and internetworking with fixed and wireless services such as ISDN and Global System for Mobile Communications (GSM). The DECT standard makes use of several advanced digital radio techniques to achieve efficient use of the radio spectrum. The combination of digital radio technology and dynamic channel selection (with additional encryption techniques), authentication, and identification procedures makes DECT radio transmissions extremely secure against unauthorized radio eavesdropping. Speech quality in a DECT system is very high thanks to applied digital radio techniques. Data applications are becoming more and more important for in-home networks. Today DECT products provide data links with up to 552 Kb/s. New modulation schemes will allow the rate to grow up to 2 Mb/s for accessing this wireless medium. Similar to HomeRF, DECT in-home digital appliances also use a CSMA/CA algorithm as the media access scheme. DECT's attractive and competitive cost position by is achieved by leveraging the same basic technology used from residential cordless phones up to complex multi-cell systems.

Figure 18.18 shows a block diagram of the DECT Cordless phone system.

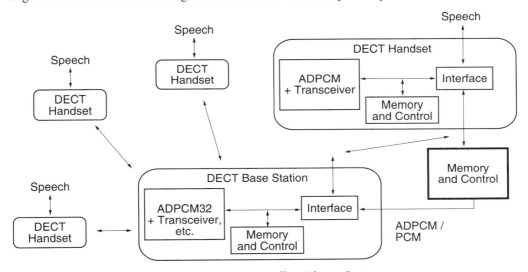

Figure 18.18: DECT Cordless Phone System

HomeRF

Two major factors are presenting a real opportunity for networking data devices within the home. The first is the massive growth of people going online and accessing the Internet. The second factor is the emergence of powerful sub-$800 home PCs. With these low-cost devices, the barrier to getting on the Internet and discovering the functionality of PCs is low enough to reach the vast majority of middle-income households. However, many consumers soon discover that their PC and Internet combination lack some key attributes in terms of mobility and convenience compared to many traditional information and entertainment options.

Several major stakeholders in the IT industry formed the Home RF Working Group with these issues in mind. At its peak there were over 85 members, with the big initiators including Compaq, Ericsson, HP, IBM, Motorola, Philips, Proxim, and Symbionics. The group's goal was to enable interoperable wireless voice and data networking within the home at an attractive price. They also worked to develop a specification for wireless communications in the home called Shared Wireless Access Protocol – Cordless Access (or "SWAP-CA").

In designing SWAP-CA, the HomeRF Technical Committee chose to reuse proven RF network-ing technology for data and voice communications. Then they simplified where appropriate for home usage. With this approach SWAP-CA inherited native support for Internet access via TCP/IP net-working and for voice telephony via the Public Switched Telephone Network (PSTN) and voice over IP. Because of this design approach the HomeRF Working Group made rapid progress in finalizing the specification and bringing it to market in a timely manner. From a technical perspective the SWAP-CA protocol was designed to operate on the 2.4 GHz ISM band. The 2.4 GHz band is an unlicensed worldwide frequency band ensuring that SWAP-CA devices can operate globally.

HomeRF's primary applications included PC-enhanced cordless telephony, mobile display appliances, and resource sharing. The HomeRF Working Group was disbanded in January 2003 due to the success of WiFi (short for wireless fidelity and marketing term for IEEE 802.11b technology).

Bluetooth

Bluetooth is a "personal area networking" specification for a low-cost, short-range radio links between mobile computers, cameras and other portable in-home devices. Bluetooth technology is the result of cooperation between leaders in the telecommunication and computer industries. It enables home networking users to connect a wide range of computing and telecommunications devices easily and simply, without the need to buy, carry, or connect cables. It delivers opportunities for rapid ad hoc connections, and the possibility of automatic, unconscious, connections between devices. It virtually eliminates the need to purchase additional or proprietary cabling to connect individual devices. Because Bluetooth can be used for a variety of purposes, it will also potentially replace multiple cable connections via a single radio link. Bluetooth is been promoted and adopted a group of companies called the Bluetooth Special Interest Group (SIG). This includes promoter companies 3Com, Ericsson, IBM, Intel, Lucent, Microsoft, Motorola, Nokia, Toshiba, and over 2000 adopter companies.

Bluetooth Technology Overview

The technology is an open specification for wireless communication of data and voice. It is based on a low-cost, short-range radio link, built into a 9-mm x 9-mm microchip, facilitating protected ad hoc connections for stationary and mobile communication environments.

Bluetooth technology allows for the replacement of the many proprietary cables that connect one device to another with one universal short-range radio link. For instance, Bluetooth radio technology built into both the cellular telephone and the laptop would replace the cumbersome cables used today to connect a laptop to a cellular telephone. Printers, PDAs, desktops, table sized handheld, fax machines, keyboards, joysticks, and virtually any other digital device can be part of the Bluetooth system. Bluetooth radio technology provides a universal bridge to existing data networks. It also provides a peripheral interface and a mechanism to form small, private ad hoc groupings of con-nected devices away from fixed network infrastructures. Designed to operate in a noisy radio frequency environment such as a home, the Bluetooth radio uses a fast acknowledgement and frequency-hopping scheme to make the link robust.

Bluetooth radio modules avoid interference from other signals by hopping to a new frequency after transmitting or receiving a packet. Compared with other systems operating in the same fre-quency band, the Bluetooth radio typically hops faster and uses shorter packets. This makes the Bluetooth radio more robust than other systems. Similar to HomeRF, Bluetooth radios also operate in the unlicensed ISM band at 2.4 GHz. Bluetooth has a maximum data capacity of only 1 Mb/s that translates to a throughput of only 780 Kb/s once the protocol overhead is taken into account.

From a security perspective, Bluetooth provides user protection and information privacy mechanisms at the lower layers of its protocol stack. Authentication is based on a challenge/response algorithm. Authentication is a key component of any Bluetooth home networking system, allowing you to develop a domain of trust between personal Bluetooth devices. For example, authentication allows only your personal notebook to communicate through your cellular telephone.

Building Blocks of a Bluetooth Solution

The role of each component in a Bluetooth based home network is briefly outlined in the following categories.

- **Personal Area Networks** – Bluetooth was originally conceived to replace the myriad of cables that are synonymous with a PC based home network. However, as Bluetooth evolved it became clear that it would also enable a totally new networking paradigm—Personal Area Networks (PANs). With PAN technology, a home networking user can organize a collection of personal electronic products (PDA, cell phone, laptop, desktop, MP3 player, etc.) to automatically work together. Over time, PANs will revolutionize the user experience of consumer electronics.

- **Piconets** – The Bluetooth system supports both point-to-point and point-to-multi-point connections. A collection of digital appliances connected to a home network via Bluetooth technology is called a Piconet. A Piconet starts with two connected appliances such as a digital set top box and cellular phone, and may grow to eight connected devices. All users participating on the same Piconet are synchronized to this hopping sequence. The range of Bluetooth (and its associated Piconet) is only 10 meters. However, there is a plan to extend this to 100 meters. This extension in range will increase the power and interference levels for appliances connected to the Piconet.

- **Scatternets** – Several Piconets can be established and linked together ad hoc, where each Piconet is identified by a different frequency hopping sequence. The resulting structure is called a Scatternet.

- **Software Framework** – Every Bluetooth system is comprised of a host-based application and a Bluetooth module (see Figure 18.19).

The Bluetooth protocol stack can be logically divided into four different layers according to their purpose in a wireless home networking environment. Each layer performs a specific, well-documented, function in much the same manner as an Ethernet stack. This well-defined architecture makes system design much easier and has enabled many implementation variations to emerge. The four layers comprise the following:

- **Bluetooth core protocols** – The main purpose of these protocols is to enable communication links between Bluetooth devices.

- **Cable replacement protocol** – The cable replacement protocol used by Bluetooth is called RFCOMM and is used to emulate RS-232 control and data transfer.

- **Telephony control protocol** – This Bluetooth protocol defines the rules for making speech and data calls across a home network.

- **Adopted protocols** – This layer comprises application-orientated protocols such as PPP, TCP/IP, HTTP, FTP, Wireless Application Protocol (WAP), etc.

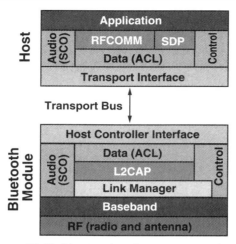

Figure 18.19: Bluetooth software protocol stack

Figure 18.20: Bluetooth RF and Baseband Controller

Figure 18.20 shows the Bluetooth RF and Baseband controller (i.e., the Bluetooth module required for Bluetooth enabling of the system).

Applications for Bluetooth

- **PC applications** – The Bluetooth Specification defines interfaces where the radio modules may be integrated into notebook personal computers or attached using PC Card or USB port. Bluetooth technology is platform independent and not tied to any specific operating system. Implementations of the Bluetooth Specification for several commercial operating systems are in development. For notebook computers, the implementation of the Bluetooth Specification in Microsoft Windows 2000 using Windows Driver Model (WDM) and NDIS drivers is being contemplated. Notebook PC usage models include:
  - Remote networking using a Bluetooth cellular phone
  - Speakerphone applications using a Bluetooth cellular phone
  - Personal card exchange between Bluetooth notebooks, handheld, and phones
  - Calendar synchronization between Bluetooth notebooks, handheld, and phones
  - File transfers (file types include, but not limited to .xls, .ppt, .wav, .jpg and .doc formats)

■ **Telephone applications** – The Bluetooth Specification defines interfaces where the radio modules may be integrated directly into cellular handsets or attached using an add-on device. The Bluetooth compliance document requires digital cellular phones to support some subset of the Bluetooth Specification. The Bluetooth contingents within the telephony Promoter companies are working with their fellow employees involved in the Wireless Application Protocol (WAP) forum to investigate how the two technologies can benefit from each other. Phone usage models include (are not constrained to):

- Wireless hands-free operation using a Bluetooth headset
- Cable-free remote networking with a Bluetooth notebook or handheld computer
- Business card exchange with other Bluetooth phones, notebook or handheld computers
- Automatic address book synchronization with trusted Bluetooth notebooks or handheld computers

■ Other applications: Usage models and implementation examples centered on other contemplated Bluetooth devices include:

- Headsets
- Handheld and wearable devices
- Human Interface Device (HID) compliant peripherals
- Data and voice access points
- Digital set top boxes
- Integrated digital televisions

Bluetooth is going to find its way into an extraordinary number of products and applications, and virtually all of them will have need to integrate a variety of different parts. Examples include:

■ Cordless base station or Pay Phone access point, which typically need a keypad, LED(s), battery charging logic, LCD, etc.

■ Access points that Bluetooth enable a set top box, DSL modem, cable modem, or Ethernet bridge.

■ Scientific and industrial equipment where Bluetooth enables a complex system.

■ Any application looking to extend or enhance Bluetooth's capability, i.e., proprietary encryption for maximum security and/or improved FEC for maximum transmission reliability.

Bluetooth chip sales are expected to top $1.7 billion by 2007. SG Cowen estimates that in 2004 total Bluetooth chipsets sales will exceed 645 million units and $2.74 billion in chipset revenues. The price per chipset is expected to reach $4.25.

Wireless LANs

Wireless local area networks (WLANs) are a rapidly emerging market. WLANs combine data connectivity with user mobility to provide a connectivity alternative for a broad range of consumers and business customers. They have a strong popularity in vertical markets such as telecommuting, SOHOs, health care, retail, manufacturing, warehousing, and academia. In these applications productivity gains are realized by using handheld terminals and notebook PCs to transmit real-time information to centralized hosts for processing. A WLAN is comparable to a cordless telephone where the user can move about and still be able to use the telephone.

Deutsche Bank AG predicts that revenues from WLAN products will exceed $2.1 billion in 2004 for the comsumer/SOHO and the enterprise markets. Deutsche Bank analysts also project that the worldwide WLAN IC unit shipments will reach 56.5 million in 2004 at a blended average selling price (ASP) of $13. The five-year unit CAGR is 41%. However, steadily declining ASPs lead to a five-year revenue CAGR of 21%.

Intel's aggressive push of Centrino on the client side and the emergence of wireless switching on the infrastructure side are the two major forces that transformed the business WiFi market in 2003. The high-tech market research firm, In-Stat/MDR, estimates that 16 million notebook PCs with embedded WiFi shipped to businesses in 2003. By 2005, they expect that WiFi will be included in 95% of notebooks as a standard feature. Hence, the extra cost of a WiFi client will be essentially transparent to the end user.

Start ups and some traditional WLAN hardware vendors have introduced "Access point/Switch" architecture as a way to ease the management, security, and configuration issues of large-scale WLAN rollouts. The WiFi business market is entering a new stage, one which promises to bring serious roll outs to horizontal businesses, not just to the tried and true verticals of education, healthcare, and retail. With laptops growing so fast across the business space, and with the majority of laptops rolling out with WiFi as a standard feature by 2005, the growth of WiFi clients is practically assured. Certainly the infrastructure market will evolve over the next 2–3 years as end users demand those solutions that best simplify installation, improve performance, and enhance and simplify security and management.

Cisco currently dominates the enterprise access point market and currently supports an intelligent access point infrastructure. Cisco has an interest in being able to support and build upon its installed base, so the company may be evaluating solutions from a handful of vendors that are developing technology designed to significantly increase performance of today's standard access point. Many of these companies, such as Bandspeed and Airgo, are relying on smart antenna technology to bolster an access point's performance, coverage, capacity, and throughput capabilities. With chipmakers Atheros, Texas Instruments, Broadcom, Intersil (now GlobespanVirata), and others focusing on IEEE 802.11 a/g solutions, and with Intel promising to roll out a dual-mode mini PCI in Centrino by 2004, the dual-mode concept has become king. Both dual-mode clients and dual-mode capable access points are shipping out in growing quantities, squashing 802.11a-only equipment.

Demand for computing and telephony mobile devices will be one of the most influential market drivers. In addition, end users demanding higher data rates and ease of use to sustain growing Internet and data applications will drive demand. However, several issues remain unsolved for the industry. Although vendors have made great strides in achieving interoperability, a common wireless standard is far from reality (today, there are seven standards). Interference from competing 2.4-GHz technologies (like Bluetooth) threatens the already crowded band. In addition, uncertainties exist with several technologies migrating to the evolving 5-GHz frequency band.

The wireless LAN is perfect for rooms in a home that are hard to cable and locations where it is not economical for home owners to implement new wire based technologies. Hence, cost savings are one of the primary drivers behind WLANs. However, there are many more advantages to be gained, including:

- Improved flexibility
- Extended reach of communication between family members
- Support transmissions over hundreds of meters

Wireless LAN Technology Overview

WLANs focus on the PHY (physical) layer, the data link layer (with the medium access control (MAC), and the logical link control (LLC) sublayers of the OSI model. The physical layer defines the electrical, mechanical, and procedural specifications that provide the transmission of bits over a communication medium or channel. WLAN PHY layer technologies used are:

- Narrowband radio, infrared (IR), OFDM (orthogonal frequency division multiplexing)
- Spread spectrum (frequency hopping spread spectrum (FHSS)
- Direct sequence spread spectrum (DSSS)

The MAC layer, which is part of the data link layer, ensures error control and synchronization between the physically connected devices communicating over a channel. It is also responsible for determining priority and allocation to access the channel.

Types of WLAN Systems

Some of the popular WLAN technologies are IEEE 802.11a, IEEE 802.11b, and HiperLAN2.

IEEE 802.11—The First Wireless Ethernet

The development of any new technology is part theory and part practice. A key issue in telecommunications is the adoption of technical standards that govern the interoperability of equipment to provide a stable environment for deployment of products and services. This does not mean that all vendor equipment will work in the exact same way. A standard sets a norm or performance expectation on the function of the technology—not its implementation. The standard that governs the wireless local area networks industry is the 802.11 family of standards that are part of the group that governs Ethernet data communications. IEEE 802.11 addresses the 2.4- and 5-GHz WLAN market. This standard is evolving and adapting to meet the needs of industry as new technology is developed to allow new product design.

The precursor to 802.11b, IEEE 802.11, was introduced in 1997. It was a beginning, but the standard had serious flaws. 802.11 supported speeds of only up to 2 Mb/s. It supported two entirely different methods of encoding—Frequency Hopping Spread Spectrum (FHSS) and Direct Sequence Spread Spectrum (DSSS)—leading to confusion and incompatibility between equipment. It also had problems dealing with collisions and with signals reflected back from hard surfaces. These defects were soon addressed, and in 1999 the IEEE 802.11b Ethernet standard arrived.

The 802.11 Extensions

- **802.11b** – The 802.11b extension of the original 802.11 standard boosts wireless throughput from 2 Mb/s up to 11 Mb/s. 802.11b can transmit up to 200 feet (61 m) under good conditions, although this distance may be reduced by the presence of obstacles such as walls. The 802.11b upgrade dropped FHSS in favor of DSSS. DSSS has proven to be more reliable than FHSS, and settling on one method of encoding eliminates the problem of having a single standard that includes two kinds of equipment that are not compatible with each other. 802.11b devices are compatible with older 802.11 DSSS devices, but they are not compatible with 802.11 FHSS devices. Also, 802.11b differs from standard 802.3 and 802.5 wired Ethernet only at OSI Layers 1 and 2, it is interoperable with standard wired Ethernet. 802.11b uses the popular CSMA/CA (carrier sense multiple access/collision avoidance) technique in its MAC. Because it is a real Ethernet standard and looks like Ethernet to applications, 802.11b is perfectly compatible with operating systems such as Microsoft® Windows®, Macintosh® OS, and Linux®. 802.11b is the most widely available wireless standard and WiFi (802.11b) networks now extend to a total of 13 million U.S. households.
- **802.11a** – Only recently available, 802.11a uses a different band than 802.11b—the 5.8-GHz band called U-NII (Unlicensed National Information Infrastructure) in the United

States. Because the U-NII band has a higher frequency and a larger bandwidth allotment than the 2.4-GHz band, the 802.11a standard theoretically achieves speeds of up to 54 Mb/s.

■ **802.11g** – This is an extension of 802.11b and operates in the same 2.4-GHz band as 802.11b. It brings data rates up to 54 Mb/s using OFDM (Orthogonal Frequency Division Multiplexing) technology. Because 802.11g is backward compatible with 802.11b, an 802.11b device can interface directly with an 802.11g access point. You may even be able to upgrade some newer 802.11b access points to be 802.11g compliant via relatively easy firmware upgrades.

Table 18.1: Key Differences Between the IEEE 802.11 Extensions

	IEEE 802.11 b	IEEE 802.11 a	IEEE 802.11 g
Popularity	Widely adopted. Readily available everywhere.	New Technology.	New technology with rapid growth expected.
Speed	Up to 11 Mbps	Up to 54 Mbps	Up to 54 Mbps
Relative Cost	Inexpensive	Relatively more expensive	Relatively Inexpensive
Frequency	2.4 GHz. Some conflict may occur with other devices like cordless phones, microwave ovens, etc.	5 GHz	2.4 GHz. Some conflict may occur with other devices like cordless phones, microwave ovens, etc.
Range (indoor)	100 to 150 feet	25 to 75 feet	100 to 150 feet
Public Access	Number of public hot spots are continuing to grow at a very fast pace, allowing wireless connectivity at airports, hotels, college campuses, and other public areas.	None	Compatible with 802.11b hot spots at the 11 Mbps rates. It is expected that the 802.11b hot spots will quickly convert to 802.11g hot spots
Compatibility	Widest adoption	Incompatible with 802.11b or 802.11g.	Interoperates with 802.11b at 11 Mbps rates. Incompatible with 802.11a.

HiperLAN2

HiperLAN2 is the fastest growing wireless technology with a proposed data rate of 54 Mb/s and a range of above 150 meters. This technology is best suited for multimedia applications. HiperLAN2 is popular in Europe due to the frequency allocation. HiperLAN2 is an OFDM-based, variable bit rate PHY layer technology operating at 5 GHz. It has FEC error control, with dynamic subchannel modulation allowing data transmission at higher rates with a strong Signal to Noise Ratio (SNR) at lower throughputs in adverse conditions. It has a generic architecture and supports Ethernet, 1394, ATM, PPP, and 3G. The HiperLAN2 data link layer/MAC provides QoS via dynamic fixed time slots. The time slotted structure allows simultaneous communication in both downlink and uplink in the same period. It is also a connection-oriented technology that allows negotiation of QoS parameters like bandwidth, bit error rate, latency, jitter, and delay requirements. This assures that other terminals will not interfere with subsequent transmissions.

WLAN products include network interface cards (or NICs/PC adapters), access points (end user-to-LAN and LAN-to-LAN), and technology bridges for communications. NICs provide an interface between the end-user device (desktop PC, portable PC, or handheld computing device) and the airwaves via an antenna on the access point. Access points act as transmitters/receivers between wired and wireless networks. They connect to the wired network via standard Ethernet cable (token ring is available, but less common), and use airwaves to transmit information to and from "connected" wireless end users. Technology bridges exist at the periphery of each product and are the most susceptible to constant change and evolution. These products need a flexible, reprogrammable and low-cost platform to accommodate for time-to-market pressures, specification changes, lack of clear direction, and short product lifecycles.

Figure 18.21 shows the wireless LAN radio, MAC, and baseband controller. A WLAN card consists of the antenna, radio/PHY, baseband controller, and the MAC. The access points are devices that provide a wireless hub or a gateway for non-wireless networks to wireless networks. They also act as the network police and perform network management. They receive, buffer, and transmit data between wireless LAN and the wired network infrastructure. Access points function within a range of 100 to several hundred feet. They also connect WLANs to other technologies such as USB and Ethernet. WLAN products will extend beyond NICs, access points, and technology bridges. They will enable every device in the home, SOHO, and enterprise with WLAN capabilities. For example, devices such as digital TV, residential gateways, set top boxes, digital modems, PC peripherals, and gaming consoles will be wireless LAN enabled.

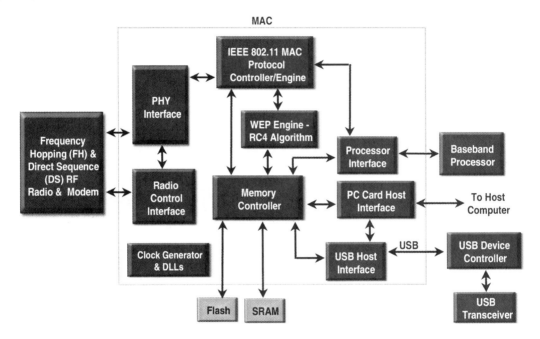

Figure 18.21: Wireless LAN Radio, MAC and Baseband Controller

Wireless Summary

A wireless home network is an intriguing alternative to phone line and power line wiring systems. Wireless home networks provide all the functionality of wireline networks without the physical constraints of the wire itself. They generally revolve around either IR or radio transmissions within the home.

Radio transmissions are comprised of two distinct technologies—Narrowband and Spread-spectrum radio. Most wireless home networking products are based upon the Spread-spectrum technologies. To date, the high cost and impracticality of adding new wires have inhibited the wide spread adoption of home networking technologies. Wired technologies also do not allow users to roam about with portable devices. In addition, multiple, incompatible communication standards have limited acceptance of wireless networks in the home.

A group called HomeRF was formed in 1997 to address these issues and develop a standard that allows cable-less connection of AV digital appliances to PCs. Since its formation, the group has developed a specification for wireless communications in the home called SWAP-CA.

What started as a purely European initiative—to develop a unified digital radio standard for cordless phones—evolved into DECT and has attracted worldwide attention. So far, the DECT standard has been adopted for use in 26 countries, and the number is growing.

Bluetooth technology is another popular solution for people who want to deploy wireless in-house networks. The Bluetooth technology facilitates real-time voice and data transmissions between devices on a home network. It eliminates the need for numerous, often proprietary, cable attachments for connection of practically any kind of communication device. Connections are instant and they are maintained even when devices are not within line of sight. While Bluetooth's technical specifications are modest in comparison to other technologies, they are acceptable for almost every application except high quality video. And they are key to achieving low cost. The combination of low cost and Bluetooth's unique functional capabilities position the technology to enable many new and innovative markets and applications, thus making it one of the highest volume emerging wireless application opportunities.

The Different Needs of Different Consumer Devices

The home networking technologies discussed above are used to network consumer devices such as consumer devices, PCs, PC appliances, etc. Each consumer device has specific applications and prefers certain home networking technologies to others. Most of these consumer products have been discussed in detail in the last few chapters. In the coming years several of these technologies will take shape and emerge as market leaders. Table 18.2 shows how different applications and consumer devices require different home networking technologies.

Table 18.2: Different Applications and Consumer Devices Require Different Home Networking Technologies

	Home Automation	Entertainment	Information	Personal Communications	Communication
Devices	– Home appliances – Security/safety systems – Utility meters	– TV sets – Set-top boxes – DVD Players – Game Consoles – VCRs – MP3 Players	– PCs – Screen phones – Printers – Modems – Routers – Hubs – Scanners	– Mobile phones – Smart phones – Handheld – Laptop – Pagers	– Corded/Cordless telephones – Fax machines
Content	Information on home processes, house environment, remote diagnostics and technical support	Rich multimedia content, electronic programming guides, impulse purchases	Discrete information on external world, shopping for household goods	Information used on the move or requiring instant action: travel, weather, local services, stock market	Information on how to reach people in time and space
Usage Pattern	Communal	Communal	Individual Shared	Individual Personal	Communal or Individual Shared
Connection to Outside World	– Power line – POTS	– Cable – DBS	– Cable modem – ADSL – POTS, ISDN	– GSM – Infrared	– POTS
Practical Networking Technology	– CEBus – X-10 – LONWorks	– IEEE 1394 (Fire Wire)	– HomeRF – HomePNA – Ethernet	– Infrared – Bluetooth	– POTS – DECT – 900MHz, 2.4GHz

Residential Gateways

The residential gateway is a device that connects or bridges an in-home network and a broadband connection. It sits at the core of the home network and enables bi-directional communication and data transfer channel among networked appliances in the home and across the Internet. As a router, it distributes Internet access among a variety of other networked devices in the home. It is the control center that functions as a bridge and links consumer devices and PCs and manages the flow of data between the outside world and devices on the network.

These gateways, also known as service gateways, media gateways, or home gateways, are the key ingredient to providing ubiquitous high speed Internet access to consumers around the globe. They provide a strategic platform and convergence point for integrating different broadband access types and several in-home networking solutions such as HomePNA, HomeRF, wireless LANs, or IEEE 1394. Another important function of the residential gateway is to serve as an access platform through which service providers can remotely deploy services to the home from the Internet. The gateway also has a security mechanism (called the firewall) to keep the network safe from intruders.

The evolution of new data broadcasting services has created the need for a special interface or gateway device that can be used to pass digital content between the Internet and a home network. These gateways provide a communications link that is required to carry information between the Internet and appliances in the home. In simple terms, the residential gateway is a device that connects the different in-home appliances (such as digital TV and digital VCR) to the necessary technology (IEEE-1394, phone lines) and to a broadband connection (such as DSL, wireless, cable, or satellite). It allows and promotes communication between devices in the home and connects these devices to the Internet through the broadband connection. It typically combines functions of a modem, router, or hub for Internet access. There are different types of residential gateways available based on the gradients of functionality required. Communication, entertainment, informational, utility management, and home automation are some of the services and delivery schemes for a single tenant residence that uses a residential gateway as the interface to in-house networks and consumer devices.

There is general deployment of high-speed Internet connections and a push by service providers to offer integrated voice, data, and video services over the same high-speed pipe to different nodes throughout the home. Hence, the residential gateway is expected to become a key integrated service enabler. Many of the home networking devices that are currently being used by consumers to access broadband networks will incorporate residential gateway functionality in the near future. In addition to these devices, a number of dedicated residential gateways such as home servers are expected to emerge.

Market Trends and Applications

Rapid implementation of integrated voice, video, and data services along with home networking and other value-added services are creating an explosive market for residential gateways. The services enabled by this residential gateway platform include broadband data, voice, video, as well as value-added services such as home networking, Internet firewalls, smart home applications (like home management and security), virtual private networking, and interactive TV.

The residential gateway is a platform for the following services, such as:

- **Entertainment services** – Video on demand, DVD downloads, video conferencing, interactive advertising channels, interactive TV games, personal video recording, console games, home videos, viewer measurement, electronic programming guide, ITV, on-demand games, sport casts, news casts, music videos, streaming audio, CD or MP3 downloads.

- **Communication services** – Voice over packet, unified messaging, voice activated interface, SMS/alerts, virtual PBX, instant messaging, chat, directories, news, finance, location services, content feeds, PIM, unified account access, scheduling collaboration, message boards, classified, greetings.
- **Smart home/Control services** – Home networking, Internet access, Internet appliances, home automation, home security, network security, power monitoring, email, family management, energy saver automation, remote control, parental control, shopping, payment gateway services, banking, asset management, home patient monitoring.

The high-tech market research firm, Cahners In-Stat, predicts the residential gateway market will rise sharply from $100 million in 2000 to $5 billion in 2005. The wide range in this projection is characteristic of the uncertain timing of market acceptance. However, the residential gateways' time has finally come due to the accelerating ramp in broadband deployments. The winning solutions will be those that are both inexpensive and sensitive to the non-technical nature of the users installing them. Media gateways will evolve over the next few years from devices that provide basic broadband access to complex integrated services gateways that enable remote management and value-added services such as home security control and video-on-demand. From a market that hardly existed in 1999, the beginning of 2001 saw actual products shipping by many different industries.

Market Drivers and Inhibiting Factors

There are a number of driving factors behind the development of the residential gateway market:
- The availability of new home networking technologies
- Increased demand for non-PC based appliances
- Intelligent homes
- TV centric applications
- Network operators expanding their service offerings
- The Internet
- Broadband connections
- International standards

The second generation of home phone line networking components is available today. Power line and wireless technologies are beginning to gather momentum and demand is increasing at an exponential rate. The availability of these new technologies combined with the explosion of non PC-based appliances is driving the demand for a single appliance that is capable of connecting in-home appliances to the public Internet–a residential gateway.

The home automation market is growing at a tremendous pace. There are about 40,000 new homes constructed in the United States each year with a value of more than $1 million. Market research firm Parks Associates estimates that about 80% are installing some form of intelligent electronic control system. The creation of a new breed of smart homes that can be managed and controlled from the public Internet can be achieved through a residential gateway. In addition to these factors, there is a large push by service providers and network operators to expand their revenue streams and move beyond their traditional service models. These large telecommunication companies are upgrading their infrastructures to facilitate the delivery of these new services to their subscriber bases. To allow them to manage these new services, the network operators are also working very closely with a number of vendors to develop residential gateways that will be capable of supporting the new services.

The Internet is also proving to be an important market driver for residential gateways. Some people believe that the Internet will be as widespread as utilities such as water and electricity. The ability to have a device like a residential gateway to control the flow of information between the internal home appliances and the Internet is seen as a crucial part of this vision. With PC costs moving down fast, people are finding it easier to have multiple PCs for different members of the home.

However, a single broadband connection is still prevalent and Internet access being driven through this one access point requires some type of residential gateway device. Industry standards are crucial for enabling mainstream consumer adoption of residential gateway products. Consortiums like the OSGi (Open Services Gateway Initiative) are working to define and promote an open residential gateway standard for connecting the coming generation of smart consumer and small business appliances with commercial Internet services. The rate of residential gateway deployments is growing at a phenomenal rate. However, there is a potential to accelerate deployments even further than the current rate if a number of inhibitors are eliminated. Hurdles that are currently inhibiting the mass deployment of residential gateways in households across the world include:

- Lack of clear business models
- Lack of customer education and mass confusion
- Lack of mechanisms for supporting residential gateways.

There is an unclear ownership model when it comes to installing residential gateways in households. Media gateways are expensive and most service providers want to get out of owning and maintaining these appliances. To implement this type of strategy, service providers will need to convince consumers to buy into the concept of purchasing a residential gateway. This is expected to be an extremely expensive business model to implement. There are many different types of consortiums and manufacturers promoting their own type of residential gateway. The wide variety of products and technologies is making life difficult for consumers who want to choose a gateway that is appropriate to their needs. Support is also major concern in residential gateway deployment. If every consumer who buys a gateway appliance also requires computing skills, then the gateways will only be suitable to a relatively small section of our society.

Characteristics of Residential Gateways

Media gateways must have a reliable and robust hardware platform. Additionally, the software that runs on a gateway appliance needs to be very reliable and not susceptible to errors. Unlike PC users, the general consumer marketplace will not put up with rebooting their residential gateways. Supporting multiple services such as voice, data, and video are absolutely essential. Another key requirement of the residential gateway is security. Functions such as secure e-commerce transactions, remote home control, and access from authorized service providers are critical. Providing quality of service to support a multiple of intelligent devices from different vendors is also extremely important.

Phased Deployment of Residential Gateways

Given the fast changing nature of the home networking industry, it is helpful to examine the different generations of residential gateways that are been deployed in households across the world. The mass deployment of gateways will come in three distinct phases:

1. Devices provide broadband access and interface to a single device such as a PC
2. Distribution of information between digital consumer devices

3. The networked home, where devices are completely networked allowing distribution of voice, video and data

Although the term residential gateway is relatively new to most people, it already exists in many of our homes. For example, most of our homes have a couple of first generation residential gateways–a set top box for receiving television and a modem that allows us to connect with the Internet. Phase two of the deployment plans involves the availability of a second generation residential gateway that includes advanced features such as broadband connectivity, home networking interfaces, and IP telephony capabilities. The third and final phase of deployment will be based on powerful residential gateways that are capable delivering video, voice, and data throughout the house. It will also supply you with other services such as home automation, energy management, security control, etc. Its hardware architecture will be modular in design. The modularity of third generation gateways will support multiple broadband and home network interfacing technologies. This support of multiple of backend and front-end interfaces will make the gateway less apt to become obsolete with technology advancements. The support for modularity will fuel the evolution of residential gateways into a type of application server that consumers will use to distribute broadband services throughout their homes.

Residential Gateway Components

The residential gateway provides a unified platform to satisfy all the needs of most consumers, providing information, entertainment, and communication. It is a centralized access point between the home and the rest of the world. Since broadband technology is relatively new and continuously evolving, the residential gateway will also evolve in its functionality. However, in its basic form it will contain and combine three distinct technological components—a digital modem, home networking chipsets, and software. Much of the configuration flexibility is brought about by a gateway's support for a range of different types of modems that provide connectivity to different types of broadband access networks. At the home networking side of the residential gateway, a chipset is available providing the interface to the particular technology running on the home network. In addition to the various types of broadband and home networking chipsets, all gateways contain computing resources that supports the software required to operate the device. The software running on the residential gateway enables the smooth inter-operation of consumer appliances and services within the home so that the complexity, distribution, and technical disparity of the system elements are hidden from the consumer.

Types of Residential Gateway

There are various types of devices that are competing for a share of the residential gateway marketplace–PC's, broadband modems, digital set top boxes, and other possible candidates.

PC-Based Residential Gateways—Home Servers

The PC-based residential gateway, also known as a home server, is conceptually the easiest path to the residential gateway. This is due to the number of PCs currently in the home, their processing power, the inclusion of a digital modem, and the presence of a platform for deployment of home networking services. Thus, combining the current PC with a digital modem provides the closest capabilities to the next generation of residential gateways. The existing PCs can perform most of residential gateway functions comfortably. However, there are certain disadvantages of the PC. The operating system has traditionally not been a robust part of the PC. The scare of the PC crashing without warning and having to reboot a gateway, which provides critical features such as security, is not acceptable. Some companies have been pushing PC-based architecture even within a separate residential gateway box such as Ericsson's E-box and IBM's Home gateway.

Modem-Centric Residential Gateways

Traditional methods to connect to the Web have hit a technological upside. Analog modems using phone lines cannot provide bandwidth beyond 56 Kb/s. Modems based on digital broadband technologies are the solution to the problem of increased Internet bandwidth access. At the broadband end there is a cable, satellite, or DSL modem termination. At the home networking side, either a HomePNA or HomeRF chipset is present which provides the home networking functionality of the particular technology. Satellite modems offer speeds of 400 Kb/s to 38 Mb/s. DSL modems offer 1.5 Mb/s to 52 Mb/s and cable-based modems can receive and process multimedia content at 30 Mb/s. Usually this gateway is sold in conjunction with services. Local cable or DSL provider subsidizes the cost of the broadband gateway by changing for extended services over 1-2 years. Usually this is an easy-to-install gateway and the service provider saves money by minimizing truck rolls. According to a recent report from Dataquest, digital broadband modem shipments are expected to reach 30 million units in the year 2004.

Today, multiple types of digital modems take advantage of the growing proliferation of broadband connectivity. In the years to come, the digital broadband modems discussed next will evolve into a residential gateway.

Advanced Digital Set-Top Boxes

Set top boxes present an interesting proposition to become a residential gateway that provides high-speed access to consumer devices in the home. Because they provide access to TV signals, they are ideal devices to provide broadband access through cable, satellite, and DSL technologies. They will not only provide traditional broadcast television, but also value-added and revenue-generating services such as pay per view system and Internet.

The set top box will have adequate quantities of memory, interface ports, and processing power. It will include a return or back channel to provide communication with a server located at the head end (e.g., CMTS–cable modem termination system). These set top boxes are capable of providing t-commerce, video-on-demand, Internet browsing, and near-video-on-demand services. The presence of a return channel further allows for broadcasts customized to the local viewing population and enables the set top box to support e-mail and local chat-style communication services.

These types of boxes have double the processing power and memory capabilities of broadcast TV boxes. They are ideal for consumers who want to simultaneously access a varied range of new multimedia and advanced Internet applications from the comfort of their homes. Set top boxes from the advanced services category bare close resemblance to a multimedia desktop computer. They can contain more than ten times the processing power of a low-level broadcast TV set top box. Enhanced capabilities in conjunction with a high speed return path can be used to access a variety of advanced services such as video teleconferencing, home networking, IP telephony, video-on-demand, and high-speed Internet TV. Additionally, users will be able to use enhanced graphical capabilities within the box to receive high definition TV signals.

Most set top boxes in this category have the capability to store video on a hard disk drive and provide the capability to record and view video simultaneously. Such receivers also come with a range of high-speed interface ports that allow them to be used as residential gateways. For cable, terrestrial, and satellite companies, set top boxes that support advanced technologies are seen as an opportunity to increase revenue streams.

Entertainment Gateway

As consumers grapple with the proliferation of digital video and audio devices in the home, as well as a variety of broadband and wireless networking options, the urge to integrate all these technologies into a single device has grown. Companies such as Microsoft and Moxi Digital, both of which recently unveiled products designed to generate buzz in this emerging category are taking up this impulse. Forrester Research predicts that more than 12 million U.S. homes will use these "entertainment gateways" in 2004, with that number rising to 25 million by 2006. These devices will be able to integrate digital cable or satellite reception, download and manage music via a broadband connection, function as both a CD and DVD player, and connect wirelessly to additional devices throughout the house.

Moxi Digital was the first out of the gate, unveiling its Moxi Media Center in early January. Essentially a home media server, the Media Center is built around a vision of media convergence in which the TV reigns supreme. When it becomes available later this year, the Moxi set top box, which uses a Linux operating system, will feature broadband and personal video recording. Like TiVo, it will allow users to record more than 60 hours of shows. It will also connect via the 802.11a wireless networking protocol to TVs and PCs in the home. Projected to cost cable and satellite operators $425 per unit, the Moxi system is designed to displace traditional set top cable boxes.

Not surprisingly, Microsoft's vision of an entertainment gateway places the PC at the center of the hub. The company is developing a Windows XP extension, dubbed Freestyle that will allow manufacturers such as Hewlett Packard and NEC to build machines that combine all the functionality of a Windows XP computer with the convenience of a digital entertainment hub. Freestyle is based on the idea of transforming a PC into a TV—without having to sit down at the keyboard.

While computer users with a TV tuner card have long been able to get television programming through their computers, Microsoft seeks to enhance the experience. Microsoft developed the software to bring the functions of a digital video recorder like TiVo to the PC, allowing a computer user to record shows on the hard drive, skip through commercials, or pause live broadcasts.

Microsoft has also unveiled software, called Mira, which allows users to roam from room to room, using a portable Web tablet to access PC programs wirelessly. Mira allows consumers to access digital music files, e-mail, or any other content on the home PC through a portable wireless tablet. This connects Windows to an intelligent display and looks like a descendant of the Tablet PC. It is a 15-inch flat panel, touch screen display that rests in a stand, but can be taken anywhere in the house to allow remote access to computer applications, media files, or the Internet. Unlike the Tablet PC, the Mira tablet lacks computer smarts. It's merely a tool to get content from the home PC through wireless home networking.

Clearly, both paths—whether PC- or TV-centric—are interesting ideas. After all, who doesn't want to get rid of a remote control or two? What's far less clear, however, is whether the average consumer is willing to pay for a dedicated entertainment gateway. So far consumers have taken a go-slow approach toward similar devices, such as TiVo's personal video recorder. Does the addition of wireless capability somehow make the gateway seem more compelling? Are consumers looking for a wireless Swiss Army Knife that promises to do a little of everything? Or do they prefer more specialized machines that perform one or two tasks extremely well? These remain open questions. Yet in the meantime, it's probably a good idea to take ambitious growth forecasts for wireless entertainment gateways with a grain of salt. Business models need to be thought through. And service providers paying for the equipment and charging for the service seems to be the most attractive option for growing this market.

Other Gateway Candidates

While residential gateways based on set top boxes, PCs, and modems seem more conceivable, gaming consoles, PVRs (personal video recorders), and utility metering devices provide an imaginative and interesting platform to evolve into potential home gateways. Companies like Microsoft and Sony believe that a high-speed gaming console could serve as the gateway or hub to the digital home of the future.

Today, PVRs provide an interface to a broadband digital network. With home networking around the corner, the role of the PVR is expected to change. PVR manufacturers will increase support for an interface that provides connectivity to a range of different home interconnection technologies. The PVR will truly evolve into the residential gateway of tomorrow.

Some of the likeliest supporters of the concept of residential gateways come from the energy-utility industry. Utility-centric gateways are an interesting part of residential gateways. In fact, the concept of a utility company installing a new gateway device in customers' homes to provision new services has been gaining ground since the early 1990s. From a utility company's perspective, service gateways provide overall cost savings by minimizing truck rolls (through automated meter reading) and providing energy management and optimization. Hardware and installation costs are recovered through provision of multiple services. The push for utility gateways will come from the service provider's end rather than the consumer.

The residential gateway, which could either be a PC, digital modem, set top box, or PVR, will be responsible for bringing broadband access into the house and distributing it within the appliances in the home. Figure 18.22 shows the different components of the gateway.

The gateway contains various flavors of silicon chips that are used to handle and process digital video and audio services. All the chips are connected to the system board and are able to communicate with each other via buses. The CPU is responsible for coordinating the different component parts of a gateway.

As more features become available to subscribers, gateways will require higher performance CPU's to keep pace with increased data throughput.

Software programs are required to operate digital set top boxes. There are three types of gateway software—operating system, middleware, and application. The operating system keeps all parts of the gateway operating together. The middleware is a layer of software programs that operates below the interactive TV applications and above the operating system. Viewers use application software to watch TV and use interactive features.

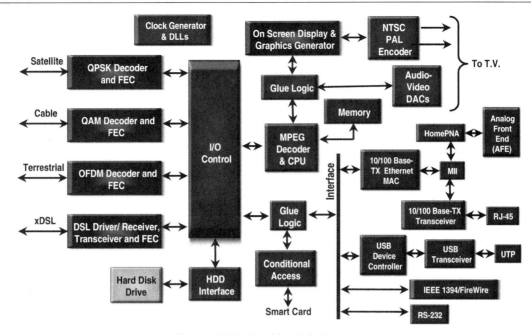

Figure 18.22: Residential Gateway

Residential Gateway Summary

The primary function of the residential gateway is to provide broadband connectivity to the home through cable, xDSL, satellite, and wireless. Residential gateways also provide home networking capabilities by distributing broadband access throughout the home using technologies such as HomePNA (phone lines) or wireless LANs.

As discussed earlier, the demand for greater Internet bandwidth is driving the need for digital Modem solutions, such as using an existing PC to provide residential gateway type services. Companies like IBM and Ericsson are strongly promoting this concept. This chapter also discussed the various classes of digital modems that are expected to evolve into the next generation of residential gateways. A cable modem is a device that allows high-speed data access—from a PC to the Internet—via a cable TV (CATV) network. It is a modem in the true sense of the word (it modulates and demodulates signals), and it delivers Internet data to the desktop at blazing speeds. It simply uses the increased bandwidth of the TV cable instead of an ordinary phone line. Cable modem services were introduced to the market much earlier than DSL services, so they gained a much larger customer base.

With the rapid increase in demand for high-speed data services, DSL modems offer telecommunications providers a technology that increases the bandwidth of the local loop without making huge investments in new fiber technologies. The convergence of television and computers is going to take a major step with the proliferation of digital TV technologies. This new environment will facilitate the broadcasting of data alongside video and audio content.

One of the more practical devices for accessing and using this new media is the set top box. The set top box, once a relatively passive device, is now evolving into a powerful residential gateway that is capable of variety of services such as automation, energy management, security, and control. Many set top box manufacturers have announced a new set top reference platform called a PVR. These new devices have two tuners and enable simultaneous recording and viewing of live TV broadcasts. The purpose of the different forms of residential gateways is to encourage consumers to use a range of new e-services.

There is an ensuing battle for eyeballs and dollars, and vendors face stiff competition in the battle for the living room. TV set top makers, satellite providers, and consumer electronics manufacturers are all trying to build the single convergence box through which consumers access their digital music, photos, and other entertainment content.

Middleware

Broken down to its simplest form, home networking middleware is a layer of software that lies on top of a consumer appliance's operating system. It provides 'hooks', or API's, to which home networking applications can be attached. Analysts are forecasting that the market for home networking middleware will be worth several billion dollars in a couple of years. To capitalize on this opportunity, a number of consortiums have been established to deliver products into this new and evolving marketplace. Current initiatives include OSGi, UPnP, Jini, HAVi, VESA, Interactive TV software providers, DVB, and OpenCable.

Introduction to Home Networking Middleware

Consumer electronic appliances such as set top boxes, PVRs, and DVD players are sophisticated and expensive digital processing systems. By connecting these consumer appliances to home networks, it is possible to share processing and storage resources between members of a family. Central to the fabric of all home networks is a software system called middleware. This software system allows connected consumer devices to exchange both control information and streaming multimedia content. "Middleware" is a relatively new term in the home networking business. Compared to an IT environment, it equates to the presentation layer of the OSI (Open Systems Interconnect) seven layer model.

Middleware is used to isolate application programs from the details of the underlying hardware and network components. In the world of home networking, we encounter many types of hardware consumer appliances from different manufacturers. A number of home networking middleware applications have evolved to provide interoperability among these diverse systems. The key middleware products are OSGi, UPnP, Jini, VESA, HAVi, OpenCable, and DVB.

OSGi—Open Services Gateway Initiative

The OSGi initiative came about in March 1999. It is now a group of over 70 companies led by giants such as Ericsson, Cisco, Nokia, Siemens, Sun Microsystems, Motorola, IBM, Nortel, Philips, Oracle, Alcatel, Lucent, Toshiba, and Texas Instruments. It was created to standardize efforts in connecting a wide array of consumer devices to the residential gateway. OSGi middleware specifications deliver an open, common architecture for service providers, system developers, software vendors, appliance vendors, and equipment manufacturers. It helps them easily develop, deploy, and manage multiple services in a coordinated fashion.

The OSGi specification is designed to compliment and enhance virtually all residential networking standards and initiatives such as:

- Bluetooth
- CAL
- CEBus
- Convergence
- EmNET
- HAVi
- HomePNA
- HomePlug
- HomeRF
- Jini technology
- LonWorks
- UPnP, 802.11b
- VESA

The specification leverages the value of existing wire line and wireless networks while providing flexibility toward cable, W-CDMA, xDSL, and other high-speed access technologies. The OSGi Framework and Specifications facilitate the installation and operation of multiple services on a single Open Services Gateway such as the set top box, cable or DSL modem, PC, Web phone, automotive, multimedia gateway, or dedicated residential gateway.

OSGi Architectural Framework

The OSGi architecture is based on requirements from several new markets. The major components of the complete end-to-end OSGi framework model include:

- **Services gateway** – The central component of the OSGi end-to-end framework is the services gateway. It enables, consolidates, and manages multimedia communications to and from the home and office networks. The services gateway can also function as an application server for a range of high value services such as energy management and control, safety and security services, health care monitoring services, appliance control and maintenance, and electronic commerce services.
- **Service Provider** – Within a home networking environment, the service provider provides a range of services to consumers. From a technical perspective, the delivery of such services is enabled through the download of a software application into the residential gateway.
- **Service aggregator** – As this computing paradigm evolves, OSGi is expecting to see the creation of new types of service providers who will offer a set of services (e.g., automatic reading of electricity, gas, and water meters) bundled together.
- **Gateway operator** – The main responsibility of this OSGi entity is to manage and maintain the residential gateway and its services. Functions of a typical gateway operator range from starting, stopping, updating, and removing services to managing the status of the residential gateway.
- **Wide area network and carrier/ISP** – The wide area network provides the necessary communications among the service gateway, the gateway operator, the service aggregator, and the service provider. This communication platform is provided and managed by a telecommunications carrier or, when the wide area network is the Internet, by an ISP.
- **Digital consumer devices and networks** – The last major piece of the OSGi framework is the local home network and the appliances attached to the services gateway.

The relationship among the major components is illustrated diagrammatically in Figure 18.23.

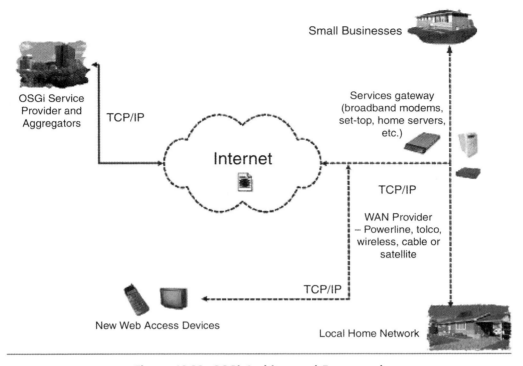

Figure 18.23: OSGi Architectural Framework

OSGi Middleware Anatomy

The major technical effort in the first release of the OSGi specification is on the Application Programming Interfaces (APIs) implemented on the gateway. OSGi is an open standard that enables multiple software services to be loaded and run on a services gateway such as a set top box, cable modem, DSL modem, PC, or dedicated residential gateway. It is focused on the residential gateway with the software environment based on Sun's Java virtual machine. The decision by OSGi to use Java as their core technology gives service providers, network operators, and appliance manufacturers a vendor-neutral application and appliance layer APIs and functions.

One of the main benefits of Java is that it is an open technology that can run on multiple platforms including residential gateways, consumer electronics equipment, household appliances, communications appliances, and computers. Technically, the services gateway is an embedded server that is attached to the broadband access network to connect external service providers to digital consumer appliances that are connected to an in-home network. In addition to supporting a Java Virtual Machine, OSGi-based gateways also include a consumer appliance access manager and a method of logging activity details. To gain a better understanding of the gateway middleware anatomy, let's take a closer look at the Java embedded server (JES) product from Sun Microsystems that complies with the OSGi specification.

Targeted at the exploding home networking market, JES is a small footprint software framework that runs in a residential gateway, providing service providers with the ability to deliver managed services to the networked home on-demand. From a technical perspective, JES consists of two primary components—a services pace framework and a number of modular in-home application services that are executed within this framework. Originally developed by Sun Microsystems, this service framework allows programmers to write Java based in-home applications as components that can be managed independently of one another and can be dynamically added, removed, executed, or updated from within a running application. Home Applications that run in the Java Embedded Server are called *application services*. The Java Embedded Server comes with a set of pre-built services that address a variety of common requirements for networked homes.

Benefits of OSGi

The main benefit of having such an organization developing API's for residential gateways include:

- **Platform neutrality** – They produce middleware that is platform independent. In other words, the OSGi software can be implemented on different types of residential gateways.
- **Security** – The OSGi specification offers several levels of system security allowing digital signing of downloaded e-services and object access control.
- **Multiple service capabilities** – The middleware provides the ability to host multiple services. By providing a single gateway platform, the gateway can become a service integration and management point in the residence, while a set of Java-based APIs provide value-added services.
- **Support for an array of home networking technologies** – The OSGi spec provides the ability to support multiple local network technologies such as emerging wired and wireless, data, and audio-video transport standards.

The OSGi specifications delineate Application Programming Interface (API) standards for a gateway platform execution environment. Open Services Gateways must support these API standards in order to conform to the OSGi specification. The APIs address service cradle-to-grave life cycle management, inter-service dependencies, data management, device management, client access, resource management, and security. Using these APIs, end-users can load network-based services on demand from the Service Provider while the Gateway manages the installation, versioning, and configuration of these services.

UPnP—Universal Plug and Play

UPnP is an industry initiative promoted by the UPnP forum that consists of 600 members. It is designed to enable easy and robust connectivity among stand-alone appliances and PCs from different manufacturers. It uses open Internet communication standards to transparently connect consumer electronic appliances to standard PCs. UPnP makes it possible to initiate and control the transfer of files and AV streams from any appliance on the in-home network. UPnP is an extension to the plug and play initiative that was introduced by Intel, Compaq, and Microsoft back in 1992. It defines a set of common interfaces that allows users to plug an appliance directly into the home network. Users can begin using a new consumer appliance without worrying about configuration settings and installing new drivers. A high level overview of the software architecture of UPnP is illustrated in Figure 18.24.

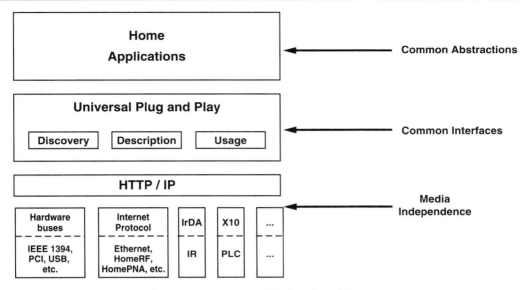

Figure 18.24: UPnP High-level Architecture

UPnP was developed within the context of existing industry standards. For instance, UPnP provides developers with a common set of interfaces for accessing services on a home network. Another advantage of UPnP middleware is its independence of the physical network media in the house. It is compatible with existing networks such as standard 10BaseT Ethernet and new networking technologies that don't require costly installation of new wiring systems such as HomePNA and HomeRF. Rather than concentrating on one particular appliance type, UPnP interconnects all types of appliances in the house, including PCs, PC peripherals, new smart home appliances, residential gateways, home control systems, and Web-connectable appliances. The result of this pragmatic and relatively simple approach from Microsoft is that implementing UPnP on your in-house network requires very little work and human intervention. UPnP is equally adaptable to both dynamic home environments and fixed, configured corporate networks.

UPnP Middleware Anatomy
The foundation blocks for the UPnP standard leverage a number of industry standards including:
- **TCP/IP** – Transmission Control Protocol/Internet Protocol
- **DNS** – Domain Name System
- **HTTP** – HyperText Transfer Protocol
- **HTML** – Hypertext Markup Language
- **UDP** – User Datagram Protocol
- **LDAP** – Lightweight Directory Access Protocol
- **XML** – eXtensible Markup Language
- **XSL** – Extensible Stylesheet Language
- **ARP** – Address Resolution Protocol

A typical UPnP middleware environment consists of the following logical components:

- **User control point** – A set of software modules that facilitates communication between it and a number of controlled appliances on a home network. Examples of consumer appliances that could function as a user control point include a standard PC, a digital set top box, and high-speed broadband modems.
- **Controlled consumer appliance** – A set of software modules that facilitates communication with a user control point. The primary difference between a user control point and a controlled appliance is that the user control point is always the initiator of the communications session. Examples of consumer appliances that could function as a controlled consumer appliance include VCRs, DVD players, security systems, and automated light controllers.
- **Bridge** – A set of software modules that allows legacy appliances to communicate with native UPnP consumer appliances.
- **Legacy consumer appliance** – Any non-UPnP compliant consumer appliance.
- **Bridged consumer appliance** – An appliance that cannot participate in UPnP at the native protocol level, either because the consumer appliance does not have sufficient hardware resources, or because the underlying media is unsuitable to run TCP and HTTP protocols. Examples of appliances that could be bridged consumer appliances are power line-controlled audio/video equipment, light switches, thermostats, wristwatches, and inexpensive toys.

Plugging a New Appliance into a UPnP Based Home Network

When a new appliance is connected to a UPnP based home network, a process called discovery is initiated. The first part of the discovery process takes place when an appliance connects into the home network. It sends out a small packet of data to other appliances on the network to identify itself. For example, the packet might say 'I am here, I am a DVD player, and you can reach me at this address.' The discovery process returns only the basic information needed to connect to the appliance into the network. More detailed information about the home networking environment is provided to the appliance in the form of a schema. A schema is a structured data definition that defines a set of values about various services operational on the network.

Benefits of UPnP

A UPnP-compliant appliance can offer a number of important advantages, including:

- **Open Standards** – Relatively simple and open protocols such as those that have been defined by the Internet Engineering Task Force (IETF) have a proven track record on the Internet. TCP/IP, for instance, allows numerous types of computing platforms to communicate reliably with each other. Because UPnP is based on standard Internet protocols, it can work with a broad range of consumer appliances, from large PCs to small consumer electronics appliances.
- **Scalability** – UPnP normally functions in small network environments. However, it is possible to scale upwards to larger networks.
- **Plug and Play** – Most home users want to plug in a device and have it work immediately with no hassles. UPnP is based on straightforward, innovative mechanisms for discovery and connectivity that provide a basis for enabling appliance services.
- **Low footprint** – Unlike traditional PC-based solutions, consumer electronic appliances have radically less systems resources at hand. Typically, they are based on a low-cost micro controller and 200 to1000 Kbytes of RAM and Flash memory. Implementing Universal Plug

and Play requires very little development work and requires only a small amount of system resources and footprint.

- **Multi-vendor and Mixed media environment** – Analysts are predicting that mixed media, multi-vendor in-home networks will be a common scenario in the future. Consequently, UPnP has been explicitly designed to accommodate these environments.

- **Smooth integration with legacy systems and non-IP appliances** – Although IP internetworking is a strong choice for UPnP, it also accommodates home networks that run non-IP protocols such as IEEE 1394-based entertainment networks. For example, a home network could use a Windows PC to host several different types of legacy appliances and use the UPnP mechanism to make these appliances discoverable to other peers on the network.

- **Non PC-centric architecture** – The configuration of a UPnP-based network can be based on peer-to-peer network architecture. This means that a home network can function without a PC. However, this doesn't mean that the PC has no role in a UPnP-based network. The PC's general-purpose nature and substantial resources will make it a valuable part of any network where it is present.

With the addition of Device Plug and Play (PnP) capabilities to the operating system, it became a great deal easier to setup, configure, and add peripherals to a PC. Universal Plug and Play (UPnP) extends this simplicity to include the entire network, enabling discovery and control of devices including networked devices and services such as network-attached printers, Internet gateways, and consumer electronics equipment.

UPnP is more than just a simple extension of the Plug and Play peripheral model. It is designed to support zero-configuration, "invisible" networking, and automatic discovery for a breadth of device categories from a wide range of vendors. With UPnP, a device can dynamically join a network, obtain an IP address, convey its capabilities, and learn about the presence and capabilities of other devices. And all this is performed automatically, truly enabling zero-configuration networks. Devices can subsequently communicate with each other directly, thereby further enabling peer-to-peer networking.

The varieties of device types that can benefit from a UPnP enabled network are large and include intelligent appliances, wireless devices, and PCs of all form factors. The scope of UPnP is large enough to encompass many existing and new scenarios including home automation, printing and imaging, audio/video entertainment, kitchen appliances, automobile networks, and proximity networks in public venues.

UPnP uses standard TCP/IP and Internet protocols, enabling it to seamlessly fit into existing networks. Using these standardized protocols allows UPnP to benefit from a wealth of experience and knowledge, and makes interoperability an inherent feature. Because UPnP is a distributed, open network architecture defined by the protocols used, it is independent of any particular operating system, programming language, or physical medium (just like the Internet). UPnP does not specify the APIs applications will use, so operating system vendors can create APIs that will meet their customers' needs.

Home API

Home API is a PC-centric system that can control a home's lights, temperature, and appliances from a PC. Home API was established to broaden the market for home control products. It was created by

Honeywell, Compaq Computer Corporation, Intel Corporation, Microsoft Corp., Mitsubishi Electric, and Philips Electronics to define and develop an open industry specification that would foster the development of computer applications for home systems and appliances. These applications will enable the control of home electronics such as televisions, VCRs, set top boxes, and other home devices including security, indoor/outdoor lighting, and temperature control systems. These new applications will enhance the entertainment, comfort, and security of consumers.

The companies are collaborating and leading an industry effort called the Home API Working Group (Home API). The goal of the group is to provide a foundation for supporting a broad range of consumer devices by establishing an open industry specification that defines application programming interfaces (APIs). These APIs are protocol and network media independent, enabling software developers to more quickly build applications that operate these devices. In addition, they will allow both existing and future home network technologies such as HAVi, Home PNA, Home RF, CEBus, LonWorks, and X-10 to be more easily utilized.

To support developers, the group intends to provide a Software Developers Kit (SDK) that will implement these APIs on the Microsoft® Windows® operating system, with availability expected in the first half of 1999. In the future, the group expects SDKs to be available for other operating systems as well.

Users will benefit from new applications in the areas of home entertainment, security, and home automation. For example, for home entertainment Home API will help simplify the set up and operation of A/V devices and home theaters. And it will be compatible with other emerging digital A/V network initiatives such as Home Audio/Video Interoperability (HAVi).

Home automation and security are other areas that can greatly benefit from Home API. House modes such as Away, Asleep, and Entertaining can be defined to control the status of the home security, sprinkler, heating/cooling, lighting, and entertainment systems according to user-defined criteria. These modes can be modified from within the home or remotely, and they can be tied to a calendar or to the weather forecast. This level of home automation can increase security and comfort, as well as save money through energy conservation.

Set up as a network-independent, Windows-based home consumer appliance discovery and control system for home automation, Home API recently combined with Microsoft's UPnP.

Jini Connection Technologies

Jini is Sun Microsystems' home networking middleware solution. It is a layer of Java software that allows appliances to plug directly into a home network without the hassle of installing drivers and configuring operating systems. In some ways the history of Jini is the history of Java. Jini is really the fulfillment of the original Java vision of groups of consumer-orientated electronic appliances interchanging information. This vision requires mechanisms that are not typically associated with desktop computers, including:

- The software infrastructure for these consumer and in-home appliances must be incredibly robust. Freezers and microwaves simply cannot fail with a message asking, "Abort, Retry, or Ignore?"
- The appliances connected to a home network have to support true, effortless "plug and play." In other words, when the devices are plugged in, they just work.
- Upgrades of software are also an important issue for home networking users. If an IT professional must be called in to upgrade all appliances on a home network, chances are the appliances simply won't get upgraded.

The vision of legions of appliances and software services working together simply and reliably had been espoused by a number of researchers and computer industry leaders before. Mark Weiser of the Xerox Palo Alto Research Center called this vision "ubiquitous computing." With this vision in mind, a group of software engineers at Sun Microsystems set out to provide a suite of technologies that realized Mark's vision. This project became known as Jini. The name Jini was chosen by the creators of the system because it is energetic and easy-to-remember word that begins with "J" and has the same number of letters as "Java". The Jini project went on at Sun, hidden from public eyes, until 1999 when the technology was officially made available to the public with a host of licensees already on board. These partners today are building Jini-enabled services and consumer appliances', including hard drives, digital cameras, handheld computers, and much more. For its part, Sun is rapidly aligning behind Jini, in much the same way it aligned around Java back in 1995.

Jini Middleware Anatomy

Jini software can run on anything with a digital heartbeat–cellular phones, digital cameras, PDAs, alarms, televisions, and even smart cards.

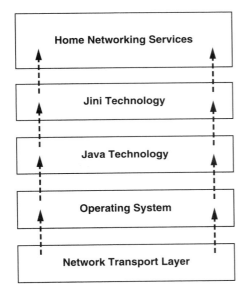

Figure 18.25:
Jini Software Architecture

As illustrated in Figure 18.25, Jini software is an infrastructure that runs on top of a Java platform to create a federation of virtual machines. Each virtual machine sits on top of an operating system that sits on top of the network. It is based on a simple model that consumer appliances with microchips should connect and work together in *communities*. A Jini community is a group of services on a home network that is available, both to each other and to the consuming applications. Each community on a home network has a unique name. The formation of these communities requires:

- No appliance drivers
- No operating systems
- No new cabling systems
- And, no human intervention

Jini technology uses a lookup service with which appliances and services register. Each appliance on the home network provides *services* that other appliances in the community may use. These appliances provide their own interfaces that ensure reliability and compatibility. Sun Microsystems is currently working with a number of manufacturers to integrate the Jini home networking technology into the next generation of digital appliances.

Plugging a New Appliance Into a Jini-Based Home Network

Similar to UPnP when an appliance plugs in to the home network, it goes through an add-in protocol called discovery and join-in. The appliance first locates the lookup service (discovery) and then uploads an object that implements all of its services' interfaces (join). For a consumer appliance to use a particular home networking service, it uses the lookup service to identify the resource. The service's object is copied from the lookup service to the requesting appliance where it will be used. The lookup service acts as an intermediary to connect an appliance looking for a service with that service. Once the connection is made, the lookup service is not involved in any of the resulting interactions between that appliance and the service.

In a Jini-enabled home network there is no central repository of drivers, or anything else for that matter. The Java programming language is the key to making Jini technology work. Consumer appliances in a network that employ Jini technology are tied together using Java Remote Method Invocation (RMI). RMI is best described as a set of protocols being developed by Sun's JavaSoft division that enables Java objects to communicate remotely with other Java objects. Jini technology not only defines a set of protocols for discovery, join, and lookup, but also a leasing and transaction mechanism to provide resilience in a dynamic networked environment.

Jini Benefits

A Jini-compliant appliance can offer a number of important advantages, including:

- **Reduced cost of ownership** – Self-managing appliances reduce the need for expert help from IT professionals, and this lowers the total cost of ownership for Jini connection technology-based systems.
- **Ease of development** – Because Jini technology is based on the Java platform, any existing development tool that can be used for Java software development can be used for Jini software development. In addition, utility classes and implementations are being developed that will be freely available which will ease the development of services and clients using Jini technology.
- **Small footprint** – The code required to implement Jini technology is so small that all types of home appliances can use it, from lamps and coffee makers to dishwashers and water heaters.
- **Appliance Agnosticism** – Jini is agnostic with regard to appliances. What does this mean for users of home networks? It means that Jini is designed to support a wide variety of entities that that can participate in a community. These entities may be appliances or software or some combination of both. If a hardware appliance is connected to a home network, Jini is flexible about how much computational power it needs to have. Jini can work with a full-blown multimedia computer capable of running multiple Java Virtual Machines and connecting with each other at gigabit speeds. It can also work with such appliances as PDAs and cell phones that may have only limited Java capabilities. In fact,

Jini is designed to accommodate appliances that are so simple they may have small amounts of or even no computational intelligence–a light switch, for instance.

- **Simplicity** – Jini technology is all about simplifying interactions with a home network. Sun has portrayed Jini as a simple way for appliances to find and use each other over a network. One of Jini's design principles is to eliminate configuration hassles and appliance drivers. Jini technology will allow users to use the network as easily as using a phone.
- **Reliability** – Communities of Jini services are largely self-healing. This is a key property built into Jini from the ground up. Jini doesn't make the assumption that in-house networks are perfect or that software never fails. Given time, the system will repair damage to itself.

HAVi—Home Audio-Video Interoperability

HAVi is a project that was started by Sony and Philips in 1996. Since then six other companies have joined–Thomson, Hitachi, Toshiba, Matsushita, Sharp and Grundig. HAVi adopted the IEEE-1394 bus standard as the underlying network technology for the HAVi protocols as well as for the transport of the real-time AV streams. IEEE 1394 is the interface of choice for digital audio and video. 1394 benefits include high-speed, flexible connectivity and the ability to link as many as 63 appliances together. No other interconnection technologies such as wireless, HomePlug, HomePNA, and USB are capable of distribution of high-speed video applications.

The HAVi middleware architecture is an open, lightweight, and platform-independent specification that allows developers to write home networking applications. It specifically focuses on the transfer of digital Audio/Video (AV) content between in-home digital appliances as well as the processing (rendering, recording and play back) of this content by HAVi-enabled appliances. However, it does not address home networking functions such as controlling lights or monitoring the climate within the house. The HAVi middleware system is independent of any particular operating system or CPU and can be implemented on a range of hardware platforms including digital products (cable modems, set top boxes, integrated TV's), internet-TVs, or intelligent storage appliances for AV content.

In the world of analog consumer electronic appliances, there exist a number of proprietary solutions for interoperability among appliances from one brand or vendor. In the coming world of digital technologies, HAVi extends this networking model by allowing communication between consumer electronic appliances from multiple brands in the home.

HAVi Middleware Anatomy

The HAVi middleware architecture specifies a set of APIs that allow consumer electronic manufacturers and software engineering companies to develop applications for IEEE 1394-based home networks. One of the main reasons HAVi selected IEEE 1394 over other transmission protocols is because of its support for *isochronous* communications.

HAVi comprises of software elements that facilitate the interoperability between different brands of entertainment appliances within the home. Interoperability is an industry term that refers to the ability of an application running on an in-home appliance to detect and use the functionality of other appliances that are connected to the home network. Table 18.3 summarizes the functions of the various architectural software elements that are present in a HAVi-based appliance. Figure 18.26 illustrates the different software elements of HAVi.

Table 18.3: Explanation of HAVi software elements

Software element	Description
Communications media manager (CMM)	This software element allows appliances to communicate over a home network based on 1394 bus technologies.
Messaging system	The main responsibility of this software element includes the passing of messages between different appliances connected to the home network
Event manager	The event manager manages events and reports the details to 'interested' software elements. An event is best described as a change in the operational state of the home network.
Registry	The registry software acts as a directory service. It allows any software object to locate another software object on the home network.
Stream manager	This element is responsible for managing real-time transfer of multimedia content between components on your home network
DCMs and FCMs	The DCM (Device Control Module) provides home networking applications with an interface to the physical appliance. The FCM (Functional Component Module) represents the functionality of an appliance. HAVi has defined these FCMs: tuner, VCR, clock, camera, AV disc, amplifier, display, AV display, modem, web proxy, and converter.
	DCM, a software element that represents a single device on the HAVi network and exposes the HAVi-defined APIs for that device. DCMs are dynamic in nature. If a device is inserted or removed from the network, a DCM for that device needs to be installed or removed respectively in the network. DCMs are central to the HAVi concept and the source of flexibility in accommodating new devices and features into the HAVi network.
	Contained within a DCM are the FCMs for each controllable function within the device. Which FCMs are present in a DCM depends on the device in question and is decided by the manufacturer. Currently, HAVi defines FCMs and corresponding APIs for functions like a tuner, VCR, disc-based storage, AV display, camera, and modem. In the future it can be expected that new FCMs will be defined for HAVi.
	The DCM Manager is responsible for installing and removing DCMs.
Resource manager	This element manages the sharing and allocation of resources on the home network.
Havlet	In addition to the above software elements specified by the HAVi architecture, devices on the in-house network may also contain a number of havlets that are specific to a home-networking environment. A havlet is typically a proprietary application that offers a user interface for controlling appliances.

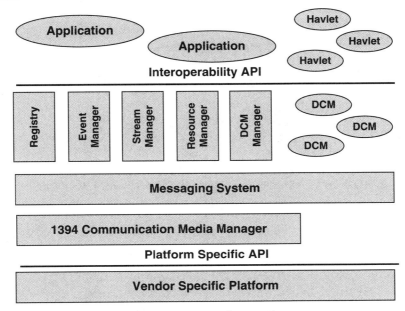

Figure 18-26: HAVi Software Elements

HAVi is a distributed software architecture, the software elements of which implement basic services such as network management, device abstraction, inter-device communication, and device user interface management. Collectively, these software elements expose the HAVi Interoperability API, which is a set of services for building portable distributed applications on the network. Software elements on different HAVi devices communicate with each other via HAVi-defined protocols to offer the desired service and the Interoperability API. The software elements on a HAVi device are implemented on top of a device- and vendor-specific platform such as a real-time operating system. Applications (which are software elements themselves) can then access these APIs transparently across the network.

Network Configuration and Appliance Classification
The underlying structure for a home network based on HAVi technologies is a peer-to-peer network where all appliances can talk to and interact with all others.

However, HAVi has been designed to allow the incremental addition of new appliances, which will most likely result in a number of interconnected clusters of appliances. Typically, there will be several clusters in the home, with one per floor or per room. Over time these clusters will be connected with technologies such as 1394 long or wireless 1394.

HAVi consumer electronic appliances can be categorized according to the degree to which they support the 1394 standard.
1. Non-1394 based appliances
2. Appliances that use 1394 but do not support the HAVi architecture
3. Appliances that use 1394 and support HAVi

Currently, most home appliances fall into the first category.

Benefits of HAVi

A HAVi compliant appliance can offer number of important advantages, including:

- **Automatic detection** – It can automatically detect other appliances on your home network.
- **Automatic registration** – Each appliance added to the HAVi network is automatically registered so that other appliances know what it is capable of. This level of functionality helps other appliances utilize the useful resources of this appliance without a need to own the same resources themselves.
- **Automatic software upgrades** – Some HAVi-compliant appliances are capable of installing new software on all appliances connected to the same home network. For instance, a HAVi Panasonic VCR can install the necessary application on a Sony TV in order to make two appliances interoperable. This reduces the need for network administration.
- **Manageability** – HAVi software takes advantage of the powerful resources of silicon chips built into modern audio and video appliances to give you the management function of a dedicated audio-video networking system.
- **Brand independence** – Entertainment products from different manufactures will communicate with each other when connected into a HAVi network. For example, a Panasonic VCR can work and share its resources with a Sony Amplifier and be controlled by a Mitsubishi TV remote control as long as all of these devices are HAVi compliant.
- **Plug and Play capabilities** – Hot Plug and Play is another exciting feature that HAVi-compliant appliances are equipped with. Here, an appliance configures itself and integrates itself into the home network without user intervention.
- **Legacy appliances** – The HAVi architecture supports legacy appliances. This plays an important role since the transition to networked devices is going to be gradual. Manufacturers are not suddenly producing only networked appliances and consumers are not suddenly replacing their existing appliances.

VESA (Video Electronics Standards Association)

This organization develops interoperability standards for appliances that connect to the Internet. In 1999, they released a draft home networking specification, making it possible for consumers to easily and remotely access digital consumer appliances such as VCRs and security systems.

The standard uses a long distance version of IEEE 1394 as the digital backbone and Internet Protocols (IP) for internetworking. External access networks such as telephone, cable TV, broadcast TV, and direct-broadcast satellite interface with the VESA Home Network (VHN) via access devices such as residential gateways, xDSL modems, or cable modems

VESA Home Network (VHN) Technical Overview

The backbone VESA home network spans the whole house so those devices located on component networks anywhere in the house can communicate with each other. The backbone provides sufficient bandwidth and quality of service for the applications and devices that communicate over it.

The component networks enable devices connected to them to communicate with each other, perhaps over a relatively short distance such as within a room. The choice of a component network is dictated by the communication needs and cost points of the device. Examples of important component networks are IEEE 1394, Fast Ethernet, Power-line CEBus, Phone Line, and RF Wireless LAN. An access-backbone Interface connects an external access network to the home network. An access-component Interface provides a similar function for an access network to connect to a component network. A POTS modem, an ISDN adapter, a cable modem, residential gateway, and a set top box are all examples of devices containing access-component Interfaces.

An appliance is a digital device connected to a network whose purpose is to provide some utility (other than network service) to the end user. Examples of End Devices are printers, TVs, audio speakers, security sensors, and HVAC controllers. A backbone-component Interface connects a component network to the backbone network. Backbone-component Interfaces may function as repeaters, bridges, or routers, and may be stand-alone, or embedded in an appliance or PC.

In August of 1992, VESA passed a standard—the VESA Local Bus (VL-Bus) Standard 1.0. This standard had a significant impact on the industry because it was the first local bus standard to be developed that provided a uniform hardware interface for local bus peripherals. Creating the VL-Bus standard ensured compatibility among a wide variety of graphics boards, monitors, and systems software. Today, several hundred companies are producing systems and peripherals based on the VL-Bus specification and more than two million VL-Bus products are shipped each month worldwide.

Since then VESA has continued to be a formative influence in the PC industry. It has been a major contributor to the enhancement of Flat Panel Display, Monitor, Graphics, Software, and systems technologies including Home Networking and PC Theatre. VESA's latest standards are described in more detail in the following sections.

Standards Initiative

VESA is developing and promoting standards in the key focus areas of displays and display interfaces. All VESA standards are created by its technical committees that consist of hardware and software professionals drawn from high technology companies around the world. Each committee is structured into technical Workgroups whose goals are to focus on specific technical requirements for developing a standards proposal. The proposals are reviewed and, when approved, submitted to the general membership for ratification. Each VESA parent member is given one vote. Current active technical committees are described next:

- **VESA Display Committee** – With the advancing technology in the display and display interface industry, particularly in the area of flat panel, LCD, and TFT displays, VESA has sought to coordinate the standards development activities of three of its most productive committees into one interactive Display Committee. These committees are the Monitor, Flat Panel Display Interface (FPDI), and Plug and Display (P&D) committees. The focused approach to technological requirements in these areas lends itself more appropriately to cohesive group activities and offers the marketplace a structured, integrated approach to developing and advancing much-needed technology. A continuation of the original committees' goals and objectives is an essential part of the new Display Committee and is summarized in the Display Committee's goals and objectives.

- **VESA Display Metrology Committee** – The mission of the Display Metrology Committee (DMC) is to specify reproducible electronic display metrology. The DMC is not in competition with other bodies that create display standards. Rather, this work details unambiguous measurement methods to supplement any display standard efforts. Performance criteria, compliance criteria, or ergonomic requirements will not be featured or pursued since such specifications should be left to the individual standards bodies outside the DMC. However, by collaborating with and accommodating other standards bodies, consistency of measurement methods and results will be achieved and any inadequacies will be clearly identified.

511

Therefore, liaisons will be maintained with ISO, CIE, IEC, EIAJ, SAE, and any other display-standards-creating body.

- **VESA Marketing Committee** – The VESA Marketing Committee was created to develop and support the total VESA organization in its effort to support VESA standards and VESA activities in the marketplace. It also promotes its member companies and their work in supporting VESA standards.

- **VESA Microdisplay Committee** – The VESA Microdisplay Committee was established in June of 1999 to develop interface, measurement, signal, and other standards for microdisplays, components, and products.

- **VESA Japan Committee** – The VESA Japan Committee was established in February 2000 to promote the development of design tools and reference guides, PlugTests, educational seminars, and other activities for the benefit of VESA member companies, particularly in Japan. Additionally, the committee serves as a conduit to other VESA committees for feedback on current standards and proposed standards.

- **DPVL Committee** – The VESA DPVL Committee was established in July 2001. The committee's efforts are to focus on defining the Digital Packet Video Link (DPVL), which is a high-level video stream protocol that enables high information content displays.

Middleware Summary

Middleware operates between the operating system and a home networking application and allows home networking users to run applications that are independent of the underlying hardware platform. It supports the seamless convergence of broadcast and home network applications. The fusion between both of these technologies facilitates the deployment of a range of new entertainment services within the home.

OSGi is a distributed, open networking architecture providing pervasive and peer-to-peer network connectivity to PCs, intelligent appliances, and wireless appliances. The OSGi specification provides a common foundation for ISPs, network operators, and equipment manufacturers to deliver a wide range of end-to-end e-services via gateway servers running in the home or remote office.

UPnP, developed by Microsoft, is a distributed open networking architecture for connecting PCs and digital consumer appliances using TCP/IP and the Internet. UPnP leverages Internet and Web components (like IP, TCP, UDP, HTTP, and XML) and enables seamless proximity networking in addition to control and data transfer among networked appliances in the home and office.

Jini, developed by Sun Microsystems, is a network and OS-agnostic Java-based proximity networking technology that enables automatic device discovery. There are currently more than 20,000 Jini licensees. Initially, an appliance first locates the lookup service in the 'discovery' phase. Next, in a process called 'join in,' it uploads an object that implements all its services. Jini is a technology that allows OEMs to develop their products based on Jini without worrying about the network.

The importance of HAVi middleware software system cannot be stressed enough. HAVi is a software standard that allows all types of digital consumer electronics and home appliances to communicate with each other. Most analysts agree that the combination of IEEE 1394 and HAVi is the best solution for distributing rich multimedia entertainment content around a home. This is due to the higher throughput required in moving audio and video content.

The VESA Home Network specification allows consumers to connect their disparate networks in the home, networks such as Ethernet, HomeRF, and other "no-new-wires" networks. The APIs used by the DVB networking standard allow in-home appliances to access a range of different Internet and home-centric applications.

Digital Home Working Group

Announced in June 2003, the Digital Home Working Group (DHWG) is a nonprofit alliance of 18 major companies from the consumer electronics, personal computer, and mobile devices industries. Heavyweights in the alliance include:

- Microsoft
- Sony
- Intel
- Hewlett Packard
- Fujitsu
- Gateway
- IBM
- Kenwood
- Lenovo
- NEC
- Panasonic (Matsushita Electric)
- Philips
- Samsung
- Sharp
- Sony
- STMicroelectronics
- Thomson
- Nokia

These companies are working closely to choose networking technologies they will promote in future products. These technologies will simplify sharing of digital content (e.g., digital music, photos, and video) among networked consumer electronics (CE), mobile devices, and PCs. The group shares a common goal of establishing a platform of interoperability based on open industry standards. It will deliver technical design guidelines that companies can use to develop digital home products that share content through wired or wireless networks in the home. Examples of these products include PCs, TVs, set top boxes, printers, stereos, mobile phones, PDAs, DVD players, and digital projectors.

Many of the companies involved in the group have a common vision of connecting products so people can use the network to access and share resources at home in the same way they share resources at work. For example, people would be able to play digital audio on their living-room stereo even though the music files are stored on a computer in the den. In addition to making it easier for consumers to access and share digital content across PCs, mobile devices, and consumer electronics products, the alliance also intends to push the mainstream adoption of new products that can be networked together. But one stumbling block to product harmonization has been the existence in the networking industry of multiple standards that perform similar functions. This defeats the purpose of having a single standard foundation for product development. The alliance has considered supporting

WiFi, UPnP, and IEEE 1394/FireWire and is developing a logo to be applied to products to indicate which are interoperable.

Companies in the DHWG are expected to release products using the chosen networking standards over the next year.

Summary—The Home Sweet Digital Home

The digital home era is here. Home networking is morphing from disparate networks to a single unified network connecting consumer devices. While each digital consumer device provides unique advantages and uses, they all require the use of the Internet and the sharing of data, voice, and video among appliances. We are seeing several value-added services being enabled by the arrival of high-speed Internet access. Because of cost, most homes will have a single access point to the Internet. Home networking technologies will be essential to provide the benefits of Internet access to all consumer devices. Middleware software will be essential for application providers to deploy services on the gateway.

The five key parts that enable the evolution of the digital home are:
- Broadband access technologies
- Residential gateway
- Home networking technologies
- Middleware
- Digital consumer devices

Broadband access provides an always on, high-speed access to the Internet. Popular technologies being deployed today include satellite, DSL, and cable. In the future, technologies such as wireless broadband, LRE, and FTTH will be coming to our neighborhoods to provide faster access to the Internet and bundled services.

Once deployed, consumers find innumerable uses for home networks. Sharing Internet access among in-home appliances has proven to be the main motivator for consumers installing a new home network. Telecommuters, knowledge workers, students, and families are some of the groups that are at the forefront of the home networking revolution.

The ultimate goal of a home network is to share voice, audio, data, and entertainment information among different digital devices in the home. Home networking allows the ability to communicate this information anywhere at any time. It will eventually bring the Internet to the hands of consumers. It will appear within many appliances in the house and help connect people across the globe. Multiple users can have access to similar data at the same time, and in different locations at home. It allows access to the home through the Internet when a user is away, thus enabling security and energy management.

Home networking provides entertainment, information, and automation. Most consumers would love to use the existing wires within their homes. Power lines and phone lines are the dominant 'no new wires' home networking technologies. Not deploying new wires means that the consumer does not have to rewire the home and the technology can be used immediately. However, using the existing wires has until now been limited.

For delivering high-speed data and video packets, Ethernet (IEEE 802.3), IEEE 1394, Optical Fiber, or USB 2.0 technologies should be used. However, all of these technologies require additional special wiring in the house. Wireless technologies such as Bluetooth, HomeRF, IEEE-802.11b, IEEE-802.11a, and HiperLAN2 can be used in areas where users cannot take advantage of the existing wires, or the cost cannot be justified for additional wiring. Wireless allows mobility but bandwidth and quality of service remains an issue.

Out of the many consumer devices, there are several that will grow into a residential gateway and be responsible for routing data, Internet, voice, and video between consumer devices. The gateway will provide Internet access to the home and will be the platform to deploy services for service providers. Some of the key gateway prospects include set top boxes, entertainment gateways, PC-modem servers, and gaming consoles.

Middleware is a software layer that provides abstraction between the basic hardware/software system and the many applications and devices that need to be supported. It is the glue that allows value-added services to be deployed on the residential gateway. Some of these middleware technologies include OSGi, Jini, UPnP, HAVi, and VESA.

Broadband access technologies, residential/media/home gateways, home networking technologies, and middleware are key to enabling the success of digital consumer devices. The provisioning of the digital home and the success of providers deploying services depends on careful execution. It can make broadband access technologies, home networking technologies, residential gateways, middleware technologies, and digital consumer devices simple and easy to use.

Some concepts of the digital home in the kitchen and bathroom have been pie-in-the-sky thoughts. Companies struggle to convince consumers that there is a benefit to having a refrigerator that automatically orders milk when the carton is empty, or a bathroom that weighs you when you step out of the shower and scrolls diet suggestions in the mirror. However, the first step in creating the digital home is in connecting consumer devices through a wireless home network (enabled with WiFi). Such networking already connects 13 million U.S. households, and this will move Web surfing from the study into any area of the home.

Digital homes will also include smart conveniences such as window shade programming for raising and lowering shades at selected times. The crisp displays available through flat panel technology are bringing digital images to every corner of the house. And the market for digital TVs is expected to exceed 10.5 million units in 2006.

Consumer devices such as digital video recorders and MP3 players are automatically synchronized with the media center PC that is typically connected to a broadband access technology. This will allow the consumer to have access to content anytime, anyplace, and on whatever device the owner desires.

Entertainment will be the gateway to the digital home. Consumers may be reluctant to trust heating, cooling, or home security to crash-prone PCs, but there is less risk in storing music, photos, and videos in a central PC. Meanwhile, the PC camp needs to boost its commitment to reliability and privacy. Also, the digital home needs easy access to entertainment content. The music industry has grudgingly embraced selling downloads, but it has surrounded them with a mix of proprietary software and complex limits on use that guarantee a poor consumer experience. Film studios are even worse and have blocked any legal online distribution of high quality video. The digital home is a huge opportunity for both consumers and vendors. Digital entertainment could be a path to much broader use of digital technology in the home. The bulk of the challenge to the industry will be providing reliability, simplicity, and privacy. Standardization and availability of solutions and products will change the digital home from a dream to reality.

Programmable Logic Solutions Enabling Digital Consumer Technology

As system vendors develop next-generation consumer products, they are faced with the challenge of providing groundbreaking solutions. Also, new features need to be integrated as the market evolves and cost continues to be a factor. In such scenarios the market demands low-cost solutions that offer flexibility, including reprogramming to accommodate bug fixes and feature additions.

The process of architecting and building next-generation digital consumer devices is filled with dilemmas. System architects are looking for solutions that can provide functionality, features, and the ability to future-proof the design, all at the lowest cost. Programmable logic has a primary advantage over custom gate arrays and standard cells by enabling faster time-to-market and shorter design cycles. By using software, developers can program their design directly into the PLD, allowing customers to make revisions to their designs relatively quickly and with lower development costs. In this chapter we will take a broad and detailed look at the many advantages of programmable logic for use in digital consumer devices. We will cover the different types of programmable logic and describe a number of products offered by Xilinx, showing how they can provide designers with lower cost and faster time-to-market solutions. We will also provide specific design information for a number of digital consumer system applications.

What Is Programmable Logic?

In the world of digital electronic systems there are three basic kinds of devices—memory, microprocessors, and logic. Memory devices store random information, such as the contents of a spreadsheet or database. Microprocessors execute software program instructions to perform a wide variety of control and computing tasks such as running a word processing program or video game. Logic devices provide specific functions to manage the interchange and manipulation of digital signals including device-to-device interfacing, data communication, signal processing, data display, timing and control operations, and almost any other function a system must perform.

Almost every electronic system contains application specific integrated circuits (ASICs). Circuits can include custom gate arrays, standard cells, and programmable logic. All these devices can compete with each other if they are utilized in the same types of applications within electronic systems. Variables in pricing, product performance, reliability, power consumption, density, adaptability, ease of use, and time-to-market determine the degree to which the devices compete for specific applications.

Fixed Logic versus Programmable Logic

Logic devices can be classified into two broad categories—fixed and programmable. As the name suggests, the circuits in a fixed logic device are permanent. Once manufactured, they perform one function or set of functions and cannot be changed. Programmable logic devices (PLDs) are standard off-the-shelf parts that offer customers a wide range of logic capacity, features, speed, and voltage characteristics. These devices can be changed at any time to perform any number of functions.

The complexity of the device defines the time required for fixed logic devices to go from design, to prototypes, and to a final manufacturing run. This time can be from several months to more than a year. With a PLD, a designer can develop, simulate, and test their designs instantly. PLDs do not require long lead times since they are already on a distributor's shelf and ready for shipment.

If the ASIC does not work properly, or if the requirements change, a new design must be developed. The up-front work of designing and verifying fixed logic devices involves substantial "non-recurring engineering" costs, or NRE. NRE represents all the costs customers incur before the final fixed logic device emerges from a silicon foundry. This includes engineering resources, expensive software design tools, expensive photolithography mask sets for manufacturing the various metal layers of the chip, and the initial prototype device costs. These NRE costs can run from a few hundred thousand to several million dollars. With PLDs, designers use inexpensive software tools to program their design into a device and immediately test it in a live circuit. PLD suppliers incur the large NRE and mask set costs when they design their programmable devices and are able to amortize those costs over their multi-year lifespan.

On the other hand, the marginal cost of producing an ASIC is usually less than that of producing a PLD because there are no wasted parts. An ASIC is optimized for a particular function. However, not all of the gates of a PLD are used once it is programmed. The general belief is that PLDs are cheaper for low-volume applications and ASICs are cheaper for high-volume applications. With the migration to advanced process technologies, the cost of PLD solutions has become comparable to ASICs, making PLDs competitive for high-volume designs. Also, PLDs allow customers to control inventory by ordering just the number of parts they need when they need them. Customers who use fixed logic devices often end up with excess inventory that must be scrapped. Or, if demand for their product surges, they may be caught short of parts and face production delays.

The PLD that is used for prototyping is the same one that will be used in the final production of a piece of end equipment such as a network router, a DSL modem, a DVD player, or an automotive navigation system. There are no NRE costs and the final design is completed much faster than that of a custom fixed logic device.

Another key benefit of using PLDs is that product designers can change the circuitry as often as they want until it operates to their satisfaction. Since PLDs are based on rewriteable memory technology, the device is simply reprogrammed to change the design. Once the design is finalized the product can go into immediate production by simply programming as many PLDs as needed with the final software design file.

ASICs sometimes have better performance because an ASIC design is optimized for a specific function. Conversely PLDs have more flexibility, which is important in prototyping during the early stages of the design cycle, where applications standards continue to emerge, or multiple variations need to be supported. PLDs can be reprogrammed even after a piece of equipment is shipped to a customer. In fact, thanks to programmable logic devices, a number of equipment manufacturers now tout the ability to add new features or upgrade products that already are in the field. They do this by simply uploading a new programming file via the Internet to the PLD and create new hardware logic in the system.

CPLDs and FPGAs

The two major types of programmable logic devices are field programmable gate arrays (FPGAs) and complex programmable logic devices (CPLDs). Of the two, FPGAs offer the highest amount of logic density, the most features, and the highest performance. The largest FPGA now shipping, part of the Xilinx Virtex™ line of devices, provides eight million "system gates" (the relative density of logic). These advanced devices also offer features such as built-in hardwired processors (such as the IBM Power PC), substantial amounts of memory, clock management systems, and support for many of the latest very fast device-to-device signaling technologies. FPGAs are used in a wide variety of applications ranging from data processing and storage, instrumentation, telecommunications, and digital signal processing. CPLDs offer much smaller amounts of logic—approximately 10,000 gates. But CPLDs offer very predictable timing characteristics and are therefore ideal for critical control applications. CPLDs, such as the Xilinx CoolRunner series, also require extremely low amounts of power and are very inexpensive. This makes them ideal for cost-sensitive and battery-operated portable applications such as mobile phones and digital handheld assistants.

In a PLD, the software is used to decide how to connect groups of transistors (gates) or groups of gates (macrocells) together in a particular way. PLDs are programmed in one of three ways:

- Connections between gates are linked by physically breaking some lines (fuses)
- Connections are made by melting some lines (anti-fuses)
- Gates are connected using software-controlled switches within the PLD

Once the connections have been made the PLD behaves as if it were a custom-designed circuit. Another way of making a customized chip (ASIC) is to have the chip designed from the beginning with fixed connections. Or have a chip that mixes groups of previously designed sections from a chip connected with fixed wires.

Types of programmable logic:

- **Simple Programmable Logic Device (SPLD)** – An SPLD is a small chip with a series of gates and macrocells that are connected to do simple functions.
- **Complex Programmable Logic Device (CPLD)** – A CPLD, which is larger than an SPLD, contains a number of pieces of circuits connected together that are each like an SPLD.
- **Field Programmable Gate Array (FPGA)** – An FPGA is generally larger than a CPLD. An FPGA has a different architecture (design) than a CPLD enabling it to do more complicated functions. FPGAs often contain memory for data storage. Some FPGAs also have specialized sub-circuits (IP blocks) that are optimized for specific functions and can be reconfigured an infinite number of times. Because of design differences from a CPLD, the number of system gates on the chip and not the number of macrocells typically measures the complexity of an FPGA.

The following tools are required to program FPGAs and CPLDs:

- Programmable Read-only Memory (PROM) chips are used together with the PLDs because the PROMs contain the PLD programming. PROM memories are nonvolatile, which means that when the power is switched off they retain the program for the PLD. In contrast, FPGAs have volatile memory on-board the chip. The PROM is needed to store all the programming information that is needed to program the PLD since they are programmed using software-controlled switches.
- Programming software is needed to program the PLDs. For example, Xilinx integrated software tools include the Foundation Series™, tailored for designers not familiar with

creating systems using FPGAs. The Alliance Series™ is tailored for designers who require high flexibility and integration into incumbent electronic design automation (EDA) environments such as Cadence, Synopsis, or Mentor design software.

■ IP (intellectual property) cores serve as plug-in modules to help shorten product development time and minimize design risks. IP cores implement predefined logic functions such as digital signal processing (DSP), bus interfaces, and peripheral interfaces.

■ IRLs reconfigure FPGAs using Xilinx Online technology to provide time-in-market.

Looking beyond a standard PLD, Xilinx has announced a system-on-chip (SOC) initiative known as the Platform FPGA. This will mix specialized sub-circuits (hard IP blocks and soft configurable IP) with generic gate arrays or macrocells making circuits that perform specific communications functions. Programmable connections between the sub-circuits and the more traditional PLD portion of the chip allow system designers to create a wide range of more powerful chips.

CPLDs provide designers with the capability to implement complex designs in a single chip. This replaces large numbers of discrete logic devices. CPLDs may be thought of as multiple interconnected SPLDs. An SPLD is a group of gates with interconnects that can be programmed to form desired logic functions. Invented in the early 1970s, SPLDs perform basic logic functions with limited internal resources. The basic building block of an SPLD is either a Programmable Array Logic (PAL) or a Programmable Logic Array (PLA). This logic feeds an output cell that may be either combinatorial or registered in nature.

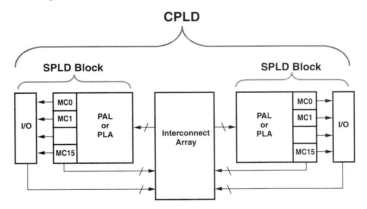

Figure 19.1: Generic PLD Architecture

CPLDs are composed of multiple SPLD blocks. The SPLD block, called a logic block, is comprised of a PAL or PLA, a macrocell, and an I/O structure. Several SPLD blocks are joined together using an interconnect array to form a CPLD. Figure 19.1 shows the generic CPLD architecture with multiple interconnected SPLDs. The CPLD was created in response to designers' demands for increased flexibility and feature sets.

The CPLD has many benefits, providing:

■ Ease-of-design
■ Low development costs
■ Faster time-to-market
■ Increased product revenue

- Longer time-in-market through field upgradeability.
- Decreased component inventory
- Reduced printed circuit board (PCB) area
- Lower cost than discrete parts

CPLDs offer a simple way to implement a design. Once a design has been described a designer can use CPLD development tools to optimize, fit, and simulate the design. Engineers who are spending a lot of time fixing old designs can be working on introducing new products and features—ahead of the competition.

FPGAs are field programmable devices that have logic structures that can be configured by the end user. FPGAs provide affordable solutions as customized very large system integrated (VLSI) chips by combining the ability to implement logic circuits and provide instant manufacturing turnaround with very low-cost prototypes. FPGAs can be programmed in minutes to offer the same functionality and performance as ASICs/ASSPs, providing a significant time-to-market advantage. The basic FPGA architecture, shown in Figure 19.2, is built on logic blocks and the interconnection between logic blocks and input/output blocks (IOBs). Logic blocks have a look-up table (LUT) where the sequential circuit is implemented. The interconnection or routing structure consists of wire segments and programmable switches that connect the different logic blocks. The IOBs provide an interface between the package pins and the internal user logic. FPGAs today primarily consist of LUTs, interconnection, and IOBs. With the growing momentum of the system-on-chip (SOC) concept, Xilinx is leading the industry in introducing features such as processors, memory and clock management functions, and multiple I/O types.

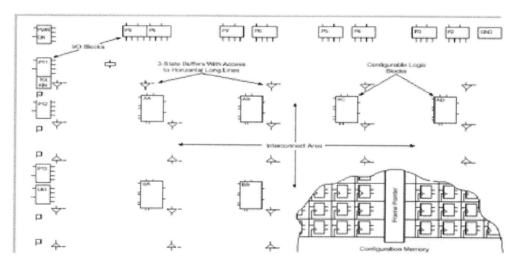

Figure 19.2: Generic FPGA Architecture

The PLD Market

Market researcher Gartner/Dataquest projects today's worldwide market for programmable logic devices is about $3.5 billion. The market for fixed logic devices is about $12 billion. However, in recent years sales of PLDs have outpaced those of fixed logic devices built with older gate array technology. And high-performance FPGAs are now beginning to take market share from fixed logic devices made with the more advanced standard cell technology.

According to the Semiconductor Industry Association programmable logic is now one of the fastest growing segments of the semiconductor business. Sales of PLDs in the last few years have increased at a faster rate than sales for the overall semiconductor industry.

According to *EDN Magazine*, a leading electronics design trade publication: "Programmable-logic devices are the fastest growing segment of the logic-device family for two fundamental reasons. Their ever-increasing logic gate count per device 'gathers up' functions that might otherwise spread over a number of discrete-logic and memory chips, improving end-system size, power consumption, performance, reliability, and cost. Equally important is the fact that in a matter of seconds or minutes users can configure, or in many cases, reconfigure, these devices at their workstation or in the system-assembly line. This capability provides powerful flexibility to react to last-minute design changes, to prototype ideas before implementation, and to meet time-to-market deadlines driven by both customer need and competitive pressures."

Over the last few years programmable logic suppliers have made such phenomenal technical advances that PLDs are now seen as the logic solution of choice for many designers. One reason for this is that PLD suppliers such as Xilinx are "fabless" companies. Instead of owning chip manufacturing foundries, Xilinx outsources that job to chip manufacturing partners like the UMC Group in Taiwan, IBM Microelectronics in the United States, and Seiko Epson in Japan. This strategy allows Xilinx to focus on designing new product architectures, software tools, and intellectual property cores while having access to the most advanced semiconductor process technologies. Advanced process technologies help PLDs in a number of key areas including faster performance, integration of more features, reduced power consumption, and lower cost. Today Xilinx is producing programmable logic devices on a state-of-the-art 90 nm, low-k dielectric, copper process on 300 mm (12 inch) wafers—one of the best in the industry.

Just a few years ago the largest FPGA was measured in tens of thousands of system gates and operated at 40 MHz. Older FPGAs were relatively expensive, often costing more than $150 for the most advanced parts. Today, FPGAs with advanced features offer millions of gates of logic capacity, operate at 300 MHz, can cost less than $10, and offer a new level of integrated functions such as processors and memory.

Just as significant, PLDs now have a growing library of intellectual property (IP) cores. These are predefined and tested software modules that customers can use to create system functions instantly inside the PLD. Cores include everything from complex digital signal processing algorithms and memory controllers to bus interfaces and full-blown software-based microprocessors. Using such cores save customers time and expense by eliminating the need to create these functions and potentially delaying a product introduction.

The value of programmable logic has always been its ability to shorten development cycles for electronic equipment manufacturers and help them get their product to market faster. Programmable logic is certain to expand its popularity with digital designers as PLD suppliers continue to integrate more functions inside their devices, reduce costs, and increase the availability of time-saving IP cores.

Xilinx Corporate Information

Xilinx is the leading provider of complete and innovative programmable logic solutions. The company's products help minimize risks for manufacturers of electronic equipment by reducing the time required to develop products and to take them to market. Customers can design and verify their proprietary circuits with Xilinx PLDs much faster than they could by using traditional methods such as mask-programmed gate arrays, ASICs, and application specific standard products (ASSPs). Because Xilinx devices are standard parts (they only need to be programmed), customers are not required to wait for prototypes or pay large NRE costs as they do for ASICs. System vendors incorporate Xilinx PLDs into products for a wide range of markets including data processing, telecommunications, networking, consumer electronics, industrial control, and instrumentation, automotive, military, and aerospace.

Founded in 1984, Xilinx pioneered the FPGA and today provides more than half of the worldwide demand for field-programmable devices. Xilinx is a publicly traded company (NASDAQ: XLNX) headquartered in San Jose, California, and employs approximately 2,612 people worldwide. Market research firm Gartner Dataquest currently ranks Xilinx as the fourth-largest ASIC supplier in the world. Xilinx now has a 50% market share of the total PLD market and holds 829 issued United States patents.

The Xilinx strategy is to provide off-the-shelf integrated circuits (ICs), software design tools, predefined system level functions such as Intellectual Property (IP) cores, Internet Reconfigurable Logic (IRL) solutions, and unparalleled field engineering support. Xilinx is continually introducing innovative products that allow customers to reduce their time-to-market and increase their time-in-market. Xilinx capitalizes on its key strengths:

- Research and development (R&D), marketing, and applications engineering
- Leading edge IC design and process technology
- Obtaining wafer capacity at competitive prices that produces the fastest, lowest cost, and most dense parts

Xilinx products address a broad application base with no single customer representing more than 10% of the company's total revenues. Xilinx products are designed primarily focused on the communications sector with 50% of the revenues generated from products such as cellular base stations, network routers, and switches. Other sales venues include computing (computer peripherals, mass storage, and high-speed servers) and consumer electronics.

Xilinx Product Solutions

Xilinx builds programmable logic ICs and develops software and cores that provide complete solutions. Xilinx designs, develops, and markets the following products to target applications for the consumer, wireless, embedded, networking, and telecom markets. Xilinx also provides application support and design services to help customers with their unique designs. However, each family of products has a unique focus:

- Programmable logic devices (Silicon solutions)
- Virtex series of FPGAs – High-performance, high-density FPGAs
- Spartan series of FPGAs – High-volume, low-cost FPGAs
- CoolRunner series of CPLDs – Low-power, low-cost CPLDs
- IP, Software, Services, and Configuration Solutions
- Software and design tools

- IP cores
- Internet Configurable Logic (IRL): Field upgradeability using the Internet

Silicon—Programmable Logic Devices (PLDs)

Xilinx offers several Silicon solutions including Virtex™ FPGAs, Spartan FPGAs, CoolRunner series CPLDs, XC9500™ series CPLDs, RocketPHY™ family of 10 Gb/s physical layer transceivers, configuration storage devices, Military and Aerospace, and IQ Automotive solutions. However, in this section we will only discuss the key product families that apply to digital consumer devices.

CoolRunner-II CPLDs

In January 2002, Xilinx introduced CoolRunner-II, a next-generation 1.8-volt family. Six devices were released with densities ranging from 32 to 512 macrocells manufactured using 180-nanometer process technology. CoolRunner-II CPLDs contain enhanced power management and system features without sacrificing performance or a cost penalty to the customer. This new class of devices is ideal for both performance-intensive applications as well as the large portable and wireless markets.

Key features include:
- Optimized for 1.8V systems
- Industry's fastest low power CPLD
- Static Icc of less than 100 µAmps at all times
- Densities from 32 to 512 macrocells
- Industry's best 0.18 micron CMOS CPLD
- Optimized architecture for effective logic synthesis
- Multi-voltage I/O operation – 1.5V to 3.3V
- Advanced system features
- Fastest in system programming (1.8V ISP using IEEE 1532 (JTAG) interface)
- On-The-Fly Reconfiguration (OFR)
- IEEE1149.1 JTAG Boundary Scan Test
- Optional Schmitt trigger input (per pin)
- Unsurpassed low power management
- FZP 100% CMOS product term generation
- DataGATE external signal control – DataGATE makes it possible to reduce power consumption by reducing unnecessary toggling of inputs when they are not in use.
- Flexible clocking modes
  - Optional DualEDGE triggered registers
  - Clock divider (by ?2,4,6,8,10,12,14,16)
  - CoolCLOCK – CoolCLOCK, another CoolRunner-II low power enhancement is implemented by dividing, then doubling, a clock signal.
  - Global signal options with macrocell control
  - Multiple global clocks with phase selection per macrocell
  - Multiple global output enables
  - Global set/reset
  - Abundant product term clocks, output enables and set/resets
  - Efficient control term clocks, output enables and set/resets for each macrocell and shared across function blocks
- Advanced design security
- Open-drain output option for Wired-OR and LED drive
- Optional bus-hold or weak pull up on select I/O pins

- Optional configurable grounds on unused I/Os
- Mixed I/O voltages compatible with 1.5V, 1.8V, 2.5V, and 3.3V logic levels on all parts - SSTL2-1,SSTL3-1, and HSTL-1 on 128 macrocell and denser devices
- PLA architecture
- Superior pinout retention
- 100% product term routability across function block
- Hot pluggable
- Wide package availability including fine pitch: Chip Scale Package (CSP) BGA, Fine Line BGA, TQFP, PQFP, VQFP, and PLCC packages
- Design entry/verification using Xilinx and industry standard CAE tools
- Free software support for all densities using Xilinx WebPACK™ or WebFITTER™ tools
- Industry leading nonvolatile 0.18 micron CMOS process
- Guaranteed 1,000 program/erase cycles
- Guaranteed 20 year data retention

CoolRunner-II CPLDs deliver the high speed and ease of use associated with the XC9500/XL/XV CPLD family with the extremely low power versatility of the XPLA3 family in a single CPLD. This allows the same parts to be used for high-speed data communications, computing systems, and leading edge portable products plus the added benefit of In System Programming. Low power consumption and high-speed operation are combined into a single easy-to-use and cost effective family. Xilinx patented Fast Zero Power™ (FZP) architecture that delivers very low power performance without the need for special design measures. Clocking techniques and other power saving features extend the users' power budget. The design features are supported starting with Xilinx ISE 4.1i, WebFITTER™, and ISE WebPACK™. Table 19.1 shows the CoolRunner-II CPLD family parameters.

Table 19.1: CoolRunner-II CPLD Family Parameters

	XC2C32	XC2C64	XC2C128	XC2C256	XC2C384	XC2C512
Macrocells	32	64	128	256	384	512
Max I/O	33	64	100	184	240	270
T_{PD} (ns)	3.5	4.0	4.5	5.0	5.5	6.0
T_{SU} (ns)	1.7	2.0	2.1	2.2	2.3	2.4
T_{CO}(ns)	2.8	3.0	3.4	3.8	4.2	4.6
$F_{SYSTEM1}$ (MHz)	333	270	263	238	217	217

Table 19.2: CoolRunner-II CPLD Family Packages and I/O Count

	XC2C32	XC2C64	XC2C128	XC2C256	XC2C384	XC2C512
PC44	33	33	–	–	–	–
VQ44	33	33	–	–	–	–
CP56	33	45	–	–	–	–
VQ100	–	64	80	80	–	–
CP132	–	–	100	106	–	–
TQ144	–	–	100	118	118	–
PQ208	–	–	–	173	173	173
FT256	–	–	–	184	212	212
FG324	–	–	–	–	240	270

Table 19.2 shows the CoolRunner-II CPLD package offering with corresponding I/O count. All packages are surface mount and over half of them are ball-grid technologies. The CMOS technology used in CoolRunner-II CPLDs generates minimal heat allowing the use of tiny packages during high-speed operation. The ultra tiny packages permit maximum functional capacity in the smallest possible area.

There are at least two densities present in each package. There are three in the VQ100 (100-pin 1.0mm QFP), TQ144 (144-pin 1.4mm QFP), and in the FT256 (256-ball 1.0mm spacing FLBGA). The FT256 is particularly important for slim dimensioned portable products with mid- to high-density logic requirements.

Table 19.3: CoolRunner-II CPLD Family Features

	XC2C32	XC2C64	XC2C128	XC2C256	XC2C384	XC2C512
IEEE 1 532	√	√	√	√	√	√
I/O banks	1	1	2	2	4	4
Clock division	–	–	√	√	√	√
Clock doubling	√	√	√	√	√	√
DataGATE	–	–	√	√	√	√
LVTTL	√	√	√	√	√	√
LVCM0S33, 25, 18, and 1.5V I/O	√	√	√	√	√	√
SSTL 2-1	–	–	√	√	√	√
SSTL 3-1	–	–	√	√	√	√
HSTL-1	–	–	√	√	√	√
Configurable ground	√	√	√	√	√	√
Quadruple data security	√	√	√	√	√	√
Open drain outputs	√	√	√	√	√	√
Hot plugging	√	√	√	√	√	√

Table 19.3 details the distribution of advanced features across the CoolRunner-II CPLD family. The family has uniform basic features and advanced features included in densities where they are the most useful. For example, it is very unlikely that four I/O banks are needed on 32 and 64 macrocell parts but very likely required for 384 and 512 macrocell parts. The I/O banks are groupings of I/O pins using any one of a subset of compatible voltage standards that share the same VCCIO level. The clock division capability is less efficient on small parts but more useful, and likely to be used, on larger ones. DataGATE, provides the ability to save power by blocking and latching inputs. It is valuable for larger parts but brings marginal benefit to smaller parts.

Architecture Description

CoolRunner-II CPLD is a highly uniform family of fast and low power CPLDs. The underlying foundation is a traditional CPLD architecture combining macrocells into Function Blocks (FBs) interconnected with a global routing matrix. In this case, the Xilinx Advanced Interconnect Matrix (AIM). Function Blocks use a Programmable Logic Array (PLA) configuration that allows all product terms to be routed and shared among any of the FBs macrocells. Design software can efficiently synthesize and optimize logic that is subsequently fit to the FBs with the ability to utilize a very high percentage of device resources. The software easily and automatically manages design changes. This exploits the 100% routability of the Programmable Logic Array within each FB. This extremely robust building block delivers the industry's highest pinout retention under very broad design conditions. This architecture will be explained by expanding the detail as we discuss the underlying Function Blocks, logic, and interconnect.

The design software automatically manages these device resources allowing users to express their designs using generic constructs without knowledge of these architectural details. More advanced users can take advantage of these details to more thoroughly understand the software's choices and direct its results. Figure 19.3 shows the high-level architecture where Function Blocks attach to pins and interconnect to each other within the internal interconnect matrix. Each FB contains 16 macrocells. The BSC path is the JTAG Boundary Scan Control path. The BSC and ISP block contains the JTAG controller and In-System Programming Circuits.

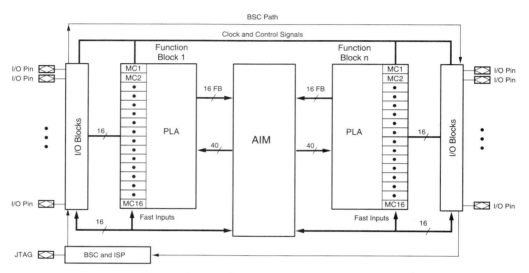

Figure 19.3: CoolRunner-II CPLD Architecture

Function Block

The CoolRunner-II CPLD Function Blocks contain 16 macrocells with 40 entry sites for signals to arrive for logic creation and connection. The internal logic engine is a 56-product term PLA. All Function Blocks are identical regardless of the number contained in the device. For a high-level view of the Function Block, see Figure 19.4.

At the highest level, the product terms (p-terms) reside in a programmable logic array (PLA). This structure is extremely flexible, and very robust when compared to fixed or cascaded product term function blocks. Classic CPLDs typically have a few product terms available for a high-speed path to a given macrocell. When needed they rely on capturing unused p-terms from neighboring macrocells to expand their product term tally. The result of this architecture is a variable timing model and the possibility of stranding unusable logic within the FB.

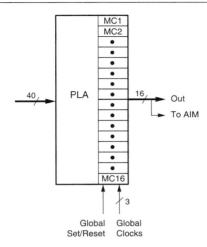

Figure 19.4: CoolRunner-II CPLD Function Block

The PLA is different—and better:

■ Any product term can be attached to any OR gate inside the FB macrocell(s).

■ Any logic function can have up to a maximum of 56 p-terms attached to it within the FB.

■ Product terms can be re-used at multiple macrocell OR functions. Once created within a FB, a particular logical product but can be re-used up to 16 times.

This works very well with fitting software that identifies product terms that can be shared. The software places as many of those functions as it can into FBs. Other than to reside in the same FB, which is handled by the software, there are no other restrictions or the need to force macrocell functions. Functions need not share a common clock, common set/reset, or common output enable to take full advantage of the PLA. Also every product term arrives with the same incurred time delay. There are no cascade time adders for putting more product terms in the FB. When the FB product term budget is reached, there is a small interconnect timing penalty to route signals to another FB to continue creating logic. And Xilinx design software handles all this automatically.

Macrocell

The CoolRunner-II CPLD macrocell is streamlined and extremely efficient for logic creation. Users can develop sum of product (SOP) logic expressions within a single function block that comprise up to 40 inputs and span 56 product terms. The macrocell can further combine the SOP expression into an XOR gate with another single p-term expression. The resulting logic expression's polarity is also selectable. The logic function can be pure combinatorial or registered with the storage element operating either as a D or T flip-flop or transparent latch. Available at each macrocell are independent selections of global, function block level, or local p-term derived clocks, sets, resets, and output enables. Each macrocell flip-flop is configurable for either single edge or DualEDGE clocking. This provides either double data rate capability or the ability to distribute a slower clock and save power. For single edge clocking or latching, either clock polarity may be selected per macrocell. CoolRunner-II macrocell details are shown in Figure 19.5.

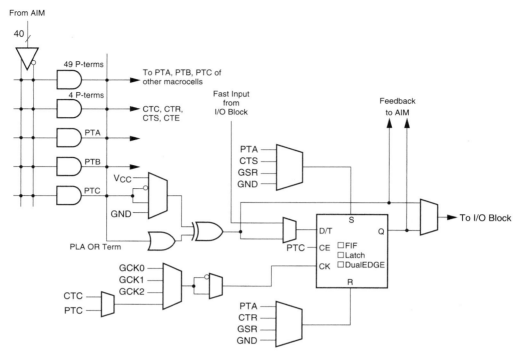

Figure 19.5: *CoolRunner-II CPLD Macrocell*

When configured as a D-type flip-flop each macrocell has an optional clock enable signal permitting state hold while a clock runs freely. Note that Control Terms (CT) are available to be shared for key functions within the FB. These are generally used when the exact same logic function would be created repeatedly at multiple macrocells. The CT product terms are available for FB clocking (CTC), FB asynchronous set (CTS), FB asynchronous reset (CTR), and FB output enable (CTE). Any macrocell flip-flop can be configured as an input register or latch that takes the signal from the macrocell's I/O pin and directly drives the AIM. If needed, the macrocell's combinational functionality is retained for use as a buried logic node. FToggle is the maximum clock frequency to which a T flip-flop can reliably toggle.

Advanced Interconnect Matrix (AIM)

The Advanced Interconnect Matrix is a highly connected low power rapid switch. For the creation of logic, the software directs the AIM to deliver up to a set of 40 signals to each FB. Results from all FB macrocells, and all pin inputs, circulate back through the AIM to provide additional connections to all other FBs as dictated by the design software. The AIM minimizes both propagation delay and power as it makes attachments to the various FBs.

I/O Block

I/O blocks are primarily transceivers. Each I/O is either automatically compliant or can be programmed with standard voltage ranges. In addition to voltage levels each input can selectively arrive through Schmitt-trigger inputs. This adds a small time delay but substantially reduces noise on that input pin. Approximately 500 mV of hysteresis will be added when Schmitt-trigger inputs are selected. All LVCMOS inputs can have hysteresis input. Hysteresis also allows easy generation of external clock circuits. The Schmitt-trigger path is best seen in Figure 19.6.

Outputs can be directly driven, 3-stated, or open-drain configured. Choice of a slow or fast slew rate output signal is also available. Table 19.4 summarizes various supported voltage standards associated with specific part capacities. All inputs and disabled outputs are voltage tolerant up to 3.3V. Figure 19.6 details the I/O pin configuration and notes that the inputs requiring comparison to an external reference voltage (SSTL2-1, SSTL3-1, and HSTL-1) are available on the 128 macrocell and other denser parts. VREF has pin-range requirements that must be observed. The Xilinx software helps designers remain within the proper pin range.

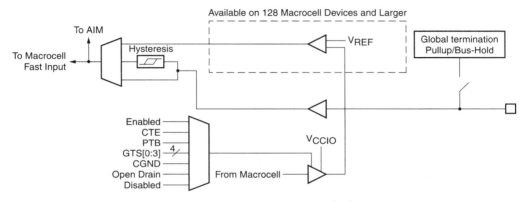

Figure 19.6: CoolRunner-II CPLD I/O Block Diagram

Table 19.4 summarizes the single ended I/O standard support and shows which standards require VREF values and board termination. VREF detail is given in specific data sheets.

Table 19.4: CoolRunner-II CPLD I/O Standard Summary

I/O Standard	V_{CCIO}	Input V_{REF}	Board Termination Voltage (V_{TT})
LVTTL	3.3	N/A	N/A
LVCM0S33	3.3	N/A	N/A
LVCMOS25	2.5	N/A	N/A
LVCMOS18	1.8	N/A	N/A
1.5V I/O	1.5	N/A	N/A
HSTL-1	1.5	0.75	0.75
SSTL2-1	2.5	1.25	1.25
SSTL3-1	3.3	1.5	1.5

Output Banking

CPLDs are widely used as voltage interface translators with the output pins grouped in large banks. The smallest parts are not banked so signals have the same output swing for 32 and 64 macrocell parts. Medium size parts (128 and 256 macrocell) support two output banks. Unless both banks are set to the same voltage the outputs will switch to one of two selected output voltage levels. The larger parts (384 and 512 macrocell) support four evenly split output banks. They can support groupings of one, two, three or four separate output voltage levels. This kind of flexibility permits easy 3.3V, 2.5V, 1.8V, and 1.5V interfacing in a single part.

DataGATE

Low power is the hallmark of CMOS technology. Other CPLD families use a sense amplifier approach to create product terms. This solution always results in drawing a residual current component. This residual current can be several hundred milliamps making them unusable in portable systems.

CoolRunner-II CPLDs use standard CMOS methods to create the CPLD architecture and deliver corresponding low current consumption. However, sometimes designers would like to reduce their system current even more by selectively disabling circuitry not being used. The patented DataGATE technology was developed to permit a straightforward approach to additional power reduction. Each I/O pin has a series switch that can block the arrival of unnecessary free running signals. Signals that serve no use, and may increase power consumption, can be disabled. Users are free to create their design, then choose sections to participate in the DataGATE function.

DataGATE is a logic function that drives an assertion rail threaded through the medium and high-density CoolRunner-II CPLD parts. Designers can select inputs to be blocked under the control of the DataGATE function. This effectively blocks controlled switching signals so they do not drive internal chip capacitances. The bus hold feature holds non-switchable output signals. Any set of input pins can be chosen to participate in the DataGATE function. Designers can use DataGATE to approach zero power in their designs.

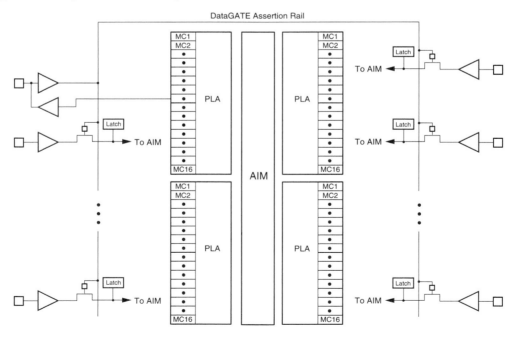

DS090_06_111201

Figure 19.7: DataGATE Architecture (output drivers not shown)

Figure 19.7 shows how DataGATE works. One I/O pin drives the DataGATE Assertion Rail. It can have any desired logic function on it. It can be as simple as mapping an input pin to the DataGATE function or as complex as a counter or state machine output driving the DataGATE I/O pin through a macrocell. When the DataGATE rail is asserted high, any pass transistor switch attached to it is blocked. Each pin has the ability to attach to the AIM through a DataGATE pass transistor and be blocked. A latch automatically captures the state of the pin when it becomes blocked. The DataGATE Assertion Rail threads throughout all possible I/Os so that each can participate if chosen. One macrocell is singled out to drive the rail and is exposed to the outside world through a pin for inspection. If DataGATE is not needed then this pin functions as an ordinary I/O.

Virtex-II Pro Platform FPGAs

The Virtex-II Pro™ FPGA family was introduced in March 2002. It consists of ten members with densities ranging from 3,000 to 125,000 logic cells (300,000 to 13 million system gates). The family has up to four IBM PowerPC™ processors, a maximum of 24 RocketIO™ multi-gigabit transceivers, up to ten megabits of embedded memory, embedded software design tools, and operating system support. Virtex-II Pro devices are delivered on 300mm wafers employing 130-nanometer copper process technology and 1.2 volts. The width of individual transistors on a chip is measured in nanometers. One nanometer equals one billionth of a meter. The Virtex-II Pro solution enables ultra-high bandwidth system-on-a-chip (SoC) designs that previously were the exclusive domain of custom ASICs. This family is enabling leading-edge system architectures in networking applications, storage systems, wireless base stations, embedded systems, professional broadcast, and digital signal processing (DSP) systems.

Key features of the Virtex-II Pro family include:

- **The power of Xtreme Processing** – Each PowerPC runs at up to 400 MHz delivering 600+ Dhrystone MIPS, and is supported by IBM CoreConnect™ bus technology. With the unique Xilinx IP-Immersion architecture, system architects can now harness the power of high-performance processors and the easy integration of soft IP into the industry's highest performance programmable logic.
- **XtremeDSP** – The Xilinx XtremeDSP solution is the world's fastest programmable DSP solution. And XtremeDSP is the industry's premier programmable solution for enabling TeraMAC/s applications. It provides up to 556 embedded 18 x 18 multipliers, 10 Mbits of embedded block RAM, an extensive library of DSP algorithms, and tools that include System Generator for DSP, ISE, and Cadence SPW.
- **An Ultimate Connectivity Platform** – The Virtex-II Pro series is the first programmable device to combine embedded processors along with 3.125 Gb/s transceivers. It addresses all existing connectivity requirements as well as the emerging high-speed interface standards. Xilinx RocketIO transceivers offer a complete serial interface solution supporting 10 Gigabit Ethernet with XAUI, PCI Express, SerialATA, and more. SelectIO™-Ultra supports 840 Mb/s LVDS and high-speed single-ended standards such as SPI-4.2, XSBI, and SFI-4.
- **The Power of Integration** – In a single off-the-shelf programmable device, systems architects can take advantage of microprocessors, the highest density of on-chip memory, multi-gigabit serial transceivers, digital clock managers, on-chip termination, and more. The result is a dramatic simplification of board layout, a reduced bill of materials, and unbeatable time-to-market.

- **Enabling a New Development Paradigm** – For the first time, systems designers can partition and repartition their systems between hardware and software at any time during the development cycle—even after the product shipped. That means designers can optimize the overall system to guarantee performance targets in the most cost-efficient manner. Designers can also simultaneously debug hardware and software at speed.
- **Industry-Leading Tools** – Optimized for the PowerPC, Wind River's industry-proven embedded tools are the premier support for real-time microprocessor and logic designs. Driving the Virtex-II Pro FPGA is lightning-fast Xilinx ISE software, the most comprehensive and easy-to-use development system available.

In addition, Virtex-II Pro EasyPath™ devices enable up to an 80% cost reduction compared to the standard FPGA device support the Virtex-II Pro family. And there is no conversion risk to the customer. This permits the customer to move to higher unit volumes without switching to ASICs. Virtex-II Pro EasyPath devices are FPGAs that have been custom tested for a specific customer application. Virtex-II Pro Series EasyPath devices leverage specialized testing methods to provide the exact same FPGA used in prototyping an application-specific form to lower unit costs in volume production. These devices are available only for the highest density members of the Virtex-II Pro family (range from XC2VP30 to XC2VP125). Customers using these devices must meet certain minimum order requirements. The EasyPath solution gives users the easiest, lowest-cost, lowest-risk, and most cost effective path for FPGA volume production. These devices are identical in every way to the customers Virtex-II Pro FPGA prototypes but are tested by Xilinx to meet individual design specifications. Therefore, they are much less expensive to manufacture and provide a total cost management solution without custom design conversion problems. The Virtex-II Pro Series EasyPath provides a no-risk, no-effort cost reduction by:

- No timing or pinout changes
- Guaranteed to work in the system just like the Virtex-II Pro FPGA prototypes
- 25-80% cost reduction
- Custom charge much lower than ASIC NRE
- Faster than an ASIC conversion.

A Virtex-II Pro Series EasyPath solution eliminates design conversion without losing feature support, part verification, design qualification, or board redesign. Designers can conserve valuable engineering resources and reduce cost reduction times.

Table 19.5 shows the Virtex-II Pro FPGA family members.

Table 19.5: Virtex-II Pro FPGA Family Members

Device	RocketIO Transceiver Blocks	PowerPC Processor Blocks	Logic Cells[1]	CLB (1=4 slices = max 128 bits)		18 X 18 Bit Multiplier Blocks	Block SelectRAM+		DCMs	Maximum User I/O Pads
				Slices	Max Distr RAM (Kb)		18 KB Blocks	MaxBlock RAM (Kb)		
XC2VP2	4	0	3,168	1,408	44	12	12	216	4	204
XC2VP4	4	1	6,768	3,008	94	28	28	504	4	348
XC2VP7	8	1	11,088	4,928	154	44	44	792	4	396
XC2VP20	8	2	20,880	9,280	290	88	88	1,584	8	564
XC2VP30	8	2	30,816	13,696	428	136	136	2,448	8	644
XC2VP40	0[2] or 12	2	43,632	19,392	606	192	192	3,456	8	804
XC2VP50	0[2] or 16	2	53,136	23,616	738	232	232	4,176	8	852
XC2VP70	16 or 20	2	74,448	33,088	1,034	328	328	5,904	8	996
XC2VP100	0[2] or 20	2	99,216	44,096	1,378	444	444	7,992	12	1,164
XC2VP125	0[2], 20, or 24	4	125,136	55,616	1,738	556	556	10,008	12	1,200

Figure 19.8 shows the generic architecture of the Virtex-II Pro family.

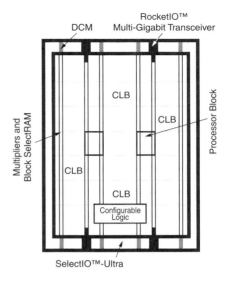

Figure 19.8:
Generic Virtex-II Pro architecture

In November 2003, Xilinx announced Virtex-II ProX platform FPGAs. These devices are the first in the industry to incorporate 10.3125 Gb/s serial transceivers. The two devices: 2VPX20 and 2VPX70, embed up to 20 Rocket IOX transceivers ranging from 2.488 Gb/s to 10.3125 Gb/s per channel. They also embed up to two PowerPC 405 processors.

Spartan-3 FPGAs

The Spartan™ FPGA families are based on the following basic philosophies:

- Use common Xilinx FPGAs architectures. This leverages all of the work in synthesis, software, and cores designed for that base architecture.
- Use the most advanced process technology available in order to combine the smallest possible die size and lowest cost. This approach is consistent with operating the transistors at a slightly larger geometry and higher voltages in order to maintain voltage compatibility with existing systems.
- Focus on total cost management throughout the entire manufacturing flow. Most of the cost savings come from areas such as wafer sort, test, assembly, factory logistics, and overhead. In the Spartan series the die cost is less than half of the cost equation for the device sizes offered.

In April 2003, Xilinx introduced the Spartan-3 FPGA family. It was the first PLD family shipping on 90-nanometer process technology. The Spartan-3 family consists of eight devices with densities up to 5 million system gates operating at 1.2V. Spartan-3 FPGAs provide the lowest cost per gate and lowest cost per I/O compared to any other programmable logic in the marketplace. These devices are programmable alternatives to gate arrays and ASICs that address a large range of cost-sensitive high volume applications. Built on four generations of proven Spartan success (Spartan, Spartan-XL, Spartan-II, and Spartan-IIE) in high-volume consumer and automotive applications, this new family of Platform FPGAs delivers unprecedented density, scalability, features, field reprogrammable functionality, and competitive price points. It also provides a platform to tackle tough connectivity, DSP, and embedded design problems.

With Spartan-3 FPGAs customers can develop logic for many new consumer, networking, and automotive applications including blade servers, low-cost routers, and medical imaging devices. Customers can get products to market faster because the system is manufactured with the same FPGA used to develop and prototype the design. This is the alternative design engineers have been

533

looking for to combat long ASIC development cycles and high ASIC NRE costs. To support an industry requirement that has grown to over $20 billion, Xilinx set the goal of low cost, unprecedented density range, and high capability for the Spartan-3 family of FPGAs that can be used both for prototyping and production of high-volume products. To meet that aggressive target Xilinx developed this Platform FPGA architecture with a small die size, a wide logic density range, and sufficient I/Os to support the mid-range of gate array and standard cell logic designs.

Spartan-3 Platform FPGAs offer a complete solution of logic, memory, and I/Os along with features such as multipliers to provide a rich fabric to support connectivity, DSP, and embedded design applications:

- **Spartan-3 Connectivity Solution** – Many design problems revolve around getting data on and off the FPGA. The best way to ease these connectivity problems is a robust I/O solution. All Spartan-3 I/O pins support the Xilinx SelectIO-Ultra functionality, dramatically increasing design flexibility. Each user I/O pin can support any of the 23 electrical interface standards. The abundant I/Os in the Spartan-3 FPGA family provide the lowest cost per parallel interconnect available in the industry. The differential I/O standards assist in achieving higher performance, lower power consumption, and lower pin count. These standards are supported by soft IP building blocks for important new interface protocols such as PCI 32/33 and PCI 64/33, RapidIO™, POS PHY Level 4, FlexBus 4, SPI-4, and HyperTransport™. To further simplify the design process and reduce costs the Spartan-3 FPGAs support XCITE digitally controlled impedance technology for both single-ended and differential I/O standards. XCITE technology provides I/O impedance matching that adapts to changes in supply voltage and temperature. XCITE technology reduces board cost by eliminating the need for most external termination resistors. And XCITE technology increases signal integrity.

- **Low-Cost Points for High-Performance DSP** – The abundant RAM resources combined with 18x18 multipliers give Spartan-3 FPGAs a significant DSP capability that can deliver up to 330 billion MACs per second. This economical and robust performance makes Spartan-3 devices a cost-effective solution for low-cost applications such as digital communications, video and imaging, and industrial control. In addition, the partnership between Xilinx and MathWorks offer MATLAB™/Simulink™ software's simple and familiar design flow to design the DSP functions.

- **Industry's Lowest Cost Soft Processor Solution** – The high feature set of the Spartan-3 FPGA family provides a low-cost platform for embedding the Xilinx MicroBlaze™ soft processor. Designers can combine the MicroBlaze processor and available peripherals with Spartan-3 logic, memory, and I/Os, to develop a custom embedded design for specific requirements. This increases on-chip integration, reduces board cost, and minimizes the risk of vendors choosing to obsolete a specific microcontroller.

- **Industry Leading Features** – Up to 1.8 Mb of block RAM can be used for large FIFOs and data storage. Abundant distributed RAM, implemented from logic cells, can be used for small FIFOs, shift registers, and constant coefficients. This combination of block RAM and distributed RAM provides a unique flexibility found only in Xilinx FPGAs. Digital clock managers (DCMs) and pre-engineered clock networks facilitate the design of high-speed systems. High-speed multipliers are embedded for DSP applications such as adaptive filters and forward error correction. Programmable SelectIO-Ultra supports 23 I/O standards and provides a low-cost bridge between different I/O standards. XCITE digitally controlled impedance technology supports increased signal integrity for both single-ended and differential I/Os. The result is a cost-optimized FPGA fabric that is flexible, scalable, and able to support many designs.

- **Xilinx Complete Development Support** – Spartan-3 FPGAs also come fully supported in the current Xilinx ISE 6.1i software design suite. With more than 150,000 installed users ISE 6.1i is the most prevalent design methodology in the industry. ISE includes a full spectrum of easy-to-use productivity options that can be inserted to fit in almost any existing corporate logic design methodology. Xilinx ISE tools increase productivity and slash design times by as much as 50% when compared to ASIC methodologies. The Xilinx ChipScope™ Pro debugging environment delivers unmatched power to uncover critical bottlenecks and optimize designs. Xilinx offers a complete hardware and software solution with the Xilinx Embedded Design Kit (EDK) for engineers using the MicroBlaze processor to develop a field programmable controller. Software tools similar to those used for IBM PowerPC designs are available to configure peripherals, develop protocols, and integrate the logic. Design can be verified in the system at speed in the Spartan-3 FPGA. This provides an easy solution for today's complex designs.

The 1.2V Spartan-3 FPGA family is specifically designed to meet the needs of high volume, cost-sensitive consumer electronic applications. As shown in Table 19.6, the eight-member family offers densities ranging from 50,000 to five million system gates. The Spartan-3 family builds on the success of the earlier Spartan-IIE family by increasing the amount of logic resources, the capacity of internal RAM, the total number of I/Os, and the overall level of performance. It also features improved clock management functions. Numerous enhancements derive from state-of-the-art Virtex™-II technology.

Table 19.6: Summary of Spartan-3 FPGA Attributes

Device	System Gates	Logic Cells	CLB Array (One CLB = Four Slices)			Distributed RAM (bits[1])	Block RAM (bits[1])	Dedicated Multipliers	DCMs	Maximum User I/O	Maximum Differential I/O Pairs
			Rows	Columns	Total CLBs						
XC3S50	50K	1,728	16	12	192	12K	72K	4	2	124	56
XC3S200	200K	4,320	24	20	480	30K	216K	12	4	173	76
XC3S400	400K	8,064	32	28	896	56K	288K	16	4	264	116
XC3S1000	1M	17,280	48	40	1,920	120K	432K	24	4	391	175
XC3S1500	1.5M	29,952	64	52	3,328	208K	576K	32	4	487	221
XC3S2000	2M	46,080	80	64	5,120	320K	720K	40	4	565	270
XC3S4000	4M	62,208	96	72	6,912	432K	1,728K	96	4	712	312
XC3S5000	5M	74,880	104	80	8,320	520K	1,872K	104	4	784	344

These Spartan-3 enhancements are combined with advanced process technology to set new standards in the programmable logic industry. They deliver more functionality and bandwidth per dollar than was previously possible. Because of their exceptionally low cost, Spartan-3 FPGAs are ideally suited to a wide range of consumer electronics applications including broadband access, home networking, and display/projection and digital television equipment. The Spartan-3 family is a superior alternative to mask programmed ASICs. FPGAs avoid the high initial cost, lengthy development cycles, and the inherent inflexibility of conventional ASICs. Unlike ASICs, FPGA programmability permits design upgrades in the field without necessary hardware replacement.

Key features of the Spartan-3 family include a Revolutionary 90-nanometer process technology.

By using a new 90 nm process technology Xilinx integrated a large logic density range with several industry leading features that reduced costs. The Spartan-3 FPGA family is the first programmable logic built with this process technology that leads to die sizes 50% to 80% smaller than any competing solution. Xilinx also uses 300 mm wafer technology to deliver the efficiency and capacity needed to produce high-volume Spartan-3 FPGAs. These 300 mm wafers deliver almost 2.5 times the number of die compared to the mature 200 mm wafers at a fraction of the cost per die.

Xilinx also implemented a dual wafer fabrication strategy using world-class foundries at both IBM and UMC to reduce production risks and increase capacity. This process technology leadership makes Xilinx the world's lowest-cost, lowest-risk FPGA provider. The Xilinx investment in design optimization, 90 nm manufacturing technology, and 300 mm wafer production delivers price points for Spartan-3 FPGAs at under $20 for a 1M-gate FPGA and under $100 for a 4M-gate FPGA. Additional features of the Spartan-3 family include:

- Very low cost, high-performance logic solution for high-volume, consumer-oriented applications
- Densities as high as 74,880 logic cells
- 326 MHz system clock rate
- Three separate power supplies for the core (1.2V), I/Os (1.2V to 3.3V), and special functions (2.5V)
- SelectIO signaling
- Up to 784 I/O pins
- 622 Mb/s data transfer rate per I/O
- Seventeen single-ended signal standards
- Six differential signal standards including LVDS
- Termination by Digitally Controlled Impedance
- Signal swing ranging from 1.14V to 3.45V
- Double Data Rate (DDR) support
- Logic resources
- Abundant, flexible logic cells with registers
- Wide multiplexers
- Fast look-ahead carry logic
- Dedicated 18 x 18 multipliers
- JTAG logic compatible with IEEE 1149.1/1532 standards
- SelectRAM™ hierarchical memory
- Up to 1,872 Kbits of total block RAM
- Up to 520 Kbits of total distributed RAM
- Digital Clock Manager (up to four DCMs)
- Clock skew elimination
- Frequency synthesis
- High resolution phase shifting
- Eight global clock lines and abundant routing
- Fully supported by Xilinx ISE development system
- Synthesis, mapping, placement and routing

Architectural Overview

The Spartan-3 family architecture consists of five fundamental programmable functional elements:

- **Configurable Logic Blocks (CLBs)** – Contain RAM-based Look-Up Tables (LUTs) to implement logic and storage elements that can be used as flip-flops or latches. CLBs can be programmed to perform a wide variety of logical functions as well as to store data.
- **Input/Output Blocks (IOBs)** – Control the flow of data between the I/O pins and the internal logic of the device. Each IOB supports bi-directional data flow plus 3-state operation. Twenty-three different signal standards, including six high-performance differential standards, are available as shown in Table 19.7. Double Data-Rate (DDR) registers are included. The Digitally Controlled Impedance (DCI) feature provides automatic on-chip terminations, simplifying board designs.
- **Block RAM** – Provides data storage in the form of 18-Kbit dual-port blocks.
- **Multiplier Blocks** – Accept two 18-bit binary numbers as inputs and calculate the product.
- **Digital Clock Manager** – (DCM) blocks provide self-calibrating, fully digital solutions for distributing, delaying, multiplying, dividing, and phase shifting clock signals.

The SelectIO feature of Spartan-3 devices supports 17 single-ended standards and six differential standards as listed in Table 19.7.

Table 19.7: Signal Standards Supported by the Spartan-3 Family

Standard Category	Description	V_{CCO} (V)	Class	Symbol
Single-Ended				
GTL	Gunning Transceiver Logic	N/A	Terminated	GTL
			Plus	GTLP
HSTL	High-Speed Transceiver Logic	1.5	I	HSTL_I
			III	HSTL_III
		1.8	I	HSTL_I_18
			II	HSTL_II_18
			III	HSTL_III_18
LVCMOS	Low-Voltage CMOS	1.2	N/A	LVCMOS12
		1.5	N/A	LVCMOS15
		1.8	N/A	LVCMOS18
		2.5	N/A	LVCMOS25
		3.3	N/A	LVCMOS33
LVTTL	Low-Voltage Transistor-Transistor Logic	3.3	N/A	LVTTL
PCI	Peripheral Component Interconnect	3.0	33 MHz	PC133_3
SSTL	Stub Series Terminated Logic	1.8	N/A	SSTL18_I
		2.5	I	SSTL2_I
			II	SSTL2_II
Differential				
LDT	Lightning Data Transport (HyperTransport™)	2.5	N/A	LDT_25
LVDS	Low Voltage Differential Signaling		Standard	LVDS_25
			Bus	BLVDS_25
			Extended Mode	LVDSEXT_25
			Ultra	ULVDS_25
RSDS	Reduced-Swing Differential Signaling	2.5	N/A	RSDS_25

These elements are organized as shown in Figure 19.9. A ring of IOBs surrounds a regular array of CLBs. The XC3S50 has a single column of block RAM embedded in the array. Those devices ranging from the XC3S200 to the XC3S2000 have two columns of block RAM. The XC3S4000 and XC3S5000 devices have four RAM columns. Each column is made up of several 18K-bit RAM blocks with each block associated with a dedicated multiplier. The DCMs are positioned at the ends of each block RAM column. The Spartan-3 family features a rich network of traces and switches that can transmit signals by the interconnecting of all five functional elements. Each functional element has an associated switch matrix that permits multiple connections to the routing.

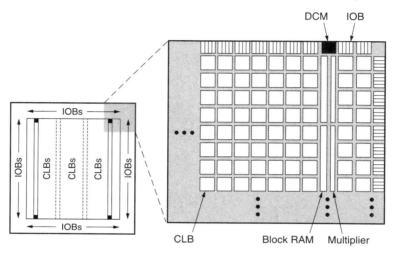

Figure 19.9: Spartan-3 Family Architecture

Configuration

Spartan-3 FPGAs are programmed by loading configuration data into robust static memory cells that collectively control all functional elements and routing resources. Prior to powering on the FPGA configuration data is stored externally in a PROM or some other nonvolatile medium either on or off the board. After applying power the configuration data is written to the FPGA using any of five different modes; Master Parallel, Slave Parallel, Master Serial, Slave Serial, and Boundary Scan (JTAG). The Master and Slave Parallel modes use an 8-bit wide SelectMAP Port. The recommended memory for storing the configuration data is the low-cost Xilinx Platform Flash PROM family that utilizes XCF00S PROMs for serial configuration and XCF00P PROMs for parallel configuration.

I/O Capabilities

Spartan-3 FPGAs offer a remarkably small die for the number of logic gates using 90 nm process technology. This smaller die has a smaller perimeter for the necessary number of I/O pads causing a potential problem for an application that demands a high I/O count. However, by using staggered pad technology, Xilinx is able to deliver two rows of I/Os on each edge of the die versus one row found in other FPGAs and ASIC alternatives. Users get more I/Os in a smaller die resulting in a low cost per I/O and low cost per gate for Spartan-3 FPGAs.

Spartan-3 FPGAs can provide functionality and performance comparable to competing ASSPs combined with the flexibility required in today's consumer applications. Spartan-3 FPGAs provide a high-bandwidth communication interface to external devices such as ASSPs, CPUs, memory, and backplanes. This makes them perfect programmable system integration and interface vehicles. This eliminates the need for custom logic, expensive clock management schemes, and various translator components including a direct interface to backplanes such as GTL I/O capability. The Spartan-3 family integrates simple ASSP functions such as PLLs, FIFOs, level translators, and others that can result in considerable cost savings. Spartan-3 FPGAs, with the low-cost, high-density, and increased feature content are ideal for evolving markets.

The dynamic consumer market is demanding that products be brought to market at competitive prices. Broadband access devices, residential gateways, home networking products, and information appliances are products that have existed for some time. Specifications and standards for these devices continue to evolve as designers develop new products. While developing ASSPs takes 12-18 months, a Spartan-3 FPGA programmed with the appropriate IP can perform ASSP functionality. A unique role for the Spartan-3 FPGA is providing an interface to the different ASSPs. This technology bridging function is crucial in the home networking market where multiple technologies and ASSPs combine with different interfaces.

Spartan-3 FPGAs facilitate and enable home networking solutions because FPGAs are reprogrammable. This offers several advantages over ASICs and ASSPs:
- No manufacturing lead time for the silicon helps reduce time-to-market
- Reprogrammability means bug fixes and updates to standards can be easily implemented ensuring longer time-in-market

Xilinx FPGAs are the superior solution featuring industry leading speed, density, power, and cost features. Xilinx FPGAs and core solutions allow designers to turn the chaos in the home networking market into an opportunity to:
- Get to market faster (no hardware lead time)
- Make changes to product features (reprogrammability)
- Update a product to a new standard faster (reprogrammability)
- Increase customer satisfaction and thereby increase market share by utilizing field upgrades to fix bugs, add new features, or fit a changed standard
- Use core solutions to speed the design time of new products

The inherent flexibility that programmable logic offers allows product customization to meet customer needs. This includes the adaptation to specification updates, feature upgrades, low risk evaluation of new markets, and the ability to remotely upgrade products already in the field.

As specifications change traditional ASSP vendors do not have the ability to provide an ASSP to address the dynamic nature of an application. The Xilinx FPGAs ability to be reprogrammed makes them ideal for this purpose. Designing an FPGA within a system can save costs and decrease board size significantly.

An example describing the value of low cost programmability is using a Xilinx FPGA instead of a PCI ASSP. A standard design might use many chips such as external PLD, external DLLs, memory, memory controllers, I/O translators, and a PCI master and slave interface PCI ASSP. A design using a Spartan-3 FPGA with a PCI LogiCORE™ IP can replace many chips with a single Spartan-3 providing lower costs and a decreased board size.

Configuration Solutions

Xilinx offers a wide range of configuration solutions for Xilinx FPGAs and programming solutions for Xilinx CPLDs and PROMs.

FPGA configuration options include:

- **Platform Flash PROMs** – Single solution for configuring all Xilinx FPGAs with densities up to 32 Mb. Platform Flash PROMs offer such features and benefits as:
- **Lowest cost per Megabit Configuration PROM**
- **Smallest area per Megabit** – VO20 and the FS48 packages reduce configuration board space.
- **One PROM family to configure all Xilinx FPGAs** – One family, from 1 Mbit to 32 Mbit, simplifies manufacturing flow and reduces inventory cost
- **In-System Programmability makes design changes easy** – Simplifies manufacturing flow and board test during development and verification by supporting on-board programming. Enables easy field upgrades with PROM re-programming.
- **Increase effective PROM density using the Xilinx patented compression technology in the high density-range of devices** – This provides for storage of up to 50% more bits and allows the use of a smaller-density, lower-cost PROM.
- **Supported in Xilinx ISE iMPACT Tool** – Lowers configuration costs.
- **System ACE** – Used for system level solutions, high density, and multiple FPGAs. Xilinx developed the System Advanced Configuration Environment (System ACE) configuration manager—a space-efficient, pre-engineered, high-density configuration solution for multi-FPGA systems. The System ACE configuration manager is a flexible two-piece configuration solution comprised of the ACE Flash™ module and the ACE Controller™ chip. The ACE Flash interface accommodates removable CompactFlash (64 Mb to more than 1 Gb) modules or the IBM Microdrive (2 Gb to 8 Gb). All have the same form factor and board space requirements.
- **Legacy PROMs** – Single solution for in-system and one-time programming configuration for densities up to 16Mb. Xilinx offers a range of one time-programmable and in-system programmable storage devices to configure Xilinx FPGAs. The 17xx family is a one time programmable solution that ranges in density up to 16 megabits. The 18xx family is in-system programmable flash PROM (programmable read only memory) that ranges in density up to 4 megabits. Xilinx PROM solutions continue to offer higher densities and low cost and targets for all FPGA designs.
- **ISE iMPACT** – A full featured software tool used to configure and program all Xilinx PLDs (FPGAs and CPLDs) and PROMs. It features a series of «wizard» dialogs that easily guide the user through every step of the configuration process:
- **Mode Shifting** – Users can quickly shift among the various programming modes.
- **Visual Feedback** – Users receive instant visual feedback on all operations and testing.
- **Output File Types** – Supports a host of output file types including STAPL and SVF.
- **Desktop Programmers and Download Cables** – Recommended for prototyping configuration of all Xilinx FPGAs using MultiPRO, PCIV, MultiLINX. This solution provides low-cost prototype solutions for all FPGAs, CPLDs, and PROMs. It also provides ease of setup, use at an Engineer's lab bench or desk, a debug environment, and in-system and stand-alone programming for single or multiple devices

CPLD and PROM programming options include:

- **Desktop Programmers and Download Cables** – Recommended for prototyping
- **Configuration of all Xilinx FPGAs** – Using MultiPRO, PCIV, or MultiLINX.
- **Xilinx HW-130** – Legacy solution for stand-alone programming of XC9500/XL/XV CPLDs, XC1700, and XC18V00.
- **J-Drive Free Download Software** – Program embedded solutions for IEEE 1532 compliant device
- **Automatic Test Equipment and Boundary Scan Tools** – Recommended for volume manufacturing
- **Embedded solutions configuration** – Xilinx has several embedded programming and download solutions using an FPGA, CPLD, or Microprocessor as a controller.

IQ Solutions

Xilinx has created a new family of devices with an extended industrial temperature range option called the IQ Solutions. These solutions provide all of the flexibility, IP cores, performance, and features needed for today's in-cabin automobile telematics systems in a programmable, extended-temperature-range product. Xilinx IQ Solutions comprise silicon, leading-edge design software, an extensive IP core library, reference designs, design services, technical support, and customer education.

The IQ Solutions silicon offering consists of all Xilinx FPGAs and CPLDs that meet the standard industrial (I) temperature range and a select group of devices that have been tested to an additional extended (Q) temperature range. IQ Solutions are ideal in automotive and industrial applications where extended temperature ranges and high quality products are required.

Q-grade devices from Xilinx have full mask set control and fabrication, assembly and test flow consistency. This provides product consistency over time. Xilinx is committed to the provision of semiconductors for the automotive industry and all of its production partners are qualified to QS-9000.

IP Cores, Software and Services

Software

Xilinx's software tools are a key part of the company's programmable logic solutions. Xilinx has a cumulative number of 175,000 users/installed software seats at customers worldwide. Xilinx offers complete software solutions that enable customers to implement their design specifications into Xilinx PLDs. Regardless of the designer's experience level these software design tools combine a powerful technology with a flexible, easy to use graphical interface to help achieve the best possible designs within each customer's project schedule. Xilinx software design tools operate on personal computers running Microsoft Windows 2000, XP and Linux operating systems, and on Solaris-based workstations from Sun Microsystems.

The ISE (Integrated Software Environment) family from Xilinx is a leading programmable design tool that delivers the lowest cost in logic design and production. ISE lowers total system costs by focusing on maximum design performance and ease-of-use to solve engineering problems. This leads to the fastest time-to-market available for logic design. ISE comes in four configurations that offer simplified design flows twice as fast as ASIC flows. Its performance dramatically exceeds competing PLD solutions. Optional design tools tackle everything from HDL simulation to embedded systems design. ISE also integrates with a wide range of third party electronic design automation

(EDA) software offerings and point-tool solutions to deliver the most flexible design environment available.

With their ISE development systems and development options Xilinx not only supports the benefits of PLDs but also offers additional cost savings. ISE 6.1i, the latest release of design software at the time of this writing, delivers a number of productivity technologies that shorten logic design flow, optimize design results, shorten implementation and verification cycles, and provide interactive design assistance. At the same time, ISE 6.1i enables users to realize even faster design performance. The end result is cost savings across the entire project. The shorter design cycles and time-to-market advantages of FPGAs and CPLDs translate to reduced engineering resources. This allows companies to make the best use of their staff when difficult economic conditions restrict the ability to hire more engineers. The fast, efficient, and highly productive Xilinx ISE software tools help get the job done in less time and make each engineer more productive.

The four ISE configurations are:

- **ISE Foundation** – Offers the most complete logic design environment for customers who desire one logic solution from a single vendor. ISE Foundation Series is a family of a fully integrated, ready-to-use Windows PC tools that support a broad range of Xilinx FPGA and CPLD design requirements including the Virtex-II Pro, Virtex-II, Spartan-3, Spartan series, and CoolRunner-II families. Available at low cost, and targeted at entry-level as well as high-end users, the Foundation Series products leverage industry standard hardware description languages (HDLs), including Verilog/VHDL. The Windows-based Foundation Series software provides access to synthesis, schematic entry, gate level simulation, verification, and implementation tools. ISE Foundation provides a great design environment for anyone looking for a complete programmable logic design solution by combining its ultra-fast runtimes, ProActive Timing Closure technologies, and seamless integration with the industry's most advanced verification products.

- **ISE Alliance** – Includes the industry's most advanced timing driven implementation tools available for programmable logic design. It is tailored for customers who want maximum design flexibility by integrating ISE into their existing EDA environment and methodology. Through its Alliance Series software, Xilinx has chosen an open system approach that allows its customers to pick the highest quality and widest variety of design and programming tools available on the market today for logic design, logic verification, high-level language design, high-speed PCB design, PCB design productivity, co-verification and co-design, and ASIC emulation/partitioning. To accomplish this Xilinx has established engineering and marketing relationships with leading third-party electronic design automation (EDA) software suppliers such as Aldec, Cadence, Celoxica, Math Works, Mentor Graphics, Model Technology, Synopsys, and Synplicity. The result has been the creation of complementary technology and tightly integrated third-party links with the Xilinx Alliance Series backend place-and-route software for FPGAs and CPLDs.

- **ISE BaseX** – Provides the industry's most cost-effective and full featured programmable logic design environment. ISE BaseX™ also includes the industry's most advanced timing driven implementation tools available for programmable logic design along with design entry, synthesis, and verification capabilities. Matching ultra fast runtimes, ProActive Timing Closure technologies, and seamless integration with the industry's most advanced verification products make ISE BaseX a great design environment for anyone looking for a complete programmable logic design solution at an affordable price.

- **ISE WebPACK** – This design suite's free Web-downloadable design and implementation modules are available for customers who only use smaller devices and a minimal set of design tools. All of the Xilinx FPGA and CPLD device families, including the newest device families CoolRunner-II, Spartan-3 and Virtex-II Pro, are supported by ISE. WebPACK offers a complete development environment with modules from ABEL and HDL synthesis to device fitting and JTAG programming. ISE WebPACK tools, a subset of their award-winning ISE Foundation design tools, provide instant access to the ISE tools at no cost. Xilinx has created a solution that allows instant productivity by providing a design solution that is always up-to-date, with error-free downloading, and single file installation. Because ISE WebPACK development tools are available for download from the Xilinx website at www.xilinx.com/ise/webpack5, users can immediately start work on designs for Xilinx CPLDs and mid-density FPGAs. This Web-downloadable design solution reduces costs by including all the required tools.

Xilinx Development System Options deliver specific benefits through optional tools that integrate seamlessly with the ISE Design Environment. These options include an HDL simulation tool, an in-system verification tool, an environment for the design of embedded programmable systems, a high-level language compiler, and a plug-in to the popular MATLAB/Simulink products from The MathWorks, Inc. The Development System Options includes the following features:

- **ModelSim Xilinx Edition (MXE-II)** – A complete HDL simulation environment that has been optimized for programmable logic design. This enables users to quickly verify source code, functional and timing models designs, and HDL source code.
- **ChipScope Pro** – The size, speed, and board requirements of today's state-of-the-art FPGAs make it nearly impossible to debug designs using traditional logic analysis methods. Flip-chip and ball grid array packaging do not have exposed leads that can be physically probed. The Xilinx ChipScope Pro solution embeds logic analyzer (ILA) and bus analyzer (IBA) low-profile software cores into their design. These cores allow you to view all the internal signals and nodes within their FPGA. This includes the IBM CoreConnect Processor Local Bus (PLB) or On-Chip Peripheral Bus (OPB) supporting the IBM PowerPC 405 located inside the industry-leading Virtex-II Pro FGPA. With the Agilent trace core (ATC) and direct interface to the Agilent FPGA Trace Port Analyzer, ChipScope Pro gives users even deeper trace memory, faster clock speeds, and more trigger options using fewer pins on the FPGA. This unique partnership from Xilinx and Agilent delivers more real-time debug power than any other solution on the market.
- **Xilinx System Generator for DSP** – Bridges the gap between the high-level abstract version of a design and its actual implementation in a Xilinx FPGA. The System Generator for DSP enables designers to develop high-performance DSP systems for Xilinx FPGAs using the popular MATLAB/Simulink products from The MathWorks, Inc.
- **The ISE Embedded Development Kit (EDK)** – An all-encompassing solution for designing embedded programmable systems design. It also supports designs of processor sub-systems using the IBM PowerPC hard processor core and the Xilinx MicroBlaze soft processor core. Hardware and software development tools combined with the advanced features of Xilinx FPGAs, especially the Virtex-II Pro the Platform for Programmable

Systems, provides users with a new level of system design. This allows users to optimize their design performance at any time during the design cycle to meet fast-changing design requirements.

- **The Forge Compiler** – Allows designers to work at a higher level of abstraction than standard HDLs. Designers can use standard Java to produce designs with fewer lines of code that are both efficient and maximize performance.
- **HDL Bencher** – Included in the ISE WebPACK toolset, the HDL Bencher™ test bench generator automatically imports the current HDL design file and by default creates an editable stimulus waveform.
- **StateCAD** – The StateCAD FSM wizard automates the state machine design process. Users can specify complex state machines to quickly meet tough product requirements. The state machines can then be automatically translated to an HDL format and included in the design flow.
- **ChipViewer** – ChipViewer is a pre- and post-fit graphical utility to assign or view pin placement and implemented logic for all Xilinx CPLD devices removing risks associated with late design process changes.
- **Xpower** – XPower is a graphical power-analysis tool. Total device power, power per-net, fitted, routed, partially routed, or un-routed designs can be easily analyzed.
- **WebFITTER** – A free Web-based CPLD design fitting tool. WebFITTER allows system designers to evaluate their designs using the XC9500 and CoolRunner series CPLDs. Fitting results are produced using the latest version of Xilinx software. Users can also receive an immediate price quote for the targeted device.

ISE 6.1i is the most cost-effective logic design suite available today. The Xilinx ISE design suite bridges the gap between today's high-performance, high-demand systems and the design productivity and ease-of-use requirements to meet and even exceed schedule demands. ISE slashes design and verification times by getting products to market ahead of the competition without sacrificing end-product performance. ISE version 6.1i delivers unique high-speed design capabilities such as new timing constraints. For example, the new clock jitter constraint has the ability to specify the true data valid window. And the addition of individual package pin flight time reporting provides more accurate place and route results for source synchronous designs. Designers using Virtex-II and Virtex-II Pro FPGAs have access to 96 local high-speed clocks that can be routed to keep clock skew within the range required for 200 MHz SDR, DDR, and QDR RAM interfaces. ISE 6.1i includes an enhanced mapping feature for ProActive Timing Closure that performs logic placement so that mapping decisions can be made based on physical location. This results in an additional 13% faster clock speed and 23% better utilization compared to ISE 5.2i. Also, for push-button flows, ISE 6.1i delivers a 16% performance improvement over ISE 5.2i.

Key advantages include:

- **Lower Design Costs and Higher Productivity** – ISE 6.1i advanced technology lowers design costs by slashing design times and gives users streamlined designer efficiency and productivity. ISE slashes time from the logic design flow with up to 2X faster design and verification times than ASIC design methodologies. ISE is still the performance leader in programmable systems with up to 20% better performance and 15% better logic utilization than the nearest competitor. This kind of performance can deliver a "virtual speed-grade" in device cost advantage. The Xilinx innovative implementation technology ProActive Timing Closure sets new standards for design performance with clock speeds of over 400 MHz using Virtex-II Pro FPGAs. Faster times save both time and money. Productivity tools such as Incremental Design slash re-implementation times letting design efforts focus where they're needed most.

■ **Support for Next Generation Logic Devices-Software Before Silicon** – ISE delivers support for upcoming leading-edge Xilinx devices so designs will be ready when the silicon becomes available. ISE 6.1i contains support for all leading-edge Xilinx devices. ISE supports enhanced Multi-Gigabit high-speed I/O transceivers and the embedded IBM PowerPC processors in the Virtex-II Pro. ISE delivers the same high productivity whether targeting glue-logic CPLDs or high-performance FPGAs. ISE is a one-tool suite delivered in four different configurations that are designed to meet budget and methodology require-ments. Upgrading software is seamless with no need to re-learn tools.

■ **Total Verification** – ISE and Xilinx software design options deliver the most complete verification suite available in logic today. ISE is also packed with advanced software technology designed to accelerate the more time-consuming parts of the design and debug logic flow. Incremental Design is a technology included in ISE that shortens design re-compile times. By locking performance for areas of the design that don't need to change, Incremental Design lets users perform re-synthesis and re-place-and-route on only those pieces of the design that have to change. This reduction in time adds up fast where debug changes are common, such as in the crucial verification cycle. The Xilinx ChipScope Pro integrated logic analyzer also delivers added productivity to the verification cycle by offering real-time debug with capabilities that aren't possible in ASIC tools or competing FPGA tools. Using small, easy-to-place software debug cores the ChipScope Pro tool allows real-time monitoring of any signal in the FPGA. This includes the IBM PowerPC 405 peripheral bus in the advanced Virtex-II Pro FPGA. Design signals are captured and brought to the outside world through the FPGA JTAG programming port. This minimizes the amount of dedicated FPGA space and I/O pins. Additional debug time can be saved since signal monitor points can be changed through the ISE FPGA editor without having to re-compile the design. The ChipScope Pro analyzer cuts verification times dramatically whether the device is on the board or in the field. XPower provides accurate device power estimates based on the design's operation. This unique «preserved hierarchy» provides speeds twice as fast for HDL simulation in ISE 6.1i and other new improvements.

■ **Redefining Ease-of-Use** – "Ease-of-use" isn't only about having good menus and a nice look-and-feel. ISE solves system and design bottlenecks and helps engineers new to pro-grammable logic breeze through designs. PACE (Pinout and Area Constraints Editor) included in ISE, simplifies pin management and area definition design bottlenecks for both FPGAs and CPLDs through a graphical interface. Architecture Wizards help simplify the design of advanced device technology like the Virtex-II Pro high-speed MGT I/O or Digital Clock Managers. Modular Design implements a "divide-and-conquer" team-design ap-proach by breaking large designs down into individual modules for easier, and faster, completion. ISE is true high-speed design with the most complete timing constraints language available. And performance technology that delivers accurate, repeatable results. Xilinx and Synopsys have defined over 100 LEDA rules for FPGA design. This helps designers get their source code right the first time. PACE also supports the ability to enter pin definitions before an HDL source design exists. This eliminates the need to complete designs before going to PCB layout. PACE also supports comma-separated value (CSV) bidirectional file transfers resulting in better integration with PCB layout design tools.

Xilinx ISE design tools have raised the industry standard for both design performance and device utilization. ISE's patented implementation algorithms allow users to achieve the fastest possible design performance. Designs can achieve better than 15% higher performance than competing solutions. This performance edge means users can potentially target a lower cost device by leveraging faster performance from the software. This allows users to meet timing goals earlier and spend less time in the design flow. Based on benchmark data, users can achieve 20% to 30% better performance in Virtex-II Pro designs using ISE than from an offering from the leading competitor. In many cases users can target a design to a slower speed grade device and still achieve targeted design performance. ISE also reduces project costs by packing more logic into Virtex-II devices allowing the design to fit in the smallest possible device.

Advanced FPGAs are not solely made of look-up tables and flip-flops anymore. Today's logic fabrics are best described as "feature rich," with processors, multiple IO standards, large memory blocks, and more. This trend requires sophisticated algorithms in both synthesis and implementation tools to provide optimal performance and logic utilization by leveraging new hardware features. Xilinx ISE development tools separate unrelated functions and assign them to different clusters (called a slice) on the fabric. This avoids conflicting placement constraint and guarantees optimal performance. As the device fills, powerful algorithms pack unrelated logic into common clusters. This gradual process ensures that the device provides the best utilization with minimal impact to design performance.

Xilinx Silicon and Software: Replacing ASICs

Xilinx has proven software and silicon leadership has accelerated the industry-wide transition from ASICs to FPGA technology. Customers are looking for more flexible, low cost solutions in the face of uncertain market conditions, skyrocketing NRE and mask costs, and shortened time-to-market windows and product lifecycles. The new software suite ISE 6.1i, coupled with Virtex-II Pro and Spartan-3 FPGAs offering breakthrough price points, device densities, and performance offer designers an ideal ASIC replacement solution. Designers can now take "push button" advantage of the world's first 90nm FPGAs—the Spartan-3 family from Xilinx with up to 5 million system gates—to dramatically reduce overall design time and costs without the verification headaches traditionally associated with ASICs.

The Xilinx solution also provides designers with a superior solution to so-called «Structured ASICs» which are plagued by long development times and high upfront costs. They also lack the flexibility, inexpensive software tools, and the robust IP library available today with Xilinx FPGAs. The inherent reprogrammability of Xilinx FPGAs enables designers to get products to market more quickly by accelerating design debug and reducing overall support costs.

Intellectual Property (IP) Solutions: System-Level Designs for FPGAs

Today, a large number of predefined cores are available to implement system-level functions directly in Xilinx PLDs. This enables customers to shorten development time, reduce design risk and obtain superior performance for their designs. These IP cores available from Xilinx and third-party partners are commonly used to perform complex functions such as DSP, bus interfaces, processors, and processor peripherals. They allow designers to cut design time and significantly reduce risk while having access to the best performing and lowest cost components available. Full information about

Xilinx cores is available online from the IP Center Internet portal of the Xilinx website. Customers can purchase a license online for the latest intellectual property cores and reference designs.

The Xilinx CORE Generator™ System is the cataloging, customization, and delivery vehicle of the IP cores targeted to Xilinx FPGAs. This tool delivers highly optimized cores that are compatible with standard design methodologies for Xilinx FPGAs. This easy-to-use tool generates flexible, high-performance cores with a high degree of predictability and repeatability and allows customers to download future core offerings from the Xilinx website. Both Xilinx and independent IP developers can design cores for the CORE Generator tool. It also serves as a cataloging and delivery system for related collateral for all designers using Xilinx.

The CORE Generator provides centralized access to a catalog of ready-made IP functions ranging in complexity from simple arithmetic operators, such as adders, accumulators, and multipliers to system-level building blocks, such as filters, transforms, and memories. Cores can be displayed alphabetically by function, by vendor, or by type. Each core comes with its own data sheet that provides detailed documentation on the core's functionality.

The CORE Generator's user interface makes it very easy to access the latest Spartan-3 IP releases and to retrieve up-to-date information. Links are also included for various partner-supplied AllianceCORE products. The use of CORE Generator IP cores in Spartan-3 designs enables shorter design times. It also helps them realize high levels of performance and area efficiency without any special knowledge of the Spartan-3 architecture. Once the CORE Generator software is installed the designer gains immediate access to dozens of cores supplied by the LogiCORE program. Data sheets are available for all AllianceCORE products. Additional separately licensed advanced function LogiCORE products are also available. New and updated Spartan-3 IP for the CORE Generator can be downloaded from the IP Center and added to the CORE Generator catalog.

The CORE Generator works in conjunction with the Xilinx IP Center (www.xilinx.com/ipcenter). To make the most of this resource, Xilinx highly recommends that whenever starting a design to first do a quick search of the IP Center to see whether a ready-made core solution is already available. A complete catalog of Xilinx cores and IP tools resides on the IP Center including:

- LogiCORE Products
- AllianceCORE Products
- Reference Designs
- XPERTS Partner Consultants
- Design Reuse Tools-IP Capture tool

LogiCORE products are IP cores designed, sold, licensed, and supported by Xilinx. These IP cores are optimized for Xilinx FPGAs and CPLDs to provide the fastest speeds possible while utilizing minimum board area. LogiCORE products include a wide selection of generic parameterized functions such as muxes, adders, multipliers, and memory cores. These are bundled with the Xilinx CORE Generator software at no additional cost to licensed software customers. Optional system-level cores such as PCI, Reed-Solomon, ADPCM, HDLC, POS-PHY, and Color Space Converters are also available as separately licensed products. The CORE Generator is commonly used to quickly generate Spartan-3 block and distributed memories. A more detailed listing of available Spartan-3 LogiCORE products is available on the Xilinx IP Center website. Types of IP currently offered by the Xilinx LogiCORE program include:

- **Basic Elements** – Logic gates, registers, multiplexers, adders, and multipliers

- **Communications and Networking** – ADPCM modules, HDLC controllers, ATM building blocks, forward error correction modules, and POS-PHY Interfaces
- **DSP and Video Image Processing** – Cores ranging from small building blocks (e.g., Time Skew Buffers) to larger system-level functions (e.g., FIR Filters and FFTs)
- **System Logic** – Accumulators, adders, subtracters, complementers, multipliers, integrators, pipelined delay elements, single and dual-port distributed and block RAM, ROM, and synchronous and asynchronous FIFOs
- **Standard Bus Interfaces** – PCI Interfaces
- **Soft processors** – 32-bit MicroBlaze and 8-bit PicoBlaze™

The AllianceCORE program is a cooperative effort between Xilinx and third-party IP developers to provide additional system-level IP cores optimized for Xilinx FPGAs. To ensure a high level of quality, AllianceCORE products are implemented and verified in a Xilinx device as part of the certification process. While optimized for Xilinx devices AllianceCORE modules are sold and supported by a network of third-party developers. Xilinx develops relationships with AllianceCORE partners who can compliment the Xilinx LogiCORE product offering. AllianceCORE products range from processors and standard peripheral controllers to datacom and telecom functions.

AllianceCORE products include customizable cores that can be configured to exact needs as well as fixed netlist cores targeted toward specific applications. In many cases partners can provide cores customized to meet the specific design needs if the primary offerings do not fit the requirements. Source code versions of the cores are often available from the partners at an additional cost for designers requiring maximum flexibility.

Xilinx offers two types of design files, XAPP application notes developed by Xilinx and reference designs developed through the Xilinx Reference Design Alliance Program. Both types are extremely valuable to customers looking for guidance when designing systems. Application notes developed by Xilinx usually include supporting design files. These are supplied free of charge but without technical support or warranty. Reference designs often can be used as starting points for implementing a broad spectrum of functions in Xilinx programmable logic.

Reference designs developed through the Xilinx Reference Design Alliance Program are developed, owned, and controlled by the partners in the program. The goal of the program is to form partnerships with other semiconductor manufacturers and design houses to assist in the development of high-quality, multi-component reference designs that incorporate Xilinx devices and demonstrate how they can operate at the system level with other specialized and general-purpose semiconductors. The reference designs in the Xilinx Reference Design Alliance Program are fully functional and applicable to a wide variety of digital electronic systems. These designs are used for networking, communications, video imaging, and DSP applications.

Xilinx established the XPERTS Program to provide customers with access to a worldwide network of certified design consultants who are proficient with Xilinx FPGAs, software, and IP core integration. All XPERTS members are certified and have extensive expertise and experience with Xilinx technology in various vertical applications such as communications and networking, DSP, video and image processing, system I/O interfaces, and home networking. XPERTS partners are an integral part of the Xilinx strategy to provide customers with cost efficient design solutions while accelerating time-to-market.

To facilitate the archiving and sharing of IP created by different individuals and workgroups within a company, Xilinx offers the IP Capture Tool. The IP Capture Tool helps to package design modules created by individual engineers in a standardized format. They can then be cataloged and distributed using the Xilinx CORE Generator. A core can take the form of synthesizable VHDL or Verilog code or a fixed function netlist. Once it is packaged by the IP Capture Tool and installed into the CORE Generator, the captured core can be shared with other designers within a company via an internal network.

Smart-IP technology is a combination of several features designed to deliver the highest performance, predictability, and flexibility when implementing IP with Xilinx FPGAs. Smart-IP technology ensures constant core performance regardless of its position in the FPGA device; maintained performance when multiple cores are integrated in the same FPGA device; and no performance degradation when migrating to larger devices. The IP is built to utilize the unique features of the Spartan-3 architecture such as dedicated multiplier or multiplexor logic. The use of Smart-IP technology means that the performance of the core is independent of core placement, number of cores used, surrounding user logic, device size, and EDA tools.

Xilinx has multiple System Generator products, which include:

- **System Generator for DSP** – A software tool for quick modeling, design, and implementation of FPGA-based DSP systems in Simulink. The tool presents a high level abstract view of a DSP system, and automatically maps it to a faithful, optimized, and synthesizable hardware implementation through Xilinx and 3rd party developed basic and system-level DSP IP.
- **System Generator for PowerPC** – A tool that automatically generates a PowerPC based system on Virtex-II Pro with user selected soft peripherals and the IBM CoreConnect bus

Xilinx is currently developing several IP cores to address the digital consumer, video, and automotive markets. The Spartan-3 family has had very robust core support since the time of its introduction. Xilinx has a range of IP cores such as memory controllers, system interfaces, DSP, communications, networking, and microprocessors. The extensive Spartan-3 IP library includes the following cores:

- **BaseBlox** – UARTs, multipliers, and DMA
- **Memory/memory controllers** – SRAM and SDRAM
- **Networking and communications** – Cell assembler, cell delineation, CRC, T1 Framer, HDLC controllers, 10/100 Fast Ethernet, UTOPIA, ATM/IP over SONET, ADPCM, IMA, SONET, OC-48, OC-192, VoIP, Reed-Solomon FEC, Viterbi FEC, xDSL modems, cable modems, satellite modems, UMTS, and W-CDMA
- **Microcontrollers and microprocessors** – 8051, 8-bit, and 32-bit RISC processors
- **DSP** – FIR filters, comb filters, FFT, color space converter, DCT/IDCT video and image processing, DES, Triple DES, MP3, QAM, QPSK, JPEG, speech recognition, programmable DSP engines
- **System I/O and standard bus interfaces** – Two-wire serial interface, CAN (Car Area Network), ISA, I2C, PCI (32- and 64-bit, 33 and 66 MHz), Compact PCI Hot-Swap, PC-104, VME, AGP, USB/USB2.0, IEEE 1394/FireWire, PCI-X 133 MHz, and other emerging high-speed standard interfaces

Xilinx Online and Internet Reconfigurable Logic (IRL)

Xilinx is pioneering a paradigm shift with the ability to remotely update FPGAs and hardware for systems already deployed in the field. The Xilinx Online program is designed to enable, identify, and promote network upgradeable systems to provide future-proofing for a system. These systems can then be upgraded, modified, or repaired after they have been deployed in the field. Many customers have been building upgradeable devices based on Xilinx technology for years. Now the explosion of networked connected devices has dramatically increased the demand for these user configurable and adaptable applications.

As an example, an engineer can reprogram and remotely upgrade a cellular base station with the latest specification and standard in a matter of minutes over the Internet using a system's Xilinx FPGA. Having the ability to remotely update hardware with new features or the latest bug fix can accelerate the time-to-market of an application. It can also extend the useful life of existing systems and significantly cut production, maintenance, and support costs. Many of today's systems have some form of communications and/or microprocessor interface built in. This simplifies the addition of remote field update capability. If designers consider remote updates during the initial specification/design process their systems can easily reap all of these benefits.

Xilinx IRL technology makes it easier to develop these systems based upon the most advanced programmable logic available.

Services (Design, Education and Support)

To extend designers technical capabilities and shorten design time, Xilinx offers a portfolio of global services that consist of education, design, and support. They also offer a personalized online technical resource at www.mysupport.xilinx.com. Xilinx is committed to helping users succeed with programmable logic designs and provides a complete and uniquely accessible array of services and training to customers with service contracts.

Education Services consist of hands-on, lab-based, and multi-day courses from fundamental to expert skill levels. They are designed to make customers proficient at high-speed logic and system design. Xilinx experts are available to provide quick problem resolution and creative and timely solutions to design challenges. They also offer design evaluation for new projects and consultation throughout the design process. Full training in design completion and methodology review is also available in addition to special application consultation.

Xilinx provides 24-hour access to a set of sophisticated tools for resolving technical issues via the Web. The Xilinx search utility scans thousands of answer records to return solutions for any given issue. Several problem-solver tools are also available for assistance in specific areas such as configuration or install. A complete suite of one-hour modules is also available at the desktop via live or recorded e-Learning. Users with a valid service contract can access Xilinx engineers over the Web by opening a case against a specific issue. Technical support on the Web is available at www.support.xilinx.com. Customers can personalize their experience at www.support.xilinx.com, through the MySupport feature. They can then access training courses, an answers database, and forums with access to an experienced Xilinx team for assistance in troubleshooting and design issues.

Design Services help shorten customers' time-to-market by augmenting their design teams with Xilinx's industry experts in FPGA design techniques and solutions. Support Services enables customer calls to get high priority response from senior application engineers with extensive design experience and a track record of solving complex problems.

Xilinx Titanium Technical Service provides on-site or off-site support for clients on a contract basis. Titanium application engineers are adept at ensuring that users start and finish designs the right way. Xilinx engineers provide design methodology coaching to make sure customers take the most efficient approach. One of the most powerful services provided by Xilinx engineers is tracing debug issues back to the design.

Emerging Standards Program (eSP)

The digitization of consumer and communications technologies is necessitating the establishment of new standards and protocols. Examples of the need for standards and protocols include:

- **Wireless communications standards** – IEEE 802.11 b/a/g and Bluetooth
- **Serial digital communications standards** – IEEE 1394 (FireWire), USB/USB 2.0, Optic Fiber, Gigabit Ethernet, 10 Gigabit Ethernet, RapidIO, Lightening Data Transport (LDT), and Infiniband
- **Streaming communications technologies** – Internet audio playing (MP3), video content distribution, and interactive conferencing

These technologies are progressing in "Internet time" and placing extraordinary challenges and pressures for new product development on system architects and ASIC/ASSP/FPGA designers. The need to develop products and introduce them to market is important. However, understanding the technology and standards to make the right decisions are paramount. The value of programmable logic solutions is in helping designers address challenges such as:

- **Dealing with multiple standards** – Often more than one alternative (standard) is available to address any given design problem. This introduces tremendous risk in selecting a solution. Programmable logic solutions, while not exempt from this risk, require less time investment and can adapt to newer architectures and standards.
- **The extensive scope and complexity of emerging technologies** – The specifications for emerging technologies are large and subject to ambiguity and error. They are new and unproven. This virtually mandates a development cycle where the implementation is cycled through numerous prototype, test, and debug iterations before it becomes stable. Programmable logic is virtually the only way to produce such designs efficiently.
- **Emerging standards need to be updated on a regular basis** – As much as the designers try, backward compatibility cannot be guaranteed. Programmable logic solutions are virtually the only platform that can remain viable over time.
- **Shrinking development cycles** – Manufacturers realize that to achieve maximum success they must get products to market as quickly as possible. Programmable logic is the superior alternative for fast time-to-market.

The Xilinx eSP initiative is a program designed and tailored to address these issues. The eSP program focuses on digital consumer convergence markets and is targeted at system architects and ASIC/FPGA designers to:

- Help them understand these technologies
- Promote the benefits of programmable solutions
- Provide support resources for Xilinx-based solution options

The eSP program assists customers shorten their time-to-market for emerging standards and protocols. The traditional time-to-product advantage of FPGAs is well known. This program further accelerates the time-to-learn and time-to-design period in product development. The eSP program provides customers and system developers in these dynamic markets with standards tutorials, white papers, application notes, system solutions, IP, and more. This helps in understanding the technology, specifications, and in making the right decisions in designing products. The eSP segments that have been rolled out include home networking, Bluetooth, digital video technologies, information appliances, wireless LANs, metropolitan area networks, and cellular technologies. These eSP segments were the industry's first Web portal dedicated to the challenges of developing products based upon emerging standards and protocols. eSP simplifies the task of designing the next product by delivering a complete set of solutions aimed at accelerating the time-to-market.

A major focus of the eSP initiative is to help designers get an overview of various specifications and to understand their complexities. The Standards Tutorials are carefully crafted to contain enough detail to do this without overwhelming the designer. The material caters to a wide audience ranging from the curious to the serious system architect. A tutorial covering the major aspects in the home networking market segment contains 2,600 carefully documented pages of information. The topics covered include market research, overviews of the various technical specifications, the hurdles faced by designers, projections on the future, system block diagrams, industry links, FAQs, a detailed glossary, and Xilinx solutions for each technology. All are available via the Internet. Similar tutorials are available for each of the other applications within every eSP segment. One other key feature of the eSP initiative is it allows users to learn about the changes to standards and protocols as they happen. Even more importantly, it helps users to understand the impact of these changes on product design. This is designed as a one-stop location for updates on all the specifications related to a specific consumer market segment.

Xilinx has put together a panel of experts to answer the most pressing questions. Each market segment explored and discussed under the eSP initiative has its own discussion forum. Some of the industry's leading experts are available to answer the toughest design challenge questions. The home networking forum has more than ten experts on the panel each with a strong understanding of the home networking industry.

The eSP also provides access to application notes, white papers, and a glossary. Users needing more information about a standard or protocol, or simply trying to determine what a term means can turn to eSP for help. The eSP website contains application notes and white papers that include in-depth discussions and market and technical analysis. They are constantly updated to reflect the latest changes in the industry. The website also contains an extensive glossary of more than 1,000 terms to quickly understand acronyms and industry term definitions.

Users trying to figure out the best way to build their next product can simply check the system block diagram pages. The eSP website has a very extensive set of system block diagrams. For example, the home networking market segment offers over 120 block diagrams that cover broadband access devices, residential gateways, home networking technologies, digital video technologies, and information appliances.

Information on intellectual property and reference boards is also contained on the eSP website. The eSP program identifies important IP required for the system architects and ASSP/FPGA designers to succeed in markets where standards and specifications are evolving. Working through the LogiCORE and AllianceCORE programs the eSP team provides a broad selection of industry-standard IP cores and solutions dedicated for use in Xilinx programmable logic. Currently developed reference designs from strategic partners are used to highlight the use of Xilinx FPGAs and CPLDs in strategic applications. Xilinx has partnered with a wide range of industry experts, ASSP manufacturers, and IP providers to develop, deliver, and support hardware reference designs for specific emerging standards.

These reference designs accelerate product development while addressing the flexibility and price constraints of targeted applications. Many of these are not just reference designs. They are complete system solutions designed to conform to all of the necessary standards and protocols. The reference designs are tested to comply with industry specifications as well as the relevant standards bodies. Because the system solutions are based upon the low-cost Spartan-II FPGA family, the reference design can be easily customized for product differentiation or to add extra features. These system solutions contain everything necessary to build and develop the final product. The reference designs include:

- Bill of materials
- Gerber files
- Software
- Software drivers
- Hardware
- VHDL or Verilog code
- Programming software
- Design tools
- IP cores
- Datasheets
- Schematics
- Applications notes
- License agreements

Xilinx Solutions for Digital Consumer Systems

Xilinx Spartan-3 FPGAs and CoolRunner-II CPLDs, in conjunction with IP and reference designs, provide different functions in digital consumer devices. Figures 19.10 through 19.39 show Xilinx solutions for different consumer devices, digital modems, residential gateways, and more.

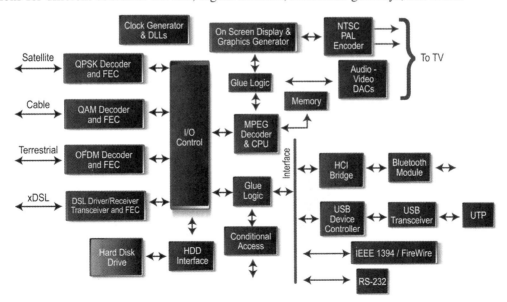

Figure 19.10: Set-top Box

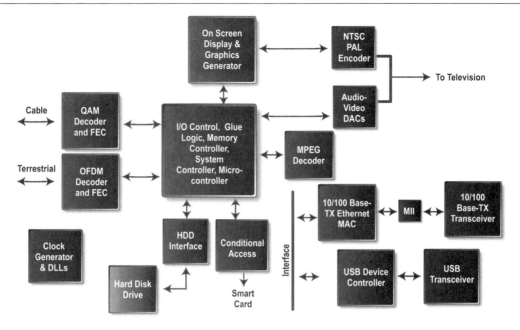

Figure 19.11: Cable Modem

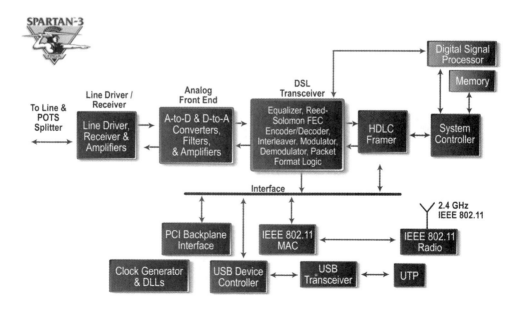

Figure 19.12: DSL Modem

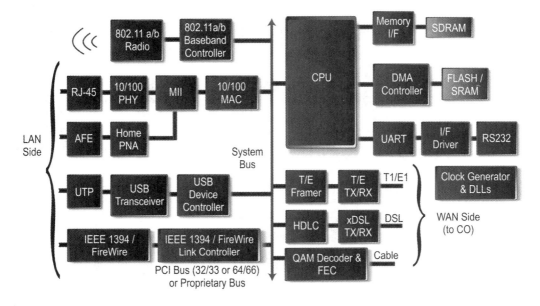

Figure 19.13: SOHO Router

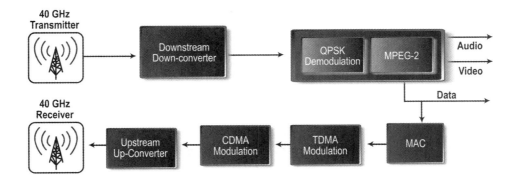

Figure 19.14: LMDS Customer Premise Equipment

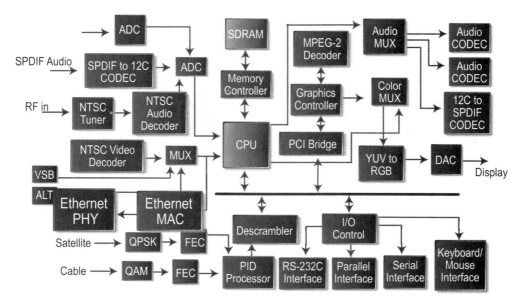

Figure 19.15: Digital TV

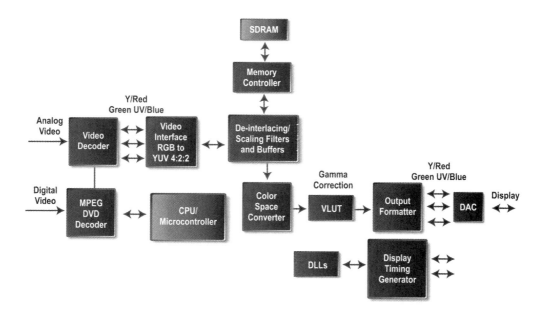

Figure 19.16: Flat Panel Controller

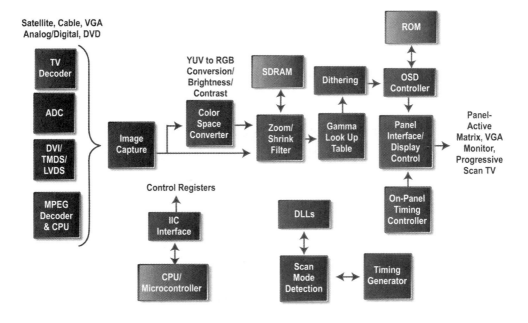

Figure 19.17: LCD Monitor

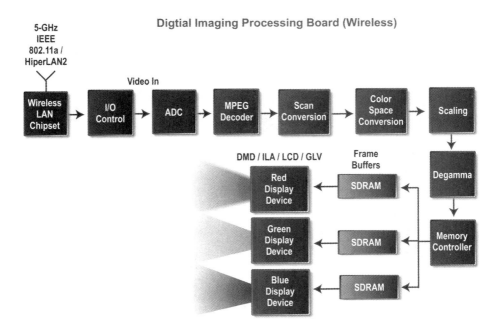

Figure 19.18: LCD Projector

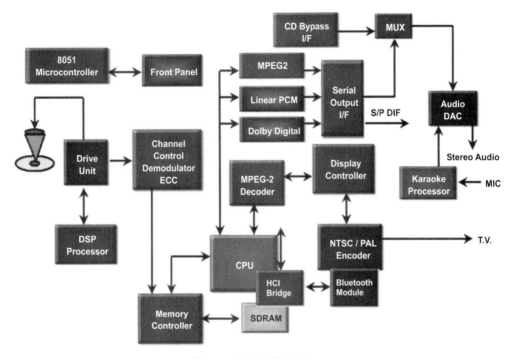

Figure 19.19: DVD Player

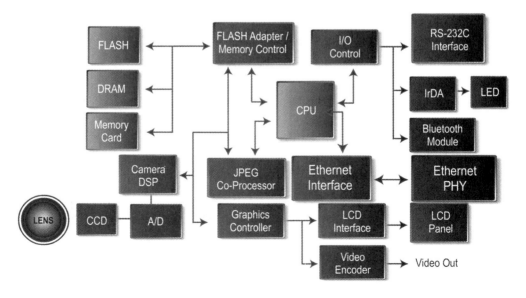

Figure 19.20: Digital Camera

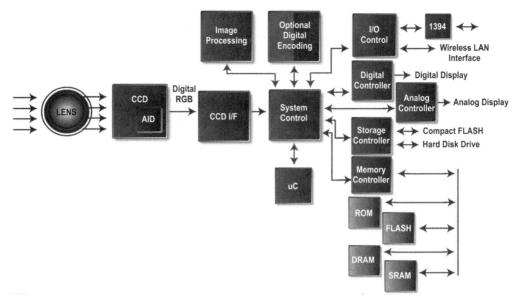

Figure 19.21: Digital Camcorder

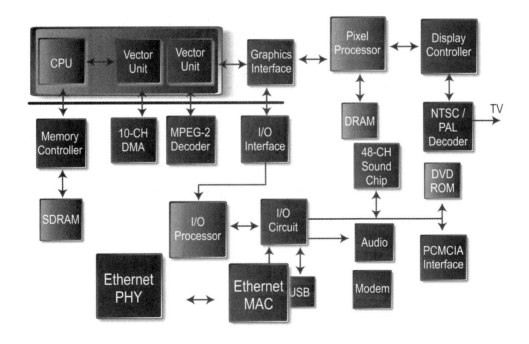

Figure 19.22: Gaming Console

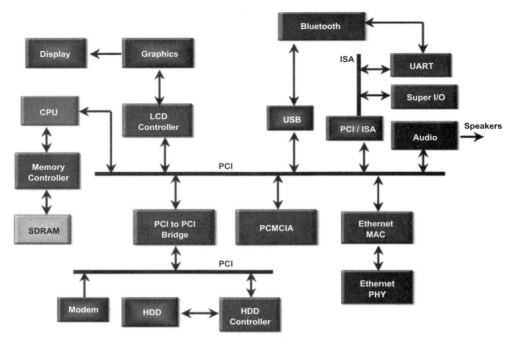

Figure 19.23: PC

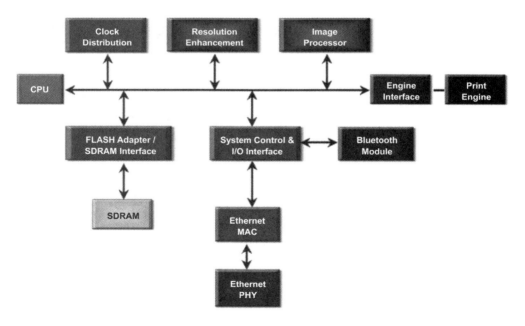

Figure 19.24: Printer

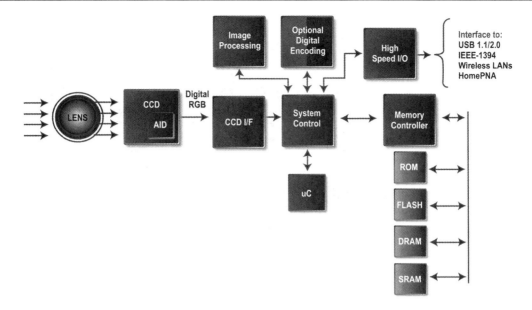

Figure 19.25: Scanner

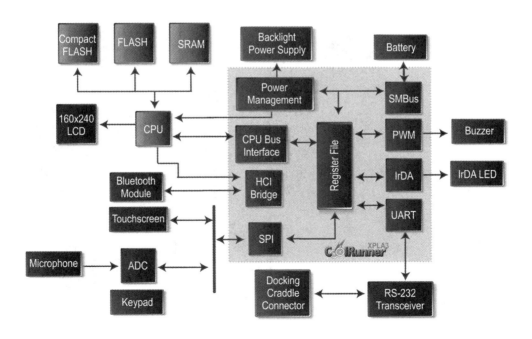

Figure 19.26: PDA

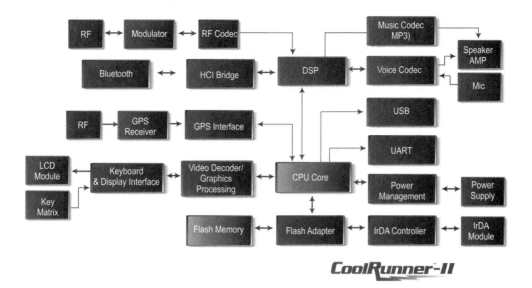

Figure 19.27: 3G Cellular Handset

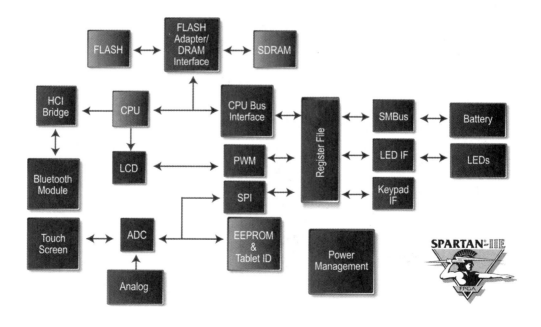

Figure 19.28: Web Tablet

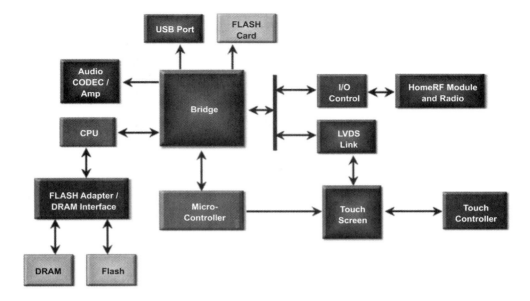

Figure 19.29: Web/Fridge Pad

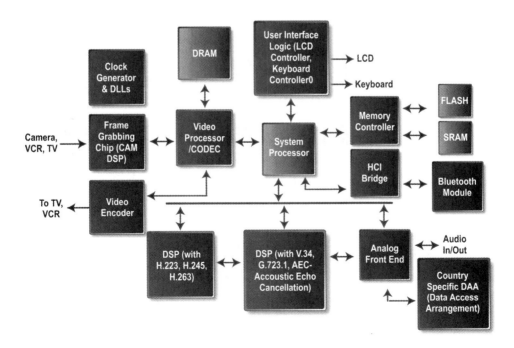

Figure 19.30: Video Phone

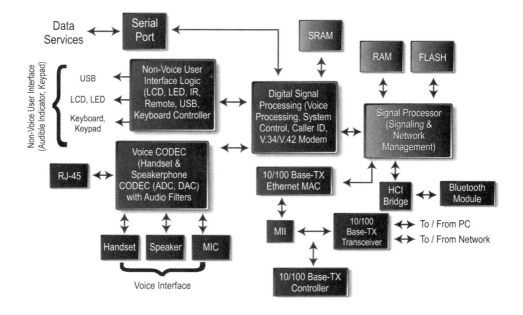

Figure 19.31: Voice over IP (VoIP) Phone

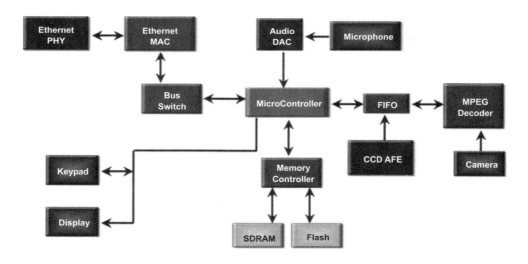

Figure 19.32: Home Security

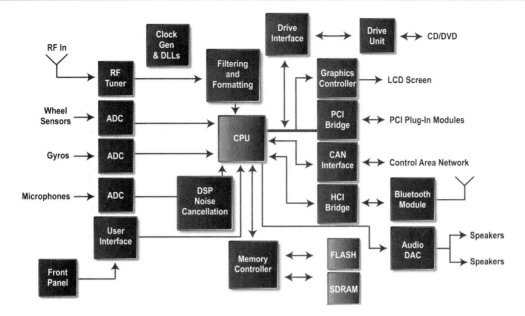

Figure 19.33: GPS Receiver

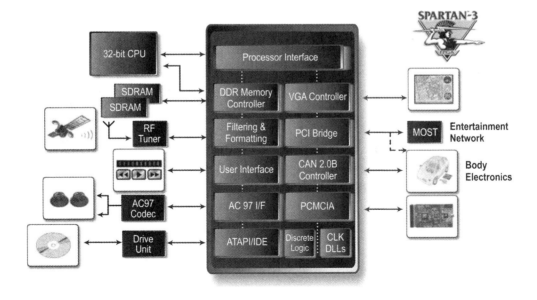

Figure 19.34: Car Multimedia (Telematics)

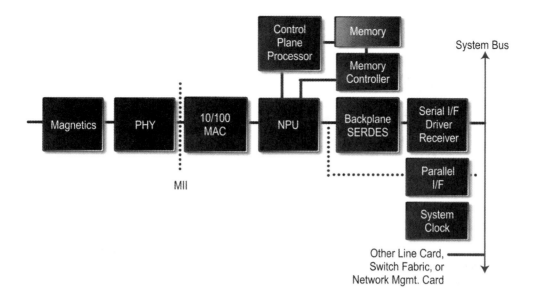

Figure 19.35: 10/100 Mb/s Ethernet Router

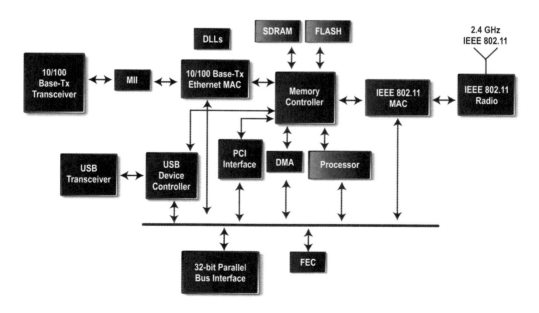

Figure 19.36: WLAN-to-Ethernet Technology Bridge

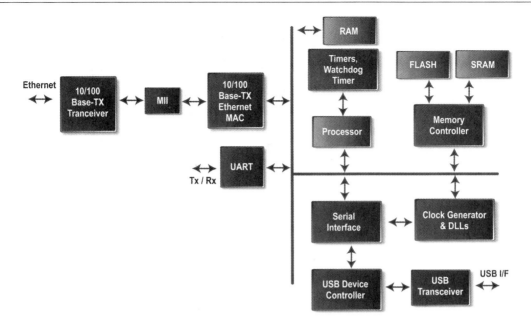

Figure 19.37: USB-to-Ethernet Technology Bridge

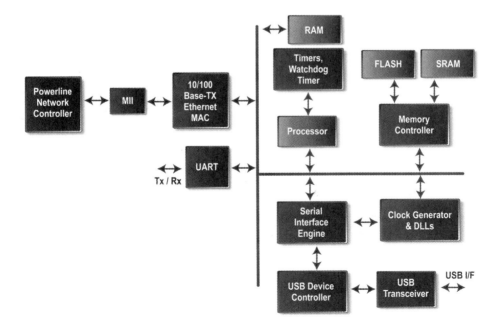

Figure 19.38: USB-to-HomePlug Technology Bridge

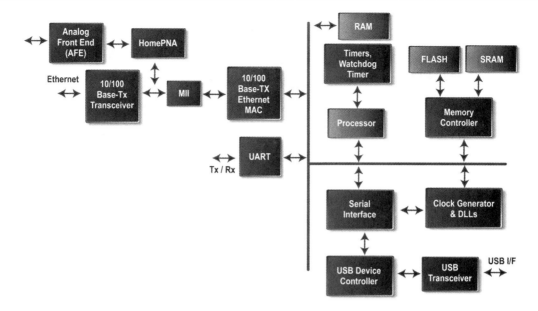

Figure 19.39: USB-to-HomePNA Technology Bridge

Addressing the Challenges in System Design

Interfacing (Bridging) and System Connectivity

Consumer and digital video products are faced with an ever-increasing list of desirable and complex connectivity options in the consumer electronics market. For instance, it may be necessary to connect internally through PCI, AGP, or LVDS. At the same time it may be necessary to integrate home networking technologies such as IEEE 1394, USB 2.0, Ethernet, HomePNA, HomePlug, IEEE 802.11b/g/a, and Bluetooth. For all these applications Xilinx programmable logic has been demonstrated to provide efficient and flexible bridging solutions to connect to a myriad of ASSPs, processors, or memories.

Component integration is the traditional strength of programmable logic. Xilinx can support chip-to-chip, chip-to-memory, and chip-to-backplane interconnection requirements. Xilinx solutions support 24 signaling standards, memory (block and distributed RAM structures for FIFOs and buffers), and a wide range of standard IP for industry standard connectivity standards such as AGP and PCI. Xilinx FPGAs and CPLDs can easily replace dual-port memories/FIFOs, clock buffers, DLLs, level translators, hot-socketing and board deskew chips, Schmitt triggers, and TTL devices. This helps reduce board area, provides overall cost savings, and higher reliability. By using fewer PCB layers and traces it also significantly reduces EMC issues.

The technology and features emerging in digital capture products make the user interface a significant challenge. And a unique user interface can be a key selling point and product differentiator in the highly competitive consumer industry. Xilinx FPGAs and CPLDs provide maximum flexibility to address this need.

Spartan-3 FPGAs provide several advantages for interfacing and system connectivity such as delivering maximum I/Os with a minimum die size. Spartan-3 FPGAs provide 784 single-ended and 344 differential pairs I/Os (high performance per pin pair-up to 622 Mb/s). The Spartan-3 FPGA family architecture gives a user more I/O pins to design with. This staggered pad technology provides two rows of I/O on each edge of the chip, while other FPGAs only have one. Spartan-3 FPGAs allow for easy density migration across similar package options. The relative position of VCC and GND remain constant across like packages, giving users the peace of mind that they won't have to re-layout their board when migrating to larger or smaller density devices. Spartan-3 FPGAs also support density migration across common packages without changing the PC board footprint. When using a FT256 package, three density members of the Spartan-3 FPGA family can be implemented, providing outstanding flexibility for design revision, upgrade, or cost optimization. And there are more I/Os to be gained with larger densities. Spartan-3 FPGA advanced interfacing options provides support for 23 different I/O standards, both single-ended and differential, giving users the freedom to choose the protocols that best suit their system needs.

Benefits of differential I/O:

- Higher I/O performance per pin pair leads to lower I/O and total costs and reduced EMI
- Low output voltage swing of approximately 350 mV allows relatively slow edge rates
- High noise immunity. Switching noise cancels between the two lines and data is not affected. The external noise affects both lines but the voltage difference stays about the same.
- Reduced power consumption

I/O termination is required to maintain signal integrity. With hundreds of I/Os and advanced package technologies external termination resistors are no longer viable. XCITE Digitally Controlled Impedance Technology dynamically eliminates drive strength variation due to process, temperature, and voltage fluctuations. Spartan-3 XCITE DCI technology highlights include:

- Series and parallel termination for single-ended and differential standards
- Maximum flexibility with support of series and parallel termination on all I/O banks
- Input, output, bi-directional, and differential I/O support
- Wide series impedance range
- Popular standard support including LVDS, LVDSEXT, LVCMOS, LVTTL, SSTL, HSTL, GTL, and GTLP
- Full and half impedance input buffers

XCITE DCI technology advantages:

- **Second generation technology** – Proven in the field and used extensively by customers
- **Lowers cost** – Fewer resistors, fewer PCB traces, and smaller board area result in lower PCB costs.
- **Absolute I/O flexibility** – Any termination on any I/O bank. Non-XCITE technology alternatives deliver limited functionality.
- **Maximum I/O Bandwidth** – Less ringing and reflections maximize I/O bandwidth.
- **Immunity to temperature and voltage changes** – Temperature and voltage variations lead to significant impedance mismatches. XCITE technology dynamically adjusts on-chip impedance to such variations reducing and improving reliability.

- **Eliminates stub reflection** – Improves discrete termination techniques by eliminating the distance between the package pin and resistor.
- **Increases system reliability** – Fewer components on board deliver higher reliability

CoolRunner-II CPLDs feature a uniform I/O structure that provides system designers with a variety of I/O options including LVTTL, LVCMOS, 1.5V I/O, HSTL Class I, SSTL2 Class I, and SSTL3 Class I. Understanding these I/O capabilities and characteristics can greatly simplify the implementation of CoolRunner-II CPLD-based designs.

New digital VCRs, gaming consoles, and set-top boxes provide the capability to store audio and video on HDDs. They also provide the ability to record and view video simultaneously. Spartan-3 FPGAs provide HDD interfaces in these products. These devices provide data buffer and disk control logic with on-chip memory for FIFOs. Programmable solutions allow support of evolving disk drive technologies that are optimized for simultaneous disk read and write. With millions of set-top boxes being shipped each year, Spartan-3 devices enable dual sourcing of multiple types of HDDs in case of a supply shortage. As the heart of evolving set-top boxes Spartan-3 FPGAs are helping revolutionize the TV experience by giving users the ability to pause live TV, view instant replay, automatically record their favorite programs, and perform advanced TV program searches.

Digital modems, set-top boxes, residential gateways, and SOHO routers have multiple front-end and multiple back-end technologies to support. Front-end technologies provide broadband (digital or analog) interface from the cable companies, telephone companies, and satellite service providers using xDSL, cable, satellite, wireless, powerline, or analog phone dial-up. With increasing numbers of consumer products being networked the home gateway requires the capability to network multiple technology islands. Product success is not only based on the pros and cons of the home networking and consumer markets but also on the geography. While DSL may be the broadband access of choice in one location cable may be the technology of choice in another. It is not cost effective for manufacturers to support multiple receivers and build individual products for cable, terrestrial, xDSL, and satellite. Slicing the market and building multiple gateway designs based on different home networking technologies is resource and cost prohibitive. This approach also prevents OEMs from second-guessing which technologies will prevail and how their products will succeed in the market.

Spartan-3 FPGAs provide the interface required to support multiple receiver and multiple home networking ASSPs. This provides a low-cost protocol translation mechanism between two disparate technologies and an interface to the system. In addition, the ASSP used is not influenced by the gateway manufacturer but by broadcaster features. This makes it nearly impossible for the set-top box or gateway manufacturer to bring a product to market in a short amount of time.

Digital Image/Video Processing

Consumer digital video applications such as flat panel displays, digital light projectors, and home video editing equipment can be split into the same three distinct processes: capture, process, and display. These following processes are complex in nature and many challenges exist in their design:

- **Capture** – The first step in digital video is signal capture—voice, video, images, and data. This process begins with some form of light to data conversion and ends with a binary data file. Examples of consumer applications that require image capture are digital camcorders, digital scanners, video screenphones, and digital still cameras.
- **Process** – Digital image processing is a key component in ensuring superior image quality. Efficient use of available bandwidth is achieved through powerful encoding techniques.

Through a myriad of other digital techniques unique and invaluable capabilities are enabled such as on screen display overlays, interactivity, and more. Virtually all digital video products incorporate some form of image processing. Examples of applications that benefit from digital image processing are digital camcorders, digital scanners, digital projectors, digital still cameras, digital TVs, DVD read/write, GPS navigation, LCD monitors, PDAs, plasma displays, telematics, video screenphones, and webpads.

■ **Display** – A display accepts a digital video data stream and converts it into the technology specific format necessary to drive that display. Before being displayed the data is scaled, colors corrected, and processing is applied to remove imperfections and adjust the image to suit the viewer's taste. Digital display technology is the result of convergence between digital television and computing. This convergence, along with the increasing availability of digital video content, has driven the development of digital transmission formats, flat panel form factors, and ever-higher resolutions and image quality.

When performing digital video processing, programmable logic enables a designer to achieve the necessary performance and still retain the flexibility to support geographically specific and continuously evolving standards. Some of the issues that must be covered include formatting and display standards, technology specific color behavior, and user perceptual preferences. These tasks are simple in principle, but are quite challenging at real-time video data rates. A key enabler of the FPGA solution is the exceptional DSP performance of Xilinx products including the embedded 18x18 multipliers, which is fundamental to all digital video processing functions. Furthermore, the integration of the Xilinx DSP toolset with MATLAB™ provides an ideal combination for these types of applications.

Several applications are pushing for higher levels of video quality as the processing limits of ASSPs and DSPs are being reached. This makes moving computationally intensive portions like motion estimation and DCT from either MPEG codec or a processor into low-cost Xilinx FPGA ideal. FPGAs are ideal for delivering the required resolution and scan rate while allowing the processor to focus on operating system tasks. Using XtremeDSP technology exploits inherent parallelism in FPGAs to perform DSP algorithms faster than any software or dedicated DSP device. FPGAs used for experimentation with hardware and software design partitioning for new techniques or enhancement algorithms leads directly to product differentiation.

Xilinx also has several image and video processing IPs such as color Space Conversion (RGB2YCrCb, YCrCb2RGB, RGB2YUV, YUV2RGB), DCT/iDCT, FIR filters, DA FIR, FFTs, MACs, MPEG-2 (SD & HD), JPEG, Huffman, Wavelet, Scaling, Rotation, and Enhancement.

Typical FPGA digital video applications include:

■ Image Scaling
■ Color Space Conversion
■ Image rotation
■ Keystone correction
■ Real-time image resizing
■ Image enhancement
■ Gamma Correction (Color and Hue Correction)
■ Half-toning
■ Contrast enhancement
■ Brightness enhancement

- Shadow enhancement
- Sharpness enhancement
- Artifact elimination
- De-contouring
- Noise reduction
- Interlacing/De-interlacing
- Motion artifact detection
- Color purity
- Color temperature
- Clearing techniques
- Dithering

Typical issues with flat panel displays include:

- Viewing angle compensation
- Black-level, contrast, and color saturation
- White saturation
- Color and gray-scale accuracy
- Motion artifacts
- Aspect ratio conversions
- Resolution/Scaling
- Interference
- Pixel refresh rate

The following table goes into more detail on functions/algorithms that are needed in a flat panel display system.

Table 19.8: Flat Panel Display System Function and FPGA Implementation Potential

Function	PDP	LCD-TV	FPGA Implementation Potential
Hardware acceleration of video algorithms	√	√	It is possible to use hardware acceleration to improve performance by offloading portions from media processors
Aspect Ratio Conversion	√	√	Can be done with high quality scaling (e.g. polyphase filter)
De-interlacing	√	√	Can use low quality de-interlacing as described in XAPP285. High quality de-interlacing requires an external frame buffer State-of-the-art 3D recursive block matching, uses Motion Estimation and Motion Compensation
Motion Estimation			Accuracy of ME can be critical for video enhancement. However, complex ME may not always be practical from an implementation point of view. (e.g. block-matching ME algorithm with full-search.)

(continued on next page)

Noise Reduction	√	√	Filtering used to eliminate different types of noise (e.g. random, salt & pepper, gaussian, impulse)
Motion Artifact Reduction	√	√	In PDPs, motion artifacts are caused by sub-field driving method for generating gray levels. Need to use motion compensation LCDs may benefit more from using methods like a flashing/scanning backlight, rather than video processing. Motion compensated interpolation is a prerequisite to counteract motion artifacts
High Quality Scaling	√	√	For example, polyphase filtering can be implemented and used for sample rate conversion.
Gamma correction	√	√	Easily done. Straight forward non-linear operation that is implemented using a LUT
Sharpness enhancement	√	√	Generally, proprietary algorithms that are key for product differentiation Usually fall into 2 categories: • Peaking - Linear operation that uses the "Mach Band" effect to improve the sharpness impression. It increases the amplitude of the high-band and/or middle-band freq using linear filtering (e.g. FIR) • Transient improvement (e.g. LTI - luminance transient improvement) Non-linear approach that modifies the gradient of the edges to enhance the sharpness
Colour Space Conversion	√	√	XAPP283, XAPP637
Video component conversion	√	√	Filter - takes advantage of parallel

Example: Partial MPEG-4 hardware implementation in Xilinx FPGAs

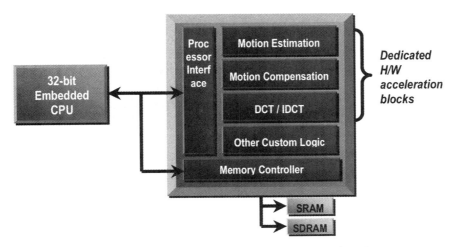

Figure 19.40: Partial MPEG-4 Hardware Acceleration

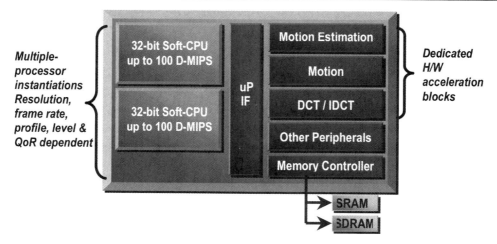

Figure 19.41: Customized MPEG-4 Implementation

Typically MPEG-4 is implemented in a processor, which excels at sequential and control tasks. Implementing MPEG-4 compression algorithms can very easily bog down a traditional processor when over 50% of processor cycles may be spent evaluating a single algorithm block (motion estimation, etc.). This necessitates the need for dedicated hardware acceleration engines such as FPGAs. Processor vendors have used inflexible and performance limited dedicated DSP blocks for hardware acceleration. These are not suitable for processing intense applications such as digital displays.

Offloading portions of MPEG-4 in the FPGA is ideal. It saves valuable processor cycles, increases quality and performance, provides potential system cost savings, and provides the ability to add more capabilities (codecs) to the system. Some functions such as motion estimation, motion compensation, DCT/IDCT, and color space conversion are ideal for implementing in Xilinx low-cost Spartan-3 FPGAs.

There are numerous encoding and decoding standards available for digital AV data including MPEG1, MPEG2, MPEG4, JPEG2000, MJPEG, and TIFF. Furthermore, many systems in a variety of industries use proprietary formats. A programmable solution can provide the flexibility and performance necessary to cope with this uncertainty of different and changing standards. As discussed previously, the high performance parallel processing of an FPGA, in conjunction with the flexibility of the microprocessor, provides a suitable platform for this type of application.

As digital video technology proliferates there is a growing need for content protection. Currently there is no unified standard and every solution is subject to the perpetual threat of being compromised. FPGAs are ideal platforms for this function with exceptional performance and the flexibility to support numerous and evolving formats.

Using System Generator for Digital Video Applications

Digital video solutions can be produced using System Generator for DSP. Using System Generator greatly shortens the path from design concept to working hardware through:

- **Simplicity** – System Generator presents designs at an appropriate level of abstraction. Signals are not just bits. They can be signed and unsigned fixed-point numbers and changes to the design automatically translate into appropriate changes in signal types. Blocks are more than stand-ins for hardware. They respond to their surroundings by automatically adjusting the results they produce and the hardware they become. Translating a design into hardware requires nothing more than the push of a button.

- **Flexibility** – System Generator allows designs to be composed from a variety of ingredients. Data flow models, traditional hardware design languages (VHDL and Verilog), and functions derived from the MATLAB programming language can be used side-by-side, simulated together, and synthesized into working hardware.

- **Speed** – System Generator simulations are considerably faster than those from traditional HDL simulators. And the results are easier to analyze. New hardware "in the loop simulation interfaces" expands this advantage both dramatically and seamlessly.

- **Power** – System Generator runs inside the Simulink framework produced by The MathWorks making available the extensive tool sets that Simulink and MATLAB provide. System Generator v3.1 includes native support for ModelSim which allows users to import HDL code and brings system level modeling capabilities (e.g., test bench creation) to the traditional hardware designer.

- **Accuracy** – System Generator simulation results are bit and cycle-accurate. Results seen in simulation mirror those that will be seen in hardware

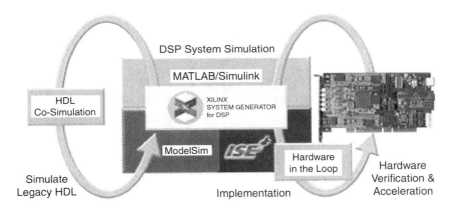

Figure 19.42: System Generator

The System Generator for DSP software enables electronic designs to be created, tested, and translated into hardware for Xilinx FPGAs. The tool extends Simulink (The MathWorks, Inc.) to support bit and cycle accurate system level simulation and automatic code generation for Xilinx FPGAs. System Generator co-simulation interfaces extend Simulink to incorporate FPGA hardware and HDL simulation into the system-level environment as naturally as other library blocks. System Generator presents a high level and abstract view of the design. It also exposes key features in the underlying silicon, making it possible to build extremely high-performance FPGA implementations.

System Generator designs are built and simulated within the Simulink block editor. An example is shown in Figure 19.43.

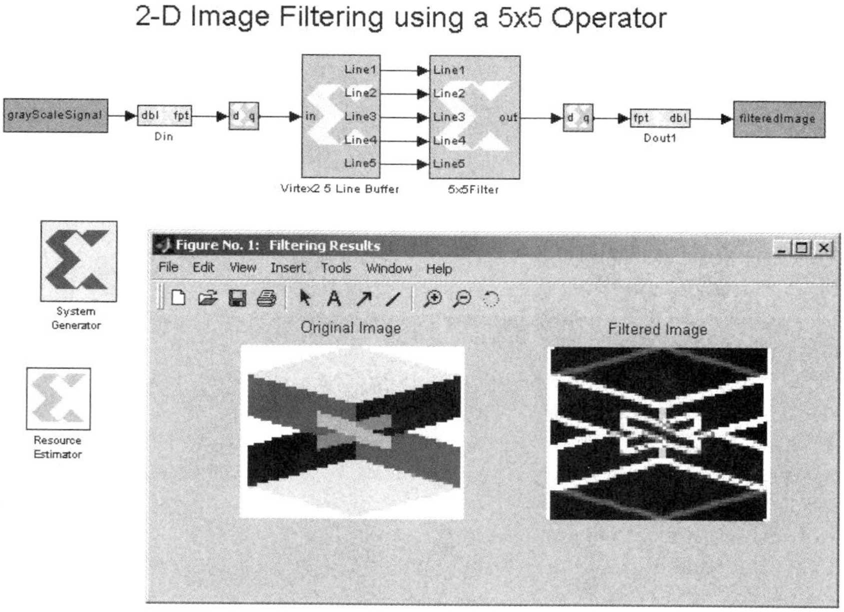

2-D Image Filtering using a 5x5 Operator

Figure 19.43: Example Showing How System Generator Designs are Built and Simulated Within the Simulink Block Editor

Each block produces results that make sense based on its setting. For example, a multiplier whose inputs are assigned fixed point numbers produces a signed result having an appropriate width and binary point position. Signal types propagate automatically, so when one block is updated System Generator adjusts downstream blocks accordingly.

The System Generator libraries include math and DSP functions, basic building blocks, high-level communication functions, memories, microcontrollers, and other functions for constructing sophisticated high-performance systems. System Generator also provides blocks for compiling MATLAB m-code into synthesizable HDL code; fast estimation of FPGA resources required by a Simulink model; and for accessing the FDATool software for filter design and analysis.

The Simulink toolkit is extensive, and Simulink blocks can be used in System Generator designs without restriction. This makes it easy to drive simulations using customized signals to analyze results with tools designed to address the problem. A black box Configuration Wizard makes it straightforward to incorporate an HDL entity into System Generator and have it co-simulated with the ModelSim simulator (Model Technology, Inc.). Hardware co-simulation interfaces will automatically encapsulate a System Generator subsystem into a Simulink object that is driven by an underlying FPGA platform. In addition to hardware acceleration of a System Generator model, this capability allows the user to verify design behavior in actual hardware.

Example: Building an FIR Filter in Xilinx FPGAs

As a simple but instructive example, let's consider how System Generator can be used to create a parametric finite impulse response (FIR) filter. An N-tap FIR filter is defined by its impulse response, a length N sequence of filter coefficients: h0, h1,..., hN-1. If x0, x1, x2,..., is a sequence of input values, where by convention we define xi = 0 for i < 0, the filter output sequence y0, y1, y2,... is defined by the convolution sum

$$y_n = \sum_{i=0}^{N1} h_t x_{nt}$$

The filter output at time *n* is computed by accumulating a sum of products of the filter coefficients with the *N* most recent input samples. In practice, all numbers must be represented with a finite number of bits. With a traditional processor numeric data is typically represented as 8, 16, or 32-bit integers, or in a floating-point representation. An FPGA has no such word length limitations. Users can create a custom data path processor having an arithmetic precision tailored to the application. System Generator supports this capability by providing an arbitrary precision fixed-point data type. Each block allows users to specify its output precision and the policy for handling quantization and overflow. The system can be modeled in the Simulink environment under a number of scenarios and the data analyzed to ensure exactly the right precision for the application.

Users can implement an FIR filter in an FPGA in many ways. A versatile approach that maps well onto an FPGA employs a multiply-accumulate (MAC) engine to compute the sum of products. As shown in Figure 19.44, a MAC unit is easily constructed in System Generator using the multiplier and accumulator blocks. The multiplier can be implemented either in the logic fabric or, for Virtex-II family FPGAs, using dedicated 18-bit x18-bit embedded multipliers. System Generator ensures the underlying IP core provides an efficient implementation. Upon reset the accumulator reinitializes to its current input value rather than zero to avoid a one-clock cycle stall.

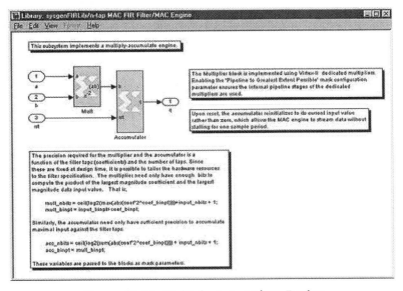

Figure 19.44: Multiply-Accumulate Engine

Although Simulink provides a graphical block editor, System Generator should not be mistaken for a "schematic capture" tool. System Generator models are fully customizable and executable in Simulink without recompilation. In fact, some blocks can be reconfigured during simulation. Xilinx blocks support Simulink's data type propagation capability. They also provide extensive error checking on their parameters and usage. System Generator's seamless integration with the Simulink tool suite allows users to customize Xilinx blocks in ways that are impossible in schematic and other visual tools.

For example, in a System Generator model we can specify the arithmetic precision of the blocks in the data path using MATLAB expressions. This makes it possible to minimize the hardware used and still avoid the possibility of overflow. For a filter with k-bit coefficients and m-bit input, we know that the output:

$$/y_n/ = /\sum_{i=0}^{N1} h_t x_{n\,t}/ \le \sum_{i=0}^{N1} /h_t x_{n\,t}/ \le \sum_{i=0}^{N1} /h_t 2^m/ = 2^m \sum_{i=0}^{N1} /h_t/$$

Thus the accumulator requires no more than m + [log2¡Æ|hi | bits. This is considerably fewer than the m+ k + [log2N] bits implied by the input and coefficient precision. The desired accumulator width is readily expressed in the MATLAB language. It may be possible to further tighten the bond when the known input values have limited dynamic range. The accumulator width just derived is a function of the input precision and the MATLAB array that stores the filter coefficients. If users want to change the input precision or change the filter itself, the same model will "right-size" the data path without any modification. The ability to use the MATLAB interpreter to customize System Generator models is unrivaled by any other design flow.

As shown in Figure 19.45, the input data buffer is implemented as an SRL16E-based addressable shift register and the filter coefficient buffer as a block memory. (Both are supplied as handcrafted Xilinx LogiCORE algorithms.) By storing the memory filter coefficients in reverse order the same address counter can be used to drive both buffers.

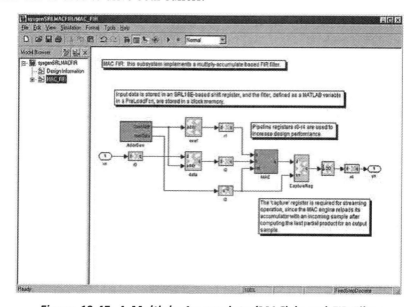

Figure 19.45: A Multiply-Accumulate (MAC)-based FIR Filter

Because there are N multiply-accumulate operations per input sample the filter must run internally at N times the data rate to supply a continuous data stream. The capture register on the output of the MAC is used to latch the accumulated sum of products. Its output is down-sampled by N to match the input data rate. This simple filter architecture is quite compact and efficient. A 64-tap non-symmetric filter with 12-bit coefficients and data requires only 110 slices in a Virtex-II XC2V250-6 FPGA. And it runs at 195 MHz, or 3 Msps (ISE 4.2i, production speeds files 1.96).

System Generator provides many ways to tailor the implementation of a design. And changes are tracked automatically and transparently. In the FIR filter, users can choose to combine dedicated embedded multipliers for the MAC engine with a block memory (BRAM) for coefficient data. Because these resources are juxtaposed in the FPGA, both can be implemented efficiently in the logic fabric. Filter throughput can be increased significantly by employing additional MAC engines. When users switch the implementation strategy, latencies can be automatically adjusted in the System Generator model to match the hardware behavior.

Soft Processing Solutions—MicroBlaze and PicoBlaze Enable System Control Solutions

Microprocessors are single chip devices typically containing the core processor technology. This includes the execution units, the register file, program counter, memory interface, interrupt controller, and in some higher performance examples the cache units and a larger peripheral set. The microprocessor requires a surplus of external components to function correctly. Typical examples are blocks of RAM and ROM used to store executable code or to provide a scratchpad memory area. Other examples include input/output (I/O) ports, timers, and serial communication ports. Figure 19.46 shows the typical layout of a microprocessor system.

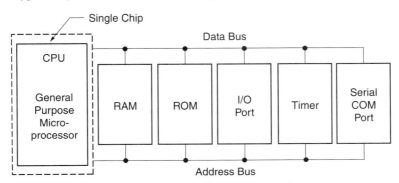

Figure 19.46: Typical Microprocessor System

Microcontrollers offer a significantly more integrated solution. Rather than enforcing a building-block platform upon the user, a microcontroller system is comprised of blocks within the boundaries of the device package. The CPU, RAM, ROM, I/Os, and peripherals are all closely integrated by localized internal connections within the device. The inputs and outputs of the microcontroller system are coupled to the other hardware blocks within the design by pins on the microcontroller device. Microcontrollers present the designer with a solution that is far easier and faster to use. Figure 19.47 shows a microcontroller implementation. A single device prevents the user from making any mistakes connecting the CPU to the memory and other peripherals. At the same time it affords them rapid development time and ease of implementation. In essence, a microcontroller provides a solution where only I/O signals need to be connected and executable code needs to be written. The flexibility of the microcontroller is reduced when compared to the use of a full microprocessor, but this is quite often an acceptable loss for the specific application. The most important issue is one of cost per unit, which often sways the decision in favor of the microcontroller solution.

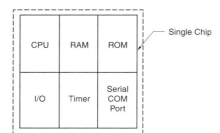

Figure 19.47: Microcontroller

Microcontrollers are used extensively in three major market areas—automotive, industrial, and consumer products. The consumer market showcases products with a heavy concentration of microcontrollers. Walk into any electronic store and users will be hard pushed to find a product that does not have some kind of microcontroller lurking beneath its cover. This includes video recorders, televisions, dishwashers, video games, set-top boxes, refrigerators, washing machines, remotely controlled lighting dimmers, telephones, answering machines, ovens, toasters, printers, and scanners. Even children's toys are equipped with microcontrollers. Because of their demand, microcontrollers are now outselling the more conventional microprocessor by around six units to one.

By the year 2005, it is expected that over eight thousand million microcontroller units will be shipped in electronic products around the world. Microcontrollers are here to stay. It is therefore important that they are able to meet our ever-increasing demands for the future. With quantities like these the individual cost of microcontrollers becomes a very big issue indeed. The only way to keep up with the demands on lower costs is to incorporate the microcontroller into existing hardware devices to form system-on-a-chip platforms. For this to succeed manufacturers need a flexible platform with an embedded microcontroller.

This flexible platform of a FPGA is an ideal base upon which to implement a microcontroller system. An embedded microcontroller takes the concept of integration one stage further by permitting the designer to embed the microcontroller system into a small section of a programmable device. The microcontroller no longer has to exist in a stand-alone package but can now be embedded deep within custom hardware. This technology offers major advantages to the designer in terms of functionality, cost, performance, circuit board area, and, most importantly, flexibility. The MicroBlaze soft processor core has taken this concept and pushed back the boundaries of what can be achieved in a programmable logic device.

The introduction of customized soft processor systems for FPGAs has offered huge flexibility and new challenges for the designer. Traditionally the designer has approached the task of processor selection by comparing the needs of their system specification to the features listed on the processor datasheet. While this may be seen as a trivial task, there are times when system specifications require unusual processor configurations. For example, the designer may desire a processor with 10 UARTs, an interrupt controller, and access to a block of external FLASH. Many off-the-shelf processors offer multiple UARTs and the other desired peripherals but they would not typically offer sufficient complexity to have other unused peripherals.

In addition to the designer paying for additional peripherals, it is often necessary that unused peripherals in this type of processor be placed into a safe mode or otherwise software disabled. An additional burden now exists for the software design team. They have to make the processor peripherals being used operate correctly and write code to disable the parts of the processor that are not being used. It is clear that purchasing an off the shelf solution for this scenario would be wasteful in terms of initial cost and engineering time during the design process. With the MicroBlaze soft processor, the designer can now start with a processor core and build the peripheral set to meet their exact requirements. Since the designers will only implement what they need, silicon waste is eliminated. Eliminating the need to write code to disable unwanted processor functionality reduces software design complexity. The creation of unusual processor configurations, which can be changed at any time to suit changes in the specification, is reduced to a simple task.

Customized processors obviously require someone to perform the customization. This is where design automation tools and intellectual property play a key role. The processor core is placed at the heart of the system and the required peripherals can then be added to the system from a catalog of IP cores. As each block of IP is added the customization process deepens by allowing the user to select the behavior and functionality of each peripheral. UARTs can be configured to operate at the correct baud rate, communicate using the desired number of data and stop bits, and employ the required parity checking. External memory controllers can be customized to insert sufficient wait states for correct and efficient memory device access. Multiple (independently configurable) banks of memory are supported from a single controller giving the designer access to SRAM, FLASH, and EPROM memory. Interrupt controllers can be configured to respond to rising or falling edge inputs or to adopt a level triggered response. Bus structures are added to connect the entire system and are configurable to meet the needs of system clock speed or silicon area.

In a variety of applications an embedded processor or controller is key to system flexibility, maintainability, and low cost. Spartan-3 FPGAs support two powerful yet flexible Field Programmable Controller (FPC) solutions shown in Table 19.9. The PicoBlaze FPC is a simple, highly efficient 8-bit RISC controller optimized for the Spartan-3 FPGA architecture. The MicroBlaze FPC is a powerful, full-featured, and high-performance 32-bit RISC processor offering high-level language and real-time operating system (RTOS) support.

Table 19.9: Embedded Processing/Control Solutions for Spartan-3 FPGAs

Function/Feature	**PicoBlaze FPC**	**MicroBlaze FPC**
Processor Architecture	8-bit RISC controller	32-bit RISC CPU
Typical Applications	Embedded control, state machines, I/O Processing	Embedded Computation and control
Memory Architecture	Harvard (separate data/code data paths)	Harvard (separate data/code data paths)
ALU/register width	8 bits (byte)	32 bits (word)
Registers	16 byte-wide	32 word-wide

(continued on next page)

	0	3
Pipeline Stages	0	3
Code Address Space	512 or 1K instructions	512 to 4G bytes
Code Storage	Block RAM (internal)	Block RAM (internal) External memory
Data Address Space	64 bytes (internal)	0 – 4G bytes
Data Storage	Distributed RAM (internal)	Block RAM (internal) External memory
I/O Address Space	256 locations	N/A
Processor Instructions	57	106
Operands per Instruction	2	3
Clocks per Instruction	2	1 to 3, 34 for integer divide
Call/Return/Interrupt Stack	31 locations (internal)	Variable size, in data memory
Interrupts	1, Expandable	1, Expandable
Maximum Interrupt Latency	4 clock cycles (46 ns at maximum clock rate)	7 to 40 clock cycles (application dependent)
Instruction Cache	N/A	0, 2K, 4K, 8K, 16K, 32, or 64K
Data Cache	N/A	0, 2K, 4K, 8K, 16K, 32, or 64K
Hardware Multiplier	N/A	32x32 + 32 in 3 cycles
Hardware Divider	N/A	Optional, up to 20% performance improvement
Hardware Barrel Shifter	N/A	Optional, up to 15X performance improvement
Hardware Debugger Support	N/A	?
LocalLink Direct Processor Interface	N/A	200 MB/sec communication

The PicoBlaze FPC is always fully embedded within a Spartan-3 FPGA using on-chip block RAM and distributed RAM for code and data storage. The MicroBlaze FPC optionally uses internal FPGA memory resources, or interfaces to external memory, to support larger code or data storage requirements. The Embedded Development Kit (EDK) for the MicroBlaze FPC includes hardware IP cores to support external Flash, SRAM, SDRAM, DDR DRAM, and ZBT SRAM memory. Similarly, the MicroBlaze FPC supports both instruction and data caches (up to 64K bytes) that increase performance when connected to external memory.

By utilizing Spartan-3 FPGAs, both MicroBlaze and PicoBlaze FPCs consume as little as $0.40 FPGA cost in high-volume applications. MicroBlaze solutions start from $1.40 in volume. Both the MicroBlaze and PicoBlaze FPCs provide significant numbers of flexible I/O at a much lower cost than off-the-shelf controllers. The peripheral set for both FPCs can be customized to meet the specific feature, function, and cost requirements of the target application. Because both FPCs are delivered in synthesizable HDL, both cores are future proof and safe from any possible product obsolescence. Being integrated into the FPGA allows both FPCs to reduce board space, design cost, and inventory.

Table 19.10: PicoBlaze and MicroBlaze Resource Requirements and Performance

Function/Feature	PicoBlaze FPC	MicroBlaze FPC
Resource Requirements		
Slices (4 slices = 1 CLB)	96	525
Block RAMs	0.5 or 1	2+
Effective cost in high-volume applications (250Ku, 2004)	From US$0.40	From US$1.40
Percent of XC3S50	13% – 25%	68%+
Percent of XC3S200	4% – 8%	27%+
Percent of XC3S400	3% – 6%	15%+
Percent of XC3S1000	2% – 4%	7%+
Percent of XC3S1500	2% – 3%	4%+
Percent of XC3S2000	1.3% – 3%	3%+
Percent of XC3S4000	0.5% – 1%	2%+
Percent of XC3S5000	0.5% – 1%	1.6%+
Performance (Spartan3 – 4 speed grade)		
Maximum clock frequency	87 MHz	85 MHz
Instructions per second	43.5M	85M
Dhrystone MIPS (D-MIPS)	N/A	68

PicoBlaze Application Development Support

The PicoBlaze FPC solution is a simple 8-bit RISC controller with an easy-to-use assembler. The PicoBlaze core can be debugged using the standard Xilinx JTAG-based interface. And a simple instruction-set simulator is also available. The PicoBlaze reference design also includes UART transmitter and receiver macros with integrated 16-byte FIFOs. The UART supports 8-bit data, no parity, with one stop bit.

MicroBlaze Application Development Support

The MicroBlaze FPC offers complete application development support. This includes a full suite of software development tools, an IP library of processor hardware peripheral functions, and in-circuit hardware debugger/emulation support.

Embedded Development Kit (EDK)

The Embedded Development Kit (EDK) is an all-encompassing solution for creating embedded programmable systems design. The EDK includes and supports the MicroBlaze soft processor core. The EDK also includes support for the PowerPC hard processor core that are only available within the Xilinx Virtex-II Pro and Virtex-II Pro X FPGA families.

Xilinx Platform Studio (XPS):

- Tools for editing software; creating hardware and software platforms
- Runs library generation, compiler tool chains, and generates implementation and simulations

Figure 19.48: XPS Tool

GNU Software Development Tools:
- C/C++ compiler for MicroBlaze and PowerPC cores (GNU gcc)
- Debugger for MicroBlaze and PowerPC cores (GNU gdb)
- Other GNU utilities

Hardware/Software Development Tools:
- **XMD** – Xilinx Microprocessor Debug engine for MicroBlaze and PowerPC cores
- **SystemACE tools**
- **Data2BRAM** – Updates internal block RAM contents without recompiling the FPGA design

Board Support Packages (BSPs):
- **Stand Alone BSP** – For non-RTOS systems (MicroBlaze and PowerPC cores)

Supported Operating Systems

Many embedded processing applications require operating system capabilities. The MicroBlaze FPC solution supports the following operating systems and real-time operating systems (RTOS):
- Micrium / uOS-II Real-Time Operating System
- uClinux Operating System
- ATI Nucleus Real-Time Operating System
- Xilinx Microkernel Libraries
- Highly modular scheduler, network stack, and file system
- Minimal resource requirements and footprint size
- Royalty-free license included with EDK purchase

Processor Peripheral IP Functions

The EDK includes the following processor IP cores that support the MicroBlaze FPC. The IP cores also include device drivers and RTOS adaptation layers. Users can add one or more IP cores to create a custom processor to meet specific application requirements.

Processor Peripherals:
- Timer/Counter
- Timebase/Watchdog Timer
- UART-Lite
- Interrupt Controller
- General-Purpose I/O port (GPIO)

Serial I/O:
- SPI Master and Slave
- JTAG UART
- 16450 UART
- 16550 UART
- I2C two-wire serial Master and Slave

Memory Interfaces:
- SDRAM controller and interface
- DDR SDRAM controller and interface
- Flash memory interface
- SRAM memory interface
- Block RAM interface

Networking Interfaces:
- Single-channel HDLC controller
- ATM Utopia L2 master and slave controller
- 10/100 Ethernet Media Access Controller (MAC) (Full and Lite versions)

In-Circuit Hardware Debugger Support
- EDK Software Debugger
- Requires MicroBlaze Hardware Debug Module
- Connects via FPGA JTAG port using Xilinx Parallel Cable IV
- Nohau In-Circuit Hardware Debugger for MicroBlaze FPC

Once the basic structure of the processor system is in place the designer can enter Phase 2 of the design process and begin allocating system level parameters to the processor design (Figure 19.49). These include the desired address map for the processor system, the selection of interrupt priorities, the allocation of the standard input, and standard output devices from the included peripheral set.

The tools can now be used to configure the system memory to contain the executable software code taken from the supplied C compiler. Memory values are allocated to the internal block RAM memory located in the FPGA core. External memory can also be configured using the various utilities supplied with the EDK.

Once the designer has completed configuring their processor system with the XPS tool, the tool performs the hard work. In a matter of seconds the hardware side of the design will be constructed and made available to the designer as a black box module. This module is then ready to be installed into the FPGA design. The tool assists the designer with this task by creating a Hardware Description Language (HDL) template.

The driving force behind all of this hard work are the Xilinx Platform Generator (PlatGen) and Library Generator (LibGen) tools. XPS will customize and synthesize each peripheral to meet the needs of the designer from the text-based Microprocessor Hardware Specification file (MHS) and the Microprocessor Software Specification file (MSS). Bus structures and banks of internal block RAM

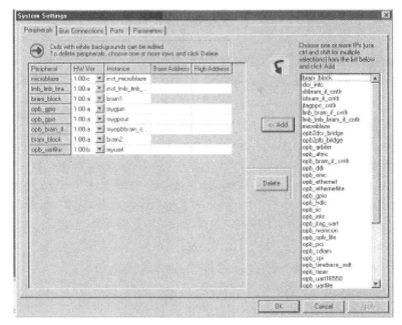

Figure 19.49: System Settings Window

are also customized and synthesized in the same manner before PlatGen connects all of the system components together. This task is transparent to the user. Users no longer need to observe the complicated and often essential design/protocol rules for the supported CoreConnect bus structures.

However, the design automation does not stop here. In the background the LibGen will also have created software libraries customized to the processor design. They contain just the software functions that can be accessed with the chosen peripheral set. Inaccessible software routines are removed from the libraries to prevent wasting valuable BlockRAM within the FPGA. The previous selections of the standard input and output peripherals are used to tailor commonly used C software functions like .printf. and .scanf., for the software design team. Device drivers are also created in C for each of the selected peripherals. By using previously determined instance names selections are created with appropriately named C functions customized for each peripheral. Interrupt control and its associated housekeeping are also handled automatically. The user simply supplies the name of the C function that should be executed when the interrupt occurs. Interrupt handler routines can be assigned to each peripheral individually and interrupt priority automatically encoded in the Interrupt Service Routine by the LibGen tool.

The completed processor system maintains totally flexibility. The designer, using the automated toolset to allow maximum flexibility and eliminate waste, can quickly reflect any change in the system specification.

The use of Xilinx Relationally Placed Macro (RPM) technology guarantees performance of the processor core. Predetermined placement of the logic elements in the FPGA guarantees effortless and repeatable performance figures. The completed system can be implemented in a wide range of the Xilinx FPGA devices. Any member of the Virtex and Spartan-3 families is a valid target for a

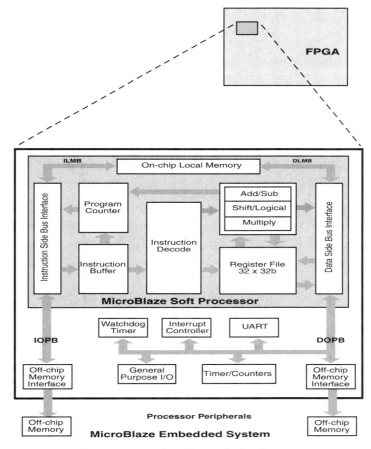

Figure 19.50: MicroBlaze Block Diagram

MicroBlaze system (Figure 19.50). The use of multiple MicroBlaze processors sharing the same bus structure allows the concepts of scalable distributed processing to become reality. Use of the Spartan-IIE device family permits low cost and high volume applications to be an ideal choice for MicroBlaze controlled systems.

Compatibility with the On-chip Peripheral Bus (OPB) from the IBM CoreConnect bus family lets the designer effortlessly include a vast range of existing IP into their MicroBlaze designs. CoreConnect compatibility also offers a simplified upgrade path if the designer wishes to migrate their processor design to the industry leading Virtex-II Pro FPGAs containing the ultra high-performance embedded PowerPC processor cores.

Microprocessor/Microcontroller Obsolescence

Obsolescence is on the mind of most design engineers. Microprocessors and microcontrollers have short life spans. Driven by consumer market trends and the ever-present need for speed enhancements, they are discontinued on short notice. Consumer products such as game consoles and mobile phones have built in obsolescence to stimulate sales of the latest and greatest products. Built-in obsolescence leads microprocessor manufacturers to chase new platform introductions and high volume sales and proliferates the obsolescence ripple effect.

Even if the design has been coded in a "portable code" such as 'C', there are always architecture specific instructions and features to hamper the migration between an obsolete processor and a next generation device. The changeover process is further complicated by different package options and I/O configurations necessitating the need for a complete board re-spin.

Xilinx offers the ability to purchase the MicroBlaze source code. This approach removes any hurdles from obsolescence and guarantees product availability. The MicroBlaze Source Code is delivered as Structural VHDL source and has been optimized for high performance in Xilinx FPGAs. Users can port the core across Xilinx product lines or even target it to an ASIC.

Using FPCs in Digital Consumer Applications

The Xilinx FPC solution allows users to create low-cost, customized processors with peripherals, memory, and logic all on a single cost-optimized programmable logic device. It can then be easily adapted to changing requirements even after the product is in the field. Xilinx provides everything users need including development tools and a wide range of IP to help get a better product, that will stay in the market longer, and get to market sooner. There is no faster, easier, or cheaper way to develop microcontroller-based products. Not only can users benefit from the flexibility, integration and upgradeability offered by programmable logic, but they can also take advantage of a processor tailored to their design needs.

The MicroBlaze soft processor core and its associated toolkit offer a new and powerful approach to embedded processor system design. Never before has this level of flexibility been available to the digital electronics industry. The combination of the high-performance MicroBlaze processor and low cost Spartan-3 FPGA make this solution extremely valuable to the modern digital designer.

Automotive Telematics Solutions

The next few years will be a proverbial minefield for automotive electronics designers. Choosing the correct data bus will be crucial to the success of integrating and testing units in production and long after the car has rolled off the assembly line. The problem is amplified for Tier 1 suppliers and aftermarket unit design companies who supply units to many OEMs. The OEM's can all opt for different data busses and protocols. The industry has seen a huge shift of design philosophy away from designing a different unit for every OEM, and every car model, to reconfigurable platforms. Reconfigurable platforms cleverly partitioned between software and reprogrammable hardware allow the designer to change the choice of system bus or interface late in the design process or even during production. The reconfigurable system concept enables different standards and protocols to be tried, tested and put on road trials. If they are found to be unsuitable, another bus interface can be loaded into the system and tested until the best configuration is found.

While this may seem unobtainable, it can be realized today by utilizing FPGAs and CPLDs. PLDs can hand the designer back control over all phases of design from prototype, pre-production, and all phases of production. This flexibility and control can be lost when developing systems based on ASICs and ASSP devices. Because generic PLD devices can be used across many projects and are not application specific, they can also alleviate over-stocking and inventory issues. Once the programmable logic based unit is on the road it can even be reconfigured remotely via a wireless communication link to allow for system upgrades or extra functions.

The reconfigurable hardware platform can be brought to market quickly by utilizing drop-in IP core blocks. For example, Memec Design recently announced the availability of a cost optimized CAN core interface. The Memec CAN core contains the complete data link layer including the framer, transmit and receive control, error core design, and flexible interface. It enables access to each internal status and frame reference. Bit rate and sub-bit segments can be configured to meet the required timing specification of the connected CAN bus. Error counters and error interrupt events report errors. The core is designed to provide a bus bit rate of up to 1 Mb/s, with a minimum core clock frequency of 8 MHz. The CAN core can provide an interface between the message filter, the message priority mechanism, and various system functions such as sensor/activator control. Alternatively, it can be embedded into a system application interfacing with the microprocessor and various peripheral functions. Another example is the LIN core available from Intelliga. The LIN core (iLIN) is a supplied reference design that uses a synchronous 8-bit general-purpose microcontroller interface with minimal buffering for the transportation of message data. In addition, the reference design includes a single slave message response filter and a software interface that allows the connected microcontroller to perform address filtration.

Instead of using a discrete device, IP cores can be used as part of a more complex design to provide the interface to the CAN or LIN bus. Reducing component count and lowering overall system cost reduces inventory, increases system reliability, and also can help reduce PCB complexity and layers. Figure 19.34 shows a generic in-car multimedia design showing the use of the CAN core coupled with PCMCIA interface, PCI bridging, IDE interface, and other functions. These functions can be modified, changed, or enhanced during the design phase or modified based on the end customer requirements. With customization taking place within the FPGA instead of at the board level one PCB can be used for many customers. This model can be extended to include modification or upgrades in the field utilizing a wireless connection to reconfigure the in-system FPGA.

Clock Management

The display driver, system timing, and integrating memory are key aspects to any display system design. The solution can also be technology and supplier specific. Xilinx FPGAs are an excellent platform for implementing these functions for two reasons. First, they are easily programmable to meet varying requirements. Second, their flexibility allows multiple versions to be implemented. This latter capability is unique and can increase the number of options with a customer's supply chain. This flexibility can more than pay for itself through increased leverage when negotiating with suppliers.

Digital Clock Managers (DCMs) provide advanced clocking capabilities to Spartan-3 FPGA applications. DCMs optionally multiply or divide the incoming clock frequency to synthesize a new clock frequency. DCMs improve system performance by eliminating clock skew. An optional DCM phase shifts the clock output to delay the incoming clock by a fraction of the clock period. The DCMs integrate directly with the FPGAs global low-skew clock distribution network.

DCMs integrate advanced clocking capabilities into the Spartan-3 global clock distribution network. Features of the Spartan-3 DCMs solve a variety of common clocking issues especially in high-performance, high frequency applications:

- Multiply or Divide an Incoming Clock Frequency. Or synthesize a completely new frequency by a mixture of clock multiplication and division.

- Condition a Clock to ensure a clean output clock with a 50% duty cycle.
- Phase Shift a clock signal either with a fixed fraction of a clock period or by precise increments.
- Eliminate Clock Skew either within the device or to external components. This improves overall system performance and eliminates clock distribution delays.
- Mirror, forward, or re-buffer a Clock Signal, often for a de-skew. Convert the incoming clock signal to a different I/O standard—for example, forwarding and converting an incoming LVTTL clock to LVDS.
- Simultaneously perform any, or all, of the above functions.

Table 19.11: Digital Clock Manager Features and Capabilities

Feature	Description	DCM Signals
Digital Clock Managers (DCMs) per device	• 4, except in XC3S50 • 2 in XC3S50	All
Digital Frequency Synthesizer (DFS) Input Frequency Range*	1 MHz to ~326 MHz	CLKIN
Delay-Locked Loop (DLL) Input Frequency Range*	24 MHz to ~326 MHz	CLKIN
Clock Input Sources	• Global buffer input pad • Global buffer output • General-purpose I/O (no deskew) • Internal logic (no deskew)	CLKIN
Frequency Synthesizer Output	Multiply CLKIN by the fraction (M/D) where M={2..32}, D={1..32}	• CLKFX • CLKFX180
Clock Divider Output	Divide CLKIN by 1.5, 2, 2.5, 3, 3.5, 4, 4.5, 5, 5.5, 6, 6.5, 7, 7.5, 8, 9, 10, 11, 12, 13, 14, 15, or 16	CLKDV
Clock Doubler Output	Multiply CLKIN frequency by 2	• CLK0 • CLK2X180
Clock Conditioning, Duty-Cycle Correction	Always provided on most outputs, Optional on CLK0, CLK90, CLK180, CLK270. 50% duty cycle ± 100ps*	All
Quadrant Phase Shift Outputs	0° (no phase shift), 90° (1/4 period), 180° (1/2 period), 270° (3/4 period)	• CLK2X • CLK90 • CLK180 • CLK270
Half-period Phase Shift Outputs	Output pairs with 0° and 180° phase shift, ideal for DDR applications	• CLK0, CLK180 • CLK2X, CLK2X180 • CLKFX, CLKFX180
Dynamic or Fixed Phase Shift Resolution	Down to 1/256th of a clock period (or ~30 to 50 ps)*	All
Number of Clock Outputs to General-purpose Interconnect	Up to all 9	All
Number of Clock Outputs to Global Clock Network	Any 4 of 9	All
Number of Clock Outputs to Output Pins	Up to all 9	All

As shown in Figure 19.51, most Spartan-3 FPGAs have four DCM blocks. The DCM blocks are located at the top and bottom of the block RAM/multiplier columns along the left and right edges. The XC3S50 has two DCMs located along the top and bottom of the block RAM/multiplier column along the left edge of the device.

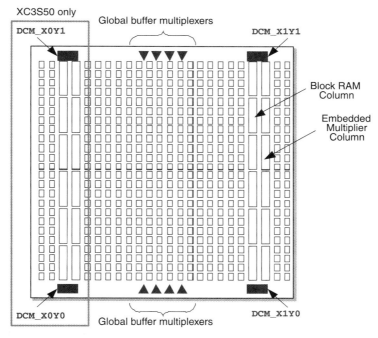

Figure 19.51: Location of the Four DCM Blocks on Spartan-3 FPGAs

The DCM blocks have dedicated connections to the global buffer inputs and global buffer multiplexers on either the top or bottom edge of the device. As shown in Figure 19.52, DCMs are an integral part of the FPGAs global clocking infrastructure. DCMs are an optional element in the clock distribution network and are available when required by the application. In Figure 19.52 a, a clock input feeds directly into the low-skew, high-fanout global clock network via a global input buffer and global clock buffer. If the application requires some, or all, of the DCM's advanced clocking features the DCM fits neatly between the global buffer input and the buffer itself as shown in Figure 19.52 b.

a. Global Buffer Inputs and Clock Buffers Drive a Low-Skew Global Network in the FPGA

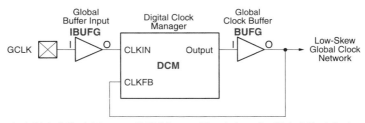

b. A Digital Clock Manager (DCM) Inserts Directly into the Global Clock Path

Figure 19.52a and b: DCMs are an Integral Part of the FPGAs Global Clock Network

DCM Functional Overview

The single entity called a Digital Clock Manager (DCM) actually consists of four distinct functional units as depicted in Figure 19.53 and described below. These units can operate independently or in tandem.

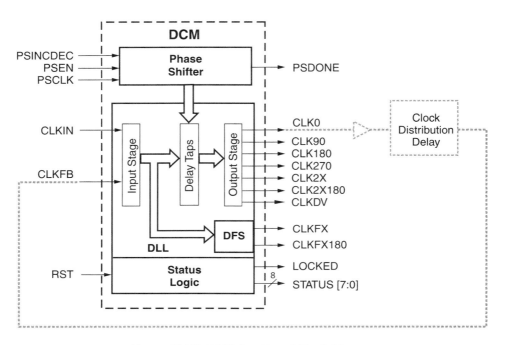

Figure 19.53: DCM Functional Block Diagram

The four distinct parts of the DCM include:

■ **Delay-Locked Loop (DLL)** – The Delay-Locked Loop (DLL) unit provides an on-chip digital deskew circuit that generates zero-propagation-delay clock output signals. The deskew circuit compensates for the delay on the routing network by monitoring either the CLK0 or the CLK2X output clock. The DLL unit effectively eliminates the delay from the external clock input port to the individual clock loads within the device. The well-buffered global network minimizes the clock skew on the network caused by loading differences. The input signals to the DLL unit are CLKIN and CLKFB. The output signals from the DLL are CLK0, CLK90, CLK180, CLK270, CLK2X, CLK2X180, and CLKDV. The DLL unit generates the outputs for the Clock Doubler (CLK2X, CLK2X180), the Clock Divider (CLKDV), and the Quadrant Phase Shifted Outputs functions.

- **Digital Frequency Synthesizer (DFS)** – The DFS provides a wide and flexible range of output frequencies based on the ratio of the two user-defined integers Multiplier (CLKFX_MULTIPLY) and Divisor (CLKFX_DIVIDE). The output frequency is derived from the input clock (CLKIN) by simultaneous frequency division and multiplication. This feature can be used with or without the DLL feature of the DCM. If the DLL is not used there is no phase relationship between CLKIN and the DFS outputs. The DFS unit generates the Frequency Synthesizer (CLKFX, CLKFX180) outputs.

- **Phase Shift (PS)** – The Phase Shift (PS) unit controls the phase relations of the DCM's clock outputs to the CLKIN input. The Phase Shift unit shifts the phase of all nine DCM clock output signals by a fixed fraction of the input clock period. The fixed phase shift value is set at design time and loaded into the DCM during FPGA configuration. The Phase Shift unit also provides a digital interface for the FPGA application to dynamically advance or retard the current shift value by 1/256th of the clock period. The input signals to the Phase Shift unit are PSINCDEN, PSEN, and PSCLK. The output signals are PSDONE and the STATUS[0] signal.

- **Status Logic** – The Status Logic indicates the current state of the DCM via the LOCKED and STATUS[0], STATUS[1], and STATUS[2] output signals. The LOCKED output signal indicates whether the DCM outputs are in phase with the CLKIN input. The STATUS output signals indicate the state of the DLL and PS operations. The RST input signal resets the DCM logic and returns it to its post-configuration state. A reset forces the DCM to reacquire and lock to the CLKIN input.

The Xilinx ISE development software simplifies applications using DCMs by including a software wizard that provides step-by-step instructions for DCM configuration. The DCM Wizard generates a vendor-specific logic synthesis file instantiating the DCM in either VHDL or Verilog syntax. The DCM Wizard generates a user constraints (UCF) file for the specific implementation. All user specifications are saved in a Xilinx Architecture Wizard (XAW) settings file.

The Spartan-3 DCM is a significant enhancement over the Spartan-II/IIE Delay-Locked Loop (DLL) function. A Spartan-3 DCM provides all the capabilities of the Spartan-II/IIE DLL plus new Frequency Synthesizer and phase shifting functions capabilities. The Spartan-3 Frequency Synthesizer multiplies an input clock by up to a factor of 32. The Spartan-II/IIE DLL has limited frequency multiplication capabilities only allow an input clock to be doubled. The Spartan-3 DCM has a wider divider range compared to Spartan-IIE DLLs.

The sixteen low-skew global clock lines and up to twelve Digital Clock Manager (DCM) circuits provide superior flexibility for high-performance clocking. The feedback in each DCM can be used to eliminate on-chip clock delay or even board clock delay. Each DCM has nine clock outputs and can drive up to four global clocks. The clock outputs provide coarse phase shifting with four-quadrant outputs. The clock can also be phase shifted with a resolution of 256 steps per clock period during configuration or phase stepped during operation. Multiples and fractions of the clock frequency are available. The FX output even provides simultaneous multiplication and division of the input frequency by any set of numbers up to 32. The DCM can eliminate the clock distribution delay since all large device flip-flops are clocked with a timing skew of less than 100ps. This eliminates concerns about internal hold time issues and guarantees short pin-to-pin input setup times as well as short clock-to-output delays.

Systems with a common (system-synchronous) clock distribution on the PCB do not experience an input hold time requirement and outputs have a specified (min) delay. I/O performance is determined by the max pin-to-pin parameters and is almost independent of chip size. Source-synchronous systems using clock forwarding have a narrow data capture window where the clock can be phase adjusted to capture data in the middle of the arriving eye pattern. The internal clock signal is delayed by the clock distribution within the chip in a system-synchronous timing diagram without DCM. This increases the output delay and makes the input set-up time less predictable. The DCM increases the total timing margin by eliminating the on-chip clock distribution delay. The receiver must recover the data by clocking close to the center of the valid data eye pattern.

Memories and Memory Controllers/Interfaces

Many applications, such as digital video applications like set-top boxes and displays, require high bandwidth memory solutions. Xilinx offers programmable logic solutions with abundant on-chip memory resources and support for off-chip data storage through external memory interfaces to meet high-speed system requirements.

For applications requiring large, on-chip memories, Spartan-3 FPGAs provides plentiful and efficient SelectRAM memory blocks. By using various configuration options SelectRAM blocks create RAM, ROM, FIFOs, large look-up tables, data width converters, circular buffers, and shift registers. Each supports various data widths and depths. The increased densities of the Spartan-3 families have improved the on-chip memory offerings and enabled many new applications. The Spartan and Spartan-XL families only had look-up table based distributed RAM elements. These are suitable for small storage elements such as DSP processing or small FIFOs. The Spartan-3 family retains the distributed RAM and adds the larger block RAM elements. These are excellent for large FIFOs and buffers such as video line buffers or packet buffers found in small office routing equipment. Wide and shallow memory structures such as SRL16 shift register logic and distributed memory are ideal for building compact DSP structures such as filters, small FIFOs, and scratch pad memories.

Spartan-3 FPGAs are also used as memory controllers to interface with the different types of memories such as SRAM, DRAM, CAM and flash.

On-Chip Memory Solutions—Look-Up Tables as Distributed RAM

In addition to the embedded 18Kbit block RAMs Spartan-3 FPGAs feature distributed RAM within each Configurable Logic Block (CLB). Each SLICEM function generator or LUT within a CLB resource optionally implements a 16-deep x 1-bit synchronous RAM. The LUTs within a SLICEL slice do not have distributed RAM.

Distributed RAM writes synchronously and reads asynchronously. However, if required by the application, it can use the register associated with each LUT to implement a synchronous read function. Each 16 x 1-bit RAM can be cascaded for deeper and/or wider memory applications. It also offers a minimal timing penalty incurred through specialized logic resources. Spartan-3 CLBs support various RAM primitives up to 64-deep by 1-bit-wide. Two LUTs within a SLICEM slice combine to create a dual-port 16x1 RAM—one LUT with a read/write port and a second LUT with a read-only port. One port writes into both 16x1 LUT RAMs simultaneously. The second port reads independently.

Distributed RAM is crucial to many high-performance applications, such as FIFOs or small register files that require relatively small embedded RAM blocks. The Xilinx CORE Generator software automatically generates optimized distributed RAMs for the Spartan-3 architecture. Similarly, CORE Generator creates Asynchronous and Synchronous FIFOs using distributed RAMs.

Distributed RAM supports the following memory types:

- Single-port RAM with synchronous write and asynchronous read. Synchronous reads are possible using the flip-flop associated with distributed RAM.
- Dual-port RAM with one synchronous write and two asynchronous read ports. Synchronous reads are also possible.

As illustrated in Figure 19.54, dual-port distributed RAM has one read/write port and an independent read port.

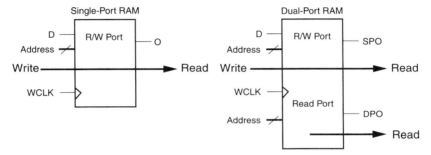

Figure 19.54: Single-Port and Dual-Port Distributed RAM

Any write operation on the D input and any read operation on the SPO output can occur simultaneously with, and independent from, a read operation on the second read-only DPO port. The operations include:

- **Write Operation** – The write operation is a single clock-edge operation controlled by the write-enable input (WE). By default, WE is active High although it can be inverted within the distributed RAM. When the write enable is High, the clock edge latches the write address and writes the data on the D input into the selected RAM location. When the write enable is Low, no data is written into the RAM.
- **Read Operation** – A read operation is purely combinatorial. The address port—either for single- or dual-port modes—is asynchronous with an access time equivalent to a LUT logic delay.
- **Read During Write** – When synchronously writing new data, the output reflects the data being written to the addressed memory cell. This is similar to the WRITE_MODE=WRITE_FIRST mode on the Spartan-3 block RAMs.

On-Chip Memory Solutions—Look-Up Tables (LUTs) as Shift Registers (SRL-16)

Spartan-3 FPGAs can configure the LUT in a SLICEM slice as a 16-bit shift register without using the available flip-flops for each slice. Shift-in operations are synchronous with the clock and output length is dynamically selectable. A separate dedicated output allows the cascading of any number of 16-bit shift registers to create size shift registers as needed. Each CLB resource can be configured using four of the eight LUTs as a 64-bit shift register.

These shift registers enable the development of efficient designs for applications that require delay or latency compensation. Shift registers are also useful in synchronous FIFO and Content-Addressable Memory (CAM) designs. The CORE Generator RAM based Shift Register module can be used to quickly generate a Spartan-3 shift register without using flip-flops . An example is using the SRL16 element(s).

The structure of the SRL16 will be described from the bottom up, starting with the shift register and building up to the surrounding FPGA structure. The LUT can be described as a 16:1 multiplexer with the four inputs serving as binary select lines. Values are programmed into the LUT serving as the data being selected (see Figure 19.55).

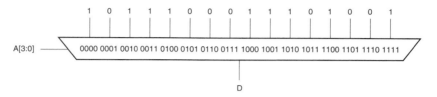

Figure 19.55: LUT Modeled as a 16:1 Multiplexer

With the SRL16 configuration, fixed LUT values are configured as an addressable shift register (see Figure 19.56). The shift register inputs are the same as those for the synchronous RAM configuration of the LUT including a data input, clock, and clock enable (not shown). A special output for the shift register is provided from the last flip-flop called Q15 on the library primitives or MC15 in the FPGA Editor. The LUT inputs asynchronously, or dynamically, to select one of the 16 storage elements in the shift register.

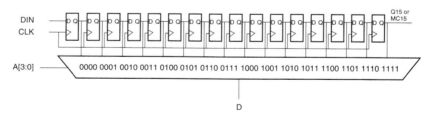

Figure 19.56: LUT Configured as an Addressable Shift Register

The address can be thought of as dynamically changing the length of the shift register. If D is used as the shift register output instead of Q15, setting the address to 7 (0111) selects Q7 as the output, emulating an 8-bit shift register. Note that since the address lines control the mux, they provide an asynchronous path to the output.

Each SRL16 LUT has an associated flip-flop that makes up the overall logic cell. The addressable bit of the shift register can be stored in the flip-flop for synchronous output or can be fed directly to a combinatorial output of the CLB. When using the register it is best to have fixed address lines selecting a static shift register length. The clock-to-output delay of the flip-flop is faster than the shift register so performance can be improved by addressing the second-to-last bit and then using the flip-flop as the last stage of the shift register. Using the flip-flop also allows for asynchronous or synchronous output to beset or reset. The shift register input can come from a dedicated SHIFTIN signal. The Q15/MC15 signal from the last stage of the shift register can drive a SHIFTOUT output. The addressable D output is available in all SRL primitives. The Q15/MC15 signal that can drive SHIFTOUT is only available in the cascadable SRLC16 primitive.

Figure 19.57: Logic Cell SRL Structure

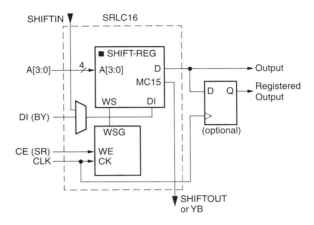

The two logic cells within a slice are connected via the SHIFTOUT and SHIFTIN signals for cascading a shift register up to 32 bits (see Figure 19.58). These connect the Q15/MC15 of the first shift register to the DI (or Q0 flip-flop) of the second shift register.

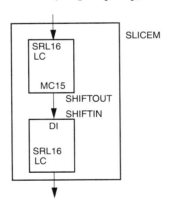

Figure 19.58: Shift Register Connections between Logic Cells in a Slice

Multiplexing together the two separate data outputs from each SRL16 allow dynamic addressing (or "dynamic length adjustment"). One of the two SRL16 bits can be selected by using the F5MUX to make the selection (see Figure 19.59).

The Spartan-3 CLB contains four slices. Each contains two LUTs but only two allow LUTs to be used as SRL16 components or distributed RAM. The two left-hand SLICEM components allow their two LUTs to be configured as a 16-bit shift register. The same cascading of SHIFTOUT to SHIFTIN available between the LUTs in the SLICEM is also available to connect the two SLICEM components. The four left-hand LUTs of a single CLB can be combined to produce delays up to 64 clock cycles (see Figure 19.60).

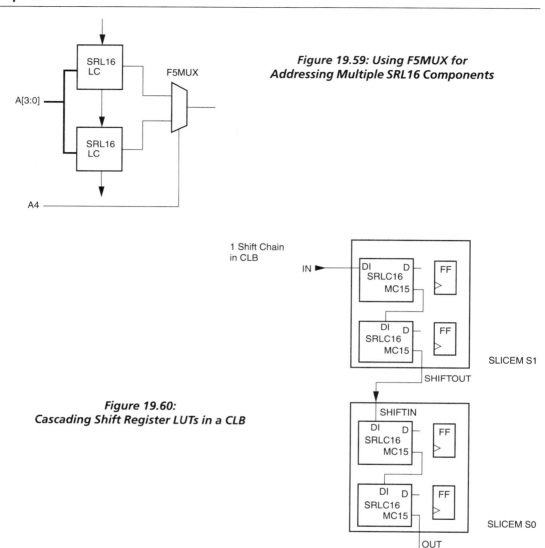

Figure 19.59: Using F5MUX for Addressing Multiple SRL16 Components

Figure 19.60: Cascading Shift Register LUTs in a CLB

The multiplexers can be used to address multiple SLICEMs similar to the description for combining the two LUTs within a SLICEM. The F6MUX can be used to select from three or four SRL16 components in a CLB. This provides up to 64 bits of addressable shift register (see Figure 19.61).

Shift Register Operations and Data Flow, with each shift register (SRL16 primitive) supporting:
- Synchronous shift-in
- Asynchronous 1-bit output when the address is changed dynamically
- Synchronous shift-out when the address is fixed

In addition, cascadable shift registers (SRLC16) support synchronous shift-out output of the last (16th) bit. This output has a dedicated connection to the input of the next SRLC16 inside the CLB resource. Two primitives are illustrated in Figure 19.62.

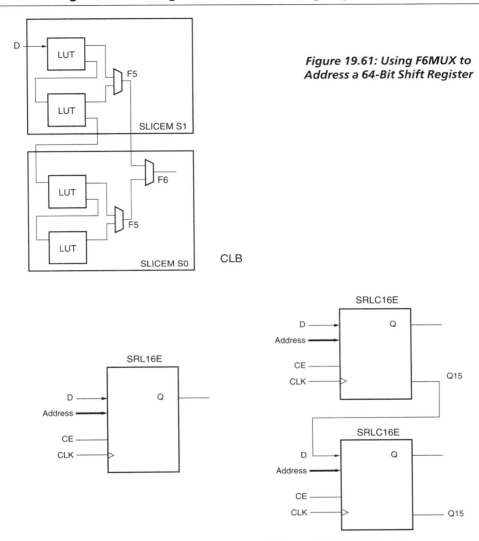

Figure 19.61: Using F6MUX to Address a 64-Bit Shift Register

Figure 19.62: Shift Register and Cascadable Shift Register

SRL16 Applications

Delay Lines: The register-rich nature of the Xilinx FPGA architecture allows for the addition of pipeline stages to increase throughput. Balancing data paths will maintain the desired functionality. The SRL16 can be used when additional clock cycles of delay are needed anywhere in the design (see Figure 19.63).

Linear Feedback Shift Registers (LFSRs): LFSRs sequence through 2n-1 states, where n is the number of flip-flops. Feeding specific bits back through an XOR or XNOR gate creates the sequence. In instances were the count sequence is not important (e.g., FIFOs), LFSRs can replace conventional binary counters in performance critical applications. LFSRs are also used as pseudorandom number generators and are important building blocks in encryption and decryption algorithms.

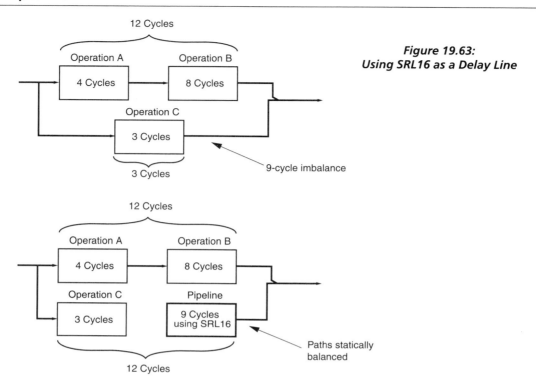

Figure 19.63:
Using SRL16 as a Delay Line

Maximal-length LFSRs need taps taken from specific positions within the shift register. There are multiple ways these taps can be made available in the SRL16 configuration. One is by addressing the necessary bit in a given SRL16 while allowing the Q15 to cascade to the next SRL16. Another is to use flip-flops to "extend" the SRL16 where necessary to access the tap points. For example, Figure 19.64 shows how a 52-bit LFSR can be implemented in one CLB with the feedback coming from bits 49 and 52. A third method is to duplicate the LFSR in multiple SRLs and address each one's different bits. Users can also generate multiple addresses in one SRL clock cycle to capture multiple bit positions. The XNOR gate required for any LFSR can be conveniently located in the SLICEL part of the CLB.

Gold Code Generator: Gold code generators are used in CDMA systems to generate code sequences with good correlation properties (see Figure 19.65). The result is a set of codes ideally suited to distinguish one code from another in a spectrum full of coded signals. Figure 19.65 shows an implementation of a Gold code generator. The logic required to initially fill the LFSR and provide the feedback can be located in the SLICEL parts of the CLB. See XAPP217 for more details.

FIFOs: Synchronous FIFOs can be built out of the SRL16 components. These provide up to 64 bits per CLB and are useful when other resources become scarce. Block RAM is the most efficient resource to use for larger FIFOs.

Counters: Any desired repeated sequence of 16 states can be achieved by feeding each output with an SRL16. Cascading the SRL16 allows even longer arbitrary count sequences. A terminal count can be generated by using the standard carry chain (see Figure 19.67).

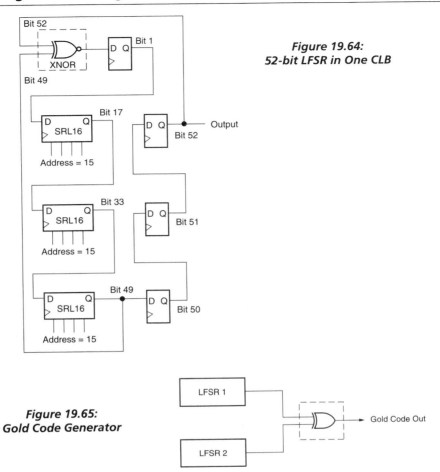

Figure 19.64:
52-bit LFSR in One CLB

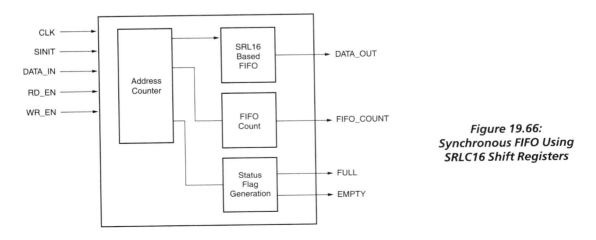

Figure 19.65:
Gold Code Generator

Figure 19.66:
Synchronous FIFO Using
SRLC16 Shift Registers

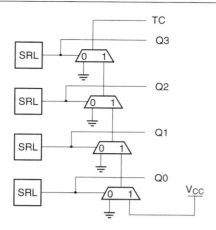

Figure 19.67:
SRL-Based Counter with Terminal Count

The SRL16 configuration of the Spartan-3 look-up table provides a space-efficient shift register. This would otherwise require 16 flip-flops. This feature will be automatically used when a small shift register is described in HDL code. However, creative consideration of the uses of the SRL16 can provide even more significant advantages in many applications.

On-Chip Memory Solutions—Block RAM

Deep and wide structures using block RAM are useful as large FIFOs for storing partial products and coefficients, real-time video line buffers, and packet line buffers. All Spartan-3 devices feature multiple block RAM memories organized in columns. The total amount of block RAM memory depends on the size of the Spartan-3 device, as shown in Table 19.12.

Table 19.12: Block RAM Available in Spartan-3 Devices

Spartan-3 Device	RAM Columns	RAM Blocks Per Column	Total RAM Blocks	Rotal RAM Bits	Total RAM Kbits
XC3S50	1	4	4	73,728	72K
XC3S200	2	6	12	221,184	216K
XC3S400	2	8	16	294,912	288K
XC3S1000	2	12	24	442,368	432K
XC3S1500	2	16	32	589,824	576K
XC3S2000	2	20	40	737,280	720K
XC3S4000	4	24	96	1,769,472	1,728K
XC3S5000	4	26	104	1,916,928	1,872K

Notes:
1. 1Kbit = 1,024 bits, per memory conventions.

Each block RAM contains 18,432 bits of fast static RAM, with16K bits allocated to data storage. In some memory configurations an additional 2K bits are allocated to parity or additional "plus" data bits. Physically the block RAM memory has two completely independent access ports— labeled Port A and Port B. The structure is fully symmetrical, both ports are interchangeable, and both ports support data read and write operations. Each memory port is synchronous with its own clock, clock enable, and write enable. Read operations are also synchronous and require a clock edge and clock enable. Though physically a dual-port memory, block RAM simulates single-port memory in an application as shown in Figure 19.68. Each block memory supports multiple configurations or aspect ratios. Table 19.13 summarizes the essential SelectRAM features. Users can cascade multiple BlockRAMs to create deeper and wider memory organizations with minimal timing penalties incurred through specialized routing resources.

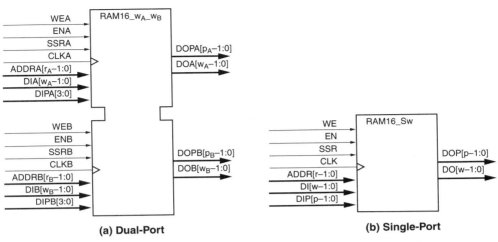

(a) Dual-Port   **(b) Single-Port**

Figure 19.68: SelectRAM 18K Blocks Perform as Dual-Port (a) and Single-Port (b) Memory

Total RAM bits, including parity	**18,432 (16K data + 2K parity)**
Memory Organizations	16Kx1 8Kx2 4Kx4 2Kx8 (no parity) 2Kx9 (x8 + parity) 1Kx16 (no parity) 1Kx18 (x16 + 2 parity) 512x32 (no parity) 512x36 (x32 + 4 parity) 256x72 (single-port only)
Parity	Available and optional for organizations greater than byte-wide. Parity bits optionally available as extra data bits.
Performance	200 MHz (estimated)
Timing Interface	Simple synchronous interface. Similar to reading and writing from a register with a setup time for write operations and clock-to-output delay for read operations.
Single-Port	Yes
True Dual-Port	Yes
ROM, Initial RAM Contents	Yes
Mixed Data Port Widths	Yes
Power-Up Condition	User-defined data, defaults to zero
Potential Applications	Local data storage, FIFOs, elastic stores, register files, buffers, stacks, circular buffers, shift registers, delay lines, waveform storage and generation, direct digital synthesis, CAMs, associative memories, function tables, function generators, wide logic functions, code converters, encoders, decoders, counters, state machines, microsequencers, program storage for embedded processor(s).

Table 19.13:
SelectRAM 18K
Block Memory Features
and Applications

The Xilinx CORE Generator system supports various modules containing block RAM for Spartan-3 devices including:

- Embedded dual- or single-port RAM modules
- ROM modules
- Synchronous and asynchronous FIFO modules
- Content-Addressable Memory (CAM) modules

Block RAM can be instantiated in any synthesis-based design using the appropriate "RAMB16" module from the Xilinx design library.

As mentioned previously block RAM is organized in columns. Figure 19.69 shows the block RAM column arrangement for the XC3S200. The XC3S50 has a single column of block RAM located two CLB columns from the left edge of the device. Spartan-3 devices larger than the XC3S50 have two columns of block RAM. They are adjacent to the left and right edges of the die and located two columns of CLBs from the I/Os at the edge. In addition to the block RAM columns at the edge, the XC3S4000 and XC3S5000 have two additional columns distributed between the two edge columns. Table 19.12 describes the number of columns and the total amount of block RAM on a specific device. The edge columns make block RAM particularly useful in buffering or resynchronizing buses entering or leaving the Spartan-3 device.

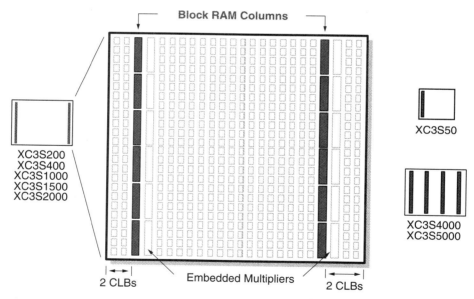

Figure 19.69: Block RAM Arranged in Columns with Detailed Floorplan of XC3S200

Immediately adjacent to each block RAM is an embedded 18x18 hardware multiplier. Co-locating block RAM and embedded multipliers improves the performance of some digital signal processing functions. Special interconnects surrounding the block RAM provide efficient signal distribution for address and data. Special provisions allow multiple block RAM to be cascaded to create wider or deeper memories.

Spartan-3 block RAM is constructed of true dual-port memory and simultaneously supports all the data flows and operations shown in Figure 19.70. Both ports access the same set of memory bits. The port's data width determines which different address schemes are used.

- Port A behaves as an independent single-port RAM supporting simultaneous read and write operations using a single set of address lines.
- Port B behaves as an independent single-port RAM supporting simultaneous read and write operations using a single set of address lines.
- Port A is the write port with a separate write address and Port B is the read port with a separate read address. The data widths for Port A and Port B can be different.
- Port B is the write port with a separate write address and Port A is the read port with a separate read address. The data widths for Port B and Port A can be different.

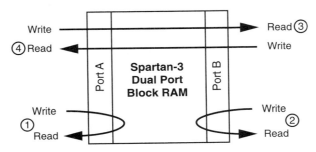

Figure 19.70: Block RAM Support Single- and Dual-Port Data Transfers

Block RAM Applications

Creating Larger RAM Structures – Block SelectRAM columns have specialized routing to allow cascading blocks with minimal routing delays. Wider or deeper RAM structures incur a small delay penalty.

Block RAM as Read-Only Memory (ROM) – By tying the write enable input Low, block RAM optionally functions as registered block ROM. The ROM outputs are synchronous and require a clock input and perform exactly like a block RAM read operation. The ROM contents are defined by the initial contents at design time. After design compilation, the ROM contents can also be updated using the Data2BRAM utility described below.

FIFOs – First-In, First-Out (FIFO) memories, also known as elastic stores, are perhaps the most common application of block RAM, other than for random data storage. FIFOs typically resynchronize data, either between two different clock domains, or between two parts of a system that have different data rates, even though they operate from a single clock. The Xilinx CORE Generator system provides two parameterizable FIFO modules. One is a synchronous FIFO where both the read and write clocks are synchronous to one another. The other an asynchronous FIFO where the read and write clocks are different.

Storage for Embedded Processors – Block RAM also enables efficient embedded processor applications. RAM performs a variety of functions in an embedded processor such as those listed below:

- Register file for processor register set. For some processors, distributed RAM may be a preferred solution.

- Stack or LIFO for stack-based architectures and for call stacks.
- Fast, local code storage. The fast access time to internal block RAM significantly boosts the performance of embedded processors. However, on-chip storage is limited by the number of available block RAMs.
- Large dual-ported mailbox memory shared with external processor or DSP device.
- Temporary trace buffers to ease and enhance application debugging.

Updating Block RAM/ROM Content by Directly Modifying Device Bitstream – In a typical design flow, the initial contents of block RAM/ROM is defined at design time and compiled into the device bitstream that is downloaded to and configures a Spartan-3 FPGA. However, for some applications the actual memory contents may not be known when the bitstream is created, or it may change later. One example is if a processor embedded with the Spartan-3 FPGA uses block RAM to store program code. To avoid re-compiling the FPGA design just to incorporate a code change, Xilinx provides a utility called Data2BRAM that updates an existing FPGA bitstream with new block RAM/ROM contents. As shown in Figure 19.71, the inputs to Data2BRAM include:

- The new RAM contents—typically the output from the embedded processor compiler/linker
- The present FPGA bitstream
- A file that describes both the mapping between the system address space and the addressing used on the individual block RAMs
- The physical location of each block RAM

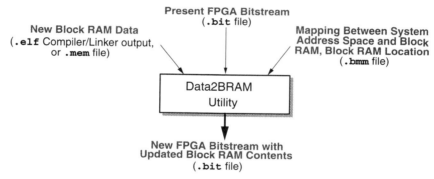

Figure 19.71: The Data2BRAM Utility Updates Block RAM Contents in a Bitstream

Two Independent Single-port RAMs Using One Block RAM – Some applications may require more single-port RAMs than there are RAM blocks on the device. However, a simple trick allows a single block RAM to behave as if it were two completely independent single-port memories, effectively doubling the number of RAM blocks on the device. The penalty is that each RAM block is only half the size of the original block, up to 9K bits total. Figure 19.72 shows how to create two independent single-port RAMs from one block RAM. Tie the most-significant address bit of one port High and the most-significant address bit of the other port Low. Both ports evenly split the available RAM between them.

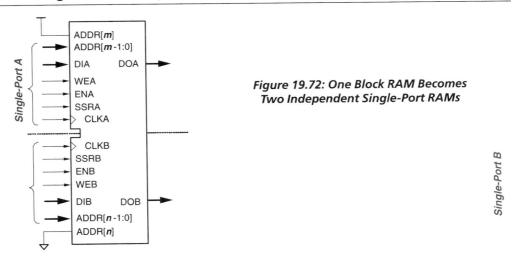

*Figure 19.72: One Block RAM Becomes
Two Independent Single-Port RAMs*

Both ports are independent, each with its own memory organization, data inputs and outputs, clock input, and control signals. For example, Port A could be 256x36 while Port B is 2Kx4. Figure 19.72 splits the available memory evenly between the two ports. With additional logic on the upper address lines, the memory can be split into other ratios.

Circular Buffers, Shift Registers, and Delay Lines – Circular buffers are used in a variety of digital signal processing applications such as finite impulse response (FIR) filters, multi-channel filtering, plus correlation and cross-correlation functions. Circular buffers are also useful simply for delaying data to resynchronize it with other parts of a data path. Figure 19.73 conceptually describes how a circular buffer operates. Data is written into the buffer. After n clock cycles, that same data is clocked out of the buffer while new data is written to the same location.

Figure 19.73: Circular Buffer

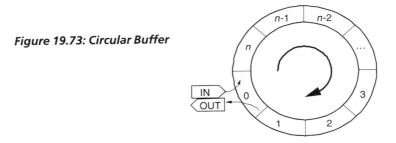

Fast Complex State Machines and Micro Sequencers – Because block RAMs can be configured with any set of initial values, they also make excellent dual-ported registered ROMs that can be used as state machines. For example, a 128-state, 8-way branch finite state machine with 38 total state outputs fits in a single block RAM as shown in Figure 19.74.

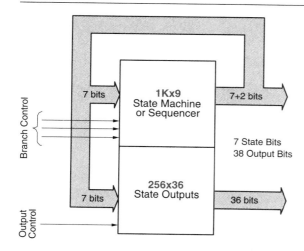

Figure 19.74: 128-State Finite State Machine with 38 Outputs in a Single Block RAM

A dual-port block RAM memory is divided into two completely independent half-size, single port memories by tying the most-significant address bit of one port High and the other one Low, similar to Figure 19.74. Port A is configured as 2Kx9, but is used as a 1K x 9 single-port ROM. Seven outputs feed back as address inputs, stepping through the 128 states. The 1Kx9 ROM has ten total address lines, seven of which are the current-state inputs and the remaining three address inputs determine the eight-way branch. Any of the 128 states can conditionally branch to any set of eight new states, under the control of these three address inputs. Port B is configured as 512 x 36 and is used as a 256 x 36 single-port ROM. It receives the same 7-bit current-state value from Port A, and drives 36 outputs that can be arbitrarily defined for each state. However, due to the synchronous nature of block ROM, the 36 outputs from the 256x36 ROM are delayed by one clock cycle. The eighth address input can invoke an alternate definition of the 36 outputs. Two additional state bits are available from the 1Kx9 block, but are not delayed by one clock.

This same basic architecture can be modified to form a 256-state finite state machine with four-way branch, or a 64-state state machine with 16-way branch. If branch-control inputs are needed, they can be combined using an input multiplexer. The advantages of this design are its low cost (a single block RAM), its high performance (125+ MHz), the absence of lay-out or routing issues, and complete design freedom.

Fast, Long Counters Using RAM – A counter is an example of a simple state machine where the next state depends only on the current state. A binary up counter, for example, simply increments the current state to create the next state. Figure 19.75 shows a 20-bit binary up counter with clock enable and synchronous reset, implemented in a single block RAM.

A 20-bit binary counter can be constructed from two identical 10-bit binary counters, with the lower 10-bit counter enabling the upper 10-bit counter every 1024 clock cycles. In this example, Port B is a 1Kx18 ROM (WEB is Low) that forms the lower 10-bit counter. The ten less significant data outputs, representing the current state, connect directly to the ten address inputs, ADDRB[9:0]. The next state is looked up in the ROM using the current state applied to the address pins. The eleventh data bit, D[10], forms the terminal-count output from the counter. In this example, the upper seven data bits, DOB[17:11], are unused.

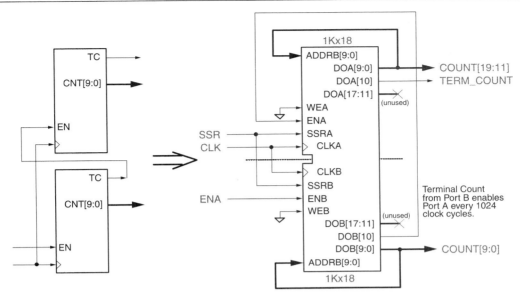

Figure 19.75: Two 10-Bit Counters Create a 20-Bit Binary Counter Using a Single Block RAM

Four-Port Memory – Each block RAM is physically a dual-port memory. However, due to the block RAM's fast access performance, it is possible to create multi-port memories by time-division multiplexing the signals in and out of the memory. A block RAM with some additional logic easily supports up to four ports, but at the cost of additional access latency for each port.

Content-Addressable Memory (CAM) – CAM, sometimes known as associative memory, is used in a variety of networking and data processing applications. In most memory applications, content is referenced by an address. In CAM applications, the content is the driving input and the output indicates whether or not the content exists in memory and. If so, it provides a reference to its location. An easy way to envision how a CAM operates is to think of an index to a book. Looking up an item, i.e., the content, first determines whether the item exists in the index. If it does, the index provides a reference to its location in the document.

Implementing Logic Functions Using Block RAM – Inside every Spartan-3 logic cell there is a four-input RAM/ROM called a look-up table, or LUT. The LUT performs any possible logic function of its four inputs and forms the basis of the Spartan-3 logic architecture. Another possible application for block RAM is as a much larger look-up table. In one of its organizations, a block RAM—used as ROM in this case—has 14 inputs and a single output. Consequently, block RAM is capable of implementing any possible arbitrary logic function of up to 14 inputs, regardless of the complexity and regardless of inversions. There are the following few restrictions, however:

- There cannot be any asynchronous feedback paths in the logic such as those that create latches.
- The logic output must be synchronized to a clock input. Block RAM does not support asynchronous read outputs. If the logic function meets these requirements, a single block RAM implements the following functions.
- Any possible Boolean logic functions up to 14 inputs may be implemented.

- Nine separate arbitrary Boolean logic functions of 11 inputs may be implemented as long as the inputs are shared.
- Various other combinations are possible, but may have restrictions to the number of inputs, the number of shared inputs, or the complexity of the logic function.

Due to the flexibility and speed of CLB logic, block RAM may not be faster or more efficient for simple wide functions like an address decoder where multiple inputs are ANDed together. Block RAM will be faster and more efficient for complex logic functions such as majority decoders, pattern matching, and correlators.

Waveform Storage, Function Tables, Direct Digital Synthesis (DDS) Using Block RAM – Another powerful block RAM application is waveform storage, including function tables such as trigonometric functions like sine and cosine. Sine and cosine form the backbone of other functions such as direct digital synthesis (DDS) to generate output waveforms. The Core Generator system provides parameterizable modules for both Sine/Cosine Look-Up Table and Direct Digital Synthesizer (DDS) modules.

Another potential application of waveform storage is in various signal companders (compressors/expanders) and normalization circuits used to boost important parts of a signal within the available bandwidth. Examples include converters between linear data, u-Law encoded data, and A-Law encoded data commonly used in telecommunications. The dual-port nature of block RAM not only facilitates waveform storage, it also enables an application to update the waveform, either with a completely new waveform or with corrected or normalized waveform data. In the example shown in Figure 19.76, Port A initially contains the currently active waveform. The application can load a new waveform on Port B.

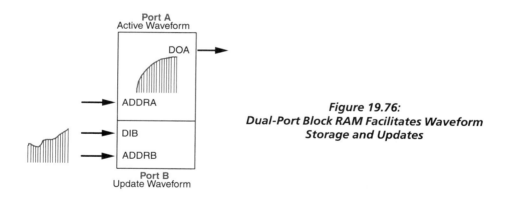

Figure 19.76:
Dual-Port Block RAM Facilitates Waveform
Storage and Updates

As in real-world engineering, sometimes it is faster to look up an answer than to derive it. The same is true in digital designs. Block RAM is also useful for storing pre-computed function tables where the output, y, is a function of the input, x ($y = f(x)$). For example, instead of creating the CLB logic that implements the following polynomial equation, the function can be pre-computed and stored in a block RAM.

$$Y = Ax3 - Bx2 + Cx + D$$

The values A, B, C, and D are all constants. The output, y, depends only on the input, x. The output value can be pre-computed for each input value of x and stored in memory. There are obvious limitations, as the function may not fit in a single logic block, either because of the range of values for x, or the magnitude of the output, y. For example, a 512 x 36 block ROM implements the above equation for input values between 0 and 511. The range of x is limited by its exponential effect on y. With x at its maximum value for this specific example, y requires at least 28 output bits.

Some other look-up functions possible in a single block RAM/ROM include:

- Various complex arithmetic functions of a single input, including mixtures of functions such as log(x) and square-root(x), are possible. Multipliers of two values are possible, but are typically limited by the number of block RAM inputs. The Spartan-3 embedded 18x18 multipliers are a better solution for pure multiplication functions.
- Two independent 11-bit binary to 4-digit BCD converters, with the block ROM configured as 1Kx18, are possible. The least-significant bit (LSB) of each converter bypasses the ROM, as the converted result is the same as the original value. That is, the LSB indicates whether the value is odd or even.
- Two independent 3-digit BCD to 10-bit binary converters, with the block ROM configured as 2Kx9 and the LSBs bypass the converters.
- Sine-cosine look-up tables are possible using one port for sine and the other for cosine, with 90 degree-shifted addresses, 18-bit amplitude, and 10-bit angular resolution.
- Two independent, 10-bit binary to three-digit, seven-segment LED output converters with the block ROM configured as 1Kx18 are possible. Leading zeros are displayed as blanks. Because input values are limited to 1023, the LED digits display from "0" to "3FF." Consequently, the logic for the most-significant digit requires only four inputs (segment a=d=g; segment f is always High).

Off-Chip Memory Interfacing Solutions

In addition to these forms of internal memory, the high-bandwidth I/Os in the Spartan-3 family can operate with external memory and can operate other ASSPs at the same high 200-MHz rate as the internal memory accesses. This transparent bandwidth across internal and external memory allows the system designer maximum flexibility so that the system can be designed for maximum cost savings.

The Spartan-3 FPGA offers unique and extensive features, including support for several IO types (LVDS, LVCMOS, LVTTL, SSTL, HSTL, GTL, and GTLP). It provides a flexible architecture that can serve as a memory controller for interfacing with different types of SRAM, DRAM, and flash. Moreover, Xilinx provides FREE VHDL source code (reference designs) for implementing the memory controllers in the low cost Spartan-3 FPGAs. Specific Spartan-3 FPGA-based memory controller reference designs available today are SDRAM, ZBT SRAM, QDR SRAM, DDR SDRAM, and CAMs. Other Xilinx FPGAs support higher-performance and next-generation memory types such as FCRAM, RL-DRAM, QDR-II SRAM, DDR-II SRAM, and DDR-II SDRAM.

Xilinx introduced a comprehensive website "Memory Corner" detailing memory and the different solutions available. It was a collaboration effort between Xilinx and major memory vendors to provide comprehensive web-based memory solutions. This website includes data sheets, application notes, tutorials, FAQs, design guidelines, and white papers. It also provides free reference designs (VHDL/Verilog) for SRAM, DRAM, and embedded FPGA memory solutions, thus making it a one-stop-shop for memory requirements.

Memory Editor Tool—Core Generator System

The Memory Editor is a tool that helps to create COE (coefficient) files to specify memory contents and initialization values for CORE Generator memory cores. Although other COE files may be used for other purposes (specifying FIR filter coefficients, for example), the Memory Editor generates COE files formatted for CORE Generator memory cores only.

A single memory is typically made up of one or more memory blocks. For each memory, the Memory Editor creates a single CGF file that defines the contents of one or more COE files. For each memory block defined in a CGF file, the Memory Editor generates a separate COE file. The Memory Editor COE Generation Format (CGF) file is a dual-purpose log and specification file. As a log file, it records the user-specified inputs that are used to generate the COE files for the memory. As a specification file, it can be used to define the contents of COE files for memory blocks. A pre-existing CGF file can be edited, saved, and then loaded into the Memory Editor and used to create a new COE file or files. The Memory Editor is accessed by selecting Tools → Memory Editor in the CORE Generator GUI.

When the Memory Editor is invoked from the CORE Generator GUI, a Memory Editor Control Panel and a Memory Contents window open (shown in Figure 19.77).

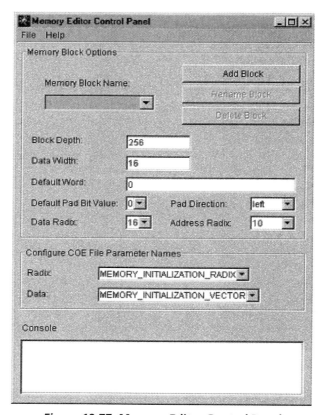

Figure 19.77: Memory Editor Control Panel

The Xilinx CORE Generator system creates distributed memory designs for both single-port and dual-port RAMs, ROMs, and even SRL16 shift-register functions. The Distributed Memory module is parameterizable. To create a module, specify the component name and choose to include or exclude control inputs. Then choose the active polarity for the control inputs. Optionally, specify the initial memory contents. Unless otherwise specified, each memory location initializes to zero. Enter user-specified initial values via a Memory Initialization File, consisting of one line of binary data for every memory location. A default file is generated by the CORE Generator system. Alternatively, create a coefficients file (.coe) as shown in Figure 19.77, which not only defines the initial contents in a radix of 2, 10, or 16, but also defines all the other control parameters for the CORE Generator system.

The output from the CORE Generator system includes a report on the options selected and the device resources required. If a very deep memory is generated, then some external multiplexing may be required. These resources are reported as the number of logic slices required. For simulation purposes, the CORE Generator system creates VHDL or Verilog behavioral models.

As shown in Figure 19.78, the Xilinx CORE Generator system provides module generators for various types of memory blocks. Choose single- or dual-port block memories, or use the higher-level functions to create FIFOs, content-addressable memories (CAMs), and so forth.

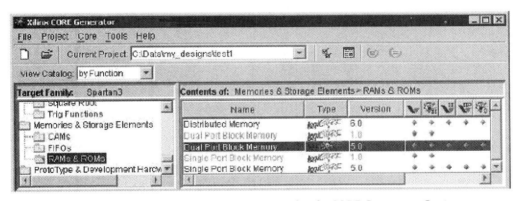

Figure 19.78: Selecting a Block RAM Function in CORE Generator System

7400 Series Replacement

We often hear that today's PC system is more powerful than the supercomputer of only two decades ago. While this is a testament to the advances of computing technology over the last 20 years, little is said about that fact that one CPLD and a little code can replace literally hundreds of discrete logic components.

Discrete logic devices, long considered the workhorse of the semiconductor industry, held a unit cost advantage over programmable logic devices for several decades. However, advances in semiconductor process technology for CPLDs have driven the costs of these devices down to the point where they now offer a highly compelling discrete logic replacement alternative. Beyond simply lower costs, CPLDs provide an unmatched solution that delivers equal, if not better performance than the fastest discrete logic device when you factor in:

- The advantages of programming flexibility
- Reduced board area

- Fast design turnaround and time-to-market
- Higher reliability

The industry seems to be getting the message as numerous products include more CPLDs and less discrete logic devices in their overall design. Driven by greatly compressed product development cycles, rapidly changing standards, and feature explosion, the prevailing trend over the last decade has been to move away from discrete logic devices in favor of CPLDs. This trend will not only continue, but also accelerate.

Discrete Logic to CPLD: The Evolution Continues

TTL Logic Comes of Age

Over four decades ago digital logic circuits were built entirely from discrete resistors and transistors. Simple logic gates, counters, and other medium scale logic functions proved to be fairly large and bulky when implemented with discrete components. Because of space constraints, integrated circuit technologies were employed to achieve higher levels of integration. What resulted was the first commercially available family of digital logic devices known as *transistor-transistor-logic,* or TTL. The classic 74-series TTL family (54-series is the military equivalent) was the first of what would eventually become a large set of families all having similar family members and functionality, but significant variations in the performance and power consumption.

Discrete TTL logic technology gained almost universal acceptance after Texas Instruments introduced their TTL 74XX family of integrated circuits in 1962 to support NASA's lunar-landing and space exploration programs. That family included:

- Logic gates (7400 quad NAND)
- Flip-flops (7474 twin D-type flops)
- Counters (74160 decade counter)
- Binary adders

all of which were implemented as TTL circuits.

Advances in discrete logic component packaging and ultra-high integration helped reduce footprint size. Designers could minimize board space component count while providing circuitry with almost "Lego Block"-like simplicity. Meanwhile, improved TTL variations were developed in the years to follow. New families were continually introduced to address a wide range of applications driven by the need for higher density and performance, lower power, and reduced costs. As a result, discrete components spawned a whole new generation of system designers.

Through the '70s, TTL variations came at a fast and furious pace. However, the rampant proliferation of multiple 7400 series families and specialized product lines that implemented specific functionality brought complexity. By the early '80s, an alphabet soup of logic family variants had been released. Variants included TTL, S, LS, AS, F, ALS, CD4000, HC, HCT, BCT, AC, ACT, FCT, ABT, LVT-A, AHC, and AHCT. This confused designers to the point where they needed a matrix scorecard to determine which family best fit each application.

The Emergence of CPLDs

Despite all the TTL diversity, a revolutionary trend began to appear in the form of programmable logic arrays. User programmable, the first generations of these chips replaced five to 10 logic gates. As logic integration improved, these devices replaced more and more standard logic functions. Today, the maximum number of gates in a CPLD is around 10,000. Devices of this size are capable of integrating in excess of 128 discrete TTL logic devices.

However, consistent with higher levels of integration, first-generation CPLDs came at a higher price. The culprit was the die costs associated with the overhead necessary for CPLD programmability. Total product cost is based upon factors including die cost, package cost, and test cost. Calculating die cost is fairly straightforward—device price is directly proportional to the number of devices that a single silicon wafer can yield. Silicon wafer pricing is a fixed cost, so the trick is to maximize the number of dice per wafer. As the size of the die increases, the number of devices per wafer decreases, and price per die increases.

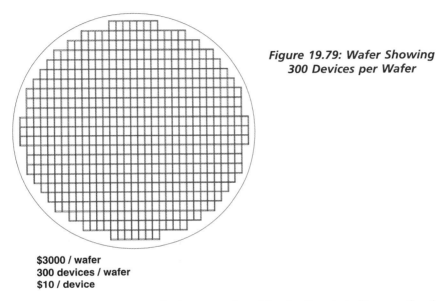

Figure 19.79: Wafer Showing 300 Devices per Wafer

$3000 / wafer
300 devices / wafer
$10 / device

For larger die sizes, costs become further exacerbated by the fact that silicon wafers have defects. The probability that a die will not yield due to a defect increases exponentially as the die size increases. Hence, the cost of a larger die increases exponentially beyond a certain point. In the past these factors lead to higher CPLD product costs when compared to discrete TTL devices.

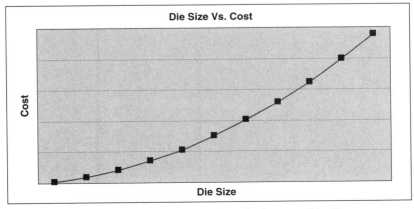

Figure 19.80: Chart Showing Die Cost vs. Die Size

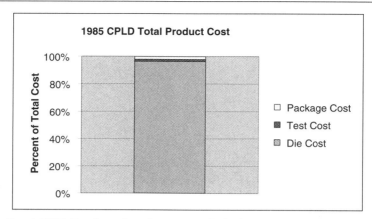

Figure 19.81: Total CPLD Product Cost in 1985 Included Die Cost, Test Cost and Package Cost

Traditionally, this high cost structure restricted the use of CPLDs primarily for pre-production prototyping or other limited volume applications—a situation that has changed dramatically.

CPLD: The Clear Discrete Logic Replacement Choice

Initially, there was a significant cost premium associated with the use of TTL devices over discrete devices. However, advances in semiconductor technologies quickly leveled the playing field in favor of TTL-based designs. This same trend is now being mirrored between CPLDs and TTL devices. In other words, the time for TTL device replacement has arrived. It is cheaper to use CPLDs rather than discrete 7400 devices.

As the diagram below illustrates, discrete devices historically had low unit costs relative to CPLDs. However, technology improvements have rewritten the cost equation, allowing CPLDs to gain equal, if not better, footing with TTL devices. The unit costs of CPLDs have been driven down to where they are equal or below that of discrete logic devices. In addition, if the cost of one CPLD device is compared to a TTL-based system that provides equivalent functionality, the CPLD wins hands down, given that multiple TTL devices are required to equal one CPLD.

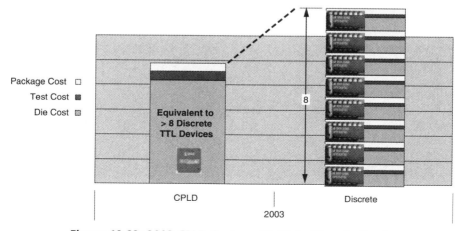

Figure 19.82: 2003 CPLD Cost vs. Multiple Discrete Devices

Factoring in the high hidden costs underlying discrete logic devices such as increased inventory, low reliability, high power consumption and EMI, prolonged time-to-market, and higher maintenance, CPLDs are the superior alternative to discrete 7400 series devices.

Discrete Logic versus CPLD Devices: Closing the Cost/Performance Gap

To support the premise that CPLDs provide a lower total cost alternative, the costs of a discrete logic component-based circuit can be compared to those of a Xilinx CPLD. The discrete logic circuit used for comparison (shown below) is a straightforward memory controller that provides functionality equal to one Xilinx 32-macrocell CoolRunner XC2C32 CPLD. To make the comparison as complete as possible, both DIP and SMT packaging for the discrete circuit are included and compared to the Xilinx CPLD with VQ44 packaging.

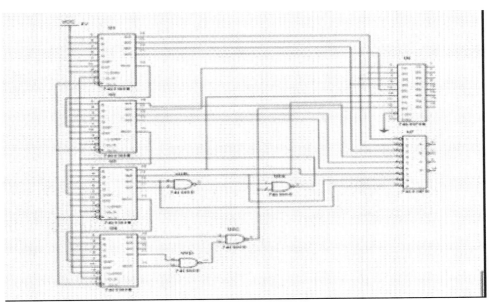

Figure 19.83: Floorplan of a CoolRunner-II CPLD Showing the Number of Discrete Devices (by part number) Implemented and Logic Required

When comparing costs between technology alternatives, it is important to consider the total system costs—the *real* cost of the solution. Accordingly, this analysis provides a detailed breakdown of all costs, starting with production costs and comparing the costs associated with total PC board area, power consumption, and the number of board layers.

Production Costs

Production costs are broken down by logic component unit costs, resistor/transistor costs, PCB material, assembly/insertion, and inventory.

Table 19.14: Discrete/CPLD Circuit Comparison

PRODUCTION COST COMPARISON			
	Discrete Circuit		Xilinx CPLD, XC2C32
Cost/Packaging	DIP	SMT	VQ44
Component costs	$1.35	$1.93	$0.90
Discrete resistors and capacitors	$2.05	$2.05	0.00
PCB material	$1.92	$1.58	$0.13
Assembly/Insertion	$0.38	$0.32	$0.03
Inventory	$1.33	$1.39	$0.26
TOTAL	**$7.03**	**$7.27**	**1.32**

Some of the assumptions in making the discrete/CPLD circuit comparison are:
- PC board quantity is 1,000/month with a two week lead time
- Total PC board area was calculated to be equal to total component area plus 30%
- Assembly/insertion costs are 20% of PCB material costs
- Inventory costs are 25% of total material costs
- Logic density is 32 macrocell
- Discrete circuit consists of:
  - (1) electrolytic capacitor (power)
  - (7) ceramic capacitors (decouple)
  - (6) ¼ watt pull-up resistors
- Standard commercial operating environment
- Power consumption data assumes frequency of 25 MHz

As Table 19.14 shows, the total cost for one Xilinx CPLD is less than one-fifth of the cost of a TTL circuit with equivalent logic density and functionality. Besides lower unit costs, this cost advantage is due to the fact that the CPLD-based design does not incur a cost penalty for auxiliary parts (resistors and transistors, extra board real estate, and assembly/insertion costs).

Although not quantified in this analysis, CPLDs gain an additional cost advantage through:
- The ability to inventory one line of CPLDs for multiple applications. This reduces the number of individual SKUs in inventory as well as the amount of scrapped parts.
- Lower availability risks and low minimum order quantities (MoQ)
- Reduced expediting costs and production delays due to parts shortages
- Avoidance of revenues lost due to lines down

Board Area Savings

When compared with discrete logic components, CPLDs requires fewer components and, thus, less board area and layers. This lowers power consumption and improves reliability. In addition, lower heat dissipation improves avoids the addition of cooling fans and heat sinks.

The component area for DIP and SMT discrete circuits is 1.298 sq. inch and 1.068 sq. inch, respectively. This leads to a PCD area utilization of the two devices of 1.6874 sq. inch and 1.388 sq. inch, respectively. The Xilinx X2C32 VQ44 device requires a 0.15 sq. inch component area and a 0.195 sq. inch PCB area. The CPLD consumes far less board space and is cheaper when measured on the PCB area cost per square inch, or cost per layer against the discrete devices. Also, power consumption of the discrete devices is 150 mW and 1750 mW for quiescent and active modes, respectively, for both DIP and SMT devices. In comparison, the CoolRunner-II consumes only 0.029 mW and 1.6 mW for quiescent and active modes, respectively. Hence, the total power cost at a rate of $0.40 per watt is $0.06 in quiescent mode and $7 in active mode for both DIP and SMT discrete devices. Comparatively, the CoolRunner-II CPLD consumes $1.16x10^{-5}$ in quiescent mode and $6.4x10^{-4}$ in active mode, a dramatic difference. Also, the reliability of discrete devices is 29.951 FIT, while for the CoolRunner-II CPLD it is 1.000 FIT. Whether measured in device costs, PCB area costs, power savings, or reliability, the CoolRunner-II device comes out on top.

Other CPLD Advantages

CPLDs also further drive down the total cost of ownership and deliver key advantages through:

- **Fast time-to-market** – Being first to market counts. In fact, every 4 weeks delay equals 14% loss of market share.
- **Reprogrammability** – CPLDs not only get to market faster, they stay in the market longer, enabling an expanded revenue stream. In addition they enable:
  - Remote bug fixes and feature upgrades that avoid costly hardware changes
  - Shorter development cycles avoiding board re-spins due to "features creep" or unnoticed bugs
  - Reduced development time being spent on rework and maintaining old designs
- **Reliability** – By employing a fewer number of devices over the discrete TTL equivalent circuits, CPLDs provide a significantly improved FIT rates, indicating a remarkable high level of reliability.
- **Electromagnetic interference** – CPLDs lower EMI levels, thus reducing the high cost and high risk of meeting EMI compliance.
- **Design security** – CPLDs offer several unique advantages over TTL devices for protecting designs.

Time-to-Market Benefits

The proliferation of electronic wireless, industrial, and communication devices, as well as ever-shrinking product lifecycles, continue to put pressure on companies to get their designs from concept to production as soon as possible. When you look at the cost of being late, it's easy to see the motivation behind being first on the market.

As the graph in Figure 19.84 illustrates, late market entry has a larger effect on profits than development cost overruns or a high product price. This is especially true in highly competitive markets and those that have short market windows. According to McKinsey & Co., even if within budget, products that are six months late earn 33% less profit over five years.

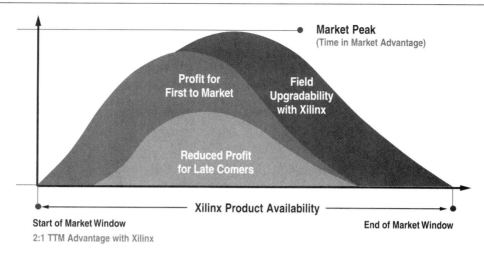

Figure 19.84: Time-to-Market Benefits
(Source: McKinsey & Co.)

Unfortunately, designs that employ discrete 7400 devices are at the mercy of several barriers that prolong time-to-market, add complexity and costs, and reduce reliability. These barriers include:

- A long manufacturing, assembly, test, and debug cycle is susceptible to delays and multiple design decisions that can directly impact board layout.
- The lack of easy-to-use design tools makes debugging and maintenance a tedious chore.
- The high number of TTL components required in discrete designs introduces availability risk. In many cases, components may be out-of-stock or even obsolete.

CPLDs, on the other hand, provide designers with numerous advantages that enable designers to react to last-minute design changes and compress time-to-market. These include component inventory reduction, faster production cycles, efficient device and board testability, and the ability to modify designs during all phases of design and production.

Programmability: The Real Advantage

While time-to-market is a major benefit, the ability to program and reprogram devices as designs change is equally important. Unlike CPLDs, once a PCB using discrete TTL devices is laid out, it cannot be altered or upgraded without ripping out components and going through board re-spins.

Another major advantage of CPLDs is quick system upgrades and bug fixes in the field. As a result, designers can easily integrate exactly the logic functionality needed without adding "cuts and jumps" on the PCB. For example, imagine a scenario where automobile manufacturers can add new functions to in-car systems by allowing consumers to dial up and purchase upgrades over the phone to reconfigure the system. In addition, CPLDs help avoid board redesigns and scrapped parts when specific programs are cancelled.

The reconfigurable nature of CPLDs also enables a single product footprint to implement multiple product personalities. That means manufacturers can standardize on a single PCB footprint and package, thus quickly leveraging economies of scale through reduced inventory overhead. Further, reprogrammability reduces the deployment of design resources for maintaining old designs, allowing engineers to focus on introducing new products and features.

Reliability

Reliability is another area that should not be overlooked when analyzing total system costs. This is especially important since TTL logic systems introduce higher complexity with more components, interconnects, layers, handling, thus lowering overall reliability. In addition, because discrete components are typically larger and require more board real estate, they increase power consumption and EMI, which further heightens failure risks.

In contrast, CPLD-based systems require fewer components and layers. The reduction in components and layers reduces PC board layout density, lowers heat dissipation, reduces EMI levels, and greatly decreases Failures In Time (FIT).

Electromagnetic Interference

Electromagnetic interference (EMI) refers to any type of interference that can potentially disrupt, degrade, or interfere with authorized electronic emissions over approved portions of the electromagnetic spectrum. EMI problems are often complex and their solutions illusive. Ideally, EMI compliance should be an integral part of the board design. Unfortunately, resolution of EMI problems is often ignored until it's too late.

EMI originates from the switching of digital circuits. Two factors are necessary for EMI to exist—a noise source and a propagation path. On a PCB, noise can come from frequency-generating circuits, component radiation, ground bounce, poor impedance control, or cable interconnects. The propagation path is the medium that carries the energy, such as free space or metallic interconnects. Antennae are the elements that both transmit and receive unwanted interference.

Concern over EMI continues to grow as the FCC and international compliance regulations clamp down harder on pollution of the electromagnetic spectrum. It is an issue that deserves the system designer's attention.

For a system designer, EMI compliance carries cost and high risk, since it can easily prolong the product design, test, and launch lifecycle. Additional techniques are required to alleviate noise levels, including employing ferrite beads, shielded enclosures, or series terminating resistors—all of which drive up costs and reduces board yield and reliability. EMI analysis is tedious and involves numerous variables, making problems difficult and expensive to fix once the board is assembled. (FCC compliance testing alone is $400 per hour.) As a result, even after deploying EMI reduction strategies, failure to meet FCC regulation standards means designers can face last-minute, trial-and-error troubleshooting exercises and board re-designs that further increase costs and prolong time-to-market.

Given the large number of components and board layers, TTL-based designs are highly susceptible to high EMI noise levels. By contrast, CPLDs significantly reduce EMI through fewer external components and other "free" features including:

- Programmable I/O slew rate
- Programmable ground
- Programmable I/O signaling
- Phase locked loops

Design Security Issues

Given the explosion of new applications in the competitive electronics market, the need to protect designs from unscrupulous competitors has never been greater. CPLDs offer several unique advantages that safeguard system designers against code theft.

Discrete logic devices are extremely susceptible to reverse engineering—which can be as simple as reading the part number directly from discrete TTL devices. But a CPLD inherently requires a user-defined bit stream that can easily prevent customer read-back.

More elaborate security schemes exploit the reprogrammable capabilities of CPLDs to keep attackers at bay by modifying the CPLD design or password on a regular basis. This seriously hampers an attacker's ability to reverse engineer the design. Exploiting the reprogramming features of CPLDs is thus a logical and efficient way to defend against copying in the market place.

Xilinx CPLD Advantages

The Xilinx CoolRunner-II CPLD family utilizes second-generation RealDigital technology to provide high performance, advanced features, and low power consumption, all at a very low price. Featuring a 100% digital core, up to 385 MHz performance, and low stand-by current, CoolRunner-II CPLDs offer a wide range of densities. It also provides abundant I/O, the flexibility to move from one density to another in the same package, and the lowest cost per I/O pin in the industry.

Learning that designers of portable products wanted even lower power than CoolRunner XPLA3 CPLDs, Xilinx added architectural features to accommodate tight power budgets. Even with FZP technology and voltage reduction, two new architectural features were added to lower overall power consumption in designs for power sensitive applications. Those features are DataGATE and CoolCLOCK. Both these features also help in EMI reduction.

When it comes to protecting designs from being damaged or copied, Xilinx CPLDs offer security measures that make designs substantially more secure than discrete logic devices products. These features include:

- Accidental overwriting as well as electrical/visual detection of configuration patterns is eliminated with four new levels of on-chip security.
- Designs can be secured during programming to prevent either or pattern theft via 'readback'. These security bits can be reset only by erasing the entire device. Bit stream protection impedes direct copying and protects intellectual property while maintaining a longer competitive edge in the market.
- Xilinx CPLDs are designed for a wide range of applied voltages to accommodate applications in the handheld, portable design world. Thus, they are much less susceptible to exposing address and data bits via externally applied voltages.
- Electrical or laser tampering causes the device to automatically lock down and erase. Even if the device is de-capped, buried interconnects make it almost impossible to trace security connections without destroying the device.

Since their introduction in the late seventies, programmable logic devices have proven to be very popular. In fact, they are now one of the largest growing sectors in the semiconductor industry. The migration to PLDs, and eventually CPLDs, has been an intriguing, four-decade evolution. It started with discrete transistors and resistors and first-generation TTL technology. However, times have changed. Just like TTL devices that replaced early generation discrete transistor-based logic designs, advances in semiconductor process technology are enabling CPLDs to offer a clear, cost-effective alternative to TTL systems. No longer held hostage to low densities and high die costs, CPLDs pack more functionality into ever-shrinking die geometries. They offer compelling benefits of on-the-spot reprogrammability, short lead times, higher performance, expanded densities, and unmatched flexibility. Xilinx CoolRunner CPLDs consume less power and offer reprogrammable flexibility with unprecedented design security–without a price premium. The result is a scalable technology that provides a great solution for many high volume applications such as PDAs, cell phones, routers, and high-speed Internet modems.

EMI Reduction in Consumer Devices

Electromagnetic Interference (EMI), which is also known as EMC (electromagnetic compliance), is interference that can be generated as a result of poor RF design. It is more often a result of over-looked digital circuit design and PCB layout. EMI is considered "black magic" because it relies on electromagnetic wave and transmission line theory and comes about through second or third order effects that generally are not a digital designer's forte.

As a quick lesson in fundamentals, EMI has two components: electric (E-field) and magnetic (H-field). Both of these run perpendicular to each other. EMI is a function of current, loop area, and frequency.

EMI is represented by the following equation:

$$EMI(v/m) = kIAf^2$$

Where,

k = constant of proportionality

I = current (A)

A = loop area (m^2)

f = frequency (MHz)

Some of the issues surrounding EMI are:

- There is an exponentially increasing amount of electronic equipment being introduced yearly, including digital consumer products such as PCs, DVD players, audio players, cell phones, and TVs.
- Equipment is moving to higher clock frequencies. For example, PCs went from the 100 MHz range in the 90's to the GHz range in this decade. And TV in the 100 MHz range moved to HDTV in the GHz range.
- There is extreme potential for unwanted radiated emissions.
- Demand on the electromagnetic spectrum exploded with greatly increased usage by cell phones, wireless local area networking, and police, emergency, and government services.
- Spurious emissions from electronic equipment caused pollution of the electromagnetic spectrum. This impacts the effectiveness and efficiency of communications and it impacts normal operation of electronic equipment.

EMI incompatibility is driving the need for compliance measures such as communications efficiency, consistent and predictable equipment operation, and safety. Hence, worldwide standards have been introduced that vary by country and provide guidance for compliance.

EMI compliance covers two components—radiated emissions (maximum energy that can be radiated at a given frequency), and susceptibility (maximum energy that must be withstood at a given frequency– conducted emissions). This is measured under specified conditions that include device type, distance, etc. EMI/EMC certification must be granted by a governing organization before a product can be sold and marketed in that geography.

EMI emission remedies include shielding, PCB layout, and signaling.

Table 19.15: National Regulations on EMC

Country	Regulation
EU	
Austria	EMV 1993 + 1995
Denmark	Law 475 and Order 475 in force July 1994
Finland	MTI Decision no 1696-93
France	Decree 95-587 + 95-283 in force 13th Mrch 1995
Germany	EMC law in force 10th December 1992
Italy	Decree 476 of 4th December 1992
Luxembourg	Rugulation of the Grand-Duche
Netherlands	Besluit van 8-14-95
Portugal	Decree law 74/92
Spain	Real Defeto no 444/1994
Sweden	Act on EMC SFS 1992; 1512, Regulation ELS
Norway	Electrical equipment regulations Jaunuary 1993
Iceland	Regulation no 146 / 1994 on EMC 28th Feb 1994
Rest of World	
Australia	Spectrum Management Agency via generic standards
Japan	VCCI, via CISPR standards
USA	FCC Rules

Shielding

Shielding is a simple, effective, but quite expensive technique. Metal or metal treated enclosures (shield) are used to intercept radiated currents. The shield is connected to signal ground with multiple connections and the grounded heat sinks can serve as shields. The proximity of the enclosure to high-speed signals determines the coupling to shield. Unless the shield is well closed, radiation will occur from the seams. Shielded cabling is used instead of PCB traces for high-speed signal transmission and is grounded at source and/or destination. In general, shielding can prove to be a brute force approach to solving EMI emissions. Its effectiveness is determined by capacitive coupling and manufacturing processes to achieve proximity. It is hard to model until the system is manufactured and adds system cost and weight. Due to these characteristics, shielding is often seen as a last resort of EMI emission remedies.

PCB layout

Multi-layer PCBs are used to counter EMI emissions for signals greater than 5 MHz or with a rise time of less than 5 ns. This technique sees a 10x to 100x EMI improvement vs. double-sided boards. A multi-layer PCB has one or more surfaces dedicated to ground and power, with a well-decoupled power plane and ground plane. This allows minimized signal and ground return path impedance levels, and minimized crosstalk between neighboring traces. And it provides characteristic impedance control along specified paths.

For example, consider the impedance of a signal trace made of copper (0.03 mm) that is 1 mm wide and 10 mm long. At a frequency of 100 MHz the impedance is four ohms. At a frequency of 300 MHz the impedance is 12 ohms. Compare this to a copper sheet where impedance between two points on an infinite plane at a frequency of 100 MHz is four milliohms. At a frequency of 300 MHz the impedance is six milliohms. At 100 MHz the impedance is about 1000 times that of a plane. And at 300 MHz it is about 2000 times.

Good layout is a key to EMI control. It is important that components are segregated according to frequency and that the high and low speed signal lines are separated. The high-speed lines are routed first and should be kept short and direct. The devices should be positioned for minimum clock runs and the clock lines should not be placed near I/O ports and should be terminated when necessary.

Improper termination of the traces will demonstrate overshoot or undershoot since the source impedance does not match the destination. The impact can result in 3 to 4 dB in radiated EMI at a given node, and multiple nodes with impedance mismatch exacerbates radiated energy.

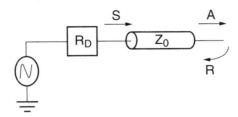

where,

S is the signal trace into the Transmission Line
R_D is the Impedance exhibited by driver source
Z_o is the Trace impedance
A is the signal accepted at the load
R is the signal reflected from the load

A series termination is the simplest form of impedance matching, where the driver impedance is matched to the trace impedance using a series resistor. There is no constant DC current compared to other schemes, and exact matching impedance can be tough using standard resistors.

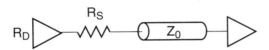

where,

$$R_D + R_S = Z_o$$

EMI improvements of up to 4 dB per node can be achieved through properly matched termination. Manually routing critical lines (clock, I/O, and bus) and paying close attention to high frequency clock trace length/cabling can provide an efficient EMI emission remedy.

Signaling

Varying aspects of signaling affect radiated emissions such as frequency, signaling technologies, and slew rate as shown in the following equation:

$$EMI(v/m) = kIAf^2$$

where,

    k = constant of proportionality

    I = current (A)

    A = loop area (m2)

    f = frequency MHz

Radiated EMI is directly proportional to the square of the operating frequency. There is a significant benefit in reducing f_{sys} (the system frequency). This can be achieved by lowering the frequency system synchronization clock by using local PLLs (phase locked loops) for clock rate multiplication and spread spectrum clocking.

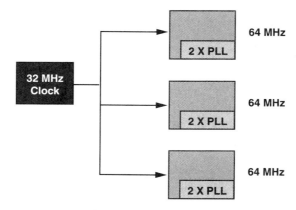

Figure 19.85: Local PLLs

A low-frequency distributed clock source with a local rate-multiplied clock reduces EMI emissions as the square of the operating frequency. This is achieved by having local PLLs.

Radiated EMI is directly proportional to the amplitude of the signaling technology. Low level signaling technologies should be considered whenever possible. Higher speed, multiplexed I/O signaling should be considered for reduced I/O count at higher frequency and lower signaling level.

A 4-dB reduction in EMI can be realized by utilizing lower voltage signaling standards. In the case of GTL and LVDS, slower edge rates provide additional EMI reduction benefits.

Table 19.16: IO Signaling Technologies and the Voltage Standards

Technology	V_{oh} max
TTL	5.5V
LVTTL	3.6V
LVCMOS	1.8V
LVDS	1.6V
SSTL	1.5V
HSTL	1.3V
GTL	1.2V

Table 19.17: I/O Signaling Standards Summary

Type	Chip to chip	Chip to Backplane	Chip to Memory	
Key Standards	LVTTL, LVCMOS	GTL, GTL+, AGP	HSTL I, III, IV	SSTL2, SSTL3
Hey Highlights	Higher voltage swing	Low voltage swing	Low voltage swing, low power, low noise, 200-400MHz	Low voltage swing, low power, low noise, SSTL3 82-166MHz, SSTL2 166-333MHz
Primary Usage	Legacy interface	Pentium CPU, backplanes	High speed SRAM, MIPS/UltraSparc-II	Synchronous DRAM interfaces (SDR & DDR)
Applications	Glue logic, ASIC chip to chip	Datacom, Pentium, add-in cards	Line cards, graphics cards, digital cameras, modems	3-D graphics cards, plasma LCD displays, DTV interfacies, Set-Top Boxes
Vendors	Most vendors	Intel, TI	Micron, IDT, Cypress, MIPS, IBM, etc.	Micron, Samsung, Toshiba, Hyundai, NEC, Siemens, etc.

LVDS (Low Voltage Differential Signaling) requires two pins per channel. It was first used as interconnect technology in laptops and displays to alleviate EMI issues and reduce connector costs and signal routing complexity. LVDS employs 8:1 multiplexing and demultiplexing on transmit and receive paths. LVDS technology has now been adopted across a broad spectrum of applications such as networking, telecom, consumer digital video, and video displays.

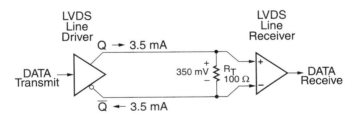

DC Parameter	Conditions	MIN	TYP	MAX	Units
Output High Voltage for Q and /Q	$R_T = 100\ \Omega$ across Q and /Q signals	–	1.38	1.6	V
Output Low Voltage for Q and /Q	$R_T = 100\ \Omega$ across Q and /Q signals	0.90	1.03	–	V
Differential Output Voltage (Q – /Q), Q = High (/Q – Q), /Q = High	$R_T = 100\ \Omega$ across Q and /Q signals	250	350	450	mV
Output Common-Mode Voltage (Q + /Q) / 2	$R_T = 100\ \Omega$ across Q and /Q signals	1.125	1.25	1.375	V
Differential Input Voltage (Q = /Q), Q = High (/Q – Q), /Q = High	Common-mode input voltage = 1.25V	100	350	–	mV
Input Common-Mode Voltage (Q + /Q) / 2	Differential input voltage = ±350 mV	0.25	1.25	2.25	V

Figure 19.86 and Table 19.18: LVDS Signaling Levels

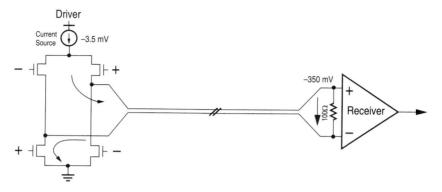

Figure 19.87: LVDS Working

LVDS uses low output swing of approximately 350 mV and slow edge rates (dV/dt) of 1V/ns. The differential (odd mode operation) magnetic fields cancel out each other. LVDS technology has "soft" output corner transitions and minimum I_{cc} spikes due to low and constant current mode operation. It has a high common mode rejection ratio (CMRR).

Hence, choose the right signaling technology for the job. Using high performance I/O typically comes at a price. Employ non-traditional signaling wherever possible. HSTL, SSTL, and GTL technologies employ lower I/O swings and dV/dt. LVDS is a clear leader for EMI management.

Another method of reducing EMI emissions at low cost is through spread spectrum clocking. The system clock is modulated such that output frequency varies slightly. It distributes energy of a single or narrow band over a much greater spectrum. The amount of EMI reduction affected by modulation profile is a percent of frequency modulation and modulation rate. Typical improvement of 6 dB can be seen.

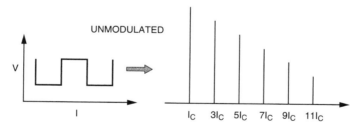

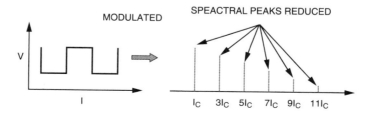

Figure 19.88: Spread Spectrum Clocking

Simultaneous switching nodes in buses such as PCI 64-bit/66 MHz generate significant EMI radiation. Redistributing energy through skewed I/Os provides meaningful EMI reduction greater than approximately 1 dB.

EMI Testing Process

The best practices in EMI testing will not guarantee compliance. While it is a good start to minimize lab test time and increase success, there are multiple variables with interactions that are hard to predict. Compliance with emissions regulations requires evaluation in a certified lab which typically requires:

- Over $200 an hour
- Several weeks
- Trial and error to identify and rectify problems
- Tools such as foil, an XACTO knife, ferrite beads, resistors, and a soldering iron

Fixes that enable (or appear to enable) compliance testing in the lab usually require PCB re-spin. Updating a PCB for inserting series termination and ferrite beads, re-routing traces, and integrating other fixes typically requires a two-week turn around. And fixing one problem may create another. Multiple PCB spins to meet compliance are not uncommon, especially for cost sensitive, high volume applications. And this is typically unacceptable. Achieving EMI compliance can have significant costs in lab test time and EMI consultation, time-to-market and opportunity costs, and increased total product cost.

Xilinx Solutions for EMI Reduction in Consumer Devices

Xilinx CoolRunner-II CPLDs and Spartan-3 FPGAs provide a wide range of features that significantly improve EMI emissions and susceptibility:

- Programmable Slew Rate
- Programmable Drive Strength
- Programmable Impedance Matching
- Programmable I/O Signaling Levels and LVDS signaling
- PLL/DLL by using the digital clock manager (DCM)
- Programmable Ground

Programmable Slew Rate

Xilinx devices provide two slew settings with a slew rate reduction of 1V/ns (2V/ns total). Since this is in-system programmable, the slew rate can be modified during any phase of product development. (See Figure 19-89.)

Programmable Drive Strength

The Spartan-3 I/O structure supports 16 output current drive strengths that are user configurable. They can be varied depending on the requirements of the application because the pull-up and pull-down drivers can be individually controlled. These programmable drive strength settings help decrease the effects of simultaneously switching outputs (SSO), reduce system noise, decrease power consumption, and improve signal integrity. (See Figure 19.90.)

Programmable Impedance Matching

Impedance matching typically requires discrete termination resistors and matched source and destination impedance values. This eliminates the over and under shoot, which is a major EMI

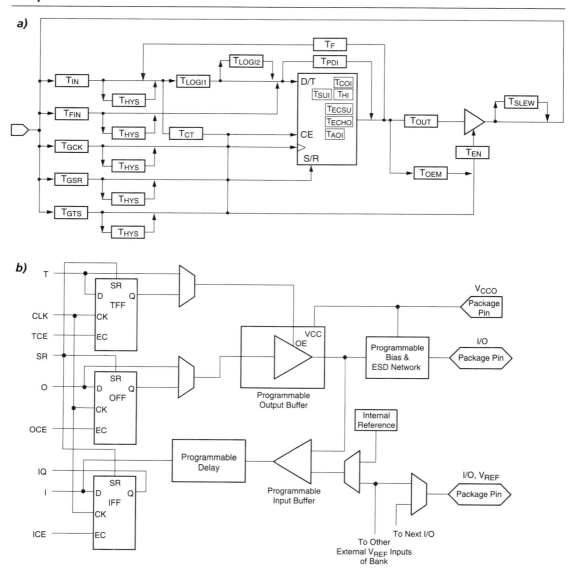

Figure 19.89: Programmable Drive Strength (a) CoolRunner-II CPLDs (b) Spartan-3 FPGAs

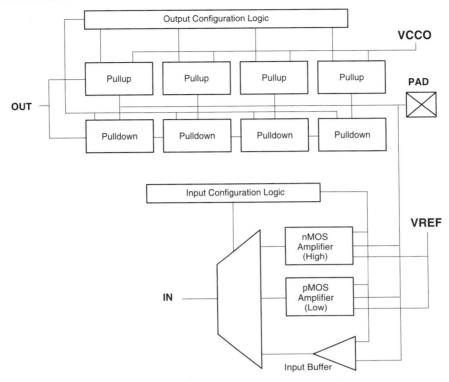

Figure 19.90: Programmable Drive Strength

emissions component. The match is limited to PCB trace modeling accuracy. Also terminating high pin count BGA devices consume board area and increases failures.

The Spartan-3 family also supports XCITE digitally controlled impedance (DCI) technology on selected I/O standards. The XCITE option is available on each user output. Here, the output impedance is matched to an external reference impedance dedicated to one of eight I/O banks within each device. This capability eliminates the need for most external termination resistors and allows high-precision impedance matching required for high speed. This provides a unique capability—Spartan-3 designs may be used in different electrical environments by matching local impedance requirements. In contrast, ASICs designed for one particular board impedance spec may not be able to work in a different board environment and cannot be tuned for adjustment.

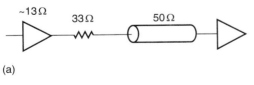

(a)

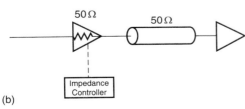

(b)

Figure 19.91: Impedance Matching
(a) Conventional IOs
(b) Digital Controlled Impedance (DCI)

Programmable I/O Signaling Levels and LVDS Signaling

The SelectIO feature of Spartan-3 devices supports 17 single-ended standards including GTL, HSTL, LVCMOS, LVTTL, PCI, and SSTL. It also supports six differential standards (including LVDS, RSDS, and HyperTransport) on all IOs shown at the beginning of this chapter. The I/Os can be configured as synchronous or asynchronous, and input or output LVDS. Two IOBs (pair) form one LVDS signal—one IOB will function as + (or P) and the other will function as – (or N). There are up to 344 differential pairs and they operate up to 622 Mb/s per differential pair.

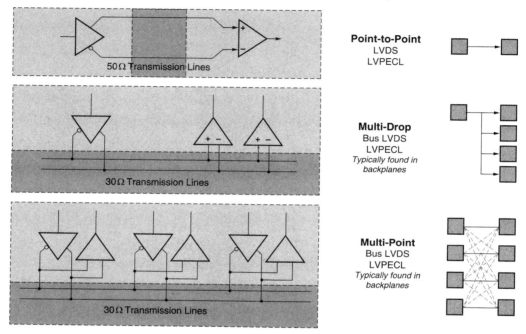

Figure 19.92: LVDS / LVPECL Configurations

Clock speeds of over 622 MHz can be distributed with ease using LVDS. Spartan-3 eliminates LVDS-to-TTL converters. Hence, this removes the 2 ns delay and skew. This leads to higher performance, lower EMI, lower cost, and fewer components. (See Figure 19.93.)

Spartan-3 LVDS can receive and convert high speed clocks with zero delay. This helps lower cost and EMI, and results in fewer components. (See Figures 19.94 and 19.95.)

Digital Clock Managers (DLL/PLL)

Spartan-3 FPGAs and CoolRunner-II CPLDs have digital clock managers (DCMs) that eliminate on-chip clock delay and board-level delay. The DCM is a robust and stable digital delay-locked loop with low jitter. The frequency can be multiplied and divided simultaneously which reduces board-level clock speed and the number of board-level clocks. It also significantly reduces EMI by attaching the f^2 term.

Spartan-3 FPGAs have eight global clock multiplexers, with eight clocks to every quadrant that are sized for target density range. There are four DCMs per device to provide clock skew elimination, clock conditioning (50% duty cycle), clock mirroring, frequency synthesis, and phase shifting.

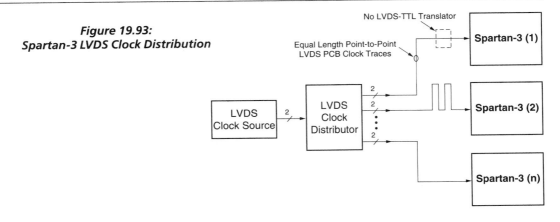

Figure 19.93:
Spartan-3 LVDS Clock Distribution

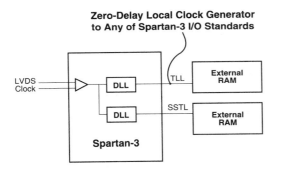

Figure 19.94:
Spartan-3 LVDS Clock Conversion

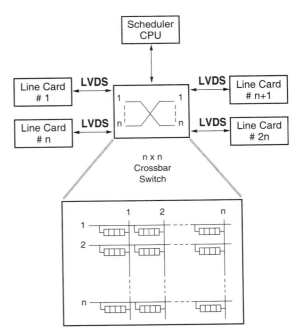

Figure 19.95: Spartan-3 LVDS
Chip-to-Chip Interconnects

CoolRunner-II CPLDs provide clock-doubling capabilities where the distributed global clock can be divided, and then doubled locally at the macrocell. This decreases I_{cc} on global clock nets and helps reduce EMI emissions. The clock doubling can be used for double data rate applications

Programmable Ground

CoolRunner-II CPLDs have the programmable ground feature that reduces device power plane impedance. The multiple grounded I/Os act as resistances (approximately 8 ohms) in parallel. The unused I/O pins can be connected to ground, which minimizes EMI power supply coupling and shields adjacent high frequency switching I/Os.

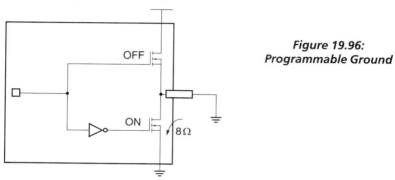

**Figure 19.96:
Programmable Ground**

EMI is an often overlooked topic and its "black magic" symptoms and solutions are typically not intuitive. EMI emissions are mostly determined at compliance testing after the PCB design is complete. Fixes traditionally require a PCB re-spin (or more) and there is no guarantee that re-spins fix the problems. Achieving EMI compliance can have significant system cost impact in direct costs such as shielding materials, PCB layers, and EMI compliance testing/consulting. It also incurs opportunity costs such as time-to-market due to PCB re-spins and EMI compliance testing time. Understanding the causes/sources and models for EMI is a good start. There are several factors affecting EMI, many of which may be inter-related. Xilinx FPGAs and CPLDs have key features that can be used as powerful tools in the battle against EMI. Most are in-system programmable and help avoid PCB re-spins, thus lowering total solution costs and reducing time-to-market costs. The following table summarizes Xilinx solutions for EMI reduction:

Table 19.19: Xilinx Solutions for EMI Reduction

Feature	CoolRunner-II CPLD	Spartan-3 FPGA
Slew Rate Control	X	X
Programmable Ground	X	
Programmable I/O	X	X
PLL / DLL	X	X
Programmable I/O Delay	X	X
Programmable I/O Drive Strength	X	X
LVDS		X
Programmable Impedance		X

Summary

Today's programmable logic continues to penetrate new markets and applications. This is due to its inherent qualities such as time-to-market, flexibility, low cost, high density, and new features. PLDs are standard components. This means that the same device type can be sold to many different customers for many different applications. As a result, the development cost of PLDs can be spread over a large number of customers. On the other hand, custom gate arrays and standard cells are custom chips for individual customers with specific applications. This means a high up-front cost to customers. Technology advances are enabling PLD companies to reduce costs considerably. This factor makes PLDs an increasingly attractive alternative to custom gate arrays and standard cells.

Xilinx is the market leader in FPGAs, CPLDs, software, IP cores, and solutions for digital consumer devices. As shown by the different system block diagrams, Xilinx products have significant advantages nd are used in many different ways. Xilinx products also provide competitive advantage through product differentiation.

Xilinx FPGAs and CPLDs address the challenges of system design for digital consumer devices by providing the following solutions:

- **System interfacing and connectivity** – With support for a high number of IO standards and available IP, Xilinx FPGAs and CPLDs are ideal for interfacing to ASSPs, memories, and processors (including DSPs and graphics processors). Xilinx FPGAs and the connectivity IP cores also provide chip-to-chip and board-to-board communication in a single FPGA. This eliminates the requirement of multiple, external, discrete components.

- **Superior digital image and video processing** – Consumers are demanding superior image quality, more access to larger displays, and higher quality source material such as DVD. Due to DSP features, Xilinx FPGAs provide far better signal processing capabilities than today's leading digital signal processors. This is true in all phases of video applications– capture, process, and display. Moving computationally-intensive functions—such as motion estimation and DCT from MPEG codecs—from multiple DSPs to a single FPGA can provide better performance.

- **Soft processing solutions for enabling system control** – Using MicroBlaze or PicoBlaze soft processors or embedded PowerPC processors, Xilinx FPGAs can act as a high performance central hub for data flow, overall system control, interrupts, high speed memory controllers, and other peripherals. This frees up the central processor and consumes less Silicon space on the board.

- **Superior clock management capabilities** – Xilinx FPGAs have digital clock managers for better clock management capabilities in complex systems. These features help reduce EMI by internal multiplication of slower PCB clocks. This also helps reduce system cost since no external clock management devices are required. Also, depending on the system, FPGAs allow the customer to choose cheaper memory types by providing a flexible IO timing budget. FPGAs with clock management capabilities also eliminate clock skew and nullify clock input and board delay in addition to internal distribution delay.

- **Internal and external memory solutions** – Xilinx offers solutions with abundant on-chip memory resources, as well as support for off-chip storage through the several IO standards in each FPGA. For buffering real-time video in displays and video applications that require high bandwidth memory, Xilinx block RAM and SRL-16 features are typically used. Xilinx FPGAs provide different types of IOs and free memory controller reference designs to interface to external memory devices such as SRAM, DRAM, CAM, and flash.

- **Discrete logic replacement** – Discrete logic devices, long considered the workhorse of the semiconductor industry, historically held a unit cost advantage over PLDs. However, silicon and packaging costs for PLDs have been aggressively driven down to the point where they now offer a highly compelling discrete logic replacement alternative. Beyond unit die costs, PLDs provide a better solution than the fastest discrete logic device when factors such as flexibility, higher performance, fast design turnaround, time-to-market, and higher reliability considered.

- **EMI and signal integrity** – Solving EMI issues is "black magic" and reducing EMI problems to meet FCC compliance is a very costly exercise. Failure to comply with FCC regulations leads to costly product redesign/shielding, board re-layout, time-to-market delays, and reduced market share and revenues. Xilinx FPGAs are "EMI friendly" and they support over 20 IO standards including single-ended interfaces, differential interfaces, and multi-gigabit transceivers. Xilinx products also have programmable output drivers, post-PCB signal integrity adjustment capabilities, and clock management using DCMs. These features provide significant cost and time-to-market reduction.

- **Feature addition** – Adding new features that help customers differentiate their products is a key FPGA use model. Functions such as encryption/decryption for preventing unauthorized access and copying and sharing can be implemented in Xilinx FPGAs. This can also be aided by using the extensive Xilinx library of IP cores such as AES, SHA, DES, and triple DES, or by using proprietary schemes.

- **Flexibility and time-to-market** – Flexibility provides the ability to improve upon existing solutions rapidly and with ease. Faster time-to-market and low cost of ownership with Xilinx products maximize profitability in the market. According to research from McKinsey & Co., products that are six months late and on budget earn 33% less profit over five years. Furthermore, every four-week delay equals 14% loss in market share. Specifications in home networking technologies, display technologies, interfacing standards, and broadband access technologies evolve rapidly and differ by geography. This requires a flexible solution such as an FPGA where minor specification and standard updates can be reprogrammed in the FPGA. Also, the splintered product volumes do not justify an ASIC, and FPGAs can be used for inventory management by programming the FPGA with geography-specific encryption or other requirements.

- **Field upgradeability** – Bugs are an unavoidable problem in the high-tech industry, with competition adding pressure to bring products to market quickly. With specifications continuing to evolve, multiple product variants increase the probability of bugs. Remaining competitive requires feature enhancement/product customization, and fixing bugs and enabling new features after product deployment requires expensive truck rolls. Bug fixes and re-spins in ASICs/ASSPs are expensive for any reason, including adding key differentiation between competitive products, future-proofing products, or enabling instant compliance to new standards. With an FPGA, companies can remotely upgrade and improve their systems via the Internet, increase the lifetime of their products, and enable product features per end-user needs. This has a large financial impact by reducing maintenance costs, providing new Internet-based revenue opportunities, increasing market share by getting in earlier and staying in the market longer, and developing customer loyalty. This feature is unique to programmable logic, and ASSPs/ASICs cannot provide such capabilities.

The eSP website (www.xilinx.com/esp) is an excellent reference for details and assistance with solutions for next-generation digital consumer devices. The eSP program helps customers shorten time-to-market for emerging standards and protocols. It also includes educational material, block diagrams, application notes, Intellectual Property, reference designs, and example implementations for different digital consumer devices, broadband access devices, and residential gateways.

Conclusions and Summary

We have recently entered a new century, one that promises to be even more technologically vibrant than the last. But this steady stream of marvels has tended to make us a bit technologically blasé. As consumers, we casually accept each new technological wonder into our daily lives even as we rage about the change from one format to the next. In fact, not only do we expect technological marvels as our right, we are impatient when they don't come fast enough to suit us.

A new era of digital consumer devices has entered the market, capitalizing on the pervasiveness of people communicating via email and the widespread use of the Internet. Access to the web has been the exclusive domain of the PC until the arrival of these appliances. However, the lack of portability, the heavy price tag, and complicated software installation associated with the PC has prevented the "big beige box" from being a visible tool outside the office or study environment in the home. Activities such as accessing email on the move, checking driving directions while on the road, and playing video games while lounging on a sofa are just examples of activities that are beyond the capability of the PC. The desire of consumers to engage in activities like these while not being tied to the PC are overwhelming, as evidenced by the explosive sales of cellular phones, telematics in cars, video game consoles, handheld PDAs, set-top boxes, and so forth.

However, consumer appliance manufacturers are in somewhat of a quandary when it comes to determining the target audience for their products. The question for these manufacturers to ask is: What is the customer profile for these products? Is it a PC-user or non PC-user?

"High-definition television pictures that are almost as sharp as film!"? Great—but when will we be able to record them? Satellite phones that allow us to communicate with anyone, anywhere and all times—why not, but why aren't the phones smaller? Powerful, full-color PDAs equipped with wireless modems and digital cameras that fit in your pocket? What took so long? But why don't they have voice recognition instead of a keyboard or handwriting recognition? Why is voice, video, and data access required from a single device and why is transmission of multimedia and Internet required over a single network? It is obvious that the emerging technologies of the 21st century still need some finishing touches. It is equally obvious that they will be here before we know it.

Could we end up with *less* of a future than we now imagine? Paranoia about piracy from the creative worlds of Hollywood and the music recording industry have put many of these digital developments on "pause" for a time. These issues could also result in devices that don't live up to their complete promise—if they result in devices at all.

Fear of the technological unknown often has been the cause of potential delays in advances and new devices. The gas company tried everything it could to stop Edison from opening his power station. The telegraph monopoly attempted to strangle the blossoming telephone industry. Nearly a century later, several Hollywood studios sued to stop the deployment of the VCR. Hollywood and the hardware makers now are engaged in ensuring the protection of HDTV video broadcasts. And the recording industry is attempting to put a technological finger in the dike of digital music downloads with its legal pursuit of sites like Napster and MP3.com.

In some cases, individual vision and persistence won the day. In other cases, Hollywood is glad it lost. In other cases, the jury is still out. In all cases, technology and history inexorably march on as they always do, despite the determined efforts of the frightened and the shortsighted to forestall change. One thing is for sure: we may be smart enough to invent these new digital technologies, but we never will be smart enough to predict how they will change us or the world. We only know that they will. This is best exemplified through the recent launch of Apple Computer's iTunes Music Store. It is stocked with hundreds of thousands of songs in 100% pristine digital quality that one can preview and own with just one click for only 99¢ each.

In this book, we have covered several emerging digital consumer devices that are part of our daily lives in the home and around us. We also covered some of the dynamics affecting consumers' lives and the business models and other requirements surrounding the success of consumer appliances. In the last chapter, we understood the use of programmable logic solutions to design digital consumer devices cheaper and better.

Landscape

As shown by the breadth of consumer devices covered in this book, the industry is dealing with "gadget" madness. Low electronic-component prices are sparking the emergence of inexpensive, multifunction, portable, computing devices. For decades, cheaper, faster, and smaller electronics have been coming out every 12 to 18 months, just as Moore's Law predicted.

The silver lining behind this for device manufacturers is that the prices for displays, memory chips, microprocessors, and many other components have dropped so much that it is now possible to produce no-compromise portable gadgets that had previously been strangled by high component costs. Gadget makers now have the opportunity to cross the traditional borders of form and function, combining in one device the features of a game player, music player, wireless email device, PDA, and cell phone. Consumers benefit by having all these features combined into one product.

Several devices such as MP3 players, digital cameras, and PDAs have common components such as flash memory. Lower cost of flash memory leads to higher availability of flash memory in the hands of the consumer, which further enables higher storage capacity of music, address books, and similar functions. Similar dynamics are true for components such as displays, processors, batteries, etc. Thus, desktop PC equivalent processors are being made available in handhelds, with a reasonable amount of battery life. That will make new and upcoming applications such as speech recognition more practical, perhaps making it possible to dictate emails in the near future.

This drop in prices will continue to permeate into wireless gadgets, making wireless email gadgets and web tablets much more affordable. Devices such as web tablets that are equipped with IEEE 802.11b wireless technology could see a higher price drop than the present prices, which are as high as the cost of a laptop. A product priced under $200 has a good chance of gaining tens of millions of customers, but one priced at $500 may easily get stuck. On the other hand, consumers have shown that they won't pay for poor products even at the lowest prices.

The success of the next hot consumer device is predicated on the manufacturers' study of what key application the appliance addresses that will motivate the consumer to shell out money. For example, wireless and portable video game players have been very successful because they provide a key application of handheld gaming. As long as there are continuous improvements, such as providing 3D entertainment and a bright display, consumers will buy this product repeatedly. And to squash rivals, companies might add still more features at low costs to products already at the high end of the feature range. For example, the addition of Microsoft's Pocket PC operating system in a PDA at a low cost helps migrate key benefits from desktop and laptop PCs to handhelds.

Consumer devices that innovate and move rapidly to introduce new products will continue to grab high market share. The Compaq iPaq PDA has taken considerable market share from the handheld market leader, Palm. The iPaq became the hottest-selling PDA because of its attractive color screen, as well as its ability to play MP3 files and low-resolution video and to keep calendars, email, and address books.

Some of the key requirements for the success of each and any of the existing or future consumer products are strategic partnerships and well-thought-out business models. While there have been several products claiming that their value will eliminate the PC, it has often led to the demise of the product category itself.

The successful business model for some digital consumer devices is going to be the one that has worked for set-top boxes and cell phones. Monthly service revenue far outweighs the hardware costs, and hence it makes sense for companies to give away the hardware and then charge for service.

The Ultimate Digital Consumer Device

Many manufacturers have already embarked on convergence by providing multiple applications in a single product, as is shown by the increasingly popular introduction of PDA-cell phone-wireless email device combination products. In a 1993 speech, Bill Gates first proposed the idea of a device that could store and access all manner of personal and public information, and dubbed it the wallet PC. This device has seen many forms in the PDA, PC companions, cell phones, wireless email clients, etc. Finally products are emerging that combine the functionality of these products.

While manufacturers such as Kyocera and Samsung have exhausted several efforts, new efforts are coming to market both from PDA vendors and cell phone manufacturers – each vying for a larger piece of the market. With computing power increasingly becoming cheaper, these Palm-phone combinations do just about anything and are slim, light, and small. The key for the success of such products is the attention to detail on the function it will perform and forming partnerships to ensure success through providing the necessary services. These products provide all the functions provided by PDAs and PC companions. They also provide cell phone and wireless email service. Many also have the capability of storing favorite MP3 audio files and digital photographs when on the road. QWERTY thumb-typing keyboards and keyboard-less models (based on Graffiti) are popular input mechanisms.

These unconventional wireless marriages will take place over the next few years, bringing together the worlds of communications, computing, and content in an easy-to-use way. The combination of a phone, PDA, digital camera, and a MP3 player will be highly successful. When not streaming Internet and video, listening to music does not impact the network. The number one thing is that consumers want to enjoy media consumption anywhere in a seamless environment that is copyright protected.

Increased network speed is enabling these multifunctional devices. The most popular features that customers are willing to pay for are color displays, short message service (SMS), wireless email, voice calls, downloadable ringers that are especially popular with the teen market, and an embedded camera on the phone.

A key to the success of these wireless convergence products is the use of Bluetooth and IEEE 802.11b technologies for short-range wireless communication, which will provide faster access to and greater benefits of the Internet. This will also provide products with the ability to synch with the PC, print on a LAN, or access the Internet. Also, GPS technology will have a big impact on the industry, making available safety features as well as targeted services where advertisements could be sent to potential customers. Other wireless trends include on-demand music, video and gaming, and personalization and customization.

Convergence is further shown through the combining of gaming and DVD players and also audio players that play multiple audio formats such as SACD, CD, DVD-Audio, MP3, etc. Other examples include the integration of the amplifier, audio and video (DVD) players with set-top boxes and digital VCR functionality into an entertainment/media gateway.

Historical Precedent—Not Everything Can and Will Converge

Another key consideration is historical success rates for combination products. A number of products have been brought to market over the years, some of which have succeeded and many of which have not. Examples of products that combine two or more functions include:

- Clock radios
- Portable stereo/CD players
- TV/VCRs
- Multifunction printers
- Telephone/answering machines
- PC/TVs
- Watches

The most successful of these combination products are the ones that clearly enhance and/or simplify the user experience, deliver a new benefit, or offer meaningful savings to a consumer that may be in the market for both products. For example, a clock radio delivers the ability to wake up or doze off to music and a TV/VCR saves space while integrating two products that naturally complement or rely on each other.

The products that successfully integrate additional functionality are able to create growth for the category or even create new categories. They also help differentiate vendors and boost price points for products that are successful. An example is the TV/VCR combo that, despite being seen by many as only moderately successful, actually has developed into a successful niche.

The less successful products are those that do not complement each other, sacrifice performance or usability of the functions relative to standalone products, or are too expensive to be perceived as a good value. A good example of an unsuccessful concept is the PC/TV, either taken as a TV tuner for a PC or as a full PC in the entertainment center.

Both PCs and TVs, arguably the most important technologies in the last half-century, have revolutionized the way we stay informed. In the late 1970s, several experiments with "interactive TV" were conducted. While the technology still needed some computing muscle, it was the public, still unexposed to the power and interactivity of the personal computer and video games, who were clearly not ready to put down their TV remote and pick up a control pad or keyboard. By the late 1980s, however, the personal computer had started invading the home, and every home with a child, or so it seemed, also had a videogame system nearly as powerful as the PC. Suddenly "Video 1" became an important switch on the TV. Since the computer monitor and the TV used the same CRT screen, and consumers now were used to typing and pushing buttons to manipulate images on a screen, why not finally combine the two? Programmers looking to get a start on a "smart" TV and convergence began in the mid-1990s to use the vertical blanking interval (VBI) to present all manner of information to the TV watcher. Electronic program guides and nascent Internet-based interactivity, such as searching for sports scores and statistics while watching a game, became available. In 1996, the set-top box-based WebTV brought the Internet to a TV near everyone, followed by AOLTV in 2000.

The problems with the PC-TV combination were immediately evident. To date, this type of product fails on each account:

- Consumers have different expectations for PCs (active interaction) and TVs (passive entertainment). The inclusion of one function into the other does not create new benefits or enhance the user experience. Plus, the PC is a single-person experience, while the TV was perceived as a group activity.
- Different technologies impair the performance of either option. TV used interlacing technology to produce an image, while the PC monitor used progressive scanning. Interlaced signal looks bad on a progressive screen and vice versa. As such, the consumer experience with the integrated function is sacrificed relative to the standalone product.
- The package solution is often more expensive than buying both a TV and a PC. For example, the Gateway Destination, which includes a large progressive format monitor for the living room, starts at $1,999. A PC of similar performance from Gateway can be purchased for about $1,200 (with monitor).
- Top-brand TV costs about $350—a sum that is $450 less than the combination product.

But thanks to the Grand Alliance's flexible HDTV solution, and the changing behavior toward their TVs by the public, the marriage between the TV and the personal computer, as well as digital audio and video recording systems, may occur in this new millennium. In the not-so distant future, consumers watching a music video could buy and download it (or the complete album) from the Internet onto a digital recorder, all without getting off their sofa. Similarly, integrating Internet functionality in a digital TV set will have less success. While consumers enjoy and want the ability to browse the Web and send email through TV sets, the evolution of different services will be enabled by a set-top box, and not by the integration of the Internet with digital TV.

Convergence also has evolved between the PC and audio. MPEG-3 or MP3 became a popular way to create concise computer files of music tracks for uploading and downloading to and from the Internet. The popularity of MP3 players is threatening the hegemony of the pre-recorded compact disc in the same way the CD threatened the vinyl record in the early 1980s. But more than that, MP3 may change the way music is distributed and sold.

Room for Many

The consumer electronics market is in excess of $100 billion in terms of sales and over $500 billion more when devices such as mobile/cell phones and PCs are included. The most popular products are DVD players, digital televisions, mobile and portable home electronics, set-top boxes (including personal video recorders), digital cameras, and video game consoles. The stand-alone DVD player sales are topping 15 million units and have reached the 25% household penetration rate faster than any other product in history. (This is followed by color TV, VCRs, PC, cell phones, cordless phones, pagers, CD players, and black-and-white TV.) From product introduction to date, DTV product sales total 6,216,831 units with a consumer dollar investment of some $10.8 billion.

The key is that there will not be just one product that takes the lion share of the market. There are enough products and product manufacturers that there is room for survival of several appliances. While niche products such as telematics and automotive entertainment devices and eBooks have greater chance for survival, there is room for emerging products such as web terminals, NetTVs, Internet-enabled refrigerators, laundry dryers, microwave ovens, toys, etc.

Components of a Typical Digital Consumer Device

In looking through several digital consumer devices, a critical question to solve is "What do all these appliances have in common?" All of these products require low cost and low-power consuming semiconductor components. The main blocks of a typical consumer device include embedded processor/CPU, memory, communications interface, user interface, software, middleware, display, keypad/keyboard, and power (battery) management.

With increased computing possible, today's processors are programmable and can handle video, audio, games, and data traffic. These processors bring more flexibility and faster time-to-market than traditional ASICs (application specific integrated circuits) and ASSPs (application specific standard products), which usually have a 12–18 month development cycle. They are also far cheaper and have lower power consumption than PC processors. These embedded processors are responsible for application execution, memory management, communications management, etc. Companies, such as MIPS, ARM, Hitachi, Motorola, Intel, Transmeta, and so on, have developed consumer appliance processors.

Typical memory types in consumer devices are flash, RAM, and ROM. Storage is a major growth area for consumer appliances as vendors try to fit increasingly large numbers of applications and data within smaller form factors while keeping costs and power requirements down. At this point, storage for mobile appliances can be split into solid state (based on flash semiconductors with no moving parts) and disk (small disk drives) and come in a variety of standards, shapes, and sizes. Compact Flash is generally associated with SanDisk, but is also offered by Hitachi, SiliconTech, Kingston Technology, M-Systems, and Lexar Media; compact flash comes on capacities ranging from 2 MB up to 160 MB and beyond. Advantages of compact flash include its small size, durability, capacity (with compression), and the fact that the controller is embedded into the unit, thereby saving the device manufacturers the cost of building it into their devices.

PC cards are based on a standard developed by the Personal Computer Memory Card International Association (PCMCIA) for integrated circuit cards that have any number of functions—storage, modems, or Ethernet connections—for notebook computers and now some mobile consumer appliances. Intel, Hitachi, Simple Technology, and SanDisk, among others, offer these flash-based cards, used for storage, which are characterized by their durability and capacity (up to 1 GB and beyond). PC cards are gradually being replaced in the mobile appliance market by smaller compact flash cards due to their higher total cost. Smart Media is used widely in the digital camera market and perhaps soon in the smart phone market. MultiMediaCard (MMC), developed by SanDisk and

Siemens and backed by Nokia, Ericsson, Motorola, and QUALCOMM is challenging flash-based smart media. MMC is expected to take a large share in the emerging smart phone market.

The SD (Secure Digital Memory) Card, developed by Toshiba, Panasonic, and SanDisk, is a modified version of the MultiMediaCard. It is compatible with the Secure Digital Music Initiative's (SDMI) standards, with 32 MB and 64 MB capacity currently available and 128 MB and 256MB available in the near future. More than 200 companies worldwide have adopted the SD Card standard, including Palm, Sharp, Casio, IBM, and others.

Developed and supported by Sony across the entire line of digital products, including PCs and portable audio, video, and multimedia devices, MemoryStick is a proprietary flash-based storage product. The Clik! drive from Iomega is a portable external disk-based storage format for notebook computers and Windows CE-based mobile appliances. The drive holds 40 MB and can be used for either notebooks or digital cameras—but not both, unless their dual-platform product is bought. The fact that it is a disk drive inherently limits its durability because of the internal moving parts. IBM MicroDrive is a mini-disk drive capable of being plugged into an industry standard CF+ Type II slot or to a PCMCIA Type II slot with an adopter. Its capacity ranges from 340 MB to 1 GB, and it can sustain a maximum data transfer rate in excess of 4 MB per second. Because of the internal moving parts, there are inherent limits to its durability.

A component that is missing on most consumer devices is the hard disk drive. Therefore, the amount of data that can be carried on an appliance is relatively limited. Flash memory cards have been adequate for most portable devices with relatively low storage needs. However, as consumers demand greater storage and applications such as email and Internet access, flash will be too expensive for a mainstream storage medium. More developments such as the IBM MicroDrive will bring disk drive to prevalence in the consumer appliance.

The communications interface depends on the type of communication technology being supported. Popular interfaces for communicating with the network and other appliances include USB, IEEE 1394, and wireless. The user interface includes the display, touchscreen, keyboard/keypad, microphone, and speaker.

The software includes the operating system and the applications that run on the consumer appliance. The popular operating system that is seen in many consumer devices is the Microsoft PocketPC due to the application support and installed PC base. Other niche operating systems, such as the Palm OS for PDAs, have been very successful in niche products. Middleware is a layer of software that lies on top of a consumer appliance's operating system. It provides 'hooks' or API's to which home networking applications can be attached.

Semiconductors enable new devices and players, but technology is increasingly becoming invisible. In the future more functionality will be available at lower price points. Consumer appliances will evolve to deliver Web content. Brands will change from 'device only' to service, solutions or customer relationship provider such as financial institutions.

The Coming of the Digital Home

While the digital home concept will be a step-by-step approach, the arrival of digital consumer devices will demand home networking and broadband access. Undoubtedly using digital technology in appliances has given rise to a number of new, exciting applications and made the existing ones a lot better.

Communications has become the cornerstone of human existence through using new and existing techniques in wireless and wired networks. The ability to have voice, video, and data access any time, anywhere is no longer a concept and is more a vision being played out step by step. This has been made possible by digital technology. The digitization of our consumer electronics and communications products is bringing utility, quality, and affordability into the hands of the consumer. While it has brought performance, reliability, lower power, and lower cost, the highest benefits of digital are seen in converging media. Exchanging content in digital format is far simpler than analog format. Voice, video, images, media, email, Internet, and data can be exchanged over landline, wireless, IP, cable, powerline networks with little difficulty once in digital format.

The Internet is permeating all facets of our communication lifestyles. The majority of future consumer devices will want access to the benefits of the Internet. This demand has affected the consumer and hence faster Internet access will continue to remain a hot and high-growth market and application for the coming years. Several technologies and business models have come about to take advantage of this new and lucrative revenue source. Also, to provide multiple consumer devices with an access to the Internet, several home networking technologies are emerging. These home-networking technologies can broadly be categorized under no new wires, new wires, and wireless – with each technology providing a set of pros and cons unique to itself.

However, the exchange and distribution of voice, data, and video requires a gateway device. This residential gateway will bring DSL, cable, and other broadband access to the home and allow access to other networked consumer appliances through home networking technologies. The gateway is often seen as an existing device such as set-top box or PC and digital modem combination that will grow to provide services such as video on demand and Internet access to consumers.

In addition, with the emergence of various appliances such as set-top boxes, MP3 players, digital TV, etc. there is an increase in digital content such as online shopping, MP3s, digitized photographs, and gaming. These combined forces are the backbone of bringing the digital revolution into the homes of the consumers.

The PC in the 1980s and the Internet in late 1990s have helped provide savings to corporations and enhance productivity. Similarly, the ability to interact and communicate over the Internet through any appliance in the home is beginning to enhance productivity, provide conveniences, and provide cost savings to the consumer. The home network improves communication and allows the sharing of expensive resources among members of a family. By the interconnection and interoperation of different home electronic appliances, entertainment devices, PC hardware, and telecommunication devices, the consumer will experience a convergence of voice, video, and data.

Key Technologies to Watch

While there are several technologies affecting the consumer market, some of the key ones that will impact the consumer electronics market include:

- Broadband (high-speed Internet) access: One of the key developments affecting the homes is the demand for broadband access. Dial-up techniques do not provide the consumer with the full benefits of the Internet. Applications such as high-definition video, CD-quality Internet radio, file sharing, web-based delivery of movies and software, online gaming, telemedicine, and home networking are just a sample of the many applications demanding faster access to Internet. There are several upcoming technologies such as LRE, PON, fixed wireless, and

FTTH. For the next few years, until the cost of these technologies are reduced, DSL, satellite, and cable access hold the highest promise of success.

■ Electronic video games: The electronic gaming industry has been around for a couple of decades, but is now in a period of massive evolution and convergence. Gaming is gaining prominence with its use of the Internet as players can use high-speed online gaming. While the primary application is gaming, these video game consoles provide access to the Internet and play DVD movies and CD audio music.

■ Digital home theater: As products converge (combining several functions and features into a single product) and digital media converges (digital voice, video and data), the home theater of the future takes unprecedented importance. This will enable communication through a single media using the TV for video, voice, data, and control of other appliances. The digital home theater includes the TV, VCR, CD players, gaming consoles, DVD players, set-top boxes, sound system (receives/amplifiers, speakers), etc. The home theater continues to offer a superlative audio and video experience to consumers with access to the Internet. In the future digital displays will become more affordable and have crisper and better quality images. Also, similar to VCRs, a hard-disk drive in the set-top box—the DVR will expand the storage options for a variety of media types. Also, multiple audio formats such as SACD and DVD-Audio will come to market.

■ Wireless technology: The dream of wireless technology one day having similar performance and quality of service as wired networks is keeping the development of wireless technology fast paced. It is about bringing a complete wired experience to wireless users, including delivery of broadband services and support for multimedia applications like streaming, image capture and wireless Internet. The different wireless technologies that will affect these developments include cellular (2.5G, 3G), broadband wireless (IEEE 802.16), wireless LANs (IEEE 802.11a, HiperLAN2), HomeRF, Infrared, DECT, GPS, satellite, and Bluetooth. We have all seen the high success of cell phones over the last few years. This will be further enhanced through the high data rates supported by 2.5G and 3G set of technologies.

Leading mobile applications include personal information management (PIM), entertainment, financial services, Internet browsing, navigation/location, m-commerce, gaming, etc. In the last mile access, digging up streets to put wires for neighborhood Internet access is a far more expensive proposition than IEEE 802.16 technology. Satellite broadband access is already one of the key broadband access technologies. Also, the wireless LAN segment is expected to grow at high rates and even challenge the cellular market in data access. It is also challenging the prevalence of 10/100 Ethernet in company local area networks. Meanwhile, low-cost Bluetooth technology will soon replace short-range corded connections and will be useful for applications such as cell phone to headphone wireless short-range connections. GPS will also be key for providing telemetry-based services.

■ Home networking: With multiple devices trying to access the Internet medium to the home, home networking technologies are much needed. Developments in this market must include consumer education and the reduction of standards and technologies targeting the market space. It not only hampers growth but also increases consumer confusion on choice of technology. Nonetheless, home networking will be a growing market and one that will see several developments over the next few years.

In Closing

The multibillion-dollar question remains—which digital consumer device will succeed and which will fail. Clearly, there is no room for all these products and businesses to survive because several of these products are going after the same dollars in the consumer's wallet. Consumer product vendors' need to look at the value of the product they are developing and what impact it will have on the market. They need to walk through strategies and details of the revenue and business models before developing such products. Whatever the case, the digital consumer device is here to stay. TVs, displays, phones, household appliances, DVD players, cameras, handhelds, etc. are growing and will continue to become pervasive. The key to success is to design and manufacture extremely easy-to-use yet sophisticated devices at price points that attract the masses.

Regardless of which digital consumer devices will dominate your home and your wallet, programmable logic solutions will be the building blocks and key technology enabler for all these systems.

Index